# Automotive Transmissions

Second Edition

Harald Naunheimer · Bernd Bertsche ·
Joachim Ryborz · Wolfgang Novak

# Automotive Transmissions

## Fundamentals, Selection, Design and Application

In Collaboration with Peter Fietkau

Second Edition

With 487 Figures and 85 Tables

Springer

Dr.-Ing. Harald Naunheimer
Vice President Corporate Research
and Development
ZF Friedrichshafen AG
Graf-von-Soden-Platz 1
88046 Friedrichshafen
Germany
harald.naunheimer@zf.com

Professor Dr.-Ing. Bernd Bertsche
Director
Universität Stuttgart
Institute of Machine Components
Pfaffenwaldring 9
70569 Stuttgart
Germany
bernd.bertsche@ima.uni-stuttgart.de

Dr.-Ing. Joachim Ryborz
Project Manager
Development Transmission
for Light Commercial Vehicle
ZF Friedrichshafen AG
Alfred-Colsman-Platz 1
88045 Friedrichshafen
Germany
joachim.ryborz@zf.com

Dr.-Ing. Wolfgang Novak
Development Engineer
Daimler AG
Mercedesstraße 119
70546 Stuttgart
Germany
wolfgang.novak@daimler.com

*In Collaboration with*
Dipl.-Ing. Peter Fietkau
Scientific Employee
Universität Stuttgart
Institute of Machine Components
Pfaffenwaldring 9
70569 Stuttgart
Germany
peter.fietkau@ima.uni-stuttgart.de

*Translator*
Aaron Kuchle
Foreign Language Institute
Yeungnam University
214-1 Dae-dong
Gyeongsan, Gyeongbuk
Korea 712-749
aaronkuchle@yahoo.com

ISBN 978-3-642-42256-0 ISBN 978-3-642-16214-5 (eBook)
DOI 10.1007/978-3-642-16214-5
Springer Heidelberg Dordrecht London New York



*Cover design*: eStudio Calamar S.L.

Printed on acid-free paper

Springer is part of Springer Science+Business Media (www.springer.com)

# Preface to the Second Revised and Expanded Edition

"Automotive Transmissions" was first published in Germany in May 1994. It was so well received that we decided to publish the book in English in 1999. Since then much has happened in the automotive and transmission sectors.

Imperatives imposed upon the development of automotive transmissions are improving driving performance, increasing driving comfort and ease of use, increasing reliability and service life, reducing weight and installation space, raising efficiency levels, profiling the brand image, reducing costs and, above all, reducing fuel consumption and pollutant emissions. Markets and market mechanisms for passenger cars and commercial vehicles differ and the emphasis placed on these requirements differs in turn. Common to all cases is that a variety of requirements leads by necessity to a conflict of goals. Approaches that can help to solve the goal conflicts are individual usage-optimised transmission solutions, higher integration of submodules, introducing more functionality and generating superordinate functions by means of networking with other vehicle components.

In the case of passenger cars, the trend toward individualised designs has caused strong segmentation with numerous vehicle classes. This has also lead to a massive diversification among transmission designs, with individual solutions and competing concepts: manual transmissions (MT), automated manual transmissions (AMT), dual clutch transmissions (DCT), conventional automatic transmissions (AT), continuously variable transmissions (CVT) and hybrid drives. The "black and white", manual vs. automatic situation existing back in 1990 no longer applies. In the case of commercial vehicle transmissions, the mechanical geared transmission with 6 to 16 speeds of either single-range or multi-range design are standard. In the heavy-duty truck segment, AMT have become successful in Europe. Their path led from semi-automatic designs right up to fully automated transmissions. Increasing integration of peripheral parts and submodules into the transmission has led to lighter, more compact and more reliable aggregates.

Electrics and electronics, actuator technology and sensor technology have played a defining role in many innovations in the area of automotive transmissions. Software is responsible for many of the functions of transmission systems, and thus for much of their customer benefit. The increase in function content and networking with other components of the vehicle leads to changes in the chain of responsibility between vehicle and transmission manufacturers.

The correct evaluation of trends in the market, in engineering and technology has taken on greater importance. The tasks now are to recognize and evaluate future demands early on, to derive new strategies and products from this basis and to develop and finally to produce these products for the market cost-effectively while maintaining a high level of quality. The goal of this book is to provide some of the tools required to do this. It intends to show the process of product development for automotive transmissions in its entirety.

The second edition integrates innovations in automotive transmissions into the systematic framework established in the first edition. Approximately 40% of the content of the second edition is either entirely new or revised with new data. As with the first edition, however, the goal is not to introduce the most current developments or to be exhaustive in details, but to provide the reader with lines of reasoning and to demonstrate approaches. Theoretical principles and concepts are explained that are of general validity and hence of enduring relevance. Therefore beside current designs, transmission systems that are no longer in production are also presented.

In order to strengthen the relation to praxis, the second edition has consolidated the knowledge of experts from different sub-disciplines. Our thanks go to them: *history*: Hans-Jörg Dach (ZF); *passenger car MT/AMT*: Christian Hoffmann (Getrag); *passenger car DCT*: Michael Schäfer (VW), Michael Kislat (VW), Michael Ebenhoch (ZF); *passenger car AT*: Christoph Dörr (Mercedes-Benz); *passenger car/commercial vehicle hybrid*: Stefan Kilian (ZF); *passenger car CVT*: Peter Schiberna (Audi); *commercial vehicle AMT*: Carsten Gitt (Mercedes-Benz); *commercial vehicle CVT*: Karl Grad (ZF); *gearing*: Franz Joachim (ZF); *operational fatigue strength*: Karl-Heinz Hirschmann (Uni Rostock); *acoustics*: Martin Hildebrand (Ford); *external gearshift system*: Andreas Giefer (ZF); *multi-plate clutches*: Dietmar Frey (ZF); *dry clutches*: Benedikt Schauder (ZF Sachs); *wet dual clutches*: Johannes Heinrich (BorgWarner); *bearings*: Oskar Zwirlein (FAG); *seals*: Werner Haas (Uni Stuttgart); *retarders*: Reinhold Pittius (Voith); *all-wheel drive*: Dieter Schmidl (Magna Powertrain), Andreas Allgöwer (Getrag), Hubert Gröhlich (VW); *electronic transmission control*: Josef Schwarz (ZF); *calculation tools*: Marco Plieske (ZF); *driving simulation*: Friedemann Jauch (ZF); *manufacturing*: Christian Wagner (ZF); *testing*: Peter Brodbeck (Porsche) – and many others who supported us with their advice and expertise.

We would like to thank the following companies for up-to-date and realistic illustrations: Allison, Audi, BMW, BorgWarner, Eaton, Ford, Getrag, Honda, LuK, Magna Powertrain, Mercedes-Benz, Opel, Porsche, Toyota, Voith and VW. Special thanks are due to ZF for all their support during the development of this book.

This English language edition could not have come to fruition without the assistance of many contributors. We are particularly indebted to Dipl.-Ing. Peter Fietkau as the manager and co-ordinator of the project, and to his assistants at the Institute of Machine Components (IMA), University of Stuttgart. We thank Springer-Verlag for their good cooperation. Our special thanks go to our families for their great patience, understanding and support during the three years spent preparing this book.

In 2002, Professor Dr.-Ing. Gisbert Lechner passed away. He was the initiator and author of the first English edition of "Automotive Transmissions". We see the second edition as a continuation of his excellent work.

Friedrichshafen and Stuttgart, May 2010

Harald Naunheimer, Joachim Ryborz
Bernd Bertsche, Wolfgang Novak

# Preface

It was in 1953 that H. Reichenbächer wrote the first book on motor vehicle transmission engineering. At that time, the German motor industry produced 490 581 vehicles including cars, vans, trucks, busses and tractor-trailer units. In 1992, production had reached 5.2 million. The technology at that time only required coverage of certain aspects, and Mr Reichenbächer's book accordingly restricted itself to basic types of gearbox, gear step selection, gear sets with fixed axles, epicyclic systems, Föttinger clutches and hydrodynamic transmissions.

Automotive engineering and the technology of mechanism design have always been subject to evolution. The current state of the art is characterised by the following interrelations:

**Environment - Traffic - Vehicle - Transmission.**

Questions such as economy, environment and ease of use are paramount. The utility of a transmission is characterised by its impact on the traction available, on fuel consumption and reliability, service life, noise levels and the user-friendliness of the vehicle.

There are new techniques which now have to be taken into account, relating to development methodology, materials technology and notably strength calculation. Examples include operational fatigue strength calculations, the introduction of specific flank corrections, taking account of housing deformation, and the need for light-weight construction.

Transmission design engineering bas been enriched by numerous variants. The manual two-stage countershaft transmission, preferred for longitudinal engines, and the single-stage countershaft transmission preferred for transverse engines now have many sub-variants, e.g. automatic transmissions, continuously variable transmissions, torque converter clutch transmissions, dual clutch transmissions, and transmissions for all-wheel drive.

The engine and transmission must increasingly be considered as one functional unit. The terms used are "powertrain matching" and "engine/transmission management". This can only be achieved by an integrated electronic management system covering the mechanical components in both engine and transmission.

The technique of systematic design developed in the 1960s, and the increasing use of computers for design, simulation and engineering (CAD) are resulting in ever-reducing development cycles. This trend is reinforced by competitive pressures. Systematic product planning is another significant factor in this regard.

It was therefore necessary to create an entirely new structure for the present book "Automotive Transmissions". Modern developments have to be taken into account. The great diversity and range of issues in developing transmissions made it difficult to select the material for this completely new version of "Automotive Transmissions", especially within the prevailing constraints. Not every topic could be covered in detail. In those places where there is an established literature, the authors have chosen to rely on it in the interests of brevity.

The purpose of this book is to describe the development of motor vehicle transmissions as an ongoing part of the vehicle development system. Only by actively taking this interaction into account is it possible to arrive at a fully viable transmission design. The aim is to highlight the basic interrelations between the drive unit, the vehicle and the transmission on the one hand, and their functional features such as appropriate gear selection, correct gear step, traction diagram, fuel consumption, service life and reliability on the other. Of course, another major concern was to represent the various engineering designs of modern vehicle transmissions in suitable design drawings.

The book is addressed to all engineers and students of automotive engineering, but especially to practitioners and senior engineers working in the field of transmission development. It is intended as a reference work for all information of importance to transmission development, and is also intended as a guide to further literature in the field.

Without the assistance of numerous people this book would not have been written. We would like to thank Dr Heidrun Schröpel, Mr Wolfgang Elser, Dr Ekkehard Krieg, Dr Winfried Richter, Mr Thomas Spörl, Mr Thilo Wagner, Dr Georg Weidner and Professor Lothar Winkler for researching and revising chapters. We also wish to acknowledge the contribution of numerous assistants and postgraduates for important work on specific aspects.

We wish to thank Christine Häbich for her professional editing. We would like to thank many employees and scientific assistants of the IMA (Institute of Machine Components) for reviewing and checking various parts of the text.

Such a book cannot be published without current practical illustrations. The publishers wish to acknowledge their gratitude to numerous companies for making illustrations available: Audi AG, BMW AG, Eaton GmbH, Fichtel & Sachs AG, Ford Werke AG, GETRAG, Mercedes-Benz AG, Adam Opel AG, Dr.-Ing. h.c. Porsche AG, and Volkswagen AG. We are particularly indebted to ZF Friedrichshafen AG who have always been most forthcoming in responding to our numerous requests for graphic material.

We are also indebted to Springer-Verlag for publishing this book. We would particularly like to thank Mr M Hofmann, whose faith in our project never wavered, and whose gentle but firm persistence ensured that the book did indeed reach completion. Dr Merkle then prepared the work for printing. We must also thank the publisher of the "Design Engineering Books" series, Professor Gerhard Pahl for his patience and advice. Our thanks especially to our families for their understanding and support.

Stuttgart,
May 1994

Gisbert Lechner
Harald Naunheimer

# Contents

# Terms and Symbols

*A formula you cannot derive is a corpse in the brain /C. Weber/*

Physical variables are related by mathematical formulae. These can be expressed in two different ways:

- quantity equations,
- unit equations.

### *Quantity equations*

Quantity equations are independent of the unit used, and are of fundamental application. Every symbol represents a physical quantity, which can have different values:

$$\text{Value of the quantity} = \text{numerical value} \times \text{unit}\ .$$

*Example:* Power P is generally defined by the formula

$$P = T\ \omega\ , \tag{1}$$

where $T$ stands for torque and $\omega$ stands for angular velocity.

### *Unit equations*

If an equation recurs frequently or if it contains constants and material values, it is convenient to combine the units, in which case they are no longer freely selectable.

In unit equations, the symbols incorporate only the numerical value of a variable. The units in unit equations must therefore be precisely prescribed.

*Example:* In order to calculate the effective power $P$ in kW at a given rotational speed $n$ in 1/min, the above equation (1) becomes the unit equation

$$P = \frac{T\ n}{9550}\ . \tag{2}$$

The unit equation (2) applies where the prescript $P$ is expressed in kW, $T$ in Nm and $n$ in 1/min.

# Designation system for steels

In several sections of this book, particular steels have been referred to according to German standard DIN EN 10027. Often there is no exact English equivalent. However it seemed important to provide an explanation of the type of steel being referred to. Therefore the basics of the specification will be explained.

The main symbol is the carbon content multiplied by 100 followed by the chemical composition of the material. The alloying elements are sorted by their alloy content, whereby the percentage of content is multiplied by a multiplier according to the following table. If there is no percentage of an element given in the specification, this means that there is just a small content of this element.

| Multiplier | Alloying element |
|---|---|
| 4 | Mn, Si, Ni, W, Cr, Co |
| 10 | Al, Cu, Mo, Ta, Ti, V, Pb, Zr, Nb, Be |
| 100 | P, S, N, C |
| 1000 | B |

Examples:

| | |
|---|---|
| 16MnCr5 | 0.16% carbon, 1.25% Cr, small content of Cr |
| 42CrNiMo4-4 | 0.42% carbon, 1% Cr, 1% Ni, small content of Mo |

# Terms and Symbols

(Only those which occur frequently; otherwise see text)

| | |
|---|---|
| $A$ | Surface area, vehicle cross-section = projection of vehicle frontal area |
| $A_R$ | Friction surface area |
| $A(t)$ | Availability |
| $B_{10}$ | System service life for a failure probability of 10% |
| $B_x$ | System service life for a failure probability of x% |
| $C$ | Basic dynamic load rating, constant |
| $CG$ | Constant gear |
| $CG_H$ | Front-mounted splitter unit constant high |
| $CG_L$ | Front-mounted splitter unit constant low |
| $CG_{main}$ | Main gearbox constant gear |
| $CG_R$ | Range-change unit constant gear |
| $D$ | Diameter, damage |
| $D_{act}$ | Actual damage sum |

| | |
|---|---|
| $D_{\mathrm{prof}}$ | Damage sum of a load profile |
| $D_{\mathrm{th}}$ | Theoretical damage sum |
| $E$ | Modulus of elasticity |
| $F$ | Force |
| $F_{\mathrm{B}}$ | Braking force |
| $F_{\mathrm{H}}$ | Manual effort, slope descending force |
| $F_{\mathrm{L}}$ | Air resistance, bearing force |
| $F_{\mathrm{Q}}$ | Shear force (transverse force) |
| $F_{\mathrm{R}}$ | Wheel resistance |
| $F_{\mathrm{S}}$ | Lateral force |
| $F_{\mathrm{St}}$ | Gradient resistance |
| $F_{\mathrm{U}}$ | Circumferential force |
| $F_{\mathrm{Z}}$ | Traction |
| $F_{\mathrm{Z,A}}$ | Available traction |
| $F_{\mathrm{Z,B}}$ | Required traction |
| $F_{\mathrm{a}}$ | Acceleration resistance, axial force |
| $F_{\mathrm{ax}}$ | Pressure force of the pressure plate |
| $F_{\mathrm{n}}$ | Normal force |
| $F_{\mathrm{r}}$ | Radial force |
| $F_{\mathrm{t}}$ | Tangential force |
| $F(t)$ | Distribution function, failure probability |
| $G_{\mathrm{R}}$ | Wheel load |
| $J$ | Moment of inertia |
| $K_{\mathrm{G}}$ | Transmission characteristic value |
| $L$ | Service life, sound level |
| $M_{\mathrm{b}}$ | Bending moment |
| $M_{\mathrm{t}}$ | Torsional moment |
| $M_{\mathrm{v}}$ | Reference moment |
| $N$ | Number of load cycles, number of oscillation cycles to failure, component service life |
| $P$ | Power, equivalent bearing load |
| $P_{\mathrm{A}}$ | Frictional power related to surface area |
| $P_{\mathrm{Z,B}}$ | Power requirement at wheel |
| $P_{\mathrm{m}}$ | Mean frictional power during synchronizer slipping |
| $Q$ | Shear force (transverse force), flow rate |
| $R$ | Reaction force, stress ratio |
| $R_{\mathrm{e}}$ | Yield strength |
| $R_{\mathrm{m}}$ | Tensile strength |
| $R_{\mathrm{p0.2}}$ | 0.2% offset yield strength |
| $R(t)$ | Survival probability, reliability |
| $S$ | Safety factor, locking safety factor of synchronizers, slip, interlock value, taper disc radius |
| $S_{\mathrm{B}}$ | Brake slip |
| $S_{\mathrm{H}}$ | Rear-mounted splitter unit high |
| $S_{\mathrm{L}}$ | Rear-mounted splitter unit low |
| $S_{\mathrm{T}}$ | Drive slip |

| | |
|---|---|
| $T$ | Torque, temperature, characteristic service life |
| $T_B$ | Acceleration torque (synchronizer), locking torque (differential) |
| $T_C$ | Clutch torque |
| $T_D$ | Drag torque |
| $T_L$ | Load torque |
| $T_M$ | Engine torque |
| $T_R$ | Friction torque, slip torque |
| $T_Z$ | Opening torque (synchronizer) |
| $U$ | Revolutions |
| $V$ | Displacement volume (oil pump) |
| $V_H$ | Total swept volume |
| $W$ | Section modulus, work, absorbable work, frictional work |
| $W_A$ | Frictional work related to surface area (specific frictional work) |
| $W_b$ | Section modulus under bending |
| $W_t$ | Section modulus under torsion |
| | |
| $a$ | Acceleration, centre distance |
| $b$ | Shape parameter, failure slope, pack length, width, fuel consumption |
| $b_0$ | Size factor |
| $b_e$ | Specific fuel consumption |
| $b_s$ | Fuel consumption per unit of distance, surface factor |
| $c$ | Rigidity, absolute speed |
| $c_W$ | Drag coefficient |
| $c_m$ | Machine capability index |
| $c_p$ | Process capability index |
| $c_s$ | Tooth rigidity |
| $c_u$ | Circumferential component of absolute speed |
| $c_\gamma$ | Meshing rigidity (average value of tooth rigidity over time) |
| $d$ | Diameter |
| $e$ | Eccentricity |
| $f$ | Deflection, frequency |
| $f_R$ | Rolling resistance coefficient |
| $f(t)$ | Density function |
| $g$ | Gravitational acceleration |
| $h_i$ | Number of stress oscillation cycles |
| $i$ | Ratio |
| $i_A$ | Powertrain ratio (from engine to wheels) |
| $i_{CG}$ | Constant gear ratio |
| $i_E$ | Final ratio |
| $i_{E,A}$ | Ratio of the axle drive |
| $i_{E,N}$ | Ratio of the hub drive |
| $i_{E,V}$ | Ratio of the transfer box |
| $i_G$ | Transmission ratio |
| $i_{G,tot}$ | Overall gear ratio, range of ratios |
| $i_S$ | Moving-off element ratio |

| | |
|---|---|
| $i_V$ | Variator ratio |
| $j$ | Number of friction surfaces |
| $k$ | Wöhler curve equation exponent |
| $k(v)$ | Characteristic value of a torque converter |
| $m$ | Gear module, mass, linear scale (converter) |
| $m_F$ | Vehicle mass |
| $m_n$ | Standard module |
| $n$ | Rotational speed, number, number of load cycles, number of bearings |
| $n_M$ | Engine speed |
| $p$ | Contact pressure, pressure, number of gear pairs, service life exponent |
| $p_{me}$ | Effective average pressure in the cylinder of a combustion engine |
| $q$ | Gradient |
| $q'$ | Gradient in % |
| $r$ | Radius, redundancy level of a system |
| $r_{dyn}$ | Dynamic wheel radius |
| $s$ | Travel, shift movement at the gearshift sleeve |
| $s_{Fn}$ | Root thickness chord |
| $t$ | Statistical variable, time |
| $t_0$ | Failure free time |
| $t_R$ | Slipping time, friction time |
| $t_S$ | Shifting time |
| $u$ | Gear ratio, circumferential speed |
| $v$ | Speed, flow rate |
| $v_F$ | Vehicle speed |
| $v_W$ | Wind speed |
| $v_{th}$ | Theoretical speed with slip $S = 0$ |
| $w$ | Absorbed work |
| $x$ | Addendum modification coefficient |
| $x, y, z$ | Co-ordinates |
| $z$ | Number of speeds, number of friction surfaces, number of teeth, number of load cycle passes |
| $z_i$ | Number of teeth gear $i$ |
| | |
| $\Delta$ | Interval, difference |
| $\Delta S$ | Wear path (synchronizer) |
| $\Delta V$ | Wear (synchronizer) |
| | |
| $\alpha$ | Meshing angle, cone angle of a cone synchronizer, viscosity-pressure coefficient |
| $\alpha_0$ | Effort ratio |
| $\alpha_{St}$ | Gradient angle |
| $\alpha_k$ | Stress concentration factor |
| $\alpha_n$ | Normal meshing angle |
| $\beta$ | Helix angle at reference circle, opening angle of dogs |

| | |
|---|---|
| $\beta_k$ | Fatigue notch factor |
| $\delta$ | Reference cone angle, degree of pump irregularity (volumetric flow pulsation) |
| $\varepsilon$ | Total contact ratio |
| $\varepsilon_\alpha$ | Transverse contact ratio |
| $\varepsilon_\beta$ | Overlap ratio |
| $\eta$ | Efficiency, dynamic viscosity |
| $\vartheta$ | Temperature |
| $\lambda$ | Performance coefficient (converter, retarder), rotational inertia coefficient |
| $\lambda(t)$ | Failure rate |
| $\mu$ | Torque ratio, torque conversion, coefficient of friction |
| $\mu_{stall}$ | Stall torque ratio |
| $\mu_H$ | Static coefficient of friction |
| $\nu$ | Speed ratio, speed conversion, kinematic viscosity |
| $\rho$ | Density, radius of curvature |
| $\sigma$ | Normal stress |
| $\sigma_D$ | Endurance strength |
| $\sigma_H$ | Hertzian stress |
| $\sigma_b$ | Bending stress |
| $\sigma_v$ | Reference stress |
| $\tau$ | Torsional stress, torque increase of a combustion engine |
| $\varphi$ | Gear step, bending angle |
| $\varphi_1$ | Base ratio change with progressive stepping |
| $\varphi_2$ | Progression factor with progressive stepping |
| $\varphi_{th}$ | Gear step with geometrical stepping |
| $\omega$ | Angular velocity |

## Subscripts

| | |
|---|---|
| 0 | Nominal or initial state |
| 1 | Pinion (= small gearwheel), input |
| 2 | Wheel (= large gearwheel), output |
| 3 | Frame |
| 1, 2, 3, ... | At point 1, 2, 3, ... |
| A | Available, related to area, powertrain, axle |
| AM | Angular momentum |
| B | Required, brake, acceleration |
| C | Clutch |
| CC | Converter lock-up clutch |
| CG | Constant gear |
| CS | Countershaft |
| D | Endurance, endurance strength, deficit, direct, drag |

| | |
|---|---|
| E | Final ratio |
| Ex | Excess |
| F | Vehicle, tooth root |
| G | Gearbox, propeller shaft |
| H | Static friction, main, main gearbox, ring gear, high (= fast), Hertzian, displacement, manual |
| IS | Input shaft |
| L | Air, load, low (= slow) |
| L, L1, L2 | At bearing point, at bearing point 1, 2 |
| M | Engine, model |
| MS | Main shaft |
| MSW | Main shaft wheel |
| N | Hub, rear-mounted range unit |
| OS | Output shaft |
| P | Pump, pump wheel, planetary gear |
| PV | Pump test |
| Q | Transverse |
| R | Reverse gear, roll, slip, friction, wheel, range-change unit, reactor, rotor (retarder) |
| Roll | Roll |
| Rot | Rotation |
| S | Sun gear, splitter unit, stator (retarder), system, lateral, shifting, moving-off element |
| Sch | Pulsating (strength) |
| St | Gradient |
| T | Turbine, drive |
| TC | Torque converter |
| U | Circumferential |
| V | Front-mounted range unit, loss, test, variator, transfer box |
| W | Alternating (strength), wind |
| Z | Traction, opening |
| | |
| a | Acceleration, axial, values at tip circle |
| abs | Absolute |
| act | Actual |
| ax | Axial |
| b | Bending |
| calc | Calculated |
| dyn | Dynamic |
| e | Effective |
| exper | Experimental |
| f | Values at root circle |
| i | Inner, input, control variable $i = 1, 2, 3, ..., n$ |
| i, j | At point i, j |
| id | Ideal |
| in | Input |

| | |
|---|---|
| j | Control variable |
| k | Control variable, notching effect |
| m | Mean, machine, number of stress classes |
| main | Main |
| max | Maximum |
| min | Minimum |
| n | Nominal, normal, n-th gear, standard |
| o | Outer, output |
| out | Output |
| p | Process |
| perm | Permissible |
| r | Radial |
| red | Reduced |
| ref | Reference |
| rel | Relative |
| res | Resultant |
| s | Surface, distance |
| spec | Specific |
| stat | Static |
| t | Torsion, time, tangential |
| th | Theoretical |
| tot | Total |
| trans | Transverse |
| u | Circumferential |
| v | Reference |
| w | Pitch circle |
| x, y, z | In $x$, $y$, $z$ direction, around $x$, $y$, $z$ axis |
| z | Highest gear, number of speeds |

# 1 Introduction

*Every vehicle needs a transmission!*

## 1.1 Preface

All vehicles, aircraft and watercraft included, require transmissions in order to convert torque and engine speed. Transmissions are distinguished in accordance with their function and purpose – e.g. selector gearboxes, steering boxes and power take-offs. This book deals exclusively with transmissions for road vehicles as well as for vehicles designed for both on-road and off-road use (Figure 1.1).

Figure 1.2 provides an overview of common transmission designs and their systematic classification. Further details can be found in Chapter 6 "Vehicle Transmission Systems". Dual clutch transmissions are assigned to automatic transmissions with various gear ratios due to their similarity with respect to control and functionality.

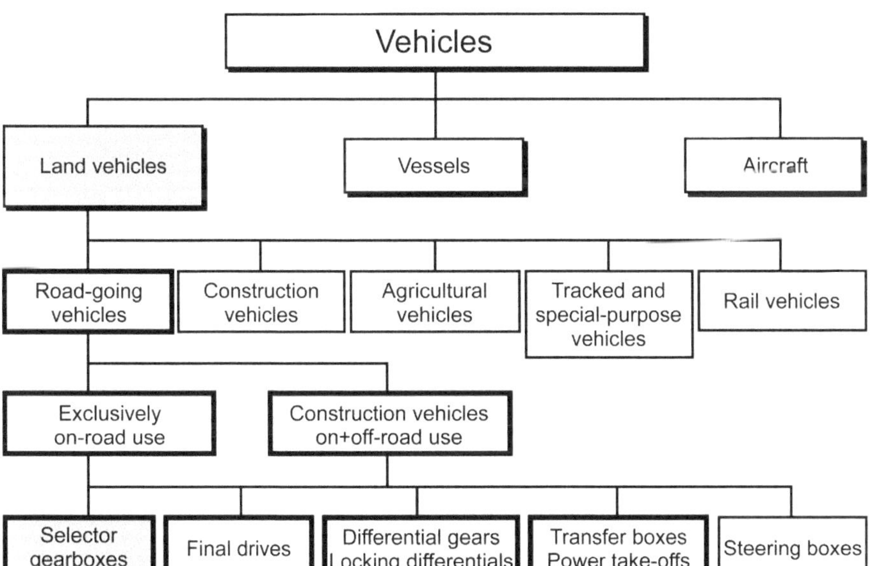


**Fig. 1.1.** Definition of the term "automotive transmission" as this book uses it

H. Naunheimer et al., *Automotive Transmissions*, 2nd ed.,
DOI 10.1007/978-3-642-16214-5_1, © Springer-Verlag Berlin Heidelberg 2011

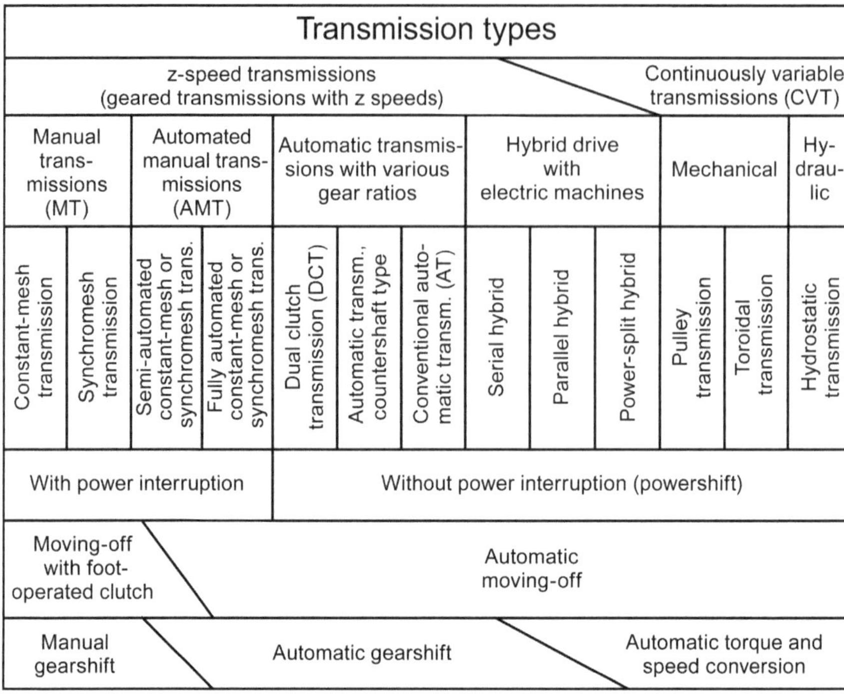


**Fig. 1.2.** Systematic classification of automotive transmission types

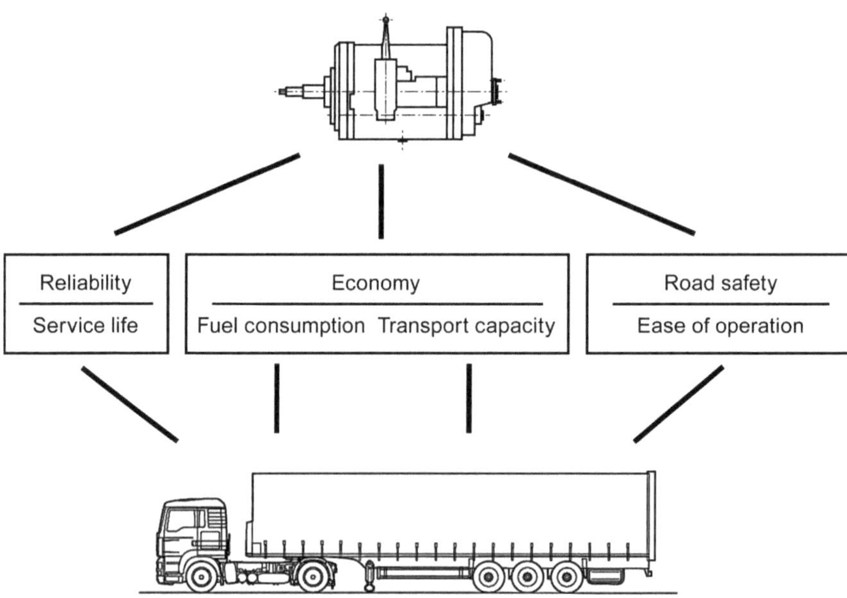


**Fig. 1.3.** The effect of the transmission on basic attributes of a vehicle

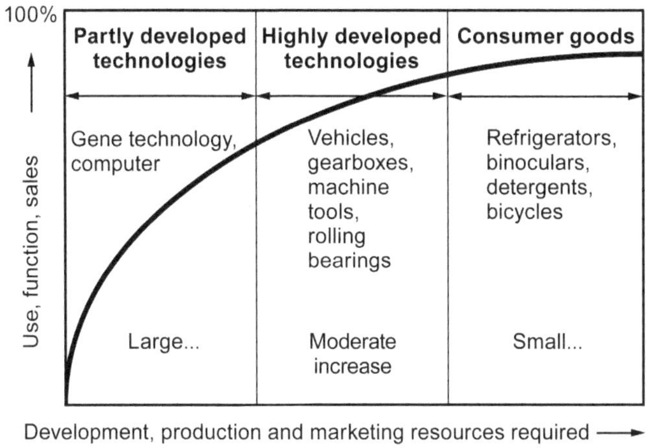


**Fig. 1.4.** Achievable increase in the practical value of a product by additional development effort

The task of a transmission is to convert the traction available from the drive unit, satisfying requirements placed on it by the vehicle, the road, the driver and the environment. Technical and economical competitiveness are essential here. In addition to the driving and transport performance of passenger and commercial vehicles, transmissions are of central importance with respect to reliability, fuel consumption, ease of operation and road safety (Figure 1.3).

**Table 1.1.** Comparison between industrial and automotive transmissions

| Transmission | Number of speeds (forward) | Ratio 1st gear/overall gear ratio | Power (kW) | Input torque (Nm) | Mass (kg) | Specific power (kW/kg) |
|---|---|---|---|---|---|---|
| Industrial 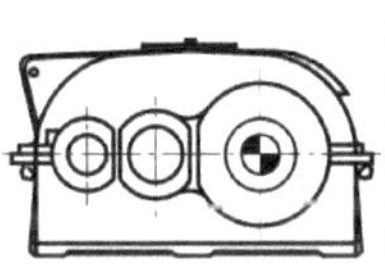 | 1 | 12.5<br>– | 330 | 2100 | 680 | 0.48<br>100% |
| Commercial vehicle (AMT) 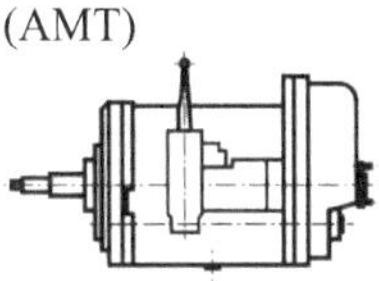 | 16 | 14.1<br>17.0 | 397 | 2600 | 266 | 1.49<br>300% |
| Passenger car (MT) 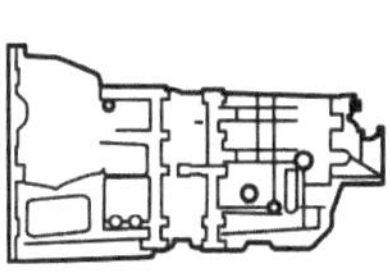 | 6 | 4.2<br>5.1 | 294 | 500 | 46 | 6.39<br>1300% |

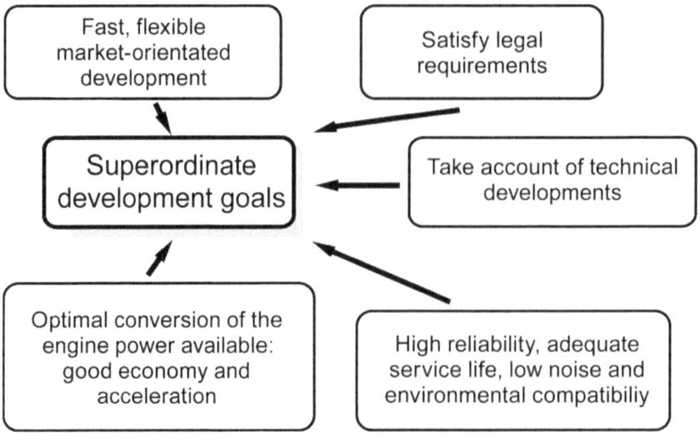


**Fig. 1.5.** Superordinate development goals for vehicle transmissions

Automotive transmissions are mass-produced products of a high technical and technological order. They are classified as highly developed technologies (Figure 1.4). What is remarkable is the specific power $P_{spec}$ in kW/kg of commercial vehicle transmissions, which is more than three times more than that of industrial transmissions (Table 1.1), despite the fact that automotive transmissions have more speeds. On the other hand, industrial transmissions have to be designed for a longer service life.

Basic innovations in the field of automotive transmissions are no longer to be expected. Instead, we are witnessing a process of gradual evolution. This process is characterized by system thinking focused on the factors Environment ⇔ Traffic ⇔ Vehicle ⇔ Engine/Transmission and by the use of electronics for operational, control and monitoring processes. The superordinate design objectives for automotive transmissions resulting from these tendencies are shown in Figure 1.5.

Vehicle transmission development must be fast and market-oriented. Customers' preferences, especially in the case of commercial vehicles, must be accommodated flexibly. Legal conditions, kW/t-regulation or emission policies for example, must be met. Furthermore, emotional aspects like driving pleasure must also be taken into consideration.

The main goal when designing an automotive transmission is an optimal conversion of the traction available from the engine into the traction force of the vehicle over a wide range of road speeds. This must be done such that there is a favorable compromise between the number of speeds, the climbing and acceleration performance and fuel consumption. Further technical and technological developments should obviously be considered – reliability and service life as well. It is also essential to have regard for environmental and social considerations.

The design of vehicle transmissions should always stay within the planning horizon for new vehicles (Figure 1.6). During the developmental phase of a vehicle, a corresponding transmission must also be created or further developed. To this end, new manufacturing technologies for mass production must also be prepared and introduced.

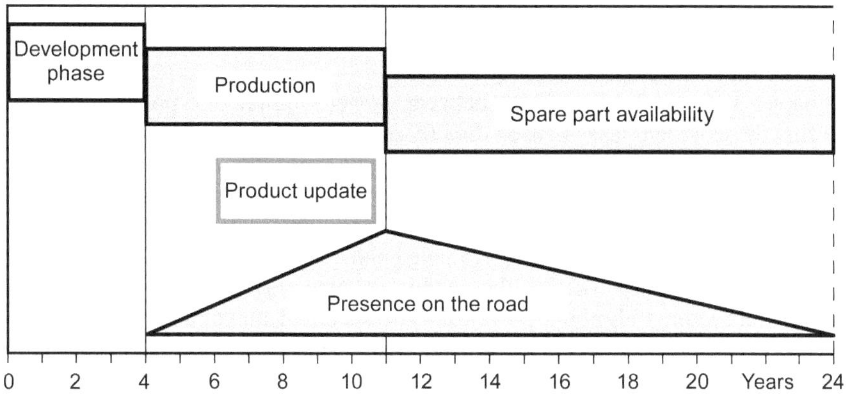


**Fig. 1.6.** Time dimensions and planning horizons in the automotive industry, from [1.1]

After the end of the production phase, it should be guaranteed that spare parts are available. For this purpose, the life cycles of additional components, including semiconductor components, have to be taken into consideration.

This book seeks to present the automotive transmission development process as a whole (Figure 1.7). It should show ways of thinking that go beyond mere component design. Regardless of which product is at hand, it is always necessary to assess the total system in which that product will later be employed. Such a system overview is indispensable and will be presented in Chapter 2.

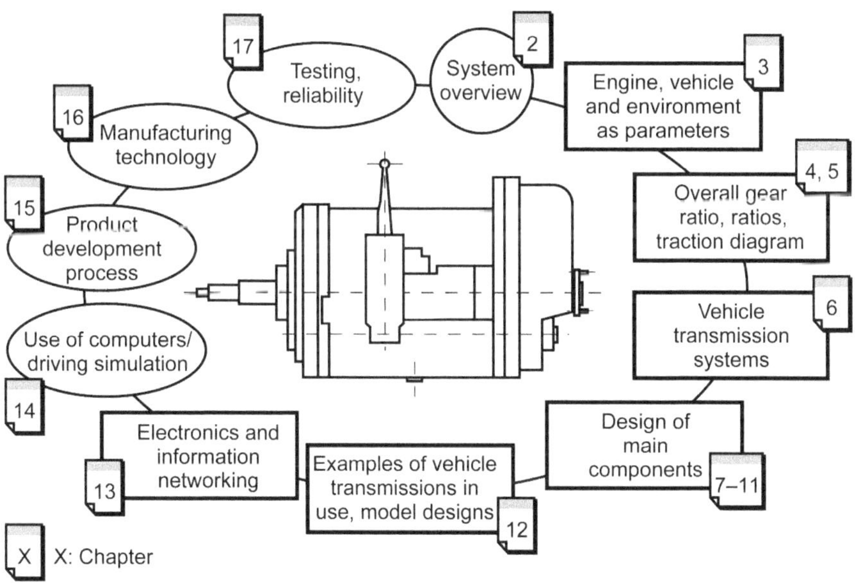


**Fig. 1.7.** The tasks involved in developing automotive transmissions, overview of chapters

Automotive transmissions are decisively influenced by the vehicle, the engine and the road profile. Without basic knowledge of these factors, meaningful developments are impossible.

Chapter 3 shows the interaction between power required and power available. The first development task focused directly on vehicle transmission is then selecting the range of ratios to be covered, the "overall gear ratio". In conjunction with selecting the number of speeds $z$, the gear ratio of the individual speeds, the resultant gear steps and the gear ratio of the final drive, the interaction of the vehicle and its transmission system can be evaluated and defined. Observing the road profile, it must be decided whether the vehicle is being sufficiently accelerated and whether the required climbing power and the specified maximum speed $v_{max}$ are reached. We can then establish at the same time whether the transmission unit also permits economical driving – driving with low amounts of fuel consumption in particular. This is dealt with extensively in Chapters 4 and 5.

Creative design, which is indispensable, is complemented by systematic design. Here, a functional analysis is carried out during the conceptual phase. Solutions for individual functions must be found, evaluated and joined together to make an overall solution, i.e. the transmission design. Chapter 6 provides the information regarding the vehicle transmission systems necessary for this.

Following this in Chapters 7 to 11 are the layout and design of the most important components of a transmission: gearwheels, shafts, bearings, synchronizers, clutches, parking locks, pumps as well as hydrodynamic clutches and converters. A treatment of all the details involved in highly developed computation and simulation methods would go beyond the scope of this book. We have instead confined ourselves to the basics of calculation methodology and operations.

In Chapter 12, the structure of various transmission designs and important detailed solutions are explained with the help of a plentiful amount of design examples. Electronic transmission controls built with microprocessors have been the standard in automatic transmissions since 1982. They are among the most complex electronic components in the vehicle and are undergoing a very dynamic development with respect to both hardware and software. Chapter 13 explores this topic and deals with their integration and interconnection with other control devices in the vehicle.

Tools and parameters for the development of automotive transmissions are handled in the latter part of the book. Chapter 14 is dedicated to calculation and simulation tools. In Chapter 15, we take a look at the product development process. Manufacturing technology has a large influence on transmission design, competitiveness and quality. Chapter 16 provides insight into the broad and innovative field of machining, assembly and final inspection.

Quality is a decisive competitive factor. The final customers are interested above all in the reliability and service life of the overall system. Methods for planning and guaranteeing quality as well as corresponding testing programs and test stations are described in Chapter 17.

Of particular concern in this book is to show the reader different approaches and to supply data as amply as possible regarding practical development work on automotive transmissions. As Dudeck put it, "The task of engineering science is, among other things, to refine complicated models to the point of simplicity". This book strives towards that aim.

## 1.2 History of Automotive Transmissions

*Knowledge of the past and of the state of the Earth adorns and nourishes the human spirit /Leonardo da Vinci/*

Learn from the past for the future! Development engineers and designers should have a grasp of the historical development of their products. Then they can estimate what progress is still possible and what technological potential the current product development has already realized. Such knowledge compliments that of systematic design (see Chapter 15).

### 1.2.1 Basic Innovations

Basic innovations are discoveries, inventions and new developments, without which products of today could not have been developed. They lead in turn to further discoveries, inventions, new developments and designs that culminate inevitably in new products (Figure 1.8).

In the course of such developments, certain phenomena should be explained and researched in order to guarantee that the product will function reliably.

**Table 1.2.** Examples of fundamental innovations in automotives and automotive transmissions

| | | | |
|---|---|---|---|
| **4000** BC | Mesopotamian vase with a picture of a cart | **1829** | *Stephenson* Rail vehicle, steam locomotive |
| **2500** BC | Wheels made of two semicircular wooden discs, presumably with leather tyres | **1877** | *Otto* Patent for four-stroke gas engine with compression |
| **2000–1000** BC | Spur gears with pin wheel gear as drive element for water scoops (Sakie, Figure 1.10), worm gears for cotton gins | **1885** | *Benz* Three-wheeler with internal combustion engine |
| **500** BC | Greek scholars discover the principles of mechanics | **1897** | *Bosch* Magneto-electric ignition |
| **200** BC | Lever, crank, roller, wheel, hoist, worm gear and gearwheel are in use | **1905** | *Föttinger* Hydrodynamic torque converter |
| **1754** | *Euler's* law of gears for gearwheels, involute toothing | **1907** | *Ford* Mass production of model T; the passenger car becomes a mass-produced item |
| **1769** | *Watt* Patent for steam engine | **1923** | *Bosch* Injection pump |
| **1784** | *Watt* Gearbox with constant-mesh engagement | **1925** | *Rieseler* Automatic passenger car transmission with torque converter and planetary gear set |

Table 1.2 is an attempt to retrace the seminal innovations in mechanical engineering that have led to the motor vehicle and thus to automotive engineering.

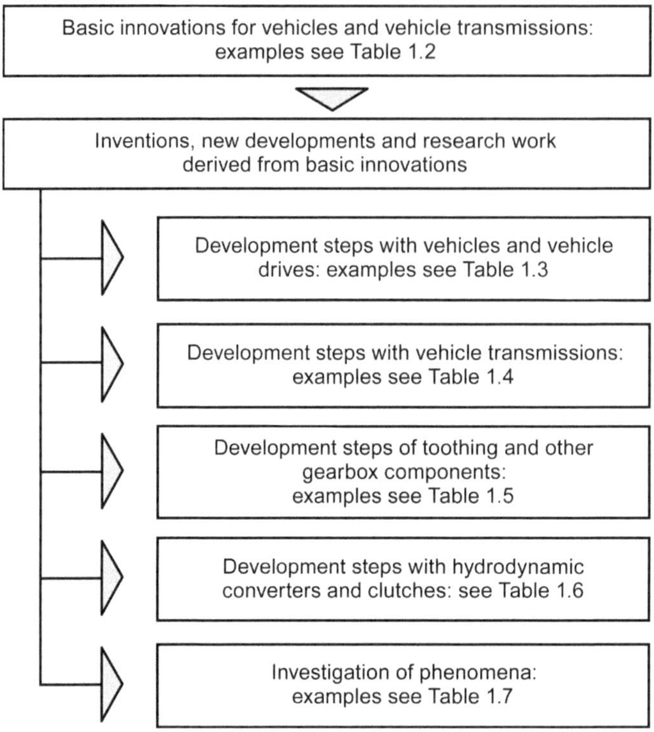


**Fig. 1.8.** Product developments are built upon basic innovations

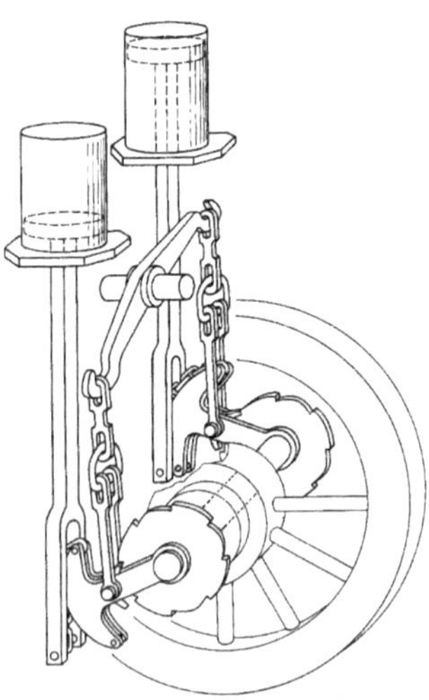

**Fig. 1.9.** Conversion of reciprocal movement into rotary movement. Twin-cylinder power unit with opposed pistons in the steam passenger car designed by Cugnot (1725 to 1804)

## 1.2.2 Development of Vehicles and Drive Units

The idea of equipping an engine with a gear unit in order to adjust speed and torque to the power output required is 100 years older than our present-day automobile with its official birth date of 1886. In the early days of the engine, the problem was how to convert the reciprocating motion of the piston into rotary movement. One possible solution is shown in Figure 1.9. The historical development of the transmission is thus closely tied to that of all engines (see Table 1.3).

**Table 1.3.** A chronology of important developments in vehicles and drive units

| | |
|---|---|
| **5000–500** BC | First technical inventions known: wheel, cart, gearwheel |
| **1500** | *Dürer* Sketch of a self-propelled vehicle |
| **1690** | *Papin* designs an atmospheric steam engine with cylinder and pistons |
| **1769** | *Cugnot* Steam vehicle with rectifier transmission |
| **1784** | *Watt* Double-acting steam engine with rotary movement and flywheel |
| **1800** | *Trevithick* Patent for high-pressure steam engine |
| **1801** | *Trevithick* Use of steam vehicle to carry passengers |
| **1801** | *Artamonow* Metal bicycle with pedal cranks |
| **1814** | *Stephenson* First steam locomotive |
| **1817** | *Drais* Steerable road wheel |
| **1832** | *Pixii* Rotating alternating current generator |
| **1845** | *Thompson* Invention of the pneumatic tyre |
| **1862** | *Lenoir* Double-acting gas piston engine |
| **1866** | *Siemens* Discovery of the dynamo-electric principle and design of an operational dynamo |
| **1877** | *Otto* Patent for four-stroke gas engine with compression |
| **1884** | *Parsons* Patent for steam turbine |
| **1885** | *Benz* Three-wheeler with combustion engine |
| **1885** | *Daimler* Motorcycle |
| **1886** | *Daimler/Maybach* Four-wheel motor car |
| **1888** | *Dunlop* Pneumatic rubber tyre |
| **1889** | *Maybach-Daimler* Steel wheeled passenger car with open 2-speed transmission |
| **1897** | *Bosch* Controlled electric magneto ignition |
| **1897** | *Diesel* Diesel engine; heavy fuel engine with compression ignition |
| **1903** | *Wright brothers* Powered flight |
| **1907** | *Ford* Introduction of mass production line |
| **1926** | *Gregoire* Constant-velocity joint. The Tracta joint opens the door to mass-produced front-wheel drive |
| **1934** | *Porsche* Project draft of the Volkswagen |
| **1935** | *Opel* designs the first frameless body for mass production vehicles |
| **1959** | Presentation of the BMC Mini, which will be the archetype for compact cars |
| **1970** | *Thyssen Henschel* Transrapid maglev monorail |
| **1979** | *Mercedes and BMW* Introduction of electronic engine control units and digital ABS |
| **1980** | *France* TGV high-speed trains |
| **1989** | *Audi* Introduction of direct injection and exhaust driven turbochargers for passenger car diesel engines |
| **1992** | After the Japanese vehicle manufacturers the European ones are introducing multi-valve engines in series production |
| **1997** | Common-rail injection in passenger car diesel engines |

**Fig. 1.10.** An early transmission system! Egyptian water wheel (Sakia) in Luxor, approximately 2000 to 1000 BC

### 1.2.3 Stages in the Development of Automotive Transmissions

Gears were doubtlessly used over 1000 years ago for enhancing human and animal labour. Similarly to bullock gear systems, still used for water supply in Egypt today, the two mating parts interlock by means of wooden pins or teeth (Figure 1.10).

The first drawings of gear systems date from the Middle Ages. Motor power was lacking and thus muscle power had to be used in its place. Human "machines" had to do the heavy work in the process. The first "vehicle transmissions" originated. In an etching by Albrecht Dürer from about 1500, the limited human power stroke is converted into propulsive force by means of a thrust crank, an angular gear and a spur gear stage.

Table 1.4 provides examples for important stages in the development of automotive transmissions. Note that all essential elements and design principles for transmissions had already been developed by 1925. Since then, further progress has pursued the goals of increasing service life and performance, reducing weight and noise reduction and optimising operability. There are four basic lines of development (Figure 1.11, see also Figure 1.2):

- mechanical z-speed transmissions (including automated ones),
- automatic transmissions with various gear ratios,
- continuously variable mechanical or hydrostatic transmissions and
- hybrid drives.

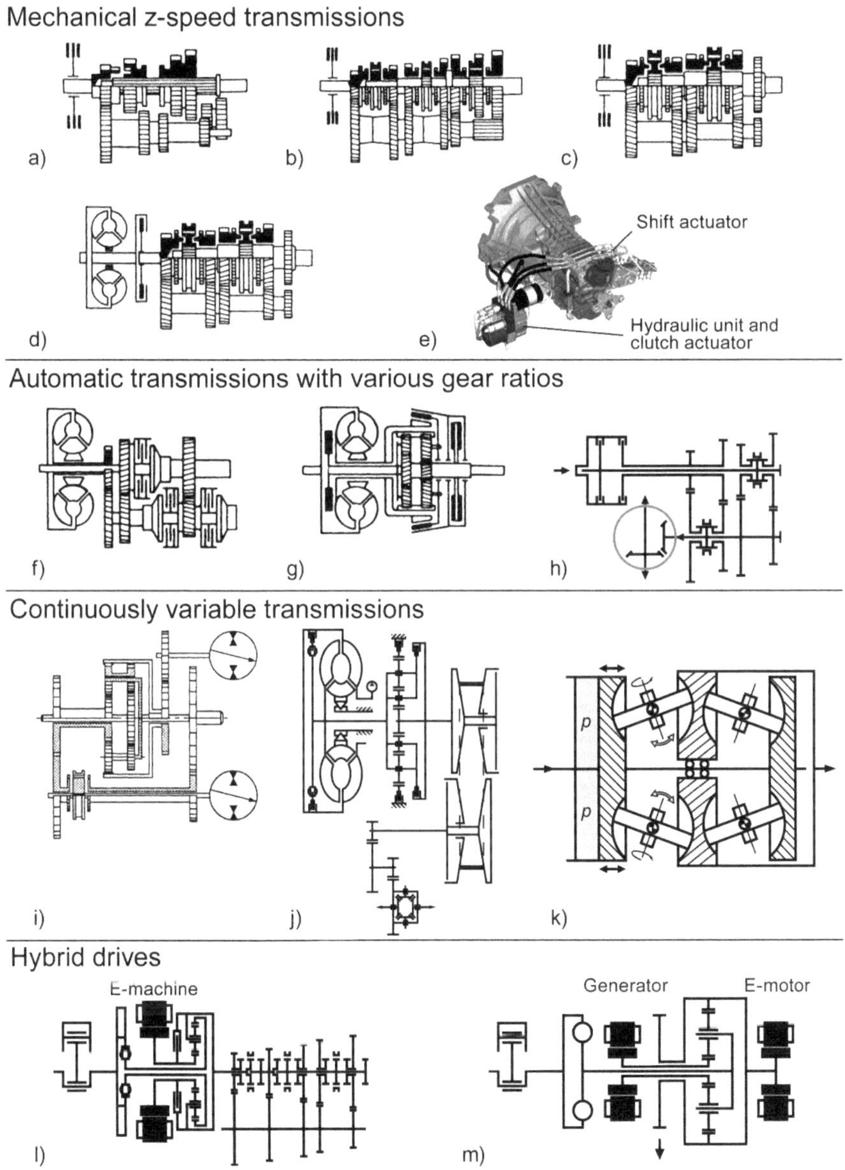


**Fig. 1.11.** Development sequence of passenger and commercial vehicle transmissions. *a* Transmission with sliding gear engagement; *b* transmission with constant-mesh engagement; *c* synchromesh transmission; *d* torque converter clutch transmission; *e* "Add-On"-automated manual gearbox; *f* countershaft-type automatic transmission; *g* conventional automatic transmission; *h* dual clutch transmission; *i* hydrostatic continuously variable transmission with power-split; *j* mechanical continuously variable transmission with taper discs; *k* friction gear, toroidal; *l* 1-E machine hybrid with z-speed transmission; *m* 2-E machine hybrid with summarising gear (power-split)

**Table 1.4.** Examples of important stages in the development of vehicle transmission

| | |
|---|---|
| **1784** | *Watt* stipulates that steam engines require additional ratios for road-going vehicles |
| | *Watt* patents variable-speed gearbox with dog clutch engagement and constant mesh of gearwheels (Figure 1.12) |
| **1821** | *Griffith* 2-speed transmission with sliding gears (Figure 1.12) |
| **1827** | *Pecqueur* First differential in a road-going vehicle (Figure 1.12) |
| **1834** | *Bodmer* Planetary transmission with stallable ring gear body using brake belt |
| **1849** | *Napier/Anderson* 2-speed belt transmission (Figure 1.12) |
| **1879** | *Selden* Patent enclosed sliding gear transmission with reverse gear and clutch (Figure 1.12) |
| **1885** | *Marcus* Cone clutch for motor vehicles |
| **1886** | *Benz* Belt-driven bevel gear differential (Figure 1.12) |
| **1889** | *Maybach-Daimler* 4-speed transmission with sliding gears (Figure 1.13) |
| **1890** | *Peugeot* Complete powertrain with sliding gear drive (Figure 1.13) |
| **1899** | *Buchet* Continuously variable belt transmission with axially adjustable taper discs |
| **1899** | *Krauser/Schmidt* Continuously variable friction gear with taper discs |
| **1899** | *Darracq - Léon - Bollée* 5-stage variable-speed belt "transmission gearbox" |
| **1899** | *Oliverson - Killingsbeck* Continuously variable belt transmission with axially adjustable taper discs |
| **1900** | *Reeves-Pulley* Continuously variable V-belt transmission with thrust links and axially adjustable taper discs |
| **1900** | *Léo* 3-speed transmission with face dog clutch engagement, integral differential and chain drive reverse gear |
| **1900** | *Lang* 3-speed geared transmission with constant-mesh wheels and draw key shifting |
| **1900** | *Diamant Speed Gear Company* Helical gear transmission |
| **1905** | *Pittler* Hydraulic drive system with hydro pump and hydro motor |
| **1906** | *Renault* Pneumatic transmission with piston compressor and piston engine |
| **1906** | *Didier* Two-stage planetary gear transmission with shifting using brake band and clutch of the planetary gear via friction plate face clutch |
| **1907** | *Renault* Hydrostatic transmission with axial piston pump and axial piston motor |
| **1907** | *Ford* Mass production of the model T with 2-speed planetary gear |
| **1915** | *ZF Soden transmission* 4-speed all constant-mesh transmission with constant-mesh gearwheels with preselector shifting and with synchronizing aids |
| **1925** | *ZF* Commercial vehicle standard gearbox with spur toothed sliding gears |
| **1925** | *Rieseler* Automatic passenger car transmission with torque converter and planetary gear set |
| **1926** | *Cotal* 3-speed planetary gear with automatic shifting via three electromagnetic clutches |
| **1928** | Development of the Trilok converter – a precondition for modern hydromechanical "conventional" automatic transmissions |
| **1928** | *Maybach* Overdrive auxiliary gearbox for reducing engine speed; shifting by means of override face dogs, and ground helical-cut gearwheels to reduce noise |
| **1929** | *ZF Aphon transmission* Helical-cut 4-speed transmission with multi-plate synchronizers |

**Table 1.4.** *(continued)*

| | | | |
|---|---|---|---|
| **1931** | *DKW F1* with driven front wheels. Transverse-mounted 2-cylinder 2-stroke engine | **1953** | *ZF* Hydromedia transmission for buses; 3-speed transmission with converter and hydraulically activated multi-plate clutches |
| **1932** | *Wilson transmission* Multi-stage planetary coupled gear with identical ring gears that are alternately fixed against the housing by means of brake belts | **1957** | *ZF* S 6-55, 6-speed commercial vehicle transmission, first fully synchronized commercial vehicle transmission |
| **1934** | *ZF All-synchromesh gearbox* 4-speed gearbox, helical cut, all speeds synchronized | **1958** | *Smith* Magnetic-particle dual clutch with rear-mounted 3-speed spur gear transmission and electrically activated dog clutches |
| **1939** | *General Motors Hydra-Matic transmission* First mass-produced conventional automatic transmission: 13 million produced; hydrodynamic clutch, 4-speed planetary transmission | **1961** | *Daimler Benz* 4-speed automatic transmission, of 2-range design with hydrodynamic clutch |
| **1939** | *ZF* 4-speed transmission, helical cut, prototypes with electro-magnetic multi-plate clutches | **1962** | *Eaton* 9-speed commercial vehicle transmission with power-split to two countershafts for a shorter overall design length |
| **1940** | *Franke* Patent on dual clutch transmissions | **1965** | *ZF* 3 HP 12, 3-speed-automatic transmission for passenger cars: converter without lock-up clutch, 3-stage planetary gear set and hydraulic actuation |
| **1948** | *General Motors* Dynaflow-transmission with polyphase converter and 2-speed Ravigneaux planetary gear set | **1967** | *VW* Semi-automatic transmission with torque converter clutch and rear-mounted 3-speed geared transmission |
| **1950** | *ZF* AK6-55 6-speed commercial vehicle transmission, all speeds with dog clutch engagement | **1970** | *ZF* 5K/S 110 GP, 9-speed commercial vehicle transmission (1+4x2) with dog clutches or synchronized and rear-mounted range-change unit in planetary design |
| **1950** | *Packard* Ultramatic transmission. Conventional automatic transmission with torque converter lock-up clutch, 2-stage 2-phase converter and 2-speed planetary gear | **1970** | Various companies develop a torque converter clutch transmission for commercial vehicles with a torque converter lockup clutch and secondary 6–8 speed transmission |
| **1950** | *Van Doorne "Variomatic"* Mass production of continuously variable V-belt transmission with axially adjustable taper discs (diameter adjustment) | **1971** | *Sundstrand "Responder"* Mass produced hydrostatic commercial vehicle gearbox with power-split through planetary gear set |
| **1952** | *Borg-Warner "Warner-Gear"-transmission* Conventional automatic transmission with Trilok converter and 3-speed planetary gear set | **1972** | *Turner* Commercial vehicle transmission with output constant gear and synchromesh on the countershaft to increase service life |
| **1953** | *Borgward* Automatic transmission with converter and 3-speed spur gear drive with electrohydraulic shifting | | |

**Table 1.4.** *(continued)*

| | |
|---|---|
| **1975** | *Van Doorne* Continuously variable passenger car transmission with steel thrust link chain and axially adjustable taper discs |
| **1978** | 5-speed passenger car gearboxes with increased overall gear ratio to reduce fuel consumption |
| **1979** | *ZF Ecosplit* 16-speed commercial vehicle transmission with integrated front-mounted splitter unit and rear-mounted range-change unit |
| **1980** | Trilok converter with lock-up clutch in automatic passenger car transmissions |
| **1983** | *Eaton/Fuller TwinSplitter* 12-speed commercial vehicle transmission with 4-speed main gearbox and two rear-mounted splitter units |
| **1985** | *Porsche* Re-discovery of the dual clutch principle as an automatic transmission for passenger cars |
| **1987** | *ZF* Semi-automation for commercial vehicle transmissions, AVS automatic preselection gear-shifting |
| **1989** | *Porsche* Automatic transmission with finger-tip control and adaptive shifting strategies |
| **1990** | Mass production of conventional automatic transmissions with torque converter, lock-up clutch, five speeds and electrohydraulic shift |
| **1991** | Renewed interest in alternative powertrain concepts: electrical and hybrid drives |
| **1996** | *Fendt Vario* Hydrostatic continuously variable power-split transmission with two driving ranges for tractors |
| **1998** | *Getrag* Automated 6-speed transmission in multi-range design for Smart compact cars |
| **1998** | *ZF AS-Tronic* Fully automated commercial vehicle transmission with 12 or 16 speeds in 2-countershaft design |
| **1999** | *Audi Multitronic* Mass production of continuously variable transmissions. Tensional link chain and wet master clutch |
| **1999** | *VW* 6-speed manual transmission for front-wheel drive with transverse engine |
| **2000** | *Toyota* Mass production of hybrid drives with the Prius 1 |
| **2001** | *ZF* 6-speed automatic transmission for standard drive |
| **2002** | *Aisin* 6-speed automatic passenger car transmission for front-wheel drive with transverse engine |
| **2003** | *VW* Dual clutch transmission with 6 speeds for front-wheel drive with transverse engine |
| **2003** | *Mercedes-Benz* 7-speed automatic transmission for standard drive |
| **2005** | *Getrag* 7-speed automated manual transmission for BMW M5 |
| **2006** | *Aisin* 8-speed automatic passenger car transmission for standard drive |
| **2008** | *VW* Dual clutch transmission with dry clutch and 7 speeds |
| **2009** | *ZF* 8HP, 8-speed automatic passenger car transmission for standard drive with optimised efficiency |

The invention of the steam engine soon brought forth the desire to adjust the available power to the intended use. The first steam-powered vehicles were driven by ratchet gears (Figure 1.9). Climbing gradients required higher ratios than driving on an even surface. In 1784, James Watt patented the constant-mesh gear with constantly meshing gearwheels still common today (Figure 1.12). Variable-speed transmission was born.

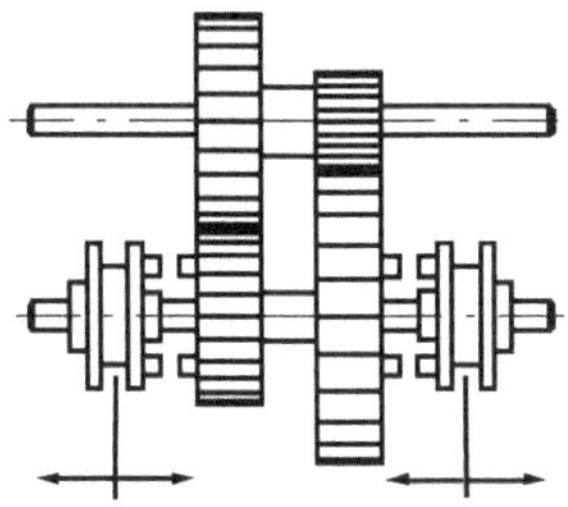

1784 *Watt patent*
2-speed gearbox with dog clutch engagement

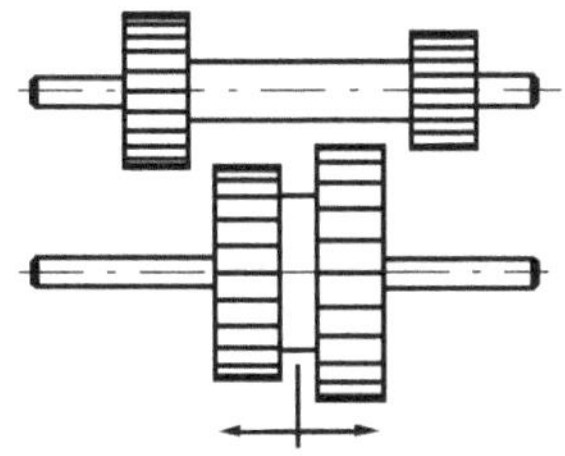

1821 *Griffith*
2-speed gearbox with sliding gears

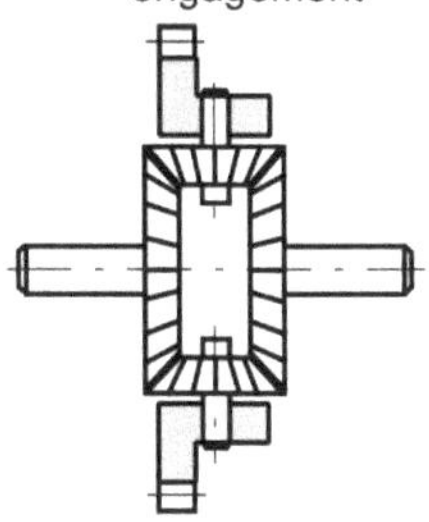

1827 *Pecqueur*
Differential gear

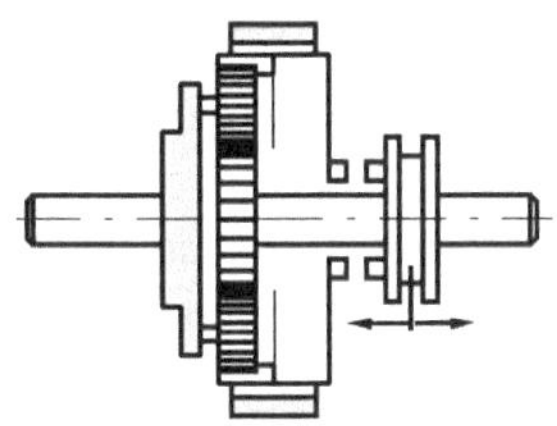

1834 *Bodmer*
Shiftable planetary gear

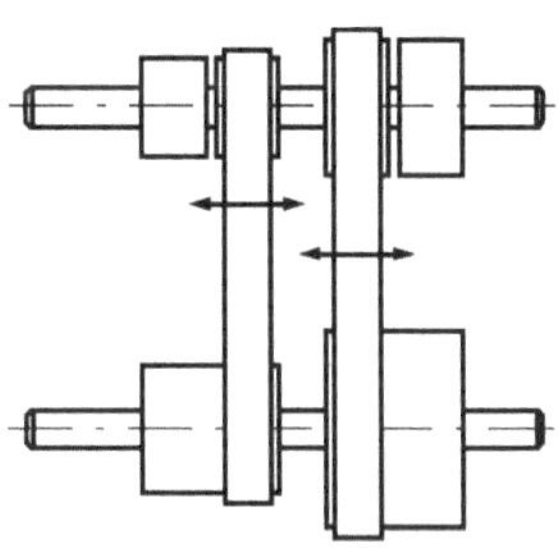

1849 *Anderson*
Shiftable belt transmission

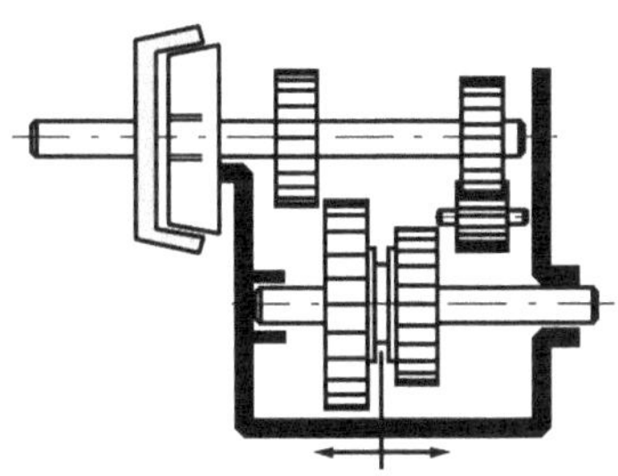

1879 *Selden*
Complete vehicle transmission with clutch, R gear and housing

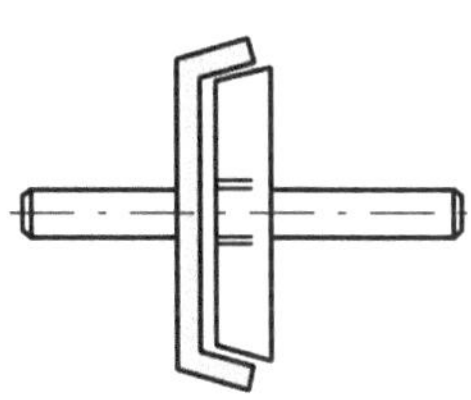

Around 1885 *Marcus*
Engaging cone clutch

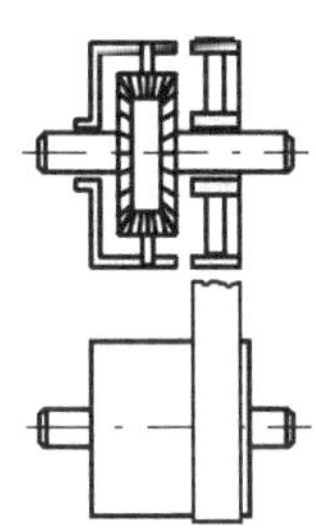

1886 *Benz*
Belt-driven bevel gear differential

**Fig. 1.12.** Early vehicle components and gears

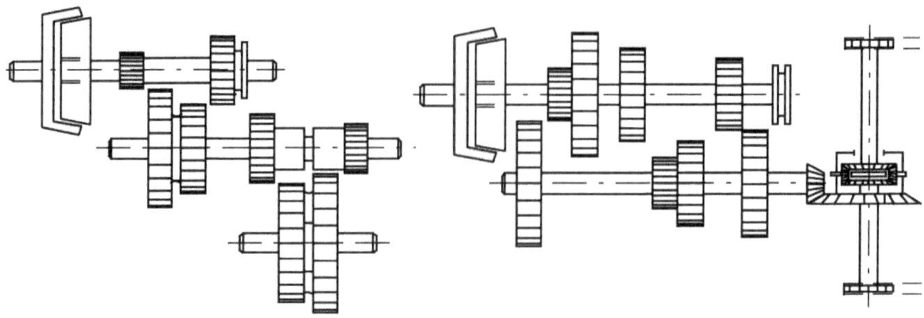


**Fig. 1.13.** Transmissions from the early days of the automobile

Actual production of road vehicles only began several decades later. In the year 1801, steam vehicle builders Evans and Trevithick solved the problem of torque adaptation, yet this still entailed interchanging a gear pair.

Already the beginning of the 19th century saw a series of important inventions (Figure 1.12). In 1821, Griffith revealed his sliding-gear transmission system, an inexpensive solution that was widely used well into the 20th century. Pecqueur managed in 1827 to equalize varying wheel speeds while cornering with the use of a differential.

In 1834, Bodmer designed a partially power-shiftable planetary gear. The change in gear ratio is achieved by disengaging the shifting dogs and tightening a brake band. As part of an overall patent for a vehicle with a piston engine, Selden patented a sliding-gear countershaft transmission with clutch and reverse gear in the year 1879.

It is remarkable how intensively research efforts already at the turn of the century were focused on the continuously variable transmission which is most ideally suited to internal combustion engines. Besides electrical and mechanical solutions, hydrostatic and even pneumatic ones were considered (Table 1.4). However, they did not gain acceptance, be it because of their insufficient power or mechanical complexity. The Föttinger torque converter (Table 1.6), invented in 1905 for ship propulsion systems, was first adapted to vehicle powertrains around 1925.

Direct drive was a crucial advancement. With it, Benz created the classic countershaft transmission with coaxial input and output still in use today. It was not yet included in Peugeot's exemplary powertrain from 1890 (Figure 1.13). The countershaft transmission design, with direct drive and four forward gears, proved a success. The basic problems of stepped gear change had been solved.

Around 1920, a further development phase began. Comfort had to be increased. The primary development goals were now making the gearshifting process easier and reducing noise by implementing ground and/or helical-cut spur gears or by reducing engine speed. Another important breakthrough was standard transmission, which was brought on the market for commercial vehicles in 1925. With it, gearboxes that were structurally identical or that varied only in their ratios and connections allowed for rational and inexpensive production. The gearbox has sliding gears.

The first facilitations in gearshifting date from the year 1915. The ZF Soden transmission included constant-mesh gearwheels, a preselector and synchronizing mechanisms. This transmission had preselection gearshifting: the driver sets a knob on the steering wheel to the desired gear and presses the pedal. The clutch is disengaged. As the shifting pedal is released, the preselected gear clicked automatically into place. The advantage of nearly effortless gearshifting could not compete with its disadvantages, such as the difficulty of adjusting the cable controls and the complex gearbox design.

In the case of a transmission from General Motors, the gearshifting process and subsequent power transmission was achieved by means of dogs with a cone synchronizer for speed synchronization between the shaft and the gearwheel. In 1928, Karl Maybach succeeded in improving vehicle running smoothness considerably by reducing gearing faults and engine speed with his auxiliary overdrive transmission and helical-ground gears. The quiet-running four-speed ZF Aphon gearbox originated at the same time. Its upper three speeds were synchronized with multiplates. In the ZF fully synchromesh gearbox for passenger cars (1934), all forward gears were already equipped with cone synchronizers.

The last conspicuous changes made to mechanical passenger car transmissions occurred after World War II, when more vehicles came out on the market, at first with rear-wheel drive, later with transverse engine and front-wheel drive – a development which has in the meantime spread to upper mid-size vehicles. For reasons of space, direct drive and coaxial design were abandoned and the engine, transmission and differential were combined into a single unit.

From about 1978, for fuel economy reasons, 5-speed transmissions with an increased overall gear ratio and finer ratio stepping gained increasing popularity for passenger cars. About ten years later, 6-speed manually shifted transmissions were also used in sports cars with longitudinal engines and rear-wheel drive. Especially in Europe, diesel engines in passenger vehicle gained in importance. Their image shifted from that of "Taxi-style" endurance machines to the active, high-torque engines. The transmission could compensate the lacking engine speed spread with more speeds. In 1999, VW began employing 6-speed manual transmissions for passenger cars with strong diesel engines mounted front-transverse. By 2005, six speeds were widely used among manually shifted vehicle transmissions.

Improvements in service reliefs all the way up to automatic gearshifting are a distinctly important line of development. From about 1956, Fichtel & Sachs furnished DKW (now Audi) with an electrically controlled, semi-automatic clutch, the *Saxomat*. This system consists of a centrifugal master clutch and a vacuum-operated gearshifting clutch. Upon contact with the gearshift lever, the gearshifting clutch is opened by a vacuum-controlled servo device. When the gearshift lever is released, air is slowly released to the servomechanism through a nozzle, thus engaging it. Pressing the accelerator pedal accelerates the air supply and thus the engaging motion. In comparison with a vehicle with a foot-activated clutch, driving comfort is significantly increased. In 1967, VW presented a semi-automated three-speed torque converter clutch transmission (TCCT) for passenger vehicles. From about 1995, a first generation of automated manual transmissions was introduced for passenger cars and light commercial vehicles up to 3.5 t. They were based on the concept of "add-ons", i.e. the attachment onto existing mass-produced transmissions of automated actuators for clutch and gearshifting. Figure 1.11e illustrates such an "add-on" version as exemplified by the MT75 5-speed transmission installed at that time in the Ford Transit. The basic concept of "adding on" actuators was retained for the second generation of AMT as well. In the third generation, from about 2008, the peripheral parts were integrated.

Already in the year 1925, H. Rieseler designed an automatic transmission comprising a torque converter and rear-mounted planetary transmission. He thereby designed a transmission, the essential components of which – torque converter with planetary gear shifted by means of clutches and brakes – are now typical for all conventional automatic transmission systems. Rieseler had made an extraordinary contribution, the advantages of which were not yet recognized by subsequent design engineers. The latter consistently sought only to replace the mechanical clutches with a fluid clutch. The conventional automatic transmission, consisting of a torque converter (some with a hydrodynamic clutch), 3- or 4-stage planetary gear set and hydraulic control, began to become established as of 1939. The manufacturing technology required for this was developed in the USA.

The first mass-produced transmission of this kind was the General Motors *Hydramatic*. These transmissions spread rapidly in the USA after World War II, capturing market shares of around 85%. In Europe on the other hand, conventional automatic transmissions for passenger cars only reached a market share of around 13%. In 1953, Borgward developed the first automatic transmission designed in Germany. It had a powershift countershaft transmission with a front-mounted torque converter used only for starting. Daimler-Benz followed in 1961 and ZF in 1965 with their own designs. Daimler-Benz still had the old design reminiscent of the *Hydramatic* transmission, with planetary gear transmission and front-mounted fluid clutch. These automatic transmissions underwent constant development aimed at fuel conservation. The slip-controlled torque converter lock-up clutch as well as transmissions of up to 8 speeds for increasing the range and improved adaptation of ratios became standard.

Under competition by dual clutch transmissions, the field of conventional automatic transmissions has been focusing since around 2003 even more on spontaneity, dynamics and fuel consumption in order to open up further potential. The dual clutch transmissions going into production prior to 2010 predominately have torque ranges larger than 300 Nm and have wet-operating clutches. Dry dual clutches are presently being developed for transmissions below 300 Nm.

The continuously variable transmission reappeared fifty years after its first development. Van Doorne's *Variomatic* was developed in 1950, and in 1958 it became the first mass-produced continuously variable transmission. The power was transmitted by rubber V-belts and V-belt pulleys, the diameter of which could be varied by axial displacement. In the *Variomatic*, centrifugal weights and a membrane acted on by vacuum achieved this adjustment of the pulleys. On the output side, the pressure is produced by a spring. In such a design with two parallel mounted belts a differential is unnecessary. The difference in rotational speed is compensated by belt slip. The rubber V-belts placed a limit on power. The permissible input torque was around 100 Nm. The transmission was therefore only suitable for small passenger cars.

Van Doorne then conceived the notion of the "steel V-belt". The thrust link chain consists of a steel belt made up of thin belts, onto which the thrust links are pushed, connected to the V-belt pulleys. This transmission, developed around 1970, was ready for use around 1975 and went into production around 1987. The Audi *Multitronic* started production in 1999 as a continuously variable transmission with a tensional link chain and wet master clutch. This transmission serves mid-size vehicles of up to 350 Nm engine torque. While continuously variable passenger car transmissions can claim a considerable market share for small cars in Japan, they appear, especially in Europe, not to have lived up to prior expectations. In the case of small cars, especially weight and costs speak against the continuously variable transmission. It thus appears that automated manual gearboxes have become more competitive in the European small car market.

Until the World War II, commercial vehicle transmissions were distinguished from passenger vehicles essentially only in dimension. This then changed fundamentally. Payloads increased as new tyres could take on heavier loads; trucks began being used not only regionally but also for long-distance transport; the motorway network was expanded etc. – all this necessitated a larger range of ratios (i.e. a greater overall gear ratio) and thus more gears.

The initial development goals for mechanical transmissions for commercial vehicles were low weight (= larger payload), noise reduction and improved ease of use with the introduction of synchronizers. One important requirement was a long service life of up to 1 million km. Initially, five to six speeds were sufficient, although front-mounted splitter units already provided a finer stepping of the overall gear ratio. The 6-speed gearbox was expanded to 12 speeds. The increase in the specific power output (kW/t) of commercial vehicles then led to the need for an increased overall gear ratio. Transmissions with nine or more speeds were developed. For better fuel economy or, alternatively, better performance, transmissions with twelve to sixteen speeds became common for heavy trucks in the early 1970s. Such transmissions are designed as multi-range gearboxes (see Chapter 6).

Due to problems concerning service life and cost, synchronizers did not become as established in commercial vehicle gearboxes as in passenger cars. While passenger cars already had fully synchronized transmissions before World War II, the first fully synchronized commercial vehicle transmission did not come onto the market until 1957 with the ZF S 6-55. But especially in Europe, more and more commercial vehicle transmissions were equipped with synchromesh to ease the gearshifting process. Other approaches to improving ease of operation were also explored. The companies Faun and Siemens began developing the Symo gearshift mechanism in 1954. In this engine-based synchromesh, the electronically controlled gear is engaged at the exact moment when the element to be engaged is synchronized. The electronics also control acceleration/deceleration during shifting. In critical situations, such as steep downhill slopes or hills, it is possible that equalization of rotational speed by the engine may not alone be sufficient, or it may not be executable if the electronic system fails. Since this situation – dangerous for the driver, the vehicle and the load – could never be completely excluded, the system never went into mass production. Around 1970, an attempt was made to make commercial vehicle transmissions semi-automatic by developing torque converter clutch transmissions (Figure 1.11d). The combination of a torque converter with a conventional separating clutch and a 6 to 16 speed transmission made moving-off with heavy tractor-trailer units easier. The torque converter increased the overall gear ratio. Transmissions of this sort are indeed in use, but have not become popular, claiming a mere 1 to 2% of the market share. The reasons for this are mostly to be sought in their high price (due to their complexity) as well as in increased fuel consumption.

Semi-automated commercial vehicle transmission designs have been coming onto the market since about 1985. Representative examples of such systems are the AVS system (automatic preselection shifting) by ZF or Mercedes-Benz's EPS system (electronic pneumatic shift). Fully automatic transmissions have become pervasive among heavy-duty commercial vehicles since about 2000. In such systems, both the moving-off and gear-changing processes are completely automated. As is the case for automated transmissions for passenger cars, commercial vehicle transmission designs have also taken the "add-on" route towards integrating peripheral parts into the transmission.

Automatic transmissions have not yet become common in trucks. This has to do with questions of economy and reliability. When commercial vehicles are exported to developing countries, the main concern is that they be easy and reliable to maintain. However, automatic transmissions are standard for city buses (Figure 1.11g). 1971 saw the first production version of a continuously variable hydrostatic power-split transmission (via a planetary gear set) for city delivery vehicles, the Sundstrand *Responder*. It proved unsuccessful however, and production was discontinued. Later attempts to utilize hydrostatic units with mechanical power-splits via planetary gears in city buses were also unsuccessful. Instead, power-split hydrostatic continuously variable transmissions have been widely used in the production of tractors of various manufacturers since about 1996.

Presently, fuel cell related research is focusing intensively on the development of electrical drives, particularly for city buses.

### 1.2.4 Development of Gear-Tooth Systems and other Transmission Components

The components of automotive transmissions are themselves in a state of evolution. We will now examine the development of components such as gearwheels, shafts, bearings, synchronizers and clutches as well as electronic controls (Table 1.5).

The most important component is the gearwheel. It would be impossible to provide historical evidence of the first gearwheels. But gear drives were used early on both for increasing human or animal power and for exploiting wind and water power. We can assume that the use of wooden gearwheels with crossed axes – similar to the bullock gear systems still in use for irrigation in Egypt today – is one of the earliest examples of the use of the gearwheel (Figure 1.10). Derived from these primeval gearwheels are mill drives and series-connected geared drives used to realize greater transmission ratios. Designs of such systems are recorded in great diversity in contemporary drawings. The use of gearwheels for the transmission of power has proven particularly fruitful in mining and mill construction. The great artist and inventor Leonardo da Vinci provided the foundations for our present-day machine component already back in the 15th century.

The scientific study of gears began at the end of the 17th century with de la Hire. Euler, Willis and Reuleaux continued this work. The law of gears as it was finally formulated by Saalschütz in 1870 states:

> *There will be uniformity of transmission of motion between two meshing gearwheels where the common normal of both tooth curves passes through the pitch point C at any contact point of the flanks.*

The creation of theoretically correct flank profiles on a mathematical and graphical basis was the prerequisite of mechanical gear technology. The development of the rolling process was groundbreaking for industrial gearwheel production (Table 1.5).

While lantern and cycloid gears had previously been the most important types of gear, today it is the involute. Because of its straight-flanked tool, which meshes on the base circle, it can be manufactured and measured precisely. Moreover, it has the characteristic of being insensitive to changes in gear centre distance.

Developments since 1980 have opened up new possibilities in gearwheel manufacturing. With numerically controlled tooth-hobbing machines, the rotary and longitudinal movements required to produce the tooth profile are controlled and synchronized electronically. In this way, arbitrary tooth profiles can be produced for special purposes, e.g. for low-noise gear pumps, which however still satisfy the requirements of the gear law.

**Table 1.5.** Chronological development of gear tooth systems and other transmission components

| | |
|---|---|
| **2000–1000** BC | Spur gears with lantern gear, worm gears. Transport of heavy loads on rollers |
| **230** BC | *Philon v. Alexandria* Multi-lever wheel with gear rack |
| **100** BC | Sun wheels and planetary gears in the astrolabe of Anticythera |
| **1300** | *Giovanni da Dondi* Astronomic clock with internal gearing and elliptical gearwheels |
| **15th C.** | Idea of helical gears<br>Sprocket wheels for link chains |
| **15th C.** | *Leonardo da Vinci* "Book of movement", "Book of gravity", "Book of moment of force", principle of virtual speeds, principle of independent super–position of movements, principle of the potential lever |
| **15th C.** | Gearwheels for transmitting movement in windmills |
| **1639** | *Désargues* Cycloid profiled gearwheels |
| **1694** | *De La Hire* Founder of gearing science, point gearing: teeth paired with points or journals, pitch circles |

**Table 1.5.** *(continued)*

| | |
|---|---|
| **1733** | *Camus* Pair gearing, teeth paired with teeth, cycloid toothing |
| **1754** | *Euler* Involute toothing |
| **1765** | *Euler* Curvature centre-points |
| **1780** | *Wasborough/Pickard* Thrust crank transmission |
| **1820** | Axial ball bearing with cage as bearing for castors |
| **1820** | *Tredgold* Beginning of gear-wheel strength calculation |
| **1850** | *Willis* Systematic classification of gears: Modules: possibility of combining any gearwheels from the same module |
| **1856** | *Schiele* Hobbing process useable with insertion of index gears |
| **1857** | Application and spread of rolling bearings in bicycles, first patented cup-and-cone bearing |
| **1865** | *Reuleaux* Description of "general gear hobbing" |
| **1869** | *Surirey* Ball bearing |
| **1872** | *Wagen-Thorn* Gear shaping method |
| **1876** | *Reuleaux* Line of action |
| **1881** | *Hertz* Theory of contact and pressure of solid elastic bodies; Hertzian stress |
| **1882** | *Bilgram* Invention of bevel gear production |
| **1883** | *Petroff/Tower/Reynolds* Hydrodynamic lubricant film theory in plain bearings |
| **1885** | *Marcus* Cone clutch for automobiles (Figure 1.12) |
| **1887** | *Grant* Gear shaping method for helical gears |
| **1890** | *Sachs* Patent on precision bicycle wheel hub |
| **1895** | *Maybach* Gate shift for automotive transmissions, grouping gears in "gates" |
| **1897** | *Pfauter* Universal gearwheel milling machine for spur gears, worm gears and helical gears |
| **1902** | *Stribeck* Work on the chief characteristics of plain bearings and roller bearings |
| **1903** | First deep groove ball bearing |
| **1907** | *SKF* Self-aligning ball bearing |
| **1908** | *Norma* First useable cylindrical roller bearing |
| **1912** | *Humphrie* Synchromesh to make changing gear easier |
| **1915** | *Maag* Gear grinder |
| **1916** | *v. Soden* Patent application for synchromesh |
| **1925** | *Gleason* Hypoid gear |
| **1927** | *ZF* Bevel grinding |
| **1930** | *Palmgren* Method for calculating rolling bearings based on the concept of service life |
| **1934** | Determination of module series |
| **1938** | *ZF* Introduction of lock synchronizer |
| **1938** | *Simmer* Patent for rotary shaft seal |
| **1956** | *Fichtel & Sachs Saxomat* Electrically controlled semi-automatic clutch comprising centrifugal master clutch and vacuum-activated gearshifting clutch |
| **1955** | *Novikov* Round-flank toothing for unhardened spur gears |
| **1982** | Transmission control of automatic transmissions with microprocessors |
| **1983** | Free tooth formation according to the law of gears using numerically controlled gear hobbing machines |
| **1997** | *Mercedes-Benz & Siemens* combine in the automatic transmission W5A 180 electronic transmission control, actuating elements, sensors and hydraulics to one mechatronic system and place it inside the transmission |
| **>2000** | System and information networking of vehicle components |

Initially, heat-treated steels were used to make gearwheels. Case-hardened steels soon became necessary in order to improve performance while simultaneously reducing weight. To reach the level of quality necessary for noise reduction, gearwheels have to be shaved after hobbing or ground after hardening. Current methods of machining after heat treatment are described in Chapter 16.

Other important transmission components such as rolling bearings, clutches and synchronizers were then developed in the second half of the 19th century and the beginning of the 20th. Since about 1995, there have been essential innovations in automotive transmissions in the highly dynamic fields of electronics, software, function development as well as system and information networking.

Finally, it should be mentioned that toothed gearing, as a means of converting torque and rotational speed, has a better power/weight ratio than other converters such as belt or chain drive, hydrodynamic or hydrostatic transmission or the electric motor.

### 1.2.5 Development of Torque Converters and Clutches

Initially, the individual components of automatic transmissions developed slowly, but this development has accelerated markedly, especially considering the complexity of such systems.

The foundation was laid by H. Föttinger in 1905, when he had a torque converter patented and a hydrodynamic clutch shortly thereafter. Föttinger constructed this torque converter for use in ships and never considered installing one in an automobile. The development of the torque converter is a good example of the systematic development of a transmission component (see Table 1.6 and Chapter 10). As an electrical engineer, Föttinger recognized the potential of combining a hydrodynamic prime mover (pump) with a machine (turbine) and first developed the idea theoretically.

It lasted almost two decades until attempts were made to apply Föttinger's torque converter and clutch to an automotive transmission. The Trilok converter devised by Spannhake, Kluge and van Santen combined the less efficient torque converter with the more efficient clutch. By mounting the reactor in the housing by means of a freewheel unit, the reactor runs freely when the reaction torque is discontinued, that is, exactly at the moment that the output torque falls below the input torque. The torque converter becomes a clutch and can thereby make use of the high level of efficiency of the fluid clutch in the high speed range. This combination has long been prevalent in automatic transmissions worldwide. In 1925, Rieseler recognized the potential inherent in the torque converter as both a moving-off and limited torque conversion mechanism for automatic vehicle transmissions. In the USA, the technology for mass production of hydrodynamic clutches and torque converters was developed shortly before World War II.

In order to bypass the slip necessary for power transmission in the Trilok converter, the pump and the turbine have been fitted with a lock-up clutch in the main driving ranges. This lock-up clutch has been slip-controlled since about 1994, thus making it possible to lock the converter even in lower gears and at low engine speeds. Developments such as the turbine torsional vibration damper or the two-damper converter have further improved the filtering of engine excitation.

**Table 1.6.** Chronology of the development of torque converters, clutches and their use in conventional automatic transmissions

| | |
|---|---|
| **1900** | Steam turbines start to replace steam engines. Ship propulsion systems require a reversible reduction gearbox approx. 1:4 for several 1000 hp between the turbine and the propeller |
| **1902** | *Föttinger* is commissioned by the "VULCAN" shipyard where he works to study this problem; the largest gearwheel transmissions deliver only 400 hp |
| **1905** | *Föttinger's* patent specification on 24 June, with the basic idea of hydrodynamic power transmission. Integration of pump and turbine to reduce losses, German Patent No. 221422 |
| **1910** | German Patent No. 238804 for hydrodynamic clutch = converter without reactor |
| **1917** | Gearwheel transmissions catch up with and displace torque converters in marine engineering. But the significance of the hydrodynamic clutch continues to increase |
| **1925** | *Rieseler*, a colleague of Föttinger, builds and tests an automatic vehicle transmission with torque converter and planetary gear unit |
| **1928** | The *TRILOK* consortium in Karlsruhe (Spannhake, previously a colleague of Föttinger, Kluge and van Sanden) develop the Trilok converter. Both phases run in a single fluid circuit, first the torque phase ($\eta_{max} = 0.8–0.9$) and then the clutch phase ($\eta_{max} = 0.98$) |
| **1939** | *General Motors* develops the first mass-produced (10 million) fully automatic vehicle transmission, the Hydramatic, with hydrodynamic clutch |
| **1948** | *Dynaflow* transmission by GMC with 4-phase torque converter |
| **1955** | *Borgward* Borgward builds the first automatic mass-produced transmission in Germany, with hydrodynamic converter and rear-mounted 2-speed transmission |
| **1961** | The first in-house development by *Daimler-Benz*. Hydrodynamic clutch with rear-mounted 4-speed 2-range planetary transmission |
| **1965** | 3 HP 12 from the gear manufacturer *Zahnradfabrik Friedrichshafen AG*: Trilok sheet metal converter with rear-mounted 3-speed Ravigneaux planetary gear set |
| **1965** | Trilok converter with lock-up clutch for commercial vehicle torque converter clutch transmission. Cast pump, sheet metal turbine |
| **1980** | Trilok converter with lock-up clutch for automatic passenger car transmission |
| **1994** | *ZF* Slip-controlled lock-up clutch in passenger car transmission 5 HP 30, lock-up also in lower gears |
| **1996** | *LuK* Turbine torsional vibration damper, closing of the lock-up clutch at low engine speeds |
| **2006** | *ZF-Sachs* Two-damper torque converter for broadband filtering of engine excitation |

### 1.2.6 Investigation of Phenomena: Transmission Losses and Efficiency

For the successful and reliable utilization of automotive transmissions, a great variety of phenomena need to be researched. Hertzian stress, tooth root strength, elasto-hydrodynamic lubrication and operational fatigue strength are just a few examples.

One representative example of historic development is the phenomenon of friction. Heat is generated in transmissions by friction. Friction occurs when tooth flanks and bearing parts make rolling or sliding contact from shifting and from circulating, flowing oil.

Heat generation in transmissions was soon a matter of great interest. Determining transmissions losses, i.e. toothing, bearing and churning losses, became increasingly important. Inquiry into the friction coefficient along the contact path became topical. An understanding of the transmission's efficiency and how this is related to design, load and speed is essential for any energy-saving measures. Table 1.7 provides an overview of research into these phenomena.

**Table 1.7.** Chronology of the development of research into transmission loss phenomena

| | |
|---|---|
| **1869** | *Reuleaux* First formulations to determine frictional work losses |
| **1883** | *Ernst* Losses in spur gears and perpetual screws |
| **1886** | *Lewis* Measurement of efficiency of worm gears |
| **1911** | *Rickli/Grob* Measuring loss in transmissions with a torque test bench. The reading is the actual loss, and no longer the input and output power |
| **1946** | *Hofer* Approximation formula supported by measurements for calculating the efficiency of a gear stage |
| **1954** | *Niemann* develops a formula for calculating efficency $\eta = 1 - \frac{P_V}{P_1}$ $\eta = 1 - \frac{i \pm 1}{7\,i\,z_1}$ |
| **1960** | *Niemann, Ohlendorf* Systematic experiments and calculations to determine transmission losses. Gear losses in the mixed friction area (power loss through dry friction), information on churning losses and bearing losses |
| **1965** | *Hill* investigates the connection between gearing geometry and efficiency; he calculates the transmission efficiency at a constant average coefficient of friction |
| **1967** | *Lechner* Scuffing resistance with spur gears made of steel. Heat generation in gearwheels. Investigation of the phenomenon of gear scuffing as a function of gearing geometry and operating conditions |
| **1971** | *Duda* Detailed analysis of the influences of tooth geometry on efficiency |
| **1972** | *Schouten* Rolling, sliding action as elasto-hydrodynamic problem |
| **1975** | *Rodermund* Elasto-hydrodynamic lubrication with involute gearwheels. Losses with variable coefficient of friction along the contact path |
| **1980** | *Lauster* Investigation and calculation of the thermal economy of mechanical transmissions |
| **1982** | *Walter* Investigation of splash lubrication of spur wheels at circumferential speeds of up to 60 m/s |
| **1985** | *Funk* Heat dissipation in transmissions under quasi-static operating conditions |
| **1988** | *Mauz* Hydraulic losses of spur gear systems at circumferential speeds of up to 60 m/s |
| **1990** | *Greiner* Investigation of lubrication and cooling of injection-lubricated spur gear systems |

### 1.2.7 Historical Overview

The development of automotive transmissions can, historically speaking, be split into four stages:

**Ca. 1784 to 1884** Recognition of the fact that the torque/speed characteristics of steam engines and internal combustion engines must be adapted to the load by means of a transmission in order to obtain the maximum power. The first solutions were variable-speed transmissions with sliding or constant-mesh gears.

**Ca. 1884 to 1914** Hunt for the correct principle for torque/speed conversion. Besides toothed gearings, a great diversity of transmission designs was attempted: chain, belt and friction gears, electric, hydraulic and even pneumatic transmissions, geared transmissions and especially continuously variable transmissions were tested. All the while, every transmission design was specially tailored for a particular vehicle.

**Ca. 1914 to 1980** Geared transmissions became more accepted because of their high power/weight ratio. The notion of standardized gearboxes that could easily be modified for use in different vehicles became established. Their development has continued through the subsequent decades up to the present time in terms of service life, reliability, noise level and ease of operation (synchromesh, conventional automatic transmission, shifting with uninterrupted traction, semi-automated transmission with electronically controlled shift aid). The number of speeds and the overall gear ratio constantly increased. Mass increases in motorization have been a crucial impetus behind the development of service reliefs for passenger cars.

**Ca. 1980 to date** The main focus of further research has been "individual" solutions tailored to particular uses (see also Chapter 2.5 "Trends in Transmission Design"). The palette of transmission designs has gotten much larger. Alternative transmission designs for passenger cars are competing with each other: manual transmissions, automated manual transmissions, dual clutch transmissions, conventional automatic transmissions, continuously variable transmissions and hybrid drives. Geared transmissions have 5–8 speeds. All-wheel technology has gained in importance. In the case of commercial vehicles, geared transmissions have 6–16 speeds and the greatest possible overall gear ratios. In the European heavy-duty commercial vehicle sector, automated manual gearboxes have become widespread. Now even commercial vehicles have attained a high level of operational comfort and can be driven by practically anyone. There are also important developments in both passenger and commercial transmission technology in the fields of electronics, software, function development as well as in system and information networking.

# 2 Overview of the Traffic – Vehicle – Transmission System

*Communication and mobility are the prerequisites of all human interaction! /Walter Koch, 1980/*

## 2.1 Fundamental Principles of Traffic and Vehicle Engineering

The interrelations between traffic and traffic engineering and the economy as a whole are as close as they are fundamental. Transport processes have a *basic economic function* similar to that of money, without which a modern economy, based on the division of labour and with complex system processes, cannot function. As an example of this interrelation, Figure 2.1 shows a constant increase in goods traffic performance, both universally and with reference to the population of Germany. The lion's share of this goods traffic takes place on the road.

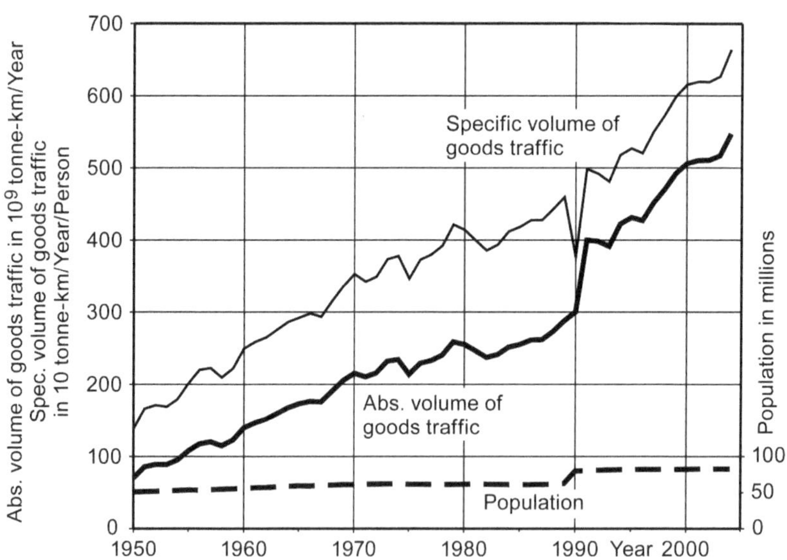


**Fig. 2.1.** Example: Development of goods traffic and population in Germany; figures for the whole of Germany from 1990 [2.5, 2.14]

H. Naunheimer et al., *Automotive Transmissions*, 2nd ed.,
DOI 10.1007/978-3-642-16214-5_2, © Springer-Verlag Berlin Heidelberg 2011

Automotive transmissions are a sub-system embedded within the transport system "road traffic". This system is characterised by the following factors:

*Person ⇔ Vehicle ⇔ Road ⇔ Traffic Volume ⇔ Goods in Transit*

There is a conflict of goals here that must be considered (Figure 2.2). If the individual wants to increase his or her own quality of life, this is only initially positive for the quality of life of society as a whole as well. Should everyone attempt to increase his or her own individual quality of life without compromise, the overall quality of life of that society will suffer. This conflict of goals is becoming especially obvious today in consideration of the traffic and environment issue.

On the subject of the **Road Traffic Transport System**, H. J. Förster writes as follows [2.7]:

> *"Since humanity, with all its wishes and needs, far outweighs all other interests, optimising the system is not necessarily the same thing as optimising transportation performance. Both people actively using the traffic system as well as others simultaneously suffer from its ill effects. Classic measures of transport effectiveness, such as transportation volume (passenger kilometres), and the cost and speed of travel, should therefore become secondary considerations. Priority has to be given to more complex human criteria such as journey quality, human satisfaction, and especially environmental impact. For goods traffic however, economic factors such as transportation volume (tonne-km), transport costs (cost per tonne-km) and journey speed (km/h) continue to outweigh considerations of social and environmental impact."*

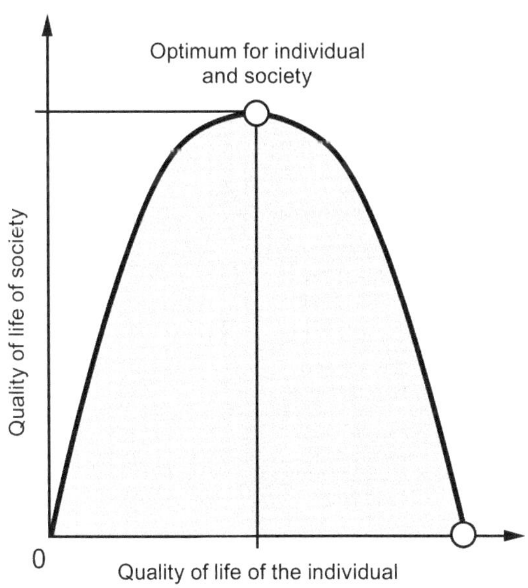


**Fig. 2.2.** Relationship between the quality of life of a society and that of an individual

### 2.1.1 The Significance of Motor Vehicles in our Mobile World

Mobility is a basic, age-old human need. Two factors influence human choice of means of transport. One is actual satisfaction of his objective needs, such as transportation performance, door-to-door access and destination attainability. The second factor is the satisfaction of perceived subjective needs such as comfort, convenience, and freedom to decide the means, destination and timing of the journey. Individual mobility by motor vehicle is also an expression of our free social order. Individual traffic is stochastic; it is neither determinable nor subject to a planned economy. Public transport is determinable. Its use can be planned.

According to Helling [2.9], we can sketch out situations and development goals for road traffic if we consider it as a black box (see Figure 2.3) and compare the costs with the results. This simplified approach leads to the task of realising the desired transport performance with minimal negative side-effects and at low resource costs. The resources required to manufacture motor vehicles are characterised as ambivalent insofar as they also add economic value and help create jobs (Figure 2.4).

The economic importance of the automotive industry is enormous, in terms of both employment and human sustenance. For example, in the year 2005, every 7th German citizen made his or her living from the automobile industry! The sales volume of the automotive industry in Germany is twelve times higher than that of the machine tool industry.

A company that mainly produces motor vehicles or products for the motor vehicle industry could find no other products with the same production volume to provide nearly as many jobs.

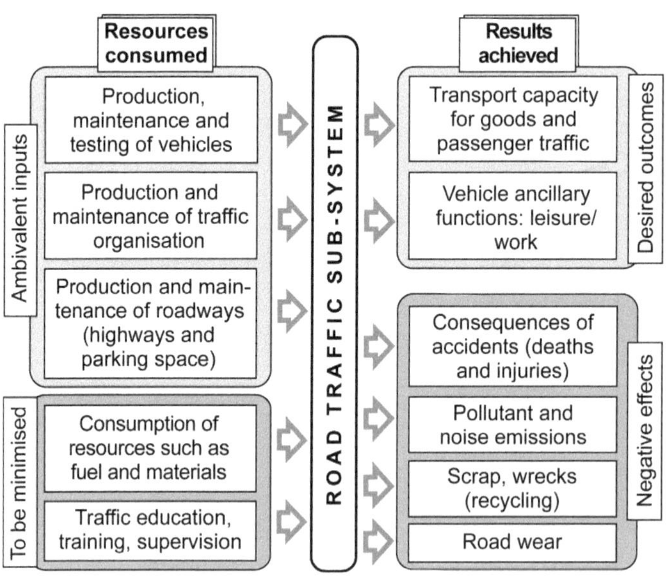


**Fig. 2.3.** Resources consumed and results achieved by the road traffic system [2.9]

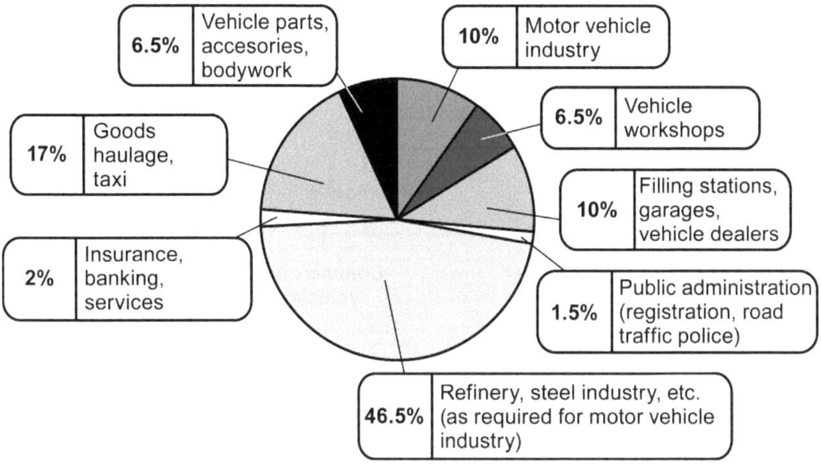


**Fig. 2.4.** Breakdown of jobs dependent on the motor transport industry in Germany

Motor vehicles have gained great significance for humanity. They should serve to improve each individual's quality of life.

There are no alternatives to motor vehicles in sight. The trend towards increasing motorisation (Figure 2.5) continues unabated despite the threat of gridlock. No alternative system presently existing or under development points towards the possibility of replacing the motor vehicle in the forseeable future.

The rekindled pioneer spirit of the railways, with high-speed trains and rail trailer shipment for long-distance haulage, is pointing in the right direction. For the time being however, the motor vehicle fulfills not only the basic human need for mobility, it also allows door-to-door transport of both people and goods.

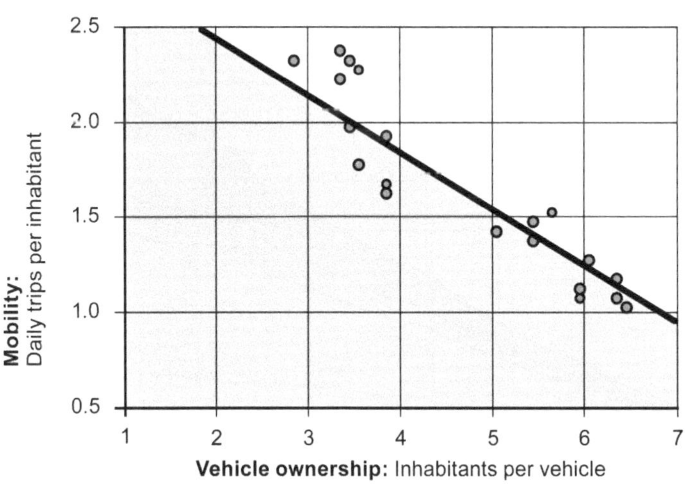


**Fig. 2.5.** Relationship between mobility and vehicle ownership [2.8]

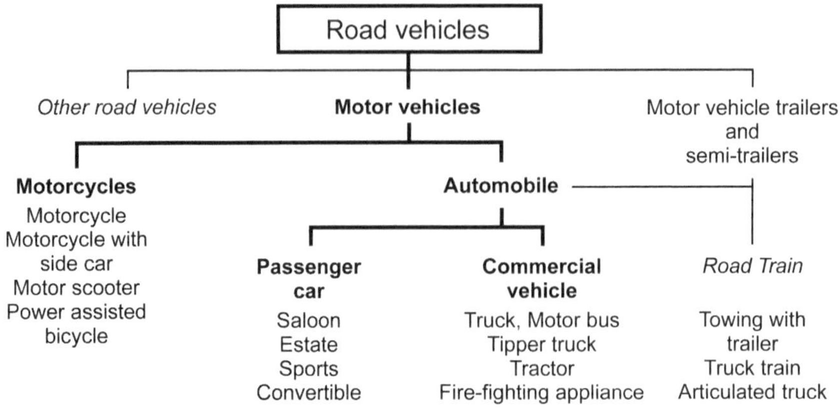


**Fig. 2.6.** Classification of road vehicles and motor vehicles to German standard DIN 70010 [2.2]

Motor vehicles play a dominant role in both passenger and freight transport. The various vehicle types are shown in Figure 2.6.

Statistics show that motor vehicles are an unhalting trend. Neither environmental destruction nor the threat of gridlock have discouraged us from our craving for mobility. Since 1946, the number of motor vehicles in the world has risen at a rate of 10% per year (Figure 2.7).

Despite declines caused by the World Wars, we can still see an average yearly increase of 9% between 1907 and the German reunification in 1990 (Figure 2.8). Within the same time period, the number of motor vehicles per capita rose from 0.00044 to 0.52345.

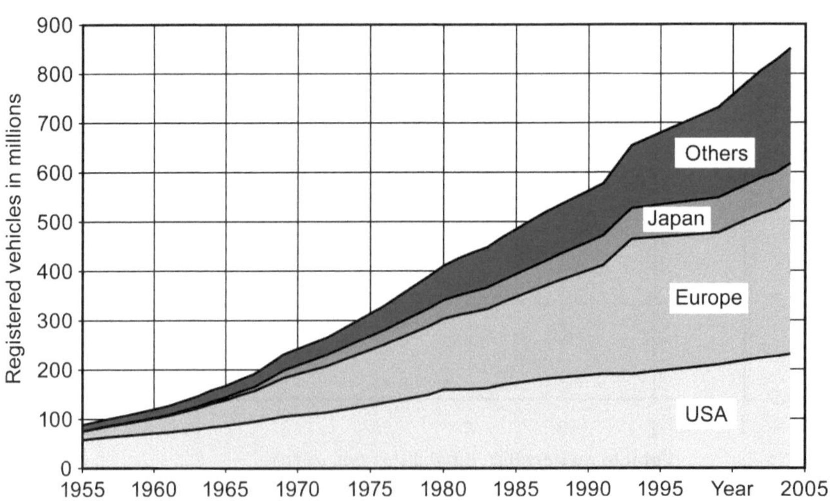


**Fig. 2.7.** Increase in the number of motor vehicles world-wide [2.16]

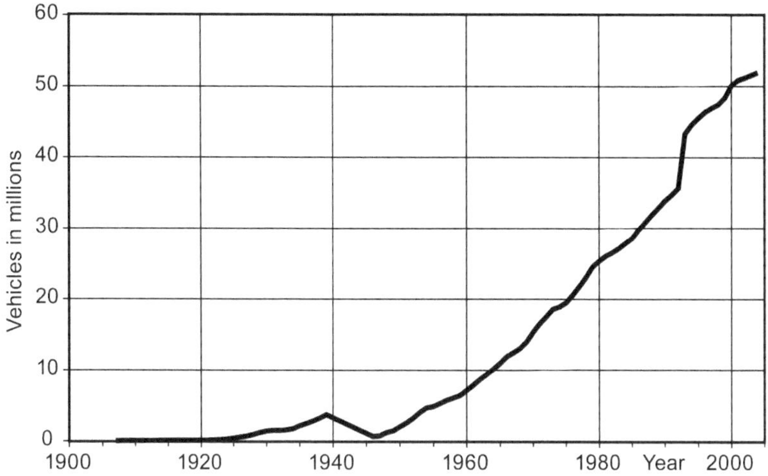


**Fig. 2.8.** Growth of the number of vehicles in Germany [2.16]

Thus, in 1990 about every second inhabitant of Western Germany possessed a motor vehicle. In 2004, the total number of motor vehicles per capita in Germany rose to 0.63008 with 52 million vehicles in total (Figure 2.9).

Even with the intensive promotion of mass transit, the demand for motor vehicles, especially in Eastern Europe and Asia, will probably continue to climb, however not to the same extent as in industrial countries previously. It is therefore imperative that the vehicles exported to and produced in these developing countries be as resource-conserving and efficient as possible.

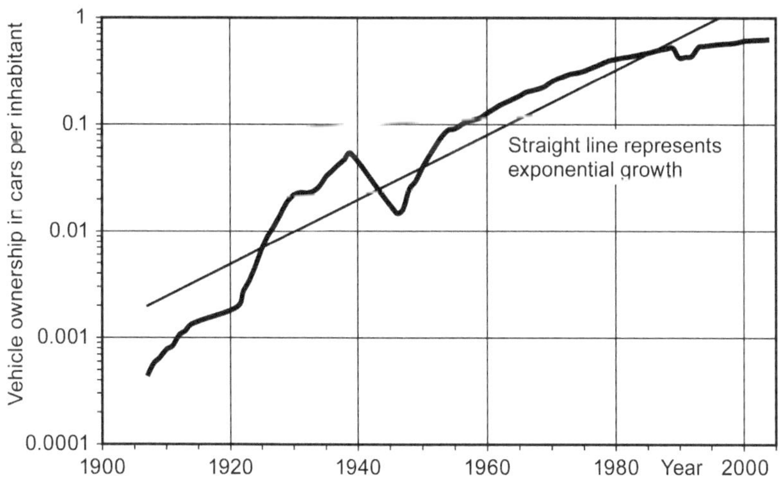


**Fig. 2.9.** Growth in vehicle ownership in Germany [2.16]

The prognosis for the **road traffic transport system** is therefore:

> *The significance of the motor vehicle will, despite some shifts within particular transport areas, not change in the longer term. On the one hand, the motor vehicle is an excellent answer to both the passenger's need for mobility and the necessity of flexibility in goods transport. On the other hand, infrastructure expenditure permits only slow and gradual change because of the complex established structure of traffic systems* [2.7].

### 2.1.2 Trends in Traffic Engineering

Traffic is the sum of all processes serving to overcome distance, comprising all relocation of persons, goods and information. The environment of the automotive transmissions as a product is determined by the traffic system. We differentiate between five categories of traffic:

- *local traffic*: urban traffic,
- *regional traffic*: traffic within agglomeration areas,
- *long-distance traffic*: traffic between agglomeration areas,
- *continental traffic*: long-distance traffic and
- *intercontinental traffic*.

Specifically for passenger and commercial vehicle transmissions, we only distinguish between local and long-distance traffic, while in the case of buses there is a three-fold distinction between city traffic, local traffic and long-distance traffic (coaches). This traffic structure also has a perceptible effect in the development and design of vehicle transmissions.

The parameters of traffic performance are extensively defined in Table 2.1. The most salient figures are the level of vehicle ownership and the passenger kilometres and tonne-kilometres per year as measures of passenger and goods traffic performance.

The term "transportation performance" defines traffic performance as relates to goods traffic (Table 2.1/5b). In every historical epoch, the gradient of increase of specific transportation performance has been greater than the gradient of population growth, i.e. goods consumption per capita and the haulage distance have always grown more quickly than the population.

Developmental trends in modern traffic engineering derive from the solutions to four main problems:

- satisfying all transportation requirements,
- increasing the environmental friendliness of the means of transportation,
- reducing primary and secondary energy consumption and
- realising the potentials of electronic communication.

The most varied means of transportation such as road, rail, canal and pipeline can be distinguised into transport systems according to their purpose or their technology.

**Table 2.1.** Measures of traffic performance

| Name | Definition | Calculation | Unit |
|---|---|---|---|
| 1. Vehicle population *MVP* | Total number of vehicles circulating in a region or state | $MVP$ | Motor vehicle |
| 2. Level of vehicle ownership *MVO* | Number of vehicles per head of population in a region or state | $MVO = \frac{MVP}{Inh}$ | $\frac{\text{Motor vehicles}}{\text{Inhabitants}}$ |
| 3. Total volume of traffic *VT* | Number of vehicle journeys in a particular period | $VT = \sum_{i=1}^{n} f_i = \sum_{i \in I} f_i$ | $\frac{\text{Journeys}}{\text{Year}}$ |
| 3a. Passenger traffic $VT_P$ | Number of car and bus trips in a particular period | $VT_P = \sum_{i \in I_{Car}} f_i + \sum_{i \in I_{Bus}} f_i$ | $\frac{\text{Journeys}}{\text{Year}}$ |
| 3b. Individual traffic $VT_{In}$ | Number of car trips in a particular period | $VT_{In} = \sum_{i \in I_{Car}} f_i$ | $\frac{\text{Journeys}}{\text{Year}}$ |
| 3c. Goods traffic $VT_G$ | Number of truck trips in a particular period | $VT_G = \sum_{i \in I_{Com.veh.}} f_i$ | $\frac{\text{Journeys}}{\text{Year}}$ |
| 4. Transport volume *TV* | Weight of transported goods in a particular period | $TV = \sum_{i \in I_{Com.veh.}} \sum_{k=1}^{f_i} g_{ik}$ | $\frac{\text{tonne}}{\text{Year}}$ |
| 5. Total traffic performance *TP* | Total number of km travelled by all vehicles in a particular period | $TP = \sum_{i \in I} \sum_{k=1}^{f_i} s_{ik}$ | $\frac{\text{km}}{\text{Year}}$ |
| 5a. Passenger traffic $TP_P$ | Km travelled by passenger vehicles multiplied by the number of occupants | $TP_P = \sum_{i \in I_{Car} \cup I_{Bus}} \sum_{k=1}^{f_i} s_{ik} p_{ik}$ | $\frac{\text{Pkm}}{\text{Year}}$ |
| 5b. Goods traffic $TP_G$ | Km travelled by goods vehicles multiplied by the weight of the load | $TP_G = \sum_{i \in I_{Com.veh.}} \sum_{k=1}^{f_i} s_{ik} g_{ik}$ | $\frac{\text{tonne-km}}{\text{Year}}$ |
| 6. Total spec. traffic performance | Traffic performance as above, but related to the number of inhabitants | $TP_{spec} = \frac{1}{Inh} \sum_{i \in I} \sum_{k=1}^{f_i} s_{ik}$ | $\frac{\text{km}}{\text{Inh. x Year}}$ |
| 7. Transport flow *TF* | Effective volume per hour of a traffic conduit | $TF$ | $\frac{m^3}{h}$ |
| 8. Specific transport flow $TF_{spec}$ | Transport flow related to the cross-sectional area required by the traffic conduit | $TF_{spec} = \frac{TF}{A}$ | $\frac{m}{h}$ |

**Observations:** Time interval considered $\Delta t = 1$ year
*MVP*: Motor vehicle $i \in 1,\ldots,MVP = I$ where $I = I_{Car} \cup I_{Bus} \cup I_{Com.veh.}$
$Inh$ : Inhabitants $j \in 1,\ldots,Inh$
$f_i$ : Number of trips of the i-th motor vehicle/$\Delta t$ $k \in 1,\ldots,f_i$
$s_{ik}$ : Journey length of the k-th journey of the i-th vehicle
$p_{ik}$ : Number of people of the k-th trip of the i-th vehicle
$g_{ik}$ : Weight of the load of the k-th trip of the i-th vehicle

A transport system consists of:

- means of transport:
  - transport medium (vehicle),
  - transport infrastructure (road, track, and rail) and
- transport organisation (operational control, administration).

Vehicle transmissions are therefore components of a transport system. Factors influencing this system are environmental preservation, market needs, legislation and individual customer demands.

### 2.1.3 Passenger and Goods Transport Systems

Traffic engineering is aimed at developing and providing reliable transport systems and means of conveyance.

We distinguish between passenger and goods traffic. The most important means of transport for passenger traffic are walking, bicycles, motorcycles, private cars, taxis (passenger cars on demand), public transport (local bus and urban rail transport), railways, airplanes and ships. Figure 2.10 illustrates how the largest percentage of passenger traffic is carried by passenger cars. Passenger traffic on public transport and railway is drastically lower. If we compare the journey distance and transportation performance of the various means of transport (Figure 2.11), we can see that there are no adequate means of transport in the ranges both between 1 to 10 km and between 100 to 1000 km.

Figure 2.12 shows travel times for various means of local passenger transport in relation to the length of the journey.

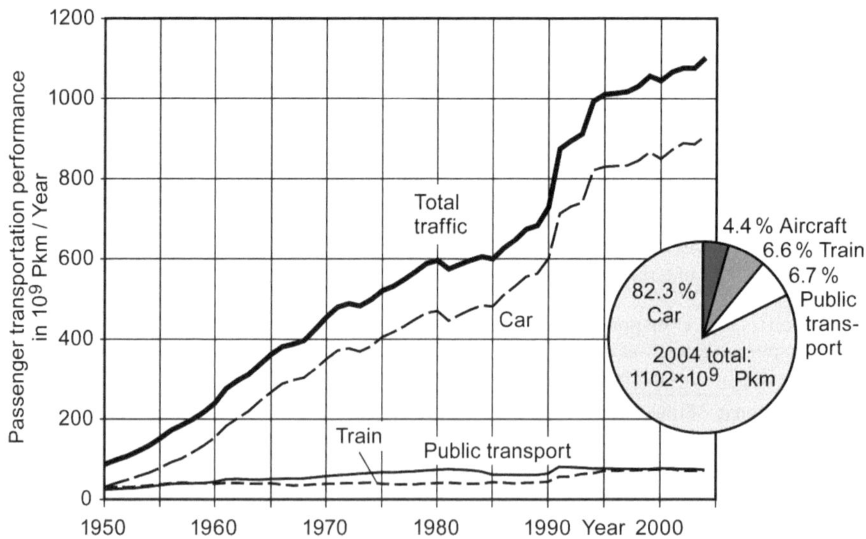


**Fig. 2.10.** Growth of passenger traffic in Germany [2.5]

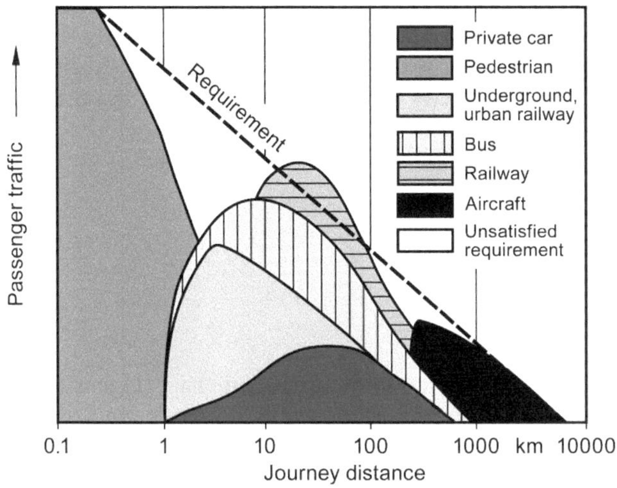


**Fig. 2.11.** Comparison of passenger transport supply and demand related to length of journey [2.9]

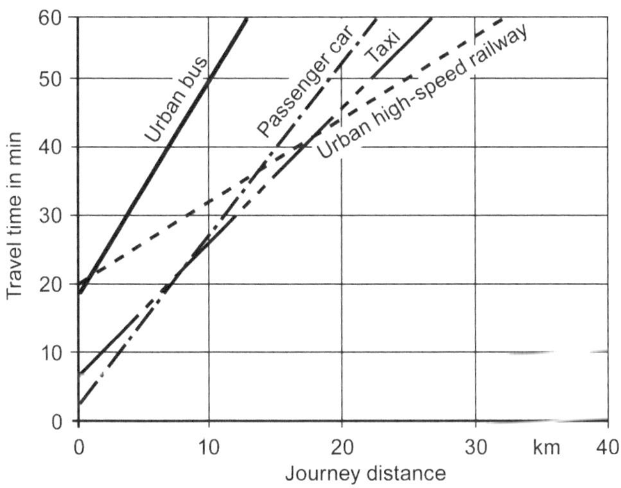


**Fig. 2.12.** Local passenger traffic journey times [2.9]

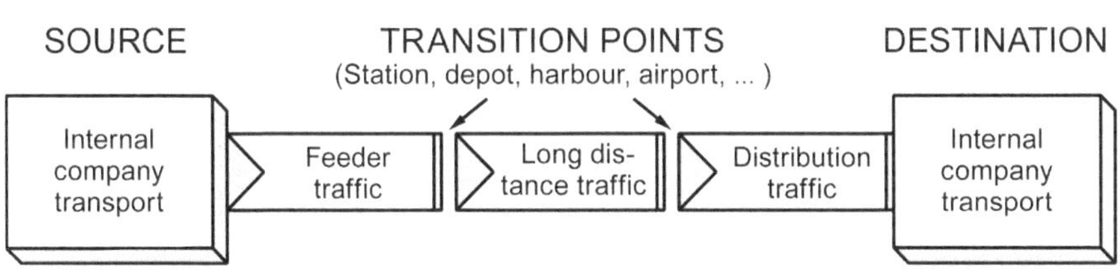


**Fig. 2.13.** Traffic distances: Diagram of a goods transport chain

Passenger cars and taxis are the fastest means of transport for short-distance passenger trips. Due to their low average driving speed, city buses are clearly slower than urban high-speed railways given equal idle times. For longer journey distances, from about 17 km, high-speed city railways provide shorter travel times than passenger cars or taxis.

For goods transport, five different means of transport are available:

- railways,
- commercial vehicles (road traffic),
- ships (canals, maritime transport),
- airplanes (airfreight) and
- pipelines.

These means of transport are often joined to form a transport chain (Figure 2.13). New approaches to limiting the amount of goods traffic on roads are urgently needed. Various characteristics such as transport speed, transport flow, space required and transport flow related to the space requirement make it possible to compare these means of transport. If we compare various means of transport with repsect to transport speed and flow (Figure 2.14), the railway performs particularly well, followed by trucks. Pipelines, however, compare unfavourably.

Of particular interest is a comparision of the specific transport flow. This indicates how fully the transportation infrastructure is being utilised relative to the cross sectional area it requires.

| **Means of transport** | **Cross-sectional profile** | **Transport speed** $v$ (km/h) | **Transport flow** $TF$ (m³/h) | **Profile surface** $A$ (m²) | **Specific transport flow** $TF_{spec} = TF/A$ (m/h) |
|---|---|---|---|---|---|
| Railway | 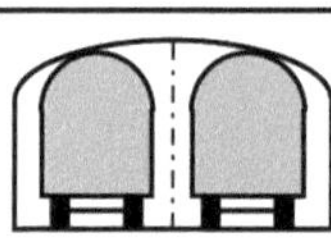 | 50 | 20 000 | 37 | 541 |
| Motorway |  | 50 | 14 500 | 115 | 126 |
| Canal | 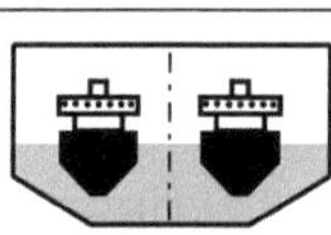 | 12 | 6 250 | 470 | 13.3 |
| Pipeline | 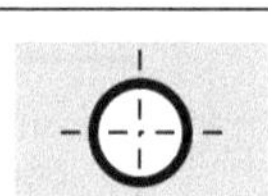 | 7.2 | 2 850 | 0.4 | 7125 |

**Fig. 2.14.** Comparison of goods transport alternatives: rail, motorway, canal, pipeline [2.8]

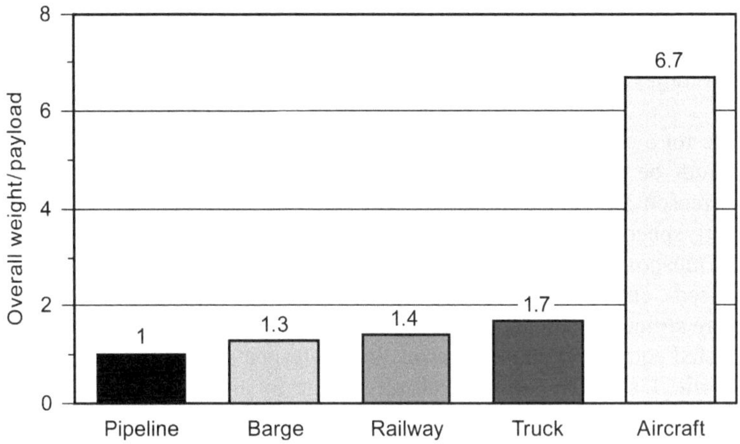


**Fig. 2.15.** Relationship between total weight and payload for various means of transport [2.8]

This comparison gives the pipeline a distinct advantage, followed by the railway and motorway. The canal ship is particularly disadvantageous in this context.

Figure 2.15 shows the efficiency of various means of transport. The ratio between total weight and payload is the most favourable for the pipeline, followed by the barge, railway and truck. In contrast, the payload ratio is clearly less favourable for aircraft.

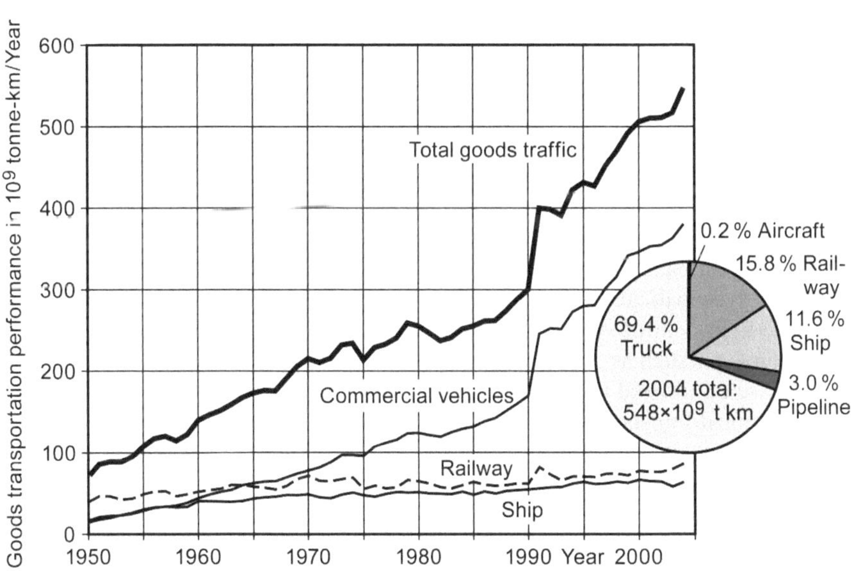


**Fig. 2.16.** Growth of goods traffic in Germany [2.5]

Commercial vehicles carry most of the annual volume of goods traffic (Figure 2.16), with rail and barge well behind. There is no immediate prospect of expanding rail freight to relieve the burden of goods traffic on our roads to a noticable extent.

A key feature for a goods transport system is door-to-door access, or target purity: can the goods be transported by the same means of transport without transhipment? The reason for the enormous increase in the number of trucks is door-to-door transport, speed and economic efficiency, and just-in-time delivery to assembly lines. Transport systems have to be assessed on the basis of satisfying transportation needs, environmental impact and energy efficiency.

First, there are structural adjustments that are necessary. In particular, road and rail must be treated equally in financial terms. In this respect, rail is at a disadvantage to road traffic. Using the road as a cheap storage facility in the just-in-time system of delivery is not economically viable in the long term as it promotes increased traffic congestion.

### 2.1.4 Alternative Transport Concepts

Innovative mass transit systems have been under consideration since around 1960. We distinguish between local transport within congested urban areas and high-speed trains for easing the burden of long-distance traffic on the roads. Prototypes of such concepts are in existence, some of them using new technologies such as maglev or air cushion technology. Some experimental tracks have been constructed, yet such concepts have not become popular. They have but few advantages compared to traditional transpost systems for the most part. This is all the more true considering the fact that the railway continues to offer further development potential.

The only one of these systems to have had success is that of the Transrapid [2.15], which went into commercial service in December 2003 in Shanghai. The Transrapis is a maglev monorail with linear motor drive. Designing and developing such systems is expensive and has to be co-ordinated internationally to make it viable. Their success depends on legislation and market acceptance.

Both conventional and innovative vehicles and transport systems for local traffic can be systematically categorised in a morphological matrix (Table 2.2) according to their control system and type of use (or availability to the user). A methodical analysis such as this helps us to develop ideas for new transport systems.

Buses are considered to have better than average development prospects for local and regional transport because of their flexibility in use, low investment costs and relatively low energy requirements. The importance of buses has been reflected in an intensive development of automatic transmissions specifically for them.

One interesting development has been that of buses that are partly guided along tracks, operating in "dual-mode". This means buses can operate both freely on conventional roads and also under guidance along special tracks.

**Table 2.2.** Classification of transport systems by control system and type of use. (Individual utility decreases from top to bottom) [2.9]

| **Control** | **Use** Individual | On demand | Planned |
|---|---|---|---|
| Free | Passenger car<br>Commercial vehicle<br>Motorcycle | Taxi<br>Demand bus | Schedule service bus |
| Dual-mode | Automated motorway | Dual-mode taxi | Dual-mode bus |
| Track-bound | Transport belt | Cabin taxi | Railway |

The benefits of such systems are reduced driver stress, tracks that are easy to build and less environmentally damaging and minimal tunnel diameters. "Demand buses" are mini-buses which can be called to bus stops. A process computer optimises transport routes within the service network and notifies the passenger of departure time.

In addition to new developments in road traffic, high-speed trains are being increasingly used that shorten rail travel times considerably. The ICE (Germany), TGV and Thalys (France), Eurostar (England and France) and AVE (Spain) form a European network of high-speed trains with a maximum speed of 300 km/h or more. These developments are characterised by high transportation performance, virtual door-to-door access, and a high level of passenger comfort.

## 2.2 The Market and Development Situation for Vehicles, Gearboxes and Components

Progressive vehicle and transmission developments must not only focus on technilogical sophistication but also be market-orientated. Vehicles and transmissions are developed cyclically and have a relatively long product and production lifecycle. Vehicle transmissions generally only require redevelopment after some 10–15 years. The transmission developer must therefore be acquainted with the market situation and be able to assess the market and changing values in society in the long term. This requires continuous observation of the market and of technological developments as well as project planning, implementation and analysis of "futuristic" projects. Incorrect product development decisions generally lead to serious financial loss.

### 2.2.1 Market Situation and Production Figures

The automotive industry is a crucial factor in the global economy. In 2004, 63.04 million motor vehicles were manufactured (Figures 2.17 and 2.18). This figure comprises 53 million passenger cars and 10 million commercial vehicles [2.16].

***Definitions:***

Passenger car: Motor vehicle designed and equipped mainly for transporting people, with a maximum of nine seats.

Commercial vehicle: Motor vehicle designed for the purpose of:
- transporting people – *Bus*;
- for transporting goods and pulling trailers – *Truck*; or just for pulling trailers – *Tractor*.
  This excludes passenger cars.

There are three main centres competing in the development of motor vehicles: Europe, the US and Japan/South Korea/China. Europe is the largest manufacturer of passenger cars (Figure 2.17). The proportion of passenger car production accounted for by small, mid-range and luxury passenger cars varies greatly in the various European producer countries. Whereas France and Italy produce mostly small and mid-range passenger cars, Germany produces a larger proportion of mid-range and luxury passenger cars. Germany produces more luxury passenger cars than the rest of Europe put together.

Every market has specific conditions essentially dictated by the socioeconomic situation of the buyers, social values, geographical factors and, not the least, by legislation. To be successful, motor vehicles have to fulfill the requirements of the respective market. This particularly applies to the transmission as the link between the road and the engine. For example, while in the US over 85% of passenger cars are equipped with automatic transmissions, it is only 13% in Europe.

For commercial vehicles of over 3.5 t gross weight, the gearbox is usually selected specifically for the particular case of application. There are often different numbers of speeds and different methods of operation (manual or automatic) available for a commercial vehicle gearbox from different manufacturers. The spectrum of types of transmission for commercial vehicles is as broad as the spectrum of applications (Figure 2.18). In the US for example, constant-mesh gearboxes are mostly used for trucks weighing over 16 t. For long stretches where no shifting is necessary, the driver is equipped with the less convenient unsynchronized constant-mesh gearbox. This is also the case in many developing countries, where driver comfort is of less concern than the longer service life of the constant-mesh transmission. In Europe on the other hand, synchromesh gearboxes dominate for heavy-duty trucks as well, accounting for about 90%.

A large proportion of European commerical vehicle production is accounted for by the class up to 3.5 t gross weight rating (Figure 2.18). These vehicles usually have 5 or 6 speed synchromesh transmissions, which are often either identical to passenger car transmissions or are modified passenger car transmissions.

Assuming that, for the sake of spare parts, 10% more gearboxes than vehicles are manufactured, the number of transmission components comprising gears and synchronizer packs can be estimated. In 2004, about 240 million gearwheels and 46 million synchronizer packs were produced for passenger car synchromesh gearboxes in Europe. For commercial vehicle constant-mesh and synchromesh transmissions, about 38 million gearwheels and 7.5 million synchronizer packs were manufactured.

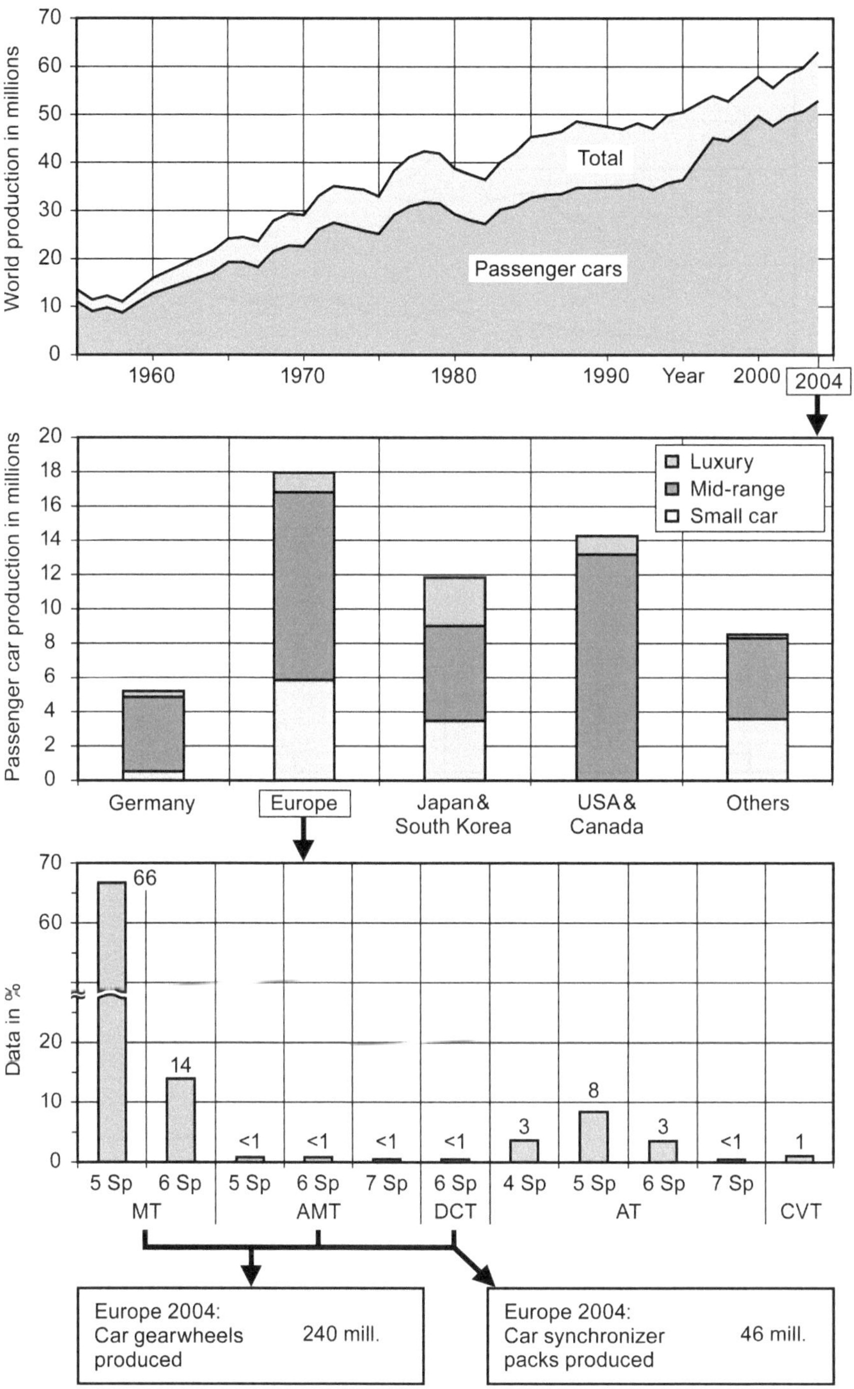


**Fig. 2.17.** Output figures for passenger cars (from [2.16])

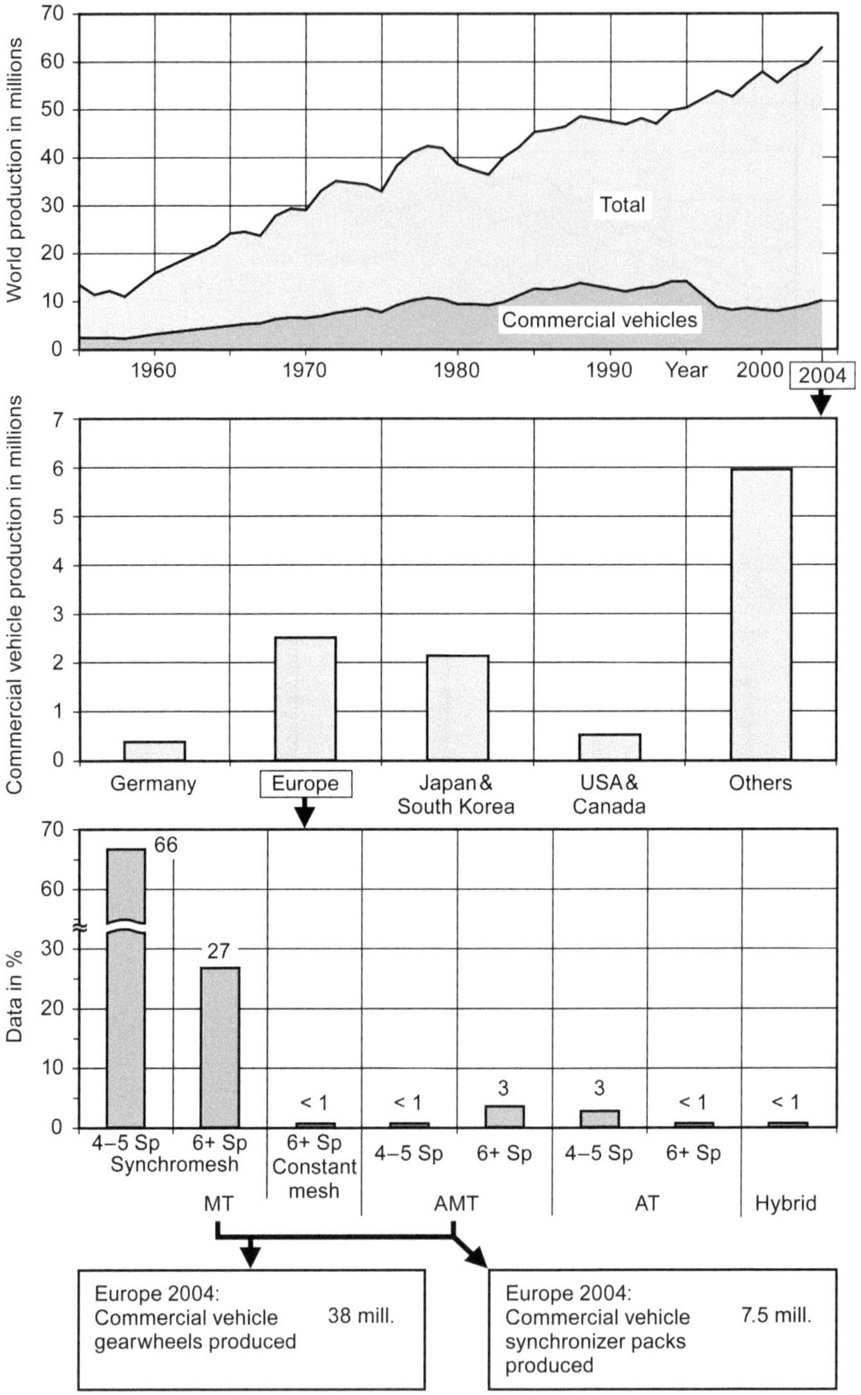


**Fig. 2.18.** Output figures for commercial vehicles (from [2.16])

**Table 2.3.** Germany's balance of trade in motor vehicles [2.16]

| | Quantity in millions | 1997 | 1998 | 1999 | 2000 | 2001 | 2002 | 2003 | 2004 |
|---|---|---|---|---|---|---|---|---|---|
| Export | Passenger cars | 2.82 | 3.27 | 3.44 | 3.46 | 3.64 | 3.62 | 3.67 | 3.67 |
| | Commercial vehicles | 0.22 | 0.24 | 0.24 | 0.27 | 0.28 | 0.25 | 0.25 | 0.25 |
| | Total | 3.04 | 3.51 | 3.68 | 3.73 | 3.92 | 3.87 | 3.92 | 3.92 |
| Import | Passenger cars | 1.95 | 2.04 | 2.17 | 2.05 | 2.09 | 1.95 | 2.03 | 2.04 |
| | Commercial vehicles | 0.22 | 0.23 | 0.23 | 0.23 | 0.21 | 0.20 | 0.19 | 0.15 |
| | Total | 2.17 | 2.27 | 2.40 | 2.28 | 2.30 | 2.15 | 2.22 | 2.19 |

These figures are indicative of the enormous economic importance of the motor vehicle. No other product supports the production of such technically sophisticated components in such quantities. There is no product in sight that could replace the motor vehicle as the engine of the economy.

Table 2.3 shows the balance of trade in recent years for German motor vehicles. In 2004, 3.92 million vehicles were exported and 2.19 million imported. Table 2.4 shows the most important independent manufacturers of automotive transmissions. Vehicle transmissions, especially those for mass-produced passenger cars, are primarily produced by the vehicle manufacturers themselves.

**Table 2.4.** Some independent manufacturers of automotive transmissions (headquarter)

| | Passenger cars | | Commercial vehicles | |
|---|---|---|---|---|
| | mechanical | automatic | mechanical | automatic |
| EU | - Getrag<br>- ZF | - ZF<br>- General Motors | - ZF | - ZF<br>- Voith |
| USA | - Transmission Technologies Corp. (TTC) | | - Eaton | - Allison<br>- Eaton<br>- Twin Disc |
| Japan | - Aichi<br>- Aisin<br>- Fuji Univance | - Aisin<br>- Jatco | - Aisin<br>- Fuji Univance | - Aisin<br>- Jatco |

### 2.2.2 Development Situation

The course of technological development has accelerated during recent years. Microelectronics is constantly finding new areas of application in vehicles, in transmissions and in their development. The pace of product development is increasingly becoming an important competitive factor for individual companies. Table 2.5 shows characteristic development times for vehicle transmissions. Current research aims to reduce these durations.

Suppliers from the Far East with longer working hours demand new development strategies from Europe. European industry must accordingly make better use of existing potential and assess its own depth of development and production [2.1].

The general principles of automotive transmission development are elaborated in Chapter 15 "The Automotive Transmission Development Process".

## 2.3 Basic Elements of Vehicle and Transmission Engineering

It is important to define both the motor vehicle and its use clearly in order to achieve a practical development of vehicle transmissions. In the following, conventions, definitions and physical foundations of automotive and transmission technologies will be outlined, forming the basis for the following chapter.

The concept of "automotive transmission" as it is used in this book comprises all components in the powertrain assembly with the exception of the engine (Figure 2.19).

**Table 2.5.** Typical development lead times for automotive transmissions

| **Development phase** | **Passenger cars** | | **Commercial vehicles** | |
|---|---|---|---|---|
| | Synchromesh transmission | Automatic transmission | Synchromesh and contant-mesh transm. | Automatic transmission |
| Concept phase | 4 | 5 | 6 | 9 |
| Design and development | 5 | 7 | 6 | 12 |
| Prototype production | 6 | 9 | 9 | 12 |
| Testing | 12 | 12 | 15 | 15 |
| Preproduction development | 9 | 15 | 12 | 18 |
| **ΣMonths** | **36** | **48** | **48** | **66** |

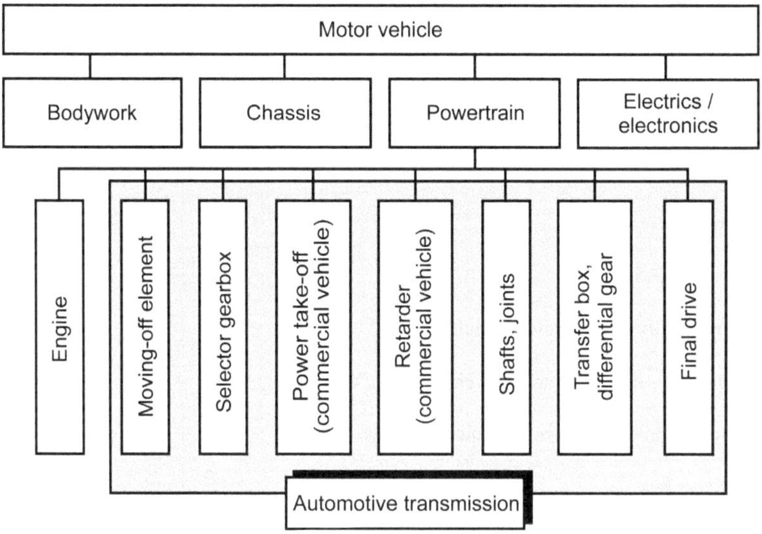


**Fig. 2.19.** Definition of the subject area of automotive transmissions

In the development of automotive transmissions, we must distinguish between variables that the designer can influence – internal factors – and those that cannot be influenced, but which still must be taken into consideration – external factors. This is shown in Table 2.6.

## 2.3.1 Systematic Classification of Vehicles and Vehicle Use

The development of a vehicle transmission conforms to the type of vehicle, its power unit and its intended use. A classification of vehicles oriented to transmission development assists systematic analysis. Table 2.7 shows a transmission-oriented classification of vehicles which has proved effective in practice. Vehicles are first divided into passenger cars, commercial vehicles, construction-site vehicles, agricultural tractors and special vehicles.

**Table 2.6.** Internal and external factors affecting the development of automotive transmissions

| **Internal factors** which can be influenced by the design engineer | **External factors** which cannot be influenced by the design engineer |
|---|---|
| • Bodywork | • Road profile |
| • Chassis | • Driving style |
| • Electrics/electronics | • Payload |
| • Engine | • Traffic conditions |
| • Automotive transmission (s. Fig. 2.19) | • Weather conditions |

**Table 2.7.** Transmission-oriented classification of motor vehicles by type of vehicle and type of use; GVW: Gross vehicle weight. *) Another feature is the number of seats

| **Type of vehicle** | | | **Type of use** | | |
|---|---|---|---|---|---|
| | | | On-road | On-/off-road (building sites) | Off-road |
| Passenger cars | | Power $P < 75$ kW | | | |
| | | Power $P > 75$ kW | | | |
| | | Vans < 3.5 t | | | |
| Commercial vehicles | Trucks | Light-duty commercial vehicles: GVW < 7.5 t | | | |
| | | Medium-duty commercial vehicles: GVW < 16 t | | | |
| | | Heavy-duty commercial vehicles: GVW > 16 t | | | |
| | Buses *) | Urban bus | | | |
| | | Interurban bus | | | |
| | | Multi-purpose bus | | | |
| | | Coach | | | |
| Agricultural tractors | | | | | |
| Construction-site vehicles | | | | | |
| Special vehicles | | | | | |

Passenger cars are split into three main groups, under 75 kW engine size, over 75 kW engine size and vans smaller than 3.5 t. The commercial vehicle category is split into buses and trucks. The truck category is further broken down by gross vehicle weight. The bus category can conveniently be broken down by stops per kilometer for urban buses, interurban buses, multi-purpose buses or coaches.

There are three basic types of use in a transmission-orientated classification of automobiles:

- *On-road use,*
- *On-/off-road use*, e.g. construction-site vehicles. This combined type of use, which is typical e.g. for dump trucks, means the transmission must provide economical propulsion both on- and off-road and
- *Off-road use*. Vehicles move predominantly off-road, possibly with occasional on-road use. This category includes tracked vehicles or extremely heavy special vehicles not permitted on normal roads, such as landfill vehicles or mining vehicles.

### 2.3.2 Why do Vehicles Need Gearboxes?

Almost all automobiles in use today are driven by internal combustion engines with cyclical combustion, working on the spark-ignition or diesel principle. The factors determining the power output and performance characteristics of internal combustion engines are explained in Section 3.3.

In addition to the many advantages of the internal combustion engine, such as high power-to-weight ratio, relatively good efficiency and relatively compact energy storage, it has three fundamental disadvantages:

- unlike steam engines or electric motors, the combustion engine is incapable of producing torque from rest (zero engine speed), see Figure 3.14,
- an internal combustion engine only produces maximum power at a certain engine speed (Figure 3.14) and
- fuel consumption is strongly dependent on the operating point in the engine's performance map (Figure 3.18).

With a maximum available engine power $P_{max}$ and a road speed $v$, the so-called "ideal traction hyperbola" $F_{Z,Aid}$ and the effective traction hyperbola $F_{Z,Ae}$ can be calculated as follows

$$F_{Z,Aid} = \frac{P_{max}}{v} \quad \text{or} \quad F_{Z,Ae} = \frac{P_{max}}{v} \eta_{tot} \,. \tag{2.1}$$

Thus, if the full-load engine power $P_{max}$ were available over the whole speed range, the traction hyperbolas shown in Figure 2.20a would result. But for the internal combustion engine the traction profile also shown in Figure 2.20a would result. The maximum traction between tyres and road is limited by the friction limit.

The problem with the internal combustion engine as a prime mover is now clear. The whole shaded area in Figure 2.20a cannot be used without an additional output converter. The output converter must convert the characteristic of the combustion engine in such a way that it approximates as closely as possible the ideal of the traction hyperbola (Figure 2.20b).

Output converter:

- *speed converter*
  ≡ mechanical or hydrodynamic clutch and
- *speed-torque converter*
  ≡ geared transmission or continuously variable transmission.

The proportion of shaded area, i.e. the proportion of impossible driving states, is significantly smaller when an output converter is used, and the power potential of the engine can be better applied.

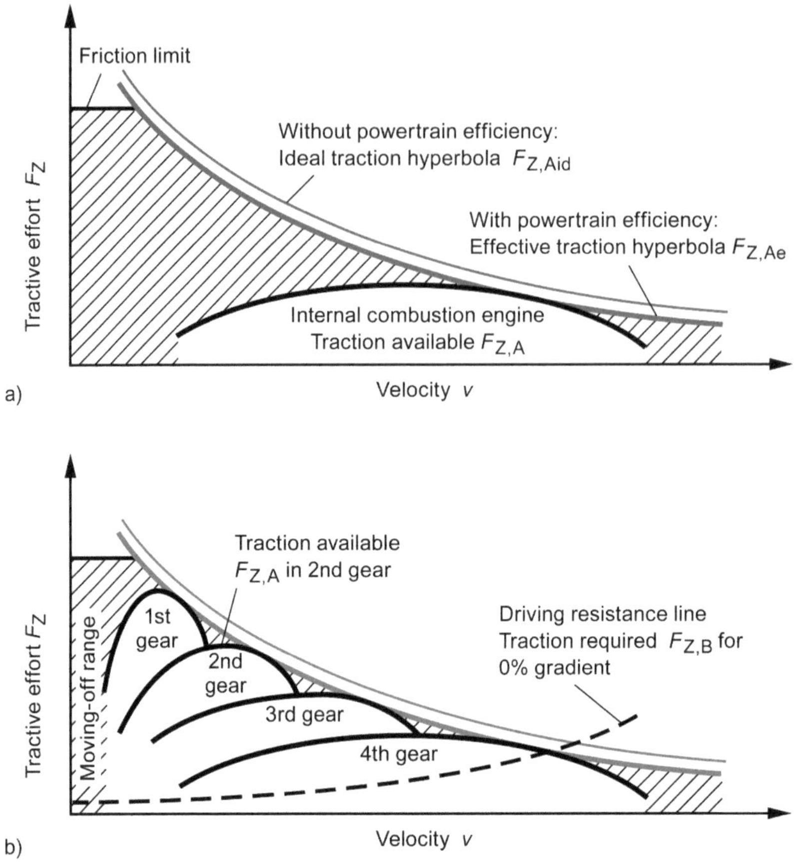


**Fig. 2.20.** *a* Secondary map of an internal combustion engine without a gearbox; *b* Secondary map of an internal combustion engine with rear-mounted 4-speed gearbox: traction diagram

Figure 2.20b shows that increasing the number of speeds as much as possible gives a correspondingly better approximation to the traction hyperbola. With continuously variable transmissions, the traction hyperbola can correspond to the traction characteristic curve over the range of ratios.

The second fundamental disadvantage of the internal combustion engine, that it does not deliver torque from rest, is overcome by means of a moving-off element (force locking clutch). The moving-off clutch (master clutch) is generally mounted between the engine and the transmission in the powertrain. See Chapter 10 "Moving-off elements".

The task of the transmission in defining engine operating points that are favourable in terms of efficiency and performance is dealt with in detail in Chapter 5 "Matching Engine and Transmission".

### 2.3.3 Main and Auxiliary Functions of Vehicle Transmissions, Requirements Profile

For transmissions to function adequately as the link between the engine and the drive wheels, it is advisable for the design engineer to consider the "vehicle transmission" as a functional whole, including gearbox and moving-off element, i.e. the whole system of adapting speed and torque, including changing gear and starting.

The four main functions of a vehicle transmission are to:

- *Enable the vehicle to move-off from rest.*
- *Adapt power flow.*
  Convert output torque $T_2$ and output speed $n_2$.
  Enable reverse motion.
- *Enable permanent power transmission.*
  Positive or force locking engine power transmission with minimal loss.
- *Control power matching.*

In addition to these main requirements there are some ancillary transmission requirements, also known as operational requirements, which substantially affect its competitiveness. The result of a survey of manufacturers and users of commercial vehicle transmissions is shown in Table 2.8.

The importance of individual ancillary requirements is displayed on a scale from 0 ≡ unimportant to 10 ≡ very important. Such a rating list of ancillary requirements is referred to as a *requirements profile.* The ancillary requirements of vehicle transmissions can be broken down as follows:

- operational reliability,
- gearbox costs,
- ease of repair,
- ease of operation,
- power matching,
- efficiency,
- installation dimensions and weight,
- customisability and
- emissions (noise, oil etc.).

Starting from the requirements profile derived from statistical surveys or empirical data, goal conflicts due to design or economic constraints can be recognised and a suitable compromise sought on the basis of the weighting allocated.

These compromises are translated into specific criteria for the development engineer, contained in the *specification,* which is presented in Chapter 15.

**Table 2.8.** A typical requirements profile for commercial vehicle transmissions. Score ≡ average assessment of importance on a scale of 0–10

| Requirements | Score | Requirements | Score |
|---|---|---|---|
| Long service life | 9.00 | Number of parts | 1.58 |
| Low repair costs | 4.89 | Power take-offs | 1.55 |
| Low production costs | 4.59 | Assembly tools | 1.55 |
| Range of ratios | 4.02 | Time to remove and replace | 1.50 |
| Gear step | 3.47 | Shift connections | 1.49 |
| Early failures unusual | 3.13 | Temperature resistance | 1.48 |
| Length | 2.92 | Crawler gear available | 1.46 |
| Long maintenance interval | 2.63 | Accessibility | 1.42 |
| Operating travel/force | 2.59 | Type of range-unit design | 1.40 |
| Low weight | 2.55 | Ratio variants | 1.39 |
| Traction constantly available | 2.47 | Method of assembly | 1.34 |
| Vibration resistance | 2.35 | Moving-off | 1.32 |
| Small number of seals | 2.33 | Spare parts procurement | 1.26 |
| Danger of operator error | 2.20 | Low power loss | 1.22 |
| Low maintenance costs | 2.19 | Clutch engagement | 1.16 |
| Overload capability | 2.08 | Low development cost | 1.16 |
| Overdrive available | 2.06 | Standard connections | 1.11 |
| Installation of wearing parts | 1.92 | Height above main shaft | 1.07 |
| Type of shift | 1.86 | Clear gearshift pattern | 1.06 |
| Owner repairable | 1.72 | Good service network | 1.00 |

### 2.3.4 Interrelations: Direction of Rotation, Transmission Ratio, Torque

The key factors in a gearbox are the direction of rotation, the transmission ratio and the torque. In order to be able to compare and assess various transmission designs and variants, we therefore need definitions to use as a standard for all considerations [2.11].

#### *Definition of Direction of Rotation*

The direction of rotation in a powertrain is defined as positive when the direction of rotation is clockwise in a right-handed Cartesian system of co-ordinates. This is as viewed against the forward direction of movement related to the vehicle, as shown in the left-hand diagram in Figure 2.21.

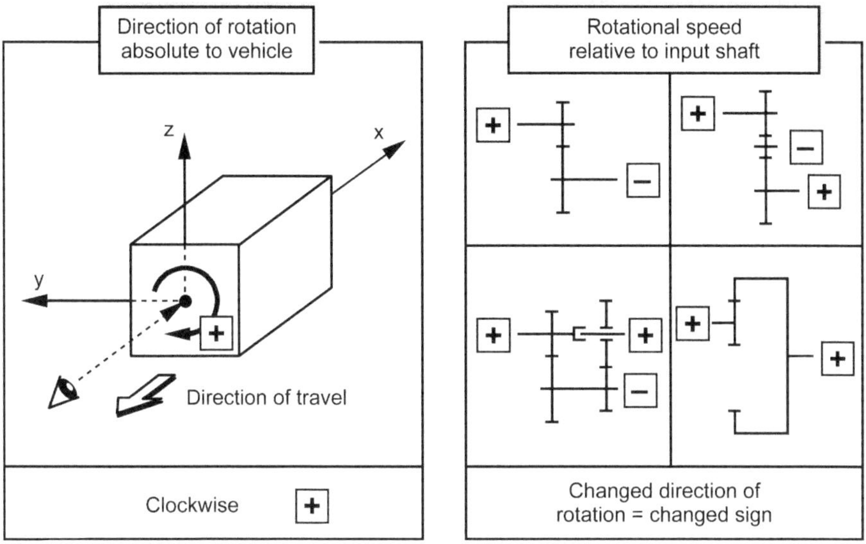


**Fig. 2.21.** Definition of direction of rotation in a powertrain [2.3, 2.4]

In the case of complicated gear plans, especially in the case of planetary gears, it is advisable to represent the speeds of rotation of the individual transmission elements with their sign and relative to each other. It is in principle of no importance which of the two possible directions of rotation is defined as positive, but normally the direction of rotation of the transmission input shaft is taken as positive (right-hand diagram in Figure 2.21).

### *Definition of Transmission Ratio*

The transmission ratio $i_G$ is the relationship between the angular velocity $\omega_1$ of the input shaft of a gearbox to $\omega_2$ of the output shaft

$$i_G = \frac{\omega_1}{\omega_2} = \frac{n_1}{n_2}. \tag{2.2}$$

The relationship between the output speed $n_2$ and the input speed $n_1$ in a powertrain component is called the speed conversion $v$ (Equation 4.2). The torque conversion $\mu$ gives the relationship between the output torque $T_2$ and the input torque $T_1$ of a powertrain component (Equation 4.3). With Equation 2.2 and the sign rules derived above, the following characteristics result for the transmission ratio:

$i_G > 0$ transmission input and output shaft rotate in the same direction,
$i_G < 0$ change of direction of rotation in the transmission,
$|i_G| > 1$ speed reducing ratio,
$|i_G| < 1$ speed increasing ratio.

In the case of continuously variable transmissions and with transmission combinations:

$i_G = \infty$ stationary output with rotating input,
$i_G = 0$ stationary input with rotating output.

The ratios inside a gearbox are designated by the gear ratio $u$. The gear ratio $u$ of a gear pair is the ratio of the number of teeth $z_2$ of the larger wheel to the number of teeth $z_1$ of the smaller wheel (pinion)

$$u = \frac{z_2}{z_1} \text{ with } z_2 \geq z_1. \tag{2.3}$$

German standard DIN 3990 specifies that in the case of spur gears the number of teeth of a wheel with external gearing is positive, and that the number of teeth of a wheel with internal toothing (ring gear) is to be taken as negative.

### Definition of Torque

Further important factors affecting a gearbox are the torque values acting on its shafts. Their directions of action must be defined by showing their sign. Here again, it is in principle of no consequence which torque direction is taken as positive.

The torque direction of the transmission input shaft is normally also defined as positive. By separating the transmission components and establishing torque equilibrium, it can be shown that the torque direction is always reversed along a free connecting shaft.

As shown in Figure 2.22, the torque direction changes along a transmission component, but the direction of rotation does not. The sign of the power $P$ absorbed (positive) or delivered (negative) at a particular point can be determined from the speed of rotation and the torque at that point in the transmission by means of Equation 2.4

$$P = T\,\omega = 2\pi\, n\, T\,. \tag{2.4}$$

Müller [2.12] proposes four key rules for speeds of rotation, torque values and power values in a transmission:

- *All parallel shafts in a transmission rotating in the same direction will have speeds with the same sign.*
- *In an "input shaft", the signs for speed of rotation and torque are the same; in an "output shaft", they are opposite to each other.*
- *"Input power" is always positive; "output power" is always negative.*
- *The two equal connecting torque values of a free connecting shaft have opposite signs at the connecting ends.*

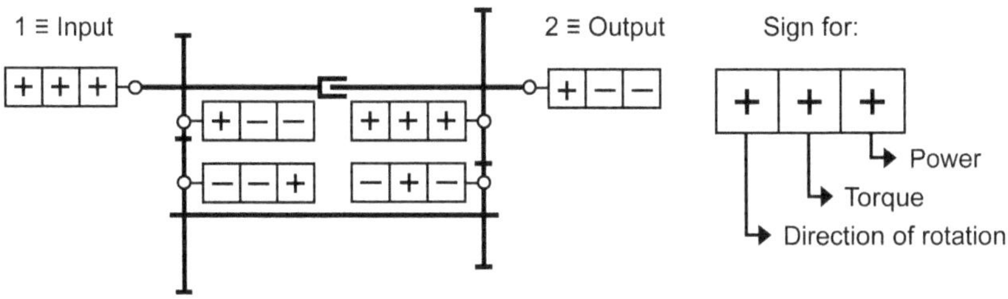


**Fig. 2.22.** Sign rules for rotational speed, torque and power

A transmission consists of at least three parts, one of which must be the "frame". This important condition is necessary to provide a reaction for the difference in force or torque between the input and output side resulting from the conversion of movement. In vehicle transmissions, the gearbox housing is the frame.

The symbolic representation proposed by Wolf [2.17], as shown in Figures 2.23 and 2.24, clearly illustrates these relations. From Equations 2.2, 2.9 and 2.10 it follows that

$$T_2 = -T_1 \frac{\omega_1}{\omega_2} = -T_1 \, i_G \,. \tag{2.5}$$

With Equation 2.7 the reaction torque $T_3$ can be calculated as

$$T_3 = -T_1 - T_2 = T_1 \left( i_G - 1 \right). \tag{2.6}$$

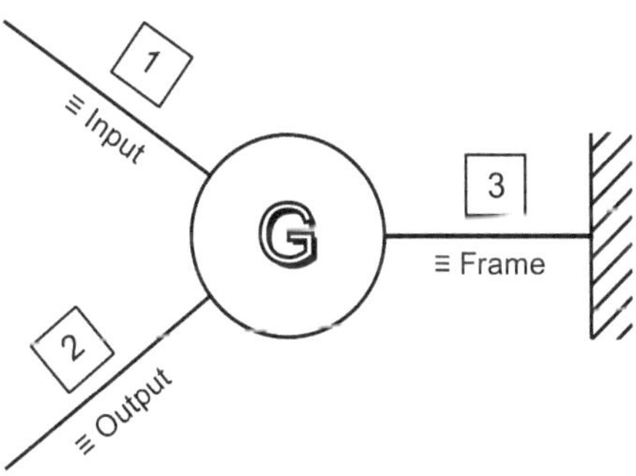


$$T_1 + T_2 + T_3 = 0 \,, \tag{2.7}$$

$$P_1 + P_2 + P_3 = 0 \,, \tag{2.8}$$

$$T_1 \, \omega_1 + T_2 \, \omega_2 + T_3 \, \omega_3 = 0 \,, \tag{2.9}$$

$$\omega_3 = \omega_{\text{frame}} = 0 \,. \tag{2.10}$$

**Fig. 2.23.** Wolf transmission symbols

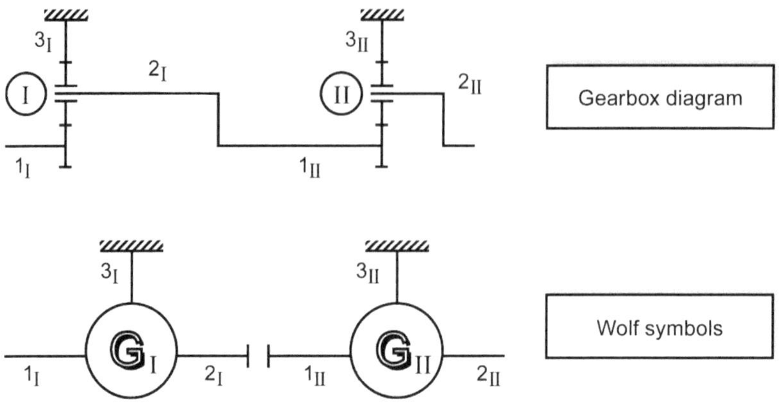


**Fig. 2.24.** Planetary coupled gear represented by Wolf symbols

Two fundamental characteristics of transmissions concerning reaction torque emerge from Equation 2.6:

- For a transmission ratio of $i_G = 1$, i.e. direct drive, the transmission takes on the function of a clutch, i.e. its frame does not have to provide any reaction torque.
- The reaction torque of the frame changes its sign, i.e. its direction, when shifting from a speed reducing gear to a speed increasing gear.

The third "frame" part can also be in the form of a second input or second output member. In these cases the term "differential drives" is used.

### 2.3.5 Road Profiles, Load Profiles, Typical Vehicle Use and Driver Types

In addition to the "internal" factors affecting the transmission (i.e. the design data of the individual vehicle sub-assemblies), the designer should be fully informed about "external" factors such as driving style, vehicle use and road type (see also Table 2.6).

This information can be acquired through field trials and customer surveys, focusing on the following criteria:

- *Road types*:
  Proportion of total mileage on various types of roads such as motorway, rural road, urban traffic or mountain road (see Table 2.9).
- *Loading*:
  Percentage distribution of journeys with numbers of passengers, cargo and trailer weight.
- *Driving style*:
  Shifting frequency, gearshift engine speed, acceleration habits in town (moving-off from traffic lights), on rural roads (when leaving built up areas) and on the motorway (overtaking).

In addition to the abovementioned criteria it is useful to classify drivers by type. Table 2.9 shows an example of driver type classification according to the percentage of roads used.

The definition of the so-called CARLOS driver (Car Loading Standard) is used to get a consistent mix of roads for the load profile. Table 2.10 shows the road mix used for determining the load profile for automatic transmissions, the CARLOS-PTA (Powertrain Automatic).

The CARLOS-PTA road mix describes percentages of road type $s_p$. For a reference route $s_{Bez}$ taken as standard for the interpretation, the respective road length results from the formula $s_U = s_{Bez} \times s_p$. Using an extrapolation factor $f_E$, the distance to be measured is $s_M = s_U / f_E$.

The following determinations apply as boundary conditions for determining the load process:

- *Style of driving*:
  Measured journey on public roads in compliance with traffic regulations. Driving style variations result from varying drivers.
- *Vehicle weight*:
  Measurements both solo (vehicle without trailer: GVW = Gross Vehicle Weight) and in combination (vehicle weight with trailer: GCW = Gross Combined Weight) in order to determine maximum stress.
- *Roads*:
  Public roads. Special off-road routes are not a part of the CARLOS-PTA road mix and must be considered seperately should the vehicle manufacturer demand it.
- *Accidental factors*:
  The traffic situation and the weather are examples of accidental factors that cannot be freely chosen. Should these factors become extreme, they are not taken into consideration so as not to distort the result.

The load patterns determined by road tests and computer simulation can be translated into load profiles using suitable classification methods. The transmission service life is then generally estimated using the Miner and Haibach damage accumulation hypothesis, based on the load profiles and the corresponding Wöhler curves.

**Table 2.9.** Breakdown of driver type in terms of powertrain loading in a passenger car [2.10]

| Driver type | Proportion of kilometres covered in % | | | |
|---|---|---|---|---|
| | Motorway | Rural road | Urban traffic | Mountain road |
| Motorway driver | 70 | 14 | 13 | 3 |
| Rural road driver | 30 | 56 | 11 | 3 |
| Urban driver | 30 | 23 | 45 | 2 |
| Mountain driver | 40 | 30 | 20 | 10 |

**Table 2.10.** CARLOS-PTA road mix for a reference route of 150 000 km [2.6]

| Road type | Proportion of road type $s_p$ (%) | Length of road type $s_U$ (km) | Length of measurement $s_M$ (km) | Extrapolation factor $f_E$ |
|---|---|---|---|---|
| Motorway | 60 | 90 000 | 900 | 100 |
| Rural road | 25 | 37 500 | 375 | 100 |
| Urban road | 40 | 60 000 | 600 | 100 |
| Mountain road | 15 | 22 500 | 225 | 100 |
| Sum | 140 | 210 000 | 2 100 | 100 |

Information on load profiles and service life is given in Section 7.4 "Operational Fatigue Strength and Service Life". The development of transmissions using computer-aided driving simulation is described in Chapter 14.

## 2.4 Fundamental Performance Features of Vehicle Transmissions

*The development process is successful only if it results in a product that sells*

Any specification for a product must be:

- suited to its intended purpose (functionally suitable),
- cost-effective (economic) and
- acceptable (environment friendly, easy to operate, and functional).

These superordinate development goals define the fundamental performance features of a transmission. Vehicle transmissions have to be designed to provide torque conversion suited to operating conditions, with low fuel consumption and at a competitive price. Specifically, this means service life appropriate to the intended use (operational fatigue strength), reliability, ease of operation, small installation space, low noise level, low weight and high efficiency (Figure 2.25).

Such performance features are often enough for a quick rough comparison of one's own designs with those of others.

### 2.4.1 Service Life and Reliability of Transmissions

Early failure and early wear impair the availability and economic efficiency of a passenger car or commercial vehicle. It is often overlooked that a slight over-rating of the transmission relative to the drive motor can achieve an enormous increase in service life. Over-rating by 10% doubles the service life, whereas the price of the transmission only increases by 10% at a first approximation (see Figure 2.30).

The standard values listed in Table 2.11 apply to service life, the most important performance characteristic of transmissions. $B_{10}$ service life refers to the service life within which 10% of the transmissions of a production batch of a given transmission type fail. With passenger car transmissions, the required $B_{10}$ service life is 150 000 km. The market requirements of commercial vehicle transmission service life are significantly higher, depending also on the particular application. In this case, the $B_{10}$ service life is between 250 000 and 1 200 000 km (Table 2.11).

The applicable load profile for the individual speeds (Section 7.4) is particularly important. It depends on the route profile, vehicle powertrain (axle ratio and dynamic tyre radius), vehicle loading (e.g. weight of load, oil sump temperature etc.) and driving style.

Transmission reliability requires thorough calculation and analysis at the design stage. See also Section 7.4 "Operational Fatigue Strength and Service Life" and Chapter 17 "Reliability and Testing of Automotive Transmissions". It must be borne in mind that a transmission is a system with mechanical, mechatronic and, if applicable, electronic components that are more or less reliability-critical.

Components such as gearwheels, shafts or bearings are now well calculable. Not only electric components of mechatronic modules (e.g. electric motors and circuit board components of transmission or clutch actuators) but also electronic components (e.g. electronic transmission control with a number of electronic components such as processors, amplifiers, capacitors, resistors etc.) must undergo an examination of their technical reliability.

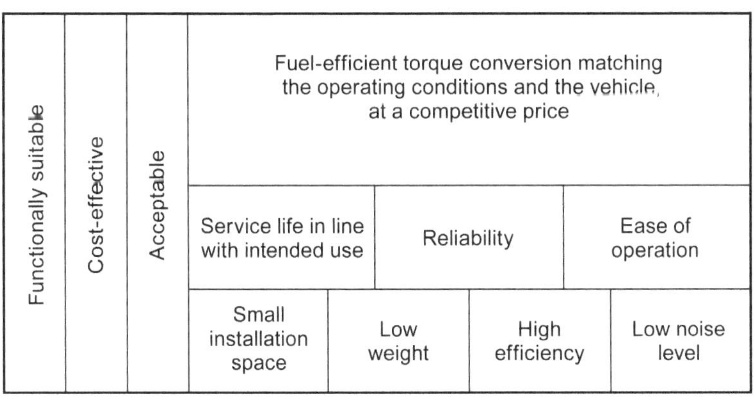


**Fig. 2.25.** Fundamental quality and performance features of vehicle transmissions

**Table 2.11.** $B_{10}$ service life of passenger car and commercial vehicle transmissions under different operating conditions

| **Operating conditions** | | **$B_{10}$ service life (km)** |
|---|---|---|
| Passenger cars | | ≥ 150 000 |
| Commercial vehicles | | |
| Light-duty commercial vehicles (and vans) | | ≥ 250 000 |
| Medium-duty and heavy-duty trucks | | |
| On building sites | (Off-/on road) | ≥ 400 000 |
| Urban traffic | (Stop and go) | ≥ 600 000 |
| Long distance | | ≥ 1 200 000 |
| Buses | | |
| Urban bus | | ≥ 700 000 |
| Interurban bus | | ≥ 1 000 000 |
| Coach | | ≥ 1 000 000 |

### 2.4.2 Centre Distance Characteristic Value

Characteristic values can be determined for transmissions and other products by using standard procedures to establish values for key basic variables of the future design quickly and without complex calculations. This method of "design using characteristic values" is ideal for use with computer-aided design. When the draft design has been transferred to the computer, the graphic and quantitative aspects need to be refined. Such characteristic values relate for example to size, mass or cost.

The centre distance $a$ of a countershaft transmission is its most important parameter. The smaller the centre distance can be with a given output torque $T_2$, the smaller the overall dimensions of the transmission. The centre distance is determined by the gear with the greatest torque multiplication $i_{G,max}$ (first gear). It is possible to gain a good impression of the order of size of a competitive centre distance before carrying out any calculation by relating the centre distances of one's own transmissions and those developed by competitors to their output torque $T_2$. This type of analysis encapsulates a great variety of production and operating experience with gearbox designs proven in practice.

Figure 2.26 is an example of such a centre distance analysis, showing the trend of the centre distance with coaxial, two-stage single-range countershaft gearboxes as well as coaxial multi-range gearboxes with one countershaft as a function of the output torque $T_2$ at the output shaft. Centre distances for passenger car gearboxes are accordingly 70–95 mm. In the case of vans and light-duty commercial vehicles, gearboxes with centre distances of 75–105 mm are used.

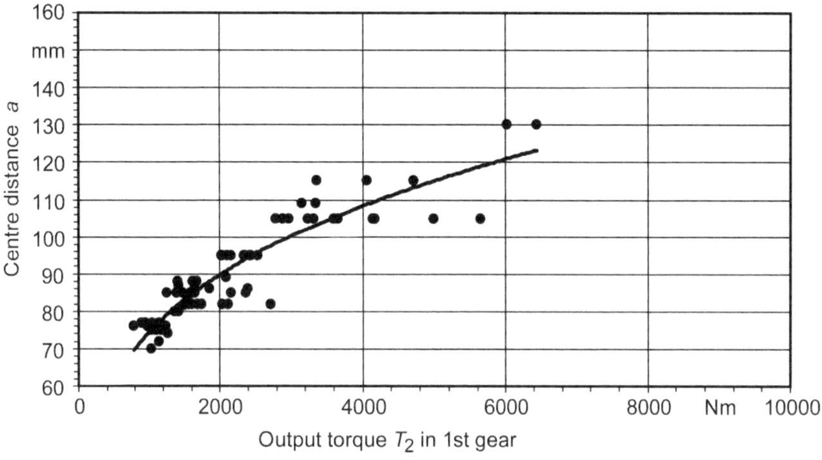


**Fig. 2.26.** Trend of centre distance *a* with coaxial two-stage single-range countershaft gearboxes for passenger and commercial vehicles as a function of the output torque $T_2$ at the output shaft

Gearboxes for medium-duty commercial vehicles have a centre distance of 100–130 mm, whereas units for heavy-duty trucks are in the range 130–160 mm. The spread is explained by differing design methods, by different applications with different load profiles, and by technical production factors. As a result of the existence of transfer lines for centre distance drill holes in the gearbox housing, it is often more economical to achieve the required service life by adapting the face widths rather than optimising the centre distance.

Another reason is in-company standardisation of centre distance intervals. By reference to Figure 2.26 and the relationship derived from it between centre distance $a$ and output torque $T_2$ in first gear, we derive for coaxial, two-stage single-range passenger and commercial vehicle countershaft transmissions

$$a = 11.483\, T_2^{\,0.271} \tag{2.11}$$

$$\text{or}\quad a = 11.483\left(i_{G,\max}\; T_1\right)^{0.271} \tag{2.12}$$

with a centre distance scope of application between 70 mm and 130 mm.

The relation of the centre distance $a$ and the output torque $T_2$ in first gear for commercial vehicle multi-range transmissions designed with a single countershaft can be seen in Figure 2.27.

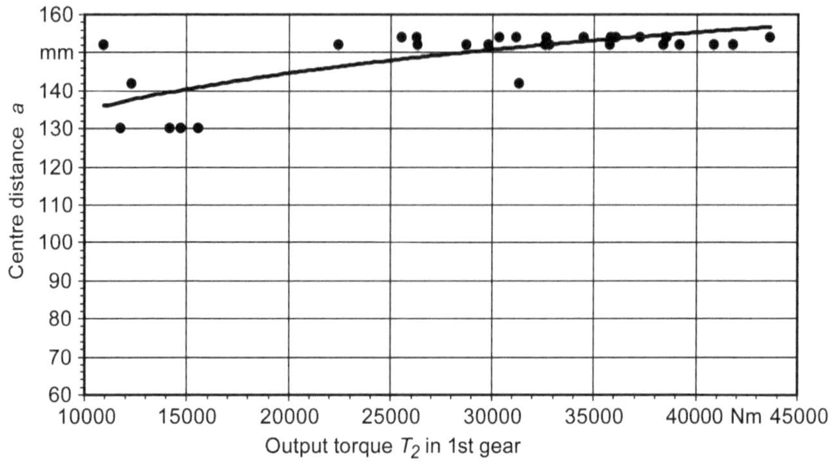


**Fig. 2.27.** Trend of centre distance $a$ with commercial vehicle multi-range transmissions with one countershaft as a function of the output torque $T_2$ at the output shaft

The associated equation in the scope of application of centre distances between 130 mm and 160 mm is:

$$a = 52.201\, T_2^{0.103} \tag{2.13}$$

$$\text{or } \ a = 52.201 \left(i_{G,\max}\, T_1\right)^{0.103}. \tag{2.14}$$

This enables centre distances in millimetres to be estimated for coaxial, two-stage countershaft gearboxes of single-range design as well as for multi-range gearboxes with one countershaft, leading to an economical gearbox size. For this purpose only the maximum transmission input torque $T_1$ in Nm and the required maximum ratio of the transmission $i_{G,max}$ need to be known. Of course such a prognosis cannot replace precise transmission design and centre distance calculation. See Chapter 7 "Design of Gearwheel Transmissions for Vehicles".

### 2.4.3 Gearbox Mass Characteristic Value

Another key performance characteristic of the transmission is its mass $m_G$. The cost of a gearbox is proportional to its weight at the first approximation. The mass of the transmission can be related to the input torque $T_1$ the maximum ratio $i_{G,max}$ and the number of gears.

Figures 2.28 and 2.29 show an analysis of a large number of transmissions that have been designed and produced. The gearbox mass $m_G$ in kg is plotted against the parameter $T_2\, z^{0.5}$ or $T_1\, i_{G,max}\, z^{0.5}$.

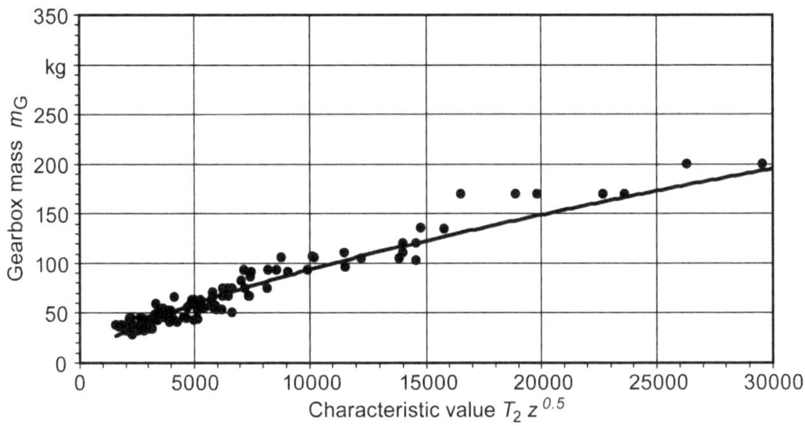


**Fig. 2.28.** Trend of gearbox mass $m_G$ for coaxial two-stage passenger and commercial vehicle countershaft transmissions of single-range design as a function of the output torque $T_2$ at the output shaft and the number of speeds $z$

The equation for coaxial two-stage passenger and commerical vehicle gearboxes of single-range design is (see Figure 2.28)

$$m_G = 0.199 \left( T_2^{\ 0.669}\ z^{\ 0.334} \right) \tag{2.15}$$

$$\text{or}\ \ m_G = 0.199 \left( \left( i_{G,\max}\ T_1 \right)^{0.669}\ z^{\ 0.334} \right). \tag{2.16}$$

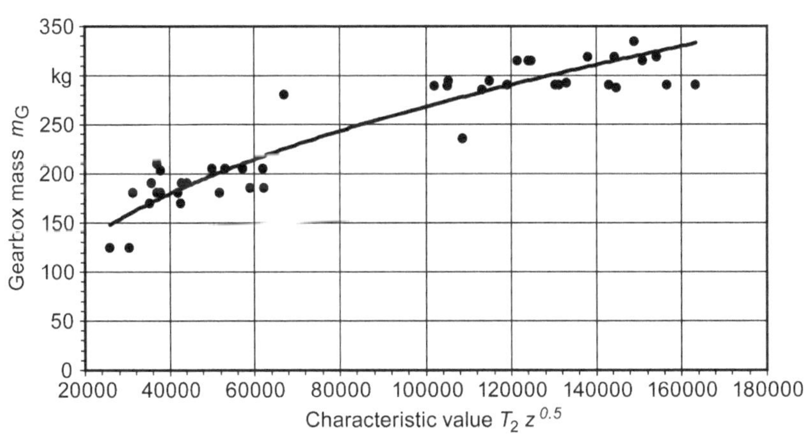


**Fig. 2.29.** Trend of gearbox mass $m_G$ for commercial vehicle multi-range gearboxes with one countershaft as a function of the output torque $T_2$ at the output shaft and the number of speeds $z$

Figure 2.29 shows the trend of the gearbox mass $m_G$ for multi-range commercial vehicle transmissions with one countershaft as a function of the parameter $T_2\,z^{0.5}$ or $T_1\,i_{G,max}\,z^{0.5}$. The equation reads:

$$m_G = 1.723\left(T_2^{0.439}\ z^{0.219}\right) \tag{2.17}$$

$$\text{or}\ m_G = 1.723\left(\left(i_{G,\max}\ T_1\right)^{0.439}\ z^{0.219}\right). \tag{2.18}$$

The diagrams in Figures 2.28 and 2.29 can be used in two ways:

- assessing the anticipated weight of a transmission, whose total ratio, input torque and number of speeds are known from the specification,
- checking whether a developed transmission is competitive in relation to its mass.

### 2.4.4 Gearbox Cost Characteristic Value

As with transmission mass, parameters can be established for transmission costs or selling price. Figure 2.30 shows an analysis of transmissions that have been built. The relative selling price *RSP* is plotted against the parameter $T_2\,z^{0.5}$ or $T_1\,i_{G,max}\,z^{0.5}$. The relative transmission selling price is based on a manually shifted passenger car transmission with an input torque of $T_1 = 350$ Nm, number of speeds $z = 6$ and a maximum transmission ratio in the 1st gear of $i_{G,max} = 5.5$, which corresponds to a relative transmission selling price of RSP = 1.

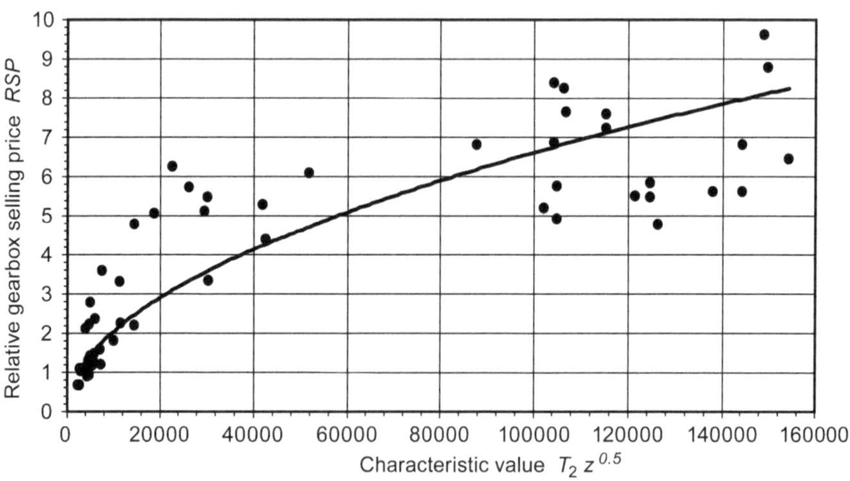


**Fig. 2.30.** Trend of gearbox selling price *RSP* as a function of the output torque $T_2$ at the output shaft and the number of speeds $z$ (Reference: manual passenger car transmission: $T_1 = 350$ Nm, $z = 6$ and $i_{G,max} = 5.5$: $RSP = 1$)

The points in this diagram can be approximated by a parabola with the equation:

$$RSP = 0.0183\, T_2^{0.512}\, z^{0.256} \tag{2.19}$$

$$\text{or}\ \ RSP = 0.0183 \left(i_{G,\max}\, T_1\right)^{0.512} z^{0.256}\,. \tag{2.20}$$

The selling price (or production cost) trend shown in the graph enables the transmission designer to estimate whether the efficiency of his transmission is worthwhile for the customer in terms of the price/performance relationship of other transmissions. It is also possible to forecast the achievable market price at the product planning stage.

The comparative cost of different types of vehicle gearboxes is also of interest. The specific design and production features and different operating conditions and safety requirements of the different types of gearboxes result in significantly different relative weight costs. The relative weight costs (EUR/kg) are shown in Figure 2.31 for eleven different types of transmissions used for different purposes.

The design basis for the relative weight costs is the passenger car synchromesh transmission with 15 EUR/kg (=100%). Commercial vehicle synchromesh transmissions have the lowest weight costs (13.5 EUR/kg at 2005 prices). Then come construction vehicle transmissions. Automated manual passenger car gearboxes are more expensive than syncromesh transmissions for passenger cars. Conventional passenger car automatic transmissions are not significantly more expensive because of the large production volume. Passenger car dual clutch and continuously variable transmissions are more expensive in comparison to conventional passenger car automatic transmissions.

On the other hand, the reliability requirements of commercial vehicle automatic transmissions for heavy load profiles and low production volumes increase weight costs considerably. Directly compared, automated manual gearboxes for commercial vehicles are significantly less expensive. Aircraft transmissions have by far the highest weight costs because of their complex weight-saving design and 100% quality monitored production.

## 2.4.5 Gearbox Noise

The transmission often generates a high proportion of noise, being the main link in the engine/powertrain/wheel chain. In addition to the general broadband noise, the meshing frequencies of gearwheel transmissions are very disturbing because of their discrete sounds. Primary noise reduction is therefore necessary at the design stage, the gear specification stage and during production, combined with quality assurance (see Section 7.5 "Developing Low-Noise Transmissions").

Because of its importance, transmission noise is subject to strict regulations (see Table 7.3), which are likely to become more stringent.

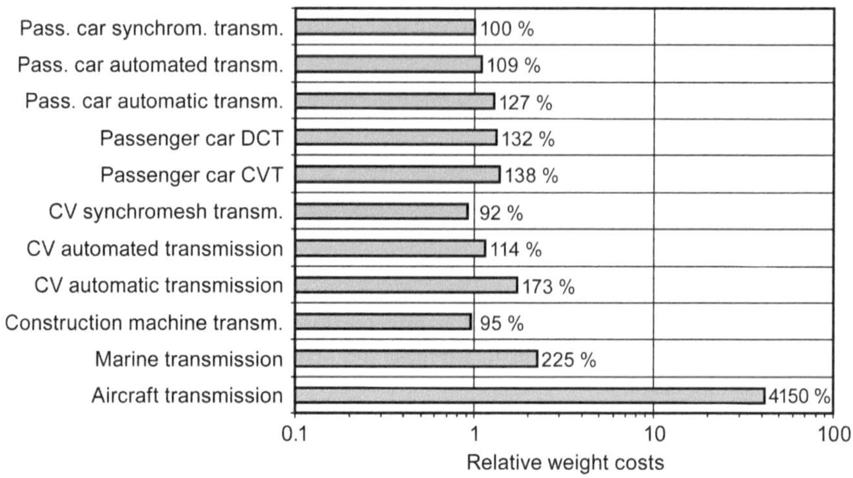


**Fig. 2.31.** Reference values for relative weight costs for various types of transmissions

### 2.4.6 Gearbox Losses and Efficiency

As part of the effort to save energy, the question of efficiency, and hence of gearbox friction losses, has received more attention. No mechanism converts torque as effectively as a gear pair in terms of production cost, converter ratio and efficiency.

Nevertheless the requirement is to determine and if possible improve the efficiency of vehicle transmissions as a function of torque, speed and other characteristic values.

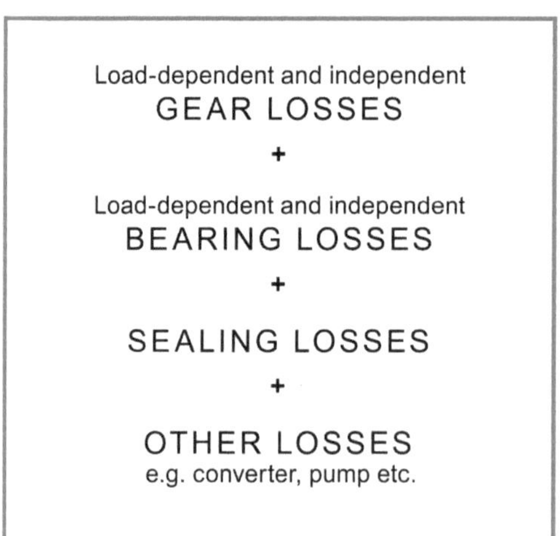


**Fig. 2.32.** Composition of losses in vehicle gearboxes

**Table 2.12.** Reference values for the efficiency ranges of gearwheels and vehicle gearboxes

| **Type of gearbox** | | $\eta$ (%) |
|---|---|---|
| Gear pair | Spur gear | 99.0–99.8 |
| | Bevel gear | 90–93 |
| Mechanical transmission with splash lubrication | Passenger car | 92–97 |
| | Commercial vehicle | 90–97 |
| Automatic transmission with various gear ratios (AT, DCT) | | 90–95 |
| Mechanical continuously variable transmission | | 87–93 |
| Hydrostatic continuously variable transmission without power-split and mechanical part | | 80–86 |

The overall power loss of a transmission is made up of various load-dependent and load-independent elements (Figure 2.32). See also Section 3.1.7 "Efficiency Map" (especially Figure 3.5).

Gearwheel-based vehicle transmissions are the most efficient of all torque/speed converters (Table 2.12), and have the best power/weight ratio. This is also why single- and multi-stage transmissions, of both single- and multi-range design, have been successful for vehicles. The use of hydrostatic transmissions, torque converters or continuously variable transmissions based on the pulley or frictional wheel principle generally leads to lower efficiencies.

## 2.5 Trends in Transmission Design

The motor vehicle market has undergone fundamental changes since about 1975. Especially for passenger cars, the trend towards individualisation has led to considerable segmentation with numerous vehicle classes. This has also led to diversification in transmission design. Instead of a "black or white" alternative between manual (MT) or conventional automatic transmissions (AT), still the case in 1990, there is in the meantime a variety of transmissions designs. These will be described in Chapters 6 and 12. Figure 2.33 shows the diversity of passenger car gearbox designs.

Despite higher costs in product development and for manufacturing plants, the trend is leading away from the "standard transmission" towards individual solutions. This entails a large number of basic designs and a multitude of adaption alternatives. It is therefore of the utmost importance that trends in the market and in technology be assessed, as companies indeed do on a systematic level. Future demands must be anticipated and evaluated in order to develop new strategies.

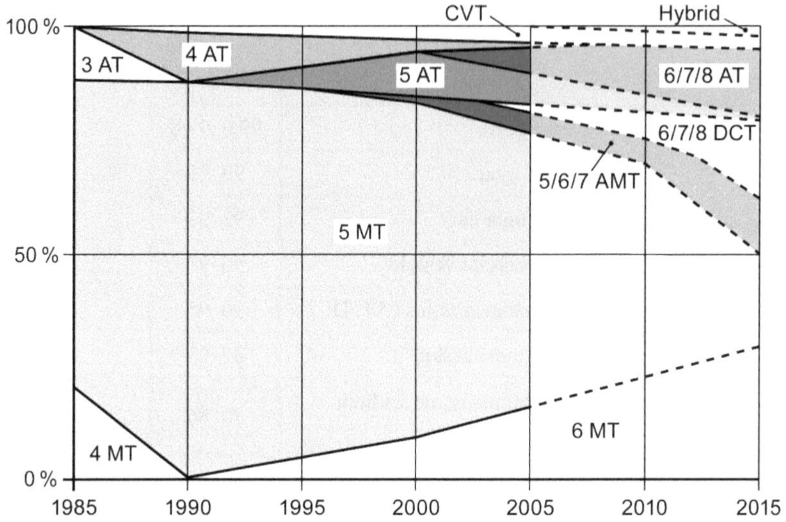


**Fig. 2.33.** Trend in the use of passenger car transmissions in Europe

Vehicle transmission requirements were described in Section 2.3.3. The goals of designing new transmissions and of further development include:

- increasing operational reliability and functionality,
- increasing ease of operation,
- increasing service life and reliability,
- reducing gearbox and repair costs,
- emphasising brand image,
  - increasing comfort,
  - increasing dynamics,
- reducing weight and installation space,
- increasing efficiency,
- reducing consumption and emissions.

Markets and market mechanisms for passenger cars and commercial vehicles differ, changing the respective emphasis on these requirements. Common to all cases is that a variety of requirements leads by necessity to a conflict of goals. The following approaches can help solve such goal conflicts [2.13]:

- development and use of individual transmission solutions,
  - varying gearbox designs, contingent on the requirements,
- increased integration on the aggregate level (Figure 2.35),
  - e.g. integration of peripheral parts and submodules into the gearbox,
- making the transmission more functional,
  - e.g. electromechanical locking differentials,
- superordinate functions by means of networking,
  - e.g. networking of information to determine the driving strategy.

In the case of passenger car transmissions, the trend towards individual solutions with alternative and competing concepts has led to a marked diversification in transmissions designs: manual transmissions (MT), automated manual transmissions (AMT), dual clutch transmissions (DCT), conventional automatic transmissions (AT), continuously variable transmissions (CVT) and hybrid drives (Hybrid). The transmission and its design underscore the character or image of the respective vehicle segment and market. As such the transmission is an essential competitive factor. Each design has its own strengths and weaknesses which depend in turn on the conditions of use. For this reason, evalutions must take place on a case-by-case basis – universally valid assertions are not possible.

Operating conditions and boundary conditions that are crucial for the suitability of a transmission design to a specific vehicle include:

- the vehicle segment,
- acceptance of the transmission design in this vehicle segment by the final customers,
- brand image,
- propulsion strategy, motor characteristics,
- powertrain configuration (e.g. front-wheel drive, all-wheel drive),
- production facility type and equipment,
- acceptance of the transmission design in different markets,
- legislation and regulations (e.g. emissions).

The coaxial design for example is well suited to standard rear-wheel drive, while the axial offset of a CVT with a taper disc variator or a gearbox with a countershaft has design-specifc advantages for front-wheel drive. The available productions facilities and associated investments also influence the evaluation of applicability.

Regional market peculiarities are a further example of boundary conditions that can demand individual solutions. For example, automated manual transmissions – with the power interruption peculiar to them – have no chance in typical automatic transmission markets like NAFTA. Figure 2.34 shows the development and prediction of market behaviour for passenger car transmissions in Europe, NAFTA and Japan. The literature varies greatly in its predictions [2.18].

The manual transmission (MT) with 5 and 6 speeds has remained dominant in Europe. Seven speeds are offered in more sporty vehicles and with automation. MTs are inexpensive and light transmissions. Their market share is declining but still remains on a relatively high level. For compact cars and light trucks smaller than 3.5 t, automated manual transmissions (AMT) will become increasingly important. Further development will improve shifting quality and reliability, thereby improving the somewhat tattered image dating from this transmission's early stages. Weight and cost speak against the use of continuously variable transmissions in compact cars. For compact and mid-range cars, the dual clutch transmission (DCT) with 6, 7 and 8 speeds is considered to have good market chances, especially for sport cars.

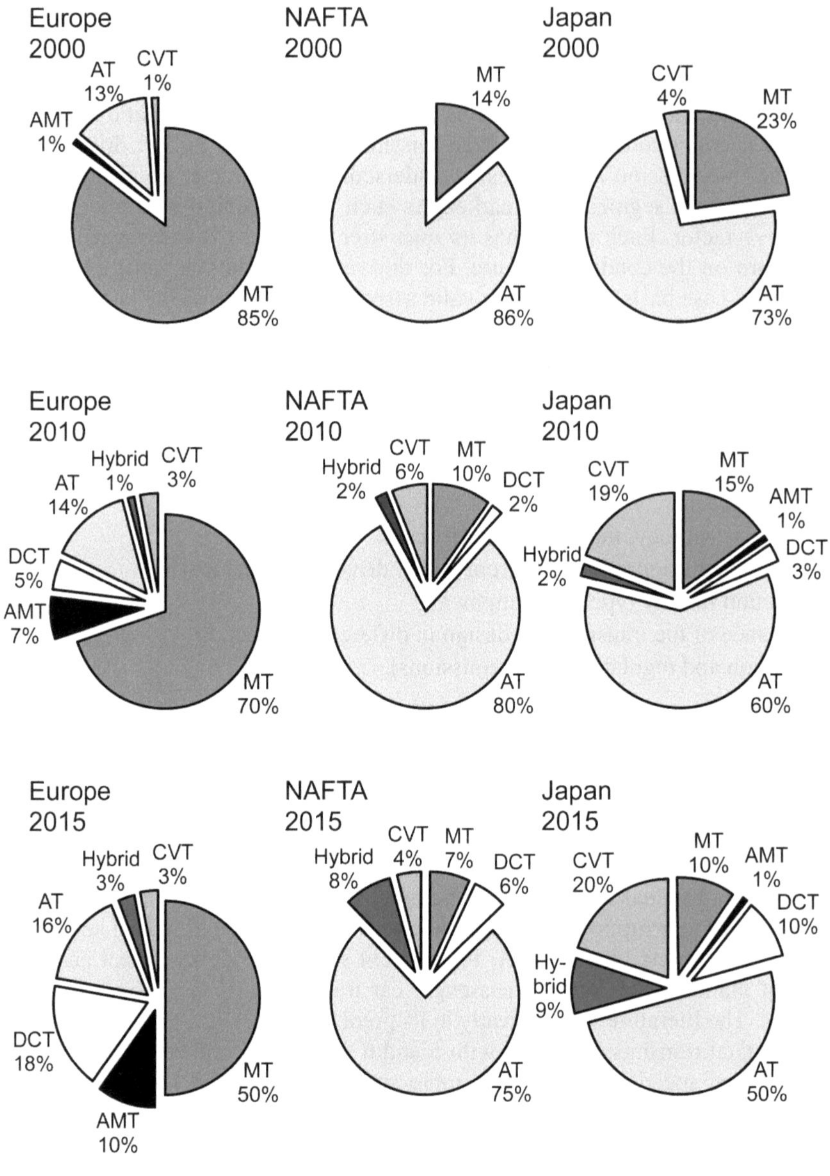


**Fig. 2.34.** Passenger car transmission designs: market shares and prediction

ATs, with torque converters and up to 8 speeds for passenger cars, will either retain or slightly expand their market share for mid-range and luxury classes. "Geared Neutral" infinitely variable transmissions (IVT) based on friction gears with torque capacities greater than 400 Nm are currently being developed.

Diesel engines continue to claim large market shares in Europe. Presently in 2007, hybrid technology is considered primarily as a boost to brand image. Hybrid drives are thus mostly found in luxury cars and in the form of mild hybrids (see Sections 6.6.5 and 12.1.5).

For passenger cars, North America and Japan are typical automatic transmission markets. Manual transmissions play a minor role there. Continuously variable transmissions (CVT) are becoming widespread especially in Japanese compact cars. The development of the market share of hybrid drives as well as their form, whether as mild hybrid or full hybrid (Table 3.7) depends greatly on legislation. There is a lot of potential predicted for them in these markets.

In the case of commercial vehicle transmissions, the mechanical geared transmission with 6 to 16 speeds of either single- or multi-range design is the standard. In the heavy-duty truck segment, automated manual transmissions have become successful in Europe. Their path led from semi-automated designs right up to fully automated transmissions (Figure 2.35). Increasing integration of peripheral parts and submodules into the transmission has led to lighter, more compact and more reliable aggregates. The number of interfaces decreases, vehicle installation is alleviated and the total system can be checked in its completion.

While in 2005, for light-duty commercial vehicles (less than 3.5 t), AMTs based on passenger car transmissions already had considerable market presence, this is still not the case for the medium-duty commercial vehicle segment (500 to 1500 Nm). Automated manual transmissions for commercial vehicles of this performance class have been on the market since 2005.

The conventional automatic transmission – with torque converter, lock-up clutch and planetary gear sets – has among commercial vehicles only become widely used in city buses. This will also remain its domain.

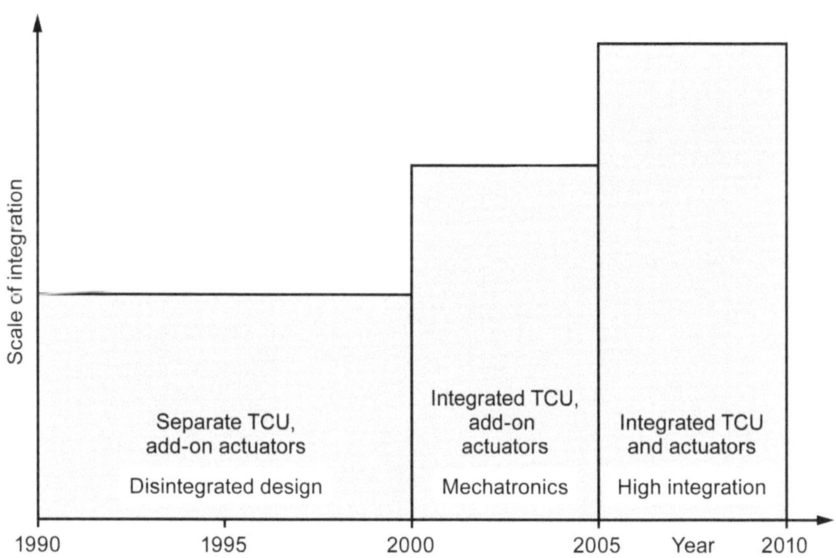


**Fig. 2.35.** Increased integration on the aggregate level using an automated manual transmission for a commerical vehicle as example

While hydrostatic continuously variable transmissions do have a market share for agricultural tractors, they have not been successful in buses. Electric drives for buses have been developed. Due to the high costs compared with manual and automatic transmissions, an extensive serial production of such drives depends heavily on legislation.

The trend towards individual transmission solutions with a large number of basic transmission designs and a variety of adaption alternatives will in the future demand a better functioning management of variants by the suppliers. The kind of partnership between the vehicle manufacturer (OEM = Original Equipment Manufacturer) and the system supplier (Tier 1) is also changing. Vehicles are being equipped with more and more complex functionalities. The increase in function content in the transmission systems and networking with other components of the vehicle lead to changes in the chain of responsibility.

Another trend is that the vehicle manufacturers, who themselves have an increased amount of tasks, expect an ever larger scope of services from their suppliers. The latter have to adapt to new contents. This means, assuming constant resources, that Tier 1 suppliers will themselves pass on tasks and responsibility for certain components to Tier 2 and 3 suppliers [2.13].

# 3 Mediating the Power Flow

*Reciprocity: Supply and Demand*

Vehicle transmissions mediate between the engine and the drive wheels. The transmission adapts the power output to the power requirement by converting torque and rotational speed. The power requirement at the drive wheels is determined by the driving resistance [3.9].

## 3.1 Power Requirement

The anticipated driving resistance is an important variable when designing vehicle transmissions. Driving resistance is made up of

- wheel resistance $F_R$,
- air resistance $F_L$,
- gradient resistance $F_{St}$ and
- acceleration resistance $F_a$.

### 3.1.1 Wheel Resistance

Wheel resistance comprises the resisting forces acting on the rolling wheel. It is made up of rolling resistance, road surface resistance and slip resistance.

Figure 3.1 shows the forces and torques acting on the wheel. The integral of the pressure distribution over the tyre contact area gives the reaction force $R$. It is the same as the wheel load $G_R$. Because of the asymmetrical pressure distribution in the wheel contact area of the rolling wheel, the point of application of the reaction force $R$ is located in front of the wheel axis by the amount of eccentricity $e$. If the wheel is unaccelerated and driven by $T_R$, then

$$T_R = F_U \, r_{dyn} + R\,e\,. \tag{3.1}$$

For a wheel rolling without drive torque and braking torque ($T_R = 0$)

$$-F_U = \frac{e}{r_{dyn}} R\,. \tag{3.2}$$

H. Naunheimer et al., *Automotive Transmissions*, 2nd ed.,
DOI 10.1007/978-3-642-16214-5_3, © Springer-Verlag Berlin Heidelberg 2011

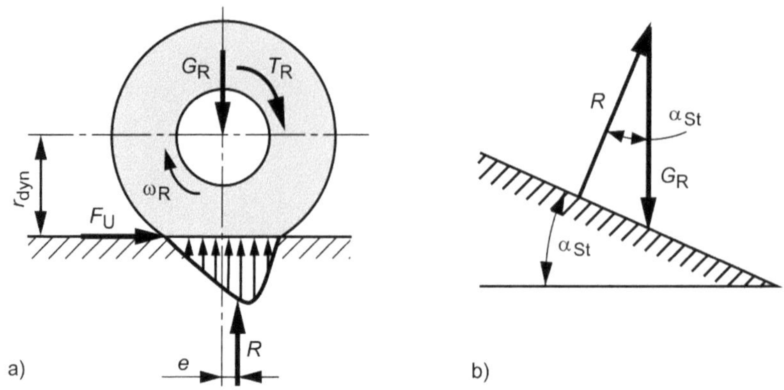


**Fig. 3.1.** Forces and torques at the wheel. *a* On the level; *b* on uphill/downhill stretch

The circumferential force $-F_U$ is equal to the rolling resistance force $F_{R,Roll}$ given these assumptions. On a level surface $R = G_R$, and therefore

$$F_{R,Roll} = \frac{e}{r_{dyn}} G_R \,. \tag{3.3}$$

Trials have revealed an almost linear correlation between the rolling resistance force $F_{R,Roll}$ and the wheel load $G_R$. The relationship is defined by the formula

$$F_{R,Roll} = f_R \; G_R \,. \tag{3.4}$$

The dimensionless proportionality factor $f_R$ is designated as the rolling resistance coefficient. From (3.3) and (3.4) it is given as

$$f_R = \frac{e}{r_{dyn}} \,. \tag{3.5}$$

Table 3.1 shows standard values for rolling resistance coefficients both on and off-road. Rolling resistance is chiefly a function of ground speed, wheel load, tyre pressure and tyre type.

Since driving resistance calculations normally assume straight running on a dry surface, and rolling resistance is anyway the dominant wheel resistance, wheel resistance is normally assumed to be equal to rolling resistance. The following formula then applies:

$$F_R = F_{R,Roll} \,. \tag{3.6}$$

When travelling up gradients/down slopes at an angle of $\alpha_{St}$ (Figure 3.1b), then

$$R = G_R \; \cos \alpha_{St} \,. \tag{3.7}$$

**Table 3.1.** Reference values for the rolling resistance coefficient $f_R$. For road speeds below 60 km/h, $f_R$ can be assumed to be constant. (See also Table 5.1)

| Road surface | Rolling resistance coefficient $f_R$ |
|---|---|
| *Firm road surface* | |
| Smooth tarmac road | 0.010 |
| Smooth concrete road | 0.011 |
| Rough, good concrete surface | 0.014 |
| Good stone paving | 0.020 |
| Bad, worn road surface | 0.035 |
| *Unmade road surface* | |
| Very good earth tracks | 0.045 |
| Bad earth tracks | 0.160 |
| Tracked tractor on acre soil | 0.070–0.120 |
| Clamp wheels on acre soil | 0.140–0.240 |
| Loose sand | 0.150–0.300 |

For the whole vehicle with a mass $m_F$, the wheel resistance $F_R$, which is considered equal to the rolling resistance, is thus given by

$$F_R = f_R \, m_F \, g \cos\alpha_{St} \,. \tag{3.8}$$

In the lower speed range, the rolling resistance coefficient can be regarded as a constant at the first approximation. The gradient angle $\alpha_{St}$ can be ignored on normal journeys with gradients/downhill slopes of less than 10%. With a gradient of 10% $\alpha_{St} \approx 5.7°$ and thus $\cos\alpha_{St} \approx 1$.

### 3.1.2 Adhesion, Dynamic Wheel Radius and Slip

There is a frictional connection between the tyres and the road surface. The transmittable circumferential force $F_U$, (Figure 3.1a), is proportional to the wheel load reaction force $R$, with a maximum value

$$F_{U,max} = F_{Z,max} = \mu_H \, R \,. \tag{3.9}$$

**Table 3.2.** Static coefficient of friction $\mu_H$ of new pneumatic tyres on road surfaces [3.1]

| Road speed | Static coefficient of friction $\mu_H$ | |
|---|---|---|
| (km/h) | Dry road surface | Wet road surface |
| 50 | 0.85 | 0.65 |
| 90 | 0.80 | 0.60 |
| 130 | 0.75 | 0.55 |

The maximum traction $F_Z$ between the tyres and the road surface is constrained by the adhesion limit (friction limit, Figure 2.20). See also Section 6.1.4 in relation to circumferential force, lateral force and Kamm circle. Table 3.2 gives static friction figures $\mu_H$ of pneumatic tyres on road surfaces.

For many vehicle dynamics calculations, the radius of the tyres is needed (Table 3.3).

**Table 3.3.** Dynamic wheel radius of common tyre sizes [3.7–3.8]

| **Size** | **Rolling circumference** (m) | $r_{dyn}$ (m) | **Size** | **Rolling circumference** (m) | $r_{dyn}$ (m) |
|---|---|---|---|---|---|
| *Passenger cars* | | | *Passenger cars* | | |
| 155/70 R 13 | 1.671 | 0.266 | 255/40 ZR 18 | 2.016 | 0.321 |
| 165/65 R 13 | 1.659 | 0.264 | 275/40 ZR 18 | 2.065 | 0.329 |
| 175/65 R 13 | 1.702 | 0.271 | 295/30 ZR 18 | 1.937 | 0.308 |
| 155/65 R 14 | 1.702 | 0.271 | 335/30 ZR 18 | 2.01 | 0.320 |
| 165/70 R 14 | 1.793 | 0.286 | 255/40 ZR 19 | 2.071 | 0.330 |
| 175/65 R 14 | 1.781 | 0.284 | 285/40 ZR 19 | 2.169 | 0.345 |
| 175/80 R 14 | 1.940 | 0.309 | 295/30 ZR 19 | 2.016 | 0.321 |
| 185/65 R 14 | 1.818 | 0.289 | 345/30 ZR 19 | 2.108 | 0.336 |
| 175/55 R 15 | 1.748 | 0.278 | 315/35 R 20 | 2.220 | 0.354 |
| 185/55 R 15 | 1.784 | 0.284 | 335/30 ZR 20 | 2.166 | 0.345 |
| 195/65 R 15 | 1.937 | 0.308 | *Vans and light-duty commercial vehicles* | | |
| 205/60 R 15 | 1.912 | 0.304 | 185/60 R 15 | 1.827 | 0.291 |
| 215/65 R 15 | 2.016 | 0.321 | 225/70 R 15 | 2.112 | 0.336 |
| 195/55 R 16 | 1.891 | 0.301 | 205/65 R 16 | 2.036 | 0.324 |
| 195/60 R 16 | 1.952 | 0.311 | 215/75 R 16 | 2.206 | 0.351 |
| 205/55 R 16 | 1.928 | 0.307 | 205/75 R 17.5 | 2.297 | 0.366 |
| 215/55 R 16 | 1.958 | 0.312 | *Trucks/Buses* | | |
| 225/55 R 16 | 1.995 | 0.318 | 215/75 R 17.5 | 2.339 | 0.372 |
| 235/50 R 16 | 1.958 | 0.312 | 245/70 R 17.5 | 2.406 | 0.383 |
| 215/45 R 17 | 1.909 | 0.304 | 265/70 R 19.5 | 2.644 | 0.421 |
| 215/55 R 17 | 2.037 | 0.324 | 305/70 R 19.5 | 2.815 | 0.448 |
| 215/60 R 17 | 2.106 | 0.335 | 275/70 R 22.5 | 2.922 | 0.465 |
| 225/55 R 17 | 2.074 | 0.330 | 295/60 R 22.5 | 2.806 | 0.447 |
| 235/65 R 17 | 2.251 | 0.358 | 295/80 R 22.5 | 3.184 | 0.507 |
| 255/60 R 17 | 2.251 | 0.358 | 315/80 R 22.5 | 3.282 | 0.523 |
| 245/45 R 18 | 2.065 | 0.329 | 495/45 R 22.5 | 3.085 | 0.491 |
| 245/50 R 18 | 2.144 | 0.341 | 13 R 22.5 | 3.428 | 0.546 |

A distinction is made between

- *static wheel radius* $r_{\text{stat}}$:
  The distance from the centre of the wheel to the road contact plane with the wheel at rest and
- *dynamic wheel radius* $r_{\text{dyn}}$:
  Calculated from the distance travelled per revolution of the wheel, rolling without slip. The dynamic wheel radius is calculated from a distance travelled at 60 km/h [3.2]. The increasing tyre slip at higher speeds roughly offsets the increase in $r_{\text{dyn}}$.

The slip between the tyres and the surface can be described as

drive slip $$S_T = \frac{\omega_R\, r_{\text{dyn}} - v_F}{\omega_R\, r_{\text{dyn}}}, \quad (3.10)$$

brake slip $$S_B = \frac{v_F - \omega_R\, r_{\text{dyn}}}{v_F}. \quad (3.11)$$

where $v_F$ is the actual vehicle speed.

### 3.1.3 Air Resistance

Air flows around the moving vehicle and through it for purposes of cooling and ventilation. The air resistance is made up of the pressure drag including induced drag (turbulence induced by differences in pressure), surface resistance and internal (through-flow) resistance.

The air resistance is a quadratic function of the flow rate. The flow rate $v$ is derived from the sum of the vehicle speed $v_F$ and the wind speed component $v_W$ in the direction of the vehicle longitudinal axis. If the wind speed direction is the same as the direction of travel of the vehicle (following wind), then the wind speed is deducted from the vehicle speed to calculate the flow rate. Driving resistance calculations normally assume calm, in which case: $v = v_F$.

Air resistance is calculated from the product of dynamic pressure $1/2\, \rho_L\, v^2$ and the maximum vehicle cross-section $A$ multiplied by the dimensionless drag coefficient $c_W$. At an air pressure of 1.013 bar, a relative air humidity of 60% and a temperature of 20°C the air density $\rho_L = 1.199$ kg/m$^3$.

The drag coefficient $c_W$ represents the special case of straight flow, i.e. the wind direction is in line with the longitudinal axis of the vehicle. Table 3.4 gives the $c_W$-values and the maximum vehicle cross-sections (projected frontal area) of some current vehicles.

**Table 3.4.** Reference values for drag coefficient $c_W$. In the case of goods trucks, the $c_W$ coefficient and the maximum vehicle cross-section are very much dependent on the particular design. *) Driver sitting and with fitting clothing

| Vehicle | $c_W$ | $A$ (m$^2$) | $c_W A$ (m$^2$) |
|---|---|---|---|
| ***Motorcycle with rider**** ⁾ | ***0.5–0.7*** | ***0.7–0.9*** | ***0.4–0.6*** |
| BMW K 1200 S*⁾ | 0.58 | 0.71 | 0.41 |
| BMW R 1200 GS*⁾ | 0.62 | 0.85 | 0.52 |
| ***Convertible*** | ***0.29–0.53*** | ***1.58–2.90*** | ***0.58–1.54*** |
| Opel Tigra TwinTop | | | |
| Roof open | 0.40 | 1.94 | 0.78 |
| Roof closed | 0.35 | 1.94 | 0.67 |
| Mercedes-Benz SLK 200 K | | | |
| Roof open | 0.37 | 1.93 | 0.71 |
| Roof closed | 0.32 | 1.93 | 0.62 |
| Mercedes-Benz SL 500 | | | |
| Roof open | 0.34 | 2.00 | 0.68 |
| Roof closed | 0.29 | 2.00 | 0.58 |
| Audi A4 Cabrio | | | |
| Roof open | 0.34 | 2.11 | 0.72 |
| Roof closed | 0.31 | 2.11 | 0.65 |
| ***Limousine/SUV*** | ***0.25–0.39*** | ***1.97–2.90*** | ***0.50–1.54*** |
| Ford Fiesta 1.4l | 0.34 | 2.06 | 0.70 |
| VW Golf V 1.4l | 0.32 | 2.22 | 0.72 |
| Mercedes-Benz B 180 CDI | 0.30 | 2.42 | 0.73 |
| BMW 320i | 0.28 | 2.11 | 0.59 |
| Audi A6 Avant | 0.31 | 2.26 | 0.70 |
| Mercedes-Benz S 320 CDI | 0.26 | 2.40 | 0.62 |
| Mercedes-Benz ML 280 CDI | 0.34 | 2.81 | 0.96 |
| Porsche Cayenne Turbo | 0.39 | 2.78 | 1.09 |
| BMW 645i | 0.29 | 2.15 | 0.62 |
| Porsche 911 Carrera | 0.28 | 2.00 | 0.56 |
| ***Van*** | ***0.35–0.40*** | ***3.1–4.2*** | ***1.1–1.7*** |
| Opel Vivaro Life | 0.37 | 3.38 | 1.24 |
| Ford Transit MWB, MJ06 | 0.35 | 4.14 | 1.45 |
| ***Bus*** | ***0.4–0.8*** | ***6.0–10.0*** | ***2.4–8.0*** |
| Setra 415 HD | 0.44 | 8.26 | 3.63 |
| ***Light trucks*** | ***0.40–0.60*** | ***4.5–6.0*** | ***1.8–3.6*** |
| ***Truck (Solo)*** | ***0.45–0.80*** | ***6.0–10.0*** | ***2.7–8.0*** |
| ***Truck with trailer*** | ***0.55–0.85*** | ***7.0–10.0*** | ***3.9–8.5*** |
| ***Articulated vehicle*** | ***0.45–0.75*** | ***7.0–10.0*** | ***3.2–7.5*** |

Drag is calculated by

$$F_L = \frac{1}{2} \rho_L \, c_W \, A \, v^2 \, . \tag{3.12}$$

The aerodynamics of vehicles with high-drag flow-impeding body, such as commercial vehicles, can be greatly improved by the use of air guiding plates.

### 3.1.4 Gradient Resistance

The gradient resistance or downhill force relates to the slope descending force (Figure 3.2) and is calculated from the weight acting at the centre of gravity

$$F_{St} = m_F \, g \, \sin\alpha_{St} \, . \tag{3.13}$$

The road gradient $q$ is defined as the quotient of the vertical and horizontal projections of the roadway (Figure 3.2).

When designing roads, gradients of more than 7% are normally avoided. Except in extreme cases, the following approximation is valid

$$\sin\alpha_{St} \approx \tan\alpha_{St} = \frac{q'}{100} \, . \tag{3.14}$$

Table 3.5 shows the maximum gradients ($q'_{max}$) of some Alpine passes.

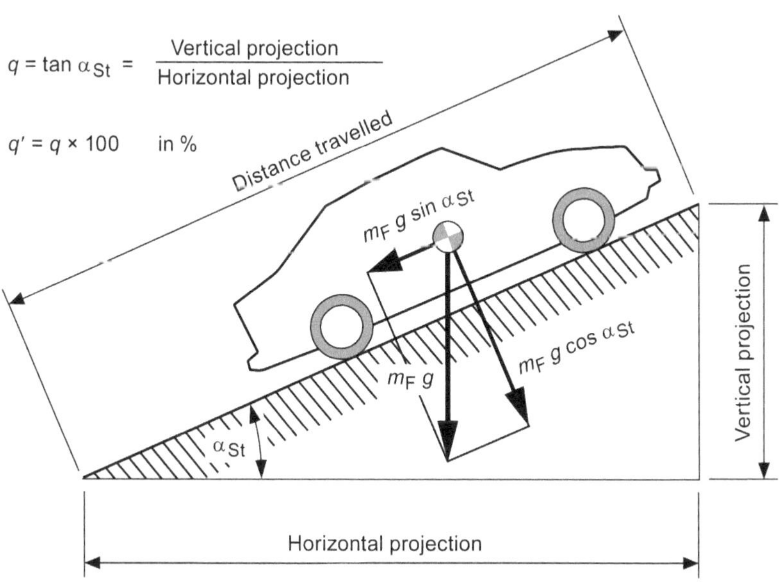


**Fig. 3.2.** Forces acting on the vehicle travelling uphill

**Table 3.5.** Maximum gradients of some passes in Europe

| **Pass** | $q'_{max}$ | **Pass** | $q'_{max}$ |
|---|---|---|---|
| *Germany:* | | *Austria:* | |
| Achenpass | 14% | Großglockner | 12% |
| Oberjoch | 9% | Timmelsjoch | 13% |
| *France:* | | Turracher Höhe | 26% |
| Col de Braus | 15% | Wurzenpass | 18% |
| Iseran | 12% | *Switzerland:* | |
| *Italy:* | | Simplon | 10% |
| Brenner highway | 12% | St. Bernardino | 12% |
| Stilfser-Joch | 15% | St. Gotthard | 10% |

### 3.1.5 Acceleration Resistance

In addition to the driving resistance occurring in steady state motion ($v$ = const), inertial forces also occur during acceleration and braking.

The total mass of the vehicle $m_F$ (translatory component) and the inertial mass of those rotating parts of the drive accelerated or braked (rotational component) are the factors influencing the resistance to acceleration:

$$F_a = m_{red,i}\, a\,, \quad \text{with} \tag{3.15}$$

$$m_{red,i} = m_F + \frac{\sum J_{red,i}}{r_{dyn}^2}\,. \tag{3.16}$$

The rotational component is a function of the gear ratio. The moment of inertia of the rotating drive elements of engine, moving-off element, gearbox, drive shaft etc., including all the road wheels (even the non-driven road wheels), are reduced to the driving axle.

This reduces to $J_{red,i}$. The acceleration resistance is frequently represented in simplified form as

$$F_a = \lambda\, m_F\, a\,, \tag{3.17}$$

where $\lambda$ is a rotational inertia coefficient, which expresses the proportion of the total mass that is rotational. Reference values for the rotational inertia coefficient for passenger cars are shown in Figure 3.3.

Since the gear ratio is a quadratic factor in determining the reduced moment of inertia, the rotational inertia coefficients for vehicles with high-ratio gears are widely spread. (Reference values for trucks as per [3.14] are: crawler gear: $\lambda \approx 10$, 1st gear: $\lambda \approx 3$, direct gear: $\lambda \approx 1.1$).

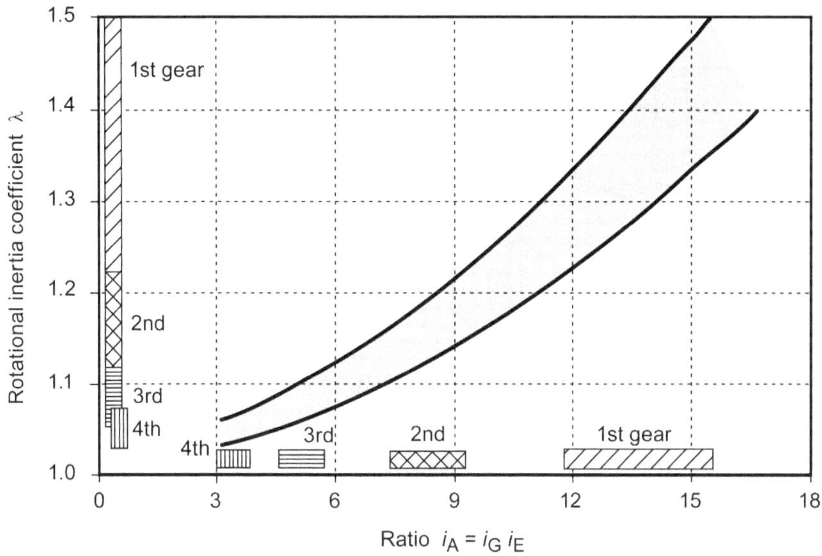


**Fig. 3.3.** Reference values for rotational inertia coefficients of passenger cars [3.9]

### 3.1.6 Total Driving Resistance

The traction $F_{Z,B}$ required at the drive wheels is made up of the driving resistance forces described above, and is defined as

$$F_{Z,B} = F_R + F_{St} + F_L + F_a \,. \tag{3.18}$$

Together with Equations 3.8, 3.12, 3.13 and 3.17, this may be expanded to

$$F_{Z,B} = m_F\, g\left(f_R\, \cos\alpha_{St} + \sin\alpha_{St}\right) + \frac{1}{2}\,\rho_L\, c_W\, A\, v^2 + m_F\, \lambda\, a\,. \tag{3.19}$$

With steady state motion ($a = 0$) and the approximations mentioned ($\cos\,\alpha_{St} \approx 1$ and $\sin\,\alpha_{St} \approx \tan\,\alpha_{St}$) this simplifies to

$$F_{Z,B} = m_F\, g\left(f_R + \tan\alpha_{St}\right) + \frac{1}{2}\,\rho_L\, c_W\, A\, v^2 \,. \tag{3.20}$$

This may be used to calculate the power requirement $P_{Z,B}$

$$P_{Z,B} = F_{Z,B}\, v \,. \tag{3.21}$$

Figure 3.4 shows the individual components of driving resistance of a mid-size passenger car, the resultant traction requirement and the power requirement.

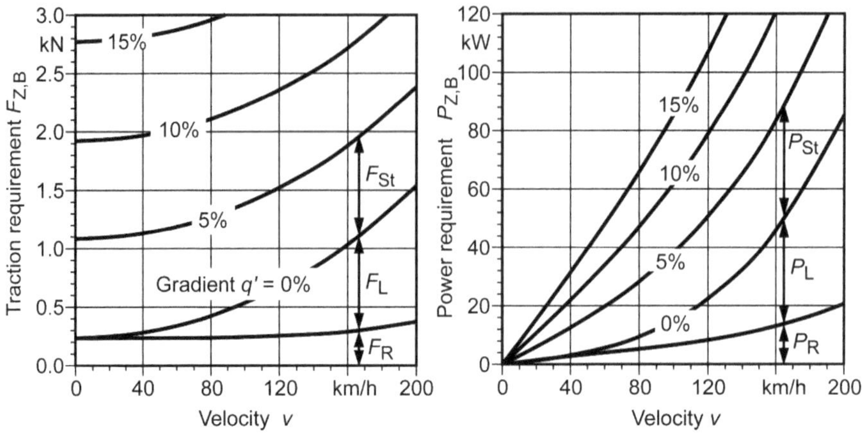


**Fig. 3.4.** Traction required and the resultant power required for a mid-size passenger car. For vehicle data see Figure 5.3

Taking into account the powertrain ratio $i_A$ and the overall powertrain efficiency $\eta_{tot}$, the traction $F_{Z,A}$ available at the wheels may be calculated from the engine characteristic curve as follows

$$F_{Z,A} = \frac{P(n_M)}{v} \eta_{tot} = \frac{T(n_M)\, i_A}{r_{dyn}} \eta_{tot} \,. \tag{3.22}$$

The traction required and the traction available for a vehicle are shown in a "traction diagram". The traction diagram is discussed in detail in Section 5.1 "Traction Diagram".

### 3.1.7 Efficiency Map

The efficiency of the engine and powertrain affects fuel consumption, emissions and driving performance. The engine efficiency is represented by the specific fuel-consumption curves. See also Section 3.3.3 "Consumption Map".

Powertrain losses can be classified according to their components into losses caused by the:

- moving-off element: e.g. torque converter,
- gearbox: e.g. gearwheel transmission, pulley transmission,
- final drive and
- auxiliary units: e.g. steering pump, oil pump in automatic gearboxes, air conditioning system, variable displacement pump in continuously variable transmissions.

Reference values for the amount of loss with different designs of vehicle transmission are given in Section 2.4.6 "Gearbox Losses and Efficiency".

To calculate the traction available or the point at which to operate an engine, it is necessary to know the powertrain efficiency $\eta_{tot}$ from the engine output shaft through to the driving wheels. In a sense the powertrain efficiency represents another driving resistance. It is made up of the efficiencies

$$\eta = \frac{P_2}{P_1} = 1 - \frac{P_V}{P_1} \tag{3.23}$$

or equivalently the power losses $P_V$ of the individual components of the powertrain:

- *gearing losses:*
  - friction losses, load-dependent,
  - churning and squeezing losses attributable to splash lubrication, load-independent,
- *bearing losses:*
  - friction losses, load-dependent,
  - lubrication losses, load-independent,
- *sealing losses:*
  - friction losses caused by rotary shaft seals at shaft exits,
  - friction losses caused by piston rings used to keep oil under pressure at the shift elements,
- *synchronizing losses:*
  - fluid friction between synchronizer ring and friction cone,
- *clutch losses:*
  - fluid friction with wet running, multi-plate clutches and brakes in automatic gearboxes and automated manual gearboxes,
- *torque converter losses:*
  - losses in the hydrodynamic torque converter,
- *auxiliary units:*
  - power to drive auxiliary units.

A further distinction is made between losses that are

- dependent on the input speed and the input torque,
- dependent only on the engine speed, including in particular pumps directly driven by the engine and
- almost independent of rotational speed and torque. For example the efficiency of the final drive is normally assumed to be constant.

Figure 3.5 shows these power losses using the example of a coaxial 6-speed manual transmission at part load in 4th gear. In this case, it is not the direct gear.

Figure 3.6 shows the efficiency of the powertrain from the engine output shaft through to the drive wheels in fourth gear of a 5-speed manual gearbox. Provision is made for a steering pump as an auxiliary unit. The level of efficiency only declines rapidly in the low load area.

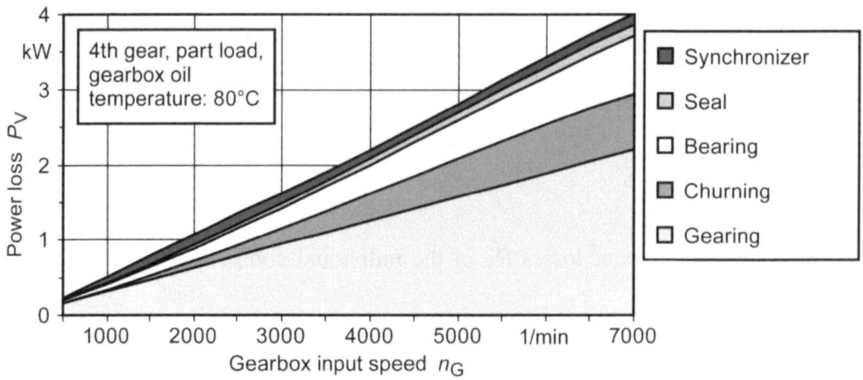


**Fig. 3.5.** Distribution and size of power losses of a coaxial 6-speed manual gearbox in 4th gear at 50% part load

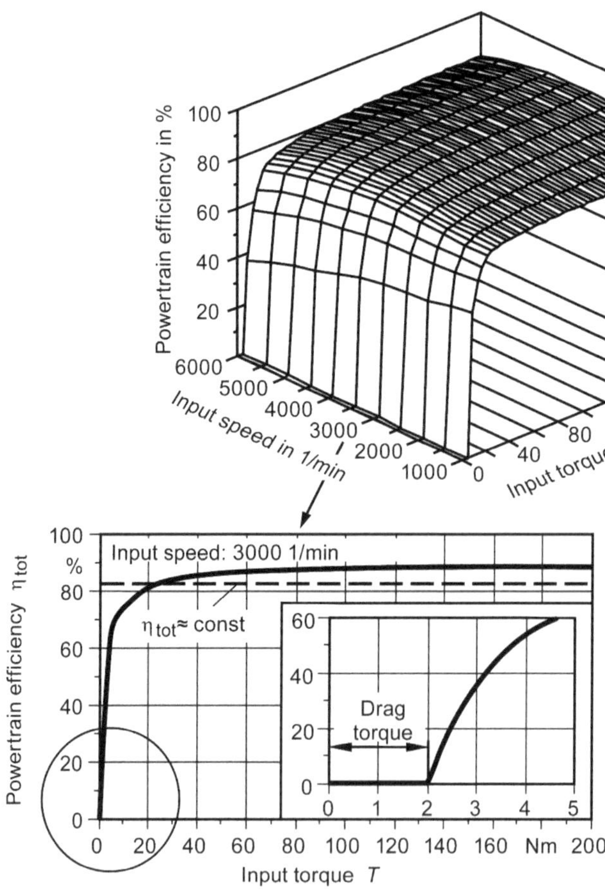


**Fig. 3.6.** 3D total powertrain efficiency map of the direct 4th gear of a 5-speed manual gearbox

In the case of manual gearboxes, a constant level of efficiency can often be assumed with sufficient accuracy. In the case of continuously variable transmissions, the part-load efficiency is significantly worse, and the drag torque is significantly higher (Figure 3.6).

## 3.2 Diversity of Prime Movers

The driving resistance described in Section 3.1 has to be overcome by the prime mover in cooperation with the other components of the powertrain.

The energy supply, which has to be carried in the vehicle, means "dead" weight and "dead" volume. Energy supplies with an energy density as high as possible are desirable. Figure 3.7 shows the working capacity at the drive wheels from various types of energy supply.

The weight of the energy accumulator is factored in, as is the transmission efficiency figure (energy at the wheel/energy of the fuel). Further important criteria in selecting transportable power storage are rapid recharging of the energy accumulator and the necessary infrastructure.

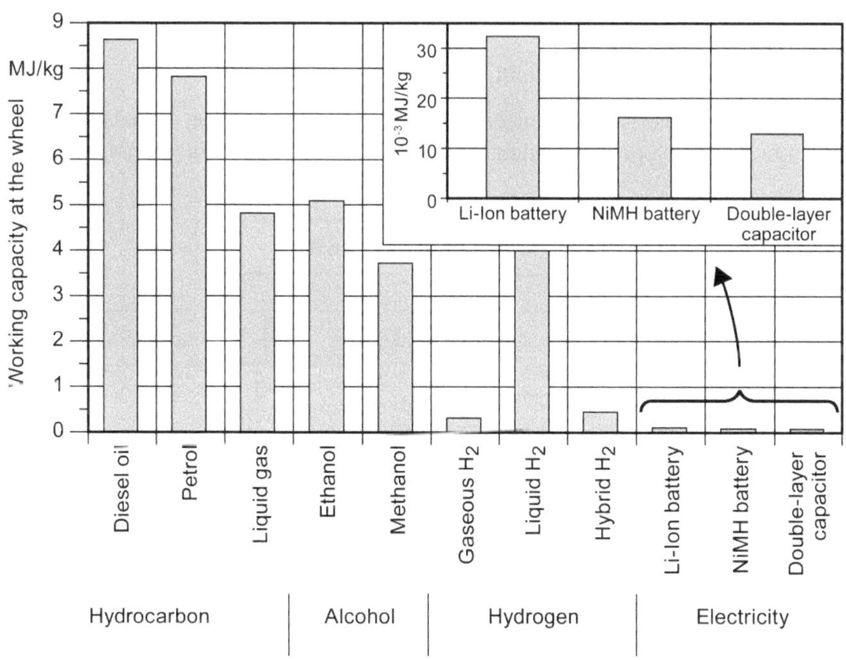


**Fig. 3.7.** The energy available from various transportable automobile energy storing systems (from [3.3, 3.6]). Mechanical energy available at the drive wheels as a function of the mass of the energy supply + energy accumulator (container). Various levels of engine efficiency with energy conversion are taken into consideration

Diesel oil and petrol perform the best in this context. The space requirement for electric batteries is about 30 times larger than for spark ignition or diesel drives with the same capacity. However, the fuel cell has in the meantime reached nearly the small storage volume of spark ignition and diesel drives if $H_2$ is extracted from methanol on the move [3.12].

### 3.2.1 Overview

The drive system of a vehicle can be made up of a variety of combinations of components for storing energy, converting energy and converting output. The prime mover used is a crucial factor in determining the assemblies and design of the associated powertrain.

Various prime movers could be used in a vehicle (Figure 3.8). They can be broken down into combustion engines and electric motors. In selecting a suitable power unit, the following factors must be considered:

- *operating performance:*
  drive characteristics, ease of control, startability, energy accumulator etc.,
- *economy:*
  specific energy consumption, specific manufacturing cost etc. and
- *environment friendliness:*
  pollutant emissions, noise, vibration etc.

The engine characteristic is a decisive technical consideration in selecting the prime mover, i.e. the power available at full load across the engine speed range.

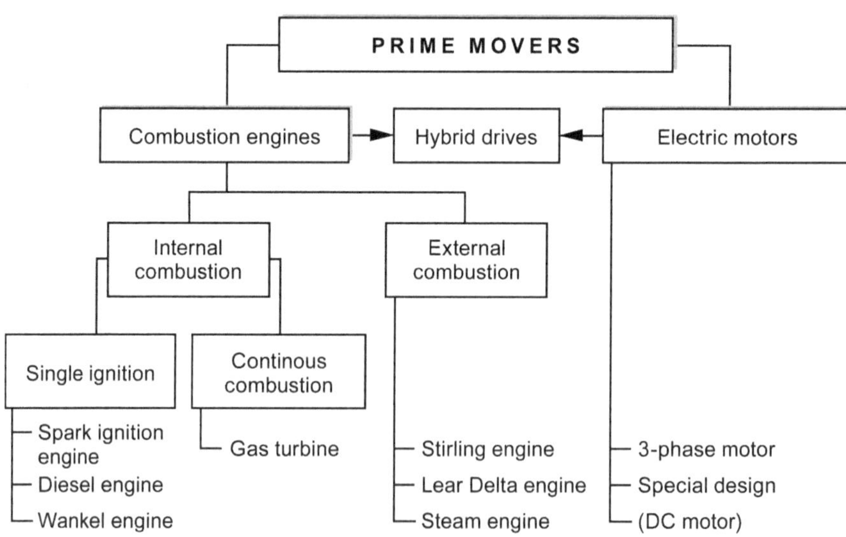


**Fig. 3.8.** Overview of prime movers for motor vehicles

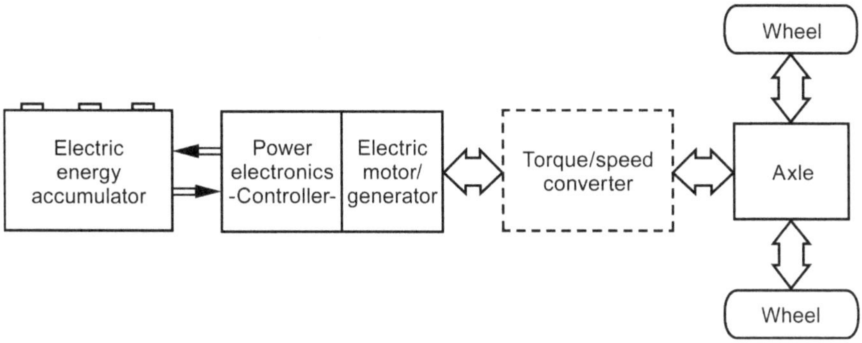


**Fig. 3.9.** Components of an electric drive

The operation of the various prime movers is not discussed here. References are made to the literature as appropriate. In the discussion that follows, the term internal combustion engine refers to a spark ignition or diesel engine.

### 3.2.2 Electric Drive with Electric Energy Accumulator

In this design, the electric motor serves as the sole prime mover. Within a certain speed range, it has the ideal engine characteristic with $P$ = const, which conforms to the ideal traction hyperbola. It can be operated from rest, i.e. motor speed zero, and apply torque. A battery is used as the electric energy accumulator. It must be charged from an energy grid before the vehicle can be propelled purely electrically (and thus emission-free) over a certain stretch of road. Then it must be reconnected to the energy grid for recharging. Figure 3.9 shows the components of an electric drive. It is not essential to have an additional speed/torque converter.

The use of passenger cars with exclusively electric drive will remain restricted to cities and their environs because of their limited range and performance and the battery charging time required.

### 3.2.3 Electric Drive with Fuel Cell

The primary energy accumulators of electric drives with fuel cells is a tank filled with a fuel (e.g. hydrogen or methanol). An electric motor is used as the prime mover (Figure 3.10).

The fuel cell, like a battery, directly converts the chemical reaction energy of the fuel into electric energy for the drive. This direct process can claim some advantages in efficiency compared to methods that take an indirect route via thermal energy conversion (i.e. internal combustion engines). Especially in part-load operation, such as in city traffic, the advantages of fuel cell drive systems become apparent. Depending on the fuel used and the method of procuring it, this drive technology shows potential for reducing pollutant emissions.

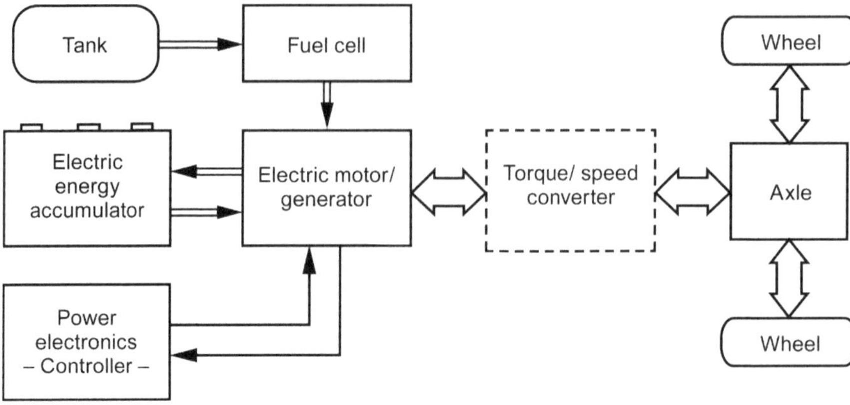


**Fig. 3.10.** Electric drive with a fuel cell

In comparison to electric drives with electric energy accumulators, the range limitation is alleviated by the fact that the system can be refuelled. However, without an additional battery, the fuel cell drive cannot recuperate any kinetic energy during braking (see Section 3.2.4), since it is difficult to operate the fuel cell in the total system reversibly.

In vehicle drive technology, the fuel cell most used is one based on a polymer electrolyte membrane (PEM). This works under temperatures less than 200°C with hydrogen or methanol as the fuel (Figure 3.11). Hydrogen can be prepared as such or obtained by reforming fuels. Systems with fuel reformation are significantly more complex.

The structure of and processes within a fuel cell will be simplified explained in the following using the example of a hydrogen-oxygen fuel cell: the fuel cell consists of two electrodes (anode and cathode) that are separated from each other by a membrane (e.g. PEM). The membrane is selective, only permeable for $H^+$ ions. The anode is supplied with hydrogen (fuel), which is there oxidised into $H^+$ ions as electrons are released.

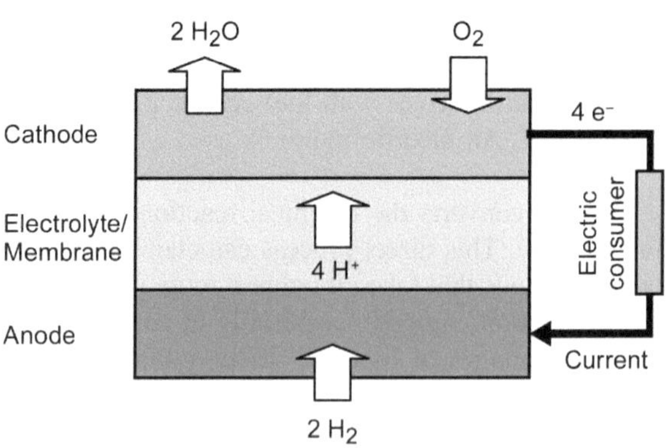


**Fig. 3.11.** Schematic structure and process in a hydrogen-oxygen fuel cell

The cathode is supplied with atmospheric oxygen (oxidising agent). As electrons are added, this is reduced to anions and reacts with the $H^+$ ions that have come through the membrane to form water or steam. Electrical potential is created between the electrodes, and an electric current can flow to an electric consumer (e.g. an electric motor).

In the case of the hydrogen-oxygen fuel cell, this electric potential is theoretically 1.23 volts in an unloaded state. The voltage of the system can be controlled by means of a series connection of several fuel cells to make a stack.

Technical difficulties still exist in the storage of the highly volatile hydrogen and in the "icing up" of the fuel cell by the freezing of exhaust steam at very low outside temperatures.

All fuel cell types are still quite high in price. This is due partially to the small numbers produced but also to the high costs of the raw materials used (precious metal catalysers, expensive polymers).

Presently in 2007, not only do open-ended technical and economical problems stand in the way of a widespread introduction of fuel cell technology to the market, there are also questions concerning the production and distribution of the necessary fuel. In the context, one interesting development is for example the regenerative production of hydrogen from water by means of electrolysis using solar technology or wind energy technology. Yet this energy conversion method is very costly: electrolysis, hydrogen transport via pipelines and liquefaction or compression for storage purposes, reconversion into electrical energy in the fuel cell. As a result, compared with direct storage of regeneratively produced electric energy in a battery, this technology has a much smaller total efficiency. Increasing amounts of research and development efforts are being undertaken with an eye to clarifying the requirements being made on the changing infrastructure (e.g. the petrol station network) [3.11].

### 3.2.4 Hybrid Drive

Hybrid drives are drives that have at least two different prime movers and energy accumulators [3.14]. Possible energy accumulators are:

- chemical energy accumulator: conventional fuel tank,
- electrical energy accumulator: – battery, high-performance capacitor,
- mechanical energy accumulator: – flywheel,
  – hydraulic accumulator.

*Electrical energy accumulators* are charged during generator operation of an electrical machine. During motor-driven operation, the stored electric energy is fed back to the powertrain. Among electric energy accumulators, the following technologies are the most competitive:

- lithium-ion batteries (Li-Ion),
- nickel-metal hydride batteries (NiMH) and
- high-performance double-layer capacitors ("ultracaps" or "supercaps").

**Table 3.6.** Energy accumulator technology – a comparison [3.6]

| | | Li-Ion battery | NiMH battery | Double-layer capacitor |
|---|---|---|---|---|
| Power density | theoretical in kW/kg@10s | 2.0–3.5 | 1.3–1.8 | 7–10 [1] |
| | used power density for hybrid vehicle layout (% kW/kg) [2] | < 30% | < 20% | < 70% |
| Energy density | theoretical (Wh/kg) [3] | 80–100 | 40–50 | 3–5 |
| | used energy for hybrid vehicle layout (DoD) [4] | 5–15% | 5–15% | up to 90% possible |
| Operation temperature | (°C) | –30 to +60 | –30 to +45 | –30 to +70 |

[1] Power density at mean discharge voltage
[2] Time-averaged boost and recuperation power obtained from drive cycles
[3] Assumption: Cells are designed power optimised
[4] To realise the required service life for automotive use, with batteries just a part of the available energy is converted per charge/discharge cycle (Depth of Discharge)

Table 3.6 provides a comparison of the most important properties of these different energy accumulator technologies.

*Mechanical energy accumulators* have so far served primarily to assist moving-off. They are used to accumulate braking energy in vehicles which must start and stop frequently, such as city buses. The mechanical energy of the flywheel or gyro accumulator is converted into electric energy and feeds the electric motor. Flywheels can pose a security risk in the vehicle however because of their high speeds and resultantly high kinetic energy.

The causes for the rapidly accelerating development of the automotive industry in the field of hybrid drive technology, which has already led to the successful market introduction of the first mass-produced hybrid vehicles, are complex: rising oil prices are putting more pressure on vehicle manufacturers to offer alternative, low-energy drive concepts. This pressure is even greater in some countries due to stiffer legislation seeking to reduce the emission of exhausts such as $CO_2$. Often, such legislation involves tax breaks for hybrid vehicles. Since hybrid vehicles suggest interesting and promising solutions to problems related to fuel consumption and emissions, the image of this drive technology (its “green image”) has improved enormously in recent years [3.13].

### *Internal Combustion Engine + Electric Drive*

The combination of combustion engine plus electric drive gives better range and availability than a vehicle with electric drive only. The electric motor’s capability of being able to reach its maximum torque even at low speeds opens up the possibility of an interesting supplement to conventional combustion engine drives (see Figure 3.12).

Even at low performance standards, electromotive torque can be relatively high. As the speed is increased, performance increases approximately linearly until the "corner speed", the point of maximum performance, is reached. Beyond this corner speed, electromotive torque decreases in accordance with a power hyperbola. In this speed range, torque and performance curves of the combustion engine typically reach their highest values, signifying that advantageous operating conditions can result from the combination of the electric motor and combustion engine [3.4–3.5]:

- temporary noise and emission-free, purely electrical driving (e.g. in congested urban areas),
- electric moving-off/manoeuvring without the combustion engine,
- recovery of kinetic energy during breaking by charging the electric energy accumulator by generator operation of the electric motor (recuperation),
- torque support of the combustion engine by the electric motor (boosting),
- combustion engine start/stop during stop-and-go traffic or when stopping at a traffic light,
- supply of the vehicle on-board power requirement by generator operation of the electric motor (eliminates the need for an alternator).

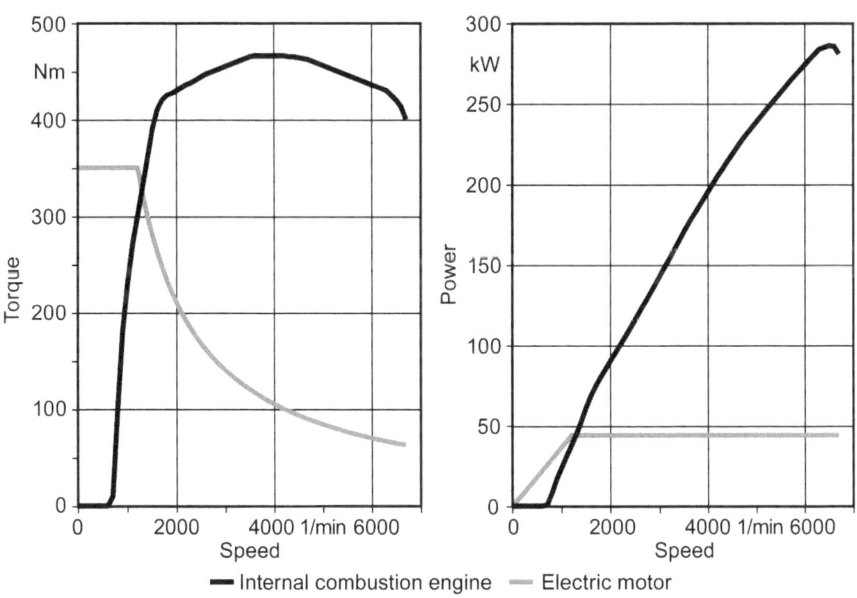


**Fig. 3.12.** Characteristic curves of torque and performance for electric motors and internal combustion engines

However, vehicles with hybrid drive have a weight disadvantage in comparison to those with only one kind of drive, since, besides the second drive, they also require a second energy accumulator.

Lower in weight and installation space requirements, three-phase AC motors have the advantage of a high power density, thus making them ideal for use in electric or hybrid vehicle powertrains. In this context, basically two different three-phase AC motor technologies are in use:

- asynchronous machines (ASM) and
- permanent-magnet synchronous machines (PSM).

Besides these, special electric motor construction types are also used:

- permanent-magnet transversal flow machines (TFM) and
- switched reluctance machines (SRM).

As high-performance drive machines, direct current motors are barely still in use in present-day motor vehicles.

The new operating conditions described above can be deduced for the functional structure of a vehicle drive as a function of performance data of the electric drive unit (electric motor and electric energy accumulator) and the associated voltage level. These conditions are potentially advantageous not only for reducing fuel consumption and emissions but also with respect to other criteria like driving dynamics and comfort. From this we derive a subdivision into different hybrid classes (Table 3.7).

Discussion of hybrid drives with internal combustion engines and electric motors involves distinguishing between these concepts:

- serial hybrid powertrain, Figure 3.13a:
  - no mechanical coupling of combustion engine and wheels,
  - mechanical gearbox not mandatory,
  - a combustion engine, in conjunction with a generator, functions solely as a electricity producer,
  - two high-performance electric machines (generator + generator/electric motor)
- parallel hybrid powertrain, Figure 3.13b:
  - both drives can be combined,
  - mechanical gearbox required, favourable derivability from existing transmission concepts ("add-on"),
  - only one electric machine is necessary. When the internal combustion engine is being used, it can run at nearly optimum efficiency. If the specific power output of the engine is higher than that required to overcome the driving resistance, the excess power can be used to charge the battery. If the specific power output of the combustion engine is lower than that required to overcome the driving resistance, the electric motor can provide support as long as the state of charge of the electric energy accumulator allows it. However: conversion losses must be taken into account!

**Table 3.7.** Hybrid classes, from [3.13]

| | Micro hybrid | Mild hybrid | Full hybrid |
|---|---|---|---|
| Power of electric motor | 2–10 kW | 4–20 kW | > 20 kW |
| Torque of electric motor | < 90 Nm | < 500 Nm | < 500 Nm |
| Voltage | 14–42 V | ≥ 42 V | 100–650 V |
| On-board power supply | • | • | • |
| Combustion engine start/stop | • | • | • |
| Recuperation | | • | • |
| Boosting | | • | • |
| Electrical driving (Duration depends on size of energy accumulator) | | | • |

- power-split hybrid drive, Figure 3.13c:
  - splitting of the combustion engine power output into mechanical and electric paths,
  - summarising gear (planetary gear sets) required for the splitting and joining of the mechanical and electric power-paths. This electric variator permits a continuous torque and speed conversion,
  - at least two high-performance electric machines are needed. However, because of their reciprocal power supply, they cannot supply their full nominal power for boost or recuperation processes.

Micro and mild hybrids are generally built as parallel hybrids. The distinction between serial, parallel and power-split hybrids pertains to full hybrids.

## 3.3 Power Output, Combustion Engine Characteristic

Internal combustion engines based on the spark ignition and diesel principle will retain their dominant position in automotive engineering for the foreseeable future.

The key features of spark ignition engines are the high power/weight ratio, good performance and low combustion noise. The disadvantages are the quality of fuel required and the high part-load consumption.

The economy of the diesel engine is based on its low consumption, especially in the part-load range, its low maintenance requirement (no ignition system), the low fuel quality required and its good gaseous emissions ratings. The disadvantages are the level of particulate emissions, higher noise than SI engines, irregular running due to higher compression, the lower engine speed spread ($n_{max}/n_{min}$), the low power output per litre, and the resultant greater weight and higher price.

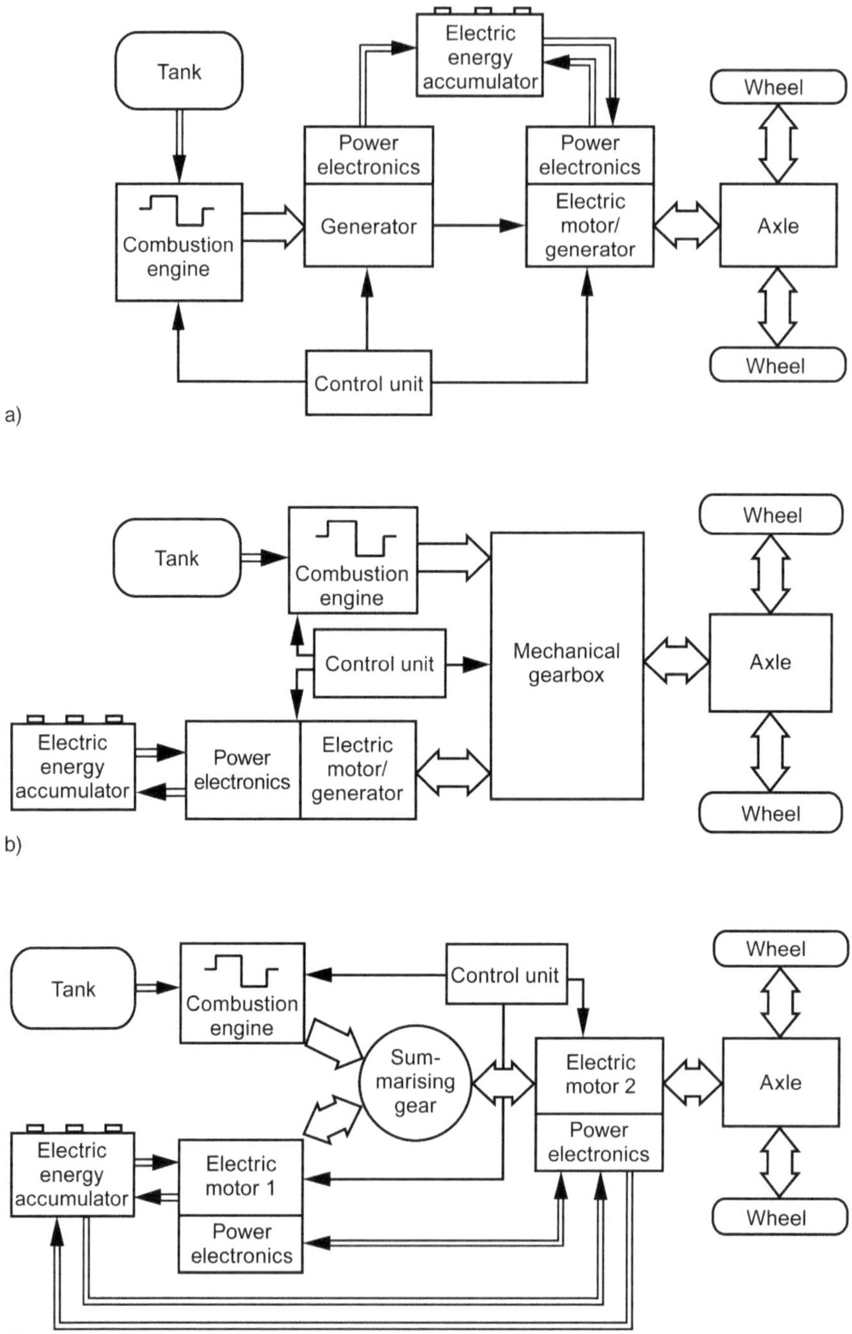


**Fig. 3.13.** Hybrid powertrains. *a* Serial hybrid powertrain; *b* parallel hybrid powertrain; *c* power-split hybrid powertrain (*a*, *b* from Höhn)

The higher capital cost means that a diesel engine only becomes economical in vehicles which do a high mileage. Almost all commercial vehicles use diesel engines.

While turbocharging has long been dominant with diesel engines, supercharging has also become widespread in spark ignition engines since 2005. One line of development is seeking to downsize to reduce fuel and emissions, another seeks to increase driving performance (sports cars). Table 3.8 shows typical torque values per litre.

### 3.3.1 Torque/Engine Speed Characteristic

There are two typical characteristic curves to describe the engine characteristic of combustion engines. One is the torque/engine speed curve at full load (100% accelerator pedal position) and the other is the corresponding full-load power curve (engine characteristic). Figure 3.14 shows the map of an internal combustion engine and the characteristic points of the full load characteristic curve. The maximum braking torque (0% accelerator pedal position) increases almost linearly with engine speed to a maximum of approx. 30% of the nominal torque $T_n$.

Various measures are used to facilitate comparison of different engines. Key variables are torque increase (torque elasticity)

$$\tau = \frac{T_{max}}{T_n} \tag{3.24}$$

and the engine speed ratio (engine speed elasticity)

$$\nu = \frac{n_n}{n(T_{max})} \,. \tag{3.25}$$

An engine is considered to have greater elasticity the greater the product $\tau\,\nu$ is. This is apparent in the form of better engine power at low and medium engine speeds, which in turn means less frequent gear changing.

**Table 3.8.** Typical torque per litre swept volume of passenger car engines.
*) complex turbocharging technique

| | Typical torque values (Nm/*l*) | | |
|---|---|---|---|
| | Without turbocharging | With turbocharging | Sport engine*) |
| Spark ignition engine | 100 | 150 | 200 |
| Diesel engine | – | 170 | 250 |

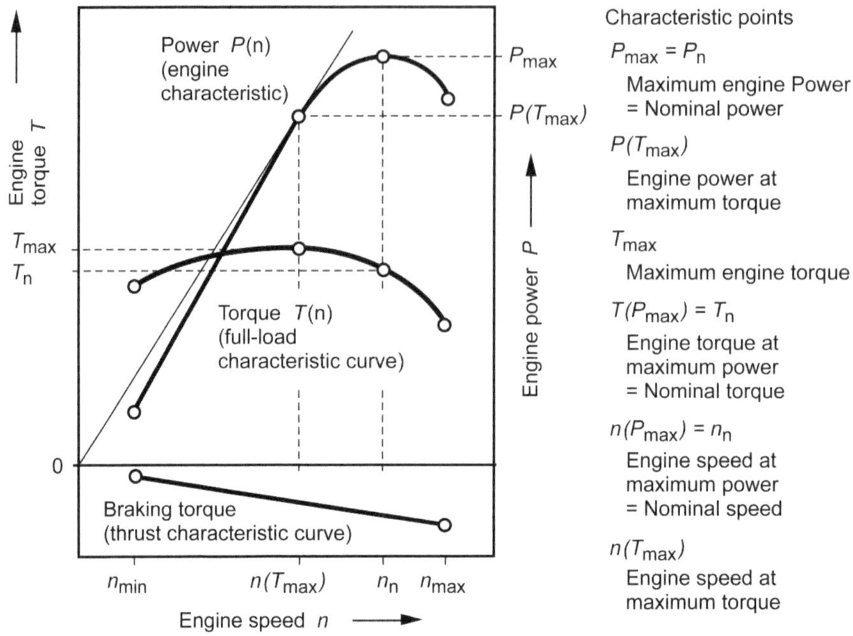


**Fig. 3.14.** Characteristic curves of an internal combustion engine

Different engine characteristics can be achieved by varying the engine design. A distinction is made in principle between three typical characteristics (Figure 3.15).

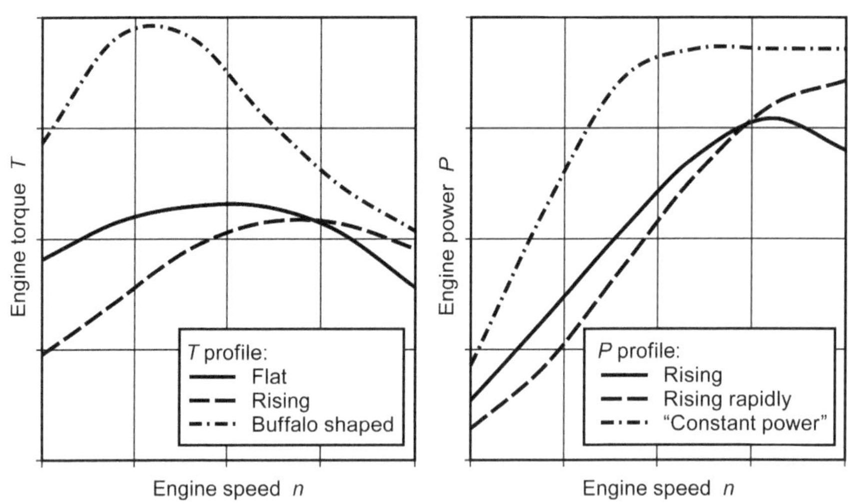


**Fig. 3.15.** Typical engine profiles

### 3.3.2 Engine Spread, Throttle Map

The spread of the engine is an important variable influencing the interaction of the internal combustion engine with the transmission (Chapter 5). In its function as a speed and torque converter, the transmission has a range of ratios: the overall gear ratio (Section 4.2.1). It is defined as the quotient of the maximum and minimum transmission ratio.

The engine spread refers to an engine's speed and torque range. Vehicles with powerful engines accordingly have a large torque spread. Diesel engines have a lower maximum speed than spark ignition engines and accordingly have a smaller engine speed spread. Figure 3.16 shows the characteristic diagram of two passenger car engines. The engine shown on the left is a spark ignition engine without supercharging; the one shown on the right is a turbo diesel engine with intercooler.

The diesel engine has a smaller speed spread than the spark ignition engine, but a greater torque spread. The transmission ratios have to be selected to accommodate this. The spread of the engine and the overall gear ratio (in combination with the stepping between gear ratios) are the main factors determining the functional characteristics of the vehicle.

The driver uses the accelerator pedal to indicate the power desired from the engine. When the accelerator pedal is fully depressed (100%) this corresponds to the engine full-load curve, and when the accelerator pedal is not depressed (0%), to the thrust characteristic curve. Figure 3.16 shows the lines for the same accelerator pedal position for the two engines. The almost equidistant pattern is typical of diesel engines. The term "throttle valve angle" is often used instead of "relative accelerator pedal position". A throttle valve angle of 90° then corresponds to the engine full-load line. Diesel engines do not have a throttle valve for preparing the mixture, so the term "relative accelerator pedal position" or "control rack travel" is used.

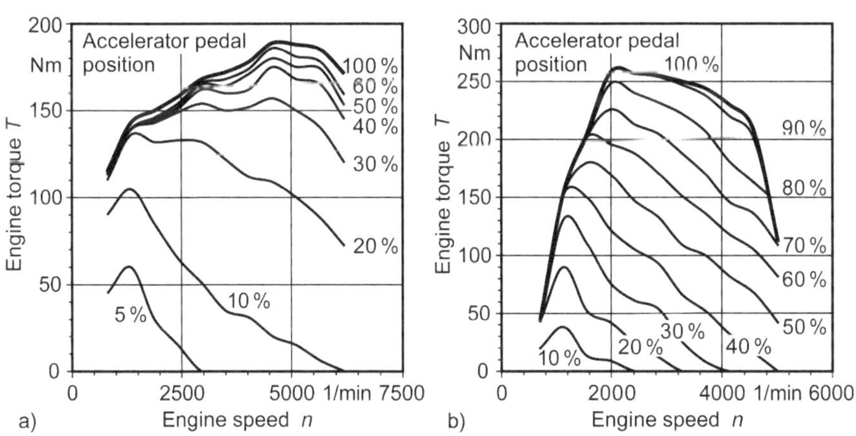


**Fig. 3.16.** Passenger car engine performance maps, accelerator pedal position (throttle map). *a* Spark ignition engine; *b* turbo diesel engine with intercooler

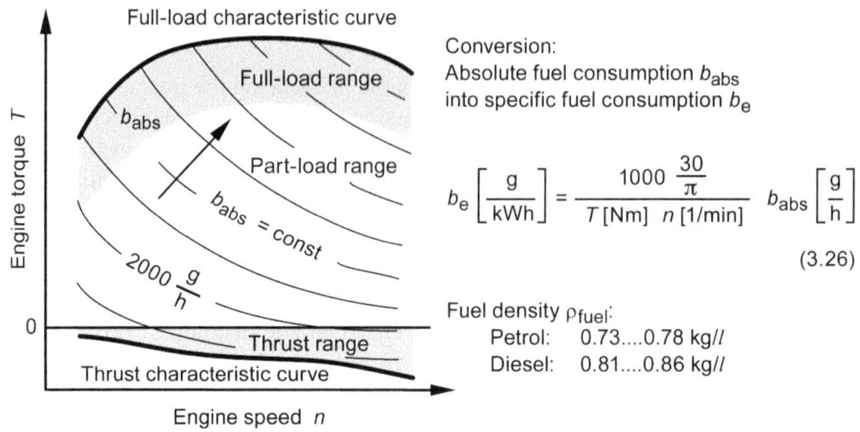


**Fig. 3.17.** Consumption map of a spark ignition engine. Absolute consumption $b_{abs}$ in g/h

### 3.3.3 Consumption Map

The fuel consumption of a stationary internal combustion engine can be represented as a function of engine speed and torque. A consumption characteristic diagram of this type is shown in Figure 3.17; the absolute consumption $b_{abs}$ is shown in g/h. It increases rapidly with the engine output. If the specific consumption $b_e$ is shown in g/kWh, the term "onion diagram" is used (Figure 3.18). There is a minimum consumption $b_{e,min}$ just below the full-load characteristic curve in the lower engine speed range. The precise position depends on the engine. In the case of spark ignition passenger car engines the minimum consumption is around 250 g/kWh, with passenger car diesel engines it is around 190 g/kWh and with commercial vehicle diesel engines it is as well around 190 g/kWh.

In the engine map the effective average pressure $p_{me}$ in the cylinder is often plotted instead of the engine torque. The following relationship applies

$$p_{me} = \frac{T_M\, 2\,\pi}{V_H\, i} \quad \text{with} \quad i = \frac{2}{\text{number of strokes}}, \tag{3.27}$$

(1 bar = $10^5$ N/m$^2$), where $V_H$ is the total swept volume in m$^3$. In four-stroke engines $i = 0.5$.

Like the engine characteristic the consumption map is an important basis for matching engine, transmission and vehicle. The transmission exploits the fuel-efficient areas of the engine performance map. Figure 3.18 shows the contour lines of constant specific fuel consumption $b_e$ (onion curves) as well as the torque/engine speed curves at a constant engine output (demand power hyperbola) $T(P$ = const). In this way the same engine output can be achieved at different torque/engine speed values – points *1* and *2* in the engine map – and thus also with different levels of fuel consumption.

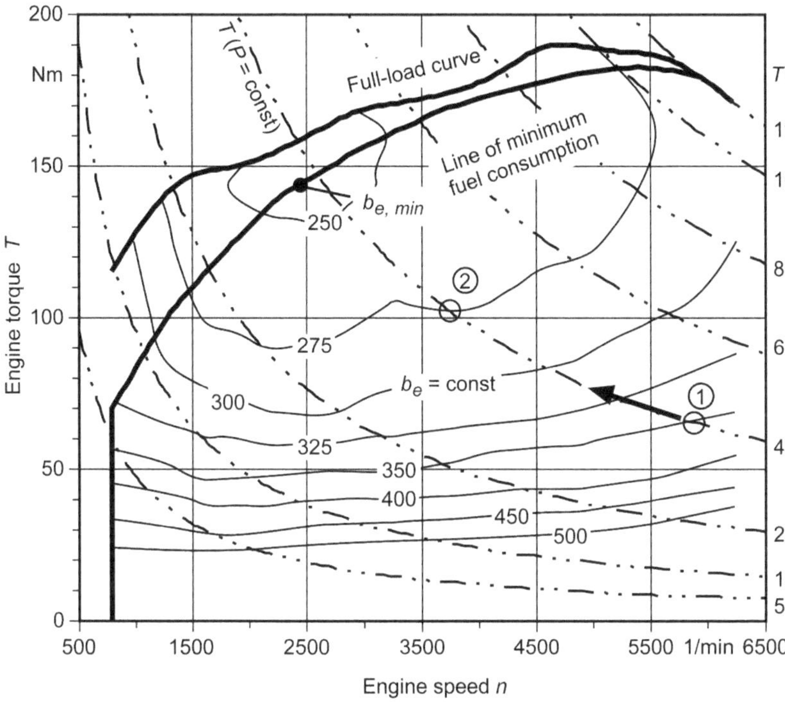


**Fig. 3.18.** Engine map ("onion diagram"), specific fuel consumption $b_e$ in g/kWh. Consumption map of a 2.0 litre spark ignition engine with 111 kW

A minimum fuel consumption point can be found on any power hyperbola. The curve passing through these points is the minimum fuel consumption curve.

# 4 Power Conversion: Selecting the Ratios

*The transmission puts the engine to work*

Chapter 3 dealt with the relationship between the power available from the engine and the power requirement arising from driving resistance. The torque/speed profile of the internal combustion engine is not suited to use in motor vehicles (see also Section 2.3.2 "Why do Vehicles Need Gearboxes?"). Output converters are needed for the final output to approximate as closely as possible to the ideal engine characteristic with $P_{max}$ = const over the entire engine speed range. Clutches serve to adapt engine speed, transmissions serve to adapt both speed and torque. The conversion ratio is determined by theoretical and practical engineering constraints, which depend in many cases on the application.

The basic design of the transmission involves first determining the maximum and minimum ratio, i.e. the "overall gear ratio" of the transmission, and then selecting the intermediate ratios. Chapters 4 and 5 deal with the selection of these key features. They are the basis for the calculation, engineering and design of components (Figure 4.1).

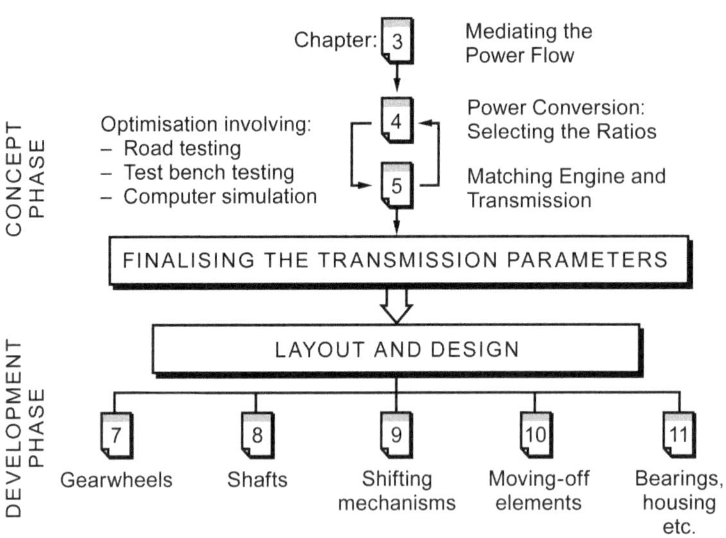


**Fig. 4.1.** The ratios selected are the key features of the transmission, and thus form the basis for subsequent development work

H. Naunheimer et al., *Automotive Transmissions*, 2nd ed.,
DOI 10.1007/978-3-642-16214-5_4, © Springer-Verlag Berlin Heidelberg 2011

## 4.1 Powertrain

In vehicles with internal combustion engines, the output conversion between the engine and the drive wheels is achieved by the combined action of the assemblies of the powertrain. Figure 4.2 shows the hierarchical structure of the various ratios in the powertrain, starting from the total powertrain ratio $i_A$. The total ratio of the powertrain is derived from the ratio $i_S$ of the moving-off element, the ratio $i_G$ of the transmission and the final ratio $i_E$,

$$i_A = i_S \, i_G \, i_E \,. \tag{4.1}$$

The ratio of output speed $n_2$ to input speed $n_1$ of a powertrain component is defined as speed conversion $v$,

$$v = \frac{n_2}{n_1} \,. \tag{4.2}$$

The torque conversion $\mu$ represents the ratio between the output torque $T_2$ and the input torque $T_1$ of a powertrain component,

$$\mu = \frac{T_2}{T_1} \,. \tag{4.3}$$

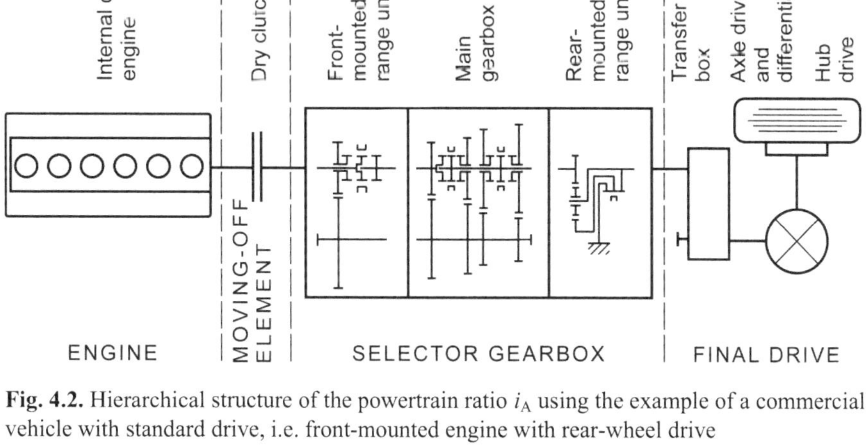


**Fig. 4.2.** Hierarchical structure of the powertrain ratio $i_A$ using the example of a commercial vehicle with standard drive, i.e. front-mounted engine with rear-wheel drive

The term ratio of $i \neq 1.0$ should only arise when there is both speed and torque conversion. In this case

$$i = \frac{n_1}{n_2}, \quad \text{if} \quad \mu > 1.0. \tag{4.4}$$

Master clutches convert only rotational speed, i.e. $i_S = 1.0$. Hydrodynamic torque converters convert both rotational speed and torque, $i_S \geq 1.0$. Torque converters are discussed in Section 10.4. The discussion below is based on a dry clutch as the standard moving-off element.

The gearbox ratio $i_G$ constantly adapts the traction available from the engine in steps – or better in an infinite number of steps – to the traction hyperbola for $P_{max}$ = const (see also Figure 2.20). Range units are fitted on the input ($i_{G,V}$) or output ($i_{G,N}$) side to increase the number of speeds where the vehicle needs a broad overall gear ratio, such as commercial vehicles and off-road vehicles. The balance between performance and fuel consumption is achieved by adjusting the final ratio $i_E$, which is especially important for commercial vehicles.

## 4.2 Total Ratio and Overall Gear Ratio

The powertrain has to offer ratios between engine speed and road wheel speed enabling the vehicle to

- move-off under difficult conditions,
- reach the required maximum speed and
- operate in the fuel-efficient ranges of the engine performance map.

The maximum ratio required $i_{A,max}$ is fixed by the first condition. The second condition gives the maximum road speed ratio $i_A(v_{max,th})$. The smallest powertrain ratio $i_{A,min}$ is given by the third condition. Figure 4.3 shows the speed spread of a transmission in a diagram of velocity against engine speed. The engine speed range (primary side) is "spread" by the transmission to the speed range of the secondary side. The operating range extends between the ratio boundaries.

Increasing speed limits and traffic density are reducing the importance of maximum speeds of passenger cars. By the same token, acceleration performance is gaining in importance.

A wide overall gear ratio is particularly important for heavy passenger cars with powerful engines and a low drag coefficient [4.1]. They need:

- a high stall torque ratio $i_{A,max}$ for moving-off and accelerating and
- a low minimum ratio $i_{A,min}$ for low engine speeds at high road speeds to reduce fuel consumption.

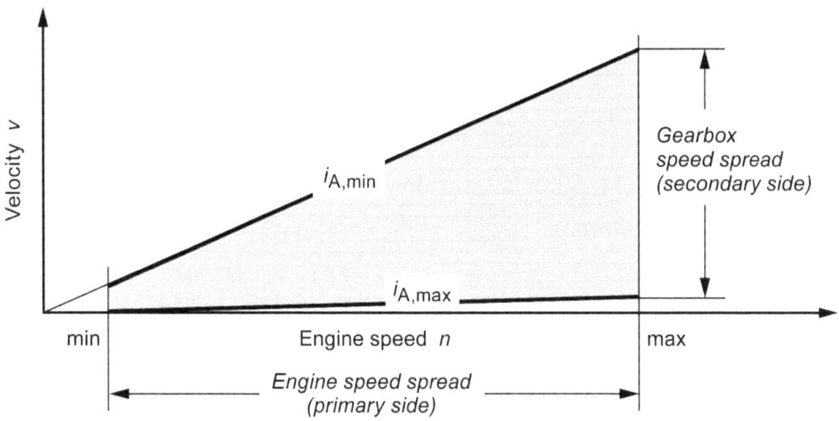


**Fig. 4.3.** Velocity/engine-speed diagram, overall gear ratio

### 4.2.1 Overall Gear Ratio $i_{G,tot}$

The overall gear ratio of the transmission, often referred to as the range of ratios, is the ratio between the largest and smallest ratio

$$i_{G,tot} = \frac{i_{G,max}}{i_{G,min}} = \frac{i_1}{i_z}, \quad \text{with the gears } n = 1 \text{ to } z. \tag{4.5}$$

The overall gear ratio depends on:

- the specific power output of the vehicle ($P_{max} / (m_F + m_{payload})$ in kW/t),
- the engine speed spread (see Section 3.3.2) and
- the intended use.

Vehicles with a low specific power output, such as commercial vehicles, need a larger overall gear ratio. The same applies for vehicles with diesel engines, which have a small engine speed spread. Reference values for overall gear ratios of various vehicles are shown in Figure 4.4.

For passenger cars in particular it is necessary to consider that:

- However great the overall gear ratio is, the transmission can only move the operating point on the demand power hyperbola (see also Figure 3.18).
- The most fuel-efficient range cannot be exploited by a passenger car with a powerful engine travelling on the level at moderate speeds since there is insufficient power required.
- The engine and all powertrain components have to fit together: powertrain matching, see Chapter 5.

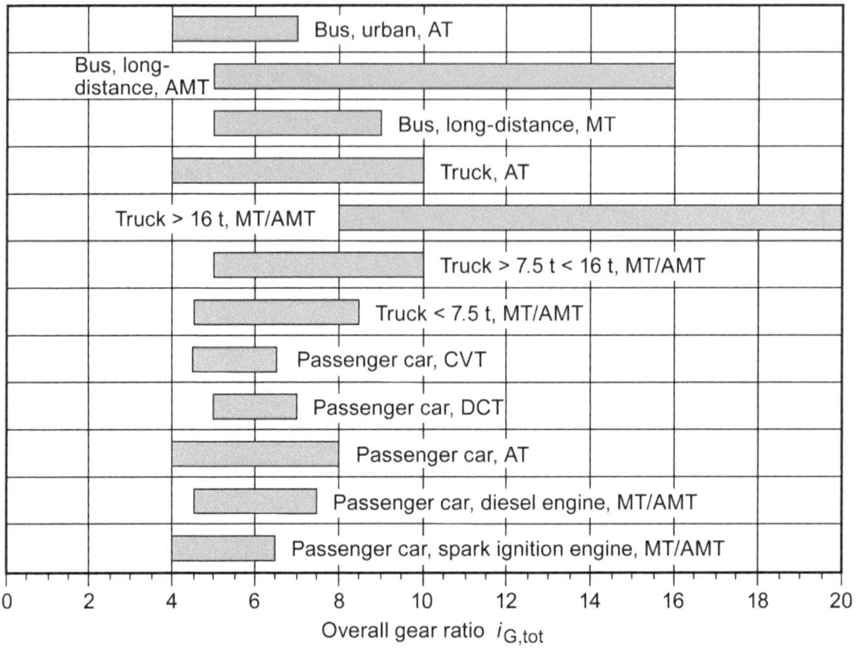


**Fig. 4.4.** Reference values for overall gear ratios for various types of vehicle. In the case of automatic transmissions, the conversion of the torque converter ($\mu_{max} \approx 2...3$) has to be added

### 4.2.2 Selecting the Largest Powertrain Ratio $i_{A,max}$

The greatest traction requirement must be known to determine the ratio of the gear with the largest torque multiplication. The friction limit – i.e. the maximum force that can be transmitted between the tyres and the road – is a physical limit and must be taken into account when establishing the traction $F_{Z,A}$ at the road wheels (see Equation 3.9)

$$F_{Z,A} \leq F_{Z,max} = \mu_H \, R\,.$$

Table 3.2 gives static friction coefficients $\mu_H$ for certain operational conditions. Air resistance may be ignored at the speeds anticipated in the lowest gear. At the drive wheels a balance must be struck between the maximum requirements of acceleration, gradient, road surface, payload and trailer load:

Maximum traction available $F_{Z,A}$ = Maximum traction required $F_{Z,B}$

$$T_{M,max} \, i_{A,max} \, \eta_{tot} \frac{1}{r_{dyn}} = m_F \, g \left( f_R \cos\alpha_{St} + \sin\alpha_{St} \right) + m_F \, \lambda \, a\,. \tag{4.6}$$

The largest ratio $i_{A,max}$, often called the stall torque ratio, depends mainly on the specific power rating (kW/t) of the vehicle. Two extreme conditions may be considered:

- The maximum gradient that can be climbed at an acceleration of $a = 0$ m/s$^2$. *Climbing performance*, Section 5.2.2 and
- The maximum acceleration on the level. *Acceleration performance*, Section 5.2.3.

The stall torque ratio for **passenger cars** and **commercial vehicles** designed for maximum gradability is, from Equation 4.6:

$$i_{A,max} = \frac{r_{dyn}\, m_F\, g\left(f_R \cos\alpha_{St} + \sin\alpha_{St}\right)}{T_{M,max}\, \eta_{tot}} . \tag{4.7}$$

The dynamic wheel radius $r_{dyn}$ of most common tyre sizes is shown in Table 3.3. Reference values for $r_{dyn}$ are: $\approx 0.3$ m for passenger cars and $\approx 0.5$ m for commercial vehicles. Reference values for the rolling resistance coefficient $f_R$ are shown in Table 3.1. A climbing performance of $q'_{max}$ greater than 50% is normally required for an unladen passenger car. This ensures that a trailer can be towed and steep ramps overcome with ease.

Acceleration performance depends not only on the stall torque ratio, but also to a significant degree on how closely the gears approximate to the traction hyperbola. The acceleration performance required depends very much on the brand image of the vehicle.

The largest ratio in **commercial vehicles** is often dictated by the vehicle's intended use. For example, building site vehicles and road sweepers have gears for extremely slow movement ($v_{crawl}$). Using the kinematic relationship

$$v = \omega_R\, r_{dyn} \tag{4.8}$$

the crawler gear in a commercial vehicle is given by

$$i_{A,max} = \frac{3.6 \frac{\pi}{30}\, n_{M,min}\, r_{dyn}}{v_{crawl}} . \tag{4.9}$$

where $n_{M,min}$ is in 1/min, $r_{dyn}$ is in m and $v_{crawl}$ in km/h. These very high-ratio gears are known as crawler gears.

### 4.2.3 Selecting the Smallest Powertrain Ratio $i_{A,min}$

Assuming there is no slip in the power transmission from wheel to road and that the (desired) maximum speed is reached at maximum engine speed, then the smallest powertrain ratio is given by

$$i_{A,min} = \frac{3.6 \frac{\pi}{30} n_{M,max}\, r_{dyn}}{v_{max}}. \tag{4.10}$$

where $n_{M,max}$ is in 1/min, $r_{dyn}$ in m and $v_{max}$ in km/h.

**Commercial Vehicles:**
The limiting factors of legal speed restrictions and diesel engine cut-off speed mean that the maximum speed will often be a design parameter when developing commercial vehicle powertrains. The design ranges for commercial vehicles in Europe arising from the maximum permissible speed $v_{max}$ are shown in Figure 4.5.

**Passenger Cars:**
There are various factors to be considered in selecting the smallest ratio. One factor is the high proportion of duty time spent in the highest gear, which can be more than 80% in the case of passenger cars. Depending on the type of design selected, a distinction is made between:

1/ $v_{max}$ – *optimum design*: $i_{A,min} = i_A(v_{max,th})$,
2/ *overrevving design*,
3/ *underrevving design*.

### 1/ $v_{max}$ – Optimum Design

In order to convert the maximum engine power installed in the vehicle into maximum performance, the required power curve $P_{Z,B}$ must pass through the point of maximum engine power available $P_{Z,Amax}$ (= $P_n$) [4.3]. This is called the design point *A*, Figure 4.6. It represents the maximum speed $v_{max,th}$ theoretically available ($q'$ = 0%; no wind). The acceleration reserve and fuel consumption in top gear are also important factors in the case of passenger car transmissions.

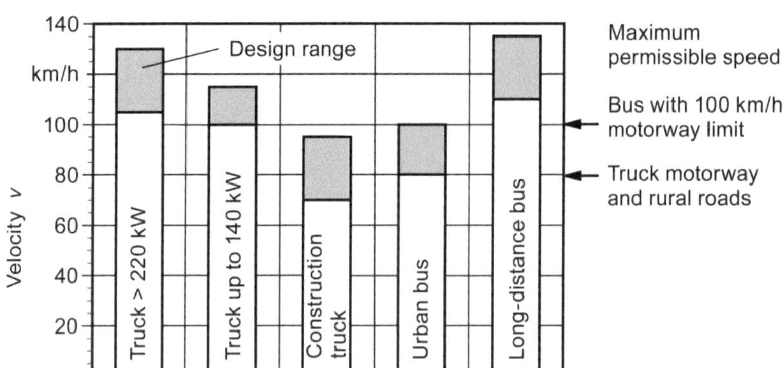


**Fig. 4.5.** Design speeds for determining $i_{A,min}$ for commercial vehicle powertrains. The maximum speed data relates to Germany

The excess power available $P_{Z,Ex}$ is a measure of acceleration reserve, and the engine speed $n_M$ serves as a measure for fuel consumption (Figure 4.6).

### *2/ Overrevving Design*

The power available and the power required intersect in the declining section of the power supply curve $P_{Z,A}$ as shown in Figure 4.6, Point *B*. The speed $v_{max2}$ which can be achieved at this point with an overrevving design is less than $v_{max,th}$.

The powertrain ratio $i_{A2,min}$ is greater than $i_{A1,min} = i_A(v_{max,th})$. This is achieved by increasing the ratio of the highest gear $i_z$ or the final ratio $i_E$. Since the engine speed is then higher for a given road speed, the operating point moves into the range of higher fuel consumption on the engine map. The high level of excess power $P_{Z,Ex2}$ makes this arrangement preferable for sports designs.

### *3/ Underrevving Design*

The power available and power required intersect on the rising section of the power supply curve, point *C*.

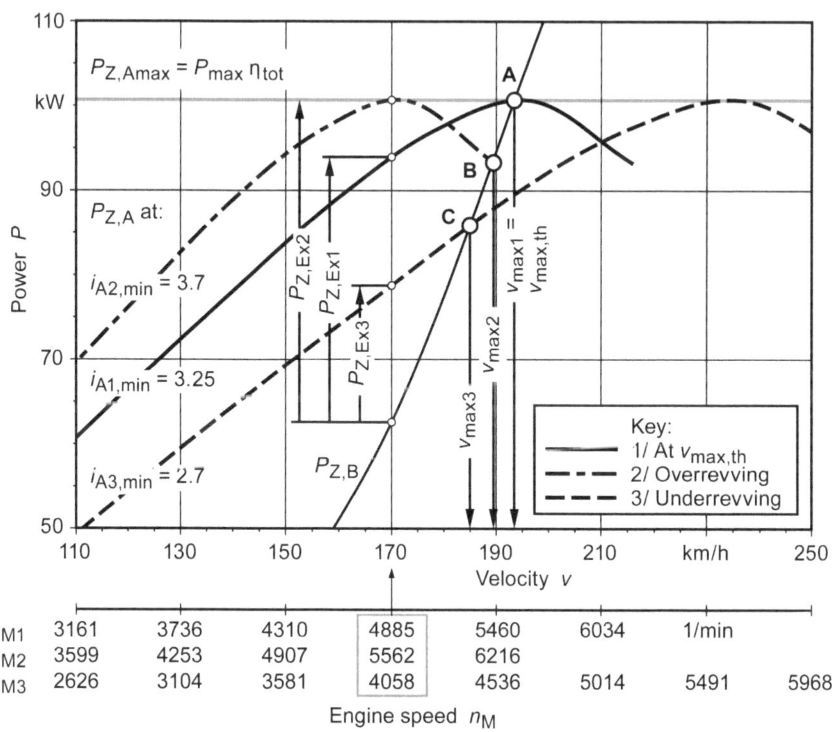


**Fig. 4.6.** Selecting the ratio in top gear for passenger cars. The excess power and associated engine speeds are highlighted for a speed of 170 km/h.
Design: 1/ at $v_{max,th}$; 2/ overrevving; 3/ underrevving

In this case the powertrain ratio $i_{A3,min}$ is less than $i_A(v_{max,th})$. The reduction in engine speed is the important feature of this design. The operating point moves into an area of improved fuel consumption. There are various approaches to reducing the powertrain ratio in the underrevving design, as follows:

- increase the overall gear ratio with the same number of speeds (Figure 4.7b),
- reduce the final ratio ("long axle design" – Figure 4.7c) and
- increase the overall gear ratio by increasing the number of gears – overdrive – (Figure 4.7d).

Figure 4.7 illustrates these approaches using the example of a passenger car powertrain with a 4-speed gearbox. The basic design is a powertrain optimised for $v_{max,th}$ (Figure 4.7a).

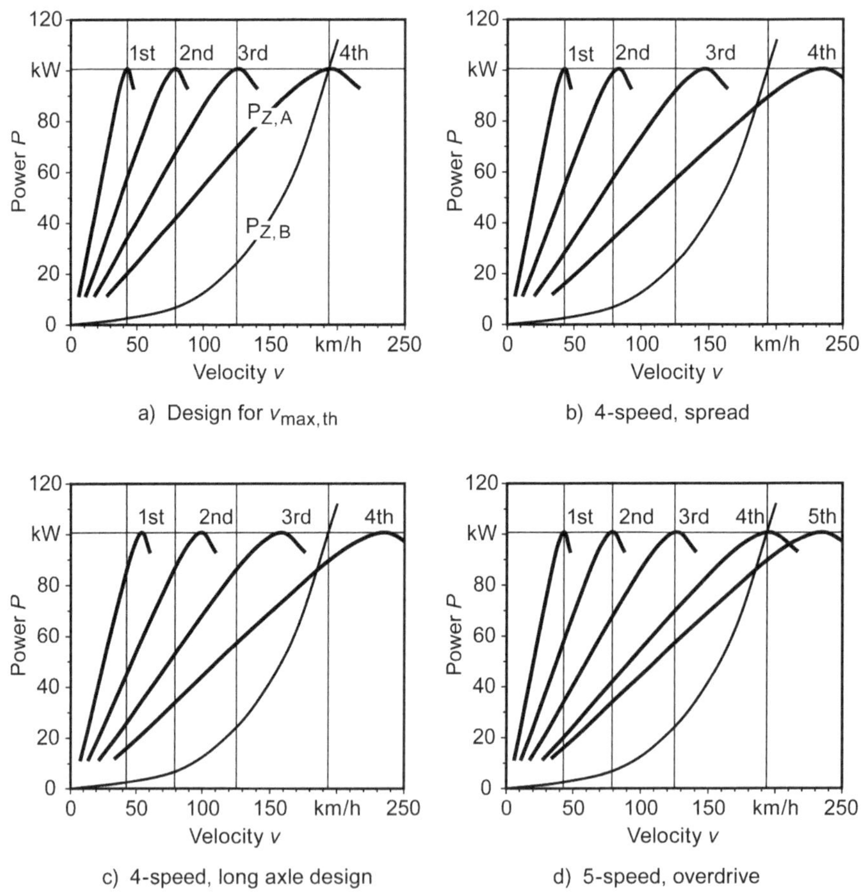


**Fig. 4.7.** Selecting underrevving powertrain ratios to improve fuel economy. Starting point: Figure 4.7*a*

The effect of increasing the overall gear ratio with the same number of gears is to create relatively large gaps in the power output, thus reducing the vehicle's acceleration performance. Increasing the final ratio ("long axle design") with the same overall gear ratio leads to a smaller stall torque ratio, and thus to reduced climbing performance and increased clutch stress when moving-off. For this reason, manufacturers now add a 5th and 6th gear (overdrive, economy gear) to the conventional 4-speed manual shift gearbox. Most top gear designs reduce engine speed by 10–20%, compared to the 4-speed design.

The fifth and sixth speed on manual passenger car gearboxes can be designed as overdrives to reduce engine speed. Alternatively, they can provide a closely stepped sports gearbox in which the increased number of gears serves to approximate more closely to the traction hyperbola and thus enhance performance.

### 4.2.4 Final Ratio

The final ratio $i_E$ selected adapts the handling characteristics and fuel consumption to the intended function, and is particularly important for commercial vehicles. For example trucks and buses operating mostly in flat terrain are fitted with longer axle designs than those operating in hilly regions. The longer axle design ($i_{E,long} < i_{E,normal}$) reduces the engine speed at normal driving speed, but also the excess traction in all gears (Figure 4.7c).

The final ratio achievable in a single stage is $2 \leq i_E \leq 7$. Larger ratios are achieved with an additional ratio stage in the final drive.

The various ways of designing the final ratio are listed systematically in Section 6.8. Section 12.3 describes some examples of final drives currently in use.

## 4.3 Selecting the Intermediate Gears

The relationship between the ratios of two neighbouring gears, the gear step $\varphi$, is given by

$$\varphi = \frac{i_{n-1}}{i_n} \leq \frac{n_{max}}{n(T_{max})}. \tag{4.11}$$

The transmission stepping should be large enough to enable the next lower gear $(n-1)$ to be engaged when the maximum engine torque is reached in gear $n$, without exceeding the maximum permissible engine speed $n_{max}$ (Figure 4.8). The following aspects should be considered when selecting the gear ratios:

- The greater the number of gears, the better the engine exploits its efficiency by adhering to the traction hyperbola. But as the number of gears increases, so does the frequency of gearshifting and the weight and size of the gearbox.

- The proportion of distance travelled in the lower gears is low, especially in the case of passenger cars.
- The proportion of distance travelled in each gear depends on the specific power output (kW/t), the route profile, the traffic conditions and driver behaviour.
- The smaller the gear step $\varphi$, the easier and more pleasant the gearshift action.
- The thermal load on the synchronizer rings is proportional to the square of the gear step.

In view of these in part contradictory aspects, compromises have to be made in designing the gearbox. Two formal methods for calculating gear steps have proved effective in practice:

- *geometrical gear steps* and
- *progressive gear steps.*

### 4.3.1 Velocity/Engine-Speed Diagram

The velocity/engine-speed diagram gives a good overview of appropriate configurations of the transmission ratios. It is often referred to as a gear plan or saw profile diagram, and has the road speed plotted against the engine speed for each gear $n$, from $n = 1$ to $z$.

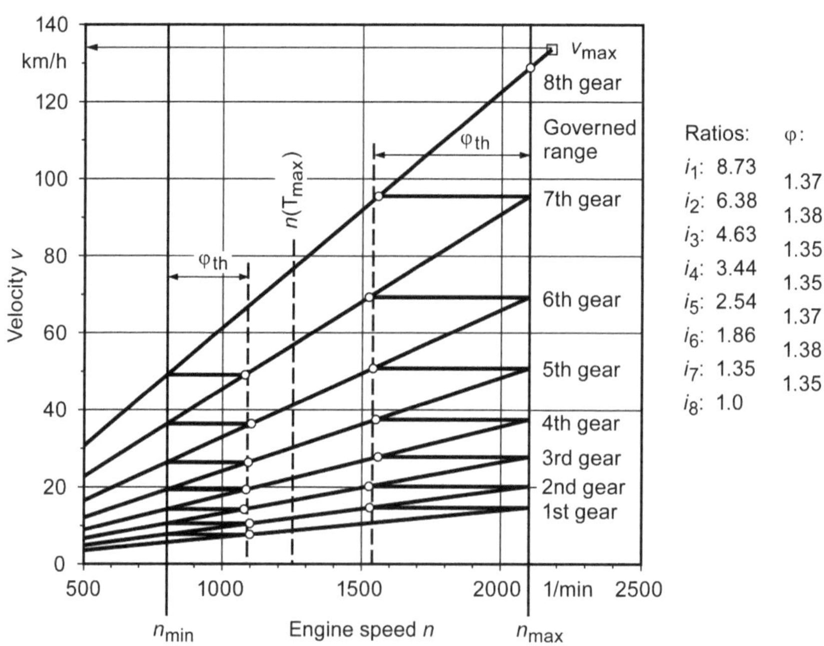


**Fig. 4.8.** Velocity/engine-speed diagram of a bus with 8-speed transmission, with range-change unit. Maximum road speed in the diesel engine governed range

Figure 4.8 shows the velocity/engine-speed diagram for a bus with an 8-speed transmission with range-change unit. The gearbox is geometrically stepped (see Section 4.3.2).

The maximum speed is reached in 8th gear in the "governed range" of the diesel engine (see also Figure 5.6). The saw profile diagram shows the earliest upshift possible without stalling the engine and the earliest downshift possible without exceeding the maximum engine speed.

### 4.3.2 Geometrical Gear Steps

In the geometric design the gear step $\varphi$ between the individual gears always has the same theoretical value

$$\varphi_{\mathrm{th}} = \sqrt[z-1]{i_{\mathrm{G,tot}}}\,. \tag{4.12}$$

The ratios of the individual gears $n = 1$ to $z$ is then given by

$$i_{\mathrm{n}} = i_{\mathrm{z}}\,\varphi_{\mathrm{th}}^{\,(\mathrm{z-n})}\,. \tag{4.13}$$

In practice the gear step will vary slightly from $\varphi_{\mathrm{th}}$ (Figure 4.8). The approximation to the effective traction hyperbola $F_{\mathrm{Z,Ae}}$ is equally good in all gears (Figure 4.9a). The difference in maximum speed between the gears consequently increases with each shift to a higher gear.

Geometrical gear steps are most common in commercial vehicle gearboxes; the lower specific power output means all the gear steps are of equal significance. Multi-range transmissions (Figure 4.8) have to be stepped geometrically to make all ratio steps the same size, preventing individual gears from overlapping (see also Section 6.7.1 Number 2/ "Multi-Range Transmissions").

### 4.3.3 Progressive Gear Steps

Progressive gear steps are used for passenger car transmissions. The higher the gear, the smaller the gear step. Figure 4.9b shows the progressive transmission stepping in the traction diagram and the velocity/engine-speed diagram (saw profile diagram). This shows clearly how the difference between the gear maximum speeds remains roughly constant with progressive gear steps. In the traction diagram the gaps between the effective traction hyperbola and the traction available are reduced in the top gears.

In the speed range relevant to passenger cars, this is reflected in improved shifting comfort (smaller $\varphi$), and in improved acceleration performance. The high level of excess power available in the lower speed range of passenger cars means that larger gaps in the traction availability are acceptable.

Given the overall gear ratio $i_{\mathrm{G,tot}}$ and the selected progression factor $\varphi_2$, the base ratio change $\varphi_1$ can be calculated thus:

$$\varphi_1 = \sqrt[z-1]{\frac{1}{\varphi_2^{0.5(z-1)(z-2)}} \, i_{G,tot}} \,. \tag{4.14}$$

The ratios $i_n$ in the gears $n = 1$ to $z$ are found to be

$$i_n = i_z \, \varphi_1^{(z-n)} \, \varphi_2^{0.5(z-n)(z-n-1)} \,. \tag{4.15}$$

Typical values are: $\varphi_1 = 1.1$ to $1.7$,
$\varphi_2 = 1.0$ to $1.2$.

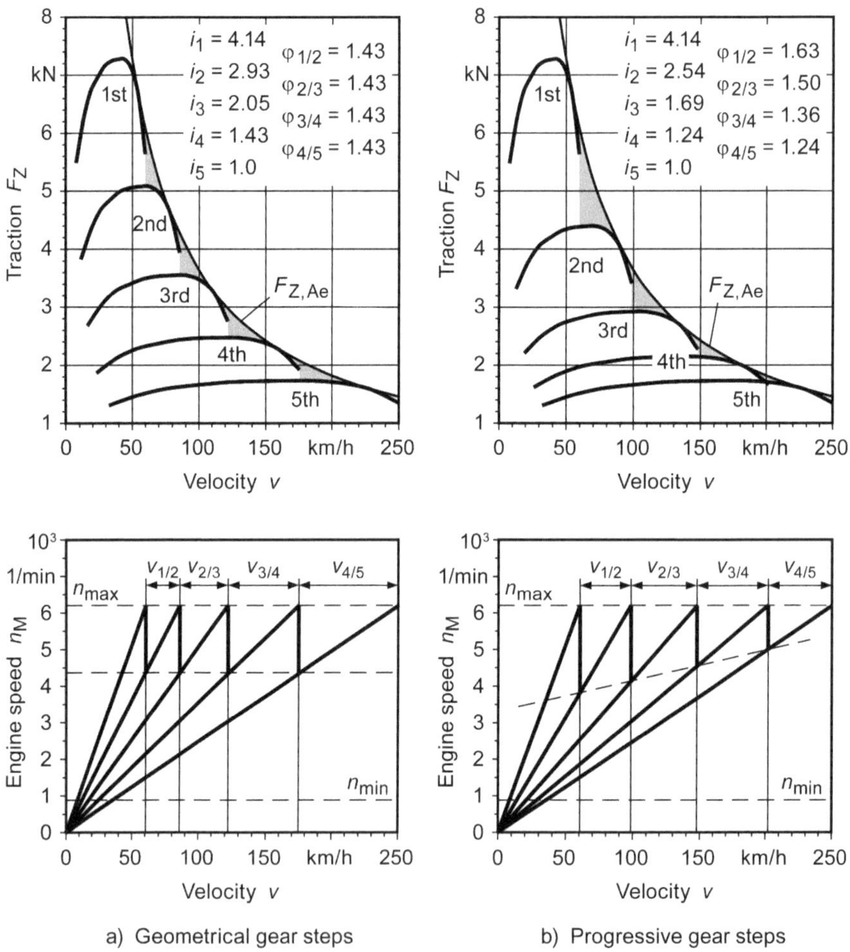


**Fig. 4.9.** Gear steps. Effects in the traction diagram and the velocity/engine-speed diagram (saw profile diagram). Ratios as shown in Table 4.1.
*a* Geometrical gear steps; *b* progressive gear steps

**Table 4.1.** *a* Geometrical gear steps; *b* transmission ratios of a 5-speed passenger car gearbox derived by calculation and ultimately built after fine tuning

| a) Gear: geom. | 1 | 2 | 3 | 4 | 5 |
|---|---|---|---|---|---|
| $i_{calculated}$ | 4.14 | 2.93 | 2.05 | 1.43 | 1.00 |
| b) Gear: progr. | 1 | 2 | 3 | 4 | 5 |
| $i_{calculated}$ | 4.14 | 2.54 | 1.69 | 1.24 | 1.00 |
| $i_{built}$ | 4.20 | 2.49 | 1.66 | 1.24 | 1.00 |

The above calculation method provides initial values to be used when selecting ratios. The gear ratios need to be adapted to the vehicle by a process of fine adjustment involving bench tests, road tests and computer simulation of field conditions. Design and manufacturing constraints have to be taken into account.

Other testing and acceptance conditions, such as fuel consumption and exhaust emissions, may also be significant in particular cases. Table 4.1 gives an example of the design of a passenger car transmission. The progression factor is $\varphi_2 = 1.1$ and the base ratio change is $\varphi_1 = 1.24$.

## 4.4 Ratio Variation in Continuously Variable Transmissions

Continuously variable transmissions are torque and speed converters whose ratio can be continuously varied without interrupting the power flow. In combination with an intelligent engine/transmission control, continuously variable transmissions make it possible to exploit the engine performance characteristic curve more fully.

Engine torque and speed may be freely selected with a continuously variable transmission, but they must be located on the current demand power hyperbola and in the operating map defined by the overall gear ratio (see also Figure 5.13 in Section 5.3.4 "Continuously Variable Transmissions"). Here the engine speed and transmission ratio are directly interrelated.

The overall gear ratio of continuously variable transmissions is normally $i_{G,tot} = 5–6$. They are normally described in terms of their adjustment range rather than the overall gear ratio. It is possible to achieve continuously variable pulley transmissions with a larger adjustment range by suitable engineering design (for example by using several power branches, see Section 6.6.6). The limited torque capacity of the chain currently restricts its use to vehicles with a transmission input torque of approximately 400 Nm.

The speed of adjustment of the transmission, and thus the change of engine speed, is a decisive factor in smooth running [4.2]. The adjustment speed is defined as

$$\dot{n} = \frac{\mathrm{d}n_\mathrm{M}}{\mathrm{d}t} = n_\mathrm{Output} \frac{\mathrm{d}i}{\mathrm{d}t}. \tag{4.16}$$

If the adjustment speed is too high, smoothness of operation suffers. The energy required to make the adjustment is partly taken from the kinetic energy of the vehicle. This can reverse the sign of the acceleration resulting in "gearshift jolt", which the driver finds uncomfortable.

If the adjustment speed is too low, the smoothness of operation is enhanced but the responsiveness of the vehicle suffers.

# 5 Matching Engine and Transmission

*Apart from design, the main factors affecting the market success of a vehicle are performance, fuel consumption, emissions and comfort*

Chapter 3 dealt with the available and the required power. Chapter 4 then elaborated the principles for selecting overall gear ratios. The purpose of this chapter is now to consider matching the transmission to the engine and the vehicle. This is a task in the field of vehicle longitudinal dynamics. The powertrain and its components are optimised by means of computer driving simulation and road and bench tests. The powertrain components – engine, moving-off element, selector gearbox, final drive etc. – must be "harmoniously" combined. This matching process is called "powertrain matching". The main optimisation criteria in this process are

- performance,
- fuel consumption,
- emissions and
- comfort.

This adaptation process has to be tackled from both sides, matching the engine to the transmission and vice versa. In practice, the characteristics of the engine dominate, and the characteristics of the transmission have to be "adapted" to match.

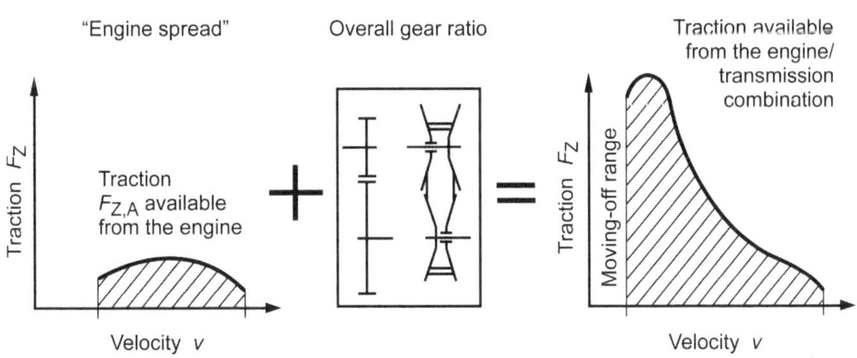


**Fig. 5.1.** Combined action of "engine spread" and overall gear ratio

H. Naunheimer et al., *Automotive Transmissions*, 2nd ed.,
DOI 10.1007/978-3-642-16214-5_5, © Springer-Verlag Berlin Heidelberg 2011

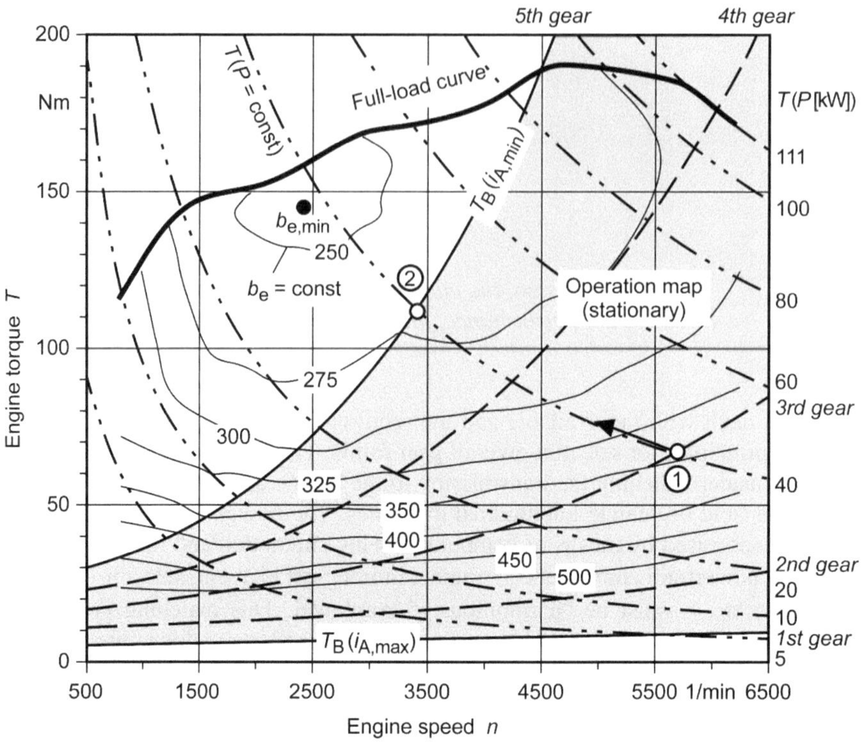


**Fig. 5.2.** Engine performance map with the characteristic onion curves of constant fuel consumption of a spark ignition engine. The driving resistance lines for minimum ($i_{A,min}$) and maximum ($i_{A,max}$) powertrain ratio define an operating map for points representing unaccelerated driving on a level surface. Vehicle, engine and transmission data as in Figure 5.3

The transmission mediates between the engine and the road surface; it adapts the traction available to the power required, ensuring the desired performance. For this purpose, the speed range of the engine is mapped to a wheel speed range or a road speed range. Similarly, the torque range of the engine is mapped to a torque or traction range at the wheels. The speed and torque range of the engine should be referred to as the "engine spread" as discussed in Section 3.3.2. Engine spread and the overall gear ratio together form a field of possible traction at the wheels (Figure 5.1). The transmission enables the most fuel-efficient operating regions of the engine to be exploited (Figure 5.2).

Fuel consumption is influenced greatly by the gear ratio and the final ratio specified, as well as the gearshift points selected. During unaccelerated driving on a level surface, there can be "discrete" operating curves (with geared transmissions), or a whole operating map (with continuously variable transmissions), between the driving resistance lines, $T_B$, for the minimum and maximum powertrain ratios. With geared transmissions the operating points lie at the intersections of the ratio-dependent driving resistance lines and the lines showing engine power available $T(P = \text{const})$. But the operating points can only lie within the specified field. The operating map shifts on uphill and downhill runs and when accelerating, because of the changing driving resistance.

In geared transmissions, the most favourable consumption area of the driving resistance curve is usually only covered by top gear. Figure 5.2 shows how an engine power of 40 kW, $T(P = 40\text{ kW})$, can be produced during unaccelerated movement on a level surface ($a = 0\text{ m/s}^2$ and $q' = 0\%$) fuel-inefficiently in 3rd gear at point *1*: $b_e \approx 350$ g/kWh, or fuel-efficiently in 5th gear at point *2*: $b_e \approx 270$ g/kWh. The ratio of the highest gear has been selected to be fuel-efficient when the driving resistance curve comes as close as possible to point $b_{e,min}$ of lowest specific fuel consumption. See also Section 3.3.3 "Consumption Map".

## 5.1 Traction Diagram

The acceleration and climbing performance in the various gears of the transmission must be checked. In the traction diagram (Figure 5.3), the traction available in each gear and the traction required at various gradients are plotted as a function of the vehicle speed using Equations 3.19 to 3.22.

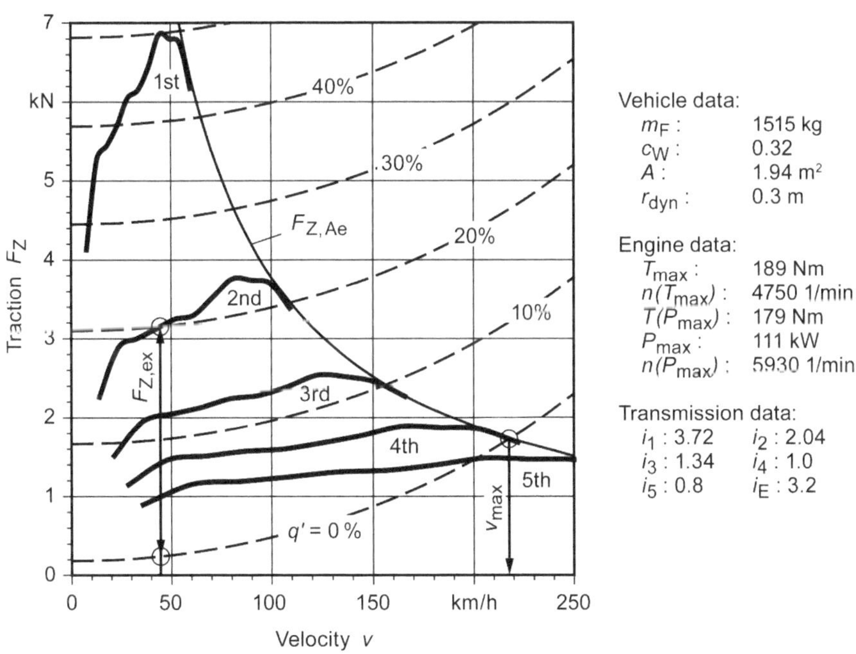


**Fig. 5.3.** Traction diagram with the demand curves for various gradients for a mid-size passenger car with a spark ignition engine as shown in Figure 5.2

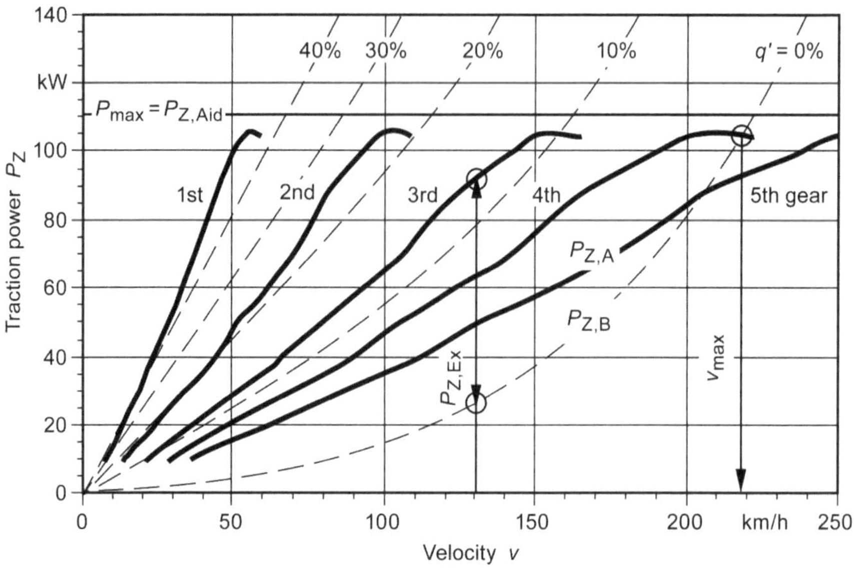


**Fig. 5.4.** Performance diagram (derived from the traction diagram, Figure 5.3)

The traction available is reduced by the powertrain efficiency, $\eta_{tot}$, which also includes the effect of losses due to auxiliary units such as servo pumps (see Section 3.1.7).

The power may be calculated by multiplying the traction by the corresponding speed, and can be shown on the performance diagram (Figure 5.4). The maximum speed, the maximum climbing performance and excess traction in the various gears can be found from the traction diagram. The excess traction $F_{Z,Ex}$ is given by the formula

$$F_{Z,Ex} = F_{Z,A} - F_{Z,B} = F_{Z,A} - F_R - F_{St} - F_L - F_a \tag{5.1}$$

$$= \frac{T(n_M)\, i_A}{r_{dyn}} \eta_{tot} - m_F\, g \left(f_R \cos\alpha_{St} + \sin\alpha_{St}\right) - \frac{1}{2} \rho_L\, c_W\, A v^2 - m_F\, \lambda\, a\,.$$

The traction diagram shows unaccelerated movement, i.e. when $a = 0$ m/s$^2$. To interpret the climbing and acceleration performance of a vehicle, the excess power available at the operating point, $F_{Z,Ex}$, can be written in two ways. Usually, to give climbing performance during unaccelerated movement

$$F_{Z,Ex} = F_{Z,A} - F_R - F_L = m_F\, g \sin\alpha_{St} \tag{5.2}$$

and to give acceleration performance during movement on the level

$$F_{Z,Ex} = F_{Z,A} - F_R - F_L = m_F\, \lambda\, a\,. \tag{5.3}$$

### 5.1.1 Deriving a Traction Diagram (Example)

Construction of the traction diagram in Figure 5.3 is explained below as an example. The procedure can be split into the following steps:

#### *A Determining the Traction Available*

1 *Specifying the initial dynamic operating parameters:*
The following calculations are based on the engine and transmission data given in Figure 5.3.

2 *Selecting some characteristic points on the full load curve:*
The full load curve of the sample engine is shown in Figure 5.2. This curve and the values at maximum torque and maximum power provide the first entries in Table 5.1.

**Table 5.1.** Design table of the traction diagram *(continued)*

| $n_M$ (1/min) | 800 | 2000 | 3000 | 4000 | 4750 | 5930 | 6200 |
|---|---|---|---|---|---|---|---|
| $T_M$ (Nm) | 115 | 150 | 170 | 175 | 189 | 179 | 170 |

3 *Calculating the associated gear-dependent speeds and tractions:*
This is to be carried out as an example for 1st gear with $i_1 = 3.72$. From Equation 4.8

$$v\,(\text{km/h}) = \frac{3.6\,\frac{\pi}{30}\,n_M\,(1/\text{min})\,r_{dyn}}{i_1\,i_E}:$$

| $v_{1stG}$ (km/h) | 7.6 | 19.0 | 28.5 | 38.0 | 45.1 | 56.3 | 58.9 |
|---|---|---|---|---|---|---|---|

The powertrain efficiency in 1st gear is assumed to be constant. $\eta_{tot} = 0.92$. Using Equation 3.22

$$F_{Z,A}\;(\text{kN}) = \frac{T(n_M)\,i_E\,i_1}{1000\,r_{dyn}}\,\eta_{tot}:$$

| $F_{Z,A1stG}$ (kN) | 4.2 | 5.5 | 6.2 | 6.4 | 6.9 | 6.5 | 6.2 |
|---|---|---|---|---|---|---|---|

4 *Entering the traction available/speed values on a diagram:*
See Figure 5.3, $F_{Z,A}$ curve for 1st gear.

#### *B Determining the Driving Resistance Lines*

1 *Determining the initial values:*
The initial values for calculating the driving resistance lines are the vehicle data given in Figure 5.3. The density of the air $\rho_L = 1.199\ \text{kg/m}^3$.

2 *Calculating the traction required at several speeds and gradients:*
Using Equation 3.20 for unaccelerated movement

$$F_{Z,B}\,[\mathrm{kN}] = \frac{1}{1000}\left[m_F\, g\left(f_R\,\cos\alpha_{St} + \sin\alpha_{St}\right) + \frac{1}{2}\,\rho_L\, c_W\, A\,\frac{v^2\,[\mathrm{km/h}]}{3.6^2}\right].$$

For gradients greater than 10%, the approximations $\cos\alpha_{St} \approx 1$ and $\sin\alpha_{St} \approx \tan\alpha_{St}$ are no longer acceptable. Entering the speed-dependent rolling resistance coefficient $f_R$ gives

| $v$ (km/h) | 0 | 50 | 100 | 150 | 200 | 250 |
|---|---|---|---|---|---|---|
| $f_R$ | 0.0124 | 0.0124 | 0.0131 | 0.0145 | 0.0200 | 0.0330 |
| $F_{Z,B0\%}$ (kN) | 0.18 | 0.26 | 0.48 | 0.86 | 1.45 | 2.29 |

3 *Entering the traction required/speed values on the diagram:*
See Figure 5.3, $F_{Z,B}$ curve for movement on the level, $q' = 0\%$.

### C Reading of Relevant Data

1 *Maximum speed:*
The maximum speed of the vehicle on a level surface is achieved in 4th gear, and is approximately 218 km/h. It is found at the intersection of the traction available line and the driving resistance line for $q' = 0\%$.

2 *Other performance data:*
See Tables 5.2 and 5.3.

## 5.1.2 Engine Braking Force

"Good braking means faster driving". This applies in particular to trucks with their large vehicle weight. Fast downhill speeds are required for trucks to achieve high average speeds and hence economic transport. Attainable downhill speeds are those which can be travelled without acceleration and without activating the service brake (friction brake). Depending on the type of braking, a distinction is made [5.11] between

- *steady-state braking:*
  preventing unwanted acceleration on downhill runs,
- *deceleration braking:*
  reducing speed and stopping if necessary and
- *braking at rest:*
  preventing undesired movement of the vehicle at rest.

In overrun conditions the internal combustion engine delivers braking torque (see Figure 3.14). The braking torque is essentially the result of pumping work in the cylinders. In the case of commercial vehicles, additional continuous service brakes, an exhaust throttle or retarder for example, can further increase the engine braking effect. See also Section 11.6 “Vehicle Continuous Service Brakes”.

The engine braking power available $F_{B,A}$ is incorporated in the traction diagram in a similar way to the traction available $F_{Z,A}$. $F_{B,A}$ is often also referred to as overrun drag. Power flow in overrun conditions is from the wheels to the engine. Whereas the traction diagram is calculated from the full load characteristic curve of the engine when power flow is from the engine to the road (Figure 5.5),

$$F_{Z,A} = \frac{T(n_M)\left(\dfrac{n_M}{n_R}\right)}{r_{dyn}}\eta_{tot} = \frac{T(n_M)\, i_A}{r_{dyn}}\eta_{tot}\,, \tag{5.4}$$

in overrun conditions the calculation is from the road to the thrust characteristic curve of the engine

$$T(n_M) = F_{B,A}\, r_{dyn}\left(\frac{n_R}{n_M}\right)\eta_{tot}\,, \tag{5.5}$$

$$F_{B,A} = \frac{T(n_M)}{r_{dyn}\,\eta_{tot}\left(\dfrac{n_R}{n_M}\right)} = \frac{T(n_M)\, i_A}{r_{dyn}\,\eta_{tot}}\,. \tag{5.6}$$

If the variation in the powertrain efficiency $\eta_{tot}$ = function (ratio, speed, torque) is taken into account in calculating engine braking force, then it must be remembered that the ratio is defined in the direction of power flow. That means that in overrun conditions “the ratio is reversed”.

The equation of motion for braking is derived from the equations for motion under power. See Section 3.1 “Power Requirement”. In deceleration braking, i.e. $a < 0$ m/s$^2$, inertial forces operate. They correspond to the acceleration resistance $F_a$. During steady-state braking $a$ is equal to 0 m/s$^2$. Braking is supported by rolling resistance and air resistance, which are given a negative sign.

On downhill runs the slope descending force $F_H$ corresponds to the gradient resistance $F_{St}$ with negative gradient, $q < 0\%$.

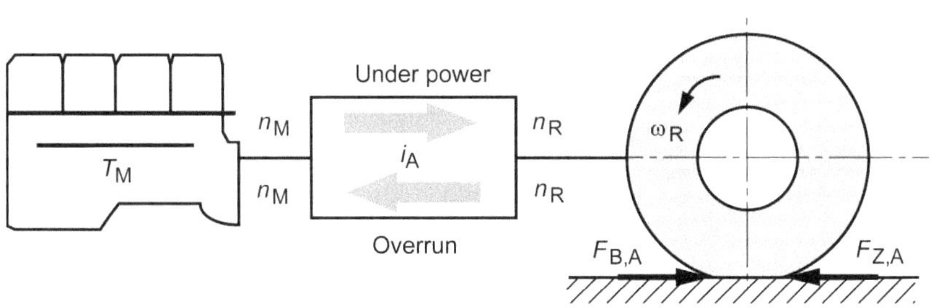


**Fig. 5.5.** Power flow under power and during overrun

Downhill, the resultant braking force requirement $F_{B,B}$ at the wheels is given by

$$F_{B,B} = F_H - F_a - F_R - F_L . \tag{5.7}$$

The braking force deficit, $F_{B,D}$, of the engine must be covered by the service brake or, in the case of commercial vehicles, by an additional continuous service brake system

$$F_{B,D} = F_{B,B} - F_{B,A}(n_M, i_A) . \tag{5.8}$$

Figures 5.3 and 5.4 are the traction diagram and performance diagram respectively of a mid-size passenger car with a spark ignition engine. Figure 5.6 shows the traction diagram of a 16 t truck. The influence of an additional splitter unit in reducing the gear steps can be seen there. The splitter unit enables a better approximation to the traction hyperbola $F_{Z,Ae}$.

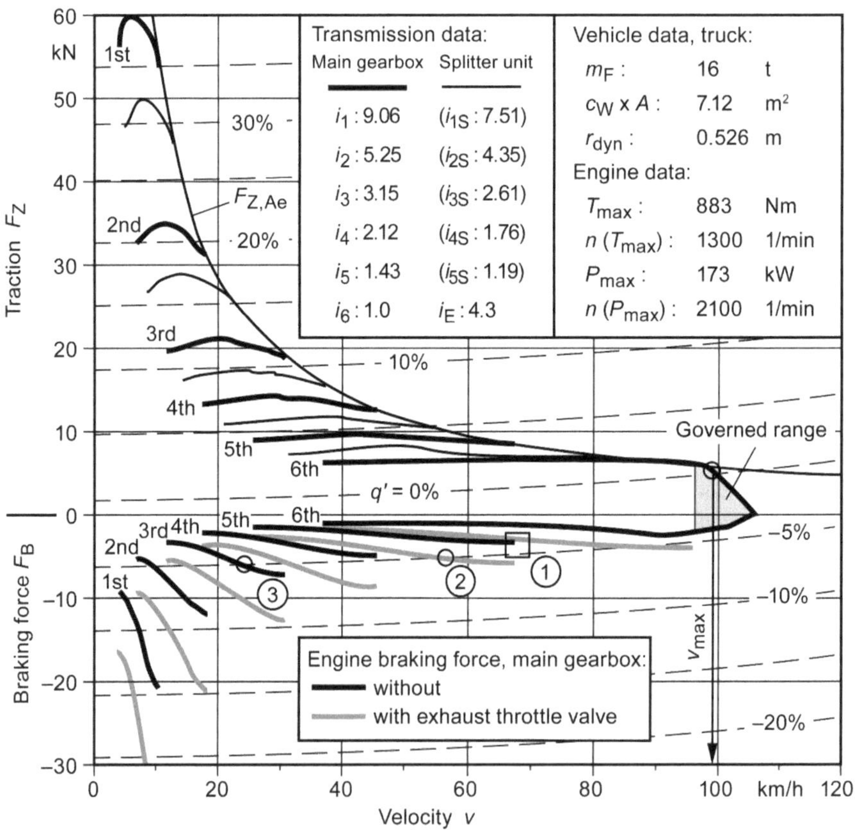


**Fig. 5.6.** Traction diagram of a 16 tonne truck with 6-speed gearbox. Engine braking curves with and without exhaust throttle valve

Besides the traction available, Figure 5.6 also shows the curve of maximum engine braking force for each gear. The 173 kW engine has an engine braking power of 57 kW at 2100 1/min. In Figure 5.6 the engine braking curves where an engine brake (exhaust throttle) is in use are represented by continuous grey lines. With an exhaust throttle the engine achieves a braking power of approximately 100 kW at 2100 1/min.

Engine braking force in 5th gear without an exhaust throttle is not adequate to prevent acceleration when travelling down a 5% slope (Point *1*); however this is made possible by using an engine brake in 5th gear (Point *2*). Without an exhaust throttle the vehicle would have to travel down the slope in 3rd gear at a lower speed (Point *3*).

### 5.1.3 Geared Transmission with Dry Clutch

Figures 5.3 and 5.6 show the interaction of a combustion engine with a geared transmission in the case of a passenger car (Figure 5.3) and a truck (Figure 5.6). In both cases the moving-off element is a conventional dry clutch. In trucks and buses with powerful engines the maximum speed is often reached in the governed range of the diesel engine, i.e. beyond the actual maximum engine speed.

### 5.1.4 Geared Transmission with Torque Converter

The converter test diagram (Figure 5.7) is necessary to determine the traction curve of a geared transmission with a torque converter of Trilok design. See also Chapter 10 "Moving-Off Elements" and Section 10.4 "Hydrodynamic Clutches and Torque Converters".

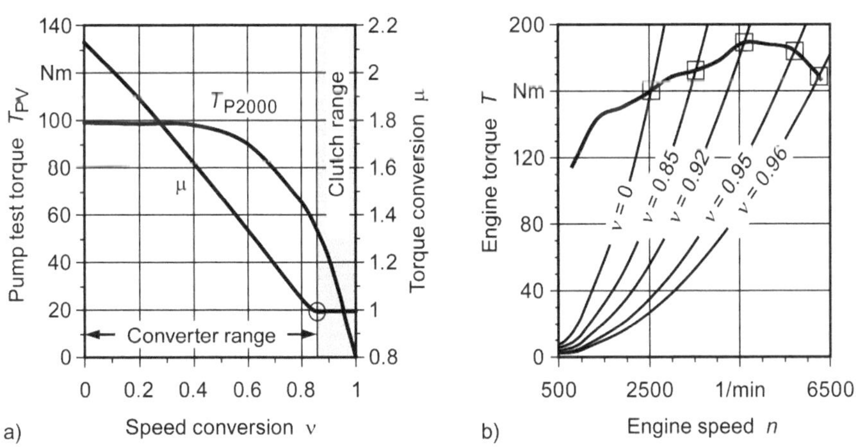


**Fig. 5.7.** *a* Converter test diagram; *b* engine performance map with torque converter parabolas. $n_{PV} = 2000$ 1/min; $\nu_C = 0.85$

In the converter test diagram (Figure 5.7a) the pump test torque $T_{PV}$ ($T_{P2000}$), determined at a pump test speed $n_{PV} = 2000$ 1/min, and the associated torque conversion $\mu$, are shown plotted against the speed conversion $\nu$.

The pump torque parabolas at constant speed conversion, the so-called converter parabolas, are added to the engine characteristic graph by using the information in the converter test diagram. The converter parabolas span a field of possible engine operating points. With the $T_{PV}$ values from Figure 5.7a and where $n_{PV} = 2000$ 1/min, the characteristic value of the torque converter then becomes

$$k(\nu) = \frac{T_{PV}}{n_{PV}^2} \tag{5.9}$$

and from this the converter parabolas may be found

$$T_P = k(\nu)\, n_P^2 , \tag{5.10}$$

where $T_p = T_M$ and $n_p = n_M$. At the intersections of the converter parabolas with the full load curve, the speed $n_P$ and the torque $T_P$ on the engine (or pump) side are now converted to the speed $n_T$ and the torque $T_T$ on the turbine side of the converter, where:

$$n_T = \nu\, n_P . \tag{5.11}$$

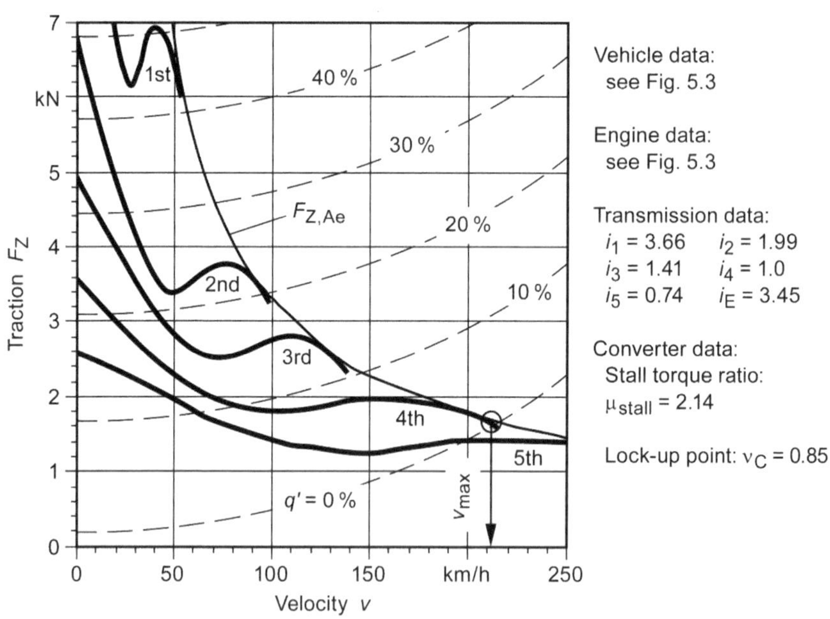


**Fig. 5.8.** Traction diagram of the vehicle as shown in Figure 5.3 with 5-speed automatic transmission and Trilok converter

For specified $\nu$-values the associated values of $\mu$ can now be read off the converter test diagram. Using these values, the turbine torque $T_T$ may be calculated as

$$T_T = \mu T_P . \tag{5.12}$$

The traction curves can now be calculated from the transmission input values $T_T$ and $n_T$, as described in Section 5.1.1. $\nu_C$ is the symbol for the speed ratio at the lock-up point. In the converter range of the Trilok converter, for $\nu < \nu_C$, $\mu > 1$. The combined action of engine and Trilok converter in a conventional automatic transmission is shown in Figure 5.8.

Figure 5.9 shows the turbine or transmission input torque curve for full and part load (lines of equal accelerator position). The lock-up point of the Trilok converter shifts in accordance with the pump parabola for $\nu_C$.

For torque transfer, Trilok converters have a differential rotational speed between the converter pump and the converter turbine even after the lock-up point. This has a negative effect on fuel consumption.

Closing the torque converter lock-up clutch (CC) – and thus the locking-up of the vibration-damping converter – is only possible if the torsional vibration induced by the engine in the powertrain does not go beyond the comfort boundary. "Hard" Trilok converters without a lock-up clutch, which have a steep $\nu_C$ parabola and are operated for long periods of time in the clutch range, are more favourable with respect to vibration decoupling than converters with a lock-up clutch. However, with respect to consumption, they are less favourable. Furthermore, hard converters with a lot of torque response are not suited to turbocharged diesel engines considering their weakness in moving-off. Trilok converters have therefore generally been equipped with lock-up clutches since the mid-1990s.

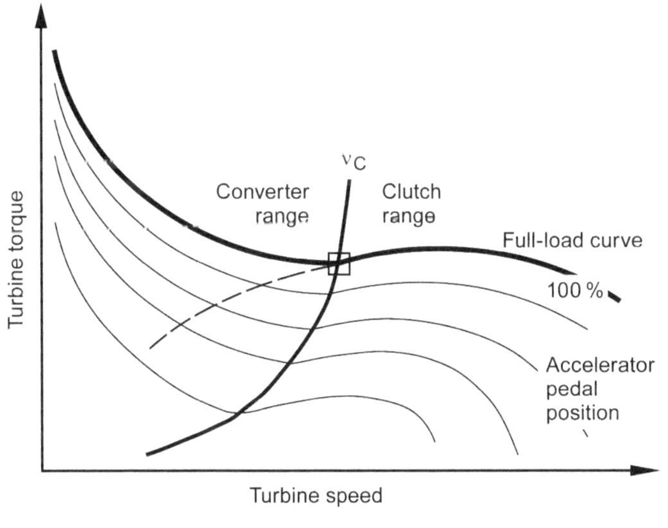


**Fig. 5.9.** Cooperation of engine and Trilok converter at full load and part load

Lock-up clutch requirements concerning reduction of fuel consumption, shifting quality, vibration decoupling and improving the vehicle's movement dynamics have increased constantly over the years. While it was sufficient until the early 1980s to support the moving-off process with the torque increase of the torque converter and then to close the clutch in the higher gears at higher speeds, torque converter lock-up clutches with controlled slip make it possible to extend significantly the driving ranges in which the clutch is active.

Modern systems with slip-controlled torque converter lock-up clutches (SCC) are already switched on in the 1st gear and operate under slip-control in all gears. This requires sophisticated control concepts as well as design measures with respect to thermal stress and vibration decoupling. See also Section 10.4.6 "Engineering Designs".

## 5.2 Vehicle Performance

The performance of a vehicle is defined by its maximum speed and its climbing and acceleration capability. The performance of a vehicle can be determined by comparing the traction available and the traction required at any point, as shown in Equation 5.1.

The procedure for determining the maximum speed, acceleration and traction of a vehicle is defined in standards (e.g. German standard DIN 70020 [5.2]). The performance data at the point of maximum engine torque, and at the point of maximum engine power are usually given in order to document the performance of a vehicle. Table 5.2 shows this for the vehicle used as an example in Figure 5.3. Table 5.3 shows some further driving conditions and consumption data for the vehicle travelling at constant speed.

### 5.2.1 Maximum Speed

The German standard DIN 70020 defines maximum speed as the average of the greatest speeds in both directions that a vehicle can maintain over a measured distance of 1 km. The main test conditions are:

- vehicle loaded with half the difference between the gross weight and the unladen weight,
- level, dry surface with good grip,
- maximum wind speed ± 3 m/s and
- the vehicle must travel along the test track in both directions without interruption.

The maximum speed may be found from the traction diagram at the intersection of the traction required curve with the traction available curve (Figures 5.3, 5.4, 5.6 and 5.8). The maximum speed in the various gears is determined by selecting the gear ratio of the individual gears of the gearbox.

**Table 5.2.** Performance data of the vehicle used as an example in Figure 5.3. Velocity $v$, traction $F_{Z,A}$, excess traction $F_{Z,Ex}$, climbing performance $q'_{max}$ and acceleration $a_{max}$ at the point $T_{max}$ = 189 Nm at 4750 1/min and $T_n = T(P_{max})$ = 179 Nm at 5930 1/min

| Gear | $v$ (km/h) at | | $F_{Z,A}$ (kN) at | | $F_{Z,Ex}$ (kN) at | | $q'_{max}$ (%) at | | $a_{max}$ (m/s$^2$) at | |
|---|---|---|---|---|---|---|---|---|---|---|
| | $T_{max}$ | $T_n$ | $T_{max}$ | $T_n$ | $T_{max}$ | $T_n$ | $T_{max}$ | $T_n$ | $T_{max}$ | $T_n$ |
| 1 | 45.8 | 57.2 | 6.9 | 6.5 | 6.6 | 6.2 | 49 | 46 | 3.7 | 3.4 |
| 2 | 83.5 | 104.3 | 3.8 | 3.5 | 3.3 | 3.0 | 23 | 21 | 2.0 | 1.8 |
| 3 | 127.0 | 158.5 | 2.5 | 2.4 | 1.8 | 1.4 | 12 | 10 | 1.2 | 0.9 |
| 4 | 170.1 | 212.4 | 1.9 | 1.8 | 0.8 | 0.2 | 5 | 1 | 0.5 | 0.1 |
| 5 | 213.2 | 266.1 | 1.5 | 1.4 | – | – | – | – | – | – |

**Table 5.3.** Driving conditions and consumption data for some constant speeds. Engine speed $n_M$, fuel consumption $b_s$ and climbing performance $q'_{max}$

| Gear | $n_M$ (1/min) at (km/h) | | | | $b_s$ ($l$/100 km) at (km/h) | | | | $q'_{max}$ (%) at (km/h) | | | |
|---|---|---|---|---|---|---|---|---|---|---|---|---|
| | 30 | 60 | 90 | 120 | 30 | 60 | 90 | 120 | 30 | 60 | 90 | 120 |
| 1 | 3112 | – | – | – | 15.1 | – | – | – | 33.7 | – | – | – |
| 2 | 1706 | 3412 | 5118 | – | 8.5 | 9.8 | 11.5 | – | 15.7 | 16.4 | 18.3 | – |
| 3 | 1122 | 2244 | 3367 | 4489 | 6.8 | 6.6 | 7.9 | 9.8 | 8.7 | 10.3 | 10.6 | 11.3 |
| 4 | 837 | 1675 | 2512 | 3350 | 6.1 | 5.7 | 6.8 | 8.4 | 4.9 | 6.4 | 6.5 | 6.0 |
| 5 | – | 1336 | 2005 | 2673 | – | 5.1 | 6.0 | 7.4 | – | 4.7 | 4.4 | 3.9 |

### 5.2.2 Climbing Performance

Climbing performance is represented by the gradient resistance as defined in Equation 3.13. Uniform speed ($a$ = 0 m/s$^2$) is assumed when determining climbing performance, so that the entire excess traction $F_{Z,Ex}$, as defined in Equation 5.2, is available to negotiate the gradient. The maximum climbing performance is given as

$$\sin \alpha_{St,max} = \frac{F_{Z,ex}}{m_F\, g} . \qquad (5.13)$$

It is normal practice to convert the angle of slope $\alpha_{St}$ into road gradient $q'$ in percent (Equation 3.14).

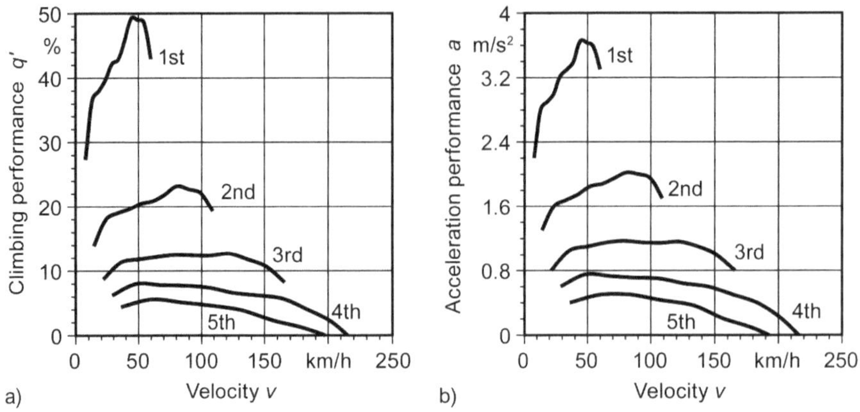


**Fig. 5.10.** *a* Dependence of climbing performance on gear; *b* acceleration of the test vehicle from Figure 5.3 dependent on gear

Excess traction as a function of speed can be read from the traction diagram for the various gears. The climbing performance in each gear can thus be calculated from Equation 5.13, and plotted in a diagram as a function of speed (Figure 5.10a).

### 5.2.3 Acceleration Performance

Maximum acceleration on the level ($\alpha_{St} = 0°$) can be derived from Equation 5.3

$$a_{max} = \frac{F_{Z,Ex}}{m_F \, \lambda_n} \,. \tag{5.14}$$

The acceleration performance in each gear may be calculated using Equation 5.14 and the gear-dependent coefficient of rotational inertia $\lambda_n$ (Figure 5.10b). In commercial vehicles the lowest gear is often given a high ratio to give the vehicle good climbing performance, even when fully loaded. The coefficient of rotational inertia can thus become very large, with the result that acceleration may be better in second gear than in first.

## 5.3 Fuel Consumption

Fuel consumption is not only a major factor determining the efficiency of a motor vehicle, but responsible utilisation of resources and pollution reduction are also becoming increasingly significant factors.

The fuel consumption of a vehicle is expressed either as

- consumption per distance travelled $b_s$ in $l$ /100 km and
- consumption per unit time $b_t$ in g/h.

It can be determined by calculation or by experiment. There are guidelines setting out the test conditions under which fuel consumption of passenger cars, trucks and buses must be measured [5.3]. Here too, the main test condition is loading the vehicle with half the difference between the gross weight and the unladen weight.

Real-world fuel consumption figures differ from those of the standard test cycles, due to the effects of driving style and road and traffic conditions, environmental factors and the condition and equipment of the vehicle. The driver has a major impact: the speeds and gears he selects determine the operating point of the engine and thus its fuel consumption.

### 5.3.1 Calculating Fuel Consumption (Example)

The specific fuel consumption $b_e$ at any momentary operating point may be read off the engine performance map with constant specific fuel consumption characteristic curves (Figure 5.2). This requires the engine speed $n_M$ and the associated engine torque $T(n_M)$. The engine speed is calculated from the road speed using Equation 4.8

$$n_M = \frac{v\, i_A}{2\,\pi\, r_{dyn}}. \tag{5.15}$$

The engine torque required $T_{Z,B}(n_M)$ is calculated from the traction required at the wheels and the powertrain efficiency, using Equation 3.22,

$$T_{Z,B}(n_M) = \frac{F_{Z,B}\, r_{dyn}}{i_A} \frac{1}{\eta_{tot}}. \tag{5.16}$$

If the engine map shows the effective mean pressure $p_{me}$ in the cylinder instead of the engine torque or engine power, then Equation 3.27 may be used to perform the conversion. The required engine power $P_{Z,B}(n_M)$ is given by

$$P_{Z,B}(n_M) = F_{Z,B}\, v \frac{1}{\eta_{tot}}. \tag{5.17}$$

The fuel consumption per unit distance can then be calculated using Equations 5.16 and 5.17:

$$b_s = \frac{b_e\, P(n_M)}{\rho_{fuel}\, v} = \frac{b_e\, F_{Z,B}}{\rho_{fuel}\, \eta_{tot}}. \tag{5.18}$$

In this example the fuel consumption at the operating points marked *1* and *2* is calculated using the engine performance map in Figure 5.2.

### *Example*

A vehicle with the data shown in Figure 5.3 is to travel on a level surface at a constant speed of 150 km/h. Given that $f_R = 0.0145$ and $\rho_L = 1.199$ kg/m$^3$, the traction requirement at the wheels is 862 N. With a powertrain efficiency of $\eta_{tot} = 0.92$ (assumed constant), Equation 5.17 gives a required engine power of approximately 40 kW. If the vehicle travels at this speed in 3rd gear, then operating point *1* in Figure 5.2 shows a specific fuel consumption $b_e \approx 350$ g/kWh. Substituting into Equation 5.18, with a petrol density of $\rho_{fuel} = 755$ g/*l* (Figure 3.17),

$$b_s = \frac{350\left(\frac{\text{g}}{\text{kWh}}\right)40(\text{kW})}{755\left(\frac{\text{g}}{l}\right)150\left(\frac{\text{km}}{\text{h}}\right)} = 0.124\left(\frac{l}{\text{km}}\right) = 12.4\left(\frac{l}{100\,\text{km}}\right).$$

Driving in 5th gear (operating point *2* in Figure 5.2), $b_e \approx 270$ g/kWh. Substituting the traction requirement directly into Equation 5.18

$$b_s = \frac{270\left(\frac{\text{g}}{\text{kWh}}\right)862\left(\text{N} = \frac{\text{Ws}}{\text{m}} = \frac{\text{kWh}}{\text{km}\,3600}\right)}{755\left(\frac{\text{g}}{l}\right)0.92} = 0.093\left(\frac{l}{\text{km}}\right) = 9.3\left(\frac{l}{100\,\text{km}}\right).$$

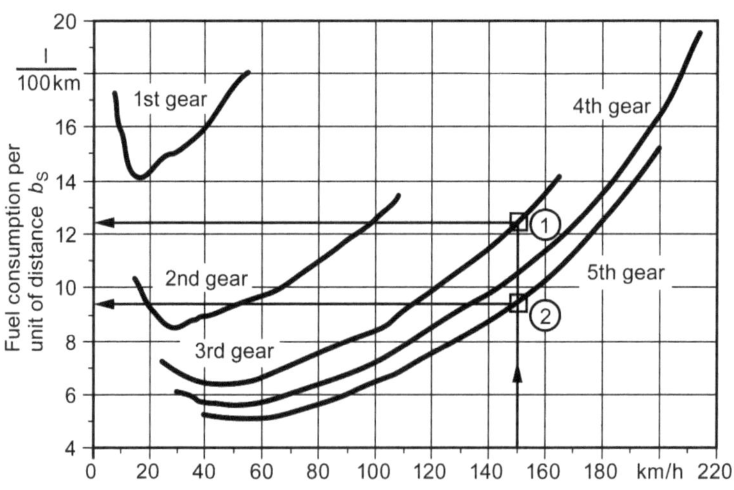


**Fig. 5.11.** Fuel consumption related to gear for the vehicle used as an example in Figure 5.3. Operating point *1,* from Figure 5.2: (150 km/h in 3rd gear) gives a fuel consumption of 12.4 litres/100 km. Operating point *2* on the other hand gives 9.3 litres/100 km in 5th gear

The driver can thus have a decisive effect on fuel consumption by his gear selection and the timing of gearshifts. Figure 5.11 shows the fuel consumption of the sample passenger car in each of the gears. In each gear there is a speed at which fuel consumption is optimal. Because air resistance increases as the square of speed, the power requirement, and thus fuel consumption, increases rapidly at high speed.

In order to use Equation 5.18 to calculate fuel consumption for driving cycles with varying speed, the cycle must be broken down into small time intervals during which acceleration is assumed to be constant. The calculation of fuel consumption for standard cycles and on free routes is an important application for driving simulation programs. See Chapter 14 "Computer-Aided Transmission Development".

## 5.3.2 Determining Fuel Consumption by Measurement

Fuel consumption is measured either on a roller test bench or in road tests. Three methods are used:

- gravimetric or volumetric measuring methods,
- flow measurement and
- determining consumption from the carbon balance of the exhaust gas composition.

Standardised cycles are normally used to assess the consumption and emissions performance of vehicles (Table 5.4). In these cycles, the most important variable is the road speed profile, with any stationary periods, plotted over time.

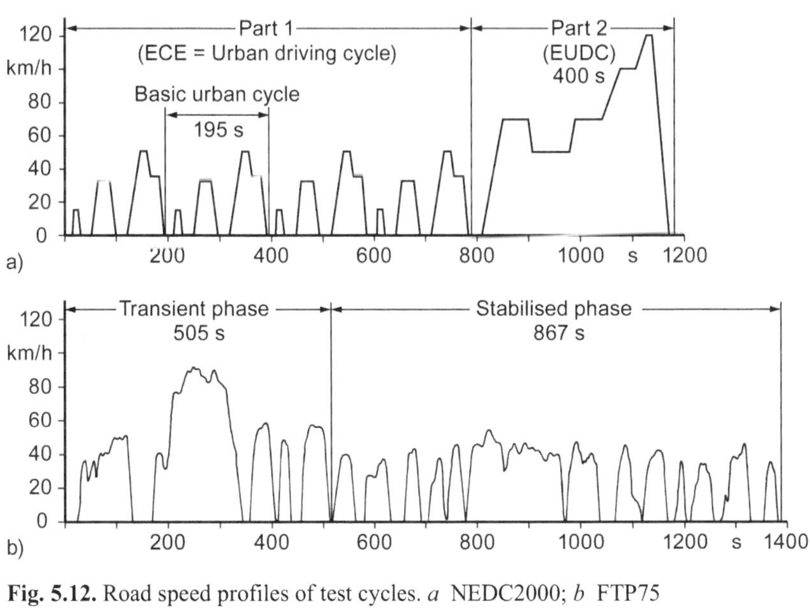


**Fig. 5.12.** Road speed profiles of test cycles. *a* NEDC2000; *b* FTP75

**Table 5.4.** Important driving cycles specified for passenger cars

| Cycle | Time (s) | Length (m) | ø-$v$ (km/h) | $v_{max}$ (km/h) | Special features | Aim |
|---|---|---|---|---|---|---|
| **NEDC2000** (EU) | 1180 | 11007 | 33.6 | 120 | • high idling proportion (28%) | • Emissions<br>• Consumption<br>• OBD test |
| **City-Cycle UDDS** (USA) | 1372 | 12068 | 31.4 | 91.2 | • Part of FTP75 | • Emissions<br>• Consumption |
| **FTP75** (USA) | 1877 | 17763 | 34.1 | 91.2 | | • Emissions<br>• Consumption |
| **Highway-Cycle** (USA) | 765 | 16444 | 77.4 | 96.4 | • Determination of fuel consumption<br>• Determination of $NO_x$ | • Consumption |
| **SC03-Cycle** (USA) | 594 | 5792 | 34.8 | 88.2 | • Start control cycle<br>• Air conditioning on<br>• SFTP cycle: single test or together with FTP75 | • Emissions |
| **US06-Cycle** (USA) | 600 | 12872 | 77.9 | 129.2 | • Load cycle (highway cycle)<br>• Hot test<br>• SFTP cycle: single test or together with FTP75 | • Emissions |
| **New York City Cycle** (USA) | 600 | 1931 | 11.4 | 44.6 | • Running loss test | • Evaporation measurement at passenger cars with SI engine |
| **CARB Unified Cycle** (USA) | 1735 | 17699 | 36.7 | 108.1 | • Optional instead of FTP75 | • OBD test |
| **10•15-Mode** (Japan) | 660 | 4160 | 22.7 | 70 | • Hot start test<br>• Until 2010 | • Emissions<br>• Consumption |
| **11-Mode** (Japan) | 505 | 4083 | 30.6 | 60.0 | • Cold start test<br>• Up to 2007 just for SI engines, from 2007 also for diesel engines | • Emissions |
| **JC08** (Japan) | 1204 | 8200 | 24.4 | 81.6 | • From 2008 planned as cold start test<br>• From 2011 planned as cold start and hot start test | • Emissions |

Figure 5.12a shows the road speed profile of the European NEDC2000 (New European Driving Cycle) from the year 2000. This driving cycle consists of 4 consecutive basic urban driving cycles of 195 s duration each – the ECE urban cycle – and of the 400 s EUDC (Extra Urban Driving Cycle). In Figure 5.12b, the profile of speed over time for the American FTP75 (Federal Test Procedure) test cycle is shown. This test program consists of a 2-phase UDDS (Urban Dynamometer Driving Schedule) urban cycle, in which the cycle is first completed (cold transition phase and stabilised phase), and the first phase (0 – 505 s) is repeated as a hot test phase after a 10 minute stationary period.

In the USA, the overall fuel consumption of all a manufacturer's vehicles, called "fleet consumption", is regulated by law. There are no such legal requirements in Europe yet. However, there is a voluntary agreement between the ACEA (European Automobile Manufacturers Association) and the EU to lower the average emission of $CO_2$ in new passenger cars to 140 g/km by the year 2008. The Japanese organisation JAMA and the Korean KAMA have made similar agreements.

Fuel consumption for trucks is normally measured on the road or calculated by computer simulation. The wide variation in equipment and fittings makes it more difficult to compare consumption between commercial vehicles than between passenger cars. Calculations are normally based on the engine's specific consumption curve at full load. The fuel consumption of the test vehicles is usually also measured under part load at a series of constant speeds, such as 70, 80 and 95 km/h. For buses, especially schedule service buses, there are company-specific duty cycle specifications.

With commercial vehicles there is a goal conflict between road speed and fuel consumption. One possible evaluation option [5.7] is the efficiency factor:

$$\text{Commercial vehicle efficiency factor} = \frac{\text{Average speed } \varnothing v}{\text{Consump. per unit distance } b_\mathrm{s}}. \tag{5.19}$$

The higher the efficiency factor, the better.

### 5.3.3 Reducing Fuel Consumption

The transmission affects fuel consumption in two ways. One factor is its own transmission losses (Sections 2.4.6 and 3.1.7), the other is providing suitable ratios for fuel-efficient utilisation of engine power. Geared transmissions are now so efficient that the former offers hardly any prospect of improvement. Transmission efficiency however remains a significant factor with continuously variable transmissions. But the main factor affecting consumption is still the driver!

The main means available for reducing fuel consumption are as follows:

- Optimising the efficiency of the internal combustion engine, in particular by reducing part-load consumption.
- Appropriate engine performance characteristics, i.e. the vehicle must be neither over-powered nor under-powered.

- Reducing driving resistance, for example rolling resistance and drag.
- Reducing the power draw of accessories such as servo pumps, air conditioning etc.
- Improving the efficiency of the transmission. This relates principally to continuously variable transmissions, which includes torque converters.
- Adaptive control of ratio selection by automatic transmissions with various gear ratios and continuously variable transmissions.
- Traffic management systems to reduce stationary periods.
- Improved driving. Intelligent control systems, which protect the driver against his own misjudgement. There are many factors involved in determining how far this "usurping" of control can go.

### 5.3.4 Continuously Variable Transmissions

Unlike geared transmissions, continuously variable transmissions offer the possibility of selecting engine operating points on the demand power hyperbola according to a pre-defined strategy. The operating point is derived from the intersection of the line $T(P)$ with the control characteristic curve. In principle any point within the operating map covered by the overall gear ratio may be selected (Figure 5.13). In steady-state travel on the level, all the operating points are inside the operating map.

Transmission efficiency is a decisive factor with continuously variable transmissions (which here relates principally to pulley transmissions). It is significantly worse than that of geared transmissions. This is partly due to the chain itself, but also the variable displacement and contact pressure pump which the design requires; it is needed to ensure that the taper discs make contact at all power levels. The efficiency of the engine and the transmission offset each other. You could say "what the right hand gives the left hand takes away".

Different control characteristics result, depending on which of the two criteria is optimised – consumption or performance (Figure 5.13). The control characteristic optimum in terms of fuel consumption is given by the line of minimum fuel consumption. A control characteristic giving optimum performance gives rise to high excess traction at every operating point.

The control characteristic chosen is bound to involve a compromise taking into account driveability. A single control characteristic is not adequate for this purpose. Adaptive strategies are required for changing the ratio in line with the driving situation. See also Chapter 13.

## 5.4 Emissions

As with consumption, a distinction can be made in the case of emissions between those caused directly by the transmission and those which result from the engine operating point to which the ratio gives rise.

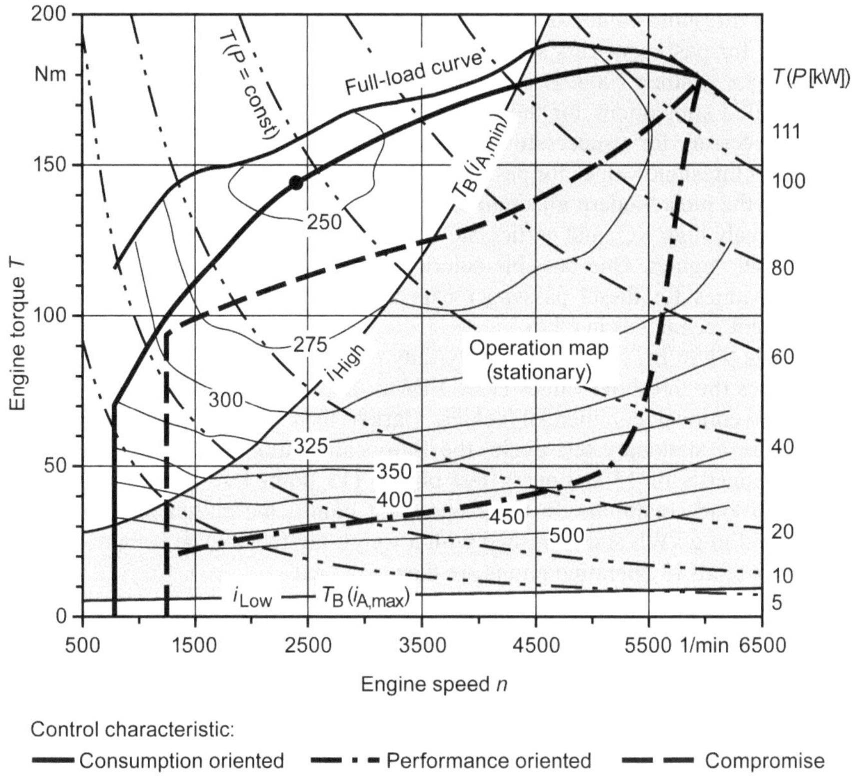


**Fig. 5.13.** Engine performance map as shown in Figure 5.2 with exemplary control characteristics for a continuously variable transmission: consumption oriented, performance oriented and a compromise solution

The causes and remedies of transmission noise emissions are discussed in Section 7.5 "Developing Low-Noise Transmissions".

Emissions from internal combustion engines are harmful to the environment. The chief pollutants are

- carbon monoxide CO,
- nitrogen oxides $NO_x$,
- unburned hydrocarbons HC and

with diesel engines also

- soot particulate.

Emissions of the greenhouse gas $CO_2$ are proportional to fuel consumption. Emissions for passenger cars are assessed and compared on the test cycles described in Table 5.4.

In the EU, there have been exhaust gas threshold values for passenger cars since the beginning of the 1970s and for commercial vehicles since the end of the

1970s. The threshold values are determined by Directive 98/69/EC. EURO IV has been valid for passenger cars since 2005 and for light-duty commercial vehicles (less than 3.5 t) since 2006. Besides the threshold values, the directive also provides detailed stipulations for the preparations, boundary conditions and test sequences necessary for a successful homologation. Table 5.5 shows the example of exhaust gas threshold values for passenger cars in accordance with EURO IV.

Despite the most modern injection technology and favourable consumption figures, relatively high $NO_x$ and particulate emissions will remain a challenge even in future diesel engines. One possible solution is the particulate filter. The particle threshold values for diesel passenger cars are determined with the appropriate measurement techniques in g/km.

Directive 99/96/EC is valid for heavy-duty commercial vehicles within the EU. It designates the threshold values EURO IV as of 2005 as well as EURO V as of 2008. Conventional commercial vehicle (larger than 3.5 t) diesel engines are measured in a stationary test cycle, the European Stationary Cycle (ESC). The ESC test consists of 13 stationary test phases (13 point test) at varying engine speeds and load conditions. In these operating points, the pollutant components are recorded in g/kWh and furnished with a weighting factor. The weighted emission values of all 13 operating points are then averaged.

Commercial vehicle diesel engines with exhaust gas aftertreatment technology are, in addition to the ESC, also subject to a dynamic test cycle, the European Transient Cycle (ETC). The ETC test consists of a series of unsteady, quickly shifting test phases. For cost reasons, exhaust gas inspection of heavy-duty commercial vehicles is executed on a roller test bench or on engine test benches.

The successful homologation is a first milestone. Of course, the goal is maintaining low emissions in real operation throughout the vehicle's entire service life. One tool for this is "On-Board Diagnostics" (OBD). OBD is a suitable means for checking emissions by monitoring the functionality of each vehicle component that is relevant to exhaust gas. If a component fails or shows a malfunction leading to a decline in exhaust gas values, the MIL (Malfunction Indication Lamp) activates on the instrument panel and an error code is recorded. The diagnostic functions must be designed such that they are executed frequently, not only during trial test, but particularly in everyday driving.

The internal combustion engine has the greatest influence on the exhaust gas behaviour of a vehicle. For this reason, a large part of on-board diagnostics is executed in the engine control unit (ECU).

**Table 5.5.** Exhaust gas threshold values for passenger cars according to EURO IV (limits for light commercial vehicles are variably higher, depending on the weight of the vehicle)

| Passenger cars | CO (g/km) | HC (g/km) | HC + $NO_x$ (g/km) | $NO_x$ (g/km) | Particulate mass (g/km) |
|---|---|---|---|---|---|
| Spark ignition engine | 1.0 | 0.1 | – | 0.08 | – |
| Diesel engine | 0.5 | – | 0.3 | 0.25 | 0.025 |

In principle however, every assembly group in the powertrain is relevant to OBD as soon as it exhibits its own control unit. This brings the transmission into the OBD's focus as well. For example, faulty regulation of the torque converter lock-up clutch of a conventional automatic transmission can have a negative influence on consumption and exhaust gas. To guarantee proper transmission functions, diagnostic routines are constantly running in the transmission control unit that monitor all actuators and sensors for functionality, plausibility and electric faults and, if necessary, activate replacement or emergency functions. Exhaust gas-relevant information are also processed and relayed by the program module "OBD management" for on-board diagnostics, also causing the MIL to activate. "OBD calibration" complies with requirements pertaining to both the vehicle and legislation.

The way that the OBD information of the transmission is processed in the control unit network depends on the vehicle. Figure 5.14 shows an example of the cooperation of the engine (ECU) and transmission (TCU) control units. Here, "OBD management" of transmission functions runs independently in the transmission, while transmission-relevant "OBD calibration" is subdivided between the TCU and the ECU. MIL activation is controlled solely by the ECU.

In the US, the forerunner with respect to exhaust gas legislation for motor vehicles is CARB, the California Air Resources Board. In the case of California's LEV legislation (LEV = Low Emission Vehicle), there are no consistent threshold values for all new vehicles such as is the case in EU directives. Rather, there are different threshold values and their gradual implementation. Vehicle manufacturers are obligated to maintain an average fleet emission standard of all vehicles sold.

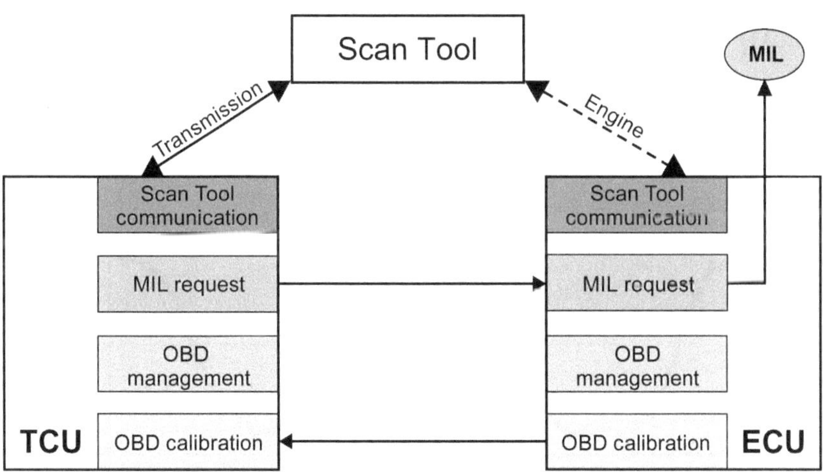


**Fig. 5.14.** Example of the cooperation of the transmission (TCU) and engine (ECU) control units during transmission-relevant OBD management. By means of an external diagnostic tool (scan tool), faults relevant to exhaust gas can be checked and removed

## 5.5 Dynamic Behaviour of the Powertrain, Comfort

Customer comfort requirements on the dynamic behaviour of the powertrain are very high, especially in the case of modern passenger cars and buses. This means paying particular attention to the vibration and thus, most importantly, the noise produced by the powertrain. The powertrain is an oscillatory system. The individual components have different masses, stiffness and damping factors. Mechanical substitute models of the powertrain can be developed using springs, dampers and inertial masses (Figure 5.15). The degree of complexity of the model relates to the purpose of the investigation.

The main source of excitation is the irregular, (considered ideally) sinusoidal speed profile of the combustion engine (Figure 5.16). The drive torque of the combustion engine pulsates with the ignition frequency and stimulates these torsional vibrations. It is not the base engine frequency (i.e. engine speed) but rather its multiple (engine order) that is decisive with respect to torsional vibration. For a four-stroke combustion engine, the half of the cylinder number comprises the decisive share of torsion oscillation. In the case of a four-cylinder, four-stroke combustion engine for example, the second order is the dominating multiple of the engine speed.

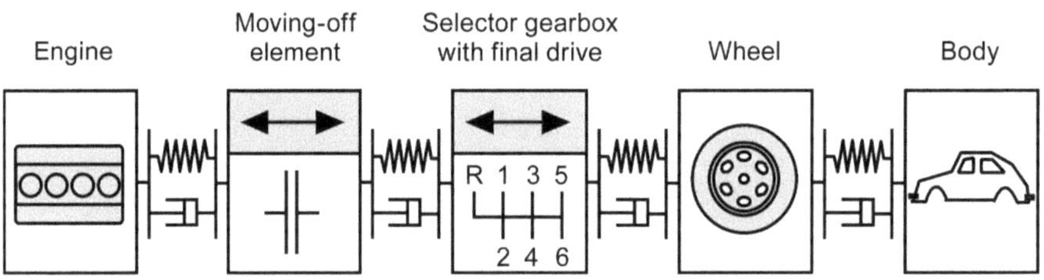


**Fig. 5.15.** Simple vibration substitute model of a powertrain

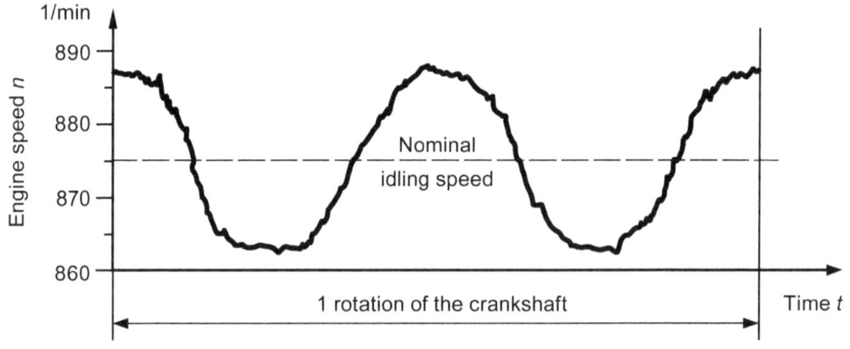


**Fig. 5.16.** Internal combustion engine idling. Irregularity of a 1.8 litre four-cylinder diesel engine during one crankshaft revolution

With the development of highly dynamic internal combustion engines with better fuel consumption and emissions performance in recent years, there has been a sharp increase in the degree of irregularity of engine running. Therefore the vibration characteristics of the powertrain is becoming more important [5.17–5.18]. See also Section 7.5 "Developing Low-Noise Transmissions".

Numerous development and optimisation tasks with respect to consumption and exhaust gas reduction as well as noise emission guidelines have led to increases in torsional vibration. This leads in turn, among other things, to further encroachments on passenger comfort due to vibrations and noise in the cabin.

For this reason, measures aimed at minimising these powertrain vibrations are indispensible. Comfort can be enhanced by tuning and optimising the powertrain with an eye to reducing torsional vibrations. To this purpose, special significance is attached to decoupling the combustion engine from the powertrain.

Known measures taken to decouple the internal combustion engine from the powertrain are:

- *torsion dampers* in the dry clutch driving plate [5.10],
- *dual mass flywheels (DMF)* are based on the principle of distributing the moments of inertia of the flywheel [5.1, 5.5, 5.12],
- *hydraulic torsion dampers* combine the DMF advantages of mass distribution with a variable damper system [5.15, 5.16],
- *absorber dampers* to reduce resonance effects [5.6],
- *speed-adaptive dampers* allow for a resonance-free compensation of the dominant torques at the crankshaft in the entire speed and load range [5.13, 5.19],
- *permanently slipping friction clutch* with electronic clutch systems: an electronically controlled adjusting mechanism filters out vibration peaks by using controlled slip,
- *integrated starter-alternator dampers* reduce vibration by arranging an electric machine on the crankshaft. However, the primary objective of this system is to provide and to manage electrical energy in motor vehicles [5.14],
- *hydrodynamic torque converters* with automatic transmissions and
- *map-controlled torque converter lock-up clutches* [5.4].

Hydrodynamic components in the powertrain (converter, retarder) have very good damping. For example in vehicles using a torque converter as the moving-off element, the torsional vibration of the engine is uncoupled from the transmission. As soon as the converter is locked-up with a lock-up clutch, the vibration problems reappear.

The transmission is not just a "passive" component in the powertrain vibration system. Gearwheel transmissions themselves also cause parametrically excited vibrations that lead in turn to noise-creation. These mechanisms are explained in detail in Section 7.5. Vibration is also caused by the power transmission function of the universal joints as used in drive shafts. Random vibrations are also introduced to the powertrain (against the direction of power flow) from the road through the wheel. The complex field of powertrain dynamics is dealt with in greater detail in the literature [5.8, 5.9].

# 6 Vehicle Transmission Systems: Basic Design Principles

*The ideal transmission design is determined by the intended application, systematic thinking and experience*

This chapter presents a systematic exposition of basic design concepts for vehicle transmissions. These principles are related to specific examples in Chapter 12 “Typical Designs of Vehicle Transmissions”.

## 6.1 Arrangement of the Transmission in the Vehicle

The important decisions at the concept phase of a new vehicle are the type of vehicle (saloon, coupé, sports car etc.) and the type of drive (front-wheel drive, standard drive etc.). The type of drive has a significant effect on handling, ride comfort, economy, safety and space available. There are numerous factors influencing the design of the transmission, both with front-wheel and rear-wheel and with all-wheel drive. There are also a lot of known alternatives for the relative position of the engine, transmission and final drive to each other. These basic configurations are set out below, based on the classification by transmission and type of use presented in Table 2.7.

### 6.1.1 Passenger Cars

The possible engine and drive configurations for passenger vehicles are shown in Table 6.1.

**Table 6.1.** Theoretical alternative engine and drive configurations

| **Configuration** | | **Drive** | | |
|---|---|---|---|---|
| | | Front | Rear | Front + Rear |
| **Engine** | Front | Front-wheel drive | Standard drive | All-wheel drive |
| | Rear | Not feasible | Rear-motor drive | All-wheel drive |

H. Naunheimer et al., *Automotive Transmissions*, 2nd ed.,
DOI 10.1007/978-3-642-16214-5_6, © Springer-Verlag Berlin Heidelberg 2011

The dominant technology for road-going passenger cars is currently *front-wheel drive* or *standard drive*. *Rear-motor drive* used to be common, but is now used mainly in sports cars. *All-wheel drive* on the other hand is firmly established in new designs. Almost every model series now includes an all-wheel drive model. Because of the variety of designs in use, all-wheel drives for passenger vehicles, both on-road and off-road, will be treated separately in Section 6.1.3.

The possible combinations of passenger car powertrain components are shown in the morphological matrix below (Table 6.2). The table also lists design variants for individual components.

Figures 6.1–6.3 show various configurations of the powertrain and its components in the vehicle. Variants commonly encountered in practice are listed below.

**Table 6.2.** Morphological matrix for passenger car powertrains

| **Parameter** | **Configurations (passenger cars)** | | | | |
|---|---|---|---|---|---|
| Position of engine | Front – longitudinal | Front – transverse | Rear – longitudinal | Rear – transverse | — |
| Driven axle(s) | Front-wheel drive | Rear-wheel drive | All-wheel drive | — | — |
| Position of engine relative to gearbox | Engine in front of gearbox | Engine behind gearbox | Engine above gearbox | Engine beside gearbox | Engine longitudinal, gearbox transverse (T configuration) |
| Position of engine and gearbox relative to final drive | Engine, gearbox and final drive as a block | Final drive separate from engine and gearbox (standard drive) | Engine separate from gearbox and final drive | — | — |
| Gearbox/final drive structurally combined | Final drive integrated in gearbox | Final drive flange-mounted to gearbox | Gearbox and final drive separate | — | — |
| Final drive | Spur gears | Bevel gears – helical bevel drive | Bevel gears – hypoid bevel drive | Worm gears | Belt drive |
| Differential lock | Unlocked | Self-locking | Manual locking | Electronic locking | — |
| Differential gear | Spur gears | Bevel gears | Helical gears | Worm gears | — |
| Gearbox design | Single-stage | Two-stage, e.g. countershaft gearbox | Multi-stage, e.g. three-shaft gearbox | — | — |

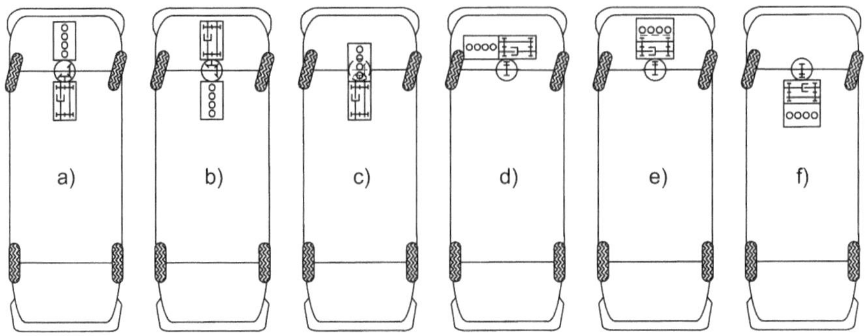


**Fig. 6.1.** Front-wheel drive. *a* Longitudinal engine in front of axle, longitudinal gearbox; *b* longitudinal engine behind axle, longitudinal gearbox; *c* longitudinal engine above axle, longitudinal gearbox; *d* transverse engine beside the gearbox; *e* transverse engine above the gearbox; *f* transverse engine behind the gearbox

*Front-wheel drive*, Figure 6.1:

- longitudinal engine in front of the axle, longitudinal transmission (Figure 6.1a),
- transverse engine beside the transmission (Figure 6.1d).

*Standard drive*, Figure 6.2:

- longitudinal engine in the front above/behind the front axle, transmission flanged to the engine longitudinally, final drive with differential at the rear axle (Figure 6.2g).

*Rear-motor drive*, Figure 6.3:

- longitudinal engine behind the axle ("rear engine") (Figure 6.3j),
- longitudinal engine in front of the axle (Figure 6.3k).

For rear-motor drive, depending on the position of the engine relative to the axle, a further distinction is made between rear-engine and mid-engine design.

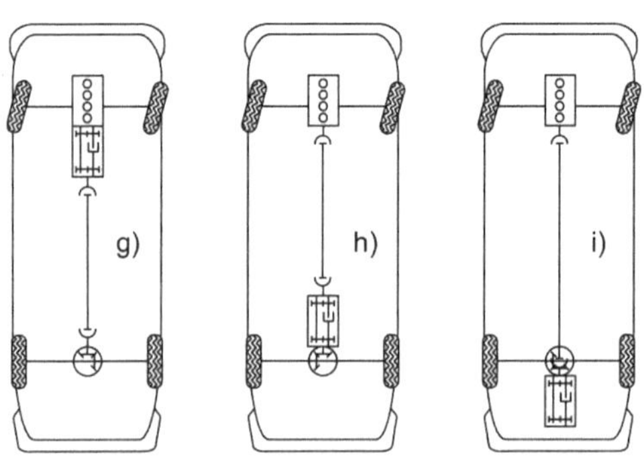


**Fig. 6.2.** Standard drive. *g* ("Standard drive") Longitudinal engine front-mounted above/behind the front axle, gearbox flange-mounted longitudinally to the engine, final drive with differential at the rear axle; *h* longitudinal engine front-mounted above/behind the front axle, gearbox mounted longitudinally in front of or *i* behind the rear axle with integral final drive and differential (transaxle principle)

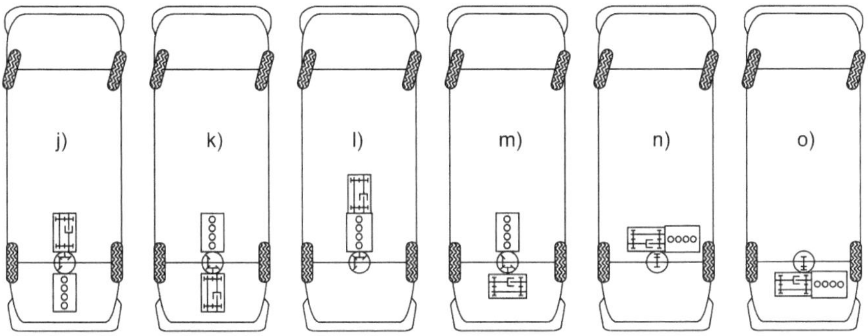


**Fig. 6.3.** Rear-motor drive. *j* Longitudinal engine behind axle ("rear engine"); *k* longitudinal engine in front of axle; *l* gearbox and engine in front of axle; *m* longitudinal engine in front of axle, transverse gearbox behind the axle (T configuration); *n* transverse engine beside the gearbox in front of axle; *o* transverse engine beside the gearbox behind the axle

*Transaxle design*: The term transaxle is a general term for *transmission + axle*. This means that the transmission and the axle drive are designed as a combined unit. The concept accordingly applies to front-wheel drive as well as to transmissions with the final drive mounted at the rear axle. The term as used in Germany is more closely circumscribed, and relates only to the configuration of front-longitudinal engine and rear transmission with final drive (Figure 6.2h and Figure 6.2i). This incongruity of terms between German and English often leads to misunderstandings.

## 6.1.2 Commercial Vehicles

The drive configurations for vans up to approximately 3.5 t gross weight are closely related to passenger car technology (see Table 2.7). The normal configurations are

- front-mounted longitudinal engine in front of/above the front axle, transmission longitudinally flanged to the engine, final drive with differential at the rear axle,
- transverse engine beside the transmission.

For trucks with an allowable gross weight of more than 3.5 t, the standard drive has become almost universal. In this arrangement, the engine torque is transmitted to the driving axle via a master clutch, gearbox and normally a cardan shaft. Pure front-wheel drive, i.e. driving the vehicle through the front axle only (steering axle), occurs only very rarely in some transport vehicle designs and special-purpose designs. This option is not viable for heavy commercial vehicles because of the unfavourable front axle load distribution (less than 40% with two-axle vehicles, less than 30% with three-axle vehicles and approximately 25% per front axle with four-axle vehicles when fully loaded), and the resultant traction difficulties.

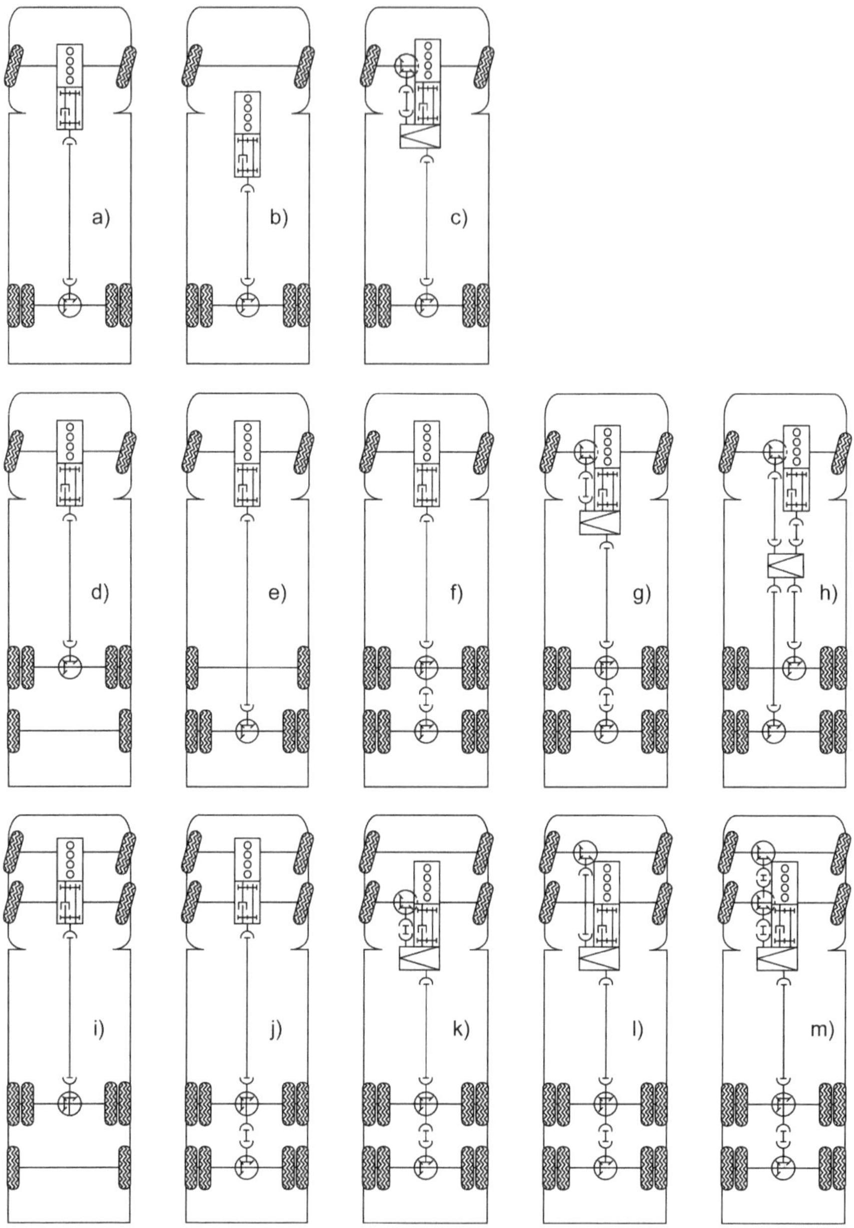


**Fig. 6.4.** Drive designs for trucks with one or more powered axles.
*a* 4 x 2; *b* 4 x 2, underfloor engine; *c* 4 x 4, all-wheel-drive; *d* 6 x 2, trailing axle; *e* 6 x 2, leading axle; *f* 6 x 4; *g* 6 x 6, with drive-through to second rear axle; *h* 6 x 6, second rear axle with direct drive; *i* 8 x 2, trailing axle; *j* 8 x 4; *k* 8 x 6, second front axle driven; *l* 8 x 6, first front axle driven; *m* 8 x 8, with drive-through to first front axle

With multi-axle vehicles, almost all non-steering axles are driven; in the case of all-wheel drive (off-road mode) all axles are driven, including the steering axle. The normal applications for two-axle and three-axle vehicles are shown in Figure 6.4. The designation of the drive variants shown in Figure 6.4 is as follows: overall number of wheels (pairs of wheels) x the number of driven wheels (pairs of wheels). For example: 4 x 2 represents a total of four wheels with two driven wheels. The engine can be located above or near to the front axle, and in some applications in the middle of vehicle (underfloor), which gives a favourable weight distribution.

Buses very often have a rear engine, either longitudinal or transverse. This allows greater freedom for designing the passenger accommodation. Figure 6.5 gives examples of drive configurations for buses. Vehicles for use off-road and on building sites are almost always equipped with all-wheel drive. The engine is located at the front or in the middle of the vehicle, and the drive axles are steering drive axles at the front, and single or double rigid drive axles at the rear. The drive is delivered through a transfer gearbox (Section 6.8.5).

**Table 6.3.** Morphological matrix for commercial vehicle powertrains

| **Parameter** | **Configurations (commercial vehicles)** | | | | |
|---|---|---|---|---|---|
| Engine and gearbox configuration | Front – longitudinal | Front – transverse | Rear – longitudinal | Rear – transverse | Underfloor |
| Number of drive axles | One | Two | Three | Four | — |
| Type of axle | Drive axle with/without drive-through | Steering axle | Driven steering axle | Trailing axle | Leading axle |
| Drive-through to second axle | Yes | No | — | — | — |
| Final drive | Bevel gears – helical bevel drive | Bevel gears – hypoid bevel drive | Worm gears | Double bevel gears | Spur gears |
| Reduction in centre drive | Single | Spur wheel reduction, shiftable | Spur wheel reduction, not shiftable | Planetary reduction, shiftable | Planetary reduction, not shiftable |
| Differential gear | Spur gears | Bevel gears | Helical gears | Worm gears | — |
| Differential locking | Unlocked | Self-locking | Manual locking | Electronic locking | — |
| Hub drive, see Figure 6.75 | Without hub drive | Spur gear reduction, external toothed | Spur gear reduction, internal toothed | Spur gear planetary reduction | Bevel gear planetary reduction |

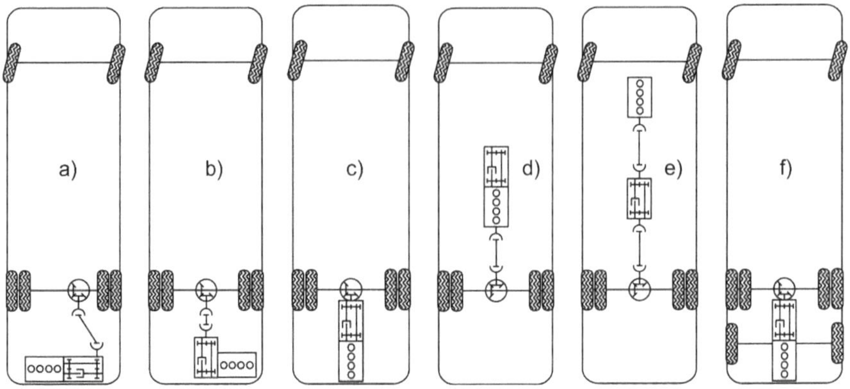


**Fig. 6.5.** Drive designs for buses. *a* Transverse engine behind axle, transverse gearbox; *b* transverse engine behind axle, longitudinal gearbox; *c* longitudinal engine behind axle, longitudinal gearbox; *d* longitudinal engine in front of axle, gearbox flange-mounted longitudinally to engine; *e* longitudinal engine in front of axle, longitudinal gearbox; *f* longitudinal engine and longitudinal gearbox, trailing axle

In contrast to passenger cars, the drive configuration of commercial vehicles has little effect on the design of the vehicle transmission. The various possible powertrain configuration options for commercial vehicles can be derived from the morphological matrix (Table 6.3).

### 6.1.3 All-Wheel Drive Passenger Cars

All-wheel drive technology has established itself increasingly in recent years, for the following reasons:

- increased climbing performance,
- increased performance (traction) through full utilisation of static friction (only possible with all-wheel drive),
- improved moving-off traction,
- more precise handling,
- increased payload and trailer load and
- identical self-steering properties under different weather conditions (dry surface, wet, ice and snow).

The principal corresponding disadvantages are greater technical complexity, increased weight (fuel consumption) and increased space requirements.

In addition to dedicated all-wheel drive designs for off-road passenger cars and special vehicles (e.g. rally vehicles), all-wheel drives have been designed based on the following drive configurations [6.4, 6.18, 6.31, 6.41, 6.45]:

- front-motor drive with longitudinal engine in front of the axle,
- front-motor drive with transverse engine beside the transmission,

- standard drive with longitudinal engine over the front axle, longitudinal transmission flanged onto the engine, final drive with differential at rear axle,
- rear-motor drive with longitudinal engine in front of the axle ("mid-engine"),
- rear-motor drive with longitudinal engine behind the axle ("rear-engine").

Figure 6.6 shows these drive configurations. Various alternative all-wheel drive designs are given in Section 12.4. Figure 6.6 shows that the all-wheel drive design is very largely dependent on the drive concept of the original vehicle. Further characteristic features are defined by the objectives when introducing all-wheel drive. For example a crucial factor for engineering complexity and thus for cost is whether the purpose of all-wheel drive is to improve handling, or merely to provide improved moving-off traction. The appropriate type of all-wheel drive can be selected from the systematic classification of all-wheel drives in Figure 6.7.

The decisive criterion is the type of link between the two axles to be driven. The power flow between the front and rear axles can be differential controlled or clutch controlled ("hang-on system"). See also Section 6.8.3 "Differential Gears and Locking Differentials" and Section 6.8.5 "Transfer Gearboxes".

### *1/ Differential-Controlled All-Wheel Drive*

In *differential-controlled* systems, torque is distributed to the front and rear axle by a planetary gear differential or a bevel gear differential. With planetary gear differentials, the drive torque can be split to the two drive axles as required, by selecting the ratio. Typical torque distribution ratios between front and rear axle are 50% : 50% to 33% : 66%.

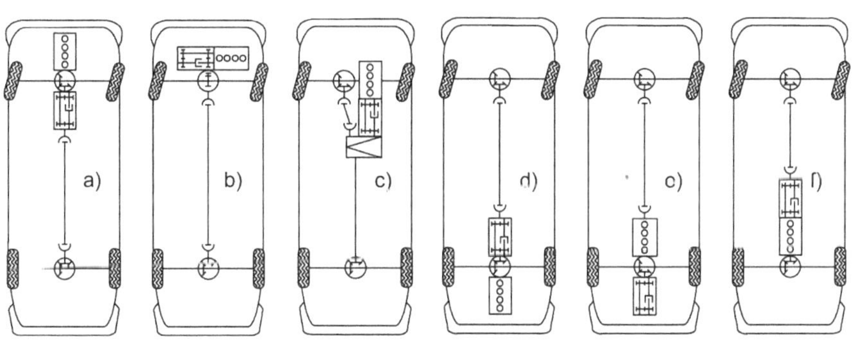


**Fig. 6.6.** Drive designs for all-wheel drive passenger cars. *a* Front-motor drive with longitudinal engine in front of the axle; *b* front-motor drive with transverse engine beside the gearbox; *c* standard drive with longitudinal engine above the front axle, gearbox flange-mounted longitudinally to the engine, final drive at rear axle; *d* rear-motor drive with longitudial engine behind the axle; *e* rear-motor drive with longitudial engine in front of the axle; *f* engine and gearbox in front of the axle

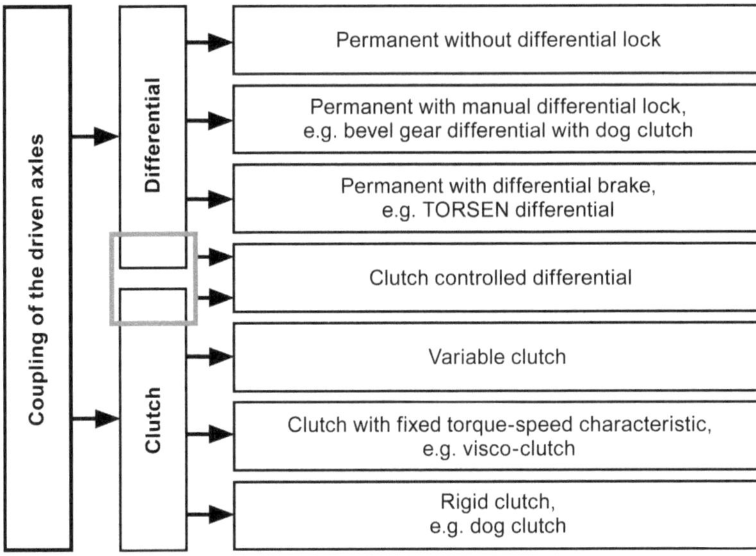


**Fig. 6.7.** Systematic classification of passenger car all-wheel drives

In bevel gear differentials, the torque distribution is fixed at 50% : 50%. Selecting a fixed torque ratio between the front and rear axles means the traction distribution is ideal for only one point, the design point.

The drive torque is thus not split in proportion to the axle load required by the current driving conditions. If the traction reserves are to be completely exhausted when there is a high degree of slip (which is theoretically only possible with variable torque distribution between the front and rear axle), the interaxle differential can be braked or locked. With a locking effect which engages as the speed difference increases (e.g. visco-lock), handling is not impaired, preventing permanent distortion of the powertrain such as can occur with positive locks. A TORSEN transfer differential (TORSEN stands for "torque sensing") acts in this regard as a self-locking differential.

### 2/ Clutch-Controlled All-Wheel Drive

*Clutch-controlled* all-wheel drives are characterised by the fact that only one axle is permanently driven. The second axle is engaged manually or automatically as required. The most inexpensive option for engaging the second axle is to use a rigid controllable clutch. This system, however, can only be used for moving-off, since distortion occurs in the powertrain with 100% lock, as in the case of the centre differential (see Figure 6.76).

The use of a visco-clutch gives another way of linking two axles, building up the coupling torque between the front and rear axles through viscous friction (slip-controlled) depending on the speed difference between the front and rear axles. The transition to all-wheel drive is gradual as the speed difference between the front axle (FA) and the rear axle (RA) increases. Permanent distortions in the powertrain are not possible. The level of coupling torque depends on the clutch characteristic selected. It can be influenced by the level, viscosity and temperature of the oil (using silicone oil with low viscosity change with temperature fluctuations). A "soft" coupling characteristic (i.e. low coupling torque with high relative rotational speed) is desirable to avoid distortions in the powertrain, whereas a "hard" coupling characteristic (i.e. large coupling torque with small relative rotational speed) is necessary when an axle is spinning. The visco-clutch is protected against destruction under high load by the *hump effect*, i.e. heat-related frictional engagement between the plates [6.6, 6.33].

The last type of clutch controlled system involves clutches with externally adjustable coupling torque (e.g. multi-plate clutches). For this type, the coupling torque can be selected to match current driving conditions. This enables torque distribution between the front and rear axles to be adapted to the changes in dynamic axle load, i.e. depending on acceleration, gradient, loading etc.

### 3/ Hybrid Form

A further variant, located between the differential-controlled and clutch-controlled systems, is all-wheel drive using an "electronically controlled" multi-plate clutch and a "lockable" differential. This solution gives a high degree of comfort with identical handling (two-wheel drive) of the basic vehicle. This "traction control system" however involves a high level of engineering complexity.

**Table 6.4.** Generations of passenger car all-wheel drives, derived from [6.18]. FA...Front axle, RA...Rear axle

| Generation | Power-split | Example of differential-controlled system | Example of clutch-controlled system |
|---|---|---|---|
| 1 | FA/RA torque distribution constant | Spur gear transfer differential | ——— |
| 2 | FA/RA torque distribution along system-dependent characteristic curve | TORSEN transfer differential | Visco-clutch, uncontrolled |
| 3 | FA/RA torque distribution along controllable characteristic curve | Transfer differential with controlled multi-plate lock | Visco-clutch, controlled |
| 4 | FA/RA torque distribution selectable | Vario drive | Multi-plate clutch controlled for FA and RA |
| 5 | Torque distribution to each wheel selectable | ——— | "Torque-Vectoring"/ "Torque-Splitting" |

**Table 6.5.** Morphological matrix for passenger car all-wheel drives

| **Parameters** | **Configurations (all-wheel drive passenger cars)** | | | | |
|---|---|---|---|---|---|
| Position of engine | Front – longitudinal | Front – transverse | Rear – longitudinal | Rear – transverse | —— |
| Concept | Manual selection | Automatic engagement | Permanent | —— | —— |
| Torque transmission FA/RA | Rigid clutch | Multi-plate clutch: controlled/ uncontrolled | Centrifugal clutch | Visco-clutch | Transfer differential |
| Torque distribution | Constant, e.g. 50% : 50% | Slip-dependent | Controlled according to wheel load | —— | —— |
| Type of interaxle differential lock | Rigid clutch | Self-locking differential | Visco-lock | Multi-plate clutch, uncontrolled | Multi-plate clutch, controlled |
| Type of interwheel differential lock | Rigid clutch | Self-locking differential | Visco-lock | Multi-plate clutch | Locking function activated by brake action |
| Transfer differential | Spur gear differential | Bevel gear differential | Other | —— | —— |
| Transfer box | Integrated in gearbox | Individual sub-assembly | Flange-mounted onto gearbox | —— | —— |
| Braking stability | Freewheel | Multi-plate clutch | —— | —— | —— |

All-wheel drives for passenger cars are often divided into different generations. Table 6.4 describes the characteristic features of the particular all-wheel drive generation, giving an example for each. Table 6.5 shows a morphological matrix for passenger car all-wheel powertrains.

Section 6.8.3 "Differential Gears and Locking Differentials" examines the necessity, function and design of the various types of differential and differential lock, including a description of the visco-clutch that acts both as a clutch and as a differential lock.

### 6.1.4 Transverse and Longitudinal Dynamics with All-Wheel Drive

Section 6.1.3 lists some of the advantages of all-wheel drive compared to two-wheel drive. To further illustrate the advantages of all-wheel drive, the longitudinal and transverse dynamics of the vehicle and/or the tyre are examined below.

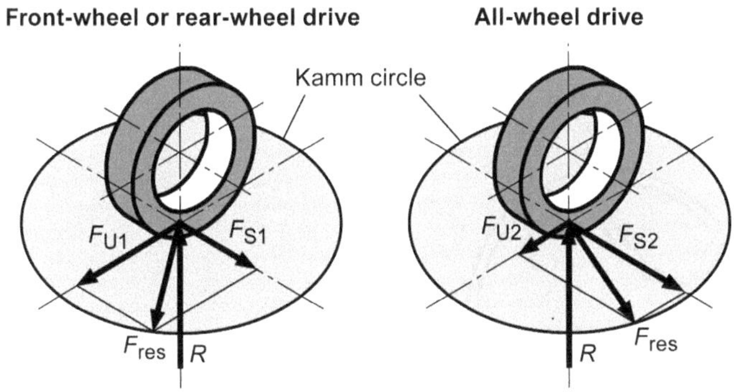


**Fig. 6.8.** Difference between two-wheel drive and all-wheel drive

This requires an understanding of the fundamental connection between circumferential force and lateral force on the wheel at high lateral acceleration. Wheel load, wheel load fluctuations, and self-aligning torque and slip angle are ignored for the sake of simplicity.

Figure 6.8 shows the conditions for a wheel at the friction limit for two-wheel drive (left) and for all-wheel drive (right). With all-wheel drive the circumferential force $F_{U2} = F_{U1}/2$, given simplified assumptions. The wheel can transmit greater lateral forces $F_S$, until it reaches the friction limit at $F_{res}$.

The maximum transmittable circumferential force $F_{U,max}$ is derived from Equation 3.9 as

$$F_{U,max} = \mu_H R . \tag{6.1}$$

The maximum lateral force $F_{S,max}$ is

$$F_{S,max} = \mu_H R . \tag{6.2}$$

When the circumferential force $F_U$ and the lateral force $F_S$ both occur simultaneously, they make up a geometrical sum (see Figure 6.8). To avoid sliding, this must not exceed $F_{res} = \mu_H R$ (Kamm circle). The Kamm circle represents the friction limit for the rolling wheel transmitting both circumferential and lateral forces at the same time. The following relationship applies:

$$F_{res} = \sqrt{F_U^2 + F_S^2} \le \mu_H R . \tag{6.3}$$

## 6.2 Transmission Formats and Designs

Completed transmissions are distinguished in terms of *format* and *design*. Transmission format relates to the morphology or external appearance of the transmission, or the configuration of input and output.

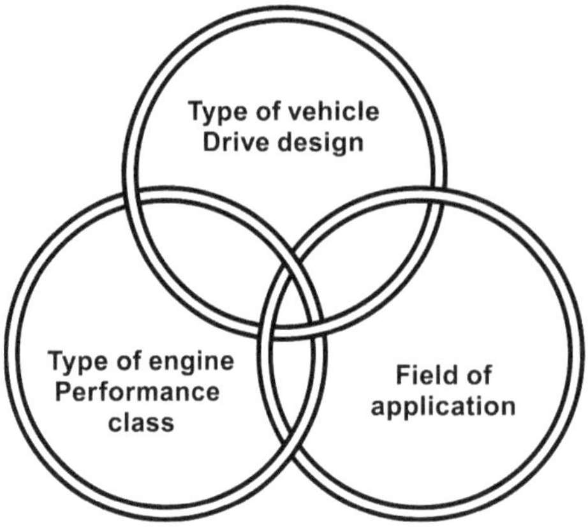


**Fig. 6.9.** Specification factors for transmission format and design

The transmission design describes how the main functions of the transmission are engineered. It relates to the internal configuration. Transmissions can thus have different designs and share the same format. The format selected for a design depends on various criteria; principally the vehicle design, the type of engine and the intended use (Figure 6.9).

### 6.2.1 Transmission Format

The format of the transmission (Figure 6.10) is determined primarily by the position of the transmission in the vehicle or in the powertrain (Section 6.1), and any additional geometrical constraints, such as space limitations. The format is also affected by assembly considerations (both as regards the transmission itself and as regards its installation in the vehicle), by gearbox housing rigidity and by noise emissions. Transmissions often comprise several individual gearboxes, which can also be housed in separate gearbox housings. In this case, the relative position of the individual housings is an important factor influencing the format of the transmission as a whole.

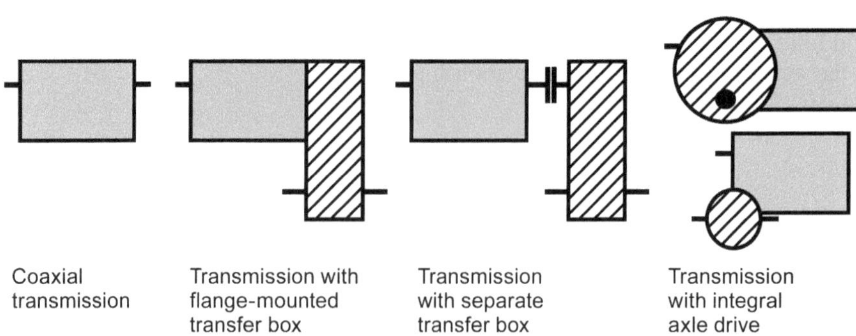


**Fig. 6.10.** Examples of different transmission formats

The format of a transmission concerns the design engineer principally when adapting or developing existing designs, for example adapting an existing transmission to a new vehicle with different dimensional constraints.

With standard drive (front-mounted longitudinal engine and transmission, rear-wheel drive, Figure 6.2g), the coaxial transmission format is used. If there are two driven rear axles, or if all-wheel drive is used, then a transfer box is needed, which may be flange-mounted directly onto the gearbox or separate from it.

For front-wheel drive layouts, a transmission format is used in which the axle drive with the differential is integrated into the gearbox. Input and output are not coaxial in this case.

### 6.2.2 Transmission Design

The transmission design is derived from the functional principles applied, to fulfil the main functions of the transmission. As already indicated in Section 2.3.3, a vehicle transmission has four main functions: "Moving-off from rest", "Changing ratio/rotational speed", "Shifting/establishing power flow" and "Operating/controlling the gearbox".

The "Moving-off from rest" function can be carried out mechanically, electromechanically or hydraulically. The "Changing ratio/rotational speed" function can be carried out using spur gears, planetary gears, hydrodynamic or hydrostatic transmissions or mechanical continuously variable transmissions. The "Shifting/establishing power flow" function can be divided into the two functional principles positive engagement or frictional engagement. The "Operating/controlling" function can be carried out by manual shifting, automation or an automatic system with associated control unit.

Their selection depends on the power to be transmitted, considering traction utilisation and ease of operation. Especially in the case of new developments, the design engineer has to decide on the design or combination of designs of the transmission.

Combinations of different designs are, in principle, always an option for carrying out the various main functions. In the last 100 years numerous possible solutions have been proposed for vehicle transmissions. These can be systematically represented in a morphological matrix (Table 6.6). The main functions are shown in the four rows of this table, and the associated solution principles applied appear in the columns. By combining these principles to form a complete transmission, you get all possible combinations of transmission designs. Not all theoretical combinations are of significance or relevance in practice.

A preliminary selection can be made by assessing the design under consideration, and other alternatives. This preliminary selection follows on from the concept phase of transmission development.

In multi-range transmissions (Section 6.7.1), these main functions can take different forms for each individual range unit. Each individual range unit must have the following main functions: "Changing ratio/rotational speed", "Shifting/establishing power flow" and "Operating/controlling the gearbox".

**Table 6.6.** Morphological matrix of solution principles for the main functions. The principles underlying a conventional manual gearbox are highlighted in grey

| Main function | Principle of operation | | | | |
|---|---|---|---|---|---|
| | 1 | 2 | 3 | 4 | 5 |
| Enable moving-off from rest | Mechanical Dry | Mechanical Wet | Electro-mechanical | Hydro-dynamic | Hydro-static |
| Change ratio/ speed | Spur gears | Planetary gears | Hydro-dynamic | Hydro-static | Mechanical continuous |
| Shift/establish power flow | Positive-engaged Sliding gears | Positive-engaged Shifting dog, synchronized | Positive-engaged Shifting dog, unsynchron. | Frictional-engaged Multi-plate clutch | Frictional-engaged Multi-plate brake |
| Operate/control the transmission | Manual shift | Automated electr./mech. | Automated electr./hydr. | Automated electr./pneum. | Automatic electr./hydr. |

Even with multi-range transmissions, only one principle is used for the main function of enabling "Moving-off". The number of functional principles and their physical principles of operation can change as technology advances.

## 6.3 Basic Gearbox Concept

Geared transmissions are categorised by their technical design or the number of ratio stages making up the individual gears:

- single-stage transmissions,
- two-stage transmissions and
- multi-stage transmissions.

The term "stage" refers here to a gear pair or the power flow from one gearwheel to another. A stage generally involves power flow from one shaft to another. Figure 6.11 shows designs of four-speed countershaft transmissions. The term "countershaft transmission" is defined in Section 6.4. Single-stage transmissions are primarily used in front-wheel drive vehicles, since they require no coaxial transmission of the power flow, unlike standard drive vehicles.

In the standard powertrain configuration (engine and transmission in the front, drive at the rear), the two-stage countershaft transmission with coaxial input and output shaft is virtually universal.

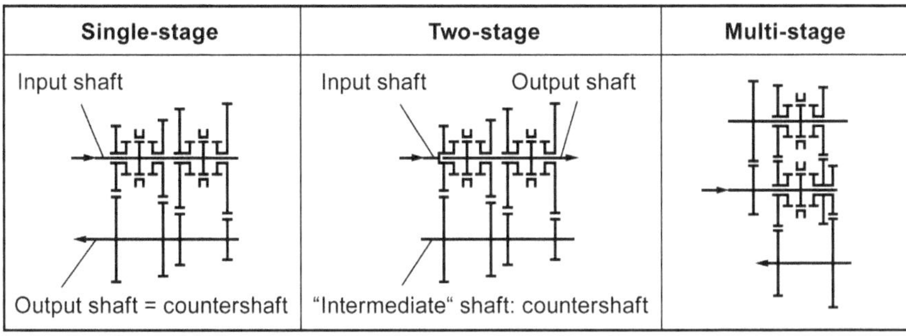


**Fig. 6.11.** Configuration of the ratio stages using 4-speed gearboxes as examples

Multi-stage (more than two-stage) transmissions are just as suitable as single-stage transmissions for front-engine front-wheel drive vehicles. The number of gear stages they have depends upon the number of gears. The multi-stage design enables short gearboxes to be constructed. Multi-stage coaxial transmissions are used principally in commercial vehicles with front- or rear-mounted range units (see Section 6.7.1).

To decide on the type of transmission for a particular application, first the basic ratio change options need to be defined. The shifting elements involved also by definition constitute part of the transmission.

## 6.3.1 Shifting with Power Interruption

The transmission is shifted without load, i.e. the power flow between the prime mover and the wheels is interrupted during the gear change operation. The vehicle coasts during the gear change operation. This can entail a loss of speed (Figure 6.12), depending on the difficulty of the terrain (gradient, high rolling resistance). In order to limit this loss of speed, the shifting operation must not take too long; the whole gear change operation must therefore be concluded in less than one second. For multi-range transmissions this means that the gear change operations in the individual range units must be carried out within 0.2 to 0.3 seconds (assuming they are in succession). This is one reason why the number of ranges in a transmission cannot be increased indefinitely, although this would lead to a reduction in the number of gear pairs needed (see also Section 6.7.1). The requirement for several individual shifting actions to occur synchronously at the various shifting points is demanding in engineering terms.

Transmissions with power interruption can be used wherever the application is such that vehicle speed does not decrease (or on downhill runs increase) significantly during the shifting process, and the shifting operation is reasonably practical for the driver. In addition to manually shifted transmissions, this also applies to automated transmissions with which the power flow is interrupted by opening the master clutch when shifting gears.

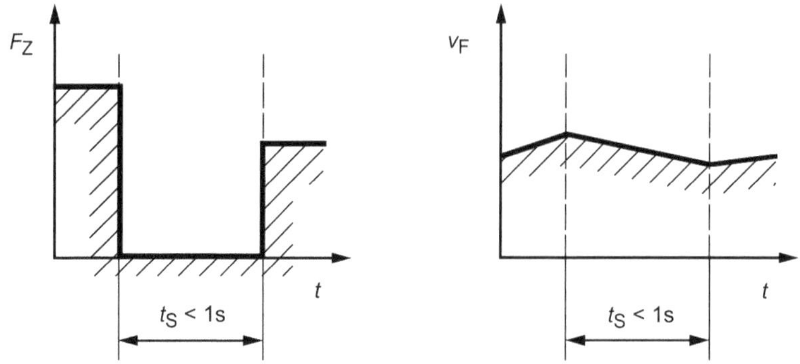


**Fig. 6.12.** Qualitative traction and velocity profile when shifting up with power interruption

In the case of automated countershaft-type truck transmissions, shifting normally involves power interruption. Vehicle acceleration forces are relatively low, the vehicle mass is high, and ride quality is not the top priority.

### 6.3.2 Shifting without Power Interruption

As in the case of shifting with power interruption, the transmission ratio is changed in steps. But in this case, the power flow is not interrupted during the gear change operation (Figure 6.13).

Such transmissions are known as frictional transmissions or powershift transmissions. The transition from one ratio to another is carried out without interrupting the power flow. The ratios can be engaged under load by means of additional braking or clutch elements. In this case the gear set which is being shifted out of is disengaged from the power flow, whilst the new gear set is engaged in the power flow. There is no reduction in road speed.

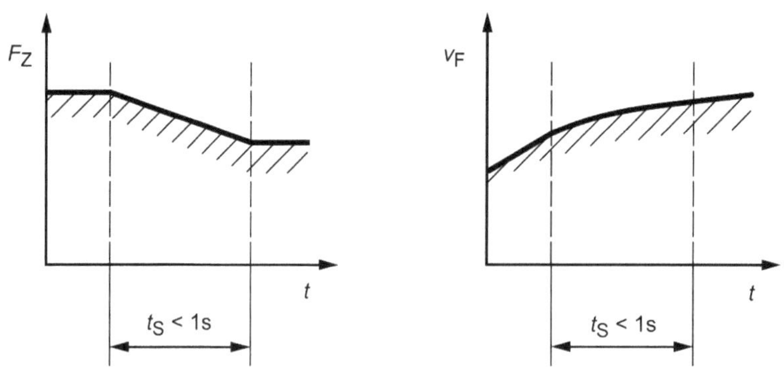


**Fig. 6.13.** Qualitative traction and velocity profile for upward powershift

Examples of this type of transmission are automatic transmissions with various gear ratios (conventional automatic transmissions, countershaft-type automatic transmissions and dual clutch transmissions, see Figure 1.2).

Powershift transmissions are well suited for fast shifting. Transmissions of this type are used in heavy vehicles, where vehicles operate in difficult terrain and in all vehicles where the driver is to be relieved of gearshifting activity. They are fitted both with manual and with automatic gear selection.

### 6.3.3 Continuously Variable Transmissions without Power Interruption

Here ratio shifting is no longer in steps, but varies continuously (see also Sections 5.3.4 and 6.6.6). The traction is adapted to the driving resistance without any intervention by the driver (Figure 6.14). This type of characteristic output conversion represents the theoretically ideal solution. Various mechanical variants are known in the form of friction gears or pulley transmissions which are based on converting the rotational speed to continuously variable diameters. In addition to the mechanical variants, there are also hydraulic solutions. The hydrodynamic converter is the best known example of this.

Hydrostatic transmissions comprising a combination of pump and motor also provide continuously variable regulation of rotational speed. Usually a hydrostatic transmission is coupled to a planetary gear to increase the overall gear ratio and to preselect different operating ranges, some with power-split.

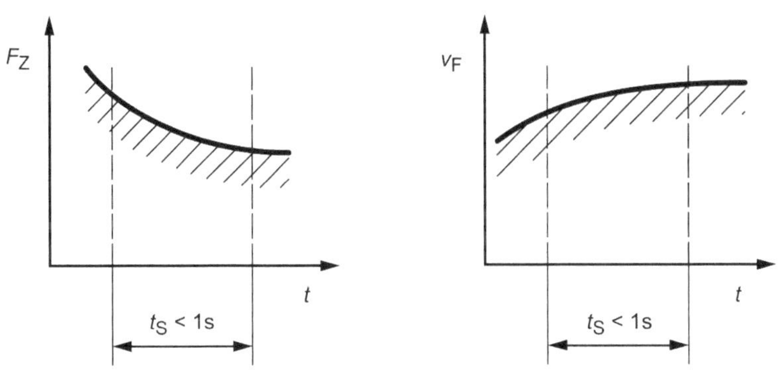


**Fig. 6.14**. Qualitative traction and velocity profile for continuously upward powershift with CVT

## 6.4 Gear Sets with Fixed Axles, Countershaft Transmissions and Epicyclic Gears

Geared transmissions are divided into:

- gear sets with fixed axles and
- epicyclic gears.

These terms relate to the axes of the gearwheels engaged in the transmission. In the case of *gear sets with fixed axles*, the positions of the axes of all the gearwheels in the transmission are fixed relative to the gearbox housing. In *epicyclic* or *planetary gear sets*, a revolving bar (spider) carries the axes of the planetary gears.

**Countershaft transmissions:** The term countershaft transmission [6.30] relates to a transmission with only one input shaft and only one output shaft, and a countershaft running in a fixed position in the housing (Figure 6.11). Countershaft transmissions are thus gear sets with fixed axles. In the case of single-stage countershaft transmissions, the output shaft and the countershaft are combined, so they could also be called "reduced" countershaft transmissions.

**Planetary transmissions:** In planetary transmissions there are always three or more planetary gears on a spider (Figure 6.15) to ensure uniform and lower stress. The number of planetary gears and the number of teeth on each have no effect on the transmission ratio; they merely reverse the direction of rotation at this point. The axes of the planetary gears thus complete a rotational movement around the main axis of the gearbox.

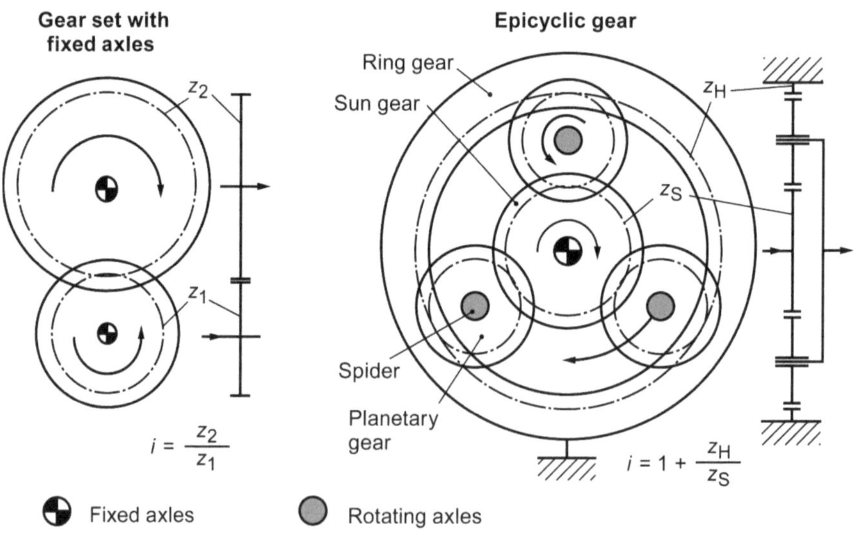


**Fig. 6.15.** Gear set with fixed axles and epicyclic gear

There are also hybrid designs combining elements of gear sets with fixed axles and epicyclic gears. Where the location of the spider is fixed, an epicyclic gear by definition becomes a gear set with fixed axles.

Planetary transmissions provide nine combinations of possible states of motion in one planetary gear set. These are derived from the fact that in principle the position of the ring gear, the spider or the sun wheel can be fixed, to act as a “frame”. The two remaining transmission components can be used as input or output of the planetary gear set.

The ratios of the individual states of motion cannot be selected independently of each other, but are defined by the numbers of teeth on the sun gear and the ring gear (Table 6.7). This table does not list the three trivial states of motion in which the transmission rotates as a block. Furthermore, not all transmission ratios are suitable for use in motor vehicles.

If none of the parts in a planetary gear set is in a fixed position, then it is referred to as a differential drive, or a summarising gearbox, transfer gearbox or differential gear. If several planetary gear sets are linked together, the result is a so-called coupled gear. This sort of gear makes it possible to achieve different ratios between input and output, depending on how the individual transmission components are linked together and which components are in a fixed position. The components are linked together by clutches, and the components are linked to the housing by brakes.

The great variety of possible ratios available in transmissions with just one planetary gear set is further substantially increased in coupled gears, but not all the ratios that can be derived from the transmission are relevant in motor vehicles. There are other important designs in addition to the simple planetary transmissions discussed here. You may wish to refer at this point to Section 6.6 and the relevant literature [6.29–6.30, 6.32].

**Table 6.7.** States of motion and ratios of a simple planetary gear set [6.29]. The number of teeth on internal geared wheels is to be entered as a positive value in formula

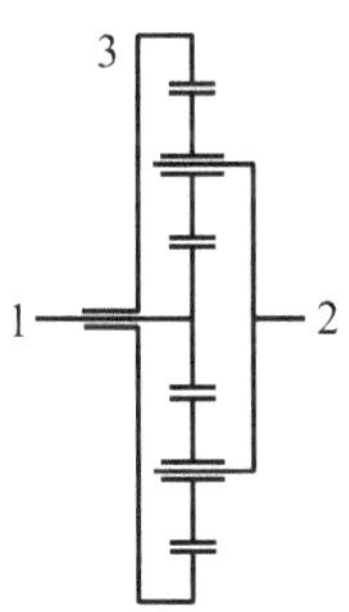


1 ... Sun gear

2 ... Spider

3 ... Ring gear

| State of motion | Type of gearbox | Input | Output | Frame | Planetary step ratio |
|---|---|---|---|---|---|
| a) | Gear set with fixed axles | 1 | 3 | 2 | $i_P = \frac{n_1}{n_3} = i_S = -\frac{z_3}{z_1}$ |
| b) | | 3 | 1 | | $i_P = \frac{n_3}{n_1} = \frac{1}{i_S} = -\frac{z_1}{z_3}$ |
| c) | Planetary transmission single drive | 1 | 2 | 3 | $i_P = \frac{n_1}{n_3} = 1-i_S = 1+\frac{z_3}{z_1}$ |
| d) | | 2 | 1 | | $i_P = \frac{n_2}{n_1} = \frac{1}{1-i_S} = \frac{1}{1+\frac{z_3}{z_1}}$ |
| e) | | 2 | 3 | 1 | $i_P = \frac{n_2}{n_3} = \frac{1}{1-\frac{1}{i_S}} = \frac{1}{1+\frac{z_1}{z_3}}$ |
| f) | | 3 | 2 | | $i_P = \frac{n_3}{n_2} = 1-\frac{1}{i_S} = 1+\frac{z_1}{z_3}$ |

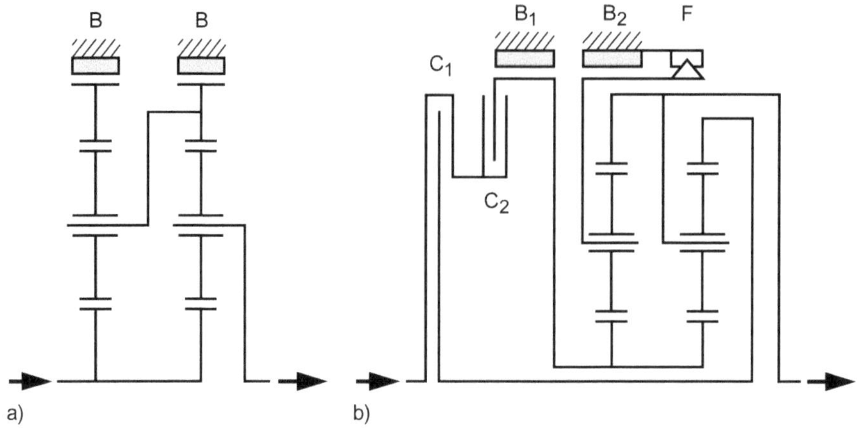


**Fig. 6.16.** Planetary coupled gear.
*a* Wilson planetary gear set; *b* Simpson planetary gear set

Conventional automatic transmissions with various gear ratios are made up of several individual planetary gear sets. The ratios of the individual gear steps cannot be freely selected independently of each other, since the same gearwheels are used for several gear steps. A section from a Wilson transmission and from a Simpson planetary gear set are shown in Figure 6.16 as an example of such a planetary transmission.

Planetary coupled gears can also be power-split. *Reactive power* can also occur in calculating the power values in the various paths. Reactive power can be envisaged as power, flowing in a circuit, which is not detectable from outside. However, it stresses the components through which it flows, and impairs the overall efficiency of the transmission. Planetary transmissions can reach very low overall levels of efficiency, which in extreme cases can even become negative. This represents transmission interlock, which in certain circumstances can be desirable if the transmission is not to be moveable from the output side.

## 6.5 Solution Principles for Part Functions, Evaluation

In the concept phase of developing a transmission, solution principles are established; see also Figure 15.13. A large number of transmissions can be created by combining the individual principles used for the main functions, as shown in the morphological matrix in Section 6.2 (Table 6.6). The number of viable alternatives is however significantly reduced when a *technical/economic evaluation* is carried out. This can be demonstrated using the example given in Table 6.8 for the main functions “Enable moving-off” and “Change ratio”. This is given as an example, and does not claim to be comprehensive.

**Table 6.8.** Example of solution principles for the functions "Enable moving-off" and "Change ratio". *0*...not possible; *1*...very unfavourable; *2*...unfavourable; *3*...moderate; *4*...favourable; *5*...very favourable

| **Function** | **Solution principles** | | | | | |
|---|---|---|---|---|---|---|
| | Gear-wheel | Pulley drive | Friction clutch | Fluid clutch | Torque converter | Hydro-static gearbox |
| Convert torque | 5 | 4 | 0 | 0 | 4 | 4 |
| Vary slip | 0 | 0 | 4 | 3 | 3 | 3 |
| Efficiency | 5 | 4 | 4 | 3 | 3 | 4 |
| Service life | 4 | 3 | 3 | 4 | 4 | 2 |
| Reliability | 2 | 3 | 2 | 4 | 4 | 3 |
| Ease of use | 4 | 2 | 3 | 3 | 3 | 2 |
| Space demand | 5 | 3 | 4 | 4 | 4 | 2 |
| Price | 5 | 2 | 4 | 2 | 2 | 1 |
| **Total points** | 30 | 21 | 24 | 23 | 27 | 21 |

A complete evaluation of all proposed solutions for the main and ancillary functions of the transmission should be carried out after the concept phase. The proper design phase can begin when this evaluation has been completed.

Table 6.8 suggests that the gear pair commends itself as by far the most cost-effective element for torque conversion. The disadvantage that this eliminates all but geared transmissions becomes a secondary consideration. Friction clutches are still the best available compromise for moving-off and for speed synchronization. The hydrodynamic torque converter also has many advantages.

## 6.5.1 Reverse Gear as Example

There are numerous designs for implementing the ancillary function of reverse gear. Figure 6.17 shows six different variants which can be developed using morphological matrices. Then the possible solutions will be assessed according to different categories.

The required reversal of the direction of rotation of the gearbox output shaft is usually achieved by inserting an idler gear into the power flow. The general rule for toothed gearing is that increasing or reducing the number of ratio stages by one reverses the direction of rotation of the output shaft.

Not all the variants shown in Figure 6.17 are of equal significance in practice. The following highly simplified assessment in Table 6.9 is intended to highlight their strengths and weaknesses.

If reverse gear does not use a gearwheel of the forward gears, the cheaper straight-cut spur gear toothing can be used for reverse, because of the relatively small proportion of time spent in reverse gear. The resultant increased noise level is acceptable.

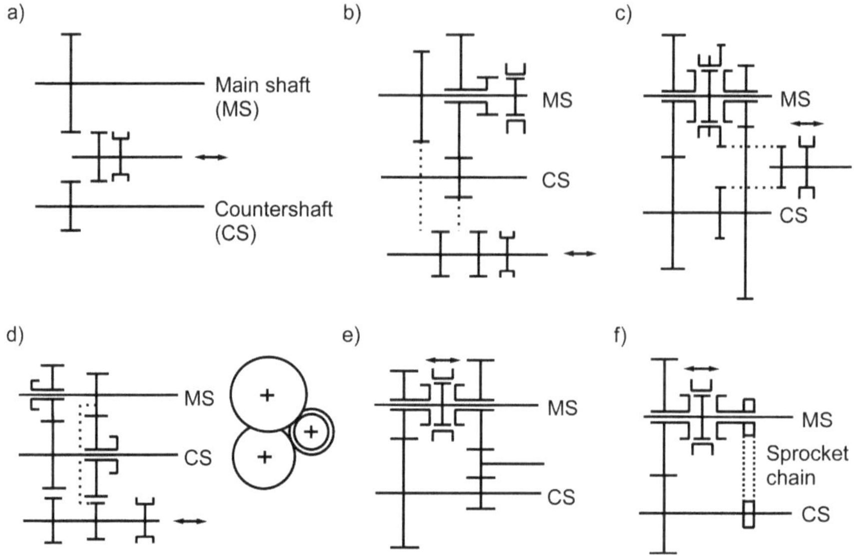


**Fig. 6.17.** Alternative reverse gear configurations. *a* An axial sliding gear is inserted between each fixed wheel of the main shaft and the countershaft ; *b* shiftable shaft with two pinions between a reverse gearwheel of the main shaft and a forward gearwheel of the countershaft; *c* the sliding gear is inserted between a fixed wheel of the countershaft and a toothed sliding sleeve of a synchronizer on the main shaft; *d* sliding shaft with two pinions between a forward gearwheel of the main shaft and a forward gearwheel of the countershaft; *e* reverse gear with intermediate pinion constantly engaged, shifting with sliding sleeve; *f* reverse gear using tooth-type chain, shifting with sliding sleeve

**Table 6.9.** Assessment of various types of reverse gear according to Figure 6.17; + advantage, – disadvantage

| **Solution** / **Evaluation criterion** | a) | b) | c) | d) | e) | f) |
|---|---|---|---|---|---|---|
| Easy to synchronize | | | | | + | + |
| Can be synchronized at rest | + | + | + | | | |
| Saves components compared to a) | | + | + | + | | |
| No ratio or toothing constraints | + | | | | + | + |
| No axial space requirement | | | + | | | |
| Sufficient shaft clearance to accomodate the toothing | | | – | | | |
| Reverse gear must be helical cut | | – | | – | | |
| Practicability | | | | – | | |

## 6.6 Passenger Car Transmissions

Passenger car transmissions are classified into the following main designs and formats (see also Figure 1.2):

- manual transmissions (MT),
- automated manual transmissions (AMT),
- automatic transmissions with various gear ratios,
  - dual clutch transmissions (DCT),
  - conventional automatic transmission (AT) (consisting of a hydrodynamic torque converter and a rear-mounted planetary transmission),
  - countershaft-type automatic transmissions,
- hybrid drives,
- mechanical continuously variable transmissions (CVT).

The following will introduce the basic design principles of passenger car transmissions according to the classification above. Table 6.10 shows in summarised form the transmissions treated in systematic sections 6.6.1 to 6.6.6 and designs sections 12.1.1 to 12.1.6.

**Table 6.10.** In Sections 6.6.1–6.6.6 as well as Sections 12.1.1–12.1.6 introduced passenger car gearboxes. *FT* front-transverse drive; *S* standard drive; *FL* front-longitudinal drive; *FLA* front-longitudinal all-wheel drive; *RL* rear-motor longitudinal drive; *RT* rear-motor transverse drive; *TCC* torque converter clutch; *CC* converter lock-up clutch

| Diagram Fig.-No. | Speeds | Characteristics | Configuration | Manufacturer | Name | Design Fig.-No. | No. |
|---|---|---|---|---|---|---|---|
| 6.18a | 4 | MT, 1-stage | FT | VW | MQ | — | — |
| 6.18b | 5 | MT, 1-stage | FT | VW | MQ | 12.1 | 1/ |
| 6.19a | 4 | MT, 2-stage | S | Getrag | 4-speed | — | — |
| 6.19b | 5 | MT, 2-stage | S | ZF | S 5-31 | 12.2–12.4 | 2/ |
| 6.20a | 6 | MT, 2-stage | S | Getrag | 286 | 12.5 | 3/ |
| 6.20b | 6 | MT, 1-stage | FT | Opel | F28-6 | 12.6 | 4/ |
| 6.21a | 6 | MT, 1-stage, 3-shaft | FT | Getrag | 285 | 12.7–12.8 | 5/ |
| 6.21b | 6 | MT, 1-stage, 3-shaft | FT | MB | FSG 300-6 | 12.9–12.10 | 6/ |
| 6.22a/b | 6 | MT, 2-stage | S | Getrag | 217 | 12.11 | 7/ |
| 6.23a | 6 | MT, 1-stage | FL | Audi | ML350-6F | 12.12 | 8/ |
| 6.23b | 6 | MT, 1-stage | FLA | Audi | ML450-6Q | 12.64 | 8/ |
| 6.24 | 3 | AMT, 1-stage, TCC | RL | VW | Built 1967 | — | — |
| 6.25a | 6 | AMT, 1-stage, multi-range | RT | Getrag | 431 | 12.13–12.14 | 9/ |

**Table 6.10.** *(continued)*

| | | | | | | | |
|---|---|---|---|---|---|---|---|
| 6.25b | 7 | AMT, 2-stage | S | Getrag | 247 | 12.15–12.16 | 10/ |
| 6.26 | 6 | DCT, principle | FT | VW | DSG | 12.17–12.20 | 11/ |
| 6.27 | 7 | DCT | S | ZF | 7 DCT 50 | 12.21 | 12/ |
| 6.30–6.31 | 4 | AT, w/o CC | FT | ZF | 4 HP 14 | — | — |
| — | 5 | AT, w/o CC | S | MB | W5A 030 | 12.22 | 13/ |
| 6.32 | 5 | AT | S | MB | W5A 580 | 12.23 | 14/ |
| 6.33 | 7 | AT | S | MB | W7A 700 | 12.24 | 15/ |
| 6.34 | 6 | AT | S | ZF | 6 HP 26 | 12.25 | 16/ |
| 6.35 | 6 | AT | FT | AISIN | TF 80-SC | 12.26 | 17/ |
| 6.36 | 5 | Countershaft-type automatic transmission | FT | MB | W5A 180 | 12.27 | 18/ |
| 6.37–6.38 | 6 | Hybrid, parallel | S | BMW/ZF | Active Transmission | 12.28 | 19/ |
| 6.39 | ∞ | Hybrid, power-split | FT | Toyota/Lexus | P310 | 12.29 | 20/ |
| 6.41 | ∞ | CVT, toroid principle | — | — | — | — | — |
| — | ∞ | CVT, chain variator | FL | Audi | Multitronic | 12.30 | 21/ |
| — | ∞ | CVT, chain variator | FT | MB | Autotronic | 12.31 | 22/ |
| 6.43 | ∞ | CVT, chain variator | FT | ZF | CFT 30 | 12.32 | 23/ |
| 6.44 | ∞ | CVT, geared neutral | — | — | — | — | — |

## 6.6.1 Manual Passenger Car Transmissions (MT)

Manual passenger car transmissions include all transmissions in which both the process of changing gear and the process of engaging the master clutch and moving-off are carried out manually by the driver. They are all fitted with spur gears. Transmissions with dog clutch engagement are not available for passenger cars. They are only offered with synchromesh. To shed light on the basic principles involved, the following will include an historical account of the development of manual transmissions. The systematics of manual transmissions will be discussed using examples ranging from the 4-speed transmissions still common in the 1990s up to the most current designs. As always in this book, all gear numbers refer to forward gears. A separate section will be devoted to deriving all-wheel drive variants from the basic principles.

Manual passenger car transmissions can be subdivided according to number of stages into further categories (see also Section 6.3). "Stage" is defined here as the power flow from one shaft to another. This subdivision relates mainly to the forward gears of the main gearbox itself, not to any integral final drives, differentials and intermediate shafts needed to drive them. This yields the following categories:

- single-stage countershaft transmissions with 4 to 6 gears and integral final drive (e.g. Figure 6.18a) and
- two-stage (coaxial) countershaft transmissions with 4 to 6 gears (e.g. Figure 6.19a).

Single-stage countershaft transmissions are used in passenger cars in which the engine is located next to the drive axle, which is to say in front-wheel drive vehicles with front engines, or in rear-wheel drive vehicles with rear engines. This applies to both normal engine configurations – transverse and longitudinal. In the case of single-stage countershaft transmissions, the final drive is usually integrated into the gearbox housing. If a very short overall transmission length is required for space reasons, the ratio can occur by means of a third, offset shaft. Figure 6.21 shows such a three-shaft gear unit.

In the transmission diagrams used in this chapter, integral final drives, where present, and reverse gears of the various transmissions are represented by "grey" lines for the sake of completeness. It should be noted that in reverse gears the shafts of the idler gears are located in a different plane to the main shafts (compare also with Section 6.5). The location and size of the idler gears are intended only to give an impression of the fundamental design.

Two-stage countershaft transmissions are used in passenger cars with standard drive. They normally contain no integral final drive components since they are generally flanged directly onto the front-mounted engine, and linked to the drive axle by a propeller shaft. An exception is two-stage transmissions mounted on the rear axle to give more even weight distribution with front-mounted engines (see also Figure 6.2h/i). Parts of the final drive are integrated in such transmissions.

The synchronizer packs are each allocated to one shift gate, and serve mostly to shift two neighbouring gears. In each shift gate there is usually first and second gear, third and fourth gear, fifth and reverse gear, or alternatively fifth and sixth gear. There are also designs which use a separate shifting element for the fifth and reverse gear, which can be unsynchronized in reverse gear.

The example of a single-stage 4-speed gearbox is the VW unit, as used for example in the early 1990s in the VW Golf (Figure 6.18a). In this gearbox, the gear pair of the first gear is located directly beside a shaft bearing. The total number of gear pairs remains the same compared to a two-stage 4-speed transmission, since although the gear pair of the *input constant gear CG* (sometimes called *head set*) (Figure 6.19) is not required, one is needed for the fourth gear. Single-stage transmissions have no direct gear. Single-stage 5-speed transmissions (Figure 6.18b) differ from single-stage 4-speed transmissions only in having an additional gear pair which is "attached onto" the housing side opposite the input side.

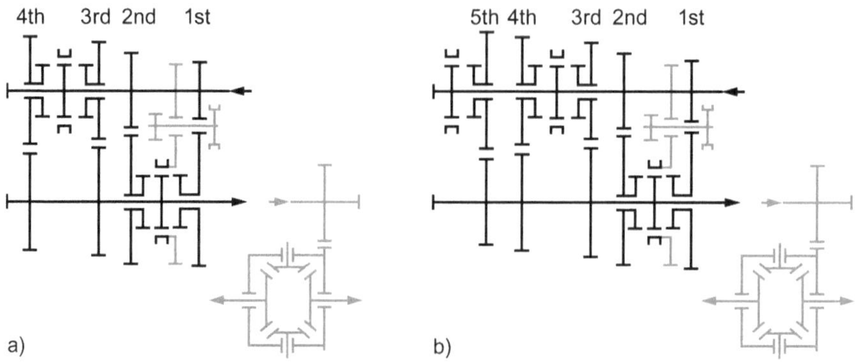


**Fig. 6.18.** *a* Single-stage 4-speed gearbox (VW); *b* single-stage 5-speed gearbox (VW), *production design Figure 12.1*

This does not require any design changes in the original gear unit. Often, 5-speed gearboxes have been developed from existing 4-speed gearboxes.

One example of the two-stage 4-speed gearbox is the Getrag gearbox in Figure 6.19a. In accordance with the design principle of placing gear pairs with high torque changes near bearings in order to minimise shaft deflection, the gear pair of the first gear is located on the gearbox output side. The fourth gear is the direct gear. In the 5-speed gearbox shown on the right in Figure 6.19b, the fifth gear is the direct gear. Frequently, the fifth gear is speed increasing (overdrive) and the fourth gear is the direct gear.

In two-stage countershaft transmissions with six gears, as in Figure 6.20a, the gears of the first and second gear are near a shaft bearing. It should also be borne in mind that such transmissions are used principally in high-performance passenger cars, and therefore have a high torque design. Figure 6.20b shows a single-stage countershaft transmission with final drive.

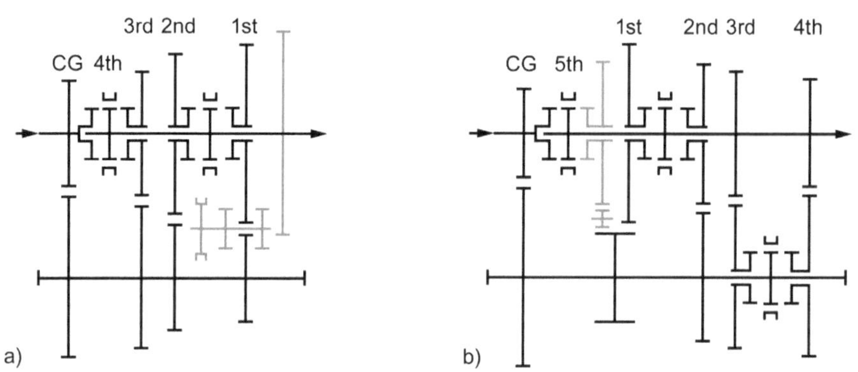


**Fig. 6.19.** *a* Two-stage 4-speed gearbox (Getrag); *b* two-stage 5-speed gearbox with direct 5th gear, "sports gearbox" (ZF), *production design Figure 12.2*

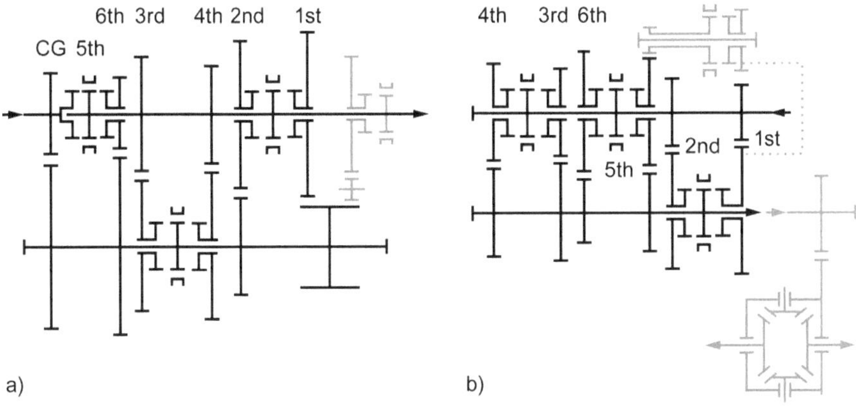


**Fig. 6.20.** *a* Two-stage 6-speed gearbox (Getrag), *production design Figure* 12.5; *b* single-stage 6-speed gearbox (Opel); *production design Figure 12.6*

Since the mid-1990s, almost all manufacturers have tended towards 6-speed manual transmissions. The existing 5-speed gearboxes are being replaced by new designs with 6 gears, primarily among vehicles with more highly powered engines. Since the length of the gearbox is of great importance when it is transversely mounted, corresponding solutions must be found for short gearboxes.

The gears of transmissions with a three-shaft design are distributed among two output shafts *OS1* and *OS2* which lie parallel to the input shaft *IS*. This allows a very short gearbox, which is necessary for transverse mounting. With this design, 6-speed transmissions have the same overall length as a 4-speed transmission. Three-shaft transmissions are single-stage countershaft transmissions with integral final drive.

Figure 6.21a shows a design optimised for minimal overall length. A fixed gear is allotted to the third and fifth gears and to the fourth and sixth gears respectively – a so-called "double-use" of the fixed gears for two gears each. The different ratios must thus be entirely effected via the respective ratios of the constants, and this with the same axial distance of the output shafts to the gearbox input shaft. A third double-use is realised in the reverse gear. The reverse gear uses the shift gear of the first gear as a reverse idler gear. This is possible because the reverse gear shares the short constant gear *CG2* with the third and fourth gears. This design is thus nearly optimal with respect to overall axial length and the number of necessary gear set parts, with a small limitation of ratio selection. As a result of the two double-uses, only five gears can be selected freely; the sixth gear is the outcome of the others [6.19].

Figure 6.21b shows a three-shaft transmission which allows all gear ratios to be freely selected and is comparable to other manual transmissions. In the figure, the first to fourth gears are on output shaft *OS1*, the fifth, sixth and reverse gears on output shaft *OS2*. Both higher starting torque ratios and higher overall gear ratios can be achieved. The transmission length is, however, somewhat greater, since only one double-use is implemented for the fourth and fifth gears. The reverse idler gear of the reverse gear has its own countershaft.

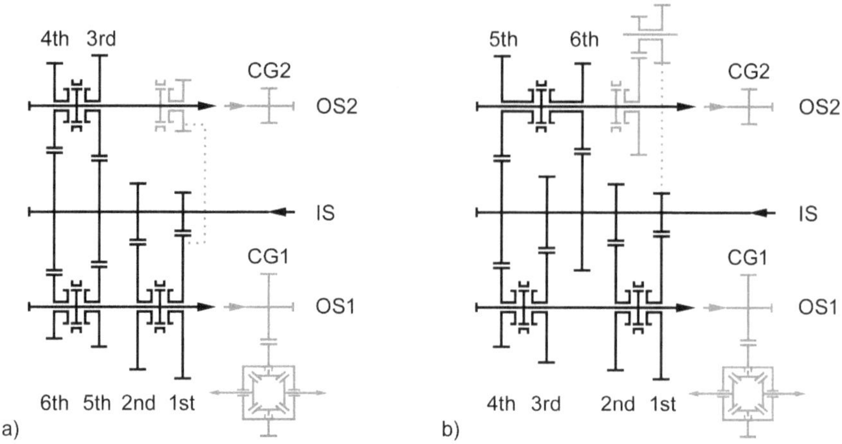


**Fig. 6.21.** *a* Single-stage 6-speed gearbox with three-shaft design (Getrag), *production design Figure 12.7*; *b* single-stage 6-speed gearbox with three-shaft design (Mercedes-Benz), *production design Figure 12.10*

New designs for standard drives have also almost exclusively six gears. These transmissions are always designed as two-stage transmissions. Examples of such transmissions, also referred to as in-line transmissions, are shown in Figure 6.22. Figure 6.22a shows the variant designed for spark ignition engines. The basic design corresponds to that of Figure 6.20a. Figure 6.22b shows the design for diesel engines. Since a greater overall gear ratio is required for diesel engines, the fourth gear was designed as a direct gear instead of the fifth gear, as for petrol engines. The fifth and sixth gears thus have a ratio smaller than 1. These transmissions are also referred to as double-overdrive transmissions. In order to receive the same gear pattern in both cases despite their varying gear configurations, small modifications of the internal gearshift system are required.

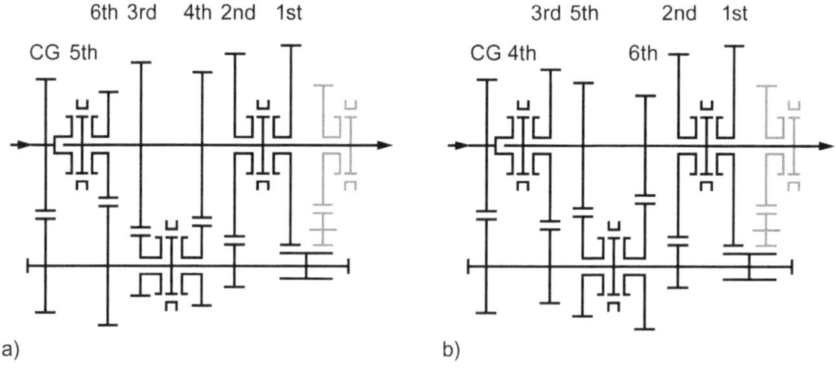


**Fig. 6.22.** Two stage 6-speed gearbox (Getrag), *production design Figure 12.11*. *a* Design for spark ignition engines; *b* design for diesel engines

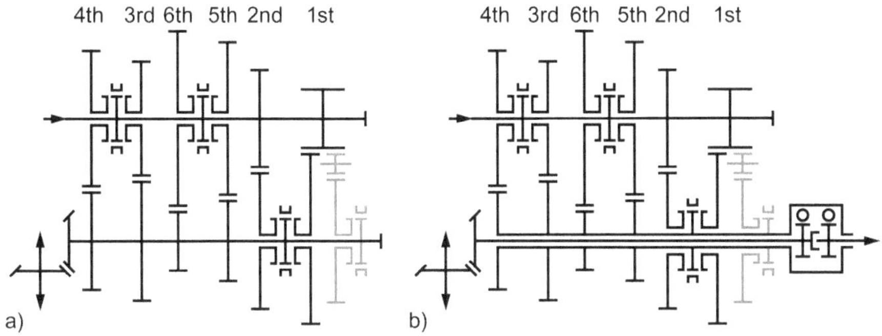


**Fig. 6.23.** Single-stage 6-speed gearbox for front-longitudinal mounting (Audi). *a* For front-wheel drive, *production design Figure 12.12*; *b* for all-wheel drive with TORSEN centre differential, *production design Figure 12.64* (bevel gear drive rotated 90°)

New designs for longitudinal transmissions with final drive are exclusively 6-speed transmissions. These are used in front-wheel and all-wheel drive vehicles (e.g. Audi A6) as well as in rear-motor drive vehicles (e.g. Porsche 911) and have a single-stage design. The integral final drive consists of a spiral gearing.

Figure 6.23a shows a transmission for front-wheel drive, Figure 6.23b for all-wheel drive. The gear set is assembled in the same way in both designs. The power flow for the front-wheel variant goes from the input shaft over the output shaft to the front axle differential. In the case of the all-wheel drive design, the flow is from the input shaft over a hollow shaft to an integral TORSEN centre differential. There, the power is distributed among the front and rear axles. Power is transmitted to the front axle via a pinion shaft mounted in the hollow shaft to the front axle differential integrated into the transmission. Power is transmitted to the rear axle via the flange-mounted cardan shaft to the rear axle differential.

## 6.6.2 Automated Manual Passenger Car Transmissions (AMT)

When manual transmissions for passenger cars began to be automated, the term "Semi-automatic transmission" was used. The term referred to the two operations "Engaging the clutch/Moving-off" and "Changing gear". One of these operations was automated in semi-automatic transmissions (see Table 6.14 "Automation levels of manual transmissions").

A typical example of an early semi-automatic manual transmission for passenger cars is the VW torque converter clutch transmission (TCCT) (1967). In this design, there is a mechanical gearshifting clutch mounted behind a hydrodynamic torque converter (Figure 6.24). The process of engaging the gearshifting clutch and moving-off is automated, while changing gears is manual.

The torque converter has three main functions to fulfil in this process: to enable moving-off in any gear, to refine the coarse stepping (three forward gears) of the manual gearbox and to damp the torsional vibration when engaging the gearshifting clutch.

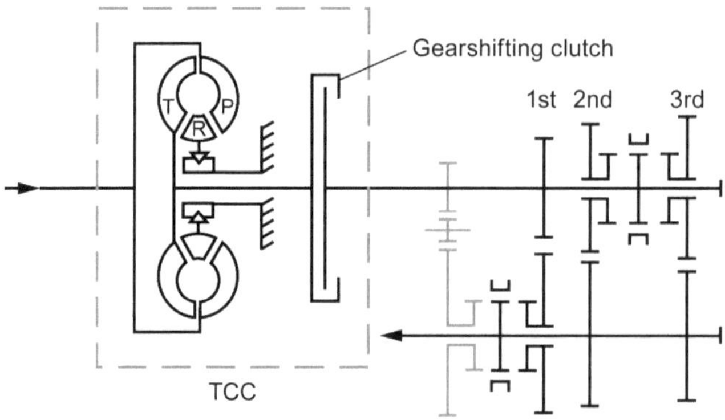


**Fig. 6.24.** Gearbox diagram: 3-speed torque converter clutch gearbox (VW 1967)

The main gearbox is a single-stage three-speed gearbox developed from a 4-speed transmission by converting what was originally first gear into a reverse gear. In practice this transmission concept had to contend with high fuel consumption. The reason for this was the constant power flow through the converter – there was no lock-up clutch – and the fact that with this transmission it was possible to move-off in second or third gear. This design therefore never became popular in passenger cars.

Semi-automatic manual passenger car transmissions have never found broad use. (Fully) automated manual transmissions (AMT) have been available on the market for passenger cars since the end of the 1990s. In AMTs, both the process of engaging the master clutch and moving-off as well as changing gears are executed by actuators which receive their control signal via shift paddles on the steering wheel, a gearshift lever or, in the case of fully automated operation, by a transmission control unit (TCU).

Automated manual transmissions combine the high efficiency of manual transmissions with the ease of operation of a fully automatic transmission. The biggest difference to automatic powershift transmissions for the user is the less comfortable gearshifting, which according to the principle of the design is subject to power interruption, as with manual transmissions. Attempts to transfer residual power with oversized synchronizing units and an incomplete opening of the clutch during the shifting process showed positive results in test vehicles, but have not led to series production.

AMTs with six gears or more and a greater overall gear ratio represent a suitable transmission design for vehicle classes and applications with an emphasis on efficiency and ease of operation, as for small passenger cars and commercial vehicles smaller than 3.5 t. The high level of efficiency of the transmission can be combined with a shifting strategy optimised for consumption. The shifting strategy of an AMT contributes significantly to reducing fuel consumption.

AMTs are further classified as add-on and integrated systems. Add-on systems furnish existing manual systems with built-on actuators. In this way, a basic gear unit can become either a manual transmission or an automated manual transmission. Integrated systems are already designed as automated manual transmissions. They cannot be used as MTs. With integrated systems, the configuration of the individual gears can be optimised for AMT. They do not have to adhere to the usual gearshift lever shifting pattern typical of MTs. For this reason, the design of the internal gearshift system is more flexible. For example, the sliding sleeves can be activated by a gearshifting drum (only serial shifting – skipping gears is not possible) or the shifting time (amount of time without power flow) can be shorted by means of an optimised gear set configuration and individual actuators for each sliding sleeve.

Systems with gearshifting drums have the advantage of only requiring one actuator; otherwise at least two actuators are required for the shifting and selecting motions. With individual actuators, each sliding sleeve is activated by one actuator. This is the most expensive, but also the technically most advanced solution.

Actuators are classified as either electrohydraulic or electromechanical systems. Hydraulic systems are generally more costly, but have advantages with respect to the maximum possible gearshift forces and the shorter shifting times associated with them. Even individual actuators are easy to realise. In hydraulic systems, linear motions are generally implemented using pistons. See also Chapter 13 "Electronic Transmission Control".

Electromechanical systems usually use rotary actuators (electric motors) and are used primarily for smaller, more cost-effective transmissions with torque capacities up to 250 Nm. The torque limitation arises from the increasing amount of power needed for clutch activation and the maximum gearshift forces in conjunction with the short shifting times. Since electric motors are used as a rule, the static power of the actuator may increase when larger actuators are used, but the dynamics of the actuators decreases due to the increased inertia. The load on the power supply also increases. Even in smaller systems, the light can be observed to flicker slightly when shifting.

Figure 6.25a shows a single-stage 6-speed gearbox for transverse mounting with integrated actuator technology developed exclusively as an AMT. This is a multi-range gearbox with three forward gears and one reverse gear as well as two selectable output constant gears (*High*/*Low*). The first to third gears are activated with the constant *Low* ($CG_L$), the fourth to sixth with the constant *High* ($CG_H$). The advantage of this design is the small overall length and the use of only three synchronizing units. Although two sliding sleeves must be activated when shifting from third to fourth gear, this is not a problem with a gearshifting drum. In the gear set, shifting proceeds from third to first gear and simultaneously from the constant *Low* to *High*.

Figure 6.25b shows a 7-speed in-line gearbox. This is a two-stage gearbox developed exclusively as an AMT. This does away with the usual restrictions on gear set design imposed by the gear pattern of a manual transmission. In order to shorten the shifting times, sequential gears, with the exception of the sixth and seventh gears, are not placed on one synchronizing unit. In this way, the shifting times can be reduced when changing gears by means of the simultaneous activation of two sliding sleeves (overlapping shifting).

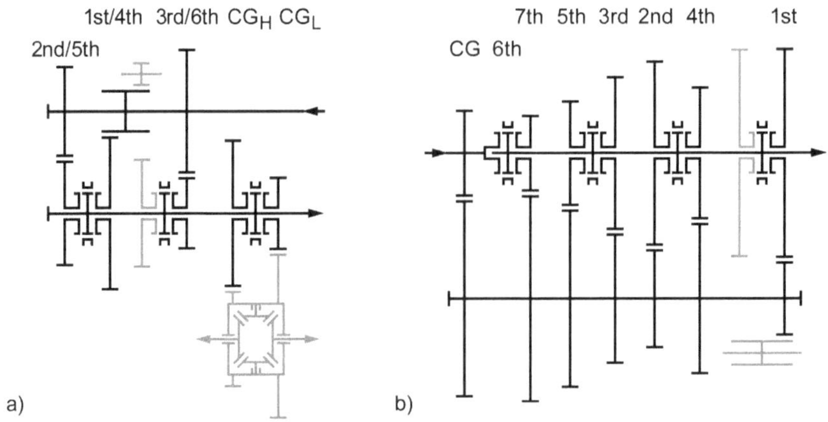


**Fig. 6.25.** *a* Single-stage 6-speed AMT with range-change unit (Getrag), *production design Figure 12.13*; *b* two-stage 7-speed AMT (Getrag), *production design Figure 12.15*

However, by no means may two gears be engaged simultaneously, since this would cause the powertrain to be blocked. Since this gearbox is designed for sports cars, the overall gear ratio is small. The sixth gear is direct drive.

### 6.6.3 Dual Clutch Passenger Car Transmissions (DCT)

Dual clutch transmissions were already being developed in the 1940s. The original intention was to furnish heavy commercial vehicles with technology which provided for driving without power interruption. Serial production was not achieved, however. In the 1980s, Porsche and Audi took up this transmission concept again and developed a dual clutch transmission for racing cars. These transmissions were not suited to serial production because the control quality of the systems was not yet sufficient.

The first DCT for passenger cars went into production in 2003. The goal of this model was to combine the advantages of manual transmissions with those of automatic transmissions. Attributes of manual transmissions are a high level of efficiency, a gear ratio spread which is freely selectable in broad ranges, as well as sportiness, driving dynamics and driving pleasure. Conventional automatic transmissions are characterised by their ease of handling when moving-off thanks to the torque converter and by automatic shifting without power interruption.

The principle of dual clutch transmission is based on the idea of two independent sub-gearboxes each connected to the engine via its own clutch (Figure 6.26). One sub-gearbox contains the odd gears (1, 3, 5…) and the other the even gears (2, 4, 6…). By dividing the gears through the dual clutch, the DCT becomes fully power shiftable. However, the dual clutch is not only implemented in a DCT for shifting; it also serves as a moving-off element.

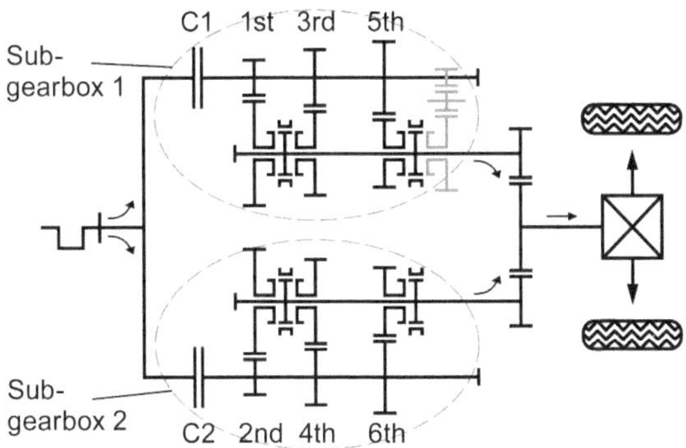


**Fig. 6.26.** Principle design of dual clutch transmissions. Division into two independent sub-gearboxes each with its own clutch *C1* and *C2*, *production design Figure 12.17* (VW DSG®)

In actual designs, the two sub-gearboxes are not arranged side-by-side, as in Figure 6.26, but rather one is nested in the other to save space. One of the two gearbox input shafts is used as a hollow shaft.

The basic functioning of dual clutch transmissions will be explained in the following on the example of the upshift from second to third gear. When a situation arises during vehicle operation which requires an upshift from the currently engaged second gear (*sub-gearbox 2*) to third gear, third gear is engaged in the free *sub-gearbox 1*. The synchronizing process of the corresponding idler gear is not noticeable to the driver. In virtue of the overlapping of closing clutch *C1* and opening clutch *C2*, the power flow is not interrupted. Once *C1* has taken over the torque, the second gear is disengaged in sub-gearbox 2, which is now free, and another gear can be preselected, if necessary. The basic process is the same for both upshifting and downshifting. See also Section 9.3.2 for more on overlapping shifting.

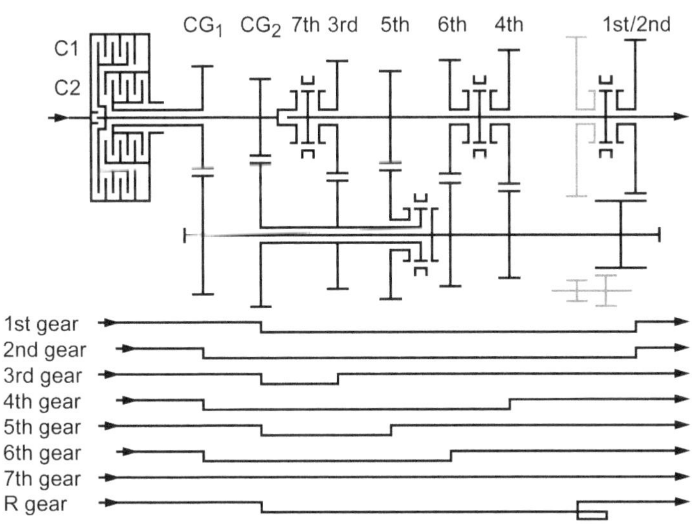


**Fig. 6.27.** Two-stage 7-speed DCT (ZF), *production design Figure 12.21*

Figure 6.27 shows a 7-speed DCT for standard drive. Because of its better performance data, the outer clutch *C1* is generally used for moving-off. In principle, however, moving-off is also possible with the inner clutch.

Through the use of a short hollow shaft on the countershaft and the multiple use of the rearmost gear pair in the first and second gear, a gear pair could be spared [6.28]. To change from first to second gear, the inner clutch *C2* only has to be closed and the outer clutch *C1* opened. Thus no gear engagement is necessary for this gear change.

The main component, i.e. the dual clutch, is classified as either a wet-running or a dry-running system. See Section 10.3 "Dual Clutches", in which the structure and function of dual clutches is explained in detail.

Like conventional automatic transmissions, wet-running systems require a hydraulic supply system for clutch activation and cooling the clutches. See also Section 11.3 "Oil Supply and Oil Pumps". An advantage of DCT over AT is that only one open clutch causes drag losses, even with wet-running dual clutch. Due to the lacking torque increase of the torque converter, dual clutch systems require a higher starting torque ratio. For this reason, these transmissions need a higher overall gear ratio. An additional gear may be required for this reason, i.e. to prevent the gear steps from getting too large.

The greatest advantage and, simultaneously, the greatest disadvantage of a DCT with dry-running dual clutch is that there is no oil in the area of the clutch. The advantage of this is a minimal drag torque for the open clutch. The disadvantage is that the frictional work and thus heat arising from moving-off and powershifting cannot be discharged by the oil, as with wet-running clutches. The limit between wet- and dry-running clutches in terms of torque capacity is approximately 300 Nm. With lower engine torques, the market tends toward dry-running systems, with higher torques, to wet-running systems.

### 6.6.4 Automatic Passenger Car Transmissions (AT)

Automatic transmissions with various gear ratios consisting of a torque converter with a rear-mounted planetary-type gearbox are known as conventional automatic transmissions (AT) or just "automatic transmissions" (see also the systematic classification of transmissions according to Figure 1.2).

The principles of planetary transmissions are introduced in Section 6.4 "Gear Sets with Fixed Axles, Countershaft Transmissions and Epicyclic Gears". Even in a single planetary gear set, planetary transmissions already provide a large number of states of motion possible by combination. The automatic passenger car transmissions treated in the following employ several planetary gear sets coupled together.

A design often used in automatic transmissions is the Ravigneaux planetary gear set (Figure 6.28). The Ravigneaux set is a so-called reduced planetary gear.

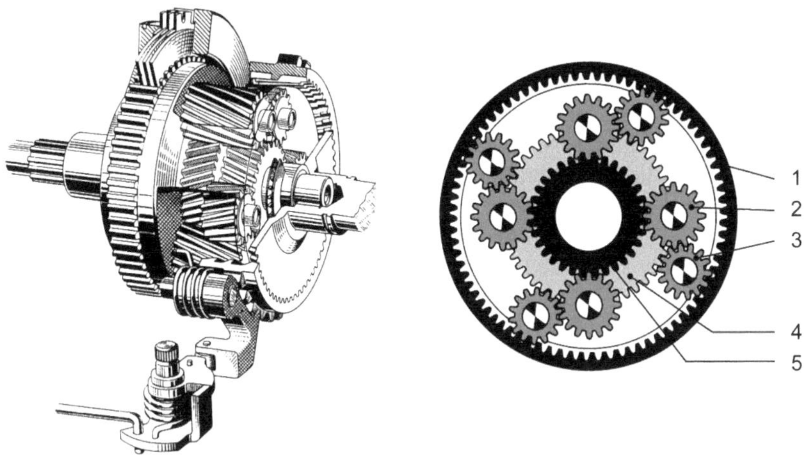


**Fig. 6.28.** Ravigneaux planetary gear set. *1* Common ring gear; *2* narrow planetary gear; *3* broad planetary gear; *4* large sun gear; *5* small sun gear [6.5]

These are planetary transmissions in which the construction resources are "reduced" since parts of the individual simple planetary gear sets are grouped together [6.30]. With this design, there are up to four practically usable forward gears and one reverse gear available.

Thanks to the distribution of the torque to several gear meshes, planetary gears have a high power density. Moreover, power transmission is not based merely on the rolling motion of the gears (rolling contact power), as with countershaft transmissions, but also on simple gear synchronization (coupling power), which purely in relation to the gear mechanism leads to a higher level of efficiency than with countershaft transmissions. Thus an effective coupling of planetary gear sets allows automatic transmissions with a large number of gears. However, the selection of ratios is restricted, since the individual gearwheels are used for several gears. The individual planetary gear sets are arranged in a row like discs. More planetary gear sets also always means a greater transmission length. This must be considered in particular in the case of transmissions for front-transverse drives. To reduce the length by designing a wide transmission, as done with multiple countershaft transmissions (e.g. Figure 6.21 "Three-shaft gearbox"), is difficult to realise.

Much of the space taken up by automatic transmissions is occupied by the clutches and brakes required to shift the gears. There are two different types of brake as standard, the belt brake and the multi-plate brake. In the belt brake, a metal belt runs once or twice around a brake drum, and brakes the drum by tightening the belt. This braking process is not as easy to control as with multi-plate brakes, since the braking action is very rapid because of the self-reinforcing physical principle involved. In view of the increasing requirement for shifting comfort, the multi-plate brake has become the most popular variant. The multi-plate brake requires more space than the belt brake (see also Section 9.3 "Layout and Design of Multi-Plate Clutches").

Since in automatic transmissions the hydrodynamic torque converter takes over part of the ratio change, they theoretically rely on fewer gear steps and a smaller overall gear ratio than comparable manual transmissions (see also Section 10.4 "Hydrodynamic Clutches and Torque Converters").

The clutches and brakes discussed above for shifting the various gear steps are hydraulically controlled by hydraulic fluid. This fluid is supplied by an oil pump. Oil supply for conventional automatic transmissions is explained in Figure 11.12, Section 11.3.1 "Oil supply". An overview of the losses in automatic transmission is given by the highly simplified block diagram of a conventional automatic transmission in Figure 6.29.

An important, if not the most important, assembly in an automatic transmission is the control unit. It is responsible for activating the brakes and clutches in the transmission. Their control has a direct influence on the "shifting quality" of the transmission as perceived by the driver. Electronic-hydraulic control units have become the standard since the mid-1990s [6.7]. Put simply, the electronics and software take care of the required intelligence, the hydraulics take care of the actuating forces. See also Chapter 13 "Electronic Transmission Control".

In the following, the functional features of a system based exclusively on a Ravigneaux planetary gear set will be examined using the 4-speed automatic transmission illustrated in Figure 6.30. The transmission went into production in 1984. This transmission is designed for use in front-wheel drive passenger cars, which would become apparent in the transmission diagram only after the planetary gear set. The components devoted to the final drive are not shown, since they have no effect on the principle of operation. The components involved in the particular gear step are shown by heavier lines.

The ZF 4 HP 14 four-speed automatic transmission consists of a torque converter with integral torsion damper *TD*. The transmission has no torque converter lock-up clutch, but works rather with power-split to improve efficiency (see Figure 10.42). There is also a crescent design oil pump linked to the pump shaft of the converter (not shown in the diagram) to provide the pressurised oil necessary to shift the gears.

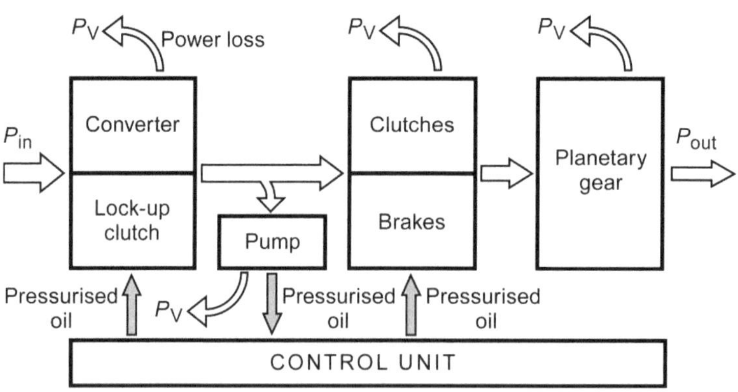


**Fig. 6.29.** Block diagram and power losses in a (conventional) automatic transmission

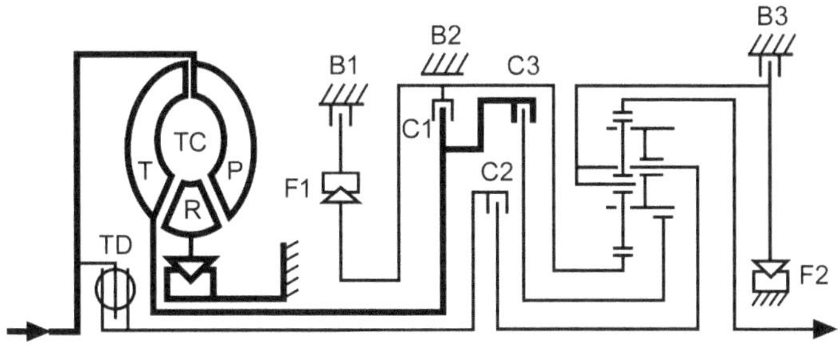


**Fig. 6.30.** Gearbox diagram: 4-speed automatic transmission with Ravigneaux gear set (ZF) in neutral. *TC* Trilok converter: *P* pump, *T* turbine, *R* reactor with freewheel; *TD* torsion damper; *F* freewheels; *B* brakes; *C* clutches

The clutches are multi-plate clutches shifted by oil pressure. The brakes are of both designs, multi-plate brakes *B1* and *B3* and belt brake *B2*.

In 1st gear, the spiders of both planetary gears are retained by the freewheel *F2*, by means of which the planetary gear set functions as a gear set with fixed axles (Figure 6.31). The input power flows through the converter and the engaged clutch *C3* to the large sun gear of the Ravigneaux set, and back out of the planetary gear set via the ring gear to the output. The effective ratio $i = 2.41$.

In second gear, the small sun gear rests against the gearbox housing by means of the freewheel *F1* and the brake *B1*. The input power flows through the converter and the engaged clutch *C3* to the large sun gear, as in the first gear. The spider of the planetary gear set now rotates, and the planetary gear set functions as a reduced planetary coupled gear. The power flows again via the ring gear to the output, and the effective ratio is $i = 1.37$.

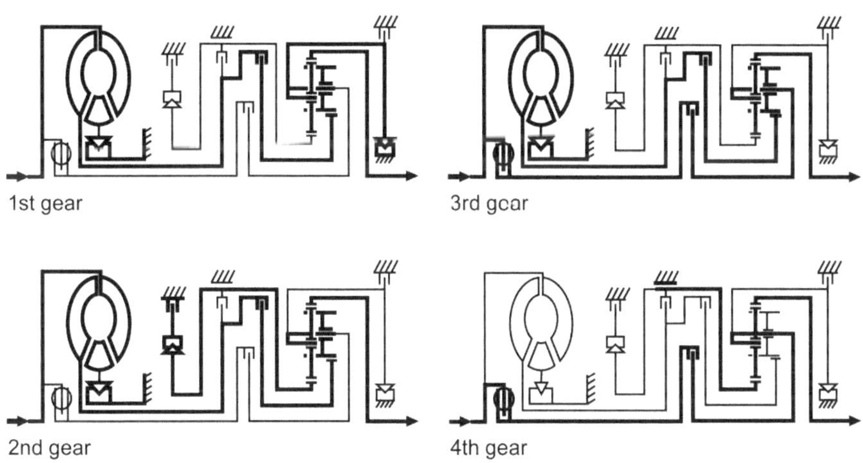


**Fig. 6.31.** 4-speed automatic transmission (ZF); power flow in the gears

The third gear is the most interesting from the point of view of its function. The transmission functions with power-split, i.e. a part of the drive power flows through the torsion damper *TD* and the engaged clutch *C2* into the planetary gear set which functions as a differential drive. The second power branch flows from the converter through the clutch *C3* to the large sun gear of the planetary gear set. Both power branches, or rotational speeds, “overlap” in the planetary gear set, and are fed to the output at the ring gear. (This operating status of the power-split is not to be confused with that of a closed torque converter lock-up clutch *CC*. In the *CC*, the impeller *P* and turbine wheel *T* of the converter are linked together, locking up the converter.) The transmission ratio in third gear depends to a small degree on the slip in the converter, and is therefore not constant. The ratio in third gear thus varies between $i = 1.0$ and 1.09.

In 4th gear the converter runs without load, and power transmission to the planetary gear set is purely mechanical through the torsion damper *TD* and the clutch *C2*. The Ravigneaux set functions as a simple planetary gear driven through its spider, and whose sun gear is supported at the housing through the brake *B2*. Output is through the ring gear. The ratio in 4th gear is $i = 0.74$, constituting an overdrive.

In reverse gear the Ravigneaux set again works as a simple planetary gear reversing the direction of rotation. The power flows through the converter and the clutch *C1* to the small sun gear. The spider is supported against the housing through the brake *B3*. The output is through the ring gear. The reverse gear ratio is $i = -2.83$.

A second example is provided by the 5-speed transmission W5A 580 by Mercedes-Benz [6.38–6.39] (Figure 6.32). This transmission, which went into production in 1995, is based on three simple planetary gear sets coupled together.

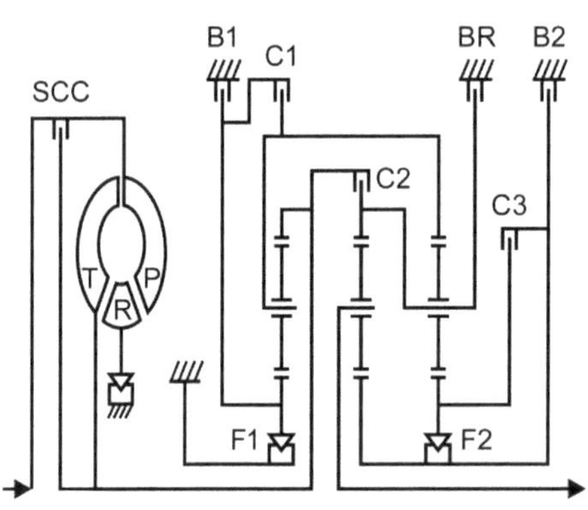


| G | Clutch | | | Brake | | | Free-wheel | | i | Gear step |
|---|---|---|---|---|---|---|---|---|---|---|
| | C1 | C2 | C3 | B1 | B2 | BR | F1 | F2 | | |
| 1 | | | (•) | (•) | • | | • | • | 3.59 | 1.64 |
| 2 | • | | (•) | | • | | | • | 2.19 | 1.56 |
| 3 | • | • | | | • | | | | 1.41 | 1.41 |
| 4 | • | • | • | | | | | | 1.00 | 1.20 |
| 5 | | • | • | • | | | (•) | | 0.83 | |
| R1 | | | (•) | (•) | | • | • | • | –3.16 | 0.61 |
| R2 | • | | (•) | | | • | | • | –1.93 | |
| N | | | • | • | | | | | 0 | Total 4.32 |

( ) = in overrun conditions

**Fig. 6.32.** Gearbox diagram and gearshift pattern of a 5-speed automatic transmission (Mercedes-Benz), *production design Figure 12.23*

The ring gear of the forward planetary gear set is driven by the turbine wheel of the torque converter. A slip-controlled torque converter lock-up clutch *SCC* can lock up the converter completely or with the necessary slip. In first gear, the sun gear is supported by means of the freewheel *F1* and the brake *B1*. The ratio acts on the ring gear in the rear planetary gear set via the planet gear carrier. The ratio is carried out here in a similar way as in the forward planetary gear set. The rear sun gear rests against the housing by means of the freewheel *F2*, the multi-plate clutch *C3* and the multi-plate brake *B2*. A ratio likewise takes place in the middle planetary gear set. The rear planetary gear carrier drives the ring gear of the middle planetary gear set, whose sun gear also rests against the housing via the brake *B2*. The middle planetary gear carrier is directly connected to the output shaft. The ratio in first gear reaches a value $i = 3.595$.

In the second gear, the brake *B1* is open and the clutch *C1* is closed. This releases the freewheel *F1* of the forward sun gear, with the forward set rotating as a block. The ratio in the rear and middle set is achieved as in first gear. The resulting ratio is $i = 2.186$.

In third gear, the multi-plate clutch *C3* is opened and the clutch *C2* engaged. This releases the freewheel *F2* of the rear sun gear, with the forward and rear planetary gear sets rotating as a block. The drive is delivered directly through the turbine *T* and the clutch *C2* to the middle ring gear. The ratio is achieved exclusively through the middle set with $i = 1.405$.

When shifting to fourth gear takes place, the brake *B2* is opened and the clutch *C3* is engaged. Now all three planetary gear sets are locked and rotate with the speed of the turbine. The ratio of the transmission in direct drive is $i = 1.0$.

In fifth gear, the forward planetary gear set is shifted as in first gear. The clutch *C1* opens and the brake *B1* closes. The rear ring gear now rotates more slowly than the turbine. Since the middle ring gear and the rear planetary gear carrier continue to rotate through the clutch *C2* with engine/turbine speed, the middle sun gear and the rear sun gear coupled via the clutch *C3* must rotate more quickly than the turbine. The rotational speed of the middle planetary gear carrier and thus of the output shaft lies between that of the ring gear and the sun gear. The speed-increasing ratio in fifth gear is $i = 0.831$.

In the first reverse gear, the forward planetary gear set is shifted as in first gear. The rear planetary gear carrier and the middle ring gear are held in place by means of the multi-plate brake *BR*. The middle sun gear and the rear sun gear connected via clutch *C3* rotate backwards, and the planetary gear carrier and the output shaft thus also rotate backwards. A second reverse gear can be engaged in analogy to the second gear.

The 7-speed automatic transmission W7A 700 introduced in 2003 by Mercedes-Benz is a further development of the gear set design of the 5-speed automatic gearbox illustrated in Figure 6.32. The forward simple planetary gear set was replaced by an inverse Ravigneaux gear set (one sun gear and two ring gears) (Figure 6.33). Through the addition of a brake *B3* and omission of two freewheels, seven forward and two reverse gears can be shown according to the gear pattern in Figure 6.33.

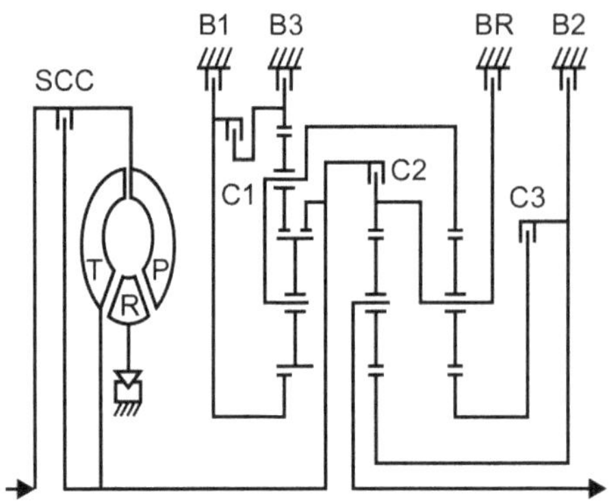


| G | Clutch | | | Brake | | | | i | Gear step |
|---|---|---|---|---|---|---|---|---|---|
| | C1 | C2 | C3 | B1 | B2 | B3 | BR | | |
| 1 | | | • | | • | • | | 4.38 | 1.53 |
| 2 | | | • | • | • | | | 2.86 | 1.49 |
| 3 | • | | • | | • | | | 1.92 | 1.40 |
| 4 | • | • | | | • | | | 1.37 | 1.37 |
| 5 | • | • | • | | | | | 1.00 | 1.22 |
| 6 | | • | • | • | | | | 0.82 | 1.13 |
| 7 | | • | • | | | • | | 0.73 | |
| R1 | | | • | | | • | • | –3.42 | 0.65 |
| R2 | | | • | • | | | • | –2.23 | Total |
| N | | | • | | | • | | 0 | 6.02 |

**Fig. 6.33.** Gearbox diagram and gearshift pattern of a 7-speed automatic transmission (Mercedes-Benz), *production design Figure 12.24*

A widespread gear set concept is that of Lepelletier. This design is based on a single planetary gear set with rear-mounted Ravigneaux gear set. In 2001, ZF used this gear set design to bring the first 6-speed passenger car transmission 6 HP 26 on the market (Figure 6.34) [6.44].

During operation in forward and reverse, the torque converter drives the ring gear of the forward planetary gear set *GS1* through the input shaft. The forward sun gear remains stationary in all gears. In first gear, the multi-plate clutch *A* is closed and the power flows over the planetary gear carrier of *GS1* to the sun of the short planetary gears of the rear-mounted Ravigneaux gear set (*GS2* and *GS3*). The brake *D* is closed and the planetary gear carrier of the Ravigneaux gear set is fixed. The ratio is $i = 4.171$.

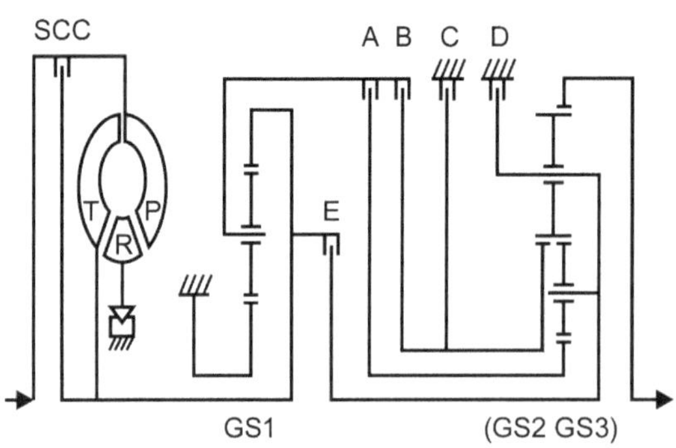


| G | Clutch | | | Brake | | i | Gear step |
|---|---|---|---|---|---|---|---|
| | A | B | E | C | D | | |
| 1 | • | | | | • | 4.17 | 1.78 |
| 2 | • | | | • | | 2.34 | 1.54 |
| 3 | • | • | | | | 1.52 | 1.33 |
| 4 | • | | • | | | 1.14 | 1.31 |
| 5 | | • | • | | | 0.87 | 1.26 |
| 6 | | | • | • | | 0.69 | Total |
| R | | • | | | • | –3.40 | 6.04 |

**Fig. 6.34.** Gearbox diagram and gearshift pattern of a 6-speed automatic transmission, principle according to Lepelletier (ZF), *production design Figure 12.25*

In second gear, the multi-plate brake *D* is opened and the brake *C* is closed. The sun of *GS2* stops moving and the planetary gear carrier rotates. The long and short planetary gears roll on of each other. The total ratio is then $i = 2.340$. When shifting proceeds from second to third gear, the brake *C* is opened and the clutch *B* is closed (see also Figure 9.33). In the Ravigneaux gear set (*GS2* and *GS3*), both sun gears are driven with the rotational speed of the planetary gear carrier of *GS1*. The Ravigneaux set rotates as a block, and the ratio is $i = 1.521$.

In fourth gear, the clutch *B* is opened and the clutch *E* is closed. In this way, the sun of *GS3* and the planetary gear carrier are driven, with the ratio reaching $i = 1.143$. When shifting from 4th to 5th gear clutch *A* is openend and clutch *B* closed. In opposition to the fourth gear the sun of *GS2* is driven. The resultant speed-increasing ratio is $i = 0.867$.

In sixth gear, the clutch *B* is opened and the brake *C* is closed, which stops the sun of *GS2*. The forward planetary gear set *GS1* is locked up. The planetary gear carrier of the Ravigneaux gear set is directly driven with the speed of the turbine, with the ratio amounting to $i = 0.691$.

In reverse gear, the clutch *B* and the brake *D* are closed. The ring gear of the Ravigneaux set now rotates in the reverse direction of the engine with a total ratio $i = -3.403$.

The 6-speed automatic transmission for front-transverse drives illustrated in Figure 6.35 is based, like the transmission described before, on the Lepelletier gear set principle. Thus the same observations regarding the power flow apply here, as well. The clutches and brakes have a different spatial configuration and the first gear has an additional freewheel. One special feature of this design is the brake *C*, which is designed as a belt brake [6.24].

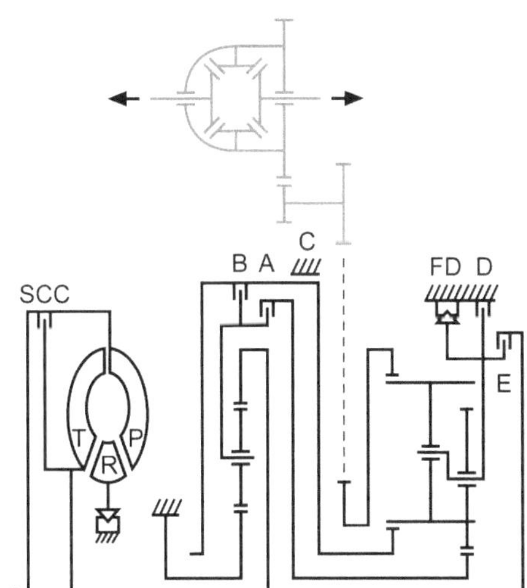


| G | Clutch | | | Brake | | Free-wheel | i | Gear step |
|---|---|---|---|---|---|---|---|---|
| | A | B | E | C | D | FD | | |
| 1 | • | | | | • | • | 4.15 | 1.75 |
| 2 | • | | | • | | | 2.37 | 1.52 |
| 3 | • | • | | | | | 1.56 | 1.35 |
| 4 | • | | • | | | | 1.16 | 1.35 |
| 5 | | • | • | | | | 0.86 | 1.25 |
| 6 | | | • | • | | | 0.69 | Total |
| R | | • | | | • | | –3.39 | 6.05 |

**Fig. 6.35.** Gearbox diagram and gearshift pattern of a 6-speed automatic transmission for front-transverse drive based on the Lepelletier gear set concept (Aisin AW), *production design Figure 12.26*

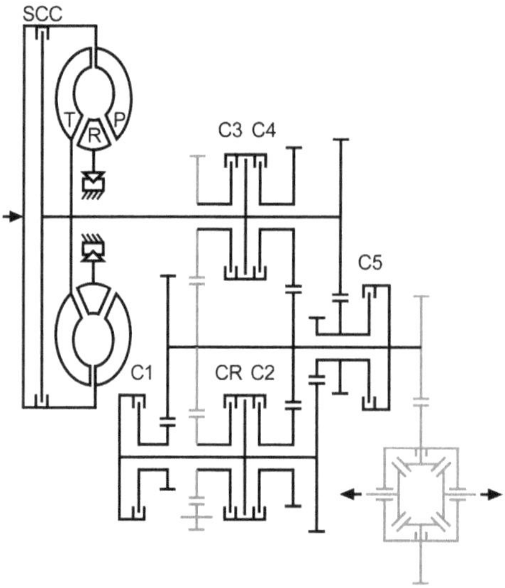


| G | Clutch | | | | | | i | Gear step |
|---|---|---|---|---|---|---|---|---|
| | C1 | C2 | C3 | C4 | C5 | CR | | |
| 1 | • | | | | | | 3.63 | 1.73 |
| 2 | | • | | | | | 2.09 | 1.59 |
| 3 | | | • | | | | 1.31 | 1.46 |
| 4 | | | | • | | | 0.90 | 1.25 |
| 5 | | | | | • | | 0.72 | Total |
| R | | | | | | • | –3.67 | 5.03 |

**Fig. 6.36.** Gearbox diagram and gearshift pattern of a 5-speed countershaft-type automatic transmission (Mercedes-Benz), *production design Figure 12.27*

### *Countershaft-Type Automatic Transmissions*

Countershaft-type automatic transmissions with various gear ratios have the advantage of allowing free choice of ratio, and comprising standard elements. The latter can in turn be advantageous with respect to manufacturing technology.

Well known examples are the Honda 4- and 5-speed automatic transmissions (Hondamatic) and the first generation transmissions of the Mercedes A-Class. In these transmissions a countershaft transmission is mounted after the converter. Figure 6.36 shows the Mercedes-Benz transmission as an example. The conventional synchromesh units are replaced by multi-plate clutch packages.

## 6.6.5 Passenger Car Hybrid Drives

In the context of vehicle prime movers, Section 3.2.4 "Hybrid Drive" discusses the principles and essential characteristics of hybrid drives. This section also contains a table (Table 3.7) which explains the current classification into micro hybrid, mild hybrid and full hybrid. Micro and mild hybrids are designed in most cases as parallel hybrids. Full hybrids have been designed as parallel hybrids and power-split hybrids.

### *1/ Parallel Hybrids*

The parallel hybrid described in the following as an example is a full hybrid. It covers the hybrid functions of supplying on-board power, engine start/stop, electric driving or manoeuvring, boosting and recuperation.

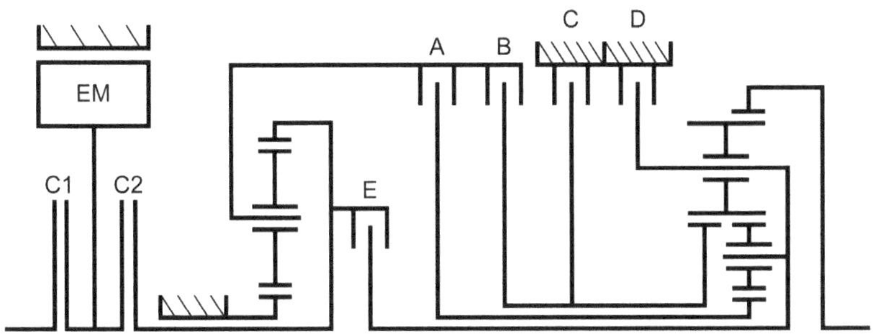


**Fig. 6.37.** 6-speed passenger car hybrid automatic transmission as a parallel hybrid with one electric machine (BMW, ZF, Continental); basic transmission, see Figure 6.34, *production design Figure 12.28*

Implementation via an "add-on" module, consisting of an electric machine *EM* with a gearshifting clutch *C1* to the internal combustion engine *CE* and a master clutch *C2* to the gear input, can be used to hybridise different basic gear units. The operational states described in the following are basic and generally applicable.

The basic gear unit in the "BMW Active Transmission" system shown in Figure 6.37 is the ZF 6-speed automatic transmission 6 HP 26 (see Section 6.6.4 with Figure 6.34; see also Figure 12.25). The hybrid module – an electric machine with the two clutches *C1* and *C2* – replaces the hydrodynamic torque converter in the converter bell and also takes over its function as moving-off element. By being mounted on the gear input shaft, the electric machine can also use the transmission ratios for characteristic conversion, like the internal combustion engine, and take over the drive or generator functions mentioned above with varying power and torque requirements. The design can be shown to accomodate the same space as a conventional automatic transmission [6.13].

An superordinate torque and power management, which will not be discussed in further detail here, must be used to make sure that, given an even charge balance, vehicle reactions are always reproducible in accordance with different driver demands. In the process, the vehicle's on-board power supply must also be guaranteed.

The hybrid functions mentioned at the beginning of this section are implemented in the different operational states represented in Figure 6.38. The directions of the arrows always indicate the direction of the power flow.

a/ Cold start of the *CE* through the electric machine via clutch *C1* (the engine starter may thus be omitted).

b/ The vehicle is stationary, the *CE* runs and loads the electric energy storage unit (electrochemical battery or double-layer capacitors) with the electric machine operated as generator via the closed clutch *C1*.

c/ If the charge state of the energy storage permits it, the *CE* can be turned off in stationary phases during stop-and-go traffic or at traffic lights (engine start/stop). The on-board power supply is taken over during these phases by the energy storage.

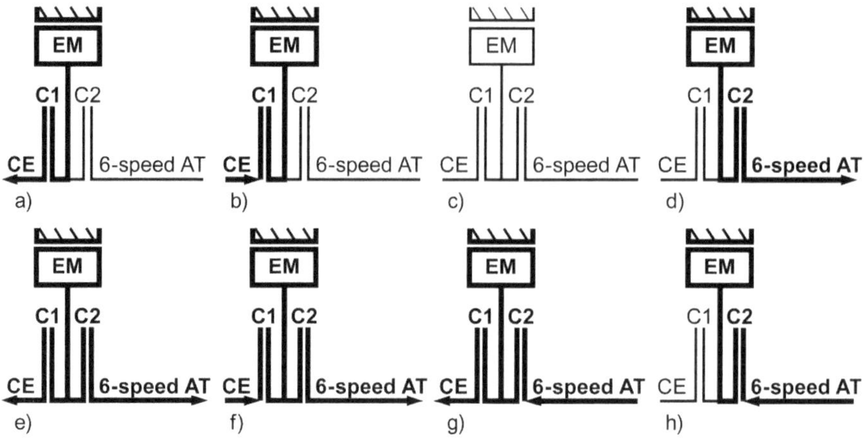


**Fig. 6.38.** Power flow in different operational states

d/ Electric moving-off or manoeuvring (*CE* is off, clutch *C1* is open)
e/ Starting up the *CE* at higher driving speeds or given a higher performance demand on the part of the driver via the slipping clutch *C1*.
f/ When the *CE* has been synchronized, the slip at the clutch *C1* is reduced and the *CE* takes over the drive power. In this operational state, the electric machine can temporarily overlay its torque with that of the engine, as e.g. when boosting or when compensating for the torque of the omitted torque increase of the torque converter.
g/ In coasting situations, e.g. driving downhill, the coasting torque can be provided by the generator operation of the electric machine. The coasting energy is thus converted into electric energy and temporarily stored in the energy storage. With a higher level of power, electric energy can be recuperated when braking.
h/ The efficiency of the conversion of kinetic energy into electric energy in coasting situations or when braking can be increased given the fact that, in these operational states, the *CE* along with its drag torque can be uncoupled and switched off due to the opening of the clutch *C1* and thus doesn't have to be additionally dragged.

### *2/ Power-Split Hybrid Drives for Passenger Cars*

The power-split hybrid powertrain described in the following can fulfill the functions of full hybrid: on-board power supply, engine start/stop, electric driving or manoeuvring, boosting and recuperation. Unlike the "BMW Active Transmission" described in the previous example, which has a conventional automatic transmission as basic gear unit and one electric machine, the Toyota/Lexus P310 power-split hybrid transmission is an assembly designed specifically for front-transverse hybrid applications and has two high-performance electric machines.

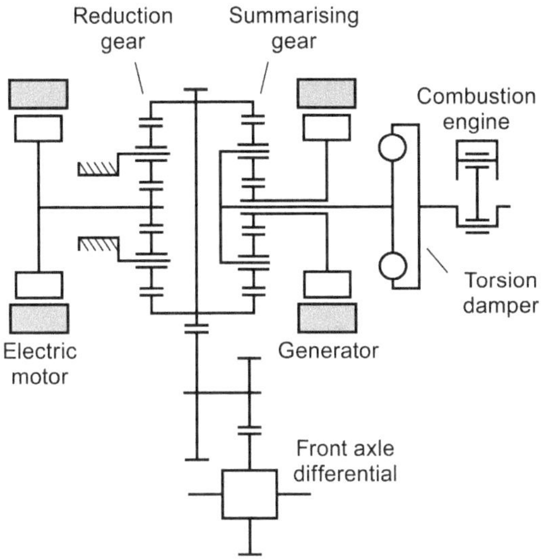


**Fig. 6.39.** Power-split hybrid gearbox (Toyota/Lexus), *production design Figure 12.29*

The structure of this hybrid transmission is shown schematically in Figure 6.39. A torsion damper with a flywheel mass is mounted between the internal combustion engine and the gearbox input to uncouple the powertrain of engine rotation irregularities. The engine acts on the spider shaft of the first power-summing/power-splitting planetary gear unit (summarising gear) via the gearbox input shaft. Here, the power of the internal combustion engine is split on the one hand into direct mechanical drive power for the vehicle (ring gear of the summarising gear) and into an electric power path (sun gear of the summarising gear) on the other. This electric power can be further conducted via an electric generator and power electronics either to an electric energy storage unit or also, via a second electric machine (the electric motor), to the vehicle drive.

Since the engine can remain switched off during moving-off and at low vehicle speeds, the spider shaft remains stationary in this case. The drive is then through the electric motor, which drives the forward gears through the ring gear. At the same time, the planetary gears of the summarising gear rotating on the stationary spider shaft set the sun gear connected to the generator in motion. In order to start the engine at increasing driving speeds, a torque is created at the sun gear though the generator. The crankshaft of the internal combustion engine begin to turn, and as soon as the engine begins to work, it transfers its power via the spider shaft both to the ring gear to drive the wheels and to the sun gear, which drives the generator (power-split).

In case of high acceleration (boosting), the engine and the electric motor function as common drive source. The electric energy accumulator makes additional energy available for the electric motor. As explained in Section 3.2.4, splitting and combining a mechanical and an electric power branch through the summarising gear allows for a continuous conversion of torque and speed. This makes the P310 hybrid transmission a continuously variable power-split transmission. However, when driving with the engine running, the engine's torque must be supported on the spider shaft of the summarising gear by the torque of the generator on the sun gear, which places high torque and power demands on the generator.

The electric motor acts on the output via a reduction gear. This gear has the function of decreasing the speed of the electric motor and increasing the torque. This torque ratio allows the electric motor to be designed more compactly. The mechanical and electric shares of the drive power are transferred to the front axle differential via the coupled ring gears of the summarising and reduction gears and an intermediate shaft with two spur gear stages [6.17].

### 6.6.6 Continuously Variable Passenger Car Transmissions (CVT)

The power available from an internal combustion engine cannot be fully exploited by the finite number of gears in traditional geared selector gearboxes. With a continuously variable transmission, the engine can be operated at the ideal operating point for economy or performance as required (see also Sections 4.4 and 5.3.4). These transmissions are referred to as CVT (Continuously Variable Transmissions). Figure 6.40 gives an overview of various CVT designs.

The transmission shown in Figure 6.39 served to explain the functioning of an electric CVT on the basis of a power-split hybrid drive. For more on purely electric drive, see also Sections 3.2.2 and 3.2.3.

The hydrodynamic torque converter per se is also a continuously variable transmission. However, it is used as a moving-off element, not as a transmission.

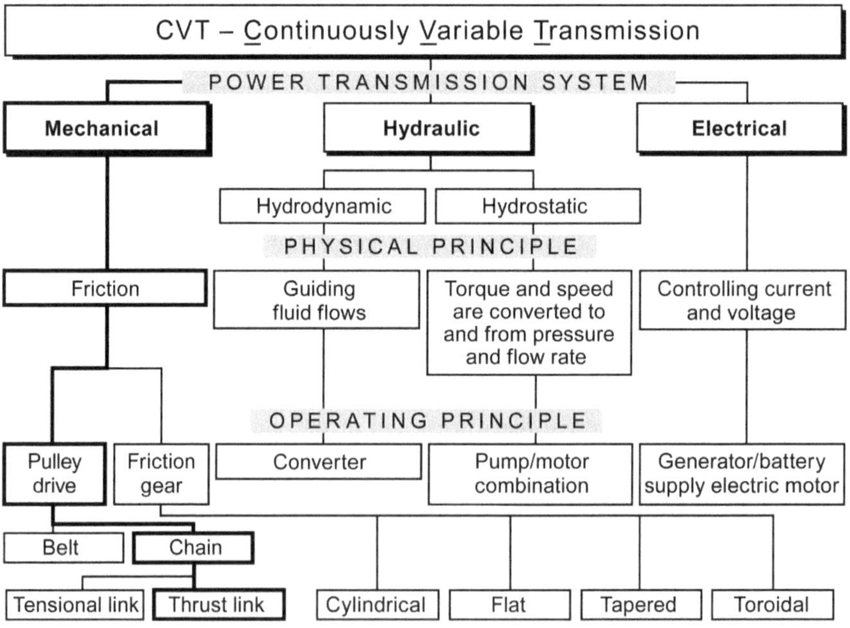


**Fig. 6.40.** Overview of CVT designs

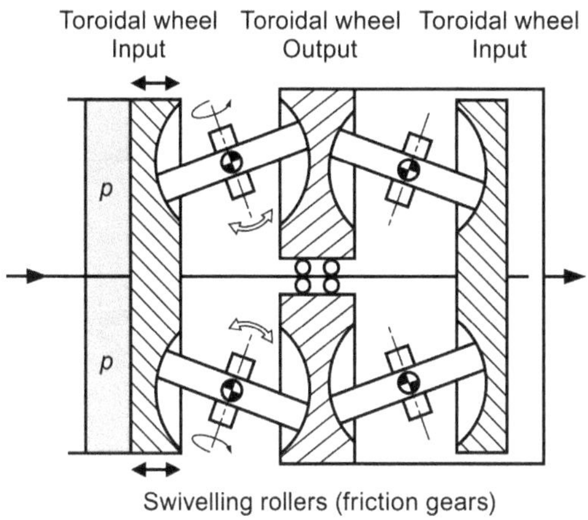


**Fig. 6.41.** Principle of a toroidal variator

Work has been done in developing hydrostatic-mechanical powershift transmissions for passenger cars which function in a continuously variable way [6.22-6.23], but they have achieved no practical significance.

With mechanical power transmission in full and half toroidal transmissions [6.10, 6.21, 6.42], the continuous variation of ratio is achieved by swivelling the friction gears (rollers). Figure 6.41 shows, in simplified form, the functioning of a full toroidal transmission. Toroidal transmissions have a higher torque capacity than pulley transmissions and lend themselves especially to coaxial powertrains.

The continuously variable transmissions used in mass-produced passenger cars are almost exclusively pulley transmissions. The following will thus be devoted to this type. The central component of the pulley transmission is the *variator*. It consists principally of taper discs and a chain. Power is transmitted frictionally through the chain, which runs between two axially adjustable taper discs. Through the axial adjustment of the taper discs, the chain runs on variable diameters, infinitely varying the ratio (Figure 6.42). The torque-related pressure of the taper discs on the chain requires a lot of attention, since excessive pressure reduces the efficiency of the chain, leading to increased power consumption, power loss by the contact pressure pump and above all an increased stress on the transmission. It is also essential to prevent the chain slipping, since this would inevitably lead to destruction of the transmission. This makes the design and reliability of the contact pressure pump, and its control, a critical factor in these continuously variable transmissions.

With chains, a distinction is made between tensional link chains and thrust link chains. Tensional link chains allow for smaller running radii and thus a greater overall gear ratio with the same centre distance. They are more efficient, since less power is required to adapt the chain to the ratio radii, and are better suited to higher torques. The extremely short pitch of thrust link chains (also referred to as thrust link belts), Figure 6.42, requires more work for this purpose. But the short pitch has many advantages at the "meshing impact" and in terms of associated noise generation.

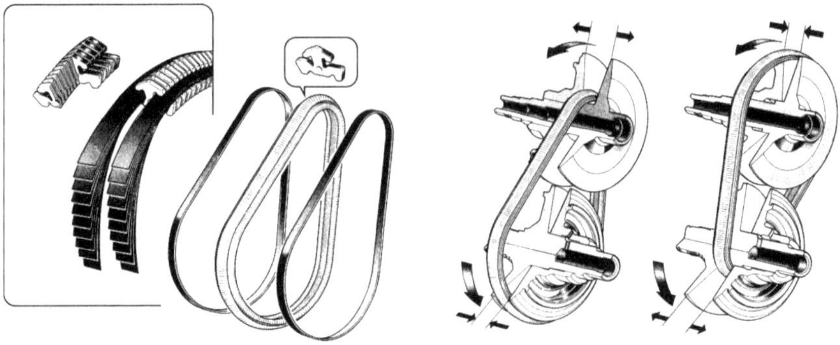

**Fig. 6.42.** Elements of a thrust link chain and principle of operation of the variator [6.43]

Figure 6.43 presents a schematic diagram of a CVT for front-transverse drive. A torque converter with a lock-up clutch serves as moving-off element. During forward driving, the planetary reversing gear set rotates as a block. In the power flow, the variator is followed by an output stage which facilitates adaption to the requirements of different vehicles. The overall gear ratio of the transmission amounts to 6.0.

In order to increase the overall gear ratio of the continuously variable transmission beyond the normal 6.0 to 6.5 of the variator, selector gearboxes with spur gears or planetary gears are front- or rear-mounted. Power-splitting is also possible. The following will describe the functioning of a CVT with power-split (geared neutral transmission) (Figure 6.44).

The total ratio $i$ of this transmission, when operated with power-split, is derived from $i_V$ and the ratio $i_G$ of the planetary gear set functioning as a differential drive, where the clutch *C1* is closed.

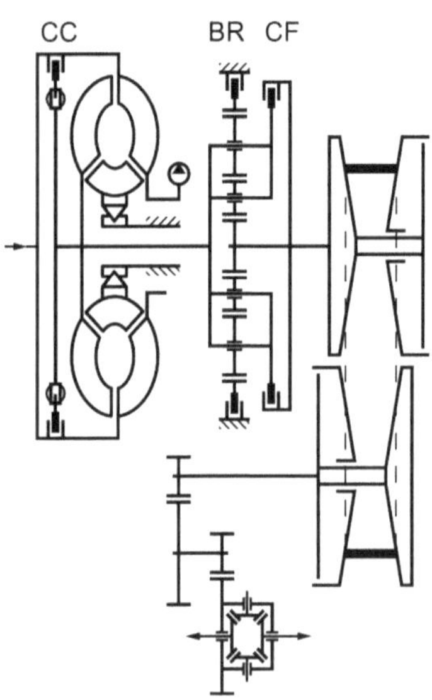


**Fig. 6.43.** CVT for front-transverse drive (ZF), production design Figure 12.32

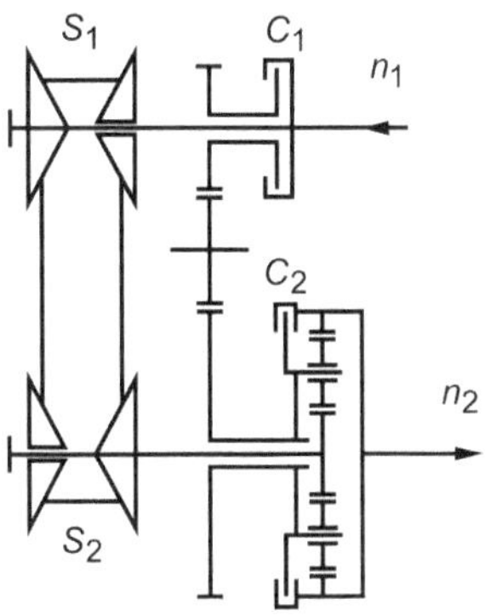


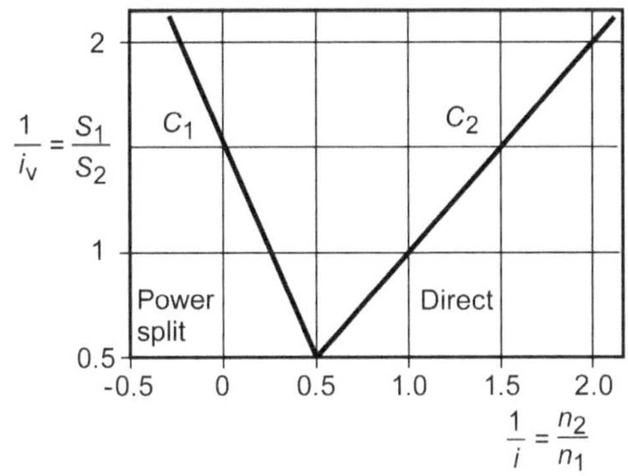


**Fig. 6.44.** Gearbox diagram: "CVT with power-split" and diagram of the ratio profile (shown as $1/i$ since $n_2 = 0$ at the geared neutral point)

Depending on the ratio $i_V$, the total ratio $i$ can also become negative, corresponding to reverse gear. The point where the sign changes is the geared neutral point. Theoretically, the transmission requires no additional moving-off element.

With the clutch *C2* engaged, the planetary gear set rotates as a block, and the equations given below apply. The profile of the total ratio as a function of the taper disc radius ratio $i_V$ and the active clutch is shown in the diagram on the right of Figure 6.44.

The overall total ratio

$$i = \frac{n_1}{n_2} = i_V \; i_{GS}, \tag{6.4}$$

(the ratio of the input speed $n_1$ to the output speed $n_2$) of a pulley transmission is made up of the ratio in the taper disc transmission (variator)

$$i_V = \frac{S_2}{S_1}, \tag{6.5}$$

(the ratio of the current taper disc radii *S2* to *S1*) multiplied by a possible ratio $i_{GS}$ of a rear-mounted or front-mounted gear stage.

## 6.7 Commercial Vehicle Transmissions

For transmissions of commercial vehicles up to approximately 3.5 t gross weight rating, the explanations in the preceding Section 6.6 "Passenger Car Transmissions" apply. The following will discuss basic design principles for commercial vehicle transmissions for vehicles with gross weight ratings greater than 3.5 t.

Table 6.11 lists common types of commercial vehicle transmissions. Depending on how their idler gears are positively locked to the shafts, manual transmissions can be subdivided into

- non-synchronized constant-mesh transmissions and
- synchronized transmissions (synchromesh transmissions)

and by the shifting system (see also Section 9.1) into

- direct shifting: gearshift lever at the transmission housing,
- indirect shifting: gearshift lever and transmission physically separated (remote shift, drive selector).

An important feature of many commercial vehicle transmissions is their *multi-range design* including many gear steps. This design is found in commercial vehicles with manual transmission (MT), automated manual transmission (AMT) and torque converter clutch transmission (TCCT). Because of their robustness, non-synchronized constant-mesh gearboxes are common in long-distance trucks in many regions of the world. Their market shares, however, are in the decline. The market share of synchronized manual transmissions, on the other hand, is as high as ever. They are used principally in local and long-distance traffic. See also Section 2.2.1 "Market Situation and Production Figures" and Chapter 9 "Gearshifting Mechanisms".

Constant-mesh and synchronized commercial vehicle transmissions are becoming increasingly automated. This trend corresponds both to an increased desire for greater ease of operation to reduce driver stress and to the fact that gearshift automation technology has become available at acceptable costs. This transmission design is becoming more prevalent worldwide in short-distance and delivery trucks, in long-distance trucks and in interurban buses and coaches.

**Table 6.11.** Market shares and applications of commercial vehicle transmissions (without hybrid and CVT)

| **Transmission type** | **Constant-mesh trans-mission MT** | **Synchro-mesh trans-mission MT** | **Automated manual trans-mission AMT** | **Torque converter clutch transm. TCCT** | **Automatic trans-mission AT** |
|---|---|---|---|---|---|
| Number of gears | 6–9–12–16 | 5–6–9–12–16 | 6–10–12–16 | 6–16 | 4–7 |
| Market share, world-wide | Decreasing | Large | Increasing | Very small | Small |
| Applications | Long-distance traffic; vehicles outside Europe | Local traffic; long-distance traffic in Europe | Local and delivery traffic; long-distance traffic; interurban buses and coaches | Heavy goods transporters; construction-site vehicles | Construction-site and delivery vehicles; urban buses |

Torque converter clutch transmissions (TCCT) fall under the category of automated manual transmissions. They are hydromechanical transmissions with power interruption and can also be fitted with an integral hydrodynamic brake (retarder) if required. Their use is restricted to heavy special-purpose vehicles.

Conventional automatic transmissions (AT) are hardly used in commercial vehicles because of their higher price, their lower reliability due to the larger number of parts, and the increase in fuel consumption. But they are common in urban buses and construction-site/distribution vehicles, where they significantly relieve driver stress.

Sections 6.7.5 and 6.7.6 discuss commercial vehicle designs for hybrid drives and continuously variable transmissions. Table 6.12 shows in summarised form the transmissions treated in systematic sections 6.7.1 to 6.7.6 and designs sections 12.2.1 to 12.2.6.

**Table 6.12.** In Sections 6.7.1–6.7.6 as well as Sections 12.2.1–12.2.6 introduced automotive gearboxes. *CS* countershaft; *TCCT* torque converter clutch transmission

| Diagram Fig.-No. | Speeds | Characteristics | Manu-facturer | Name | Design Fig.-No. | No. |
|---|---|---|---|---|---|---|
| 6.45a | 4 | MT, 1-range gearbox | — | — | — | — |
| 6.45b | 6 | MT, 1-range gearbox | ZF | S 6-66 | 12.33 | 1/ |
| 6.46 | 5 | MT, direct drive/ overdrive gearbox | — | — | — | — |
| 6.47 | 5 | MT, output constant gear | — | — | — | — |
| 6.48–6.50 | — | MT, multi-range gearbox | — | — | — | — |
| 6.51–6.52 | — | MT, multi-stage gearbox | — | — | — | — |
| 6.53 | 9 | MT, 2-range gearbox, direct | ZF | 9 S 109 | 12.34 | 2/ |
| 6.54 | 16 | MT, 3-range gearbox | ZF | 16 S 221 | 12.35–12.36 | 3/ |
| 6.55 | 12 | MT, 2-range-gearbox, 2 CS | Eaton | TSO-11612 | 12.37 | 4/ |
| 6.56 | 16 | MT, 3-range-gearbox, 2 CS | Eaton | RTSO-17316A | 12.38 | 5/ |

**Table 6.12.** *(continued)*

| | | | | | | |
|---|---|---|---|---|---|---|
| 6.58 | 6 | AMT, 1-range gearbox | ZF | eTronic 6 AS 380 VO | 12.39 | 6/ |
| 6.59 | 16 | AMT, 3-range gearbox | MB | PowerShift G241-16K | 12.40 | 7/ |
| 6.60 | 16 | AMT, 3-range gearbox, 2 CS | ZF | AS-Tronic 16 AS 2230 TD | 12.41–12.44 | 8/ |
| 6.61 | 16 | TCCT, 3-range gearbox | ZF | WSK 400 + 16 S 221 | 12.45 | 9/ |
| 6.62 | 12 | TCCT, 3-range gearbox, 2 CS | ZF | TC-Tronic 12 TC 2740 TO | 12.46 | 10/ |
| 6.63 | 5 | AT | Allison | 2000 | 12.47 | 11/ |
| 6.64 | 6 | AT, retarder | ZF | 6 HP 602 C | 12.48 | 12/ |
| 6.65 | ∞ | Serial hybrid, variants | — | — | — | — |
| 6.66 | ∞ | Serial hybrid | ZF | EE Drive 1 | 12.49 | 13/ |
| 6.67 | ∞ | CVT, hydrostatic unit, power-split | — | — | — | — |
| 6.68 | ∞ | CVT, hydrostatic unit, power-split | ZF | Eccom | 12.50 | 14/ |

### 6.7.1 Manual Commercial Vehicle Transmissions (MT)

The following will describe the structure of constant-mesh and synchronized transmissions for commercial vehicles with single- and multi-range designs. The concepts discussed apply equally to automated manual commercial vehicle transmissions (AMT) (see Section 6.7.2).

#### *1/ Single-Range Transmissions*

In the case of 4-speed to 6-speed gearboxes, the single-range design with input constant gear is standard (Figure 6.45). They are designed so that the ratio of a particular gear is derived from the individual ratios of two gear pairs.

The first gear pair, the input constant gear *CG*, stays engaged in all gears with the exception of direct gear, and drives the countershaft at a constant ratio. When shifting into another gear, only the ratio of the second gear pair changes. Such a single-range transmission with input constant gear is referred to as a two-stage countershaft transmission, or simply countershaft transmission.

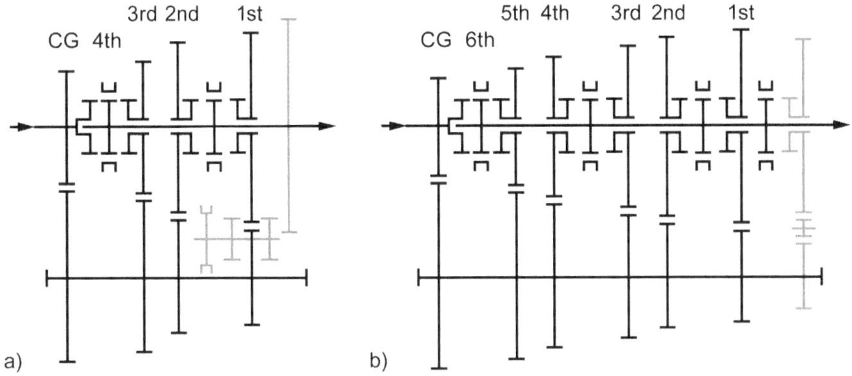


**Fig. 6.45.** Single-range design. *a* 4-speed gearbox; *b* 6-speed gearbox, *production design Figure 12.33*

There are numerous possible gearwheel configurations for any given number of gears. In general the emphasis is on locating high torque conversion in the vicinity of bearings, to minimise shaft deflection.

In commercial vehicles, these transmissions are called direct drive gearboxes or overdrive gearboxes (Figure 6.46), depending on whether in countershaft transmissions the top gear is a direct gear with a ratio equal to one or a ratio less than one (speed increasing ratio).

From a design standpoint, an advantage of the overdrive gearbox is that the potential gearbox input torque is higher in comparison to similarly large direct-drive gearboxes. With the exception of the different ratio of the constant gear and of the highest gear, the same gear pairs can generally be used. Moreover, the overdrive design potentially saves on fuel because of the resulting engine speed reduction.

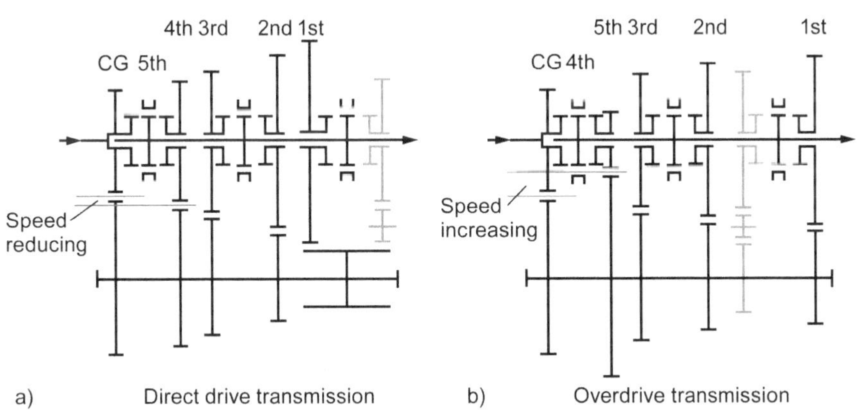


**Fig. 6.46.** 5-speed countershaft transmission. *a* Direct drive transmission – top gear is direct drive; *b* overdrive transmission – top gear is speed increasing

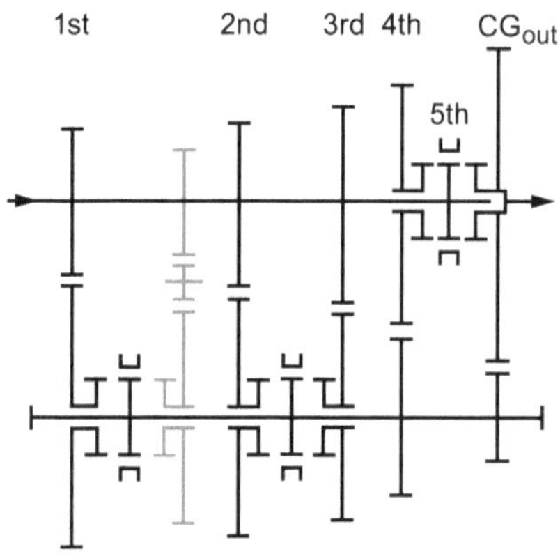


**Fig. 6.47.** 5-speed gearbox with output constant gear $CG_{out}$

Alternatively, such a transmission can also be fitted with an output constant gear $CG_{out}$ (Figure 6.47). This means that the constant ratio is located behind the gear pairs for the individual gears.

Such a configuration has the following advantages:

- The reduced moment of inertia, which is crucial for synchronizer stress and is a function of the square of the gear step, is less with transmissions with an output constant gear.
- There is less deflection of the shafts on which the gear pairs for the individual gears are mounted.

These advantages have two countervailing disadvantages:

- Gearwheels, countershaft and bearings rotate at higher speeds than in gearboxes with an input constant gear.
- The output constant gear must be of more robust design, since there are already high torque levels. Like the input constant gear, it must be designed for endurance strength since it is always in the power flow, with the exception of direct drive.

### 2/ Multi-Range Transmissions

In the case of multi-speed transmissions the task is to provide as many gear steps as possible with as few gear pairs as possible. Multi-range design is suitable for transmissions with more than 6 gears, and can be coaxial or non-coaxial. Multi-range transmissions are constructed by combining single-stage, two-stage or multi-stage single transmissions (Figure 6.48).

A single transmission which is self-contained by design is called a *range unit*. The system boundaries are, however, floating and cannot always be precisely defined. Both in a front-mounted splitter unit and in a front-mounted range-change unit, the second constant can also be used as a main gearbox gear pair (see the power flows in Figure 6.54).

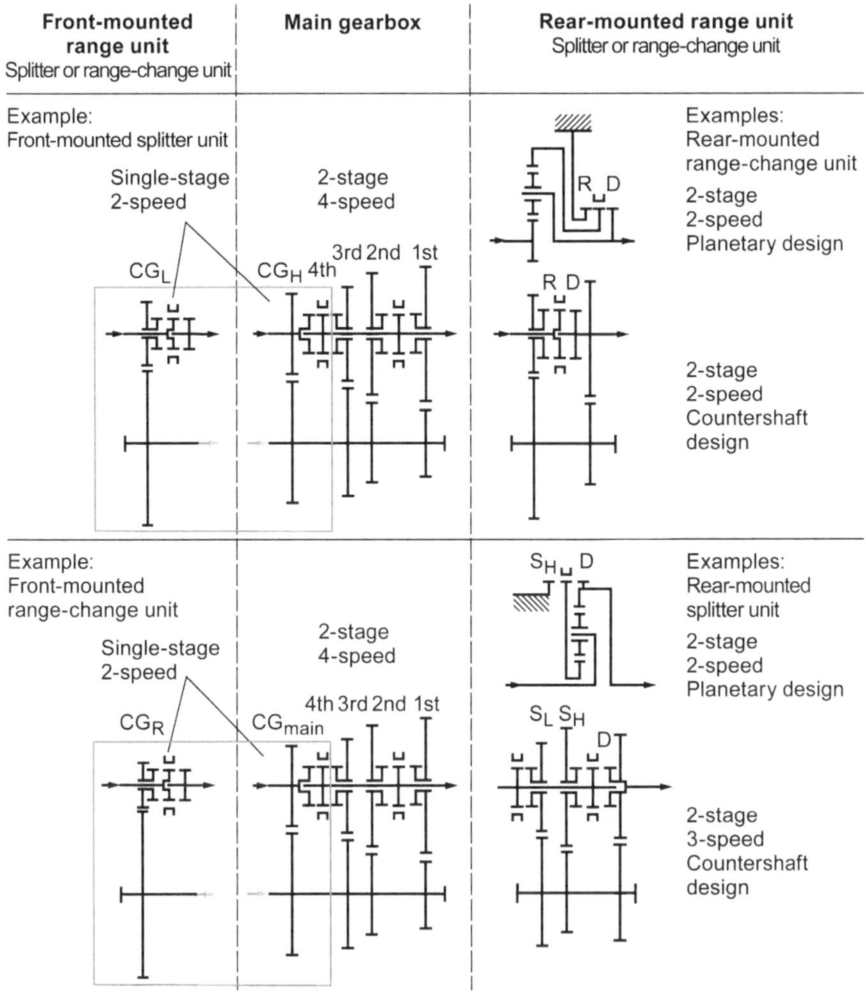


**Fig. 6.48.** Combination of two-stage main gearbox with single-stage or two-stage front-mounted or rear-mounted range units. $CG_H$ front-mounted splitter unit constant high; $CG_L$ front-mounted splitter unit constant low; $R$ range; $D$ direct; $CG_R$ range constant; $CG_{main}$ main gearbox constant; $S_H$ rear-mounted splitter unit high; $S_L$ rear-mounted splitter unit low

In the splitter unit, a distinction is made between the "High" position (fast) and the "Low" position (slow):

- The splitter unit can be speed-reducing or speed-increasing.
- The range-change unit is always speed-reducing.

The appropriate design must always be selected for each range unit. It is easy to arrange to link a countershaft transmission to a planetary gear transmission.

With front-mounted and rear-mounted range units, a distinction is made between

- *splitter unit*: compressing the gear sequence and
- *range-change unit*: expanding the gear sequence.

Figure 6.49 shows the effect of the various ranges on the gear sequence using the example of the three-range gearbox as shown in Figure 6.54 (see also the traction diagram of a commercial vehicle with front-mounted splitter unit in Figure 5.6).

The logarithm of the ratio is shown on the right of each graph in Figure 6.49 to illustrate the geometrical gear steps in multi-range transmissions.

### Splitter Unit: Compressing the Gear Sequence

A splitter unit always leads to a compression of the gear sequence (6.49a). The splitter unit can be fitted before or after the main gearbox (Figure 6.48). The gear step in the splitter unit is less than that of the main gearbox (Figure 6.49a) (half as large with geometrical gear steps). The number of gears in the main gearbox is multiplied by the number of gears in the splitter unit. The splitter unit is normally fitted with two gears.

In practice, front-mounted splitter units are almost always used. The reason for this is that front-mounted splitter units only have a small ratio change of approximately 1.1 to 1.2. This means that the rear-mounted main gearbox is loaded either with only a slightly higher torque, or in the case of a speed-increasing ratio, even with a lower torque than without a splitter unit. If the splitter unit is rear-mounted to the main gearbox (bottom of Figure 6.48), it must be designed for the highest torque multiplication reached in the main gearbox. That is a more expensive solution than a front-mounted splitter unit.

### Range-Change Unit: Expanding the Gear Sequence

The function of a range-change unit is to expand the gear sequence. This is achieved by the ratio step in the range-change unit being as big as the range of ratios in the main gearbox, multiplied with the gear step in the main gearbox (Figure 6.49b). The gear sequence with the range-change unit engaged follows smoothly on from that of the main gearbox. Overlaps in the ratios of individual gears are avoided by using geometrical gear steps (Section 4.3.2). Range-change units are always speed-reducing. The torque multiplication in the range-change unit amounts to approximately $i_R$ = 3–4. If the range-change unit was designed to be front-mounted, the high torque values would pass through the main gearbox. The range-change unit is therefore always fitted at the output end of the main gearbox. The range-change unit can be of countershaft design or a compact planetary gear unit.

### Mixed Gear Sequence

Gear sequences can also be created with one or more range units, in which the gears do not all follow each other with the same gear step (geometrical stepping). Non-geometrical stepping can lead to part of the gears no longer being useable because the gear step to a neighbouring gear is too small.

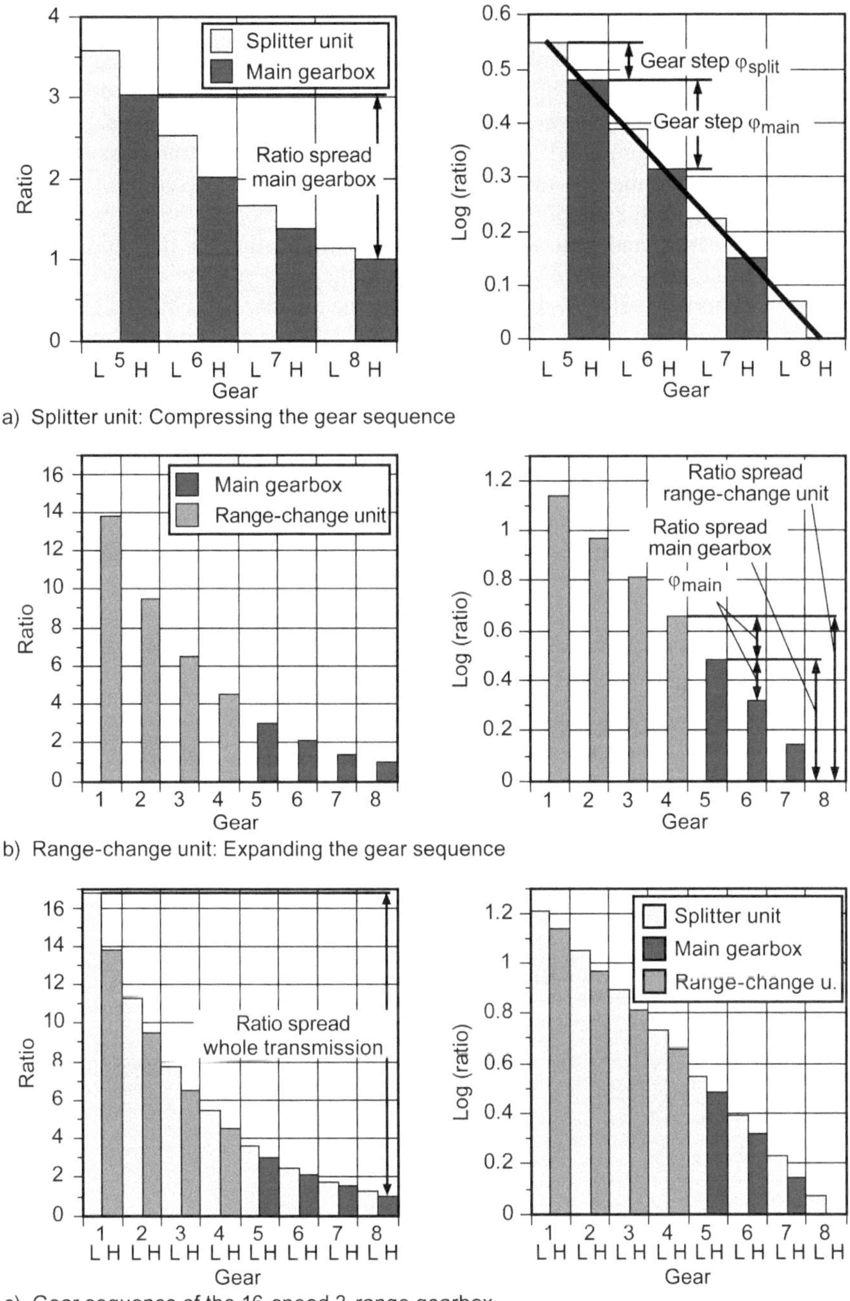


**Fig. 6.49.** Compressing and expanding the gear sequence with splitter unit and range-change unit. Based on the example of the 16-speed commercial vehicle gearbox ZF 16 S 221 shown in Figure 6.54. *L* low (slow); *H* high (fast)

### Evaluation

Table 6.13 shows various combinations of range units. Splitter units and range-change units can be used singly or in combination as front-mounted or rear-mounted range units. However, some of these possible combinations are not viable in practice. A simplified evaluation of these multi-range transmissions was carried out to determine the most suitable design. This was based on a 4-speed main gearbox, with the simplifying assumption that the centre distances in each unit are the same, and that for the torque multiplication in the splitter unit $i_{split} = 1.2$, in the range-change unit $i_R = 4$, and in the main gearbox $i_{G,H} = 3.5$.

The characteristic value $K_G$ is determined by the number of gear pairs per range unit, an investigation of the individual torque conversions, and the total ratio achievable. The resultant values represent a relative measure of the physical dimensions of the transmission, and are not absolute indications of size. They serve rather to compare the transmissions with each other. The smaller the transmission characteristic value $K_G$, the smaller the dimensions of the transmission. Front-mounted and rear-mounted range units, splitter units and range-change units were tried in different configurations. The results are shown in Table 6.13. The various possible configuration variants can be assessed in terms of their dimensions using this table.

The combinations reviewed can be subdivided into the following three categories:

- main gearbox with splitter unit (*S1* and *S2*),
- main gearbox with range-change unit (*R1* and *R2*),
- main gearbox with splitter unit and range-change unit (*RS1* to *RS4*).

**Table 6.13.** Various combinations of range units and their characteristic values $K_G$. *F* Front-mounted; *R* rear-mounted; *S* splitter unit; *R* range-change unit; *Main* main gearbox

| No. | Combination | | | | | Gearbox characteristic value $K_G$ |
|---|---|---|---|---|---|---|
| | FS | FR | Main | RS | RR | |
| S1 | X | | X | | | 3.17 |
| S2 | | | X | X | | 5.63 |
| R1 | | X | X | | | 2.89 |
| R2 | | | X | | X | 1.93 |
| RS1 | X | X | X | | | 3.15 |
| RS2 | | X | X | X | | 5.74 |
| RS3 | | | X | X | X | 2.27 |
| RS4 | X | | X | | X | 2.05 |

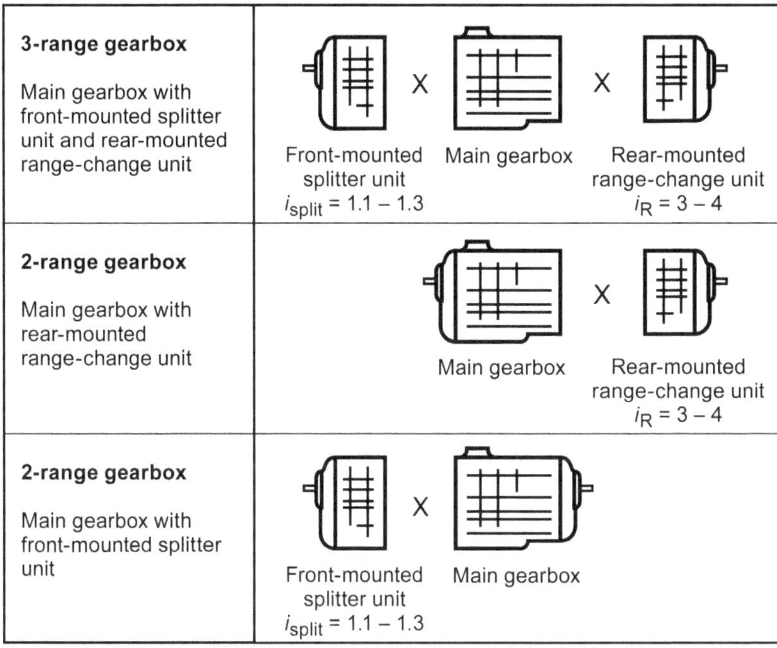


**Fig. 6.50.** Conventional configurations with two-range and three-range gearboxes

This reveals the combinations front-mounted splitter unit (*S1*), rear-mounted range-change unit (*R2*) and front-mounted splitter unit with rear-mounted range-change unit (*RS4*) as the most favourable in their respective categories in terms of physical dimensions.

Figure 6.50 shows common configurations of two- and three-range gearboxes. A high degree of flexibility can be achieved using the modular principle of two or three individual transmissions flanged together. Multi-range transmissions are in principle possible in passenger cars as well.

In multi-range transmissions, shifting times become extended, since several junctions have to be shifted in some gears. The overall shifting time should also be less than a second under unfavourable conditions.

Multi-speed transmissions can be designed with a small number of gear pairs if several junctions are shifted simultaneously when changing gear. Theoretically,

$$z = 2^{(p-1)} \tag{6.6}$$

gears can be produced with $p$ gear pairs.

Equation 6.6 applies when all gearwheels can be shifted and each gearwheel has its own shaft. Such a transmission, in which the power is transmitted several times from one shaft to the other in individual gears, is also known as a *multi-stage transmission*. In multi-stage transmissions the range units are reduced to individual gear pairs.

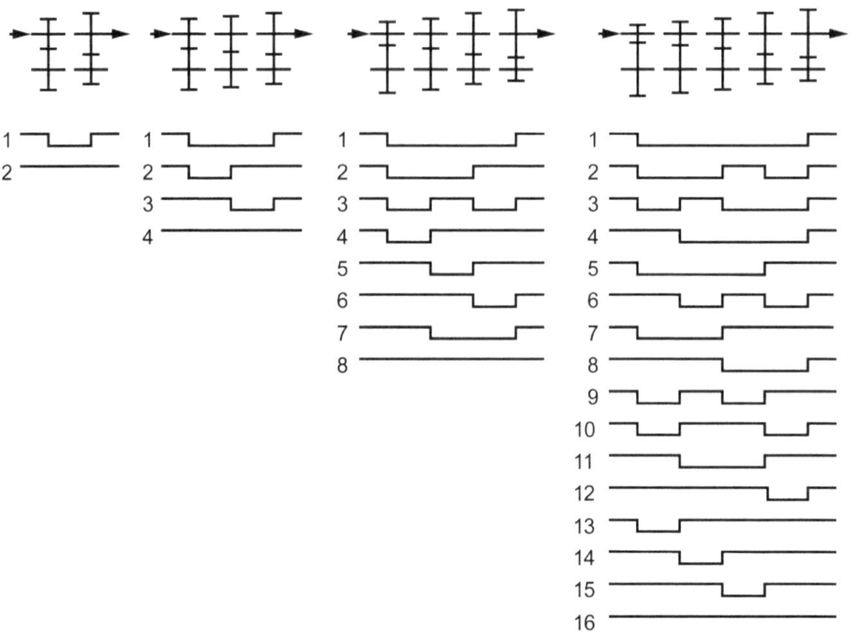


**Fig. 6.51.** Gearbox diagrams and power flows of coaxial multi-stage transmissions

In addition to the high level of engineering complexity for the shaft junctions, several junctions have to be shifted at the same time when changing gear. Multi-stage transmissions with up to 5 gear pairs and 16 gears are shown in Figure 6.51. Shifting two or more junctions at the same time can lead to high shifting times.

Depending on the design, the following transmissions can be designed using $p = 6$ gear pairs (Figure 6.52):

- single-range countershaft transmission, Figure 6.52/1: 6 forward gears,
- multi-stage transmission, Figure 6.52/2: 32 forward gears,
- two-range splitter gearbox, Figure 6.52/3.1: 10 forward gears,
- two-range range-change gearbox, Figure 6.52/3.2: 8 forward gears,
- three-range splitter/range-change gearbox, Figure 6.52/3.3: 12 forward gears.

The multi-stage transmission is of no practical interest because of the many junctions to be shifted. If a splitter unit and a range-change unit are combined with a 4-speed main gearbox, this results in a 16-speed transmission. Here the overall gear ratios of the three range units are selected in such a way that all 16 selectable combinations of gears are arranged in steps useful for the driver (Section 4.3.2 "Geometrical Gear Steps") (Figure 6.49c).

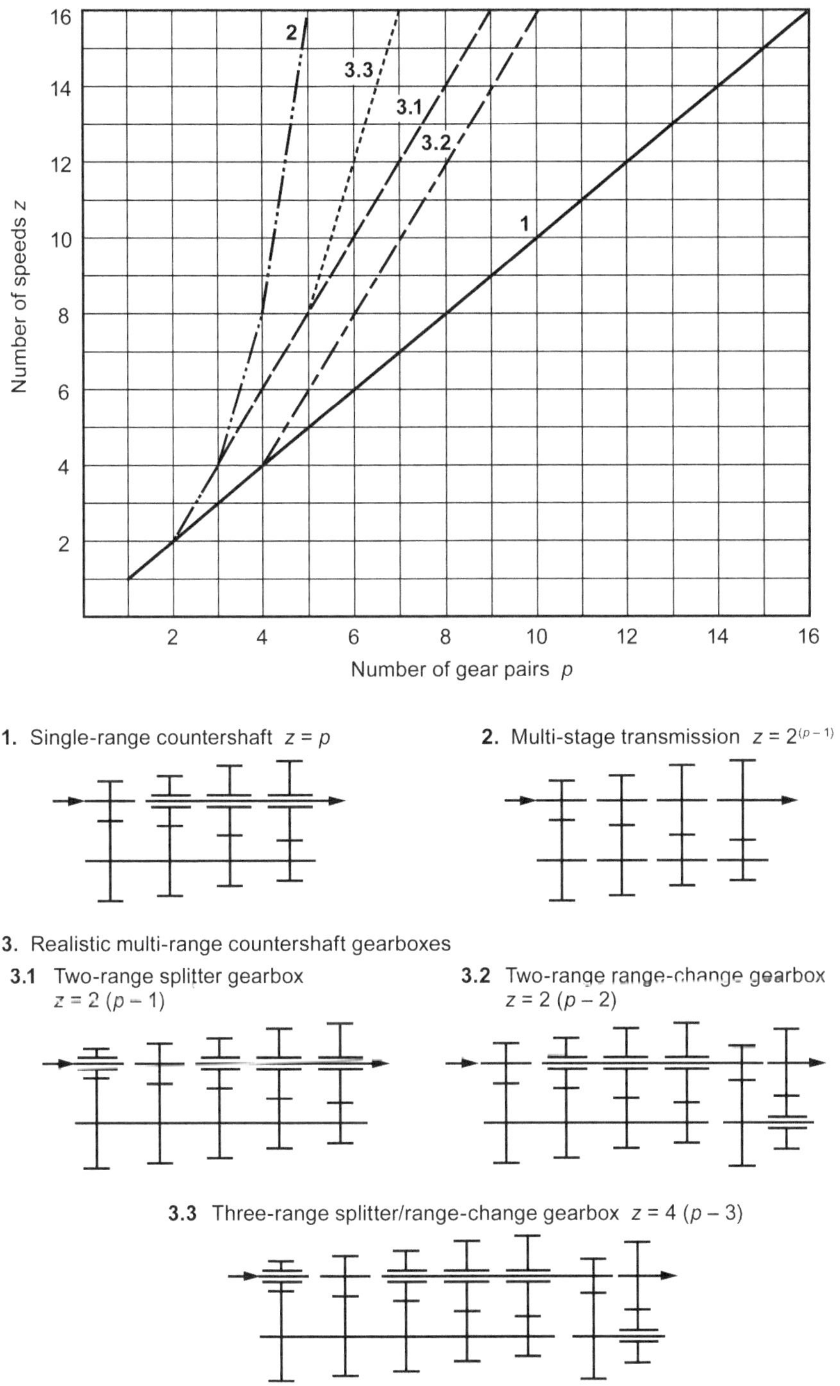


**Fig. 6.52.** Effect of the range-unit design on the number of gearwheels and speeds

### 3/ Practical Design of Two- and Three-Range Transmissions

The normal designs are two- and three-range transmissions with up to 16 gears (2 x 4 x 2) [6.26] (see also Figure 4.2 "Hierarchical structure of the powertrain ratio $i_A$"). A larger number of gears is in principle possible, but in practice no longer relevant since it involves excessively frequent gearshifting by the driver.

The ZF 9 S 109 9-speed commercial vehicle transmission (Figure 6.53) can serve as an example of a two-range type gearbox with a 4 x 2 design. The main gearbox is a 4-speed countershaft-type transmission with an additional moving-off gear stage, commonly known as a crawler, which is only used for moving-off and manoeuvring. A two-speed range-change unit in planetary design is mounted on the gearbox output side.

With a direct-drive transmission, the crawler is designed with a ratio of approximately 13; with an overdrive transmission, the ratio of this moving-off gear is at 10.3.

By means of a two-speed splitter unit, a two-range gearbox can be represented as a 16-speed gearbox with a three-range design of 2 x 4 x 2. The crawler is left out in this case. The overall gear ratio of this 16-speed transmission is approximately 13.5 in both the direct drive and overdrive designs.

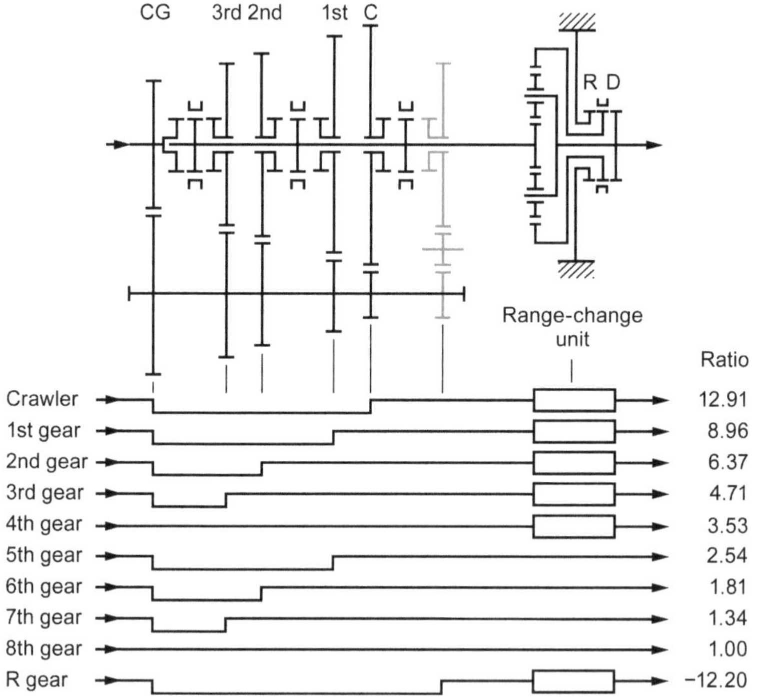


**Fig. 6.53.** Gearbox diagram, power flows and ratios of a 9-speed two-range gearbox (ZF), 4 x 2 + crawler = 9 speeds in direct drive design [6.48], *production design Figure 12.34*

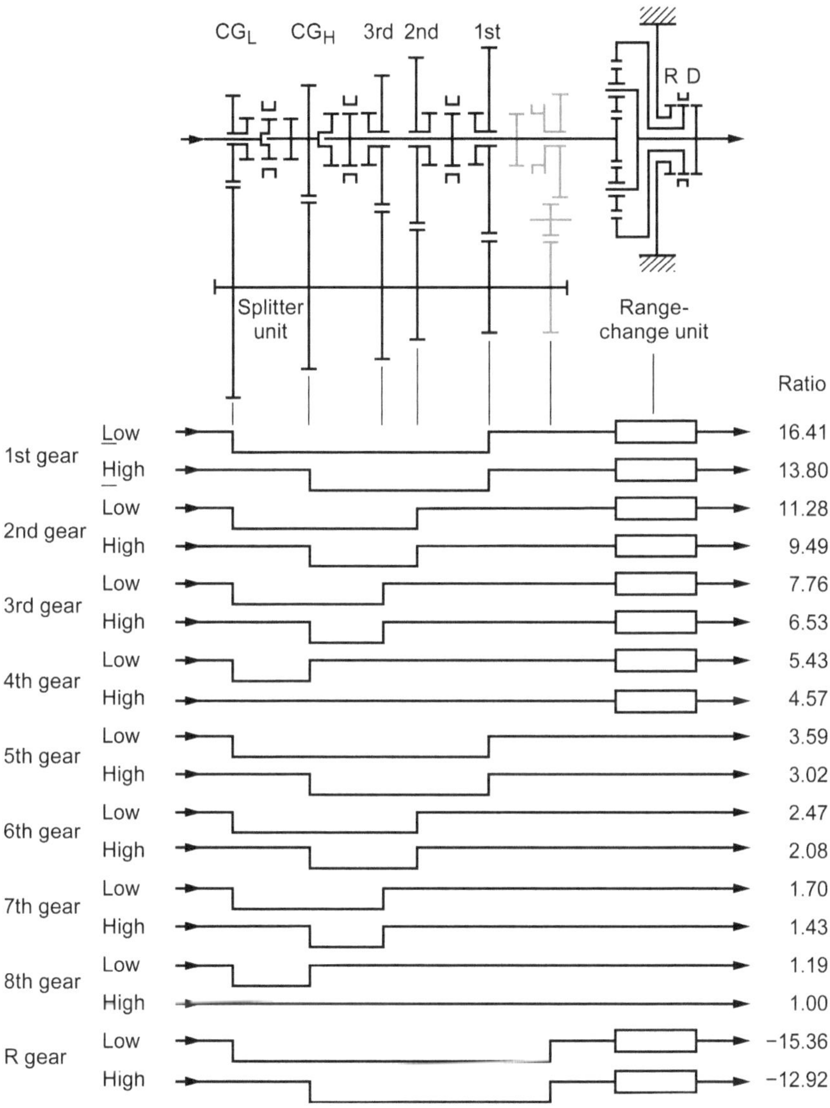


**Fig. 6.54.** Gearbox diagram, power flows and ratios of a 16-speed three-range gearbox for heavy-duty commercial vehicles in direct drive design (ZF), 2 x 4 x 2 = 16 speeds, *production design Figure 12.35*

The following principle applies to the design of commercial vehicle transmissions:

> *The transmission must be designed in such a way that the largest possible number of gear pairs is acted on with a small change in ratio, and the smallest possible number of gear pairs with a high change in ratio.*

The planetary design of the range-change unit in particular ensures compactness, bearing in mind that the range-change unit must have a large gear step, which is easy to achieve in a planetary design. The short overall length also ensures minimum shaft deflection in range-change units subject to high torques.

Figure 6.54 shows an example of a three-range gearbox with a 16-speed design (2 x 4 x 2) for heavy-duty commercial vehicles. The ZF gearbox 16 S 221 from the Ecosplit Series with direct drive design consists of a front-mounted two-speed splitter unit of countershaft design, a main countershaft gearbox with four gears and a reverse gear and a rear-mounted two-speed planetary range-change unit. Thus the transmission has a total of 16 forward gears and two reverse gears. The main feature of this transmission series is the large overall gear ratio of approximately 17. The ratios from 16.41 in first gear to 1.00 in direct gear or from 13.80 to 0.84 with the overdrive design allow an optimal exploitation of the power offered by the commercial vehicle engine [6.49].

The countershaft transmissions discussed heretofore had only one countershaft located in the power flow. The transmission diagram shown in Figure 6.55 of the Eaton Twin Splitter transmission has a 4-speed main gearbox with two countershafts as well as a three-speed rear-mounted splitter unit also fitted with two countershafts. The ratio lies between 10.90 and 0.78 in the overdrive design and between 14.05 and 1.00 in the direct drive design, with a resulting overall gear ratio of approximately 14 [6.8]. The power transmitted is split between both countershafts of the main gearbox, and flows back to the main transmission shaft. The power-split enables the gearwheels to be approximately 40% narrower than in a conventional countershaft transmission. The transmission is physically shorter, but wider. Short transmissions are advantageous especially in tractors. The shorter the transmission, the more favourable the proportions (the deflection angle resulting from the vertical offset and the longitudinal distance to the final drive) for the propeller shaft connected to the transmission.

In order to ensure uniform loading (load compensation) of the gearwheels in both branches of the power-splitter, the main shaft does not run in radial bearings but is merely radially guided. It is centred between the two countershafts when under load. Since the main shaft is not capable of absorbing large axial forces, straight-cut spur gears are used. To still achieve good running characteristics, gearwheels with a high contact ratio and high contact gearing are used.

The main gearbox of the twin splitter has four forward gears and one reverse gear. The rear-mounted splitter unit has three gears: one direct gear $i_D = 1.0$, one speed increasing gear $i_{S,H} = 0.78$ and one speed reducing gear $i_{S,L} = 1.24$. This gives 12 forward gears and three reverse gears. Rear-mounted splitter units are not usually used, because of the face widths required (see also Table 6.13).

It is nevertheless used in this case because of the low overall face width resulting from the power-split. The main gearbox is constant-mesh, the rear-mounted splitter unit is synchronized.

One example of a three-range manual transmission with two countershafts in each range is the Eaton S Series transmission shown in Figure 6.56.

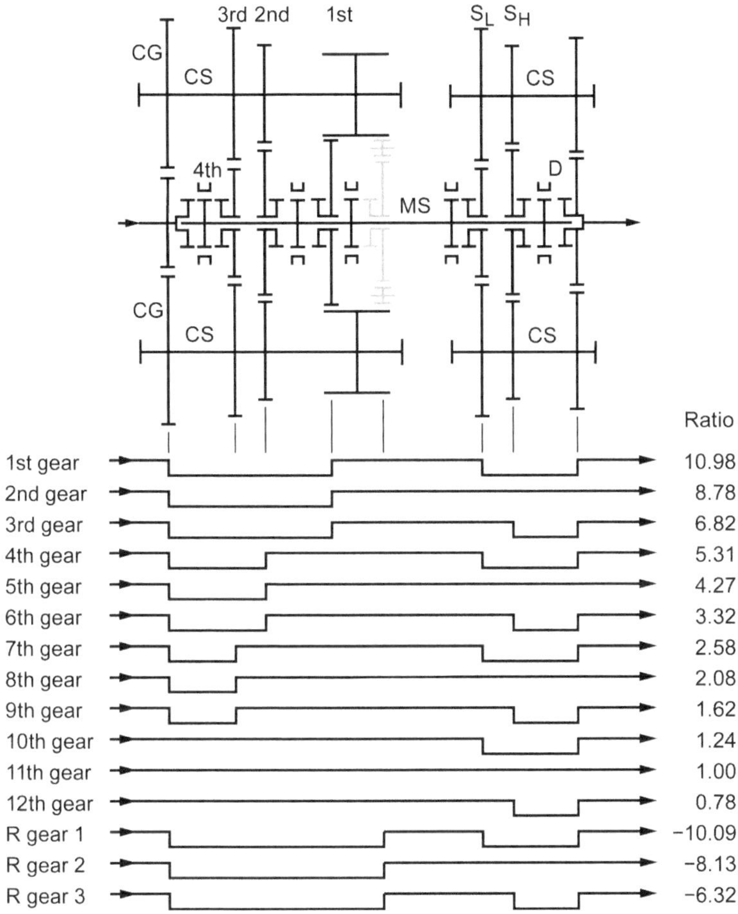


**Fig. 6.55.** Gearbox diagram, power flows and ratios of the Eaton Twin Splitter gearbox in overdrive design; 4 x 3 = 12 speeds: 4-speed main gearbox in two-countershaft design; 3-speed rear-mounted splitter unit in two-countershaft design; *CG* constant gear; *CS* countershaft; *MS* main shaft; $S_L$ rear-mounted splitter unit constant low; $S_H$ rear-mounted splitter unit constant high; *D* direct, *production design Figure 12.37*

The transmission consists of a front-mounted two-speed splitter unit, a 4-speed main gearbox and a two-speed rear-mounted range-change unit. Thus the transmission has a total of 16 forward gears and two reverse gears. Because all three units each have two countershafts, high torque absorption is possible with a short gearbox. But the transmission is wider than a one-countershaft design. With a ratio between 17.58 in the first gear with the direct drive design and of 14.45 in first gear and 0.83 in the highest gear with the overdrive design, the overall gear ratio reaches a value of 17.4 [6.9].

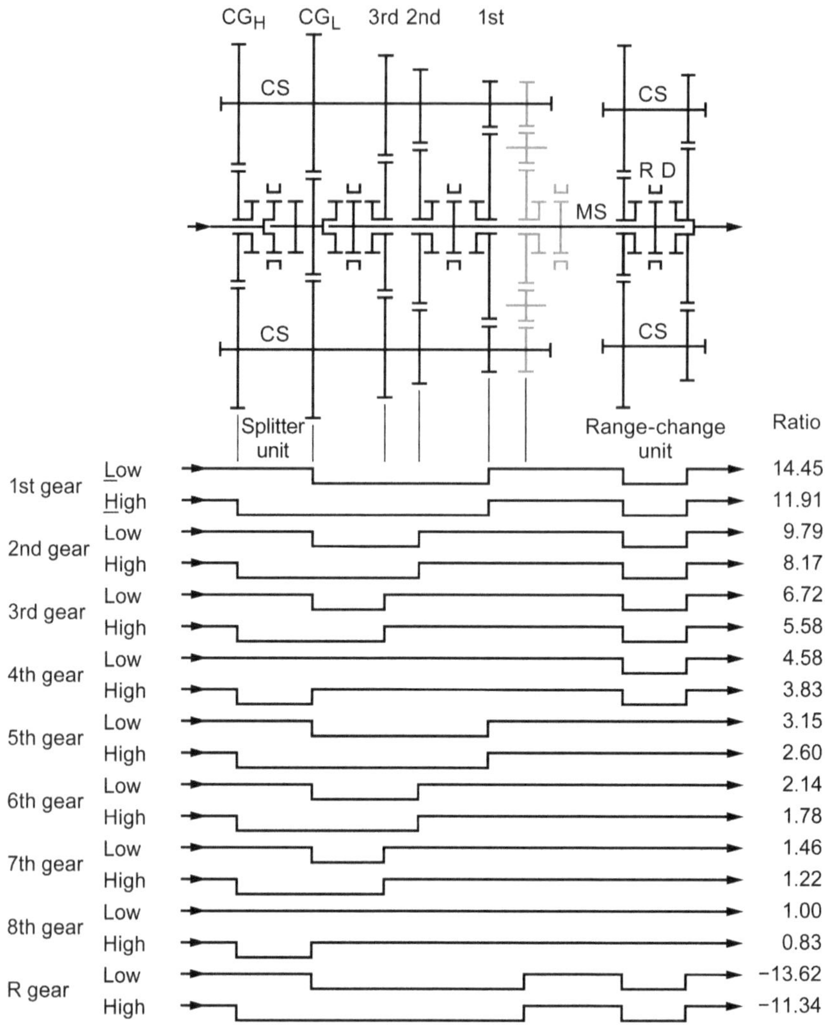


**Fig. 6.56.** Gearbox diagram, power flows and ratios of a 16-speed gearbox (2 x 4 x 2) in overdrive design (Eaton). Two-countershaft design of all range units: 2-speed splitter unit; 4-speed main gearbox; 2-speed range-change unit; *CG* constant gear; *CS* countershaft; *MS* main shaft, *production design Figure 12.38*

This transmission is also the basis for an automated design. In this case, the 4-speed main gearbox is designed without synchronization, with the other two range units being synchronized. In this design, going under the name of Eaton SAMT (Semi Automated Manual Transmission), the transmission allows a fully automated gear change without the intervention of the driver. Depending on the traffic situation, one can choose between a fully automated or semi-automated driving mode. The driver only uses the clutch when moving-off, stopping and manoeuvring [6.40].

### 6.7.2 Automated Manual Commercial Vehicle Transmissions (AMT)

Manual passenger car and commercial vehicle transmissions can have different levels of automation. Depending on the design, the moving-off process, the engagement of the gearshifting clutch and gear selection can be automated. None of these processes is automated in manual transmissions, one is automated in semi-automated manual transmissions and, in the case of fully automated transmissions, all of these processes are automated. This results in the breakdown shown in Table 6.14 listing the various degrees of automation of transmissions from manual transmissions (automation level 0) through to fully automated transmissions (automation level 4).

#### *1/ Semi-Automated Manual Commercial Vehicle Transmissions: Automation Level 2*

In automation level 2 the driver just engages the desired gear by activating the gearshift lever, then the clutch engaging action and moving-off take place automatically.

#### *2/ Semi-Automated Manual Commercial Vehicle Transmissions: Automation Level 3*

In automation level 3 the driver selects the gear or follows an automatic shift (gear) recommendation. By activating the clutch the driver triggers an automatic shift into the recommended or selected gear. Such systems have become familiar in the market under the names "Automated Preselectors (APS)" and "Ecoshift". Transmissions with an automation level of 3 and lower are no longer designed and manufactured.

**Table 6.14.** Automation levels of manual transmissions

| Automation level | Moving-off method | Shifting clutch action | Gear selection method |
|---|---|---|---|
| 0 | Foot-activated master clutch | Foot-activated clutch operation | Manual activation of a shift lever |
| 1 | Foot-activated master clutch | Automated clutch operation | Manual activation of a shift lever |
| 2 | Automated master clutch | Automated clutch operation | Manual activation of a shift lever |
| 3 | Automated master clutch | Gear change initiated by foot-activated clutch operation | Manual gear preselection from keypad |
| 4 | Automated master clutch | Automated clutch operation | Automated gear selection and engine management |

### 3/ Fully Automated Manual Commercial Vehicle Transmissions: Automation Level 4

Automated manual commercial vehicle transmissions have had level 4 automation since the end of the 1990s. This level of automation is characterised by an automated moving-off element, automated clutch engagement in gearshifting, automated gear change and data communication between the engine control unit and the transmission control unit. Through this total clutch automation, the clutch pedal can be disposed of entirely, thus allowing for a two-pedal system in commercial vehicles with acceleration and brake pedals. In addition to the automatic mode, a manual mode enables the driver to intervene at any time.

As already discussed in Section 6.6.2 "Automated Manual Passenger Car Transmissions", a distinction is made among the actuating elements between add-on systems for adding onto existing manual transmissions and integrated systems. In the case of integrated systems, the transmission is developed for purely automated implementation. The automation of manual transmissions offers many advantages, the most important of which are:

- a high level of efficiency comparable to a manual transmission,
- improved driving comfort as a result of relieving the driver of engaging the clutch and shifting,
- improved driver alertness in road traffic,
- reduction of life-cycle costs through:
  - decreased clutch wear,
  - decreased fuel consumption through the use of an optimised driving strategy (selection of shift programmes),
- increased protection of components through the prevention of shifting misuse (transmission and clutch protection),
- transmission control without rods and cables by means of a drive selector, which means:
  - reduced noise in the cabin, as there is no mechanical connection between the gearshift lever and the transmission,
  - optimised packaging, as the rod shifting system and the clutch pedal are left out, which means a more simple and cost-effective assembly.

A disadvantage of automated manual transmissions is power interruption, but since the shifting times of modern commercial vehicles are short, this is acceptable for road traffic.

#### 3.1/ Structure of Automated Manual Commercial Vehicle Transmissions

The structure of manual commercial vehicle transmissions with single- and multi-range design is explained in Section 6.7.1. The automated clutch engagement and the automated gear change are facilitated by the use of different actuating elements (actuators). These are the clutch actuator *CA* and the transmission actuator *TA*. Figure 6.57 shows the system structure of an automated manual commercial vehicle transmission.

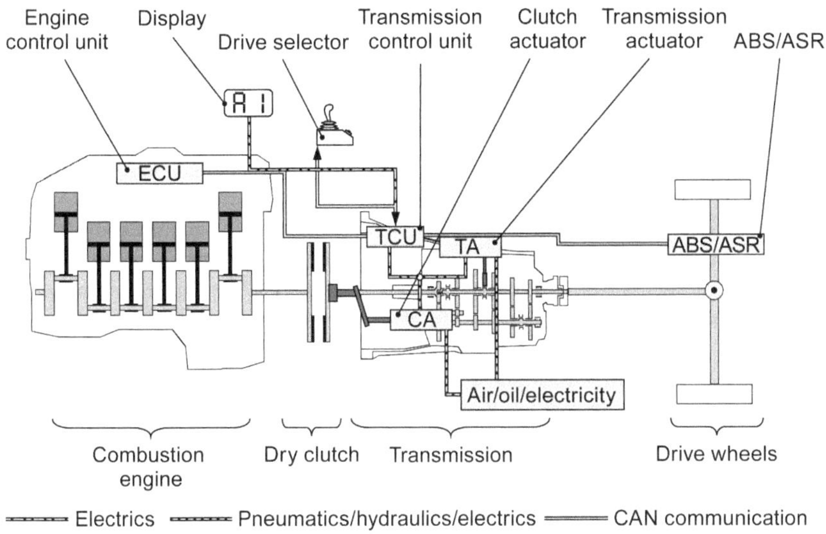


**Fig. 6.57.** System structure of an automated manual transmission

The key differentiating factor in the design of the actuating elements is the type of energy required for their functioning. The following types of actuating elements are used in automated manual commercial vehicle transmissions:

- electropneumatic actuators,
- electrohydraulic actuators and
- electromechanical actuators.

The selection of actuator essentially depends on the type of energy currently or potentially available in the vehicle. Electropneumatic actuators are suited to heavy-duty trucks, as they are equipped with a compressed air system. If no compressed air is present, either an electrohydraulic or an electromechanical system is used, depending on the required properties of the actuating element.

Actual and reference data are exchanged via the data bus (CAN bus = Controller Area Network) between the engine control unit (ECU) of the diesel engine and the transmission control unit (TCU), as well as connected sub-systems such as the drive selector, the display, the ABS/ASR and the sensors. See also Chapter 13 "Electronic Transmission Control".

Automated manual commercial vehicle transmissions are available on the market as "AS-Tronic" (ZF), "eTronic" (ZF), "Telligent EAS" or "PowerShift" (Mercedes-Benz), "Sprintshift" (Mercedes-Benz), "I-Shift/Geartronic" (Volvo), "Opticruise" (Scania) and "SAMT B" (Eaton). The following will explain some examples of these transmissions.

### 3.2/ Examples of Commercial Vehicle AMTs

The ZF 6 AS 380 VO eTronic 6-speed automated manual transmission for light-duty trucks and vans is based on the 6 S 380 VO overdrive manual transmission.

This single-range transmission has an input constant gear which, except for the direct gear (fifth gear), is always in the power flow (Figure 6.58).

The automation of this transmission is achieved by means of electromechanical actuators, mounted externally as add-on components on the gearbox housing. An electromechanical clutch actuator opens and closes the clutch. The electromechanical transmission actuator has two electric motors: one for selecting motion and one for the shifting motion [6.50].

The PowerShift transmission G241-16K by Mercedes-Benz is an example of a manual transmission for heavy-duty commercial vehicles which is automated by electropneumatic actuators. The geometrically stepped 16-speed constant-mesh gearbox is designed as a three-range transmission (Figure 6.59). A two-speed splitter unit is located in front of the 4-speed main gearbox. It "compresses" the gear sequence. The splitter unit and the main gearbox have a single-countershaft design. The two-speed planetary range-change unit is mounted behind the main gearbox. It doubles the gear sequence.

The transmission has a brake mounted on the countershaft which is responsible for adjusting the speed for upshifting. Downshifting is achieved by means of an increase in engine speed. The electronic transmission control unit is mounted together with the transmission actuator and the clutch actuator on the transmission exterior.

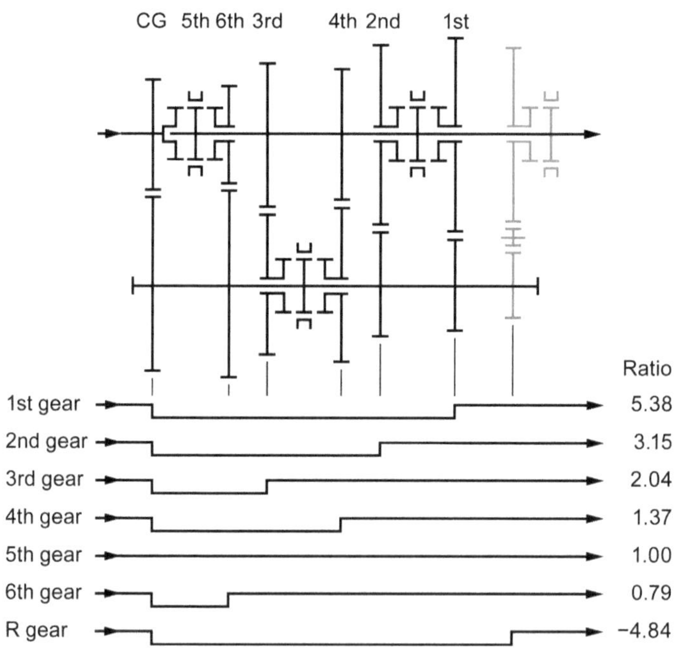


**Fig. 6.58.** Gearbox diagram, power flows and ratios of the automated 6-speed gearbox for vans and light-duty trucks in single-range design and electromechanical gearshift system (ZF), *production design Figure 12.39*

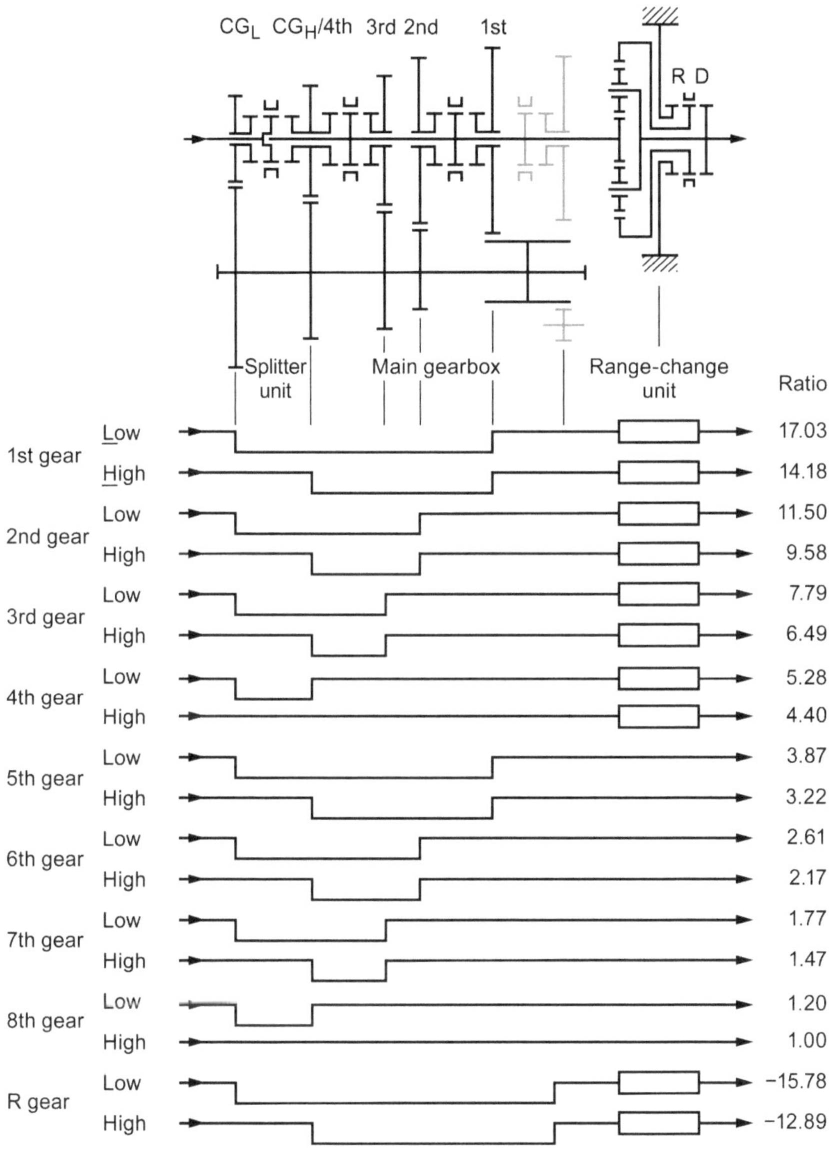


**Fig. 6.59.** Gearbox diagram of an automated 16-speed three-range gearbox 2 x 4 x 2 (Mercedes-Benz); constant-mesh gearbox with electropneumatic gearshift system and a countershaft brake, *production design Figure 12.40*

Another automated transmission for heavy-duty commercial vehicles is the ZF AS-Tronic Series three-range transmission. The unsynchronized 3- or 4-speed main gearbox has two countershafts. The two-speed splitter unit, which also has two countershafts, and the two-speed planetary range-change unit are synchronized. Due to this modular structure, 10-speed, 12-speed and 16-speed transmissions with two reverse gears can be realised [6.15–6.16, 6.25, 6.35, 6.46]. Figure 6.60 shows the gearbox diagram and power flows of the ZF 16-speed transmission AS-Tronic 16 AS 2230 TD.

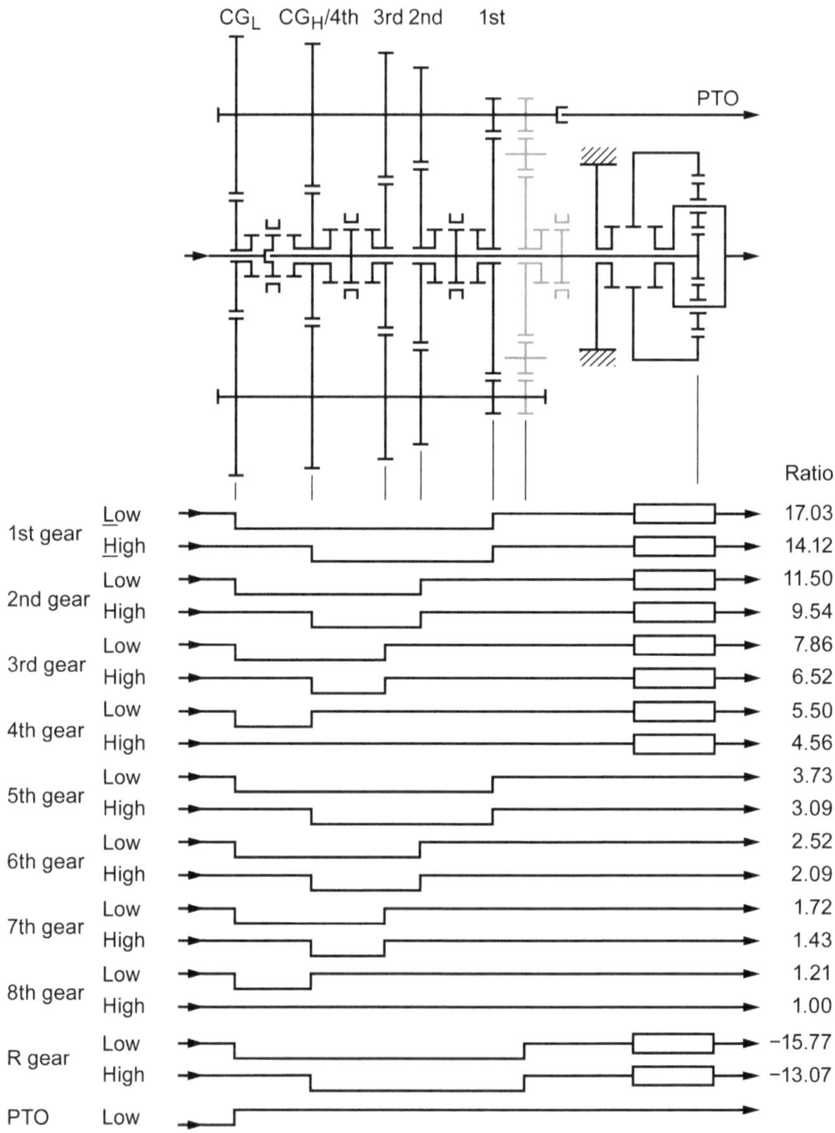


**Fig. 6.60.** Gearbox diagram and power flows of an automated 16-speed three-range gearbox in direct drive design (ZF); electropneumatic gearshift system and countershaft brake, *PTO* power take-off, *production design Figure 12.41*

The main gearbox is designed with constant-mesh gearshift and is synchronized by means of engine control and a gearbox brake. The gearbox brake mounted on one of the two countershafts is used during upshifting to reduce speed.

Like the gearbox brake, all automation components (clutch actuator and transmission actuator) are activated electropneumatically. In this transmission, the actuators are combined together into modules and integrated in the gearbox housing. The transmission has a fully automated dry clutch. This allows for a two-pedal system in the vehicle.

### 6.7.3 Commercial Vehicle Torque Converter Clutch Transmissions (TCCT)

Torque converter clutch transmissions fall under the category of automated manual transmissions. They have a very small market share. The special feature of this transmission design is the torque converter clutch unit. It consists of a hydrodynamic torque converter, a torque converter lock-up clutch, a primary retarder and a dry-running gearshifting clutch.

The torque converter clutch transmission was developed in the 1960s. The goal was to facilitate difficult moving-off processes for very heavy vehicles while retaining the manual gearbox. In Figure 6.61 the ZF 16 S 221 Ecosplit 16-speed manual transmission is located behind the torque converter clutch *TCC*. The moving-off function in this transmission is taken over exclusively by the torque converter; the gearshifting clutch is foot-activated to interrupt the power when changing gears.

A further development of the "ZF Transmatic" mentioned earlier is the "ZF TC-Tronic" with the model name 12 TC 2740 TO. The fully automated transmission consists of a torque converter clutch unit *TCC* and an automated 12-speed transmission with an electropneumatic shifting system.

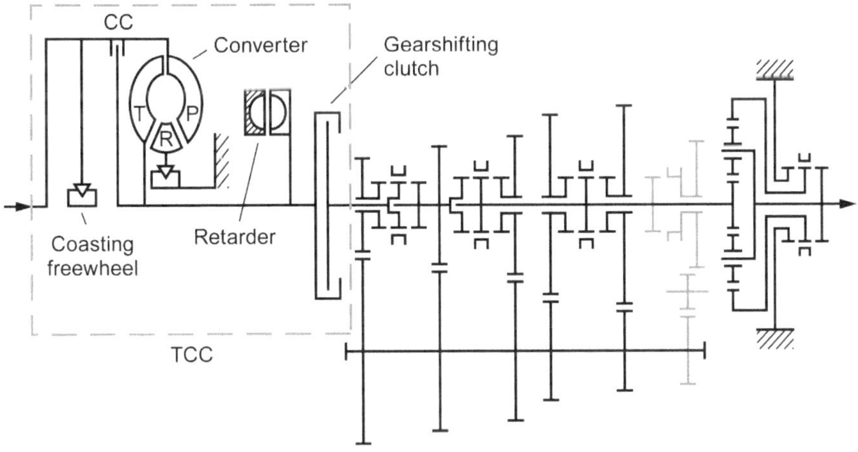


**Fig. 6.61.** Gearbox diagram of a 16-speed commercial vehicle transmissions with torque converter clutch *TCC* (ZF), *production design Figure 12.45*

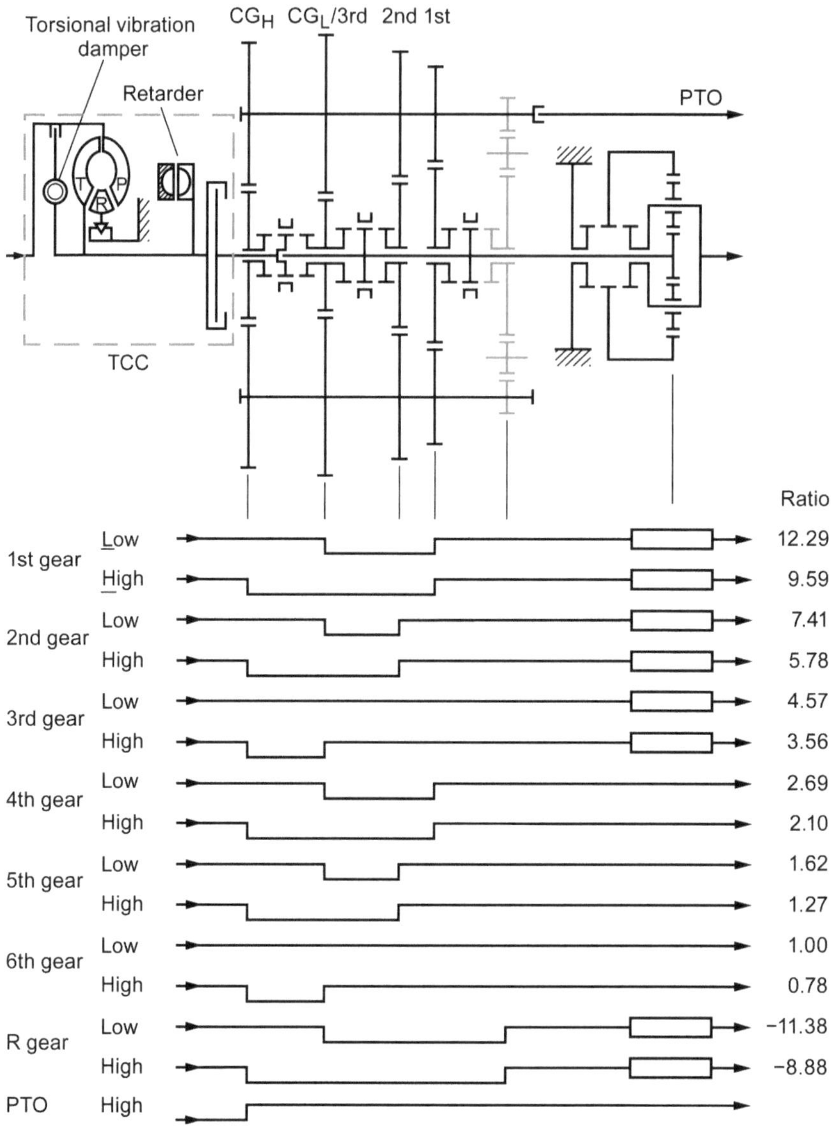


**Fig. 6.62.** Gearbox diagram of a 12-speed commercial vehicle gearbox in overdrive design with electropneumatic gearshift system and torque converter clutch (ZF), *production design Figure 12.46*

The manual transmission mounted behind the torque converter clutch unit is based on the automated 12-speed transmission 12 AS 2740 TO with overdrive (Figure 6.62). This transmission is designed as a three-range transmission (2 x 3 x 2). The two-speed splitter unit and the three-speed main gearbox have two countershafts, and the two-speed range-change unit has a planetary design.

While the main gearbox is constant-mesh, the splitter and range-change units are synchronized. See also Figure 6.60. The transmission and clutch actuators are activated electropneumatically, as is the gearbox brake mounted on one of the two countershafts.

### 6.7.4 Automatic Commercial Vehicle Transmissions (AT)

Conventional automatic transmissions for commercial vehicles are similar in design to those made for passenger cars. They are exclusively designed as hydrodynamic torque converters with power-shiftable planetary gears. See also Section 6.6.4 "Automatic Passenger Car Transmissions". Conventional automatic commercial vehicle transmissions differ from those made for passenger cars not only in terms of their layout and thus their design, but also, depending on the particular design, because they have additional components, such as retarders or power take-offs.

An example of an automatic commercial vehicle transmission for light-duty trucks, distribution vehicles, pickup trucks and small buses is the Allison 5-speed automatic transmission of the 1000/2000/2400 Series with an input torque up to 750 Nm. Figure 6.63 shows the gearbox diagram and the power flow in the individual gears.

The transmission consists of a hydrodynamic torque converter with a lock-up clutch and a planetary gear system. The so-called Polak planetary gear set is composed of three single planetary gear sets. The planetary gear carriers are respectively connected to the following ring gears. These planetary gear sets are controlled by two multi-plate clutches and three multi-plate brakes (see Section 9.3 "Layout and Design of Multi-Plate Clutches").

In this transmission, the clutch *A* is closed in the first four gears, while the clutch *B* is closed in fourth and fifth gear. The brakes *D*, *E* and *F* connected to the ring gears of the planetary gear sets are activated according to the respective gear. In fourth (direct) gear, clutches *A* and *B* are closed and planetary gear set III rotates as a block. In fifth gear, the clutch *B* and the brake *D* are closed. In reverse gear, the shifting elements *D* and *F* are activated. Theoretically, a sixth gear is possible if the clutch *B* and the brake *E* are closed. With a ratio ranging from 3.51 in first gear to 0.74 in fifth gear, the overall gear ratio of the transmission amounts to 4.74 [6.1–6.2, 6.20].

The ZF automatic transmission 6 HP 602 C, with an input torque of up to 1600 Nm, is used for urban buses and special-purpose vehicles. Figure 6.64 shows the gearbox diagram and shifting elements of this double-overdrive transmission. This is a 6-speed transmission with a torque converter, a lock-up clutch, and integrated retarder and a Wilson planetary gear set consisting of a system of three coupled single planetary gear sets. The transmission has three multi-plate clutches and three multi-plate brakes.

In the first four gears, clutch *A* is closed, and in fourth, fifth and sixth gears, clutch *B* is closed. The brakes *D*, *E* and *F* connected to the ring gears are activated according to the gear to be engaged.

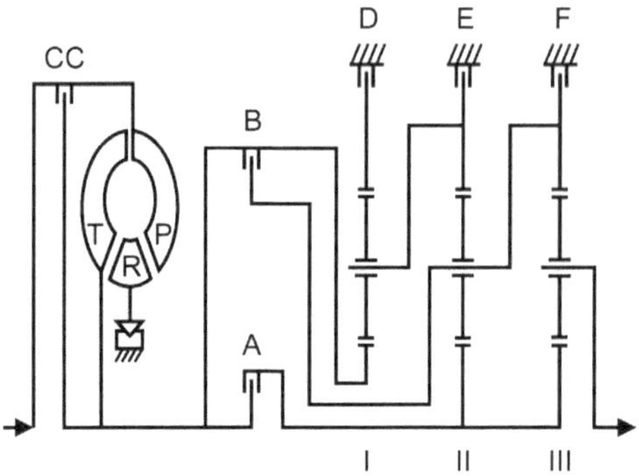


| G | Clutch | | Brake | | | i |
|---|---|---|---|---|---|---|
| | A | B | D | E | F | |
| 1 | • | | | | • | 3.51 |
| 2 | • | | | • | | 1.90 |
| 3 | • | | • | | | 1.44 |
| 4 | • | • | | | | 1.00 |
| 5 | | • | • | | | 0.74 |
| R | | | • | | • | –5.09 |

**Fig. 6.63.** Gearbox diagram and gearshift pattern of a 5-speed automatic gearbox for light-duty trucks, delivery vehicles, pickup trucks and small buses (Allison), *production design Figure 12.47*

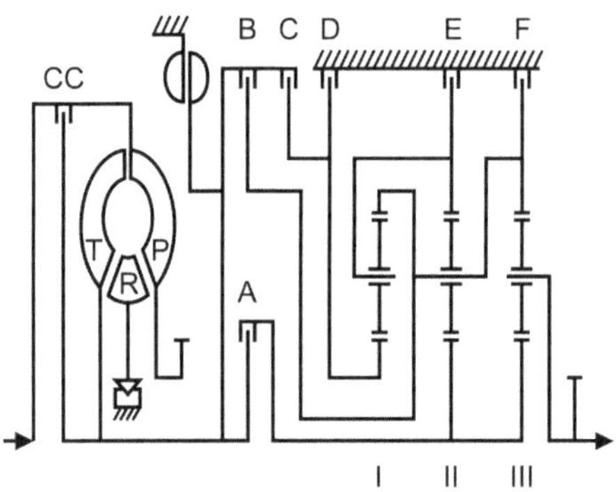


| G | Clutch | | | Brake | | | i |
|---|---|---|---|---|---|---|---|
| | A | B | C | D | E | F | |
| 1 | • | | | | | • | 3.43 |
| 2 | • | | | | • | | 2.01 |
| 3 | • | | | • | | | 1.42 |
| 4 | • | • | | | | | 1.00 |
| 5 | | • | | • | | | 0.83 |
| 6 | | • | | | • | | 0.59 |
| R | | | • | | | • | –4.84 |

**Fig. 6.64.** Gearbox diagram and gearshift pattern of a conventional 6-speed automatic gearbox (ZF), *production design Figure 12.48*

In fourth (direct) gear, clutches *A* and *B* are closed and planetary gear set III rotates as a block. Reversing the rotational direction in reverse gear is achieved by the clutch *C* and the brake *F*. With a ratio ranging from 3.43 in first gear to 0.59 in sixth gear, the overall gear ratio of the transmission amounts to 5.81 [6.11–6.12, 6.27, 6.47].

### 6.7.5 Commercial Vehicle Hybrid Drives

Serial hybrid powertrains have a practical relevance for commercial vehicles. They have no mechanical coupling of the engine to the wheels (see Section 3.2.4). The serial hybrid drive has the advantage of being highly flexible in terms of the selection of an electrical energy supply source. The following are possible:

- *Diesel generator set*: This design exhibits the features of the serial hybrid powertrain described in Section 3.2.4.

An optional electrical energy accumulator (electrochemical battery or double-layer capacitors) can be used in the intermediate voltage circuit between the generator and the traction motor(s). The energy storage unit can be used for intermediate storage of energy recuperated during braking or for carrying out purely electric vehicle operation without the combustion engine.

- *Fuel cell:* In this case, the supply of electric power carried out by the traction motor(s) is taken over by a fuel cell (see Section 3.2.3). A hybridisation with an additional battery is recommendable to ensure a system capable of recuperation.
- *Overhead contact line:* The supply of electric power carried out by the traction motor(s) can also be taken over by an overhead contact system, e.g. for city bus applications. With such applications, a small diesel generator may be used to enable manoeuvring operation without recourse to the overhead contact system.

Basic system components of a serial hybrid drive:

- *Generator:* preferably a permanent-magnet three-phase synchronous motor (PSM) flanged to the diesel engine; a special design is the transversal flux motor (TFM).
- *Feed-in:* rectifier for converting the three-phase current of the generator into direct current.
- *Intermediate voltage circuit:* electrical energy supply for the traction motors.
- *Traction inverter:* pulse-controlled traction inverter for controlling the three-phase AC traction motor with either ASM (asynchronous motor) or PSM design.
- *Control electronics:* driving electronics and hybrid management.

For all the drive configurations named, the electric traction motors can be arranged as follows:

- The central motor acting directly on a conventional axle drive; no ratio range. Application: city buses (Figure 6.65a). Possible alternative: design with tandem motors and summarising gearbox (Figure 6.65c).
- The central motor acting on a conventional axle drive via a gearbox assembly for speed adjustment, preferably a planetary gearbox; fixed ratio. Application: city buses (Figure 6.65b). Possible alternative: design with tandem motors with two planetary gearboxes and a summarising gearbox (Figure 6.65d).
- Single-wheel drive with traction motors in direct proximity to the drive wheels (wheel hub drive) as final drive with gearbox assemblies for speed adjustment, preferably planetary gearboxes, often with two-stage design with no ratio range. Application: city buses, commercial vehicles, special-purpose vehicles (Figure 6.65e).
- "In-hull drive": single-wheel drive with traction motors, e.g. mounted at the centre of the vehicle, which acts on the drive wheels via propeller shafts. Two-stage selector transmission possible (shifting when stationary or during operation, with or without powershifting). Application: special wheeled and tracked vehicles (Figure 6.65f).

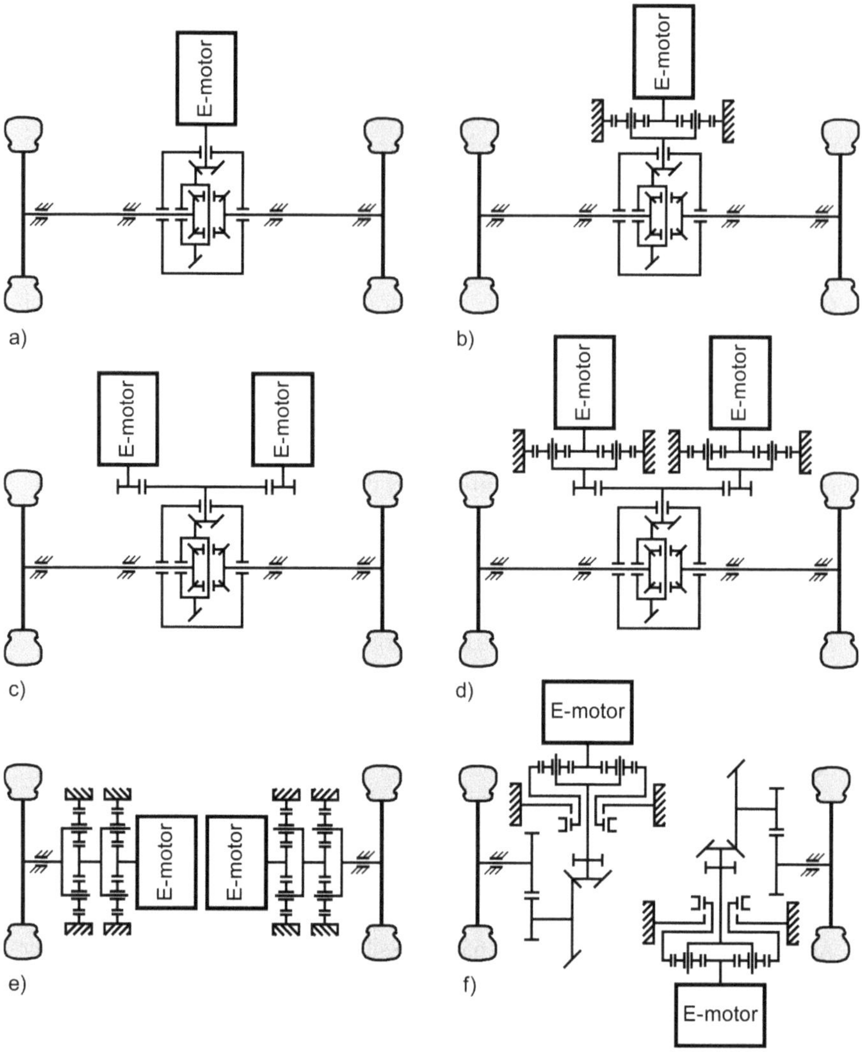


**Fig. 6.65.** Schematic design types for traction motor applications. *a* Direct central motor; *b* central motor with planetary gear set; *c* tandem motors with summarising gear; *d* tandem motors with planetary gear set and summarising gear; *e* wheel hub drive with planetary gear set; *f* "in-hull drive" with selector gearbox

Figure 6.66 shows the electric drive axle EE Drive 1 from ZF. A wheel hub drive was designed for applications in city buses which can be used for all three of the abovementioned configurations with electric drive: serial hybrid powertrain with *diesel generator set, fuel cell drive* and electric drive with a connection to an *overhead contact system*. The following will focus on the EE Drive 1 serial hybrid drive design.

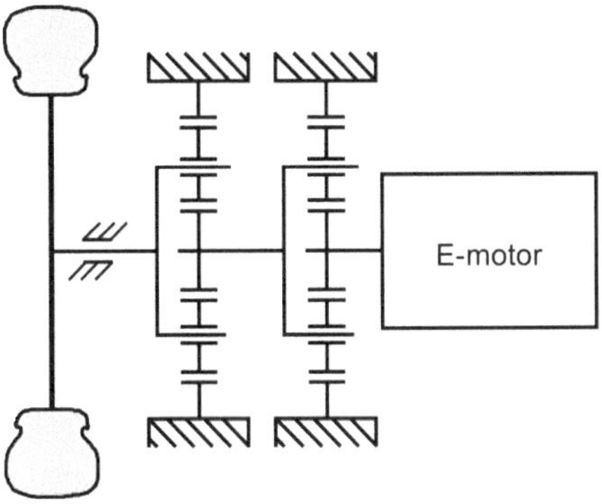


**Fig. 6.66.** Electric drive axle EE Drive 1 (ZF); schematic diagram of a single traction motor with wheel hub drive, *production design Figure 12.49*

The diesel generator set serves as a pure power source. Because there is no mechanical coupling, the engine torque does not act directly on the wheels. The electric drive is only responsible for vehicle motion and can principally execute the functions of driving and braking by reversing the direction of the energy flow. In the case of braking, kinetic energy of the vehicle can be fed back by means of the generator activity of the traction motor to the traction battery via the intermediate direct current circuit (recuperation). Additionally, the motor activity of the generator against the compression work done by the combustion engine can eliminate excess braking energy without wear. Finally, the electric energy from the intermediate direct current circuit can be converted via a brake resistor into heat (wear-free permanent brake).

In the schematic diagram of the wheel hub drive in Figure 6.66, one electric motor mounted concentrically with the wheel axis acts per driven axle on the left wheel and one on the right wheel, respectively, via a two-stage planetary gearbox. The drive for each of which is though the sun gear and the output via the spider. Both ring gears are fixed to the housing (see Section 6.8.4 "Hub Drives for Commercial Vehicles").

### 6.7.6 Continuously Variable Commercial Vehicle Transmissions (CVT)

Mechanical continuously variable transmissions are currently not in prospect for commercial vehicles. Pulley transmissions as used in passenger cars (up to 350 Nm) have neither the necessary torque capacity nor the necessary robustness for use in commercial vehicles. There are test vehicles with toroidal variators (friction gear) which have higher torque capacity than chain converters. Especially with agricultural machines and mobile work machines, however, continuously variable hydrostatic transmissions in combination with geared transmissions represent a sensible and thus widespread form of drive.

While continuously variable hydrostatic transmissions boast a simple, comfortable operation and high automatability, their overall level of efficiency is moderate at best [6.36].

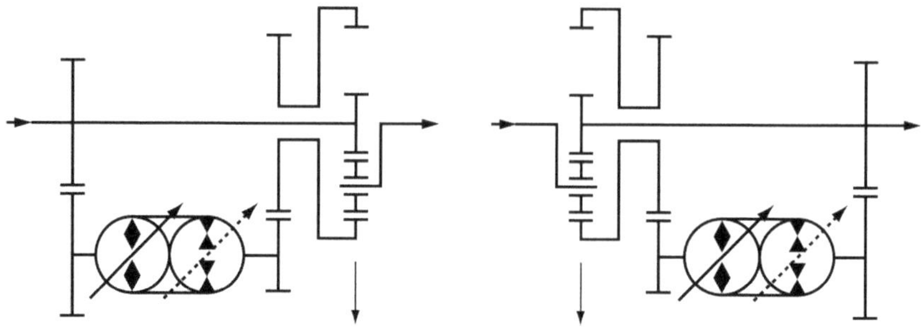


**Fig. 6.67.** Basic designs of continuously variable drives with hydrostatic power-split. *a* Input connected; *b* output connected (according to [6.36])

Purely hydrostatic continuously variable transmissions never became a popular choice for drives with a large overall gear ratio which transmit the total installed engine power and/or have high transport shares, as e.g. in agricultural tractors. Instead, tractors are increasingly being fitted with continuously variable transmissions based on hydrostatic power-splitting [6.14].

In continuously variable transmissions with hydrostatic power-split, the drive power of the engine is split into a mechanical branch and a hydrostatic branch and then combined again. The continuous ratio change is a result of the altered specific displacement volume of the hydrostatic displacement machine. Transmission designs work predominately with adjustment of the hydrostatic pump (primary adjustment). The hydrostatic power fraction is concept-dependent and can reach up to 100% of the input power; even areas of operation with reactive power flow are used. Figure 6.67 shows the two most important basic designs of drives with hydrostatic power-split. In Figure 6.67a, the power is split with a constant speed ratio and recombined in a planetary gear with a constant torque ratio. In Figure 6.67b, the exact reverse is the case. Depending on the basic design and the way the connection shafts of the planetary gearbox are connected, the hydrostatic power fraction takes different paths over the total ratio [6.37].

The transmission system shown in Figure 6.68 is a continuously adjustable hydrostatic transmission with power-split. It consists of the following components: planetary coupled gear, hydrostatic unit, reversing gear, electrohydraulic control, sensor system and electronic control unit. The total ratio of the transmission can be continuously adjusted within a range of $\infty$ to +/–0.58.

This overall range is subdivided into four fixed transmission ranges which are adjusted via the planetary gearbox and are each transmitted again with the constant gear of the countershaft reversing gear. The planetary gearbox consists of four planetary gears. The front differential drive is designed as a three-stage coupled gear with five shafts (planetary gears $P_1$, $P_2$, $P_3$).

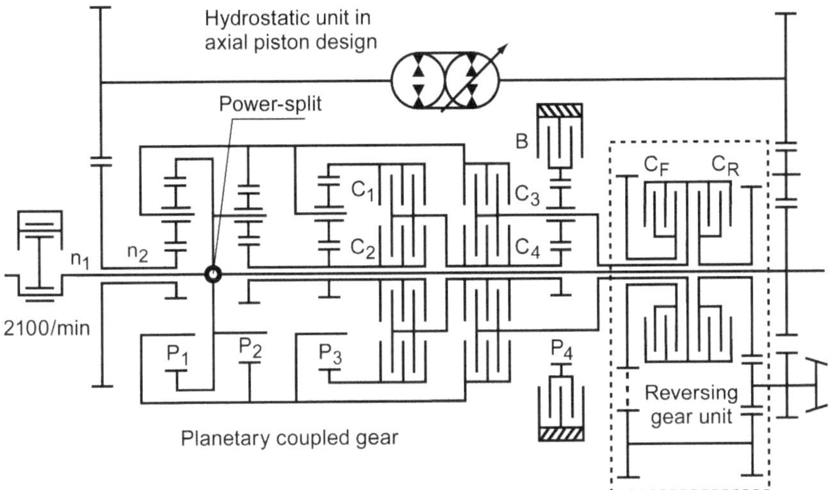


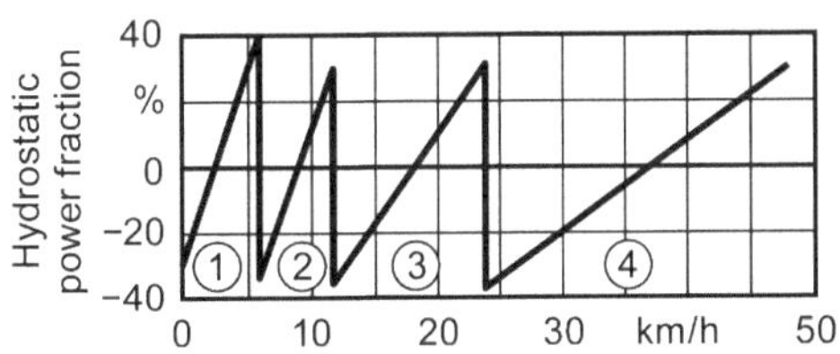


| v [km/h] | $C_1$ | $C_2$ | $C_3$ | $C_4$ | B | $P_1$ | $P_2$ | $P_3$ | $P_4$ |
|---|---|---|---|---|---|---|---|---|---|
| 0–6 | ● | | | | ● | ● | ● | ● | ● |
| 6–12 | | ● | | | ● | ● | ● | | ● |
| 12–24 | | ● | ● | | | ● | | | |
| 24–48 | | ● | | ● | | ● | ● | | |

**Fig. 6.68.** Continuously variable, hydrostatic-mechanical gearbox with power-split for the use in agricultural tractors (ZF)

Among the five shafts are two input shafts, the engine-speed dependent gearbox input shaft and the continuously variable hydrostatic input shaft, and three output shafts which can be connected to the fourth planetary gear set $P_4$ via the clutches $C_1$, $C_2$ and $C_3$. The fourth planetary gear set $P_4$ has the function of a range unit acting as reduction step in the first two transmission ranges 1 and 2 and rotates additionally as a block in ranges 3 and 4 [6.34].

A ratio range can be defined in each transmission range. Within the transmission range, the continuously variable ratio is merely dependent on the ratio of the hydrostatic input speed $n_2$ to the gearbox input speed $n_1$ and on the rotational direction of the hydrostatic input shaft $n_2$ with respect to the constant rotational direction of the gearbox input shaft $n_1$. The speed ratio $n_2/n_1$ can have any value between +1 and –1 in every transmission range.

The final ratio at the end of every transmission range is equal to the initial ratio of the transmission range which follows. Thus no speed difference arises in the respective shifting elements during the automatic shifting without power interruption. The drive power is power-split forwards and backwards in all transmission ranges and transmitted through the hydrostatic unit and through the planetary gear set. There is a ratio in every range in which the hydrostatic power becomes zero ($n_2 = 0$). In this state, the drive power is transmitted on a purely mechanical basis. The gearbox efficiency achieves its respective maximum value in this position.

## 6.8 Final Drives

Figure 4.2 shows a hierarchical classification of individual ratios in the powertrain. Figure 6.69 shows the components derived from this classification which fall under the category “final drive”:

- axle drive,
- differential gear, locking differential,
- hub drive (commercial vehicles) and
- transfer gearbox (in case of multiple driven axles).

The different final drive designs result from the position of the engine relative to the direction of travel, the position of the engine relative to the transmission, and the ratio allocation between the transmission, transfer box, axle drive and hub drive.

Because of the major differences in the final drive designs, a distinction is made below between passenger cars and commercial vehicles. Examples of the most important production designs are shown in Section 12.3 “Final drives”.

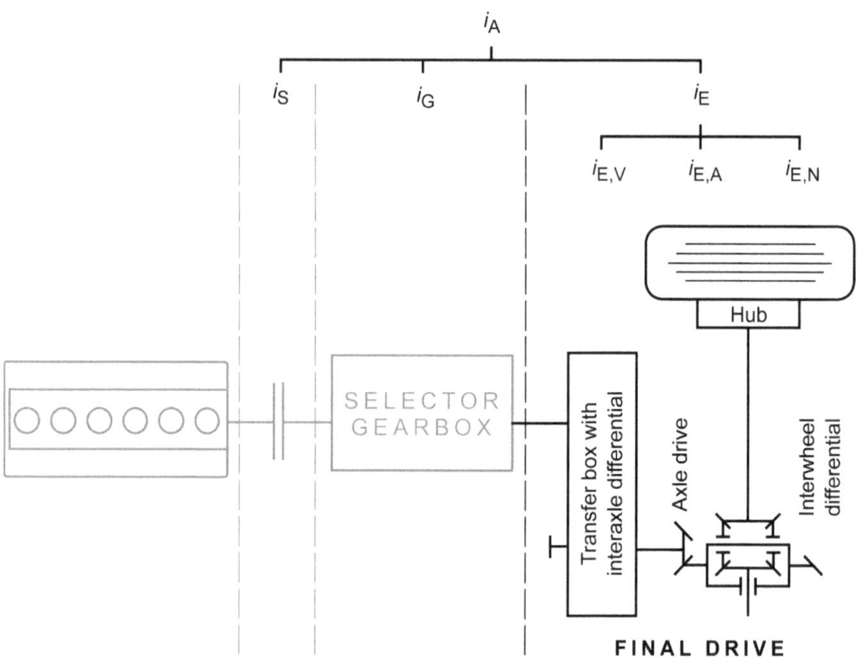


**Fig. 6.69.** Components of the final drive and classification of individual ratios

### 6.8.1 Axle Drives for Passenger Cars

The basic axle drive designs shown in Figure 6.70 can be derived from the numerous possible configurations of assemblies in the powertrain:

- spur gear axle drive,
- bevel gear axle drive of helical bevel or hypoid design and
- worm gear axle drive.

Other options for the final transmission are belt drives (DAF Variomatic) and chain drives (motorcycles).

#### *1/ Spur Gear Axle Drive*

Spur gear axle drives are now common because of the popularity of vehicles with transverse front-mounted engines. The axle drive is driven either directly by the output shaft of the transmission, or by idler gears. It is normally favourable for the differential cage drive if the engine and transmission are mounted side by side, with the disadvantage of having drive shafts of unequal length to the wheels. The reasons for their popularity are the compactness and low production cost of spur gears, normally helical cut. There are also maintenance advantages in combining transmission and axle drive, since in most cases the lubrication system is the same.

#### *2/ Bevel Gear Axle Drive*

In powertrains where the engine is longitudinally mounted, and in all-wheel drives, the power flow to the wheels has to be turned through 90°. A bevel gear axle drive is one of a wide variety of means of achieving this. The axle drive can be integrated in the transmission housing (transaxle design), or designed as an independent assembly, as in vehicles with standard drive.

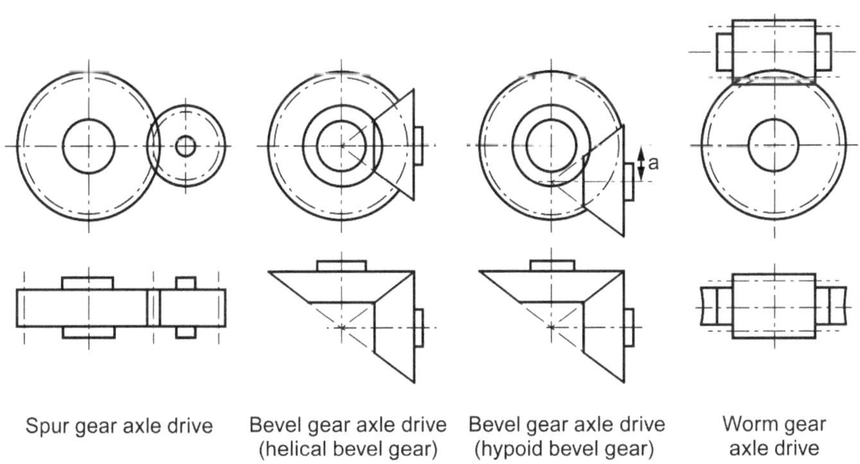


**Fig. 6.70.** Schematic view of the formats for passenger car axle drives

In the case of bevel gear axle drives a distinction can also be made between helical bevel drive and hypoid drive, according to the engagement of bevel gear and crown gear (Figure 6.70). In passenger cars, hypoid drives are usually used, in which the bevel drive pinion engages below the axial centre of the crown gear. This offset makes the diameter of the bevel drive pinion larger, and the crown gear can be smaller for the same load than in helical bevel drives in which the axes intersect. The sliding friction between the tooth flanks, which contributes substantially to reducing noise, creates very high surface pressure forces which demand pressure-resistant oil (hypoid oil) to lubricate the axle drive (see also Section 11.2 "Lubrication of Gearboxes, Gearbox Lubricants"). The offset also enables the propeller shaft to be mounted lower, reducing the size of the transmission tunnel.

### 3/ Worm Gear Axle Drive

There are now no more axle drives with worm gear axle drive in production. This type of drive was used in some Peugeot models in the 1970s. They are now rarely used, mainly because of the difficulty and expense of manufacturing the worm and the worm gearwheel. But the worm gear axle drive does have significant advantages. It offers large multiplications in a compact space. The worm can be located below or above the worm gearwheel. The former case means a low centre of gravity and low-mounted propeller shaft, eliminating the obtrusive propeller shaft tunnel. Mounting the worm above the gearwheel gives the vehicle good ground clearance, a particular advantage for off-road vehicles. If worms are used in multi-axle drives, then a continuous propeller shaft coupled to the axles can be used. Worm gear drives are superior to all other types of drive as regards quiet running, because meshing is sliding, and there is always a film of oil between the frictionally engaged tooth flanks. Since there are large axial forces with worm gear drives, the worm bearing requires careful consideration.

### 4/ Transmission Ratios

The ratio of the powertrain $i_A$ in passenger cars is derived from the transmission ratio $i_G$ and the final ratio $i_E$ (see also Section 4.1). In most passenger car transmissions the transmission ratio $i_G$ of the largest gear is fixed at $i_G \approx 0.7–1.0$.

The ratio of the powertrain $i_A$ is fixed by selecting the axle ratio $i_{E,A}$ to match the power, the desired maximum final speed etc. For individual passenger car axle drives the following ratios are typical:

- spur gear axle drive $i_E = i_{E,A} \approx 3.0–5.5$,
- bevel gear axle drive $i_E = i_{E,A} \approx 2.5–5.0$,
- worm gear axle drive $i_E = i_{E,A} \geq 5.0$.

The small ratios occur in powerful passenger cars and sports cars, whilst the large ratios occur in low-powered small passenger cars and all-wheel drive vehicles. The specification of the axle drives relates to the input torque available, i.e. the output torque of the transmission.

**Table 6.15.** Un-weighted evaluation of passenger car axle drive designs.
++ very good; + good; 0 satisfactory; – bad; – – very bad

| Feature | Axle drive | | | |
|---|---|---|---|---|
| | Spur gear | Bevel gear | | Worm gear |
| | | Helical bevel drive | Hypoid bevel drive | |
| Quiet running | 0 | 0 | + | ++ |
| Manufacturing cost | ++ | + | 0 | – – |
| Bearings | ++ | 0 | 0 | 0 |
| Lubrication | ++ | ++ | 0 | 0 |
| Efficiency | ++ | ++ | ++ | + |
| Service life | ++ | + | + | ++ |
| Load capacity | 0 | + | + | ++ |
| Space requirements | + | 0 | + | + |
| Total | 11+ | 7+ | 6+ | 8+ |

#### *5/ Comparison of Different Designs*

Table 6.15 provides a comparison of the main basic designs of passenger car axle drives: spur gear, bevel gear and worm gear axle drive. The comparison criteria are quiet running, manufacturing cost, arrangement of bearings, lubrication, efficiency, service life, load capacity and space requirements.

### 6.8.2 Axle Drives for Commercial Vehicles

In commercial vehicles the axle drive can be of single-stage or multi-stage design. There are designs in which the ratio of the "axle" is split between the axle drive ($i_{E,A}$) and the hub drive ($i_{E,N}$). In this case, the term centre gearbox is used instead of axle drive. The commercial vehicle final drive accommodates the axle drive bevel gears or the worm drive, the differential gear unit and, in the case of multi-stage axle drives, spur gear sets or planetary sets and the drive-through to the next axle. As mentioned above, axle drives can be subdivided into single-stage and multi-stage axle drives. Some designs of axle dives are given in Section 12.3.2.

#### *1/ Single-Stage Axle Drives*

Depending on the type of drive, the single-stage axle drive (Figure 6.71) is divided into:

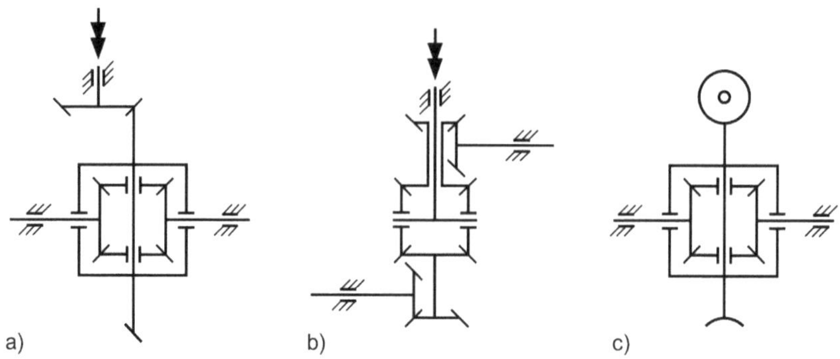


**Fig. 6.71.** Schematic view of the single-stage axle drives for commercial vehicles. *a* Bevel gear drive; *b* double bevel gear drive; *c* worm gear drive

- bevel gear drive, Figure 6.71a,
- double bevel gear drive (split bevel drive), Figure 6.71b, and
- worm gear drive, Figure 6.71c.

The spur gear axle drive is used only in commercial vehicles of up to 3.5 t gross weight rating, and will not be discussed further here.

### 2/ Multi-Stage Axle Drives

In the case of multi-stage axle drives (Figure 6.72) there are several designs:

- countershaft, front-mounted, Figure 6.72a1,
- countershaft, top-mounted, Figure 6.72a2,
- two-speed with spur gear countershaft, Figure 6.72c,
- two-speed with planetary gear, Figure 6.72d.

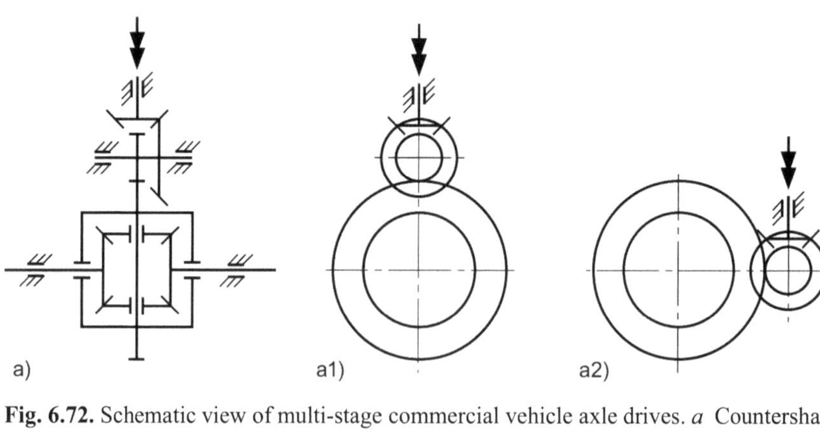


**Fig. 6.72.** Schematic view of multi-stage commercial vehicle axle drives. *a* Countershaft; *a1* front-mounted; *a2* top-mounted

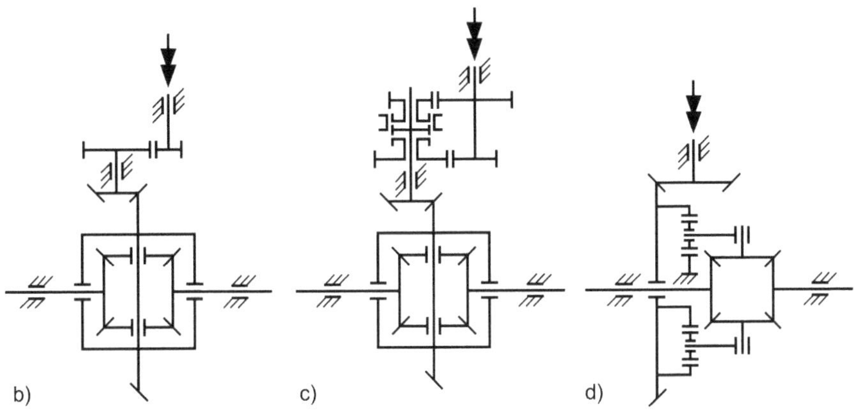


**Fig. 6.72.** *(continued) b* countershaft; *c* two-speed with spur gear countershaft; *d* two-speed with planetary gear (top: engaged; bottom: not engaged)

The term "top-mounted" refers to the fact that the propeller shaft sits higher than the drive shafts to the wheel hubs. In this design it is easy to arrange a drive-through to a second driven axle. If the propeller shaft and the drive shafts are at the same height, the drive of the axle drive is thus directly from the front, hence the term "front-mounted".

### 6.8.3 Differential Gears and Locking Differentials

Single-axle drive is the minimum for passenger cars and commercial vehicles, for reasons of directional control and traction. The engine power must therefore be distributed to a left and a right driving wheel, in the simplest case by means of a non-split wheel drive shaft as shown in Figure 6.73. But when cornering, the outer wheel covers a greater distance than the inner wheel, which with a rigid drive causes tyre abrasion and high wear and stress to the powertrain resulting from distortion.

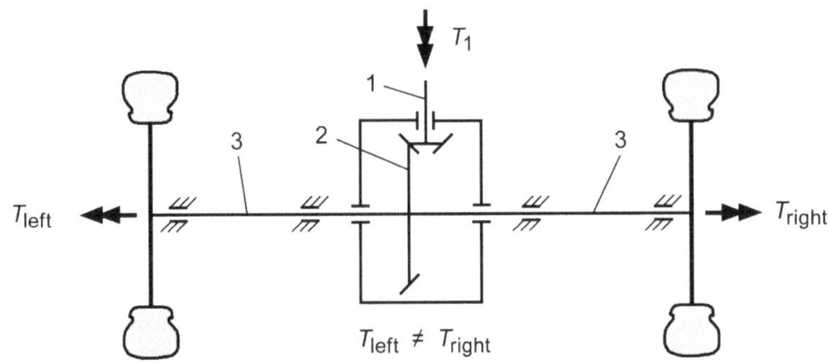


**Fig. 6.73.** Principle of a rigid, non-compensating differential gear.
*1* Drive; *2* bevel gear; *3* axle shaft

### 1/ Principles of Differential Gears

In order to avoid distortions in the powertrain and tyre wear due to a lack of speed compensation when cornering, a transmission is required which, as opposed to a rigid direct drive without split drive shaft, allows a free speed and force compensation. This transmission must provide a transverse drive torque split 50% : 50% to the left and right driving wheel. "Transverse" relates to the direction of travel of the vehicle.

Differential gear units of this type are also necessary between the powered axles of vehicles with more than one powered axle. They are only non-essential when travelling very slowly or on loose surfaces, or when there is a sufficiently small gap between the axles concerned. Interaxle compensation can be represented with asymmetrical torque distribution, depending on the traction potential of the axles and the handling required. The usual figures for all-wheel drive passenger cars are ratios of 50% : 50% or 33% : 67% for the front and rear axles respectively.

In automotive engineering, the interaxle differential for restricting the power to different drive axles is generally referred to as a *transfer box* (see also Section 6.1.3 "All-Wheel Drive Passenger Cars" and 12.4 "All-Wheel Drives, Transfer Boxes"). The interwheel differential for splitting the power to the driving wheels of an axle is also referred to as a *differential gear unit*. The interwheel differentials are usually integrated in the final drive of the powertrain. (In normal usage, these terms refer not only to the differential unit proper, but also to its drive and associated transmission, and any wheel reduction gears.)
Differential gears are subdivided into

- interwheel differential = differential gear unit:
  *Transverse split* of the power to the driving wheels of an axle,
- interaxle differential = transfer box:
  *Longitudinal split* (as seen in the direction of travel) of the power to multiple driven axles.

According to [6.30], differential drives with two or more degrees of freedom *F* can be used for the requirements described. These are generally simple or coupled planetary gears for overlaying speeds and power. Any epicyclic system with three freely moveable, coaxial shafts and which is positively actuated by two input or output movements could be used as a differential for vehicles.

In [6.3] Altmann describes ways of depicting differential gears by means of gearwheel pairings, of which the following types are common:

- bevel gear differentials,
- spur gear planetary differentials (straight differentials),
- worm gear differentials.

Straight differentials are usually used as interaxle differentials because of the possibility of asymmetrical torque distribution, and bevel gear differentials are standard for interwheel compensation. The worm gear differential (TORSEN differential) is used in both types. Some selected examples are discussed in Section 12.3.3.

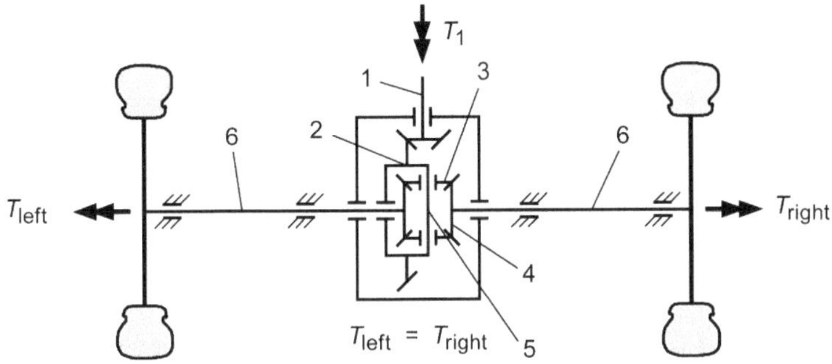


**Fig. 6.74.** Principle of a rear-axle gear unit with bevel gear differential. *1* Drive; *2* differential cage; *3* differential bevel gears; *4* axle bevel gears; *5* differential shaft; *6* axle shafts

In contrast to Figure 6.73, Figure 6.74 shows no rigid drive-through, but rather split axle shafts with an interposed bevel gear differential. The torque $T_1$ introduced through the drive *1*, for example a helical or hypoid bevel drive, is transmitted via the differential cage *2* and the differential shaft *5* to the differential bevel gears *3* and from there to the axle bevel gears *4*. The differential bevel gears act like a balance beam, establishing a torque equilibrium $T_{left} = T_{right}$ between the left and right output sides. The following applies for the rotational speeds:

- rotational speed of the outer wheel when cornering, or the one with less grip: $n_a = n + \Delta n$,
- rotational speed of the inner wheel when cornering, or wheel with more grip: $n_i = n - \Delta n$.

where $n$ is the input speed of the crown gear and $\Delta n$ is the speed difference between the output speed of the outer wheel when cornering and the input speed of the differential. When travelling in a straight line, the differential cage *2*, the axle bevel gears *4*, the axle shafts *6* which are torsionally locked to the axle bevel gears, and the differential bevel gears *3* inside the cage rotate as a block. There is no relative movement between the differential shaft *5* and the differential bevel gears mounted on it. When cornering, the outer axle shaft has to rotate faster than the inner one; axle bevel gears and differential bevel gears move on rolling contact. This enables speed compensation between the wheels.

### 2/ The Need for Differential Locking

The design of conventional differential gears described in the preceding section has two important advantages for automotive engineering:

- the rotational speeds of the drive wheels can be adjusted independently of each other according to the different distances travelled by the left and right wheels and

- the drive torque is symmetrically distributed to both drive wheels, without any yawing moment.

These two advantages are however offset by a serious disadvantage. When the frictional potential of the two drive wheels are different, the propulsive forces transmitted to the road surface for both drive wheels depend on the smaller frictional potential of the two. This comparison relates in this case to interwheel compensation in the axle drive, but applies analogously to interaxle compensation between different powered axles. This means, for example, that a wheel standing on ice will spin and the other wheel standing on asphalt cannot transfer more torque than the one that is spinning. The vehicle therefore cannot move-off. In order to overcome this disadvantage of conventional differential gears, the compensating action has to be inhibited in critical driving conditions. This can be effected in different ways:

- By means of a differential lock. It can be activated manually or automatically by mechanical, magnetic, pneumatic or hydraulic means, and blocks any compensating action 100% by locking the differential unit. This makes the axle rigid again, with all the consequent advantages and disadvantages. The use of such a traction control system is thus appropriate where there is inadequate traction for one wheel or one axle, and should preferably be automatic and temporary.
- By using self-locking differentials, also known as limited-slip or locking differentials. These are differentials with a compensating action that is deliberately tight and restricted. This enables them to transmit torque to one wheel or axle even when the other wheel or axle is spinning because of poor grip. This means losing the advantage of power transmission without yawing moment. The free adaptation of both wheel speeds to the different distances travelled by the two tracks is restricted. The axle shafts are more stressed because of the torque redistribution. Locking differentials are divided into load or torque controlled, and speed or slip controlled. The former lock the differential action as a function of the torque applied, the latter as a function of the speed difference of two out of three of the differential gear shafts.
- Using externally activated differential brakes. These systems are usually processor-controlled and hydraulically or electromechanically activated, and the degree of locking can be varied within a wide range, often from 0 to 100%, as a function of the driving conditions, normally being unlocked or only slightly locked. The advantage of such systems is that the control can inhibit the compensating function to match the driving conditions. This largely avoids negative effects on handling in situations in which a locking differential produces an unwanted locking effect.
- By combinations of the above solutions.

### *3/ The Interlock Value*

The interlock value $S$ is a key design variable, representing the degree of inhibition of the compensating action.

The interlock value for the degree of inhibition of the interwheel compensation $S_{trans}$ is defined as follows:

$$S_{trans} = \frac{\text{Locking torque } T_B}{\text{Propulsive torque } T} = \frac{T_{right} - T_{left}}{T_{right} + T_{left}}. \tag{6.7}$$

The interlock value for the degree of inhibition of the interaxle compensation $S_{longitudinal}$ is derived analogously thus:

$$S_{longitudinal} = \frac{\text{Locking torque } T_B}{\text{Propulsive torque } T} = \frac{T_{front} - T_{rear}}{T_{front} + T_{rear}}. \tag{6.8}$$

By definition the interlock value $S$ is in the range 0 to 1, and is often expressed as 0% to 100%. An interlock value of 0% describes a loss-free, non-locking differential gear; a figure of 100% represents a rigid direct drive.

In front-wheel drive passenger cars the interlock values must be kept low (maximum 17%), because of undesirable effects on the steering. The interlock values of locking differentials in rear-wheel drive passenger cars are between 25% and 50%, and in commercial vehicles up to 75%. If a locking differential with fluid clutch is used, the interlock value can reach 100% with large differences in rotational speed and high thermal stress, the so-called "hump effect".

The way the interlock works is that with a locking differential where $S = 50\%$, a maximum of 75% of the drive torque can be fed to the wheel with the better grip, with at least 25% going to the wheel which is tending to spin. The difference between these two figures is $S = 50\%$; the interlock value is as it were the "distribution figure" related to the total torque transmitted, i.e. $T_{left} + T_{right}$. In other words, the higher the interlock value, the more torque is channelled through the differential brake as braking torque or locking torque $T_B$ rather than distributed by the differential. The interlock value is thus also a measure of the power distribution between the differential and differential brake.

An example is set out below to illustrate the limited use of locking differentials as a traction control system. The example relates to a vehicle standing on a surface which is slippery on one side where $\mu_{left} < \mu_{right}$, equipped with a locking differential with an interlock value $S_{trans} = 0.3$. A maximum of $T_{left} = 25$ Nm can be transmitted to the road surface with the left wheel. It follows from Equation 6.7 that:

$$T_{right} = T_{left} \frac{1 + S_{trans}}{1 - S_{trans}}. \tag{6.9}$$

Thus a torque of $T_{right} \approx 46.4$ Nm can be transmitted to the right wheel regardless of the engine torque available. The total torque transmitted amounts to only $T \approx 71.4$ Nm. This calculation shows the limited potential of locking differentials, since depending on the driving resistance (gradient etc.) this torque could be insufficient for propulsion.

Interlock values of slip-related and load-related locking differentials have to be regarded differently: a purely load-related self-locking differential has a fixed, unchanging interlock value. This means that whatever the amount of input torque, the percentage determined by the nominal interlock value is always "diverted".

A purely slip-dependent locking differential produces a braking torque independent of the input torque, as a function of the speed difference arising. This gives rise to higher transient interlock values with small input torque, and smaller transient interlock values with large input torque. The effects of such locking differentials on performance and traction can thus only be controlled by the braking torque profile, which depends only on the difference in rotational speed.

There are many designs and principles of operation for locking differentials and self-locking differentials. Examples of some of these designs are given in Section 12.3.3. The main types currently in use are:

- load-dependent self-locking differential with multi-plate clutches,
- load-dependent self-locking differential with worm gears (TORSEN),
- slip-dependent self-locking differential with fluid clutch,
- electronically controlled (automatic) locking differentials with pressurised multi-plate clutches and
- cam self-locking differentials.

#### *4/ Alternatives to Self-Locking Differentials*

Self-locking differentials can always only represent a compromise between improving traction and driving stability on the one hand, and disadvantages in terms of steering responsiveness and possible distortions of the powertrain on the other hand. The main purpose in developing further systems is therefore only to lock a differential when absolutely necessary.

Through the use of electronically controlled multi-plate clutches, the drive torque can be freely selected and distributed between the axles and, where applicable, the individual wheels according to demand (ToD – Torque on Demand). This allows a selective influence on the driving dynamics and handling of the vehicle. Section 12.4 describes an example of such an all-wheel drive system introduced under the rubric "torque vectoring".

An economical means which makes intervening with the differential unnecessary is brake intervention. If a wheel begins to spin, it is slowed down through its service brake, which necessary causes more torque to be transmitted through the standard differential to the other wheel. This system allows for improved traction when moving-off and cornering without compromising driving stability, especially on road surfaces with highly variable friction coefficients for each wheel.

### 6.8.4 Hub Drives for Commercial Vehicles

The necessary ratio of the "axle" in commercial vehicles can be distributed either just in the axle drive, or between the axle drive and the hub drive.

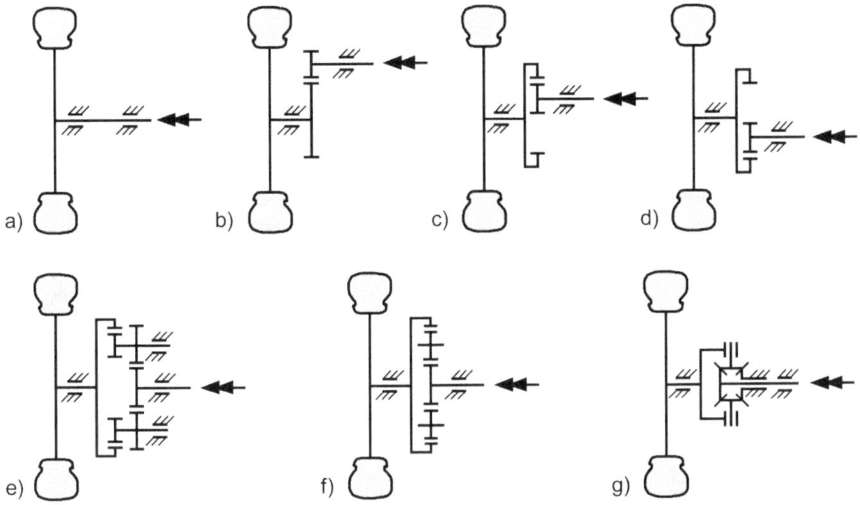


**Fig. 6.75.** Schematic view of the hub drives. *a* Without wheel hub drive; *b* spur gear countershaft with external toothing; *c* spur gear countershaft with internal toothing, above the wheel axis; *d* spur gear countershaft with internal toothing, below the wheel axis; *e* spur gear planetary drive, double planetary drive; *f* spur gear planetary drive; *g* bevel gear planetary drive

By increasing the torque directly at the hub drives, the axle drive and drive shafts to the wheel hubs can be smaller. The hub drives include the following designs:

- without wheel hub drive, Figure 6.75a,
- with external toothed spur gear countershaft with drive above, below or at the same height as the wheel axis, Figure 6.75b,
- with internal toothed spur gear countershaft, Figure 6.75c, d,
- spur gear planetary drive, Figure 6.75e, f,
- with bevel gear planetary gear (fixed at $i_{E,N} = 2$ ), Figure 6.75g.

Figure 6.75 shows a schematic view of possible hub drives. In combination with the single-stage axle drive, which is small because of the additional ratio, an external toothed spur gear countershaft with drive above the wheel axis (Figure 6.75b) creates a gantry axle. It is mainly used for off-road vehicles that require a lot of ground clearance beneath the axles. Mounting the drive below the wheel axis is favourable for low-frame vehicles.

### 6.8.5 Transfer Gearboxes

As explained in the preceding section, in the case of vehicles with several driven axles (all-wheel drive passenger cars and commercial vehicles), the engine power must be *longitudinally distributed* to the individual drive axles.

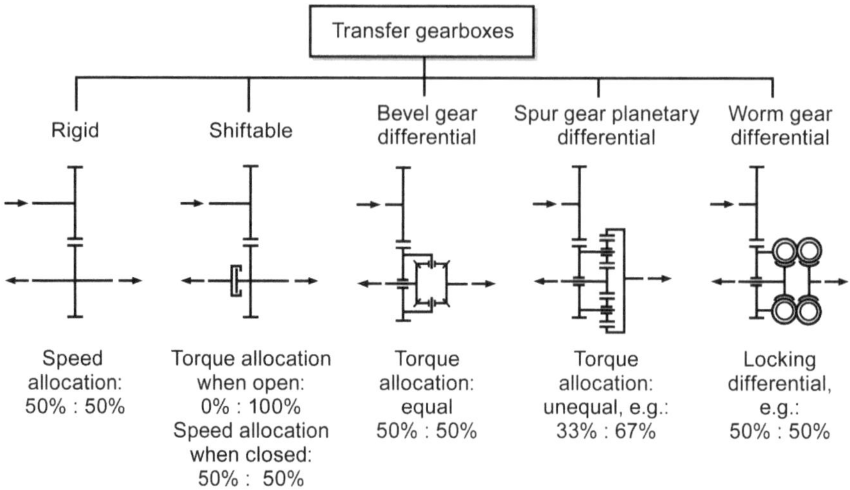


**Fig. 6.76.** Transfer gearboxes categorised by their design

In order to get by the selector gearbox (cf. Figure 6.69) and, where applicable, the engine by means of the propeller shaft, a centre distance must be created between the input and output shafts. Transfer gearboxes can be classified structurally into five groups (Figure 6.76).

To avoid serious distortion in the powertrain when cornering, rigid power distribution can only be used for axles that are close together (tandem axles). Front-axle drives (or rear-axle drives) that can be engaged as required are in use in vehicles which require all-wheel drive only part of the time, in poor traction conditions. The distortions are partly offset by wheel slip.

For passenger cars and commercial vehicles with permanent all-wheel drive, the only viable option is a transfer box with differential. The differential makes it possible to equalise the speed and the forces between the power axles. With a bevel gear differential, the torque is split equally between the front and rear axles. In straight differentials the split is unequal.

All-terrain passenger cars and commercial vehicles are also fitted with transfer boxes including a range-change unit. In the transfer box there is a choice between off-road and on-road modes. The high torque multiplication of the off-road gear thus only comes into effect after the main gearbox. For heavy-duty commercial vehicles there are also main gearboxes with integral transfer box. Existing transfer box designs are shown and discussed in Section 12.4.

## 6.9 Power Take-Offs

In commercial vehicles there is often a need for power to supply auxiliary units. The power flow can either be switched entirely to a power take-off, or can be split into a vehicle drive branch and a power take-off branch.

Power take-offs are commonly used for pumping water or mud or for driving hydraulic pumps, winches, fire-fighting ladders, crane superstructures, or sweepers (see also Figure 11.18 and Section 12.2.2, 8/). Power take-offs can be divided into three groups:

- clutch-controlled power take-offs,
- engine-controlled power take-offs and
- drive-controlled power take-offs.

### 1/ Clutch-Controlled Power Take-Offs

In clutch-controlled power take-offs the power-split to the auxiliary unit is located after the master clutch. Power flows to the auxiliary unit only when the master clutch is engaged. They are for example coupled to the countershaft of the main gearbox by means of a dog clutch. Clutch-controlled auxiliary units can be operated with the vehicle stationary (transmission in neutral), or with the vehicle moving. Since they are connected to the countershaft, they constitute an additional load on the transmission, and especially the synchronizers.

The power take-off can be in the form of an axial extension of the countershaft (Figure 6.77, variant 1), or an additional gear stage at another point in the transmission housing (Figure 6.77, variant 2).

### 2/ Engine-Controlled Power Take-Offs

Engine-controlled power take-offs are located on the engine side of the clutch, viewed in the direction of power flow (Figure 6.78). This is achieved using a hollow shaft through which the main transmission drive-shaft passes. The power take-off is thus independent of frictional engagement of the drive clutch.

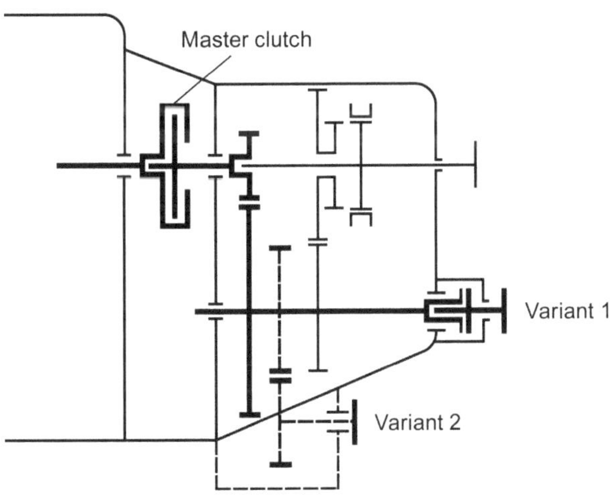


**Fig. 6.77.** Clutch-controlled power take-off

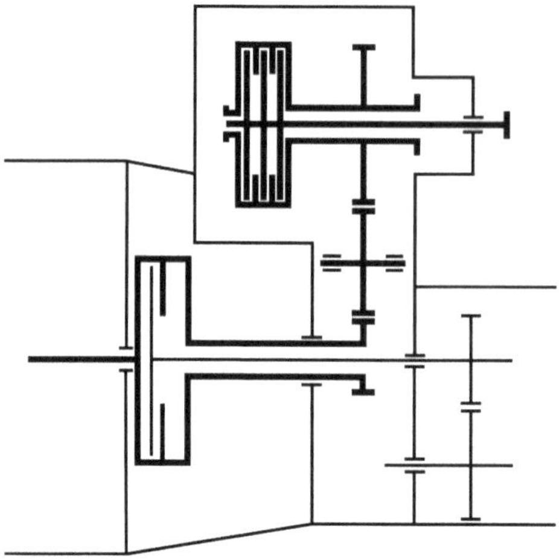

**Fig. 6.78.** Engine-controlled power take-off

Since the power flow does not go through the main gearbox, a much greater power flow can be achieved with an engine-controlled power take-off than with a clutch-controlled one. This design can function equally with the vehicle stationary or moving.

### *3/ Drive-Controlled Power Take-Offs*

Drive-controlled power take-offs are connected to the output shaft of the gearbox and are thus active as soon as the drive wheels rotate.

# 7 Design of Gearwheel Transmissions for Vehicles

*Gearwheel calculation: Global standard – much empiricism and some theory*

The declared aim of this book is to present a complete picture of the development process for vehicle transmissions. Chapters 3 to 5 showed how the ratios are selected – the fundamental design decision. In Chapter 6 some basic design concepts were introduced. Chapters 7 to 11 consider the layout and design of important components.

This does not involve the use of sophisticated calculations such as the German standard DIN 3990 gearwheel calculation, but an attempt is made to present the fundamentals of calculation methodology and calculation procedures. The aim is to equip the design engineer to quickly design important gear components "by hand". Such an approach is required for example for feasibility studies, where a quick draft is needed. For this purpose "flow charts" are displayed at suitable points throughout the following chapters, with algorithms for manual calculation.

By far the greatest proportion of vehicle transmissions are gearwheel transmissions. They still deliver the highest power-to-weight ratio for converting speed and torque and the best efficiency. The transmission flow is normally between parallel shafts, using straight-cut or, most commonly, helical-cut spur gears. The questions of *centre distance*, *transmission mass* (largely determined by *face width*), *service life* and *noise* have already been examined in Section 2.4 "Fundamental Performance Features of Vehicle Transmissions". Formulae are given below relating to these points, including measures for reducing transmission noise.

## 7.1 Gearwheel Performance Limits

The starting point for gearwheel design calculations is their performance limits, i.e. causes of failure, as well as noise considerations and bearing forces. The performance limits of a gear pair are basically determined by four different types of damage:

- tooth failure,
- macropitting and micropitting,
- gear scuffing (hot scuffing) and
- wear (cold scuffing).

H. Naunheimer et al., *Automotive Transmissions*, 2nd ed.,
DOI 10.1007/978-3-642-16214-5_7, © Springer-Verlag Berlin Heidelberg 2011

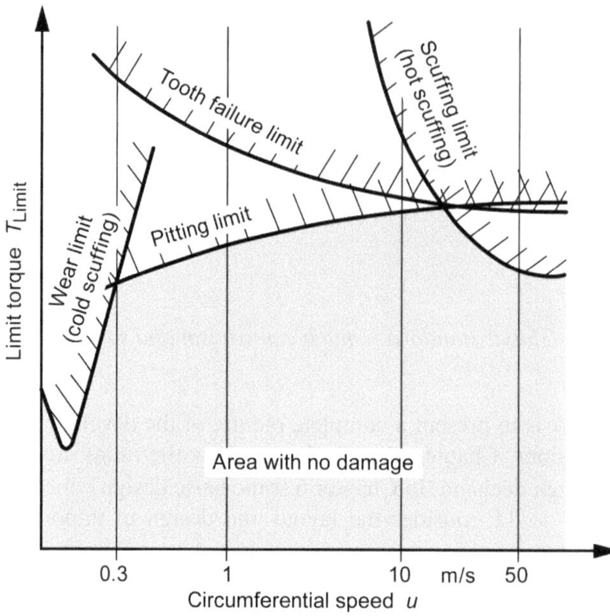


**Fig. 7.1.** Load capacity limits (limit torque) of case-hardened gearwheels and lubricants with additives

These damage types limit the load capacity of the gearwheels (Figure 7.1). The major factors affecting the performance limits indicated above are:

- operating conditions (type of load, tooth forces and additional forces, circumferential speed, temperature),
- selection of materials,
- gear geometry,
- manufacturing accuracy,
- surface treatment and surface roughness and
- selection of lubricant (chemical and physical characteristics).

**Fig. 7.2.** Brittle overload failure of a helical-cut spur gear [7.40]

**Fig. 7.3.** Vibration fatigue failure of a straight spur gear [7.40]

### 7.1.1 Causes and Types of Damage

#### *1/ Tooth Failure*

Tooth failure is where the whole tooth or part of a tooth breaks off. A distinction is made between overload failure and vibration fatigue failure (fatigue fracture). Overload failure is the result of a brief, drastic overload of the gear pair as shown in Figure 7.2.

The tooth of a gearwheel is normally subject to pulsating load. Intermediate gears are an exception, being exposed to alternating load. The maximum bending stress occurs at the tooth root. If the level of stress is frequently or occasionally in excess of the vibrational resistance of the gearwheel, this can lead to fatigue failure.

The vibrational resistance of the gearwheel is to a large extent determined by the tooth root design, surface roughness, surface strengthening in the tooth fillet, and heat treatment. Figure 7.3 shows vibration fatigue failure in a straight spur gear.

#### *2/ Macropitting*

Damage to the tooth flank by pitting is indicated by the appearance of pin holes and extended flank spalling, mostly below the pitch circle. It is a symptom of material fatigue at the tooth flanks. Depending on the assumption made, the causes can be surface cracks resulting from slip/roll stress or incipient cracks resulting from high shear stresses in the area below the tooth flank surface.

Hertzian stress is used to assess pitting load capacity, and is the basis for calculating surface stress. It is an important characteristic value for tooth flank meshing stress. But it is no more the sole cause of pitting than is the corresponding shear stress occurring below the surface [7.5].

**Fig. 7.4.** Macropitting with tip fracture due to flank surface fatigue [7.40]

Pitting only occurs in lubricated transmissions. Resistance to pitting is affected by hardness, oil viscosity, oil temperature, specific sliding, flank profile defects, surface roughness and circumferential speed. An area of spalling with pittings of different sizes is shown in Figure 7.4.

### *3/ Micropitting*

In the case of micropitting, visual inspection indicates grey zones on the worn tooth flank. Observation with a scanning electron microscope shows that these are fine surface pittings caused by fatigue (Figure 7.5). Micropittings are influenced by oil viscosity, oil additives and the surface structure of the flank. As a rule, it does not lead to gear failure; however continual material removal from the flank can lead to deterioration in transmission behaviour (noise).

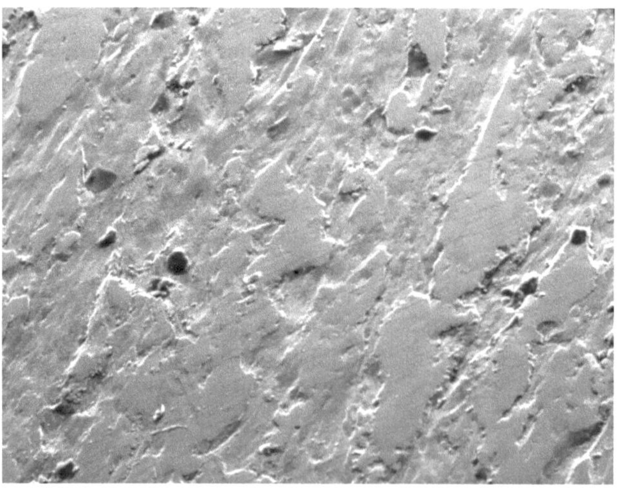

**Fig. 7.5.** SEM-image of a micropitting zone. Scale-like surface pittings [7.18]

#### 4/ Scuffing (Hot Scuffing)

Two different types of failure can occur when tooth flank lubrication fails, depending on the circumferential speed – wear (cold scuffing) and scuffing (hot scuffing). Wear occurs mostly at low circumferential speeds below 5 m/s and when unsuitable lubricants are applied. It is purely wear, and seldom occurs in vehicle transmissions.

Hot scuffing arises when the lubricant film breaks down because of high temperatures or excessive stress. This leads to metal-to-metal contact, local welding, and flaking of the tooth flanks. This gives rise to damage such as that shown in Figure 7.6. This is due to both physical and especially chemical processes. The physical phenomena are describable by elasto-hydrodynamic lubrication theory. The chemical processes occur in extremely thin layers and under high pressures, and are very complex [7.26]. A distinction must therefore be made between two different types of lubricant film: *elasto-hydrodynamic lubricant film* and *chemical protective film* resulting from the chemical reactions of rim zone material and the additives. On this subject see also Section 11.2 "Lubrication of Gearboxes, Gearbox Lubricants". We differentiate between two levels of scuffing:

- *Scoring:*
  Individual scoring or clusters of scoring appear in the sliding direction of the tooth flanks, varying from minor to serious. Typical of additive-treated oils and circumferential speeds < 30 m/s and
- *Scuffing:*
  This occurs as individual fine lines (scuffing lines), as clusters (heavy scuffing) or as areas across the full face width (scuffing zones). The main feature of the scored areas is a matt appearance. Typical for non-additive-treated and additive-treated oils at circumferential speeds > 30 m/s.

The scuffing process is critically affected by the gearwheels heating up, the critical temperature being the "tooth flank constant temperature" (tooth mass temperature), that is, the temperature which the tooth flank is constantly exposed to even when not engaged.

**Fig. 7.6.** Scuffing across the whole contact pattern of a straight-cut spur gear [7.40]

Gearwheels run more frequently in the mixed friction range, but the proportion of hydrodynamic lubrication along the contact path is high. With run-in gearwheels and circumferential speeds > 4 m/s, it is more than 60% even with high stress levels and often even 80–95% [7.25] (see also Figure 11.7).

The scuffing process is started by a breakdown of the protective chemical film on the tooth flank. The strength of this protective film depends on the temperature of the tooth flank. Stressing of the protective chemical film is determined by Hertzian stress. A thicker film of lubricant produced by higher lubricant viscosity can prevent scuffing. EP (Extreme Pressure) additives in oil to improve bearing capacity are of particular significance.

Gearwheels for vehicle transmissions are now almost without exception designed so that the "pitting" performance limit is critical. The design loads are now well established, so module and face width can be precisely selected so as to eliminate tooth failure, which is particularly serious since it causes immediate transmission failure. Gear scuffing is prevented by using suitably additive-treated gear oil.

Gearwheels are in principle case-hardened. Exceptions are ring gears and planetary gears in planetary gear units for passenger cars. Some of these are carbonitrided. For price reasons, the gearwheels are often shaved and possibly honed after hardening; ground gearwheels are used especially in low-noise gear units. Corrections in the form of profile bearing and transverse crowning, tip/root reliefs or flank line angle corrections are now state of the art. On the subject of manufacturing methods see also Section 16.3.

The outline calculations for tooth failure, pitting and scuffing to German standard DIN 3990 are set out below. A procedure for approximating centre distance and face width is then presented in Sections 7.2 and 7.3, based on the pitting load capacity calculation.

Conventionally, a component or assembly is calculated in three steps. The first step is to carry out an initial calculation determining all the principal dimensions. Correctly selecting the safety factor is of great significance here. When the dimensions have been determined, a more exact calculation for a particular type of load can be made. The final step is an operational fatigue strength calculation taking into account the actual load profile encountered.

Newer methods in automotive transmission manufacturing proceed as follows [7.2]: depending on the route, vehicle and driver, varying stresses result for each gear. For versatile operation, the transmission must be designed such that there is sufficient service life for all gears on all road types. That is, the transmissions respectively the individual gears must have tolerance for the respective unfavourable or statistically relevant stresses.

The load profile for transmission testing is also designed for these stresses. Generally, the testing time of a single-stage load profile test for case-hardened gearwheels is determined by the Wöhler curve for pitting.

Figure 7.7 provides an illustrative example of a particular design strategy: if the damage $D$ for a specific gear load profile is calculated, the acceptable boundary stress or compression stress can be computed with the corresponding Wöhler curve. For this, a suitable toothing must be designed.

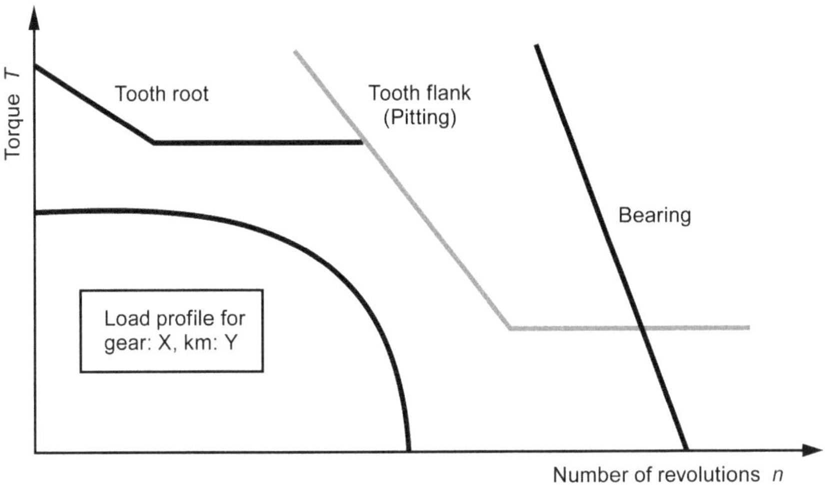


**Fig. 7.7.** Calculation of the damage $D$ using the load profile according to [7.2]

In this way, a toothing of optimal weight that is perfectly suitable for the load profile can be designed, within short development times and without iteration loops.

### 7.1.2 Calculating the Tooth Root Load Capacity

To calculate the tooth root load capacity, the maximum (local) stress of the root area must be checked. The tooth is most at risk when the normal force $F_n$ along the line of action (with its components $F_r$ and $F_t$) acts at the tooth tip or the tooth's outer contact point (Figure 7.8) (see also Figure 8.7 "Forces acting at the tooth flanks"). The tooth forces cause compression stress, bending stress and shear stress in the tooth. Investigations have shown that bending stress is generally the only stress that is critical for calculation purposes ($\sigma_v \approx \sigma_b$). The tooth cross-section to which the bending stress relates is the product of the face width $b$ and the root thickness chord $s_{Fn}$. The root thickness chord $s_{Fn}$ is determined by two tangents at the tooth root fillet at an angle of 30°. The effective root bending stress $\sigma_F$ is determined in German standard DIN 3990 [7.5] from a nominal value multiplied by various parameters

$$\sigma_F = \frac{F_t}{b\, m_n} Y_{Fa}\, Y_{Sa}\, Y_\varepsilon\, Y_\beta\, K_A\, K_V\, K_{F\beta}\, K_{F\alpha}\,, \tag{7.1}$$

where:

$F_t$ nominal circumferential force at the reference circle in N,
$b$ face width in mm,
$m_n$ standard module in mm,
$Y_{Fa}$ form factor to DIN 3990, Part 3, Page 13,

$Y_{Sa}$ stress correction value (notch stress concentration factor) DIN 3990, Part 3, Page 2,
$Y_\varepsilon$ contact ratio to DIN 3990, Part 3, Page 38,
$Y_\beta$ helical overlap to DIN 3990, Part 3, Page 39,
$K_A$ application factor,
$K_V$ dynamic load factor to DIN 3990, Part 1, Page 16–17,
$K_{F\beta}$ longitudinal load distribution factor to DIN 3990, Part 1, Page 19 and
$K_{F\alpha}$ transverse factor to DIN 3990, Part 1, Page 45.

The permissible tooth root strength $\sigma_{FG}$ is determined in accordance with German standard DIN 3990 as

$$\sigma_{FG} = \sigma_{F,lim}\, Y_{ST}\, Y_{NT}\, Y_{\delta,relT}\, Y_X\,, \tag{7.2}$$

where:

$\sigma_{F,lim}$ tooth root endurance strength value to DIN 3990, Part 5, Pages 4–10,
$Y_{ST}$ stress correction factor to DIN 3990, Part 3, Page 4,
$Y_{NT}$ service life factor to DIN 3990, Part 3, Page 40,
$Y_{\delta,relT}$ relative support figure ( = f($Y_{SA}$) ) to DIN 3990, Part 3, Page 44,
$Y_X$ tooth root size factor to DIN 3990, Part 3, Page 50.

The quotient of tooth root strength $\sigma_{FG}$ and existing root bending stress $\sigma_F$ forms the safety factor $S_F$

$$S_F = \frac{\sigma_{FG}}{\sigma_F}\,. \tag{7.3}$$

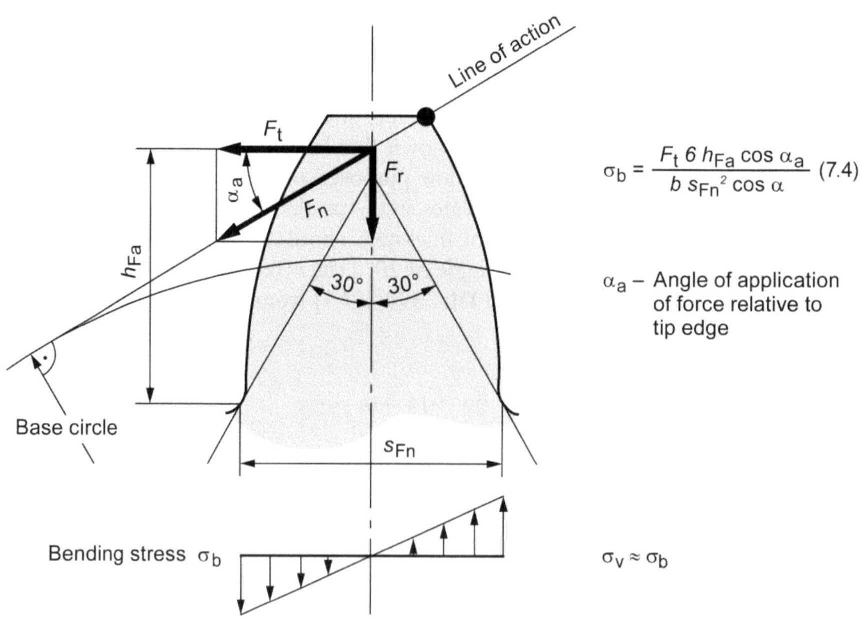


**Fig. 7.8.** Bending stress at the tooth root with force acting on the tip

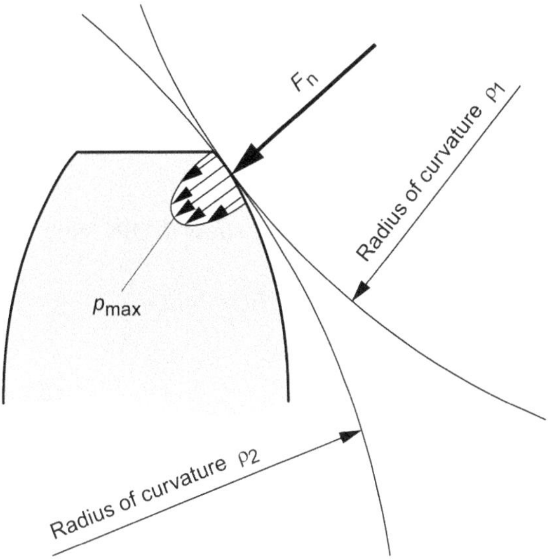


Hertzian stress:

$$p_{max} = \sqrt{\frac{F_n \, E}{5.72 \, \rho \, b}} \quad , (7.5)$$

where $E_1 = E_2 = E$ or

$$E = \frac{2 \, E_1 \, E_2}{E_1 + E_2} \quad \text{and}$$

substitute radius of curvature

$$\rho = \frac{\rho_1 \, \rho_2}{\rho_1 + \rho_2} \, .$$

**Fig. 7.9.** Stresses at the tooth flank

### 7.1.3 Calculating the Pitting Load Capacity

The bases for the calculation of pitting are the equations developed by Hertz to calculate the compression stress of two cylindrical rollers (Figure 7.9). If two rollers in contact along their common contour lines are subjected to the normal force $F_n$, they undergo flattening at the contact line. The distribution of contact pressure is unequal, peaking in the centre of the flattened surface. The Hertzian equation of roller pressing applies only for purely elastic deformation with rollers at rest. Equation 7.5 from Figure 7.9 gives only an approximation of the actual compression relations at the tooth.

Resistance to pitting is derived in accordance with German standard DIN 3990 [7.5] as the quotient of the tolerable surface stress $\sigma_{HG}$ and the existing Hertzian stress $\sigma_H$. Both values are in turn derived from a nominal value and the corresponding parameters

$$\sigma_H = Z_{B/D} \, Z_H \, Z_E \, Z_\varepsilon \, Z_\beta \sqrt{\frac{F_t \, (u+1)}{d_1 \, b u}} \sqrt{K_A \, K_V \, K_{H\beta} \, K_{H\alpha}} \, , \quad (7.6)$$

where the terms in Equation 7.6 have the following meanings:

- $Z_{B/D}$ single pinion contact factor $Z_B$, wheel contact factor $Z_D$ DIN 3390, Part 2, Page 8,
- $Z_H$ zone factor to DIN 3990, Part 2, Page 6,
- $Z_E$ elasticity factor to DIN 3990, Part 2, Page 8,
- $Z_\varepsilon$ contact ratio to DIN 3990, Part 2, Page 8,

$Z_\beta$ helical overlap to DIN 3990, Part 2, Page 10,
$F_t$ nominal circumferential force in N,
$b$ contact face width in mm,
$d_1$ reference circle diameter of the pinion (small gearwheel) in mm,
$u$ gear ratio $z_2/z_1$; $| z_2/z_1 | \geq 1$,
$K_A$ application factor to DIN 3990, Part 1, Page 55,
$K_V$ dynamic factor to DIN 3990, Part 1, Pages 16–17,
$K_{H\beta}$ longitudinal load distribution factor for surface stress to DIN 3990, Part 1, Page 19,
$K_{H\alpha}$ transverse factor to DIN 3990, Part 1, Page 45.

The pitting boundary strength $\sigma_{HG}$ is derived from

$$\sigma_{HG} = \sigma_{H,lim} \, Z_{NT} \, Z_L \, Z_R \, Z_V \, Z_W \, Z_X , \tag{7.7}$$

where the expressions in Equation 7.7 have the following meanings:
$\sigma_{H,lim}$ endurance strength value to DIN 3990, Part 5, Pages 4–9,
$Z_{NT}$ service life factor to DIN 3990, Part 2, Pages 11–12,
$Z_L$ lubricant factor to DIN 3990, Part 2, Page 13,
$Z_R$ roughness factor to DIN 3990, Part 2, Page 15,
$Z_V$ velocity factor to DIN 3990, Part 2, Page 14,
$Z_W$ material mating factor to DIN 3990, Part 2, Page 16,
$Z_X$ size factor for surface stress to DIN 3990, Part 2, Page 1.

The numerical safety factor for surface stress (against pitting) is determined accordingly from Equations 7.6 and 7.7 as

$$S_H = \frac{\sigma_{HG}}{\sigma_H} . \tag{7.8}$$

### 7.1.4 Calculating the Scuffing Load Capacity

Based on the hypothesis that the lubricant film is broken down by high surface temperatures caused by high stresses and high sliding speeds, two methods of calculation are proposed in German standard DIN 3990:

- *The Flash Temperature Method:*
  This describes varying local contact temperature along the contact path.
- *The Integral Temperature Method:*
  This gives a weighted average of surface temperature along the contact path.

The calculation is specified in German standard DIN 3990 (with reference to special considerations for vehicle transmissions see [7.19]). The permissible boundary temperature for the lubricant used is determined in a standardised scuffing test following DIN ISO 14635 [7.19]. Scuffing safety is defined as temperature safety and is calculated from the ratio of permissible integral or flash temperature to the occurring temperature.

## 7.2 Estimating Centre Distance

Besides the face width, the centre distance is the crucial parameter in automotive transmissions. It is important to obtain an approximation of this value at the start when designing gearwheel transmissions.

Pitting is the critical performance constraint. In order to derive an equation for calculating the centre distance $a$, it is therefore necessary to start with Hertzian stress at the pitch circle $\sigma_H$ (Equation 7.6)

$$\sigma_{H} = Z_{B/D}\, \sigma_{H0} \sqrt{K_{A}\, K_{V}\, K_{H\beta}\, K_{H\alpha}}\ . \tag{7.9}$$

With the base surface stress $\sigma_{H0}$

$$\sigma_{H0} = Z_{H}\, Z_{E}\, Z_{\varepsilon}\, Z_{\beta} \sqrt{\frac{F_{t}\,(u+1)}{d_{1}\, b\, u}}\ , \tag{7.10}$$

the torque to be transmitted at the pinion shaft $T_1$

$$T_{1} = \frac{F_{t}\, d_{1}}{2} \tag{7.11}$$

and the face width diameter relationship $b/d_1$ the following results

$$\sigma_{H} = Z_{B/D}\, Z_{H}\, Z_{E}\, Z_{\varepsilon}\, Z_{\beta} \sqrt{\frac{2\,T_{1}\,(u+1)}{d_{1}^{3}\, \frac{b}{d_{1}}\, u}} \sqrt{K_{A}\, K_{V}\, K_{H\beta}\, K_{H\alpha}}\ . \tag{7.12}$$

The diameter $d_1$ in Equation 7.12 is replaced by

$$d_{1} = \frac{2\,a}{1+u}\ , \tag{7.13}$$

the surface stress $\sigma_H$ is replaced by the permissible compression stress $\sigma_{H,perm}$

$$\sigma_{H,perm} = \frac{\sigma_{H,lim}\, Z_{NT}\, Z_{L}\, Z_{R}\, Z_{V}\, Z_{W}\, Z_{X}}{S_{H}}\ . \tag{7.14}$$

Now this can be solved for the centre distance $a$, giving

$$a = \sqrt[3]{\frac{T_{1}\,(u+1)^{4}}{4\frac{b}{d_{1}}u}}\ \sqrt[3]{\frac{\left(Z_{B/D}\, Z_{H}\, Z_{E}\, Z_{\varepsilon}\, Z_{\beta}\, S_{H}\right)^{2}}{\left(\sigma_{H,lim}\, Z_{NT}\, Z_{L}\, Z_{R}\, Z_{V}\, Z_{W}\, Z_{X}\right)^{2}}}\ \sqrt[3]{K_{A}\, K_{V}\, K_{H\beta}\, K_{H\alpha}}\ . \tag{7.15}$$

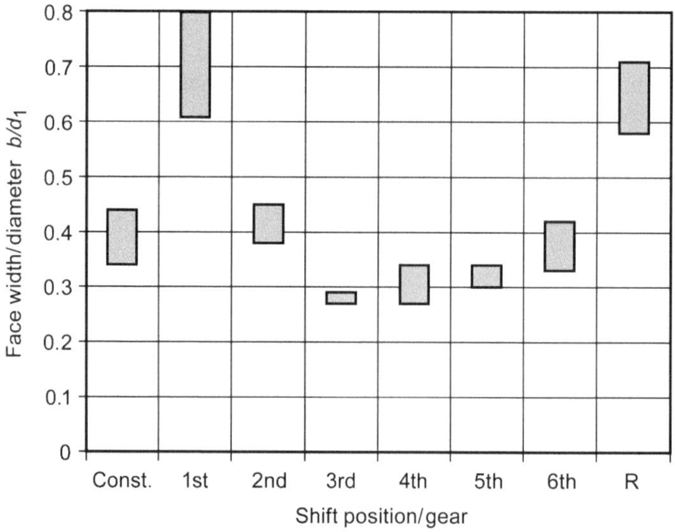


**Fig. 7.10.** Face width diameter relationship $b/d_1$ of existing 5- and 6-speed passenger car and commercial vehicle selector gearboxes

The estimate is made for the *gear with the highest torque multiplication*. For the face width diameter ratio $b/d_1$ an established practical value is to be used. In order to minimise uneven face wear, a different ratio is selected for each gear. An evaluation of existing vehicle transmissions in respect of the ratio $b/d_1$ is shown in Figure 7.10.

For the individual factors in Equation 7.15, the following values should be used to arrive at an estimate:

$b/d_1 = 0.65$,
$K_A = 0.65$ for passenger cars to DIN 3990, Part 41, Page 28,
$K_A = 0.85$ for trucks to DIN 3990, Part 41, Page 28,
$K_V, K_{H\alpha}, K_{H\beta} = 1$,
$Z_H = 2.25$ ( for $\alpha_n \approx 20°, \beta \approx 15°, (x_1+x_2)/(z_1+z_2) \approx 0.015$ ),
$Z_{B/D} = 1$,
$Z_E = \sqrt{0.175\,E} = 189.8\sqrt{\text{N/mm}^2}$ for steel/steel,
$Z_\varepsilon = 0.95$,
$Z_\beta = 0.95$,
$Z_{NT}, Z_L, Z_R, Z_V, Z_W, Z_X = 1$,
$\sigma_{H,\lim} = 1800$ N/mm$^2$ DIN 3990, Part 41, Page 29 (material 16 MnCr5),
$S_H = 1.2$.

Calculating the centre distance with *first gear* engaged, this gives the following *approximation equation* (centre distance $a$ in mm; torque $T_1$ in Nmm)

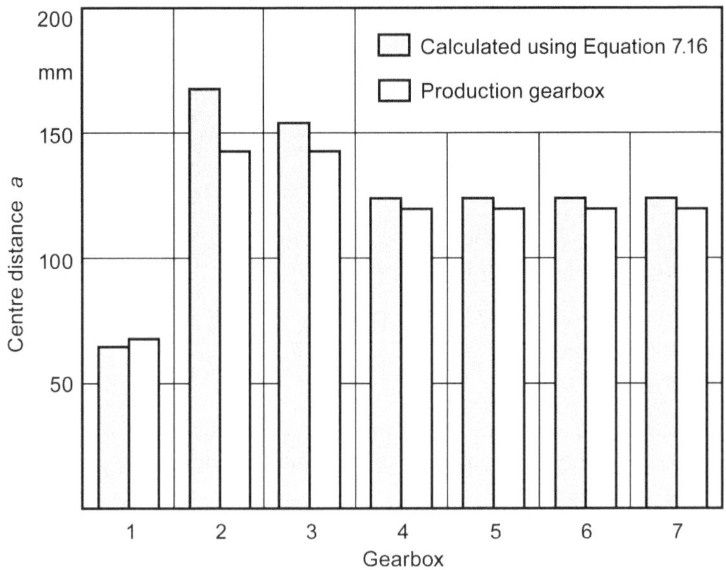


**Fig. 7.11.** Comparison of theoretical and actual gearbox centre distances

$$a = K_a \sqrt[3]{\frac{T_1 (u+1)^4}{u}} \ . \tag{7.16}$$

$K_a$ = 0.255 for passenger cars,
$K_a$ = 0.278 for trucks,
$T_1$ torque at the shaft on which the pinion of the first gear is mounted; e.g. for a two-stage countershaft transmission $T_1 = i_{CG}\, T_G$, where $i_{CG}$ is a constant ratio; $T_G$ transmission input torque,
$u$ gear ratio of the first gear pair, $|u| \geq 1$.

Figure 7.11 shows a comparison of the theoretical centre distance calculated using Equation 7.16, with the centre distance of actual transmissions.

## 7.3 Estimating Face Widths

Having estimated the centre distance it is possible to derive the pinion diameter $d_1$ from Equation 7.13, and given the transmission ratio, the wheel diameter $d_2$. The face width $b_{1,1}$ of the pinion *of first gear* is calculated from the preselected face width/diameter ratio of 0.65 as

$$b_{1,1} = 0.65\, d_{1,1\text{st}} \ . \tag{7.17}$$

**Table 7.1.** Endurance or fatigue strength $\sigma_{\mathrm{H,lim}}$, face width constant $K_{\mathrm{b}}$

| Gear | 1st gear | 2nd gear | | Other gears/constant | |
|---|---|---|---|---|---|
| $\sigma_{\mathrm{H,lim}}$ (N/mm$^2$) | 1800 | 1600 | | 1500 | |
| Face width constant $K_{\mathrm{b}}$ | – | Passenger car | Truck | Passenger car | Truck |
| | | 0.108622 | 0.142044 | 0.123588 | 0.161614 |

For the remaining gears $n = 2, ..., z$ the required face width $b_{1,\mathrm{n}}$ of the pinion is derived by equating Equation 7.12 with Equation 7.14

$$b_{1,\mathrm{n}} = \left(Z_{\mathrm{B/D}}\, Z_{\mathrm{H}}\, Z_{\mathrm{E}}\, Z_{\varepsilon}\, Z_{\beta}\right)^2 \frac{2\,T_1\,(u_{\mathrm{n}}+1)\,S_{\mathrm{H}}^2\,K_{\mathrm{A}}\,K_{\mathrm{V}}\,K_{\mathrm{H\beta}}\,K_{\mathrm{H\alpha}}}{d_{1,\mathrm{n}}^2\,u_{\mathrm{n}}\,\sigma_{\mathrm{H,lim}}^2\left(Z_{\mathrm{NT}}\,Z_{\mathrm{L}}\,Z_{\mathrm{R}}\,Z_{\mathrm{V}}\,Z_{\mathrm{W}}\,Z_{\mathrm{X}}\right)^2}\,. \tag{7.18}$$

With the factors determined as shown in Section 7.2 for the estimate, it is possible to arrive at the following simplified equation ($T_1$ in Nmm; $d_{1,\mathrm{n}}$ in mm; $b_{1,\mathrm{n}}$ in mm)

$$b_{1,\mathrm{n}} = 427\,800\,K_{\mathrm{A}}\,\frac{T_1\,(u_{\mathrm{n}}+1)}{d_{1,\mathrm{n}}^2\,u_{\mathrm{n}}\,\sigma_{\mathrm{H,lim}}^2}\,, \tag{7.19}$$

assuming the endurance strength value $\sigma_{\mathrm{H,lim}}$ as shown in Table 7.1, and using $K_{\mathrm{A}} = 0.65$ (passenger car) or $K_{\mathrm{A}} = 0.85$ (truck) as the application factor.

The pinion face widths $b_{1,\mathrm{n}}$ for the *second and subsequent gears* can be calculated from Equation 7.19 and the values in Table 7.1 using the following *unit equation* ($T_1$ in Nmm; $d_{1,\mathrm{n}}$ in mm; $b_{1,\mathrm{n}}$ in mm)

$$b_{1,\mathrm{n}} = K_{\mathrm{b}}\,\frac{T_1\,(u_{\mathrm{n}}+1)}{d_{1,\mathrm{n}}^2\,u_{\mathrm{n}}}\,. \tag{7.20}$$

$K_{\mathrm{b}}$ from Table 7.1,
$T_1$ torque at the shaft on which the pinion is mounted;
e.g. in the case of a two-stage coaxial countershaft transmission:
case 1: constant: $T_1 = T_{\mathrm{G}}$;
case 2: pinion on countershaft: $T_1 = i_{\mathrm{CG}}\,T_{\mathrm{G}}$;
case 3: pinion on main shaft ($i_{\mathrm{gear}} < 1$): $T_1 = i_{\mathrm{CG}}\,i_{\mathrm{P}}\,T_{\mathrm{G}}$;
$i_{\mathrm{CG}}$ constant gear; $i_{\mathrm{P}}$ gear pair; $T_{\mathrm{G}}$ transmission input torque,
$u_{\mathrm{n}}$ gear ratio of the gear pair of the $n$-th gear, $|u| \geq 1$.

## 7.4 Operational Fatigue Strength and Service Life

The goal of an automotive transmission design that has operational fatigue strength is reliably to calculate that fatigue strength for a certain period of use in accordance with the expected load. Economy throughout the entire period of use must also be considered as a boundary condition. This requires utilisation of the material that is as strong as possible with respect to fatigue, as well as low manufacturing and operational costs.

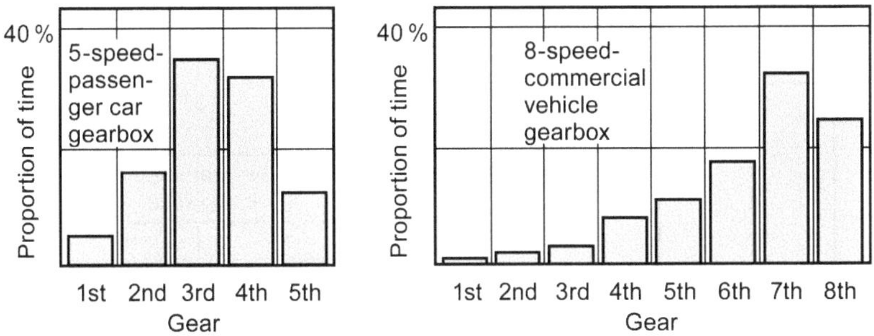


**Fig. 7.12.** Proportion of time the various gears of a 5-speed passenger car and an 8-speed commercial vehicle transmission are engaged when travelling on a mountainous rural road (example)

The service life of a vehicle transmission depends on the service life of the individual components and their collaboration. With transmissions with various gear ratios, where the respective gears are engaged in the power flow for varying periods of time (Figure 7.12), individual useful/service life calculations must be made proportionally. The load and hence the stress on and service life of all transmission components depends essentially on the driver, the vehicle and the road. Considered over time, we see individually variable load curves. These lead ultimately to wear and fatigue failures at different times. Predicting the corresponding service life is the subject-matter of operational fatigue strength calculation [7.4].

If we want to estimate quantitatively the service life of a component, we must compare stress with its resistance to stress. This can be done in two ways: mathematically by using so-called damage accumulation hypotheses or experimentally.

Computational service life estimation can only be executed for so-called A-components as they are defined in Section 17.2.2 “Qualitative Reliability Analysis”. Figure 7.13 shows the basic method. The input values in the calculation are the load in the form of load profiles and the stress resistance in the form of Wöhler curves. Since the Wöhler curves usually are determined for a constant ratio of maximum to minimum stress, i.e. a fixed stress ratio $R$, for stress curves with variable stress ratios $R$ these stresses have to be transformed to oscillations with the same damage but with the stress ratio $R$ of the Wöhler curve [7.15, 7.24]. Damage theory assumes that, the increase in load cycles is associated with progressive damage to the component. In damage accumulation, every load cycle is assigned a defined consumption of service life. If the service life supply of the component is used up, it fails, i.e. its service life has ended. Because of statistical spread of the input data, deviations caused both by the transference of Wöhler curves to components of different geometries under different load conditions as well as by unsharpness associated with damage accumulation, we designate the process of computational service life prediction and the result as service life estimation.

Experimental evidence of the operational fatigue strength of components and vehicle transmissions comes from test machines and transmission test benches in the laboratory, presented in Section 17.3 “Testing to Ensure Reliability”.

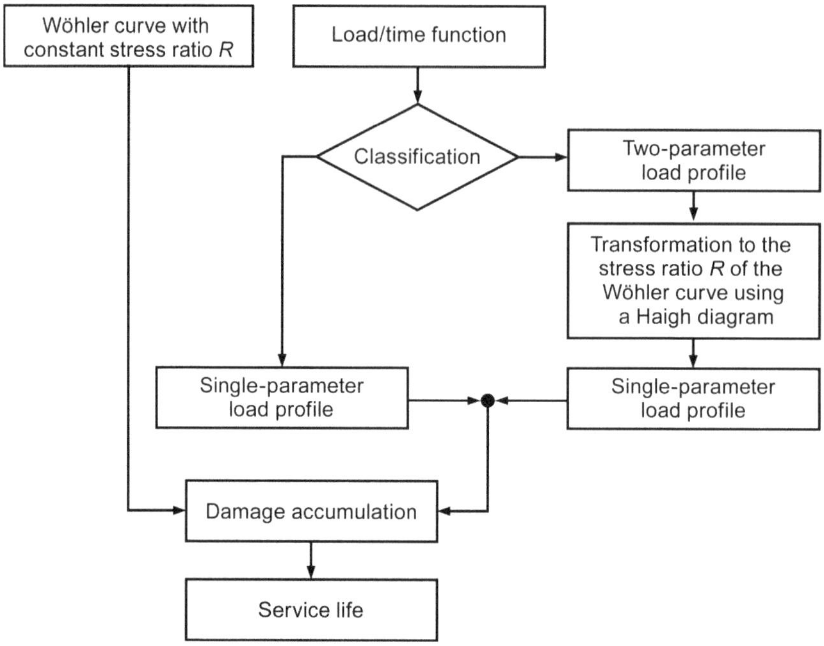


**Fig. 7.13.** Components required to calculate operational fatigue strength

The most expensive, yet closest to reality are road tests on test roads. Figure 7.14 shows a service life determination from a test bench parallel to a mathematical calculation. The starting information is the load/time function, which is classified or utilised directly to control the test bench. Frequently, the test program is obtained by retransformation of component load from the frequency range into the time range. By accelerated lifetime tests, the acceptable number of vibration cycles and as a result the test times can be reduced. The original damage effect must be preserved and the multi-stage routine sufficiently mixed in the test sequences.

The acceptable number of stress oscillation cycles until failure can be directly compared with the computational values. The results gained from the test bench are also the basis for the relative service life calculation. If load profiles are used which correspond in their distribution to real loads but have different stress levels, the computational and experimental result is the service life curve [7.15].

### 7.4.1 The Wöhler Curve

The load capacity of a component is contingent on the material and its condition, the shape and surface of the component as well as on environmental conditions. It is determined by boundary conditions specified in DIN 50100 [7.7] in a single-stage test; the result is represented in the form of a Wöhler curve. In the single-stage test, the stress ratio $R$ is held constant for all stress levels.

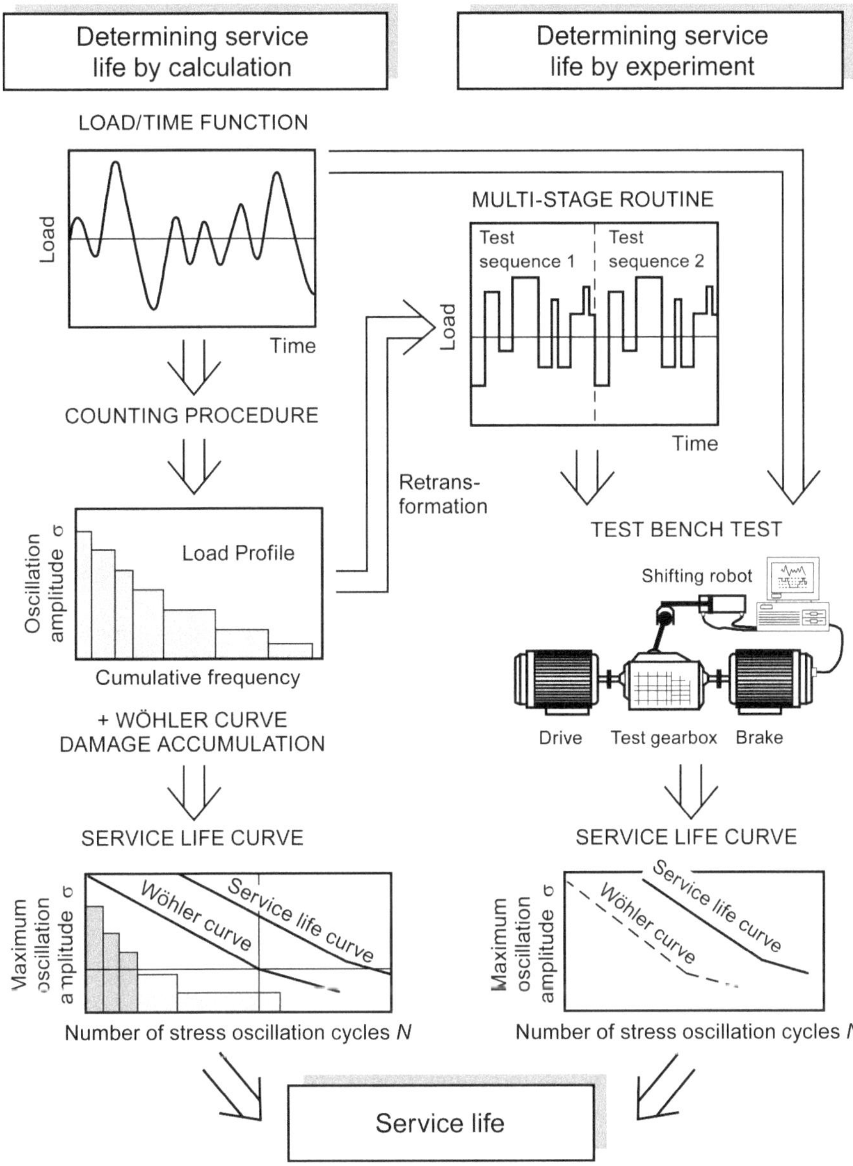


**Fig. 7.14.** Various ways of determining component service life. In contrast to the Wöhler curve, which is determined by the single-stage test, the service life curve is based on a load profile [7.15].

The simplest test parts are bar-shaped. The influence of notches is investigated on notched test bars. In the case of gearwheels, experiments are made with standardised test gearwheels.

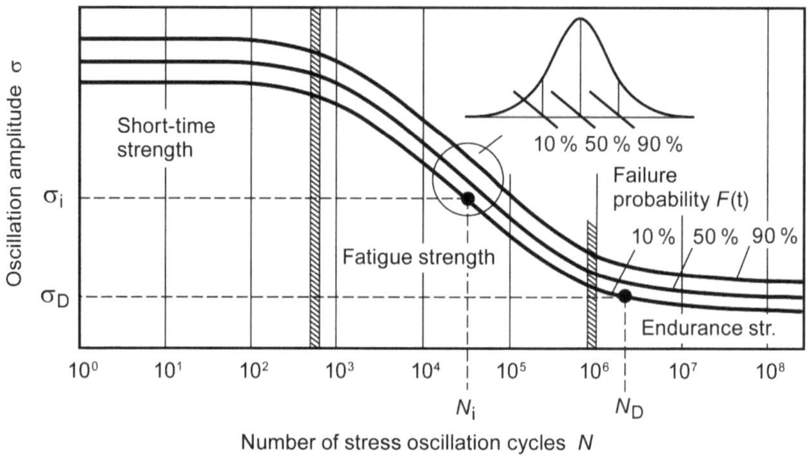


**Fig. 7.15.** Wöhler curve of permissible material stress

The Wöhler curve describes the acceptable number of vibration cycles $N_i$ up to the point of test part failure for different stress amplitudes. The typical curve shape is shown in Figure 7.15.

The resulting numbers of oscillation cycles to failure are random variables, i.e. they are spread around an average value. The most common are Wöhler curves for 10% failure probability. However, 1% and 50% curves are also common. The Wöhler curve for 10% failure probability allows us to estimate the $B_{10}$ service life of a component. This is the service life at which on average 10% of the components have already failed.

The fatigue zone of the Wöhler curve can be described as a straight line in the log co-ordinate system with the following equation

$$N_i = N_D \left( \frac{\sigma_i}{\sigma_D} \right)^{-k} . \tag{7.21}$$

The exponent $k$ determines the gradient for the fatigue strength zone. The exponent $k$ for gearwheels depends on the failure under consideration and the surface hardening process, and has values ranging from $k = 4$ to $k = 16$ (Figure 7.16). As the exponent rises, the characteristic graph becomes flatter. Even small differences in stress have a great influence on service life, which leads to a wide spread of component service life in practical use.

### 7.4.2 Load Profile and Counting Procedure

The starting values for determining the load profile is the load, which attacks a component as force or torque in one or more directions. To determine operational loadings, a load/time or a load/distance chart is needed.

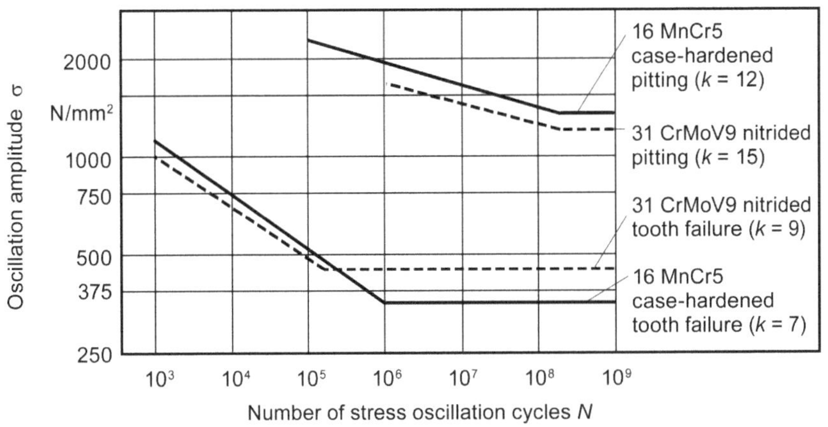


**Fig. 7.16.** Wöhler curves (10% probability of failure) for various materials and failures (examples)

The loadings can be determined by

1/ road tests on defined routes and
2/ numerical driving simulation.

By means of various classification methods (which could also be called counting procedures), the load curve is converted to a frequency distribution of load cycles. In all methods, the load amplitude range is divided into classes and the occurring loads assigned to a specific class according to the counting procedure employed. Information such as sequence and frequency of the load/time function are lost in the process. However, such information has no formal importance since such quantities have no influence on the subsequent service life calculation using damage accumulation.

In practice, single-parameter and two-parameter classification methods are used [7.39]. Single-parameter methods are applicable when the load can be sufficiently described by one parameter, e.g. in the case of purely alternating or pulsating load. Figure 7.17 shows the level crossing counting described in DIN 45667 [7.6] as an example of a single-parameter counting procedure. Here, crossings of the class limits in the positive or negative direction are counted and added. The level crossing counting provides an overview of the extreme values of the load/time function, but not of their averages and amplitudes. Further single-parameter procedures are the range counting and range pair counting methods as well as time at level and level distribution counting. The last two procedures are especially suitable for gearwheel and bearing service life calculation.

In the time at level counting method, the number of associated rotations is counted in the individual torque classes. If the torques and speeds at the transmission input or output are determined, the values can be converted to the individual gearwheels by means of the ratios. The dynamic behaviour of the transmissions is generally not taken into consideration in this context. A pulsating load thereby results for every tooth at every revolution. Intermediate gears are loaded alternately.

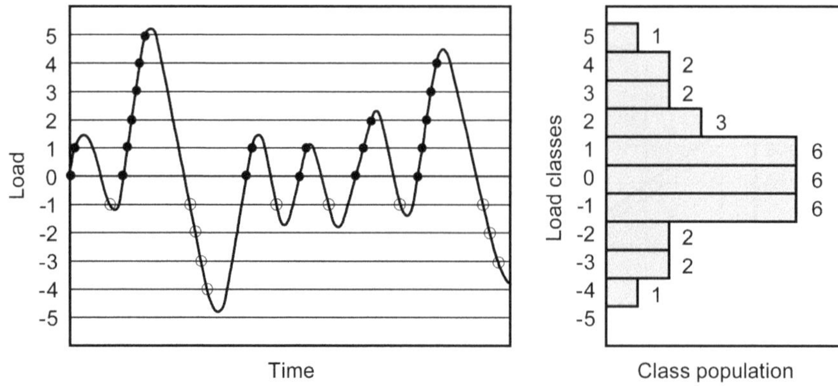


**Fig. 7.17.** Example of single-parameter counting procedure: level crossing counting according to German standard DIN 45667 [7.6]

In the case of level distribution counting, the signal value is scanned and counted at equal intervals. The frequency of counts per class is a measure for the resting time in this class. For small intervals, the result of the count practically corresponds to that of time at level counting.

Two-parameter procedures consider two quantities in load profile formation, e.g. the average value and the amplitude or the minimum and maximum value in the load/time curve [7.3]. The result is accordingly put into a matrix. Two-parameter load profiles are required, for example, for service life calculations of gearbox shafts. Procedures that are of interest in praxis are from-to counting and especially the Rainflow method. In from-to counting, the beginning and end of each load cycle flank is stored in the transition or Markov matrix, which gives an overview of the characteristics of the load/time function. In the preferred Rainflow counting method, closed hysteresis loops are formed from the load cycles, providing a good representation of the damage process. Open hysteresis loops are stored as residuum. The hysteresis loops are recorded in a matrix, for which there are various forms of represenation. In the Rainflow procedure, even small oscillations are registered. They can also be simply filtered out.

Simple algorithms allow for online classification of fast load cycles. The results of most single-parameter counting procedures can be derived from the easily manageable Rainflow matrix. This is necessary when, as is common, only Wöhler curves with a constant stress ratio $R$ exist (Figure 7.13). Conversion to a single-parameter profile with the stress ratio of the Wöhler curve takes place on the basis of the Haigh's endurance strength diagram [7.15, 7.24]. The goal is to obtain equal-damage oscillation cycles.

Load profiles allow us to ascertain loads in the form of varying stresses at different component locations. To determine the stresses, the geometry and deformation behaviour of the component, among other things, must be known. For machine components such as gearwheels, the occurring stresses are often ascertained by means of so-called nominal stresses and additional corrective factors – components of more complex design require FE methods. FEM programs supply (as do measurements with strain gauges) direct local stresses of the component. These stresses are used with the Wöhler curve to calculate service life by means of damage accumulation.

In order to design powertrain components with operational fatigue strength, representative load profiles are required [7.28] that have been obtained from load/time functions reflecting real driving operation. They can be obtained from road tests and numerical driving simulation.

#### *1/ Road Tests*

Road tests should be representative for later customer use of the vehicle. The test routes are characterised by their height profiles, their gradient and road speed distributions. The level of load involved in road tests provides as a rule an "accelerated testing effect". Customer surveys are also used to decide the types of road and loadings used in these tests. The influencing variables impinging on a transmission load profile can be divided into three groups [7.28]:

*1. Vehicle:* Engine performance map, loading, transmission ratios,
*2. Driver:* Driving style, characterised by frequency of shifting, gearshift engine speed and accelerating and braking behaviour,
*3. Road type:* Distribution of total mileage over motorways, main roads, rural roads, urban and local traffic, mountain roads, typical gradients and maximum speeds (see also Tables 2.9 and 2.10).

The costs of directly measuring a component can be reduced by determining the load at one location in the power flow of the powertrain. Thus, the vehicle's torque and speed are often measured at the transmission input as well as recording the gear in use. These values are then transferred by calculation to the components of the powertrain.

#### *2/ Determining Load Profiles by Computational Driving Simulation*

Since transmission stress depends on many different stochastic variables, substantial resources are required for statistically validated road tests. Road tests have to be repeated whenever design changes are made to the powertrain. One way of limiting the number of costly road tests is computational driving simulation (see also Chapter 14 "Computer-Aided Transmission Development").

Standardised load profiles such as CARLOS-PTM/PTA (Car Loading Standard for Powertrain with Manual/Automatic Transmission, see Table 2.10) or tests specific to the individual manufacturers offer other test alternatives [7.21].

### 7.4.3 Damage Accumulation Hypothesis

Vibrational loads damage the material as soon as that stress exceeds a certain limit. The damage accumulation hypothesis describes the effect of damage on a component proceeding from the load capacity described by the Wöhler curve and the stress based on the load profile.

Miner's [7.27] linear damage accumulation hypothesis assumes that, in the case of vibrational stress, the ratio of actual work absorbed $w$ to the maximum possible work absorbed $W$ until failure is a measure for the degree of the damage thus far incurred. This ratio is equal to the ratio of the number of load cycles $n$ to the ultimate number of cycles $N$ in the single-stage test:

$$\frac{w}{W} = \frac{n}{N} . \tag{7.22}$$

Every load cycle on the same load level thus causes a constant portion-damage, i.e. the damage sum increases linearly with the number of load cycles. According to the hypothesis, the same work $W$ until failure must be exerted above the endurance strength limit at all load levels. Then the portions of work or damage portions of different load levels can be accumulated.

Practically speaking, this means that the individual frequencies $n_i$ of the oscillation amplitudes $\sigma_i$ (divided into $j$ classes with an classification method) are correlated with the maximum number of oscillation cycles permissible at this amplitude $N_i$ (Figure 7.18). The resultant damage portions are summarised as damage sum.

$$\frac{w_1}{W_1} + \frac{w_2}{W_2} + \ldots + \frac{w_j}{W_j} = \frac{n_1}{N_1} + \frac{n_2}{N_2} + \ldots + \frac{n_j}{N_j} = \sum_{i=1}^{j} \frac{n_i}{N_i} . \tag{7.23}$$

In accordance with the Palmgren-Miner [7.27, 7.31] service life calculation, the damage sum $D_{prof}$ of a load profile generated from accumulation of the damage portions is obtained with

$$D_{prof} = \sum_{i=1}^{j} \frac{n_i}{N_i} . \tag{7.24}$$

According to Palmgren-Miner, a component will fail if the damage sum becomes $D_{prof} = D_{th} = 1$. This allows us to compute service life with Equation 7.24. In order to obtain an accurate estimate, knowledge of the actual damage sum $D_{act}$ before component failure is required. In 90% of all cases, the actual damage sum is under $D_{th} = 1$ (up to $D_{act} = 0.1$). Service life is overestimated if there are large fluctuations in mean load. Service life is underestimated if there is compressive stress due to mean load or residual compressive stress in the component. Accurate results can be obtained with the help of the relative service life calculation.

In the case of repeating load cycles with cycle number $h_i$, the damage sum is obtained by multiplying $h_i$ with the number of passes $z$. This corresponds, for example, to the $z$-th time passing through the test route used to determine the load profile.

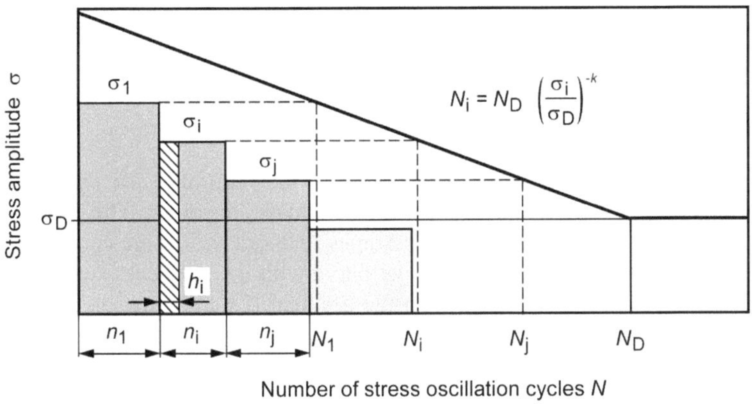


**Fig. 7.18.** The Palmgren-Miner damage accumulation hypothesis

Here,

$$D_{\text{prof}} = z \sum_{i=1}^{j} \frac{h_i}{N_i}. \tag{7.25}$$

The permissible number of load cycles $N_i$ assigned to the individual frequencies $n_i$ is determined by means of the Wöhler curve equation (Equation 7.21). Inserting Equation 7.21 into Equation 7.25, the following results

$$D_{\text{prof}} = \sum_{i=1}^{j} \frac{z\, h_i}{N_D \left( \dfrac{\sigma_i}{\sigma_D} \right)^{-k}}. \tag{7.26}$$

Solving this equation for $z$ with the actual damage sum $D_{\text{act}}$ we obtain

$$z = \frac{D_{\text{act}}}{\sum\limits_{i=1}^{j} \dfrac{h_i}{N_D} \left( \dfrac{\sigma_i}{\sigma_D} \right)^{+k}}, \tag{7.27}$$

the number $z$ of possible load cycle passes that the component can withstand until failure based on the load profile. In addition, observe the subsequent calculation example [7.17]. Inserting Equation 7.27 into Equation 7.25, we finally obtain the component service life $N_{\text{th}}$ in load cycles for $D_{\text{act}} = D_{\text{prof}} = D_{\text{th}} = 1$

$$N_{\text{th}} = N_D \frac{\sum\limits_{i=1}^{j} h_i}{\sum\limits_{i=1}^{j} h_i \left( \dfrac{\sigma_i}{\sigma_D} \right)^{+k}}. \tag{7.28}$$

It is therefore sufficient in Equation 7.28 if the load cycle numbers of a load cycle are inserted. For $z = 1$, i.e. if no repeating load cycles are used, $n_i = h_i$.

### *The Miner-Haibach Damage Accumulation Hypothesis*

It has been shown in practice that even stresses below the endurance strength limit must not be ignored where there are previous damage portions. The Miner-Haibach damage accumulation hypothesis assumes that stresses in the endurance strength range also cause damage. The Wöhler curve with the gradient exponent $k$ in the fatigue strength range continues with the gradient ($2k$-1) below the endurance strength limit (Figure 7.19). The continuation results approx. as angle bisection line between the extended fatigue strength and the endurance strength line.

For stress amplitudes above the endurance strength limit $\sigma_i \geq \sigma_D$, Equation 7.21 applies where $i = 1, ..., j$; below the endurance limit $\sigma_i < \sigma_D$ the following applies:

$$N_i = N_D \left(\frac{\sigma_i}{\sigma_D}\right)^{-(2k-1)} \quad \text{for } i = j+1, \ldots, j+n. \tag{7.29}$$

The Miner-Haibach service life calculation is carried out as above, but taking into account the stresses below the endurance strength limit

$$N_{th} = N_D \frac{\sum_{i=1}^{j+n} h_i}{\sum_{i=1}^{j} h_i \left(\frac{\sigma_i}{\sigma_D}\right)^{+k} + \sum_{i=j+1}^{j+n} h_i \left(\frac{\sigma_i}{\sigma_D}\right)^{+(2k-1)}}. \tag{7.30}$$

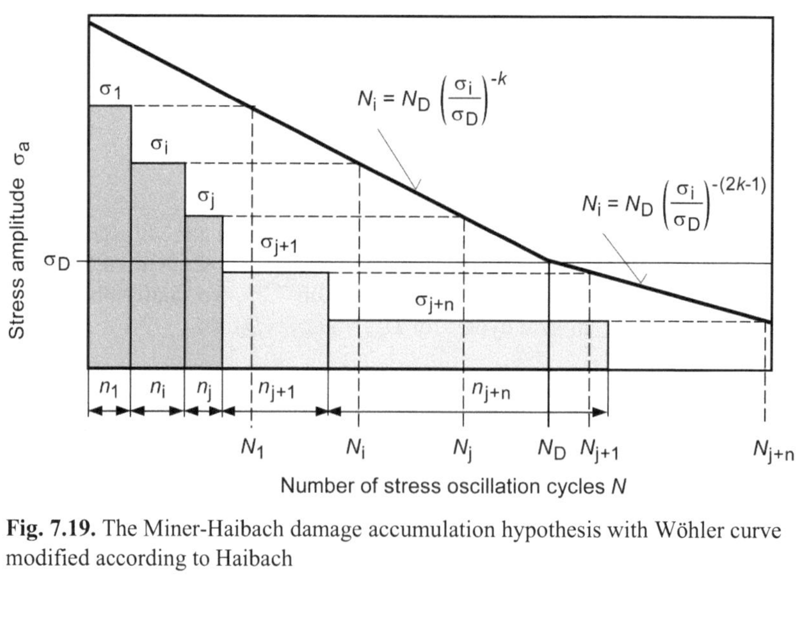


**Fig. 7.19.** The Miner-Haibach damage accumulation hypothesis with Wöhler curve modified according to Haibach

### *Relative Service Life Prediction*

In practice, the prediction of service life by calculation is frequently unreliable. This is partly due to a marked spread of the damage sum on failure. Prediction quality can normally be made more accurate by complementing purely computational methods with experimental analyses.

When designing components in automotive and aeronautical engineering the *Relative Palmgren-Miner Rule* is frequently used. Here, the load cycle $N_{exper}$ at the end of the service life is determined by means of practical experiments staying close to expected conditions. The actual damage sum $D_{act}$ is determined from $N_{exper}$, as obtained experimentally, and the calculated values ($N_{calc} = N_{th}$) as follows

$$D_{act} = \frac{N_{exper}}{N_{calc}\left(D_{th} = 1.0\right)}. \tag{7.31}$$

From this we obtain the component service life in load cycles with the Relative Palmgren-Miner Rule

$$N_{act} = D_{act}\, N_{calc}\left(D_{th} = 1.0\right). \tag{7.32}$$

In this way, experiences and analogies of the previous model or of transmissions with similar operational conditions can be exploited. False conjectures with respect to the Wöhler curves effect the result much less than when service life is determined with pure computation. However, uncertainties concerning how much the calculated load profile corresponds to the actual test load profile still remain.

Whereas the relative Palmgren-Miner rule makes use of operational fatigue strength tests, the *Modified Relative Palmgren-Miner Rule* makes use of operational experience [7.4]. This involves determining the actual service life achieved in operation. The ratio of actual to computationally predicted service life then represents the actual damage sum $D_{act}$, by which the service life predicted by calculation with Equation 7.32 for a similar component (a new design, for example) must be corrected.

### *Example of Service Life Calculation for a Gear Pair in a Vehicle Transmission*

The record of the transmission input torque $T_G = f(t)$ is available for a 6-speed commercial vehicle transmission (Figure 7.20), from which the number of revolutions $U_G$ in the various classes of the transmission input torque can be determined (see Table 7.2). On the 1 km reference route, the 5th gear under observation was only used once. It accounts for 21.5% of the distance travelled.

*1* The gearwheel load profile of the pinion of 5th gear is calculated from the gearbox input load profile, ignoring the dynamic behaviour of the transmission.

   *1.1* Converting the transmission input load profile into the load profile of the wheel $5_{MSW}$.

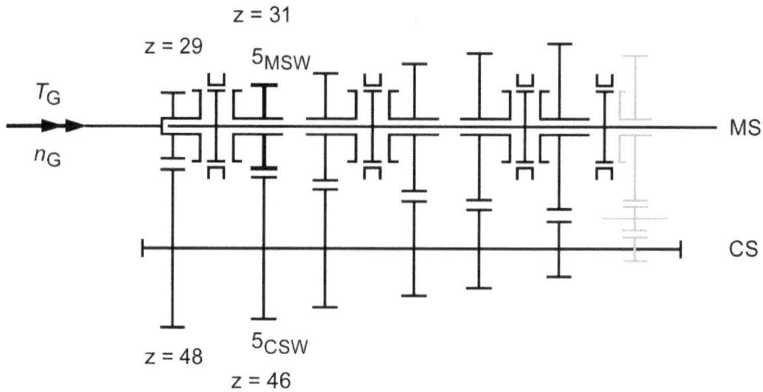


**Fig. 7.20.** Diagrammatic view of a two-stage 6-speed countershaft-type commercial vehicle transmission. *MS* main shaft; *CS* countershaft; *MSW* main shaft wheel; *CSW* countershaft wheel

Gear ratios: constant ratio $u_{CG} = \frac{48}{29}$, gear pair 5 $u_5 = \frac{46}{31}$.

This gives the torque at the countershaft: $T_{CS} = u_{CG}\, T_G$.

*1.2* For a reference torque $T_{ref} = 1400$ Nm at the countershaft, the contact pressure at gear pair 5 has been calculated for the driving flank in accordance with German standard DIN 3990: $\sigma_{H,ref} = 978$ N/mm$^2$. Conversion to any required torque

$$\sigma_H \sim \sqrt{T} \quad \text{and thus} \quad \sigma_H = \sigma_{H,ref} \sqrt{\frac{T_{CS}}{T_{ref}}} .$$

*1.3* Converting the transmission input speeds into revolutions of the wheel $5_{MSW}$, i.e. into load cycles

$$U_{5,MSW} = \frac{U_G\, u_5}{u_{CS}} .$$

*2* The service life of the driving flank of pinion $5_{MSW}$ of 5th gear as regards pitting is calculated using the Miner-Haibach damage accumulation hypothesis, Equations 7.24 and 7.25, with $D_{prof} = D_{th} = 1.0$ and $h_i = U_{5,MSWi}$

$$D_{prof} = \sum_{i=1}^{j} \frac{n_i}{N_i} = \sum_{i=1}^{j} \frac{z\, U_{5,MSWi}}{N_i} = 1.0$$

whereby

$$z = \frac{1}{\sum_{i=1}^{j} \frac{U_{5,MSWi}}{N_i}} = \frac{1}{\text{Damage sum per cycle}} .$$

**Table 7.2.** Initial values (shaded grey) and results of the calculation

| Class $i$ | $T_G$ (Nm) | $T_{CS}$ (Nm) | $\sigma_H$ (N/mm$^2$) | $U_G$ | $U_{5,MSW}$ | $N_i$ Figure 7.21 | $\frac{U_{5,MSW}}{N_i}$ |
|---|---|---|---|---|---|---|---|
| 4 | 1500–2000 | 3310 | 1504 | 4 | 3.6 | $6.3\times10^6$ | $5.71\times10^{-7}$ |
| 3 | 1000–1500 | 2482 | 1302 | 21 | 18.8 | $3.2\times10^7$ | $5.88\times10^{-7}$ |
| 2 | 200–1000 | 1655 | 1063 | 251 | 225.0 | $3.3\times10^8$ | $6.82\times10^{-7}$ |
| 1 | 0–500 | 828 | 752 | 67 | 60.1 | $1.0\times10^{12}$ | $6.01\times10^{-11}$ |

*2.1* The acceptable load cycles are derived from the Wöhler diagram modified after Haibach, Figure 7.21 (see Table 7.2).

*2.2* Calculating the number of load cycles. The number of load cycles is derived from the damage factors calculated in Table 7.2 as

$$z = \frac{1}{5.71\times10^{-7} + 5.88\times10^{-7} + 6.82\times10^{-7} + 6.01\times10^{-11}} = 543165\,.$$

*2.3* The mileage life of 5th gear is then

$$L_{5\text{th gear}} \leq z \times s_{\text{ref}} \times \text{Proportion}_{5\text{th gear}}\,.$$

For a reference route $s_{ref}$ of 1 km and a proportion of 21.5% in 5th gear, the resultant mileage life is $L_{5\text{th gear}}$ = 116 780 km. The mileage lives of the gear pairs needed in the reference route must be known to determine the service life of the transmission.

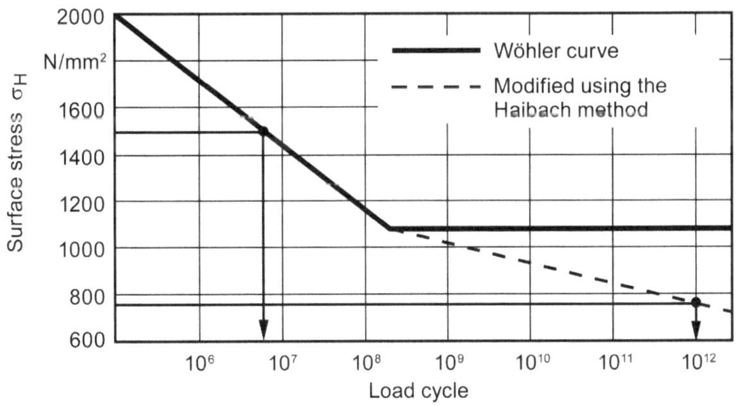


**Fig. 7.21.** Wöhler curve for pitting. 10% probability of failure.
(Ordinate simplified, not logarithmic)

## 7.5 Developing Low-Noise Transmissions

Noise reduction is an important objective when developing vehicles. Vehicle noise not only impairs the wellbeing and health of driver and passengers, but also subjects the environment to traffic noise. With respect to statutory exterior noise levels for accelerated passage, the transmission is, besides the engine (including intake and exhaust systems) and tyres, a significant source of noise only in the case of commercial vehicles. In cars, the transmission contributes only indirectly to exterior noise (DIN ISO 362 [7.8]) through engine speed as a result of a longer or shorter transmission ratio, hardly ever as an independent source of noise. Table 7.3 shows the development of legal limits on noise emissions for various types of motor vehicle.

As noise reduction measures relating to other sources within the vehicle have proved successful, so there has been increasing pressure to reduce transmission noise. It is generally not so much the absolute level of noise, as its particular character that distinguishes transmission noise from the other sources of noise in the vehicle. Some transmission noise phenomena do not in themselves constitute too great a source of noise pollution since they are only audible under certain operating conditions, like rattling and clattering. But they often lead to complaints, since the customer (wrongly) believes the vehicle to be defective. Non-specialists often mistake transmission noise for engine noise. When seeking to reduce transmission noise, it is not sufficient just to make improvements to the transmission itself. As with all vehicle noise problems, the body and other components involved in transmitting and radiating sound have to be taken into account. See also Section 5.5 "Dynamic Behaviour of the Powertrain, Comfort".

**Table 7.3.** European vehicle noise emission limits in dB(A) for accelerated passage according to DIN ISO 362 [7.8]

| | | **Noise emission limits in dB(A)** | | | |
|---|---|---|---|---|---|
| **Type of vehicle** | | **1970** | **1977** | **1994** | since **1995** |
| Passenger cars | | 82 | 80 | 77 | 74 |
| Vans, light-duty trucks | < 2 t | 84 | 81 | 78 | 76 |
| | > 2 t; ≤ 3.5 t | 84 | 81 | 79 | 77 |
| Trucks | > 3.5 t; < 75 kW | 89 | 86 | 81 | 77 |
| | > 7.5 t; ≤ 150 kW | 89 | 86 | 83 | 78 |
| | ≥ 150 kW | 91 | 88 | 84 | 80 |
| Buses | < 150 kW | 89 | 82 | 80 | 78 |
| | > 150 kW | 91 | 85 | 83 | 80 |

### 7.5.1 Transmission Noise and Its Causes

Sources of transmission noise can be divided into four categories (Table 7.4). The noise phenomena are considered below in the order of their significance as sources of noise in modern vehicle transmissions.

#### *1/ Whine*

The rolling contact noise of gear pairs under load – transmitting power – can be described as *whine (howling, squealing* or *singing)* [7.29]. This running noise has several causes:

- *Meshing impact:*
  Meshing impact is the consequence of pitch errors or geometrical errors like concentricity deviations between shaft and toothing as well as variations from the Law of Gears due to deformation of gears, shafts or the housing under load (Figure 7.22). The profile corrections made to avoid these meshing impacts are however only effective for a certain load range.
- *Parametrically excited vibration:*
  Parametrically excited vibration arises from tooth rigidity changing with the meshing position (Figure 7.23). The amount of this vibration depends on gearing geometry and speed. If the stimulation frequency (speed multiplied by number of teeth and harmonic) is close to a natural frequency of the gear pair, the resonance creates particularly large oscillation amplitudes, and hence particularly loud noises. These vibrations arise even with perfect gear teeth.

**Table 7.4.** Transmission noise and its causes

| Transmission noise | Cause |
|---|---|
| **1/ Whine** | Vibration of gearwheels under load:<br>– meshing impact,<br>– parametrically excited vibration and<br>– rolling contact noise. |
| **2/ Rattling/Clattering** | Vibration of loose parts, caused by torsional vibration of the powertrain:<br>– idler gears and<br>– synchronizer rings. |
| **3/ Clonk** | Knocking noise during beginning loading of components with clearance (gearwheels, joints, shaft-hub connection etc.) |
| **4/ Shifting noise** | Scraping and grating of the selector teeth when the synchromesh is not functioning properly |
| **5/ Bearing noise** | Running noise of rolling bearings; especially when they are damaged |

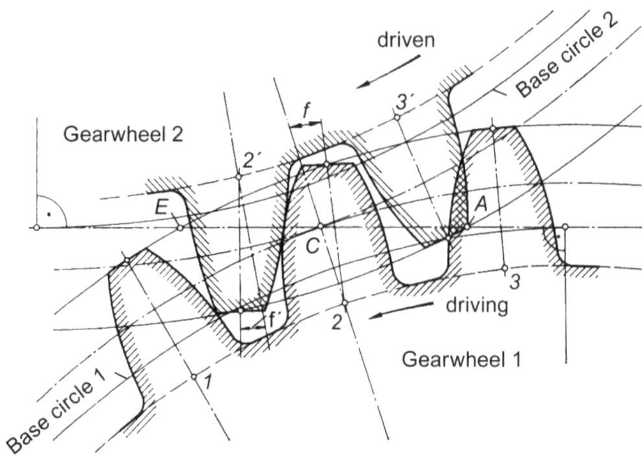


**Fig. 7.22.** Meshing impact resulting from deformed teeth [7.29]

- *Rolling contact noise:*
  Rolling contact noise occurs due to the "washboard effect" arising from inadequate surface quality. Even where production tolerances are adhered to, noise can occur if vibration is generated by certain tooth flank surface structures arising in production.

Reverse gear in many vehicles still uses straight-cut gearwheels, which are not optimised in terms of noise generation; rolling contact noise is particularly noticeable in this case.

Classical transmission noise can be reduced with modern technology to the point that it can barely be heard inside the vehicle if at all. The enormous manufacturing cost associated with increased expectations is however a problem.

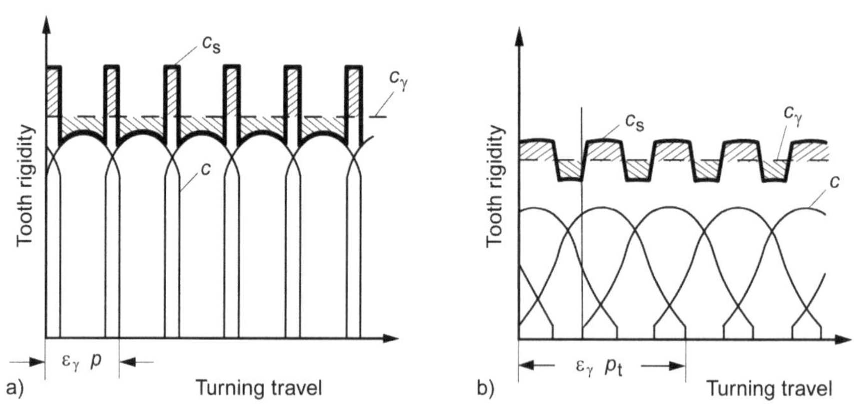


**Fig. 7.23.** Overall tooth rigidity pattern $c_s$, meshing rigidity $c_\gamma$ (average value of $c_s$ over time). *a* Straight spur gearing; *b* helical gearing [7.29]

### *2/ Rattling/Clattering*

Gearwheels and gearshift components vibrating within permitted functional and production limits when not under load produce rattling and clattering noises [7.30, 7.37–7.38], caused by torsional vibration in transmission shafts. If the amplitude of the torsional vibration exceeds a certain value, the idler gears that are not currently engaged lift away from their driving flank, and vibrate backwards and forwards within their backlash. This amplitude depends on the moment of inertia of the idler gear, the drag torque, which has a decelerating effect on this part, and the level of the torsional acceleration causing the stimulation. Torsional vibration can also be induced in synchronizer rings and sliding sleeves within their clearance. The real cause of this noise is the impacts when the loose parts encounter the clearance limits. If they occur in the transmission's neutral position, the phenomenon is called gear rattle at idling speed, and if they occur in motion in a gear, it is called traction or coast rattle or clattering. Figure 7.24 shows the possible vibration elements in form of gearwheels based on the example of two 5-speed transmission.

Torsional vibration in the transmission shafts, which gives rise to these noises, is caused by irregularities in engine speed due to the finite number of cylinders. (See also Figure 5.16 "Internal combustion engine idling".) Almost all measures to improve fuel consumption and emission standards in internal combustion engines cause irregular running. Noise from loose parts occur only under certain operating conditions.

Gear-rattle at idling speed is particularly pronounced with direct-injected diesel engines due to high excitation. Gearboxes rattle under power at low speeds. Beyond a certain degree of irregularity, coast rattle also occurs.

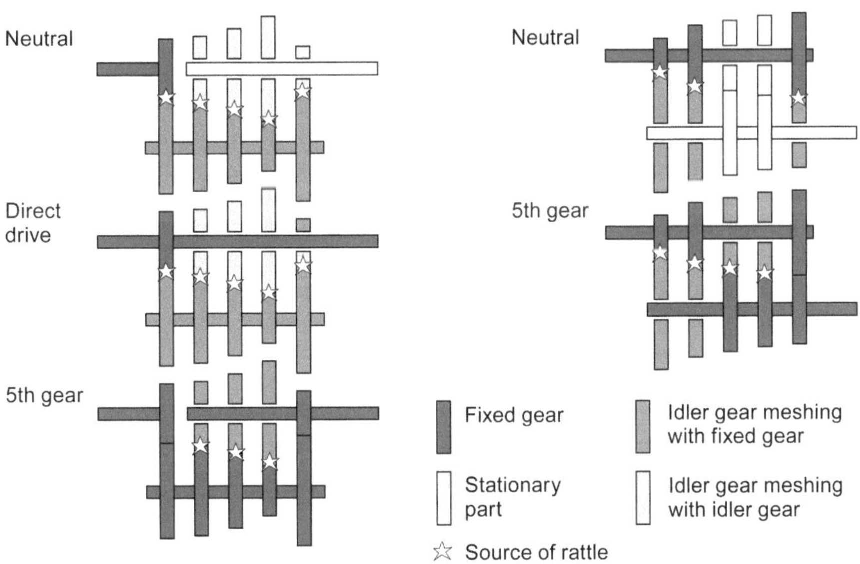


**Fig. 7.24.** Sources of rattle and clatter in coaxial and non-coaxial gearboxes

The flanks of the gearwheels and the engaging gears of the current gear and of the axle drive can also lift with very small traction or thrust loads, causing rattling noise.

Rattling and clattering are kept to an acceptable level by minimising the backlash and restricting centre distance tolerances [7.11] as well as by measures to unlink the torsional vibration of the engine from the transmission. But here again economic constraints are soon encountered. A decisive factor in minimising rattle and clatter is powertrain tuning [7.13, 7.32, 7.35], especially correct design of torsion dampers in the clutch plate, dual mass flywheels and dampers, to keep the torsional vibration amplitudes of the powertrain within certain limits in all operating situations.

### 3/ Clonk

In addition to low-frequency longitudinal jerking (about 2–8 Hz) of the vehicle as a reaction to a sudden load change (load reversal, tip in – back out), higher frequency, metallic-sounding noises (about 300–6000 Hz) occur, known as "clonk", that arise when the flanks of active components knock against each other [7.36]. For analysis, the complete powertrain from flywheel to hub must be taken into consideration. Since the significant parameters – torsional backlash, moments of inertia, rigidity and dampings – are either difficult or impossible to change, usually an optimisation is attempted by means of the diameter of the drive shafts, rigidity characteristics of the engine mount or the clutch damper. Clonk can also be positively influenced with the help of engine calibration, yet this results in somewhat less agile vehicle response characteristics.

### 4/ Shifting Noise

If the synchronizer is not functioning correctly, engagement noise will be audible when shifting [7.14]. Torsional vibrations, especially from torsional backlashes in the powertrain, can cause the sliding sleeve to mesh in the engaging gears of the idler gear prematurely, before speeds have been synchronized. This then causes grating or scraping noises. Other possible causes of this type of noise include concentricity deviation, pitch errors, tooth profile abnormalities and the shifting style of the driver [7.1]. Engagement noises can, objectively speaking, be seen as a mere comfort problem; however they can give customers the impression of lower product quality.

### 5/ Bearing Noise

Bearing noise is normally barely perceptible. Noise only arises with tightly fitted or damaged rolling bearings, increasing rapidly as the damage increases. The character of these noises can provide evidence for an early structure-borne noise damage diagnostics, which is being used more and more often in industrial transmissions. This noise phenomenon should thus be regarded rather in a positive light, since a timely discovery helps to avoid secondary failures.

### *6/ Specific Noises in Automatic and Continuously Variable Transmissions*

In the case of both automatic transmissions with various gear ratios and continuous variable transmissions, we again encounter the same basic types of noises. Because of the high damping of the torque converter in a non lock-up condition, the torsional vibrations of the engine penetrate to the corresponding gear pairs only in a very weakened state, exciting them only slightly, so that audible rattling does not occur. If the lock-up clutch is closed on the other hand, traction or coast rattling can occur.

One noise phenomenon that often appears in automatic transmissions is oil pump whine. Whine sounds coming from the actual gearbox toothing can even be masked by this, only becoming noticeable if the pump level is reduced. The masking potential of whine sounds can be determined by formula or from diagrams [7.41].

One of the main causes of noise in continuously variable transmissions is the striking of the input belt against the discs, predominantly audible as whine. Extensive research on designing the thrust link belt, the discs and their surfaces as well as on the influence of speed and load are described in [7.20]. However, the oil pump is also a dominant source of noise in the case of this gearbox design as well.

Idle drumming is another automotive noise phenomenon, not directly linked with the transmission, that is often accompanied by strong vibrations in the vehicle. This occurs if the gear selector lever is in the "D" or "R" position while this vehicle is stationary. Because of the already relatively high torques transmitted from the torque converter, the elastic rubber bearings of the engine mount can already reach their progression range, which leads to correspondingly high forces being introduced into the car body. These forces can themselves excite structural areas or sheet metal areas like the tailgate or the spare wheel trough, so that strong, very unpleasant pressure fluctuations of 20 – 40 Hz occur inside the vehicle [7.23]. Measures directed at transmission design include the implementation of a so-called "neutral idle", which reduces the initial loads on the engine mount by partially opening the clutch. Other measures, such as increasing or reducing idle speed if possible, or still other measures directed at the car body or involving the use of dampers, are also in use. Regarding mechanical dampers for torque converters, see Figures 10.37 and 10.38.

### 7.5.2 How Noise Reaches the Ear

The main source of noise is the flanks of the gearwheels. The level and character of this noise is influenced in many different ways by the path it then takes to reach the ear.

The high-level airborne noise generated within a normally sealed gearbox produces no significant level of noise outside since it does not have enough energy to excite the housing significantly. The vibration of the gearwheel bodies is transmitted to the shafts of the gearbox as structure-borne noise – directly in the case of fixed gears, and through the respective bearings in the case of idler gears. The input and output shafts transmit the vibrations outside the gearbox. But most of this vibration is transmitted to the housing through the shaft bearings. The bending vibrations of the shafts in particular excite the housing.

If the excitation frequency is close to a natural frequency of the housing, the vibrations are further amplified, and correspondingly high noise levels occur. Part of the noise from the gearbox housing is radiated as airborne noise, and the rest is transmitted as structure-borne noise to the body via the gearbox mounting (Figure 7.25). Drive shafts can also play an important role in transferring structure-borne sound from the gearbox via the axle into the body. For this reason, they should not be neglected in any noise path analysis, i.e. a calculation of the interior vehicle noise to be expected including excitation and transfer paths.

The noise radiated by the gearbox is further influenced by the insulating characteristics of the body, or, more rarely, by a special sound reduction shell. The dynamic behaviour of the engine-gearbox unit, the gearbox mounting and the car body as a whole also have a significant effect on the intensity with which structure-borne noise is transmitted or damped. "Modal alignment" plays an especially important role here, i.e. frequency-adjustment of the components among themselves with their system resonances while separating them as much as possible from the excitation frequencies.

When developing new transmissions or improving existing vehicles, not only the transmission itself has to be taken into account, but also the various routes the noise takes and the ways of controlling this.

### 7.5.3 Assessment Criteria

The subjective perception of this noise by the driver, passengers and people outside the vehicle is crucial in assessing transmission noise.

When a vehicle is under development, improvement measures are frequently evaluated by trained test drivers subjectively rating particular noise phenomena [7.12]. One rating scale frequently used is the "ATZ Evaluation" (Table 7.5).

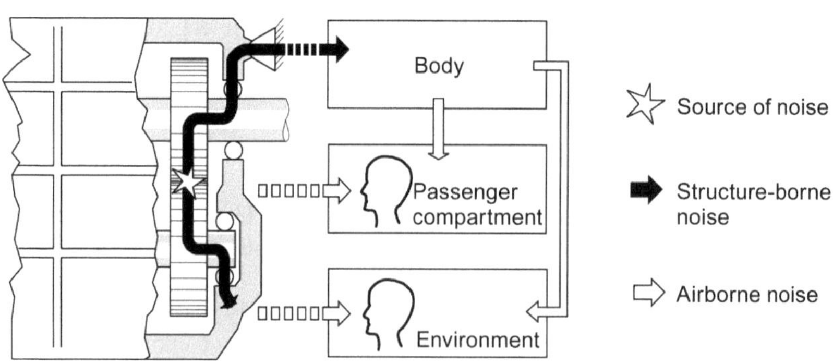


**Fig. 7.25.** Passage of airborne and structure-borne noise

**Table 7.5.** Evaluation table for transmission noise according to ATZ

| **Grade** | **1** | **2** | **3** | **4** | **5** | **6** | **7** | **8** | **9** | **10** |
|---|---|---|---|---|---|---|---|---|---|---|
| Attribute evaluation | not acceptable | | | | border case | acceptable | | | | |
| Customer satisfaction | very dissatisfied | | | | slightly dissatis-fied | pretty satisfied | | very satis-fied | outstandingly satisfied | |

The evaluation team must repeatedly compare their ratings to ensure reasonably uniform and consistent evaluation. To this end, it is best to select a reference condition or a reference vehicle with which one can always orient oneself. Noise can also be recorded digitally with an "artificial head" to be replayed at a later time for examination purposes.

Subjective judgement is, however, insufficient for a detailed analysis of causes. Objective data is required to evaluate and exactly to compare developmental measures for transmissions. Usually noise or vibration data is used, which can be measured both under operational conditions as well as with synthetic stimulations. Depending on stimulation and detection, the data can be further processed in modal, operational vibration or noise path analyses etc., thus providing valuable information.

It is helpful for many investigations to carry out a CAE-based analysis, e.g. with FEM. The effect of improvements can be gained in a significantly faster way than with prototype experiments. The requirement is that one owns a model that has been experimentally validated. Because of the existence of many nonlinearities such as backlash, friction etc., this is not always the case. The applicability and validity of CAE-models must therefore be carefully tested from case to case.

Human hearing is sensitive not only to the energy level of the source, but also very much to the frequency distribution of the noise. The frequency sensitivity of hearing has been comprehensively studied for pure sounds (sinusoidal sound pressure profile over time). The "A evaluation filter" is based on the results of these investigations. The lower frequency range is given a lower weighting since hearing is less sensitive to lower frequencies (Figure 7.26).

This evaluation function has bccn agrccd intcrnationally, and is vcry casy to use, particularly since it is incorporated in all standard sound ranging equipment. But unlike pure sound, transmission noise contains a mixture of many frequencies. The A evaluation is therefore not necessarily applicable. More complex methods of evaluation are needed, which, although known, are not widely used because of their complexity in use. But if similar noises (i.e. noises with a similar spectral distribution) are measured under the same conditions, the A evaluation levels can be meaningfully compared.

The decibel unit is so configured that 1 dB roughly corresponds to the difference threshold of human hearing. When adding up the levels of several sound sources, it is necessary to add the sound intensities of the individual sources, which are proportional to the acoustic power.

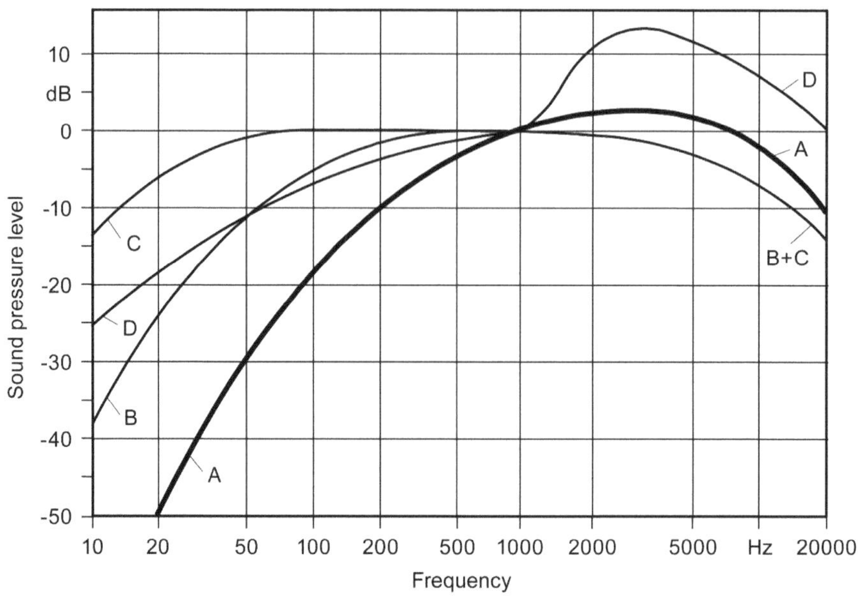


**Fig. 7.26.** Frequency evaluation for sound pressure level. Curves for evaluation filters A, B, C and D

The total of two levels of equal volume is thus 3 dB more than that of the individual levels. The result is that when two levels with a difference of more than 6 dB are added, the quieter one has practically no effect, since the total is less than 1 dB more than the louder source alone.

Airborne sound is often measured in special chambers. A distinction is made between anechoic chambers and reverberation chambers, which are used to suit the particular measuring task. Anechoic chambers are used to simulate free field conditions, reverberation chambers to simulate a sound field that is as equally diffuse as possible.

Structure-borne sound is measured as well as airborne sound. Acceleration sensors are usually used for this purpose, since they are simple to use and to attach to the desired point on the gearbox. Of particular interest in the investigation of the propagation of structure-borne sound are usually accelerations at interfaces, e.g. the elastic mounting.

Frequency analysis is an important tool for analysing transmission noise, especially when allocating a noise component to a particular transmission component. The gear pairs responsible for particular peaks in the noise spectrum can easily be identified from the speeds of rotation and number of teeth. If the amplitude frequency curves recorded at various fixed speeds are plotted sequentially over the whole speed range of the transmission in the form of a waterfall or Campbell diagram, natural vibrations can easily be distinguished from speed-related vibration. This characteristic is clearly apparent in Figure 7.27.

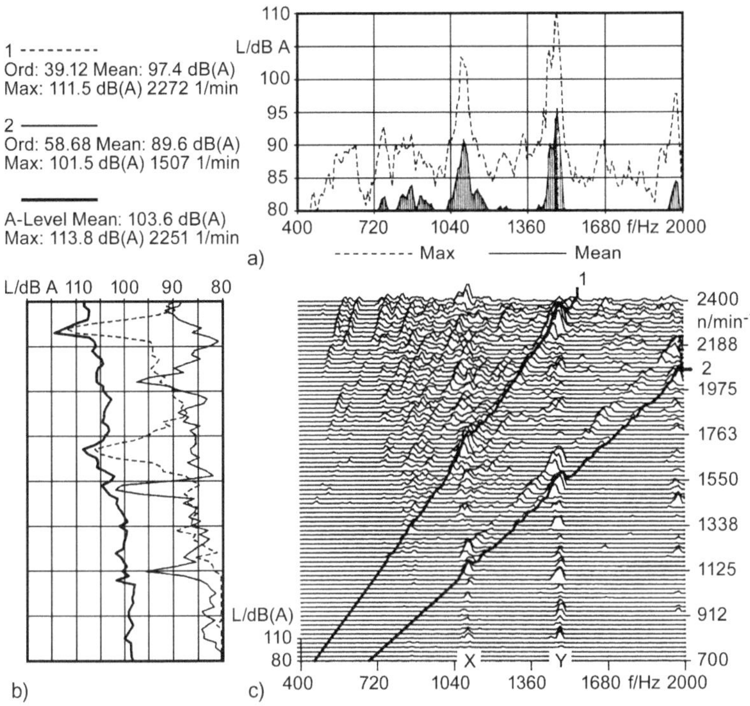


**Fig. 7.27.** Frequency analysis of noise from a truck gearbox. *a* Maximum and mean noise level as a function of frequency; *b* noise level as a function of speed for lines 1, 2 ("order cuts") and A-level; *c* "waterfall diagram"

It is clear that for example at frequencies of 1096 Hz (Figure 7.27c: *X*) and 1490 Hz (Figure 7.27c: *Y*) vibrations occur that are unrelated to rotational speed, whereas the vibration patterns marked (1) and (2) are speed-related. Figure 7.27b shows the noise level of the two salient lines 1 and 2, the "order cut" as a function of speed. The curve shape is obtained by cutting along the order of particular interest in the waterfall diagram and then looking vertically upon this sectional plane. Along with maximum noise, the total level gives an indication of how dominant the contribution of individual orders is. Figure 7.27a shows the maximum or average noise level as a function of frequency. From the wealth of information provided by the measured time signals, we obtain, besides a frequency and order analysis, parameters that describe certain signal characteristics [7.16].

Such signal parameters are normally used also for transmission quality control as well as for monitoring transmission test benches [7.10, 7.33]. For this purposes, structure-borne sound signals are almost always evaluated that can be measured with acceleration sensors on the housing. Evaluation is performed in both time and frequency ranges and is concentrated mostly on analysing the gear orders and their harmonics. A combination of parameters leads to an accurate diagnostics.

### 7.5.4 Countermeasures

The development objective "low-noise transmission" has to be considered right from the planning and design stage of a new transmission, because it is considerably more difficult and expensive effectively to reduce the noise emissions of an existing unit [7.22]. For this reason, so-called target setting is carried out already in the concept phase. Here, target values both for excitation and for transfer are set in order to obtain a certain level of interior vehicle noise. There is also an analogous process for exterior noise or vibrations perceptible to the driver. This is shown schematically in Figure 7.28.

A distinction can be made between active measures to reduce the generation of noise, and passive measures affecting its propagation (Table 7.6). Active measures thus only ever affect a particular type of vibration or noise.

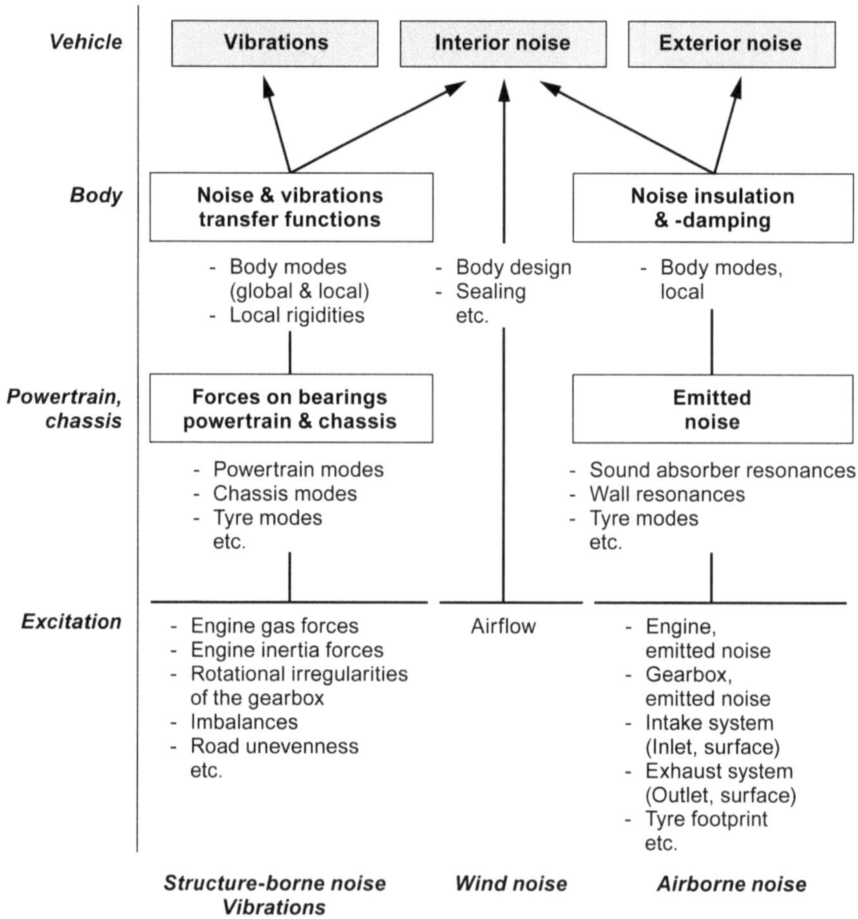


**Fig. 7.28.** Target setting process in concept phase of vehicle development

Active measures to reduce the vibrations of power-transmitting gear pairs affect the geometry and production quality of the toothing. A large transverse contact ratio and overlap ratio (high-contact gearing and helical gearing) reduces the irregularity of the resultant tooth rigidity and moderates the meshing impact.

The meshing interferences due to load and consequent deformation are reduced firstly by profile corrections (tip relief, transverse crowning), and secondly by making gear bodies, shafts, bearings and housings as rigid as possible and separating their natural frequencies to prevent excessive dynamic deformation. The production quality of the toothing is the main factor in rolling contact noise. The speed the transmission runs at has a major effect on the noise emitted by the gearwheels under load, but the load itself has little effect. But neither of these operating parameters can usually be changed.

**Table 7.6.** Active and passive measures for reducing transmission noise

| **Active measures/reduction of noise generation** | |
|---|---|
| **Internal** | **External** |
| **Vibration of gearwheels under load** | |
| *Toothing geometry:*<br>– Helical gears ($\varepsilon_\alpha + \varepsilon_\beta > 2.5$)<br>– High contact gearing ($\varepsilon_\alpha > 2$, not: 2.5)<br>– Profile correction (tip relief)<br>– No integral gear ratios<br>*Toothing quality:*<br>– Tolerances (IT 7 to IT 4)<br>– Machining processes:<br>grinding, shaving, honing (surface) | *Operating status*<br>(cannot generally be changed):<br>– Speed<br>– Torque |
| **Loose part vibration** | |
| – Backlash constriction<br>– Arrangement of idler gears<br>– Synchronizer type<br>– Drag torque increase<br>Direct measures:<br>bracing idler gears or synchronizer rings (e.g. mechanical or magnetic) | *Reduce transmission shaft torsional vibration:*<br>– Decouple engine: torsion damper in clutch plate or dual mass flywheel<br>– Reduction of resonance rise by dampers |
| **Passive measures/reduction of sound propagation** | |
| *Reduce structure-borne sound within the transmission:*<br>– Bearings and bearing support<br>*Reduce housing vibration:*<br>– Housing design (ribbing)<br>– Housing material<br>*Reduce transmission of structure-borne sound to bodywork:*<br>– Concept of transmission and engine mounting<br>– Separation of excitation frequencies and natural system frequencies<br>– High isolation using a very flexible mounting and a high dynamic connection stiffness to bodywork<br>*Encasing the transmission* | |

The extent of loose part vibration is affected by three parameters in the transmission itself. The backlash and the moments of inertia should be as small as possible, and the drag torque on the individual loose parts should be as large as possible. Since there are many functional constraints, like functioning of the transmission at low ambient temperatures, high efficiency etc., there is little potential for noise reduction here. Rattling and clattering can be significantly reduced, or even eliminated, if the amplitude of the torsional vibrations of the transmission shafts is reduced accordingly. If engine irregularity is decoupled, this creates great potential for noise reduction.

It is not yet possible to offer an universal prescription as to how to reduce shifting noise; each transmission requires individual matching and tuning. Apart from synchronization components themselves, the dynamic behaviour of the whole powertrain during gearshifting must be considered.

Passive measures in the gearbox relate in particular to the transmission of structure-borne sound through the shafts from the gear teeth to the housing. An isolation element (rubber) as soft as possible should be used as low-pass filter for the high-frequency vibrations, without there being any unacceptable deformation under load. The design of the gearbox housing is of particular importance. Noise-intensive forms of natural vibration in the transmission walls must be avoided. Stiffening measures should involve taking into consideration the fact that external ribs may be less expensive to produce, but they increase the surface of sound radiation. The housing material also has a major effect on sound radiation. The light alloys in most common use today have much poorer damping characteristics than, for example, cast iron.

Structure-borne noise transfer to the car body can be minimised by adjusting the entire powertrain. Modal alignment, in which housing, mounting, bearing and body resonances have to be attuned to each other such that no reciprocal amplification can take place, plays an essential role here. Figure 7.29 shows an example of the outcome of a modal analysis calculation. The example oscillation form is a so-called "powertrain bending", a bending of the engine-transmission unit, the frequency of which is co-determined by the rigidity of the transmission. The frequency of this vibration mode is optimised such that it is clearly outside the dominant stimulating engine orders. However, significantly more modes must be considered for a detailed modal alignment.

In the optimisation of resonance frequencies, masses and rigidities, there are almost always conflicts with other vehicle attributes, for example with installation space (package), body rigidity (safety/crash), driving dynamics and comfort as well as durability due to rubber bearing rigidities and the closely associated service life of the isolation element of the engine/transmission mount.

Airborne noise radiation can be nearly completely prevented by encasing the gearbox, but there are penalties in terms of weight, heat dissipation and price.

The effectiveness of particular measures must finally be verified taking the total vehicle into consideration. Yet improvements on the system or component level can be carried out on test benches or by means of CAE-based computations, e.g. FEM. In this context, the use of so-called hybrid models, which integrate experimentally determined quantities such as the noise transfer function into the calculation, making a more realistic prediction possible, has also proved effective.

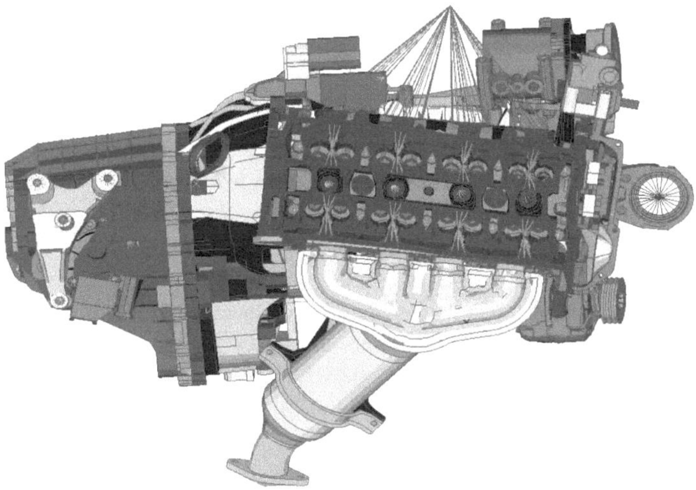

**Fig. 7.29.** Powertrain bending (lateral), result of a calculated modal analysis (FE)

# 8 Specification and Design of Shafts

*The size of a gearbox is largely determined by the diameters of the transmission shafts*

The specification and design of transmission shafts is of special significance in the layout of vehicle transmissions. Shaft diameters are a key factor in determining the centre distance of a gearbox, and thus its size. Strength *and* resistance to deformation must therefore be carefully considered during the design process.

## 8.1 Typical Requirements in Vehicle Transmissions

### 8.1.1 Configuration of Shafts in Vehicle Transmissions

Figure 8.1a shows the shaft configuration (input shaft, countershaft and main shaft) of a two-stage countershaft transmission for standard drive. Figure 8.1b shows the shaft configuration for a single-stage countershaft transmission intended for front-transverse mounting.

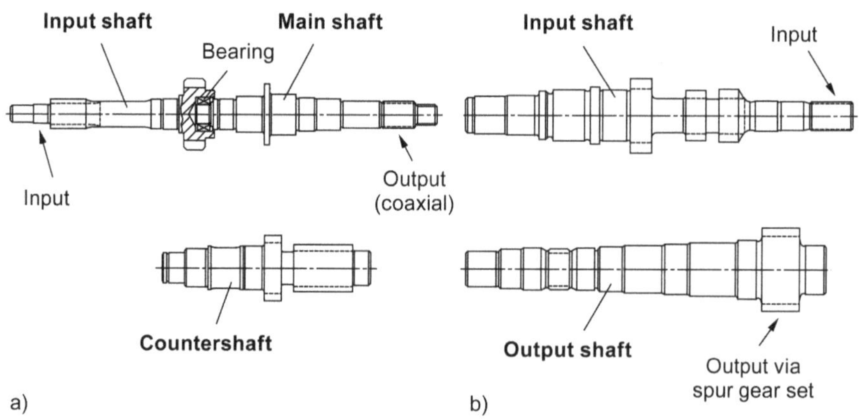


**Fig. 8.1.** Characteristic shaft configurations in vehicle transmissions. *a* Two-stage coaxial countershaft transmission; *b* single-stage countershaft transmission for front-transverse mounting

H. Naunheimer et al., *Automotive Transmissions*, 2nd ed.,
DOI 10.1007/978-3-642-16214-5_8, © Springer-Verlag Berlin Heidelberg 2011

## 8.1.2 Designing for Stress and Strength

The process of designing transmission shafts for operational fatigue strength is described in Section 7.4 "Operational Fatigue Strength and Service Life". The following factors have to be borne in mind when determining the load profiles for transmission shaft design, the "design load profiles":

- moving-off and gear-changing processes are determined largely by the driver and are consequently subject to wide variation; but they have a crucial effect on the service life of the transmission shaft, as do driver errors such as when the foot slips off the clutch pedal causing a violent engaging jolt and
- the load profile derived from a load/time function depends on the classification procedure used (see also Section 7.4.2. "Load Profile and Counting Procedure"). This means that, for example, load profiles determined for gearwheels using the level crossing counting cannot simply be used to determine the operational fatigue strength of transmission shafts. The information on the amplitude and mean amplitude of the individual oscillations is lost. Two-parameter counting procedures such as the Rainflow method are therefore preferable for operational fatigue strength design of shafts; they take into account the amplitude and mean value of each oscillation.

There are three approaches to designing transmission shafts:

1. *Preliminary specification of the shaft diameters (Section 8.3.7):*
   This involves an initial estimate of the shaft diameters required for a given loading.

2. *Design for endurance strength (Section 8.3.8):*
   The design is based on the maximum anticipated loading; the transmission must be capable of sustaining that load fatigue resistant. The maximum engine torque $T_{max}$ is used in the calculation. A transmission that is completely fatigue resistant under continuous maximum load is generally over-engineered.

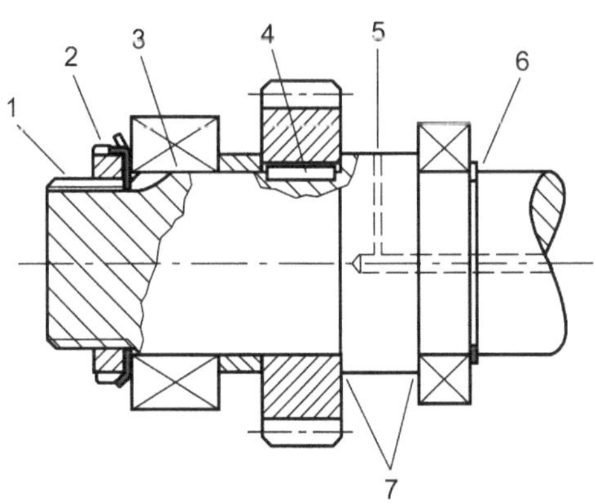


**Fig. 8.2.** Typical notches on transmission shafts.
*1* Thread;
*2* groove for locking plate;
*3* bearing seat;
*4* feather keyway;
*5* transverse drill hole;
*6* groove for circlip;
*7* shoulders

3. *Specification of operational fatigue strength (Section 8.3.9):*
   Transmission shafts are designed for a finite service life based on established load profiles. For this purpose it is necessary to establish design load profiles; in the case of gear transmissions the load classes in each gear must be taken into account as well as the proportion of mileage covered in the various gears.

The shaft configuration typical of vehicle transmissions is particularly unfavourable from the point of view of strength. Large distances between bearings cause large bending moments, and there are many notches as a result of shoulders, grooves, collars, bearing seats etc. (Figure 8.2).

### 8.1.3 Deflection

Vehicle transmissions have long shafts with long distances between bearings, and are usually subjected to asymmetrical loads. This results in large deflections $f$ and large bending angles $\varphi$ (Figure 8.3).

The resultant inclination of the teeth causes a one-sided contact pattern, i.e. the active width of the driving face is reduced, increasing stress on the teeth (Figure 8.4).

To avoid this, the shaft deflection must be checked very accurately additionally to the strength calculations, preferably taking into account the deformation of the housing and bearings.

### 8.1.4 Vibration Problems

Vibrations in the powertrain present particular challenges in transmission design. Although the stiffening effect of bearings and hubs and also the effects of idler gears and synchronizer units can hardly be measured, it is essential to analyse the vibration behaviour of the transmission, since the stress peaks and deformations caused by dynamic effects can be substantial.

The dynamic behaviour of a transmission must always be considered in combination with the powertrain as a whole. Vibration analysis can be carried out on test benches and also by computer simulation. There are numerous mathematical models available [8.1, 8.2].

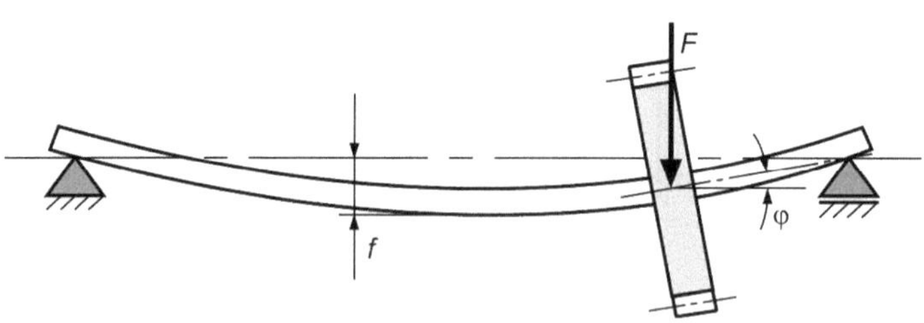


**Fig. 8.3.** Defelection $f$ and bending angle $\varphi$ in shafts with large distances between bearings and asymmetrical loading

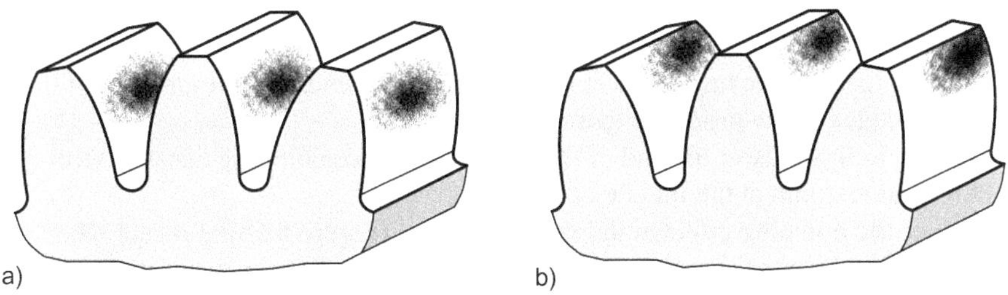


**Fig. 8.4.** Contact pattern. *a* Uniform contact pattern; *b* one-sided contact pattern

They are based on multiple-component systems with linear or non-linear linking. One significant factor with these models is the "critical speeds". Unbalanced rotating masses generate forces that cause vibration. Transmission shafts with rotating gearwheels, synchronizers etc. have several critical speeds that result in bending vibration resonance. High critical speeds are preferred, as they give quiet running and a long service life.

A distinction is made between:

- *Torsional vibration:*
  Low-frequency and high-frequency vibration principally caused by irregularities in the power flow from internal combustion engines (see also Figure 5.16) and
- *Bending vibration:*
  Higher-frequency transmission shaft vibrations. This can for example be caused by tooth meshing (see also Figure 7.23 "Overall tooth rigidity pattern").

## 8.2 General Design Guidelines

The problems specific to vehicle transmissions described in the previous section lead to three main requirements for transmission shaft design:

*1/ Avoid notches!*
*2/ Reduce bending moments!*
*3/ Increase critical speeds!*

To satisfy these requirements, the following design principles should be followed:

- reduce the distance between bearings by means of compact overall structural design,
- locate heavily stressed gearwheels close to bearings to reduce deflection and bending moment, and to achieve high critical speeds,
- keep diameter transitions between shafts below the ratio $D/d \approx 1.4$; the transitions should preferably be of conical design or with a large radius of curvature rather than with shaft shoulders,

- specify splined connections or oil press connections rather than feather key connections,
- smooth rectangular ring-grooves by using relief notches or by rounding off the inside edges of the groove (Figure 8.5a),
- locate circlips only at the end of the shaft if at all possible; use distance sleeves for axial restraint at the middle of the shaft,
- reduce the notching effect at the shaft shoulders (Figure 8.5b): *1* locate relief notches at the transition by means of rounded axial grooves; *2* use large rounding radius; *3* use radial stress relief notches; *4* use additional notches in the transition zone,
- shafts with a mounted hub must be made thicker at the wheel seat; specify a large transition radius and reduce the thickness of the hub towards the end (Figure 8.5c),
- transverse drill holes should be smoothed by relief notches at the mouth of the drill hole, by increasing the shaft diameter and using larger transition radii and by post-moulding the edges of the drill holes with a smooth thrust piece (Figure 8.5d),
- gradual power diversion using relief notches (Figure 8.5e),
- balance shafts precisely in order to minimise centrifugal forces and associated bending vibrations and
- reduce the moment of inertia of components mounted on the shaft in order to reduce deflections and increase critical speeds.

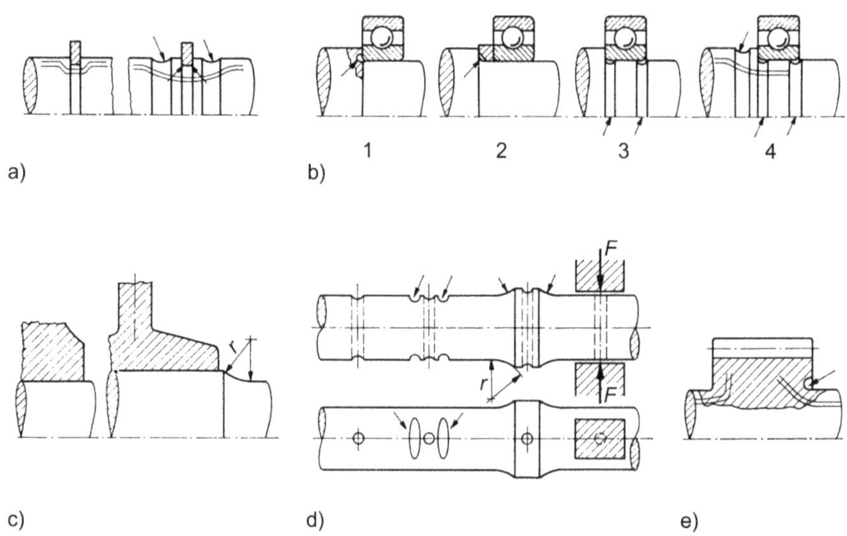


**Fig. 8.5.** Several design approaches to reduce the notching effect [8.6]

## 8.3 Transmission Drive Shaft Strength Design

The following model strength calculation is based on the transmission drive shaft (input shaft) shown in Figure 8.1b. The resultant load ratios vary depending on the gear engaged, so the following calculation has to be carried out to determine the most highly stressed point and the greatest deflection *for each gear* and also for the load conditions *under power and in overrun conditions*.

The calculation shown is restricted to statically determinate cases, i.e. to two-bearing transmission shafts. Shafts secured at multiple points are statically indeterminate and require very extensive computation which is beyond the scope of manual calculation in the case of graduated diameters (see also Section 8.4).

### 8.3.1 Loading

When designing a transmission, the designer initially works on the basis of the maximum engine torque $T_{max}$. (Engaging jolts can give temporary rise to transmission input torque values more than twice as high). If the calculation relates to some other shaft than a transmission input shaft, the effective torque in the shaft must be determined from the ratio of the respective gears.

Applied external forces, such as tooth forces and bearing forces, are treated as point loads. The forces are determined in the co-ordinate system of the powertrain, the $x$ axis corresponding to the shaft axis. This gives rise to the system of forces at the drive shaft shown in Figure 8.6.

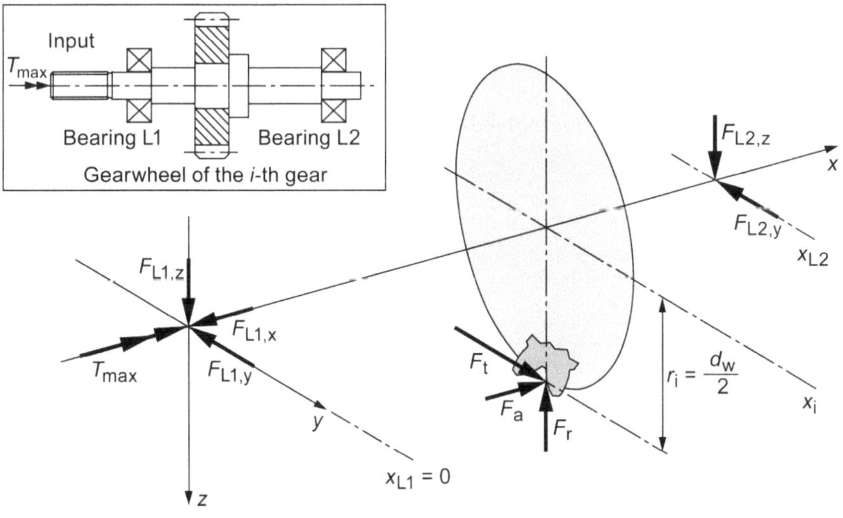


**Fig. 8.6.** Loading of the transmission input shaft of the theoretical model (single-stage countershaft transmission)

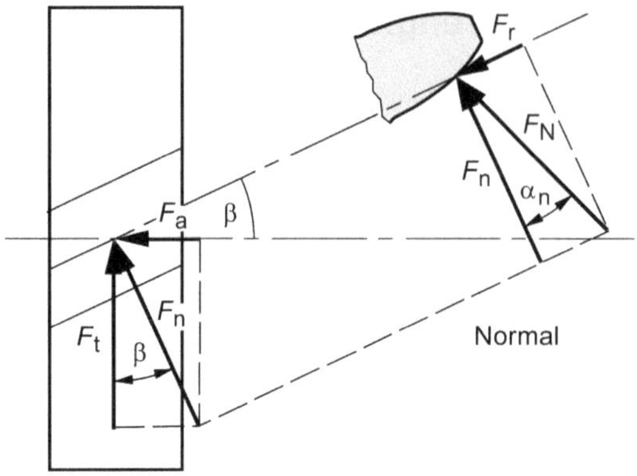


$$F_t = \frac{2\,T_{max}}{d_W} \quad (8.1)$$

$$F_r = F_t \frac{\tan\alpha_n}{\cos\beta} \quad (8.2)$$

$$F_a = F_t \tan\beta \quad (8.3)$$

**Fig. 8.7.** Forces acting at the tooth flanks of a helical-cut spur gear

The first step is to determine the forces at the tooth flanks, which depend on the gear currently selected and on the type of gear teeth. In the case of helical gearwheels there are axial forces as well as tangential and radial forces. Figure 8.7 shows the system of forces at the tooth flanks of a helical spur gearwheel.

The tangential, radial and axial forces are calculated using the equations listed in Figure 8.7. In the case of spur gears, the radius to the point of application of the force depends on the pitch circle diameter $d_w$

$$r_i = \frac{d_w}{2}\,. \quad (8.4)$$

The equations for the forces $F_t$, $F_r$, $F_a$ and for $r_i$ for straight and spiral-toothed bevel gears are summarised in Table 8.1.

If there is spatial loading, i.e. $F_t$ and $F_r$ are not located on the specified coordinate planes, they must be resolved into their $y$ and $z$ components (Figure 8.8).

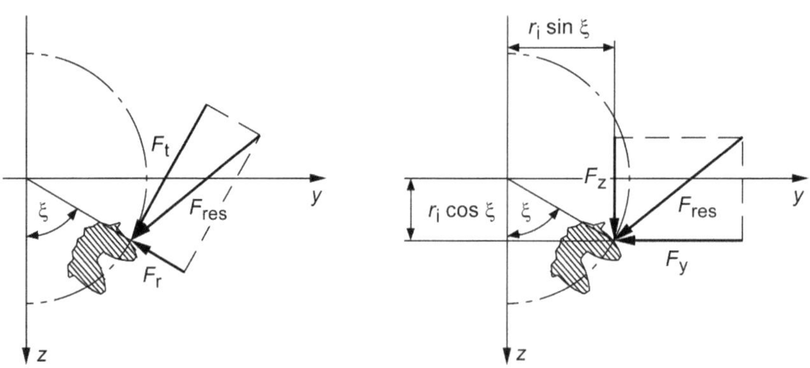


**Fig. 8.8.** Conversion of tangential and radial forces in $y$ and $z$ components in the case of spatial loading

**Table 8.1.** Forces acting on straight-cut and spiral-toothed bevel gears

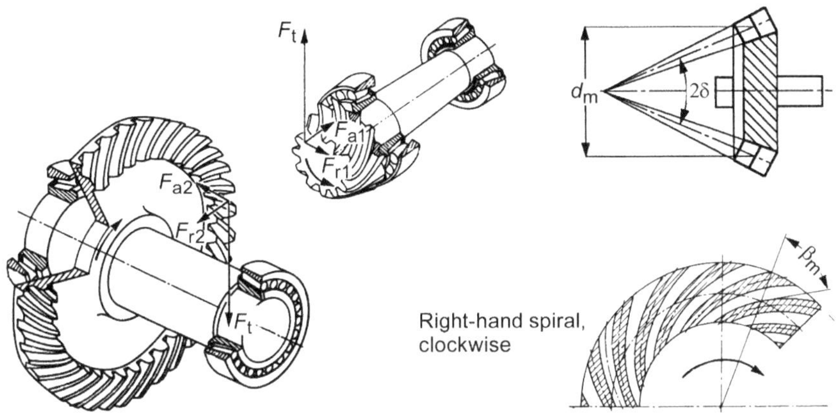


| Spiral direction/direction of rotation driving gear | Axial force | Radial force |
|---|---|---|
| Right-hand spiral, clockwise or Left-hand spiral, anti-clockwise | **Driving gear** $F_a = \frac{F_t}{\cos\beta_m}(\tan\alpha_n \sin\delta + \sin\beta_m \cos\delta)$ | **Driving gear** $F_r = \frac{F_t}{\cos\beta_m}(\tan\alpha_n \cos\delta - \sin\beta_m \sin\delta)$ |
| | **Driven gear** $F_a = \frac{F_t}{\cos\beta_m}(\tan\alpha_n \sin\delta - \sin\beta_m \cos\delta)$ | **Driven gear** $F_r = \frac{F_t}{\cos\beta_m}(\tan\alpha_n \cos\delta + \sin\beta_m \sin\delta)$ |
| Right-hand spiral, anti-clockwise or Left-hand spiral, clockwise | **Driving gear** $F_a = \frac{F_t}{\cos\beta_m}(\tan\alpha_n \sin\delta - \sin\beta_m \cos\delta)$ | **Driving gear** $F_r = \frac{F_t}{\cos\beta_m}(\tan\alpha_n \cos\delta + \sin\beta_m \sin\delta)$ |
| | **Driven gear** $F_a = \frac{F_t}{\cos\beta_m}(\tan\alpha_n \sin\delta + \sin\beta_m \cos\delta)$ | **Driven gear** $F_r = \frac{F_t}{\cos\beta_m}(\tan\alpha_n \cos\delta - \sin\beta_m \sin\delta)$ |

where

$d_m$ … average reference circle diameter (at face width centre)

$r_i = \frac{d_m}{2}$ … radius to the point of application of the force

$F_t = \frac{2\,T_{max}}{d_m}$ … circumferential force

$\beta_m$ … helix angle (spiral angle) at the reference cone in tooth centre

$\delta$ … reference cone angle of the examined gearwheel

$\alpha_n$ … meshing angle normal

$$F_y = F_t \cos\xi + F_r \sin\xi \ , \tag{8.5a}$$

$$F_z = F_t \sin\xi - F_r \cos\xi \ . \tag{8.5b}$$

When balancing moment levels, it must be taken into account that the effective lever arms are reduced in this case (Figure 8.8).

### 8.3.2 Bearing Reactions

The bearing forces $F_{L1}$ and $F_{L2}$ are calculated from the external forces. To determine the bearing reactions it is necessary to balance the forces ($\Sigma F = 0$) in the $x$, $y$ and $z$ directions and also to balance the moments ($\Sigma M = 0$) about the $y$ and $z$ axes relative to the origin of the co-ordinates.

In addition to the axial bearing forces, the resultant radial bearing forces are also of significance when calculating the bearings (Section 11.1.2)

$$F_{L1} = \sqrt{F_{L1,y}^2 + F_{L1,z}^2} \ , \tag{8.6a}$$

$$F_{L2} = \sqrt{F_{L2,y}^2 + F_{L2,z}^2} \ . \tag{8.6b}$$

### 8.3.3 Spatial Beam Deflection

In the general case of spatial beam deflection it is important to determine accurately the sign of the section reactions at the beam section. Figure 8.9 shows the method of determining the sign used for calculating transmission shafts.

In the $xz$ plane (Figure 8.9a) all the section reactions (shear force (transverse force) $F_{Q,z}$ and bending moment $M_{b,y}$) act in the direction of the positive co-ordinate axes, i.e. $M_{b,y}$ rotates about the $y$ axis in a positive sense.

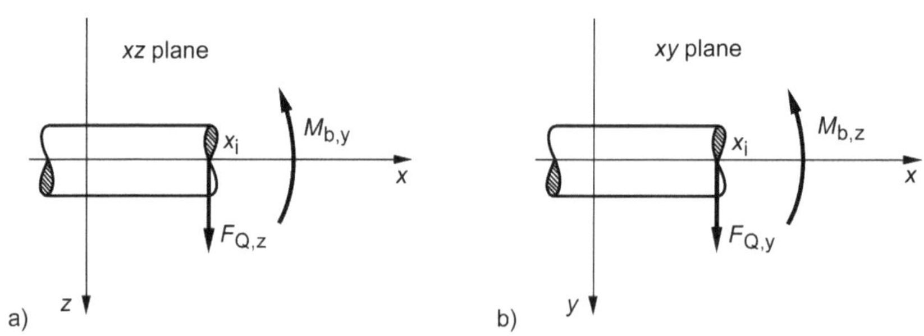


**Fig. 8.9.** Determining the sign for the section reactions at the positive beam section at the point $x_i$. *a* $xz$ plane; *b* $xy$ plane

In the *xy* plane on the other hand (Figure 8.9b) only the shear force $F_{Q,y}$ acts in the positive co-ordinate direction. The bending moment $M_{b,z}$ rotates about the *z* axis in a negative sense, but retains the same orientation to $F_{Q,y}$ as $M_{b,y}$ to $F_{Q,z}$ in the *xz* plane. But it is precisely this orientation that is important for the relation between the shear force and the bending moment.

In engineering mechanics, it is normal practice to use a closed right-hand system for the section reactions. Specifically the bending moment $M_{b,z}$ in the *xy* plane (Figure 8.9b) would be defined as positive against the direction shown.

There is thus no longer a corresponding relationship between the *xz* and *xy* planes, i.e. separate derivations must be carried out for both planes, some with changing signs. That is in a sense the price that has to be paid for a coherent scientifically exact representation of a spatial system.

However, determining the sign in the way chosen here has the decisive advantage that the *xz* and *xy* planes can be treated in exactly the same way as regards the relation between shear force and bending moment. But the following should be borne in mind:

- A positive sign in the shear force and bending moment diagram calculated means positive in relation to the sign as previously determined!
- Equations relating the shear forces, bending moment, deflections or directions of stresses are only valid if the sign as previously determined is taken into account. This is particularly important when performing comparisons with other literature or with the results of calculation programs.

Another problem with establishing the shear force and bending moment diagrams subsequently is the application of individual forces and moments. This is why there are jumps at the points $x_i$ in the shear force and bending moment diagrams. These discontinuities can be overcome by introducing the "Heavyside step function" by means of which non-analytical functions can be represented as "closed":

$$\{x - x_i\} = \begin{cases} 0 & \text{for } x \le x_i \\ (x - x_i) & \text{for } x > x_i \end{cases} . \tag{8.7a}$$

In particular, a unit step function may be defined as

$$\{x - x_i\}^0 = \begin{cases} 0 & \text{for } x \le x_i \\ 1 & \text{for } x > x_i \end{cases} . \tag{8.7b}$$

### 8.3.4 Shear Force and Bending Moment Diagrams

In order to calculate the stresses in the transmission shaft, the bending moments and torsional moments must be known at every point on the shaft.

To determine the *bending moment diagram* $M_b(x)$, the shear force diagram $F_Q(x)$ must first be calculated. The bending moment diagram can then be calculated by integrating the shear force diagram as

$$M_\mathrm{b}(x)=\int F_\mathrm{Q}(x)\,\mathrm{d}x+C\,. \tag{8.8}$$

Equation 8.8 applies both in the $xz$ plane and in the $xy$ plane because of the sign determination in Figure 8.9. The boundary conditions used to find the integration constant $C$ result from the sum of all individual moments to the left of the point $x_\mathrm{i}$. The action of individual moments must be taken into account, such as the *pitching moment* created by the axial force of a helical gearwheel (Figure 8.10). This changes the sum of all the individual moments, and there is a discontinuity in the bending moment profile at point $x_\mathrm{i}$. This must be described using the Heavyside step function, Equation 8.7.

In the $xz$ plane the shear force diagram according to Figure 8.6 is given by

$$F_{\mathrm{Q,z}}(x)=-F_{\mathrm{L1,z}}+F_\mathrm{r}\left\{x-x_\mathrm{i}\right\}^0\,. \tag{8.9}$$

The bending moment diagram can now be calculated from the shear force diagram by integration. It should be noted that there is a discontinuity of $F_\mathrm{a}\,r_\mathrm{i}$ in the bending moment diagram at point $x_\mathrm{i}$ resulting from the axial force present

$$M_{\mathrm{b,y}}(x)=-F_{\mathrm{L1,z}}\,x+F_\mathrm{r}\left\{x-x_\mathrm{i}\right\}^1-F_\mathrm{a}\,r_\mathrm{i}\left\{x-x_\mathrm{i}\right\}^0\,. \tag{8.10}$$

An analogous procedure applied to the $xy$ plane results in

$$F_{\mathrm{Q,y}}(x)=F_{\mathrm{L1,y}}-F_\mathrm{t}\left\{x-x_\mathrm{i}\right\}^0 \tag{8.11}$$

resulting in

$$M_{\mathrm{b,z}}(x)=F_{\mathrm{L1,y}}\,x+F_\mathrm{t}\left\{x-x_\mathrm{i}\right\}^1\,. \tag{8.12}$$

In the subsequent stress calculation, only the magnitude of the resultant bending moment is significant

$$M_\mathrm{b}(x)=\sqrt{M_{\mathrm{b,y}}^2(x)+M_{\mathrm{b,z}}^2(x)}\,. \tag{8.13}$$

The *torsional moment diagram* $M_\mathrm{t}(x)$ in the transmission shaft is derived from the torque equilibrium around the $x$ axis as

$$M_\mathrm{t}(x)=-T_\mathrm{max}+F_\mathrm{t}\,r_\mathrm{i}\left\{x-x_\mathrm{i}\right\}^0\,. \tag{8.14}$$

The shear force and moment diagrams can also be represented graphically (Figure 8.11). This shows clearly how the bending moment diagram (Figure 8.11b) is derived from the shear force diagram (Figure 8.11a) by integration. The torsional moment is constant between the power input and power output points, and has magnitude $T_\mathrm{max}$ (Figure 8.11c).

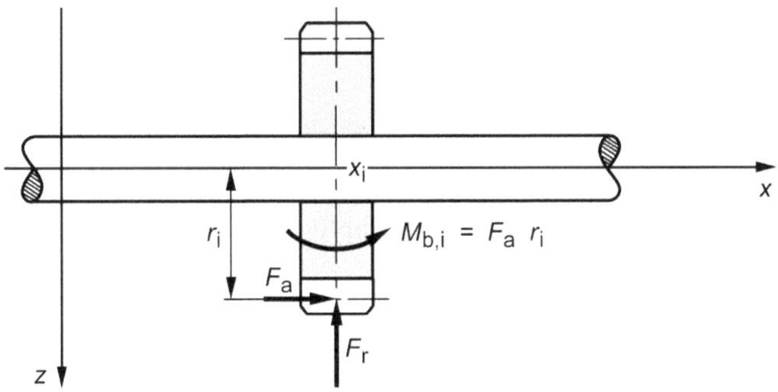


**Fig. 8.10.** Generation of pitching moment with axial forces by helical gearing

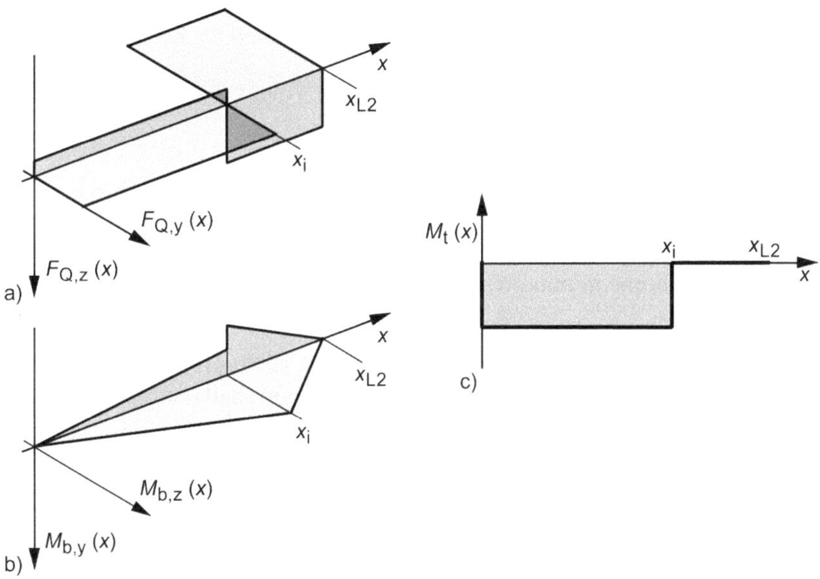


**Fig. 8.11.** Representation of the qualitative shear force and moment diagrams of the sample calculation shown in Figure 8.6. *a* Shear force diagram $F_Q(x)$; *b* bending moment diagram $M_b(x)$; *c* torsional moment diagram $M_t(x)$

### 8.3.5 Critical Cross-Section

The greatest reference stress acts in the critical, most highly stressed cross-section. It must therefore be shown that the shaft can bear the resultant stress at this point with the required safety margin. For a successful design, it is very important to check this critical cross-section. If the location of the critical cross-section cannot be precisely predicted, several cross-sections have to be analysed.

The following criteria can be used to determine the location of critical cross-sections:

- peaks in the bending moment and torsional moment diagrams (Figure 8.11 point $x_i$),
- small shaft diameters and
- notches.

*The critical factor is often the peak stresses at the notches!* Thus the critical cross-section does not necessarily have to coincide with the peak in the moment diagrams.

### 8.3.6 Stresses

In calculating the effective stresses in transmission shafts it is general practice to ignore the normal stress caused by axial forces and the shear stress caused by shear forces. The only significant stresses are those caused by bending moment and torsional moment.

#### *Nominal Stresses*

The maximum stress value $\sigma_{max}$ occurs in the notch root. Depending on the notch acuity, this value exceeds the stress $\sigma_n$ that would prevail in an un-notched shaft with the same root cross-section. $\sigma_n$ is the symbol for nominal stress. The following points must be borne in mind when calculating the nominal bending stress and torsional stress:

- The absolute value of the resultant bending moment $M_b$ from Equation 8.13 determines the nominal bending stress, since transmission shafts rotate and have a circular cross-section.
- If the critical cross-section is located at a discontinuity in the moment diagram, then the moment with the greatest magnitude must be used in calculations.

Nominal bending stress:

$$\sigma_{b,n} = \frac{M_b}{W_b}, \tag{8.15a}$$

Nominal torsional stress:

$$\tau_{t,n} = \frac{M_t}{W_t}, \tag{8.15b}$$

where $W_b$ section modulus under bending and
$W_t$ section modulus under torsion.

#### *Notch Stresses*

The increased stress in the notch root causes the nominal stresses to rise by the stress concentration factor $\alpha_k$ under static stress, and by the fatigue notch factor $\beta_k$ under dynamic stress. Thus under static stress $$\sigma_{\mathrm{b,max}} = \alpha_{\mathrm{k,b}}\, \sigma_{\mathrm{b,n}}\,, \tag{8.16a}$$

$$\tau_{\mathrm{t,max}} = \alpha_{\mathrm{k,t}}\, \tau_{\mathrm{t,n}}\,, \tag{8.16b}$$

under dynamic stress $$\sigma_{\mathrm{b,max}} = \beta_{\mathrm{k,b}}\, \sigma_{\mathrm{b,n}}\,, \tag{8.16c}$$

$$\tau_{\mathrm{t,max}} = \beta_{\mathrm{k,t}}\, \tau_{\mathrm{t,n}}\,. \tag{8.16d}$$

Of course, a transmission shaft will always be subject to dynamic stress, but if the $\beta_k$ values are difficult or impossible to determine, the calculation can also be performed using the stress concentration factor, $\alpha_k$, justifiably erring on the side of safety. Under dynamic stress the stress concentration factor $\alpha_k$, which applies under static stress, does not fully impact on the stress increase, thus

$$1 \le \beta_k \le \alpha_k\,, \tag{8.17}$$

where $\beta_k = 1$ means insensitivity to notches and
$\beta_k = \alpha_k$ full notch sensitivity.

**Table 8.2.** Determining the stress concentration factor $\alpha_k$ from the shaft geometry

| Notch type \ Load type | Bending $\alpha_{k,b}$ | | | | | Torsion $\alpha_{k,t}$ | | | | |
|---|---|---|---|---|---|---|---|---|---|---|
| 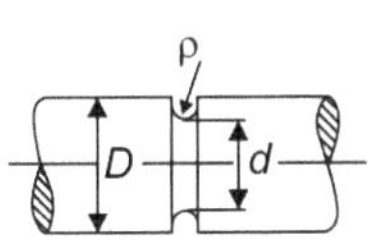 | $\rho/d$ \ $D/d$ | 1.01 | 1.03 | 1.15 | ∞ | $\rho/d$ \ $D/d$ | 1.01 | 1.05 | 1.20 | ∞ |
| | 0.05 | 1.60 | 1.90 | 2.40 | 2.60 | 0.05 | 1.30 | 1.55 | 1.75 | 1.85 |
| | 0.10 | 1.40 | 1.60 | 1.87 | 2.00 | 0.10 | 1.20 | 1.35 | 1.46 | 1.52 |
| | 0.20 | 1.27 | 1.40 | 1.55 | 1.60 | 0.20 | 1.14 | 1.22 | 1.29 | 1.32 |
| | 0.30 | 1.20 | 1.30 | 1.35 | 1.40 | 0.30 | 1.12 | 1.17 | 1.20 | 1.22 |
| 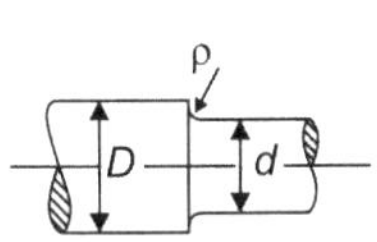 | $\rho/d$ \ $D/d$ | 1.01 | 1.10 | 2.00 | 6.00 | $\rho/d$ \ $D/d$ | 1.09 | 1.20 | 2.00 | 6.00 |
| | 0.05 | 1.55 | 1.89 | 2.19 | 2.40 | 0.05 | 1.26 | 1.55 | 1.75 | 1.80 |
| | 0.10 | 1.36 | 1.58 | 1.73 | 1.88 | 0.10 | 1.17 | 1.32 | 1.45 | 1.51 |
| | 0.20 | 1.20 | 1.37 | 1.44 | 1.50 | 0.20 | 1.11 | 1.16 | 1.25 | 1.31 |
| | 0.30 | 1.13 | 1.27 | 1.30 | 1.33 | 0.30 | 1.08 | 1.11 | 1.17 | 1.22 |
| 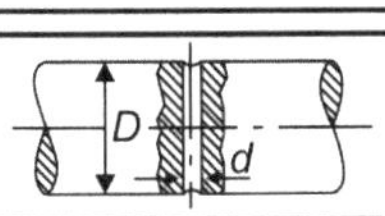 | $d/D$ | 0.01 | 0.10 | 0.20 | 0.40 | $d/D$ | 0.01 | 0.10 | 0.20 | 0.40 |
| | | 2.70 | 2.35 | 2.02 | 1.80 | | 1.90 | 1.65 | 1.50 | 1.40 |

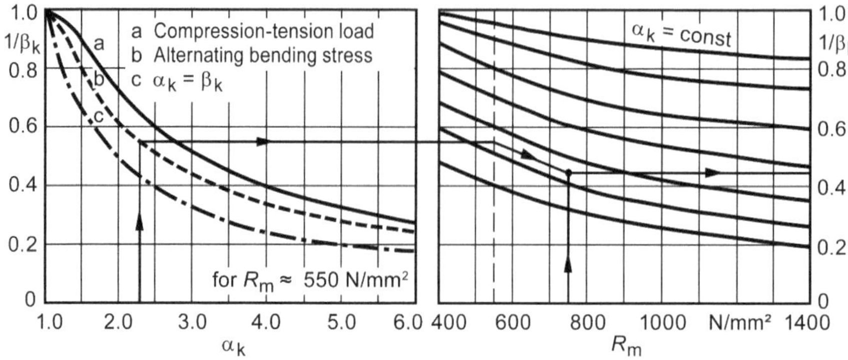


**Fig. 8.12.** Determining the fatigue notch factor $\beta_k$ using the Rühl method [8.7]. Sample reading: $R_m = 750$ N/mm$^2$, $\alpha_k = 2.3 \Rightarrow \beta_k = 2.22$

The static stress concentration factors $\alpha_k$ can be determined from the shaft geometry, and are set out in Table 8.2 for various classes of stress. There are various methods of calculating the dynamic fatigue notch factor $\beta_k$ [8.4, 8.7–8.9, 8.11]. Standard values for $\beta_k$ are also given in [8.5] and [8.6].

The fatigue notch factor $\beta_k$ can be determined as a function of the stress concentration facor $\alpha_k$, the tensile strength $R_m$ and the type of stress using the Rühl method [8.7]. Figure 8.12 shows on the left the dependence of $1/\beta_k$ on $\alpha_k$ as revealed by experiment. This diagram only applies to steel with $R_m = 550$ N/ mm$^2$, so $1/\beta_k$ for steels with different $R_m$ must be adjusted by a factor taken from the right hand side of Figure 8.12. If the diagram gives $\beta_k > \alpha_k$ then, as a consequence of Equation 8.17, let $\beta_k = \alpha_k$.

### *Reference Stress*

When designing transmission shafts reference stress arising from bending and torsional stress is calculated on the basis of the shear strain energy theory as

$$\sigma_v = \sqrt{\sigma_{b,max}^2 + 3\left(\alpha_0\, \tau_{t,max}\right)^2}\,, \tag{8.18}$$

with the "effort ratio"

$$\alpha_0 = \frac{\sigma_{perm} \text{ according time profile of } \sigma}{\sigma_{perm} \text{ according time profile of } \tau} = \frac{\sigma_{perm}(\sigma)}{\sigma_{perm}(\tau)}\,, \tag{8.19}$$

which takes into account the influence of various load scenarios for $\sigma_b$ and $\tau_t$. The torsional stress component $\tau_t$ is transformed to the time profile of the bending stress component $\sigma_b$, using $\alpha_0$. Assuming the same safety margins for $\sigma_b$ and $\tau_t$, common instances are

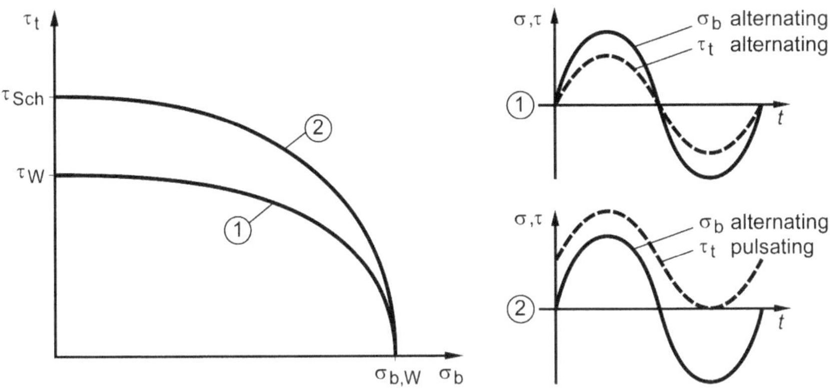


**Fig. 8.13.** Boundary lines of acceptable stress components

$$\left.\begin{array}{l}\sigma_b \text{ alternating} \\ \tau_t \text{ alternating}\end{array}\right\} \quad \alpha_0 = 1, \tag{8.20a}$$

$$\left.\begin{array}{l}\sigma_b \text{ alternating} \\ \tau_t \text{ pulsating}\end{array}\right\} \quad \alpha_0 = \frac{\sigma_{b,W}}{\sigma_{b,Sch}} \approx 0.7\,. \tag{8.20b}$$

The effort ratio $\alpha_0$ is thus derived from the permissible stresses of two types of load with different time profiles. The influence of the effort ratio $\alpha_0$ can be represented graphically by entering the pairs of permissible stress component values in a $\sigma_b - \tau_t$ plot (Figure 8.13).

### *Reference Moment*

When specifying the shaft diameter for a known stress, the "reference moment" $M_v$ is often used

$$M_v = \sqrt{M_{b,max}^2 + 0.75\left(\alpha_0\, M_{t,max}\right)^2}\,. \tag{8.21}$$

## 8.3.7 Preliminary Specification of the Shaft Diameter

A rough preliminary design of the transmission shafts is required along with a determination of the centre distance for the first outline design of a transmission. The minimum diameter of a solid shaft can be estimated using the equation $\sigma_b = M_b/W_b$ and Equation 8.21. Where $W_b = \pi d^3/32$ and $\sigma_v = \sigma_{b,perm}$, then

$$d_{min} = 2.17\sqrt[3]{\frac{M_v}{\sigma_{b,perm}}}\,. \tag{8.22}$$

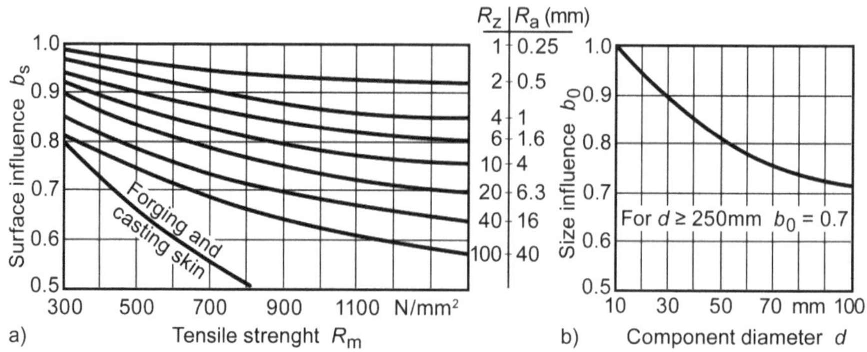


**Fig. 8.14.** For circular cross-sections: *a* surface effect $b_s$; *b* size effect $b_0$

### 8.3.8 Designing for Endurance Strength

The reference stress $\sigma_v$ calculated using Equation 8.18 depends on the maximum engine torque $T_{max}$ (see Section 8.3.1). If the shaft is to be designed for endurance strength, $T_{max}$ must be tolerated for the entire service life. The strength requirement is therefore

$$\sigma_v \leq \sigma_{b,perm} = \frac{\sigma_{b,W}\, b_s\, b_0}{S_D}, \tag{8.23}$$

where $\sigma_{b,W}$ endurance strength under alternating bending stresses,
$b_s$ surface effect (various finishes, Figure 8.14a),
$b_0$ size effect (stress gradient etc., Figure 8.14b),
$S_D$ safety factor for endurance failure.

The rare operating conditions in which the transmission input torque is greater than the maximum engine torque $T_{max}$ (e.g. clutch engaging jolts) are ignored when designing for endurance strength. With such a design ($\sigma_{b,perm}$ significantly smaller than $R_m$) and ductile materials these are normally tolerable. The stress peaks are dissipated by local plastic flow. But a strength analysis is required if extremely severe and frequent impacts are anticipated.

### 8.3.9 Designing for Operational Fatigue Strength

Considerations are different when designing for a finite life. In this case the maximum engine torque $T_{max}$ does not have to be endured indefinitely; it is sufficient if the transmission shaft does not fail during the required service life under the stress of a given load profile. The calculations required to determine the service life are extremely extensive, and are therefore normally carried out by computer.

An example follows to illustrate how the calculation can be performed manually. The requirements for the calculation are that:

- The "design load profile" for the transmission concerned is available.
- The proportion of revolutions of the transmission input shaft related to the cycle as a function of the currently selected gear $i$ and the stress class $m$ is known.
- In case a shaft other than the transmission input shaft is to be calculated, then the revolutions of this shaft must be calculated relative to the cycle with the ratio of the currently selected gear.
- In the critical cross-section, the reference stress $\sigma_v$ must be determined for each gear at load $T_{max}$.

The reference stress $\sigma_v(T_{max})$, which arises when the transmission input torque $T_{max}$ is applied, was calculated in Section 8.3.6, Eq. 8.18. If the load $T_G$ at the transmission input is now varied, $\sigma_v$ varies proportionally. Therefore the following relationship applies:

$$\sigma_v(T_G) = \sigma_v(T_{max})\frac{T_G}{T_{max}}. \tag{8.24}$$

$\sigma_v(T_G)$ is thus a reference stress value in the critical cross-section, and depends on the gear selected, $i$, and the load $T_G$ of the $m$-th stress class.

The service life calculation proceeds in a similar manner to the Palmgren-Miner damage accumulation hypothesis described in Section 7.4.3. In contrast to the gearwheel service life calculation, transmission shafts are subject to a torque that varies with the gear currently selected. As a consequence, the distance travelled in the various gears must be taken into account as well as the distance covered in the various stress classes.

The number of possible load cycles is given by Equation 7.27 on the basis of the design load profile with $D_{act} = 1.0$,

$$z = \frac{N_D}{\sum_{i=1}^{j}\sum_{m=1}^{n} h_{im}\left(\frac{\sigma_{v,im}}{\sigma_D}\right)^{+k}}, \tag{8.25}$$

where
$\sigma_{v,im}$ reference stress in the $i$-th gear, under load $T_G$ of the $m$-th stress class given by Equation 8.24,
$h_{im}$ proportion of transmission shaft revolutions related to the design basis cycle,
$\sigma_D$ endurance strength,
$N_D$ number of oscillations at which endurance strength is reached,
$k$ exponent of the Wöhler curve equation (see Section 7.4.1),
$i$ index of gears 1 to $j$,
$m$ index of stress classes 1 to $n$.

**Table 8.3.** Materials commonly used in automotive shafts, and their characteristics [8.3] (figures in $N/mm^2$)

| Description | | $R_m$ | $R_e$; $R_{p0.2}$ | $\sigma_{b,W}$ | $\sigma_{b,Sch}$ | $\tau_{t,W}$ | $\tau_{t,Sch}$ |
|---|---|---|---|---|---|---|---|
| Heat-treated steel | 25 CrMo4 | 800...950 | 530 | 430 | 730 | 300 | 450 |
| | 34 Cr4 | 750...900 | 550 | 425 | – | – | 690 |
| Case-hardened steel | 16 MnCr5 | 900...1400 | 640 | 520 | 770 | 370 | 520 |

### 8.3.10 Common Shaft Materials

The characteristics of normal shaft materials most important for strength design are summarised in Table 8.3.

## 8.4 Calculating Deformation

Confirming that the deflection and bending angles of transmission shafts do not exceed the permissible limits is as important as analysing their strength.

The strength design computation shown in Section 8.3 is easy to program. This method is no longer relevant to calculating the deflection of transmission shafts with graduated diameters. For simpler cases the Castigliano method or the Mohr graphical method could be used. But the loads that occur in reality are normally more complex, so that special programs are required to calculate the deflection curve.

One common method is the *beam deflection transfer matrices method* [8.1]. Both the load profile and the deflection profile can be calculated using this method. Only the principles of this method will be given here.

The starting point is to divide the beam into sections in which the load $q_i$ and the flexural rigidity $E_i\ I_i$ are constant (Figure 8.15). For each section, the relation between deflection $f$, bending angle $\varphi$, bending moment $M_b$ and shear force $Q$ is described by the linear system of differential equations

$$\frac{df}{dx} = \varphi(x); \quad \frac{d\varphi}{dx} = \frac{M_b(x)}{E_i\ I_i}; \quad \frac{dM_b}{dx} = Q(x); \quad \frac{dQ}{dx} = q_i(x) = konst. \tag{8.26}$$

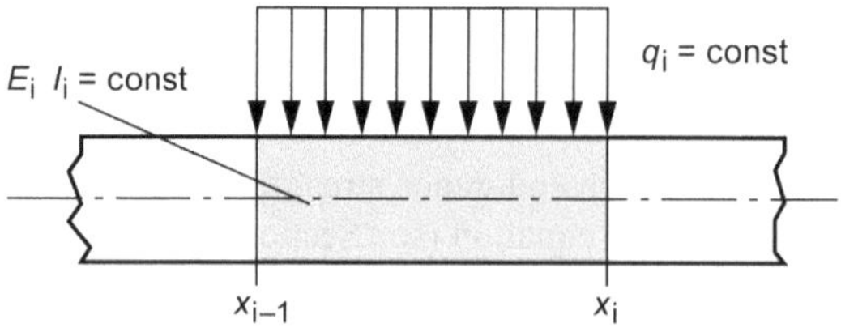


**Fig. 8.15.** Beam section for calculation using transfer matrices

**Table 8.4.** Reference values for permissible deflection and bending angles for shafts of gearwheel transmissions

| Shafts | Deflection | Bending angle |
|---|---|---|
| General rule for gearwheels | $f_{perm} \leq 0.01\, m_n$<br>$m_n$ standard module | $\tan\varphi_{perm} \leq \dfrac{2\, d_w}{10^4\, b}$<br>$d_w$ pitch circle diameter,<br>$b$ face width |
| Reference values for gearing | $f_{perm} \leq 0.02 - 0.06$ mm | $\tan\varphi_{perm} \leq 0.005$ for spur gears<br>$\tan\varphi_{perm} \leq 0.001$ for bevel gears |

A system of equations describing the values at point $x_i$ as a function of the values at point $x_{i-1}$ can be found by definite integration. The relationships can be easily represented in matrix form, hence the name "transfer matrices". There are some problems when introducing the boundary conditions, but the problem can be reduced to a simple programmable form by overlaying different solutions. Finally, the deflection $f$, bending angle $\varphi$, bending moment $M_b$ and shear force $Q$ can be calculated at any point along the shaft.

This method also enables statically indeterminate cases to be calculated by specifying appropriate boundary conditions. For example, with a three-bearing shaft, the deflection at the three bearings must be set equal to zero.

For comparison with the permissible deformations, the resultant deflections and bending angles must be calculated from the components in both planes:

$$f = \sqrt{f_y^2 + f_z^2}\,, \tag{8.27}$$

$$\tan\varphi = \sqrt{\tan^2\varphi_y + \tan^2\varphi_z}\,. \tag{8.28}$$

Toothed gearing systems are very sensitive to shaft deformation; tilting in particular can easily lead to canting of gearwheels or to edge pressure in the bearings. The requirements for permissible deflections and bending angles are correspondingly high (Table 8.4).

## 8.5 Flow Chart for Designing Transmission Shafts

Figure 8.16 shows a summary flow chart for calculating transmission shafts, referring to the equations and tables in Chapter 8.

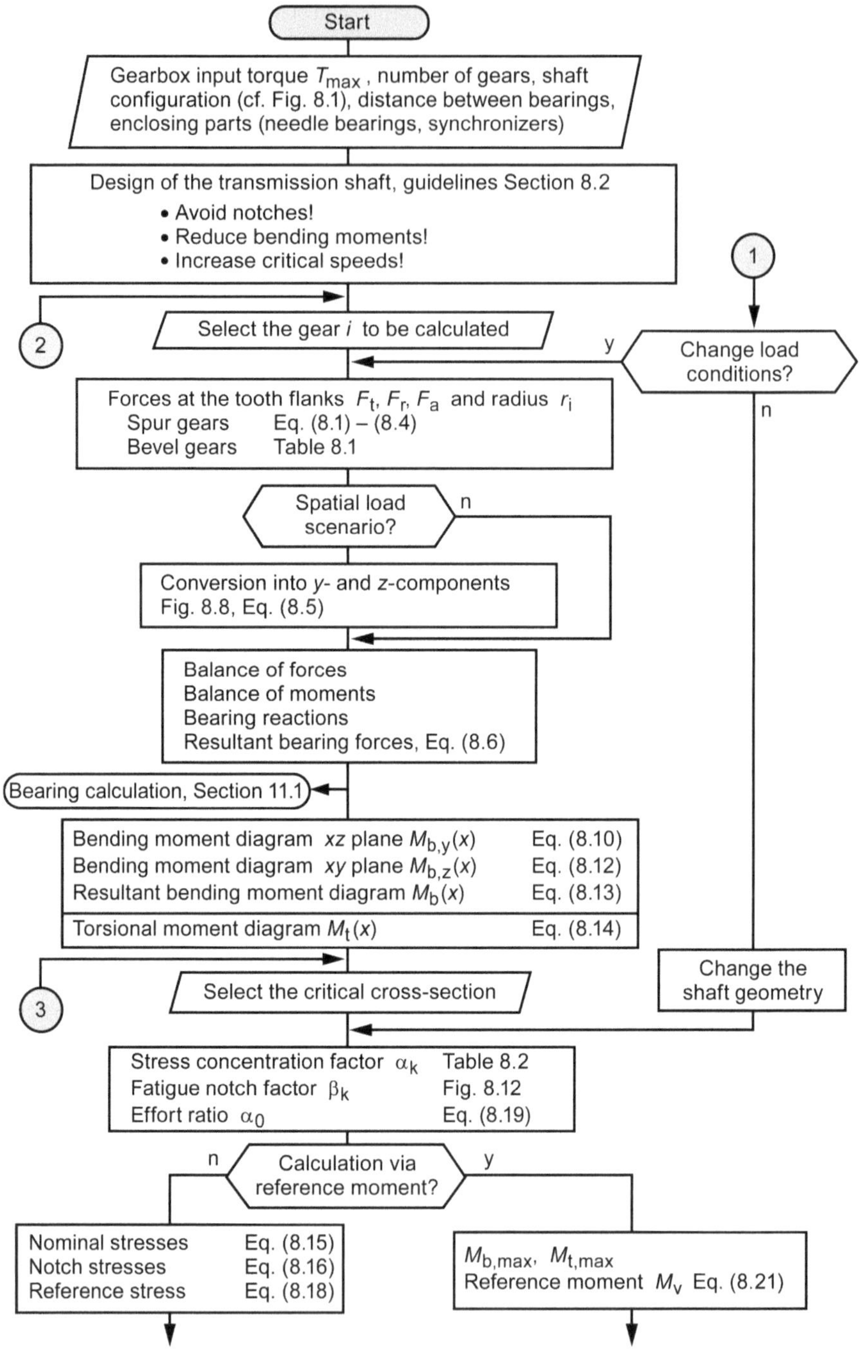


**Fig. 8.16.** Flow chart for designing transmission shafts

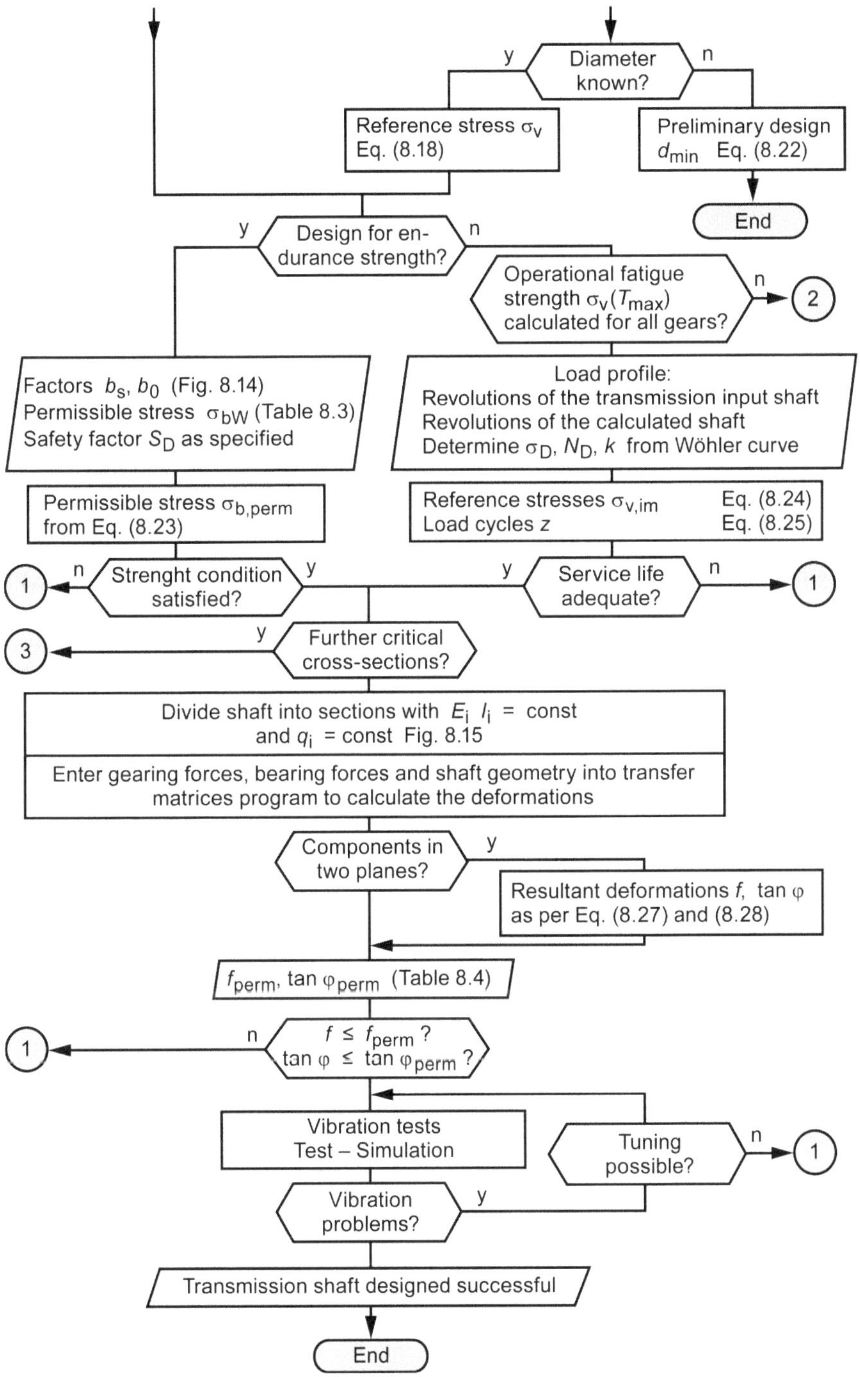


**Fig. 8.16.** *(continued)*

# 9 Gearshifting Mechanisms

*Changing connections*

Vehicle transmissions require devices to match the ratio, and thus the power available, to the prevailing driving conditions. "Power matching" is one of the four main functions of a vehicle transmission. In manual gearboxes, changing gear is controlled and carried out by the driver. Depending on the amount of automation, in all other gearboxes electronics and actuator systems take over this function partially or completely. Certain transmission functions, such as Neutral, Reverse, and Park are however still controlled by the driver using a shifting device.

The gearshifting mechanism thus plays an important role in the interface between driver and vehicle. Its handling has a major influence on perceived comfort. The components used in a gearshifting mechanism depend largely on whether shifting gear involves interrupting the power flow. Other factors are the type of vehicle (passenger car or truck), the type of drive (front-wheel or rear-wheel drive) and the operating conditions. In the following discussion, a distinction is made between

- *internal shifting elements*:
  shifting elements inside the transmission, such as selector bars, swing forks, synchronizers, multi-plate clutches and
- *external shifting elements*:
  shifting elements outside the transmission, such as gearshift levers, gearshift gate, linkage, four-bar linkages and cable controls.

The concept "shift-by-wire" pertains to the substitution of mechanical gearbox operation with electronic operation (see Section 9.1.3). Synchronizers are a decisive assembly group in gearboxes that shift with power interruption (MT, AMT) and are also used in dual clutch transmissions (DCT). This important internal shifting element is dealt with thoroughly in Section 9.2. Then, in Section 9.3, we focus on multi-plate clutches as essential components when shifting without power interruption. The parking lock is an assembly group whose function is to prevent vehicles that have no mechanical coupling between the engine and the output from rolling away (AT, DCT and CVT). This is the topic of Section 9.4.

Figure 9.1 shows internal shifting elements for engaging gearwheels into the power flow. A distinction is made between positive locking clutches (e.g. dog clutch) and friction clutches (e.g. multi-plate clutch).

H. Naunheimer et al., *Automotive Transmissions*, 2nd ed.,
DOI 10.1007/978-3-642-16214-5_9, © Springer-Verlag Berlin Heidelberg 2011

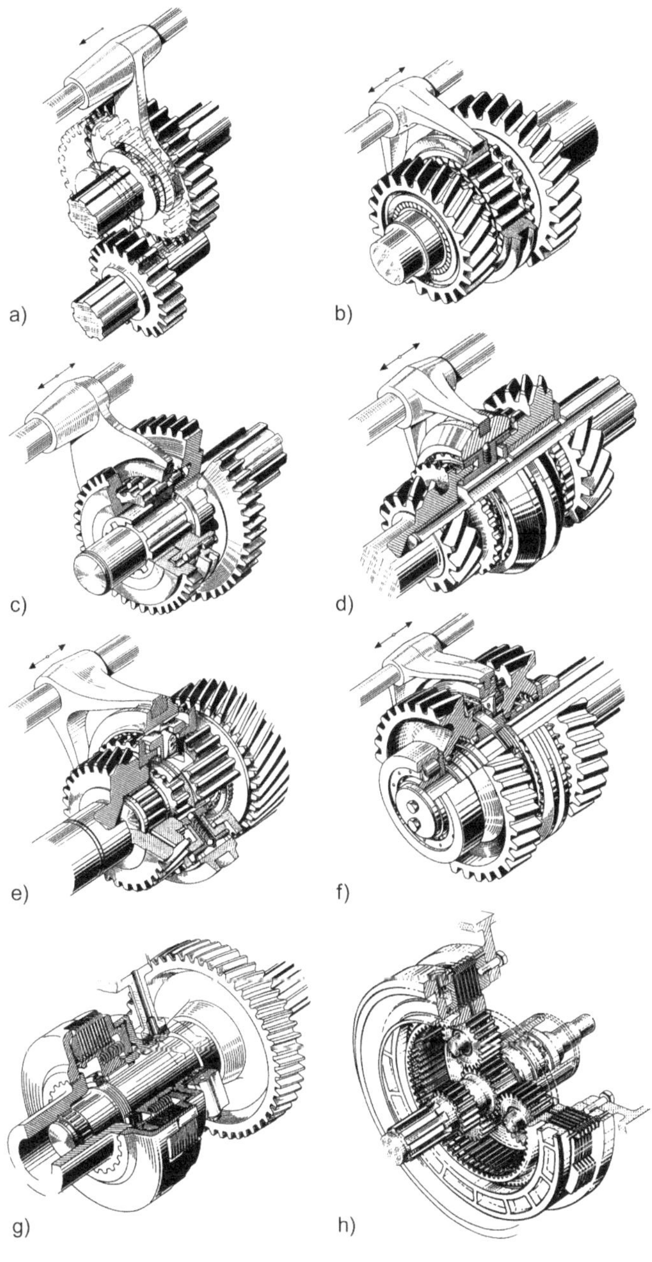


**Fig. 9.1.** Internal shifting elements in automotive transmissions. *a* Sliding gear; *b* dog clutch engagement; *c* pin engagement; *d* synchronizer without locking mechanism; *e* synchronizer with locking mechanism; *f* servo lock synchronizer mechanism (Porsche system); *g* hydraulically activated multi-plate clutch for powershift transmission; *h* hydraulically activated multi-plate brake for planetary gear [9.17]

## 9.1 Systematic Classification of Shifting Elements

The variety of designs and combinations of internal and external shifting elements is virtually unlimited, so this section will restrict itself to describing the basic principles. Sections 12.1 and 12.2 explain some typical examples of existing designs. The morphological matrix gives an overview of shifting elements (Table 9.1).

**Table 9.1.** Morphological matrix for shifting elements

| **Parameters** | **Configurations (shifting elements)** | | | | |
|---|---|---|---|---|---|
| **External shifting system** | Linkage | Multi-bar linkage | Cable control | Shift-by-wire | — |
| **Example** | Single bar actuation [9.12] | VW Golf III, Figure 9.4 | Shift and select cable MB A-class [9.8] | Automatic transmission ZF 6 HP 26, Figure 12.25 | — |
| **Generate shifting force** | Mechanical, manual effort | Electro-mechanical | Electro-hydraulic | Electro-pneumatic | Electro-magnetic |
| **Example** | Shift lever | Transmission actuator, Figure 12.39 | Automatic transmission, Figure 9.1h | Commercial vehicle range unit, Figure 12.34 | Electro-magnetic clutch |
| **Gear selection and shifting force transformation** | Selector bars, levers | Ball joint, four-bar linkage | Selector shaft, turning shaft | Gearshifting drum | — |
| **Example** | 3-bar shift mechanism, Figure 9.3 | VW Golf III, Figure 9.4 | Figure 12.3 (ZF) | Smart (Getrag), Figure 12.14 | — |
| **Shifting** | Shift fork | Swing fork | Piston | — | — |
| **Example** | Figure 12.9 (MB) | Figure 12.1 (VW), Figure 12.3 (ZF) | Conventional automatic transmission, Figure 12.23 (MB) | — | — |
| **Frictional connection** | Single/Multi-cone | Spreader ring | Multi-plates | Belt | Sprag |
| **Example** | Cone synchronizer | Porsche synchronizer | Clutch, brake, Figure 6.30 | Brake, Figure 6.35 | Freewheel (AT), Figure 6.30 |
| **Positive engagement** | Dogs | Pins | Sliding gear | Draw key | — |
| **Example** | Figure 9.1b, d, e, f | Figure 9.1c | Figure 9.1a | Motorcycle gearbox | — |

### 9.1.1 Shifting Elements for Transmissions with Power Interruption

Shifting by hand entails more than bringing gearwheels into the power flow. An exact and smooth-running operation of the gearshift lever is needed. This involves the interaction of external shifting with internal shifting elements such as detent devices, guides and synchronizers (Section 9.2). Passenger cars must have attributes such as short shift strokes, a fluid shifting process, a crisp, sporty shifting feel and low shifting forces.

#### *1/ Internal Gearshift Systems for Manual and Automated Transmissions*

The simplest type of shift system is *sliding gears* (Figure 9.1a). The gearwheels are not constantly meshed but are shifted into the power flow as needed. Sliding gears are used for reverse gear in both passenger car and commercial vehicle transmissions.

Unsynchronized *constant-mesh transmissions* are often found in commercial vehicle transmissions. The constant-mesh gear pairs run on rolling bearings and have a positive locking connection to the transmission shaft via a sliding dog sleeve (gearshift sleeve) (Figure 9.1b). The gears are prevented from disengaging (gear dropout) by undercut dogs (Figure 9.2).

Gearshifting is always made up of a selecting movement and a shifting movement. The selecting movement selects the gearshift sleeve to be shifted, and the shifting movement moves the gearwheel into the power flow. Figure 9.3 shows an example of this for a *synchromesh gearbox* with direct shifting (gearshift lever on the gearbox housing, e.g. in commercial vehicles) by three selector bars. The gearshift lever *1* and the ball joint *2* serve to select the gear and transmit the manual effort.

The path where the gearshift lever can move a gearshift sleeve is known as the gate. When a gate is selected, the selector finger *3* of the gearshift lever engages in the grooves of one of the individual selector bars *4* (Figure 9.3). The selector bar is shifted axially with a longitudinal movement of the gearshift lever, thus changing gear.

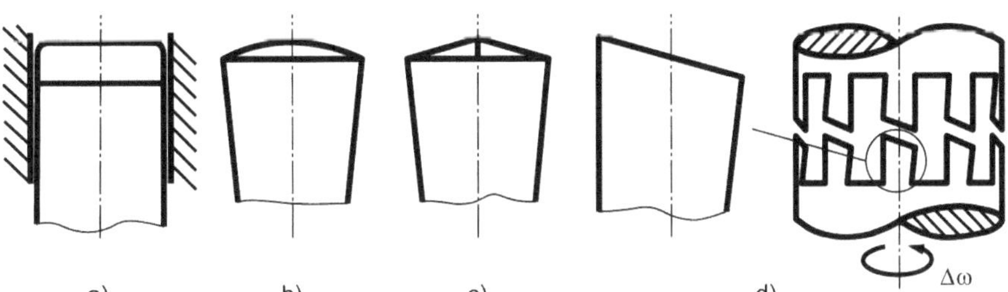


**Fig. 9.2.** Dog shapes in unsynchronized mechanisms. *a* Fuller dog; *b* ZF dog; *c* Berliet dog; *d* deflector dog (Maybach override dog): as long as there is relative movement, the bevelled deflector surfaces prevent engagement

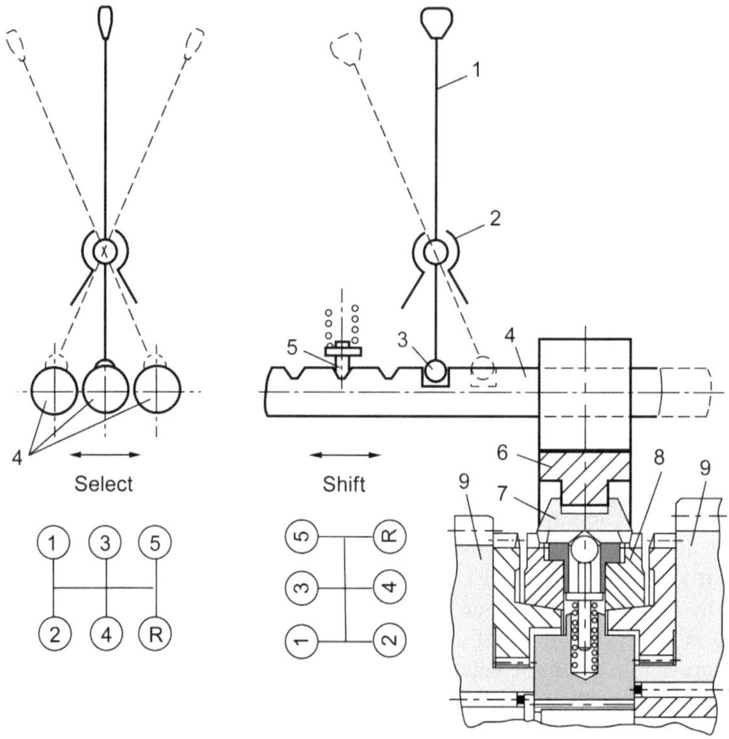


**Fig. 9.3.** Direct shifting of a 5-speed gearbox with three selector bars. *1* Shift lever; *2* ball joint; *3* selector finger; *4* selector bar; *5* detent; *6* shift fork; *7* gearshift sleeve; *8* synchronizer; *9* idler gear

A shift fork *6* engages in the gearshift sleeve *7*. Since each shift fork can shift either of two opposing idler gears *9*, three shift positions (two end positions and one middle position) of the selector bar *4* are secured by a detent *5*. The shift forks can both be moved axially as shown, or swivelled around a fixed pivot. This is referred to as a *swing fork*. By selecting appropriate lever proportions, gearshift effort can be reduced at the expense of increasing the shift stroke (see also Figure 12.3).

### 2/ External Gearshift Systems for Manual and Automated Transmissions

Kinematics is needed to describe the breakdown of gear changing into selecting and shifting movements. In order to avoid load changing reactions and vibrations in the gearshift lever and gearshift housing, the external gearshift system is suitably decoupled from the gearbox and the car body. The reverse gear is safeguarded against operating errors with a locking device, depending on the philosophy of the vehicle manufacturer. Figure 9.4 shows the external shifting elements, and some of the internal shifting elements of a passenger car with a transverse-mounted gearbox.

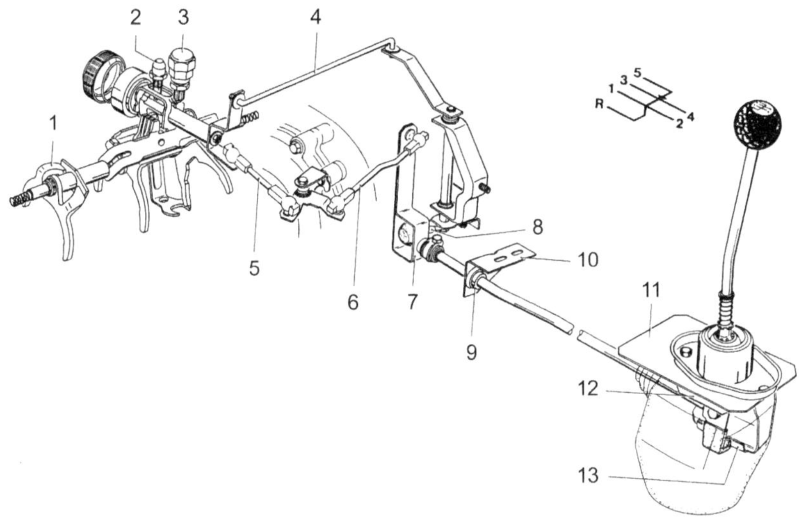


**Fig. 9.4.** Shifting mechanism of a 5-speed gearbox for front-transverse mounting (VW Golf III). *1* 5th gear shift fork; *2* selector shaft detent; *3* 5th gear lock; *4* connecting bar; *5* front selector rod; *6* rear selector rod; *7* selector lever; *8* relay lever; *9* selector bar bearing bush; *10* bearing plate; *11* shift lever bearing housing; *12* 5th gear end stop; *13* 1st/2nd gear end stop

The remote shifting is mechanical, using a four-bar linkage. The shifting arrangement shown was produced in series by VW in the Golf III until the mid-1990s.

Instead of costly linkage kinematics, it is popular to use cable controls, especially for passenger cars with transverse-mounted gearboxes. Shifting and selecting forces are transmitted from the gearshift lever in the shift housing to the internal gearshift system [9.3, 9.8].

The BMW 3 Series (model year 2005) and the Mercedes-Benz SLK (model year 2004) are examples of external gearshift systems with linkage activation [9.12]. Single-bar activation is used as the external gearshift design for these longitudinal-mounted passenger car gearboxes. They allow for short shift strokes and high precision without redirecting.

In commercial vehicle multi-range transmissions an additional control is needed to shift the gears in the splitter unit and/or range-change unit. Often a switch is fitted in the knob of the gearshift lever, which controls a pneumatic valve (see Figure 12.36). Table 9.2 shows the twelve possible conditions under which shifting may occur, highlighting those that are critical for heavy-duty commercial vehicles with respect to the duration of changing gears.

In heavy-duty commercial vehicles, gearshifting is often servo-assisted to reduce the effort required by the driver; the existing compressed air system is then used to activate the final control elements. The shifting sequence is electronically controlled. Driver effort is reduced to varying degrees depending on the degree of automation of shifting.

**Table 9.2.** The twelve possible gearshift states. ○ is non-critical and ● is critical with regard to changing gear for heavy-duty commercial vehicles

| | Shifting up | | Shifting down | |
|---|---|---|---|---|
| | Power | Overrun | Power | Overrun |
| Level | ○ | ○ | ○ | ○ |
| Uphill | ● | ○ | ● | ○ |
| Downhill | ○ | ○ | ○ | ● |

Automated manual transmissions (AMT) are currently popular in Europe for medium-duty and heavy-duty trucks [9.24]. (See also Figure 2.35 and Table 6.14 "Automation levels of manual transmissions".) Automation of manual gearboxes is an important aspect of the shifting of technology towards shift-by-wire (Section 9.1.3 "Shift-by-Wire").

### 9.1.2 Shifting Elements for Transmissions without Power Interruption

Automatic transmissions following the definition given in Figure 1.2 (dual clutch transmissions, countershaft-type automatic transmissions and conventional automatic transmissions) are designated as powershift transmissions. The gear to be shifted is brought into the power flow frictionally and without power interruption. The process of power shifting is explained in Sections 6.3.2 and 9.3.2.

The following remarks on internal shifting elements, particularly on external gearshift systems, can be applied to mechanical continuously variable transmissions (CVT) as well.

#### *1/ Internal Gearshift Systems for Automatic Transmissions*

Important internal shifting elements in automatic transmissions include:

- multi-plate clutches and
- freewheels.

Multi-plate clutches in automatic gearboxes are generally externally shifted clutches bathed in pressurized oil. (For more information on the layout and design of multi-plate clutches, see Section 9.3.) Freewheels are clutches activated by direction. A freewheel allows only one direction of rotation; in the other direction, it is blocked. Freewheels connect shafts to each other or against parts of the housing. In connecting shafts, it is removed from the lock by the higher relative speed and then rotates freely.

Freewheels grip independently at equal speeds. The gripping process is a step function with respect to the increase in torque and is only damped by elasticity. With respect to freewheel design, the particular manufacturers should be consulted. There are a great number of designs, varying in terms of the kind of clamp device and spring used. Important designs include:

- roller freewheel,
- ball freewheel and
- sprag freewheel.

With respect to the timely delivery of the torque (synchronous speed), a freewheel is absolutely unbeatable. For this reason, rollout shifts are easier to master with freewheels than with externally activated multi-plate clutches. On the other hand, freewheels are relatively large and are not tolerant to overloading. Thus, flanking protective functions should be provided in the software whenever possible.

Among existing designs, freewheels are used to support downshifts in the lower gears (e.g. Figures 6.32 and 12.23). Another significant area of application is the Trilok converter. In this case, the freewheel connects the reactor with the housing (see Section 10.4.6 as well as Figure 10.32).

### 2/ External Gearshift Systems for Automatic Transmissions

The external gearshift system is the interface between the driver and the transmission.

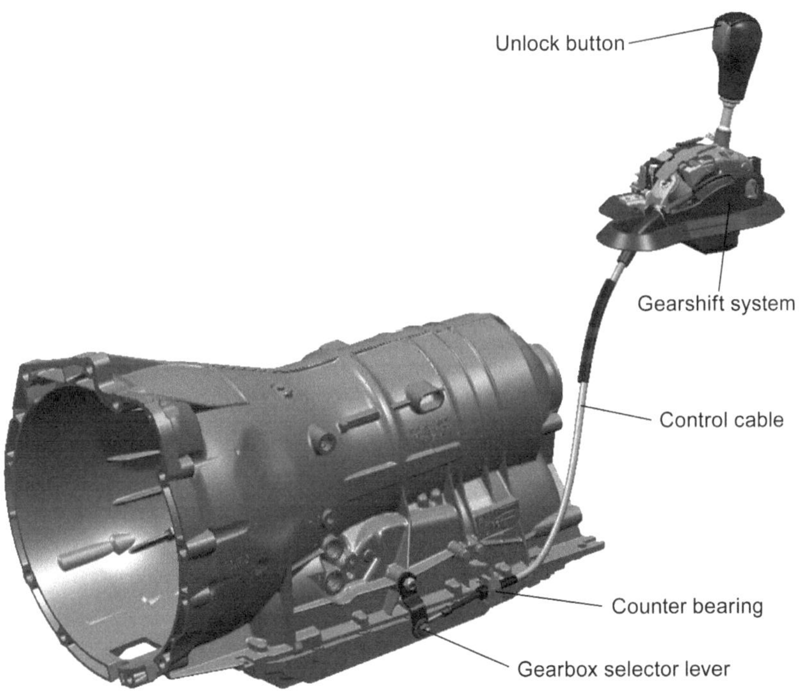


**Fig. 9.5.** External gearshift system of an automatic transmission: an AT for standard drive

In automatic transmissions, the wishes of the driver are transferred to the transmission usually by cable control (mechanical gearshifting, Figure 9.5), but also using electric signals (from tip-shifting to complete shift-by-wire). In passenger cars, the shift lever is customarily placed in the centre console, but also in the dashboard or on the steering column.

The bases for the design of automatic shift activation systems are two important safety regulations of the NHTSA (National Highway Traffic Safety Administration) in the Unites States. FMVSS (Federal Motor Vehicle Safety Standard) 102 contains passages relevant to the shift pattern and display. FMVSS 114 comprises shift system-relevant regulations on anti-theft devices and on the prevention of accidents caused by unauthorised use or the rolling away of parked vehicles with automatic transmissions.

Highly varied shifting patterns with two basic principles have been developed taking account of FMVSS 102. The "straight" path, in which all shift positions lie on a line, requires a locking system to prevent an unintentional gearshift from P (Park) or from N (Neutral) to R (Reverse) or from R to P. The lock is realised by means of a pull rod or a push rod in the selector lever tube and a gate. The selector lever can be released with an unlock button in the gear knob.

The second basic principle, the "labyrinth" path with laterally shifted positions, may have the same shifting sequence, but the selector lever can only be released by means of movement to the side when shifting.

Display of the current and possible selector lever positions required by FMVSS 102 occurs by means of corresponding symbols in the gate area (cover panel, display) and often also in the display of the dashboard.

The so-called key lock or interlock system originated from the requirements of FMVSS 114 for passenger cars. Here, a logical connection between the ignition lock and automatic gearshift activation has been made. On the one hand, the selector lever can only be moved from P if the ignition key releases it; on the other hand, the ignition key can only be removed if the selector lever is locked in P (parking lock engaged in the gearbox). In this way, rolling away and unauthorised movement of the vehicle is prevented.

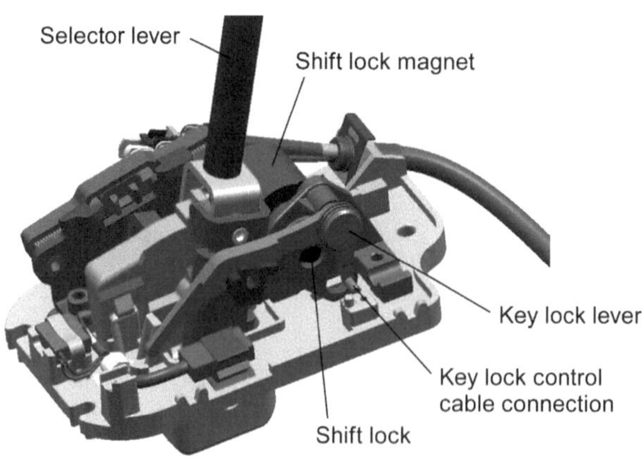


**Fig. 9.6.** Example of a key lock and shift lock system

There are two common key lock systems on the market. The mechanical variant connects the locks in the ignition lock and gearshift system with a cable control, while the electric variant has one locking magnet in both the ignition lock and gearshift system. The locking magnet is operated by means of sensor signals (Figure 9.6). Since electric key lock systems have to lock without electricity, they are usually supplemented with an emergency release so that the vehicle can still be moved in case of emergency.

The shift lock system in automatic gearshift systems has become the safety standard, serving as an additional safeguard against unintentional starting of the vehicle. Customarily, the selector lever is fixed in position P with an electric actuator, while release is only possible by activating the brake. In European vehicles, the shift lock function usually also takes hold in position N.

### 9.1.3 Shift-by-Wire

Mechanical devices such as linkages or cable controls have determined the connection between gearshift activation and the transmission for a long time. However, the increasing number of electronic systems in vehicles has also brought about changes in gearshift technologies. The concepts of "shift-by-wire" and "e-shifting" refer to the substitution of mechanical gearshift activation with electronics.

One driving force behind this change in technology is the automation of manual transmissions (AMT). Since the start in 1996, shift-by-wire for AMT has been pursued by almost every automobile manufacturer. A further motivation is the possibility of a freer design of the vehicle interior, since mechanical assembly restrictions originating from the transmission can be omitted. In the year 2001, e-shifting in conventional automatic transmissions was realised in mass production in the BMW 7 Series.

Gearshift activation controls can be built much smaller. Relevant forces and manual handlings of the gearshift process can be defined for the driver according to movement-physiological and haptic points of view. Other functions can be realised automatically, e.g. the automatic activation of the parking lock if the door is opened or the ignition key removed at negative seat occupancy detection.

The acoustic linkage between the transmission and the passenger cabin, eliminated by the omission of mechanical connections, makes the job of vehicle acousticians easier. Assembly becomes easier, and the necessary amount of adjustment work is reduced. A floor opening is no longer required, eliminating sealing problems and improving the crash behaviour of the vehicle.

However, there are also risks involved in shift-by-wire technology. In mechanical gearshift activation (e.g. cable control systems), the selector lever position always agrees with the gearbox position because of mechanical coupling. The gearbox positions are stable, making the selector lever positions stable as well. Because of this mechanical coupling, the driver is always informed about the actual gearbox position.

Electric gearshift systems eliminate any mechanical linkage between the selector lever and the gearbox. This causes a problem when shifting gears with stable selector lever positions: in case of error, the selector lever no longer agrees with the gearbox position, and the driver receives false information. In order to avoid this situation, gearshift activation with stable positions can be blocked or "guided" to the actual gearbox position. That is, in case of not allowed operations, the chosen position is not locked, the control element moves back, or in case of error a true position is taken automatically.

A less expensive solution is gearshift activation with a mono-stable selector lever here the mono-stable position is always the true position, and the driver has to continually read off the current gearbox position from a display controlled by the transmission.

Reaching the same level of availability and safety as mechanical systems will only be possible at considerable cost. Sensor redundancy, intrinsic safety of the electronics and backup communication in the vehicle network make such systems more expensive. Moreover, most automatic transmissions also require a mechanical emergency release of the parking lock.

Since the start in 1996, highly varied concepts and shift patterns have been developed for AMT shift-by-wire. There are systems modelled on automatic gearshift activation with gear locks, shift locks and key locks as well as systems that rely solely on the electronics with respect to operating errors. With the first shift-by-wire gearshift activation systems for automatic passenger car transmissions, more focus was put on the safety of the systems. The operation and function of the parking lock must also be designed in compliance with legal requirements.

## 9.2 Layout and Design of Synchronizers

This section deals with the transmission synchronizer, a crucial internal shifting element. It is a decisive assembly group in transmissions with power interruption (MT, AMT). Synchronizers are also found in dual clutch transmissions (DCT) shifting under load. The synchronizing process of the preselected idler gear takes place in the load-free branch of the transmission.

### 9.2.1 Synchronizer Functional Requirements

All passenger cars with manual gearboxes have synchromesh. In Europe, almost all commercial vehicles with manual transmission are also fitted with synchromesh gearboxes to improve road safety (a gear can be engaged at any time) and ease of use. The life of the synchronizer becomes critical in determining the system service life in the case of large transmissions, where there are high input torques and large masses to be synchronized.

Rotating shifting dogs can only be positively locked without "grating" if they have the same circumferential speeds. A synchronizing mechanism is therefore required to match the circumferential speeds of the parts to be connected in 0.1 to 0.3 seconds with the application of a minimum of force, and to prevent premature locking by blocking the shift movement. A gearwheel transmission with multi gears may be synchronized in the following ways [9.16]:

- synchronizing mechanism for each individual gear,
- central synchronizer for the whole transmission (Section 9.2.6) and
- speed synchronization by the prime mover (Section 9.2.6).

It is technically possible to dispense with synchronizers when

- there is a small gear step between the gears ($\varphi < 1.15$) or
- the masses of the gears are small, for example in a motorcycle gearbox.

Synchromesh is frequently omitted in commercial vehicles for reasons of economy and to improve transmission reliability. Transmissions without synchromesh are more robust.

A mechanical synchromesh unit as shown in Figure 9.7 frictionally matches the different speeds of the transmission shaft *6* (and the gearshift sleeve *5* rotationally fixed to it) and of the idler gear to be shifted *1*. When their speeds have been synchronized, the elements are positively engaged. This synchromesh unit incorporates a frictionally engaged clutch and a positive locking clutch (see also Figure 10.3 "Systematic classification of moving-off elements").

### 1/ The Gear Changing Process

This section describes the gear changing process using as an example a notional vehicle with a 2-speed coaxial countershaft transmission (Figure 9.8).

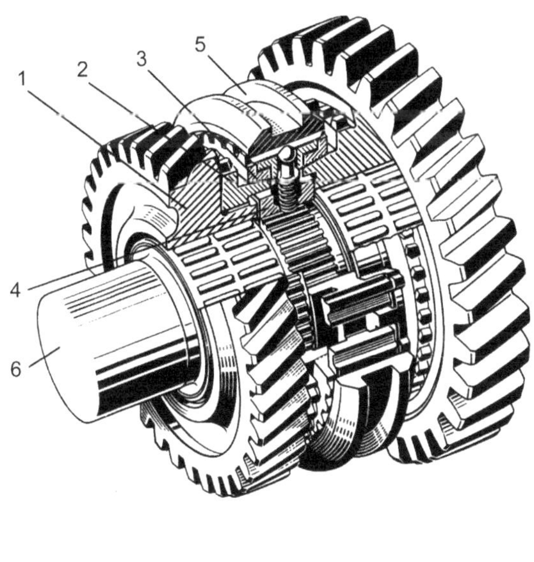


**Fig. 9.7.** Single-cone synchronizer (ZF-B), see also Figure 9.12.
*1* Idler gear with needle roller bearings;
*2* synchronizer hub with selector teeth and friction cone;
*3* synchronizer ring with counter-cone and locking toothing;
*4* synchronizer body with internal toothing for positive locking with the transmission shaft and external dog gearing for the gearshift sleeve;
*5* gearshift sleeve with internal dog gearing and ring groove;
*6* transmission shaft

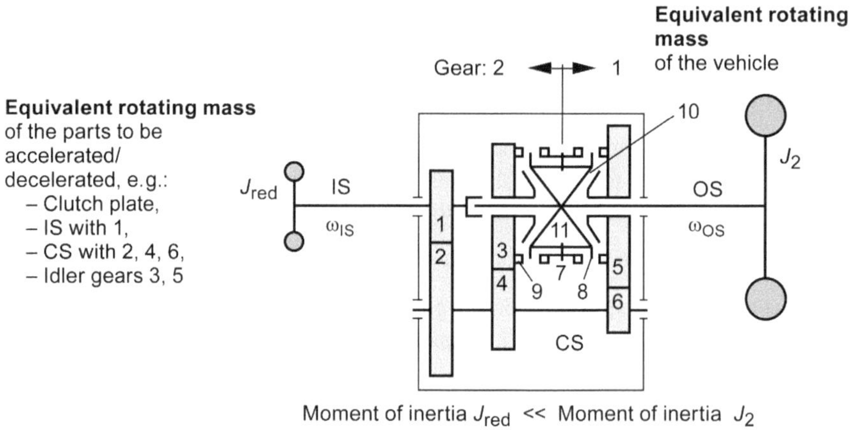


**Fig. 9.8.** Gear changing process. *1, 2, 4, 6* Fixed gears; *3, 5* idler gears; *7* gearshift sleeve with dogs; *8* locking mechanism; *9* selector teeth; *10* friction surfaces (cone and countercone); *11* synchronizer body; *IS* input shaft; *OS* output shaft; *CS* countershaft

As the speed of the vehicle $v$ in second gear drops, there is a particular angular velocity $\omega_{IS}$ at the input shaft *IS*. When the master clutch is fully engaged, $\omega_{IS} = \omega_M$. It is not possible to shift down into first gear until $\omega_M$ in first gear after shifting is less than $\omega_{M,max}$ (see also Figure 4.8).

The moment of inertia $J_2$ of the vehicle is significantly greater than the combined moment of inertia $J_{red}$ of the masses to be synchronized. The fall in angular velocity of the output shaft *OS* during the shifting period (slipping time $t_R$) may thus be ignored initially, i.e. $\omega_{OS}$ = constant. This simplification is no longer acceptable for more precise models, for example changing gear on a hill.

The shifting process is initiated at time $t_0$ (Figure 9.9). The gearshift sleeve *7*, rotationally fixed to the output shaft, rotates with an angular velocity $\omega_{OS}$ and the idler gear *5* to be engaged rotates with an angular velocity $\omega_{5,0}$.

The angular velocity difference to be adapted is $\Delta\omega_i = \omega_{OS} - \omega_{5,0}$ (Figure 9.9). After the response delay $t_1 - t_0$ the synchronization process starts at time $t_1$. The idler gear *5* and the masses connected to it are accelerated when shifting down. The angular velocity $\omega_5$ of the idler gear *5* to be shifted increases according to a certain function until a velocity $\omega_{OS}$ is reached, when the idler gear is synchronized with the gearshift sleeve *7*.

During the slipping time $t_R = t_2 - t_1$, the friction surfaces *10* slip at a relative angular velocity $\omega_{rel} = \omega_{OS} - \omega_5$. A similar result applies when shifting up from first to second gear. When the angular velocities are synchronized, a locking device *8* unlocks the shift movement and the gearshift sleeve *7* may be positively connected to the selector teeth *9* without "grating".

### 2/ Main Functions and Ancillary Functions

Table 9.3 shows the main functions and ancillary functions of synchronizers, along with possible mechanical solutions.

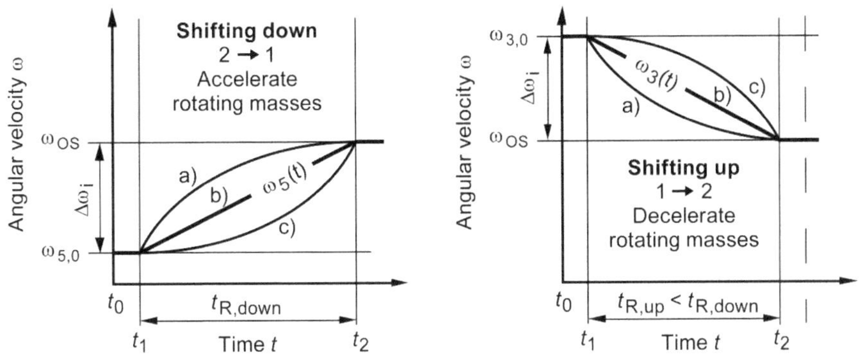


**Fig. 9.9.** Basic angular velocity curve during synchronization. $\omega$ increases or decreases depending on the gearshift effort and coefficient of friction curve according to a particular law. Ideal $\omega$ curve: *a* degressive; *b* linear; *c* progressive

**Table 9.3.** Main and ancillary functions of synchronizers

| Main functions | Comments | Mechanical solution |
|---|---|---|
| 1/ Adapt speed, accelerate or decelerate masses | • Low slipping time $t_R$, Table 9.4 | Internal energy transfer, using the energy accumulator $J_2$ (vehicle), power flow through friction clutch |
| 2/ Measure speed difference, determine synchronous speed | • Reliable functioning under all operating conditions | Speed comparison using friction, as a function of relative speed |
| 3/ Locking the positive engagement until speeds are synchronized | • Forcing the gears before speeds are synchronized should be difficult or impossible | Friction lock mechanism with differential speed dependent effect |
| 4/ Establish positive engagement and power flow | • Shift stroke *s* as short as possible<br>• Ensure positive locking, prevent gear dropout | Dog clutch with undercut toothing |
| **Ancillary functions** | **Comments** | **Mechanical solution** |
| 5/ Shifting comfort | • Shifting force profile<br>• Shifting force cycle | • Table 9.4 |
| 6/ Reliable operation under all conditions | • Low temperatures<br>• Quick shifting | • Arctic temperatures<br>• $t_R \leq 0.1$ s |
| 7/ Overload capacity | • Operator error | • Abuse test |
| 8/ Service life | • Adequate mechanical and thermal specification | • Passenger car ≥ 150 000 km<br>• Commercial vehicle up to 1 200 000 km |

**Table 9.3.** *(continued)*

| | | |
|---|---|---|
| 9/ Performance | • Synchronizable masses<br>• Slipping time<br>• Performance limits | • Design<br>• Permissible stress values, Table 9.7 |
| 10/ Costs | • Development/production<br>• Replacement of wearing parts | ——— |
| 11/ Weight/space constraints | • Reduce space required<br>• Shorten shift stroke | ——— |

### *3/ Speed Synchronization with Slipping Clutch*

The friction surfaces of mechanical synchronizers involved in speed synchronization are of flat, conical or cylindrical design. Systems using friction cones are common in both passenger car and commercial vehicle transmissions (Figure 9.10). The gearshift effort applied by the driver through the gearshift lever, shift fork and gearshift sleeve is amplified by a cone.

The following section concentrates on cone synchronizers. Cone synchronizers may be divided into

- internal cone synchronizers:
  - single-cone synchronizers, e.g. Borg-Warner system,
  - multi-cone synchronizers,
- external cone synchronizers.

Synchronizers using friction cones are a special case of friction clutches with planar friction surfaces, so the same theoretical basis applies.

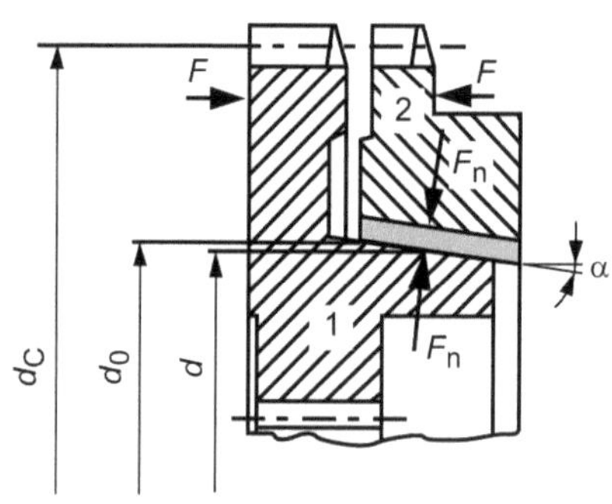


$F$ Gearshift effort
$F_n$ Normal force
$\alpha$ Cone angle
$j$ Number of friction surfaces
$d$ Effective diameter
$d_0$ Nominal diameter
$d_C$ Clutch diameter
1 Synchronizer hub
2 Synchronizer ring with counter friction surface

| Single-cone | Multi-cone | Multi-plate |
|---|---|---|
| $j = 1$<br>$\alpha = 6.5 – 8°$ | $j \geq 2$<br>$\alpha = 7 – 12°$ | $j \geq 2$<br>$\alpha = 90°$ |

**Fig. 9.10.** Common formats of mechanical synchronizers, dimensions

The normal force $F_n$ on the friction surfaces is derived from the gearshift effort $F$ as

$$F_n = F \frac{1}{\sin\alpha}. \tag{9.1}$$

The friction torque $T_R$ is derived from the gearshift effort $F$ and the coefficient of dynamic friction $\mu$

$$T_R = j\,F \frac{d}{2} \frac{\mu}{\sin\alpha}, \tag{9.2}$$

where $j$ is the number of friction surfaces and $d/2$ the effective radius. The simplification $d/2 = d_0/2$ is frequently used in practice. In order to prevent the cones from self-locking the following must apply for the cone angle $\alpha$:

$$\tan\alpha > \mu. \tag{9.3}$$

Equation 9.2 provides some starting points for design measures to increase performance and reduce gearshift effort. Multi-cone synchronizers accordingly offer less gearshift effort, or greater torque capacity, than single-cone systems. Multi-plate synchronizers have planar friction surfaces. There is no shift effort multiplication as there is with cone systems. The power handling capacity increases, and the shift effort decreases with the number of friction surfaces, but the overall length of the unit increases.

#### 4/ Synchronizer Dimensions

Figure 9.11 shows the main dimensions of a synchronizer.

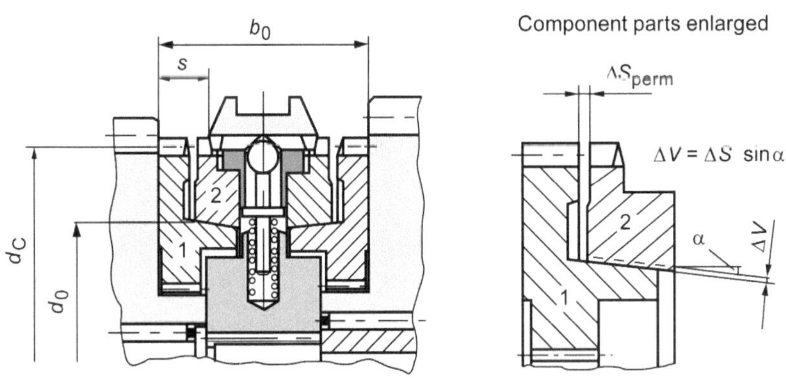


**Fig. 9.11.** Dimensions. $b_0$ Overall pack length; $d_0$ nominal diameter; $d_C$ clutch diameter; $\Delta S$ wear path; $\Delta S_{perm}$ permissible wear path incl. clearance; $s$ shift stroke at the gearshift sleeve; $\Delta V$ wear at the synchronizer ring; $\alpha$ cone angle

The wear on the friction surfaces is usually the factor that determines service life. The shift movement $s$ at the gearshift sleeve is approximately 7.5–13 mm. The permissible wear $\Delta S_{perm}$ is generally between 1 and 1.5 mm. The wear reserve of the synchronizer unit is calculated by subtracting the operating clearance from the permissible wear. The maximum wear $\Delta V_{max}$ is of the order of 0.15 mm per friction pairing for cone synchronizers.

### 9.2.2 The Synchronizing Process

Single-cone synchronizers based on the "Borg-Warner" system are widely used in manual transmissions. The various phases of the synchronizing process are shown in Figure 9.13, based on the ZF-B synchronizer (Figure 9.12, "B" standing for Borg-Warner system).

The synchronizer body *4* is fixed to the transmission shaft. The synchronizer ring *3* is guided by stop bosses in the synchronizer body. These are narrower than the grooves in the synchronizer body, which allows the synchronizer ring a certain amount of freedom to twist radially.

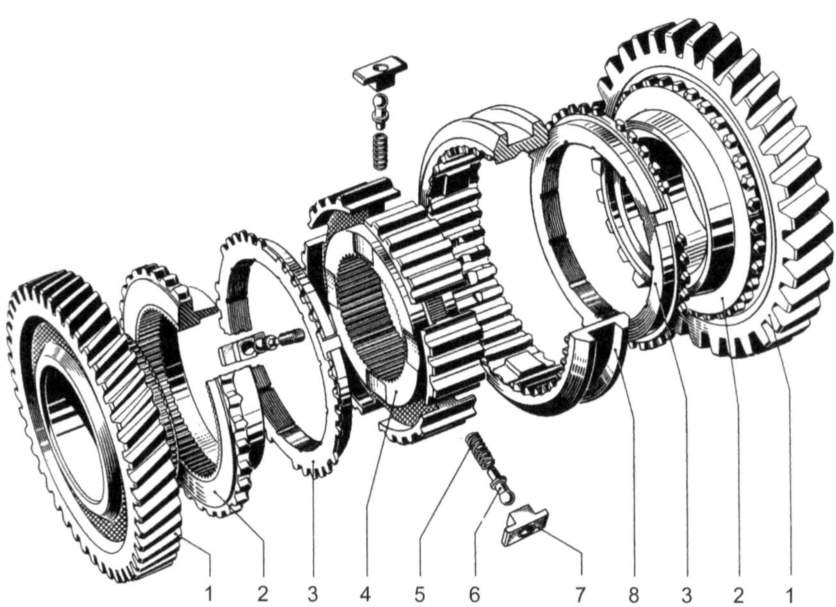


**Fig. 9.12.** Borg-Warner system single-cone synchronizer (ZF). *1* Idler gear running on needle roller bearings; *2* synchronizer hub with selector teeth and friction cone; *3* main functional element, synchronizer ring with counter-cone and locking toothing; *4* synchronizer body with internal toothing for positive locking with the transmission shaft and external toothing for the gearshift sleeve; *5* compression spring; *6* ball pin; *7* thrust piece; *8* gearshift sleeve with internal dog gearing

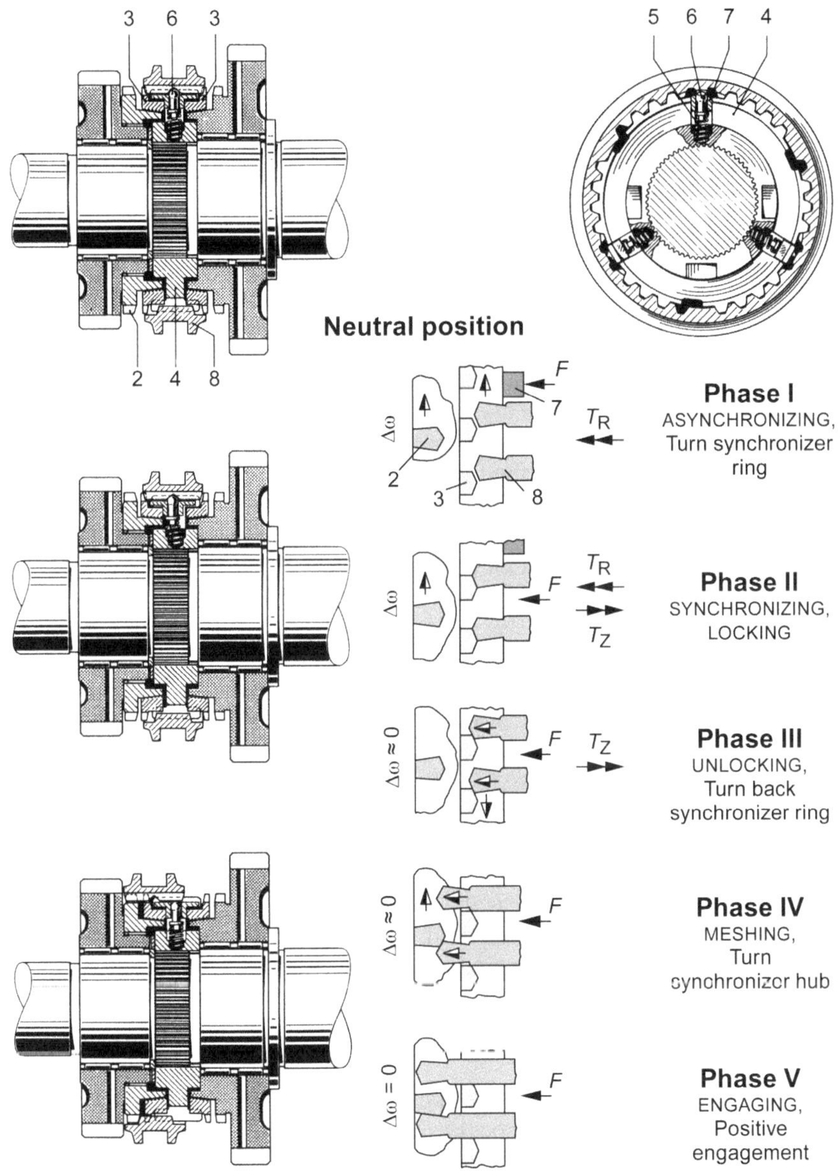


**Fig. 9.13.** Synchronizing process. Arrows with half-filled tips indicate the direction of movement, the torque arrows indicate the moments acting on the synchronizer ring. *2* Synchronizer hub with selector teeth and friction cone; *3* main functional element, synchronizer ring with counter-cone and locking toothing; *4* synchronizer body with internal toothing for positive locking with the transmission shaft and external toothing for the gearshift sleeve; *5* compression spring; *6* ball pin; *7* thrust pieces; *8* gearshift sleeve with internal dog toothing

Before the shifting process starts, the gearshift sleeve is held in the middle position by a detent. The gearshift force $F$ triggers the axial movement of the gearshift sleeve *8,* which causes the ball pins *6* to act on the thrust pieces *7* to press the synchronizer ring *3* with its counter-cone against the friction cone of the synchronizer hub *2*. The speed difference between the gearshift sleeve *8* and the synchronizer ring *3* relative to the idler gear *1* causes the synchronizer ring to turn until the dogs contact the groove walls. This first phase of the synchronizing process is known as "asynchronizing" (Figure 9.13).

The gearshift sleeve is moved further. This brings the bevels of the internal dog gearing of the gearshift sleeve *8* and the external dog gearing of the synchronizer ring *3* into contact. The main synchronization action starts, Phase II. The gearshift force is applied to the synchronizer ring via the thrust pieces *7* and the dogs *8*, the force being divided between them. The gearing torque $T_Z$ arising at the bevels acts so as to open the locking device. $T_Z$ is smaller than friction torque $T_R$ that acts to close the locking device. In the slipping phase the gearshift sleeve cannot be shifted. In the literature the gearing torque $T_Z$ is frequently referred to as the index torque $T_I$, and the friction torque $T_R$ as the cone torque $T_C$.

When speed synchronization has been achieved, the friction torque tends towards zero, Phase III. The unlocking process starts. The gearing torque becomes greater than the friction torque, and acts via the bevels to turn back the synchronizer ring. The gearshift force decreases rapidly in this phase. Throughout the axial movement of the gearshift sleeve, the spring-loaded ball pin slides along the inclined grooved surface. This presses it against the spring *5* into the thrust piece, until it is covered by the gearshift sleeve.

During shifting, the gearshift sleeve toothing encounters the bevels of the selector teeth of the synchronizer hub *2*. In this Phase IV the ball pin is covered. The synchronizer ring is pressed against the friction cone of the synchronizer hub only by residual pressure via the thrust pieces. This residual pressure arises from the friction between the moving gearshift sleeve and the thrust pieces (with ball pins). The gearshift sleeve toothing twists the synchronizer hub relative to the synchronizer ring. The shift movement is enabled. The gearshift sleeve positively engages the power flow between the gear pair and the transmission shaft, Phase V.

### *Shifting Comfort*

Proper sequential co-ordination of the above functions is important to ensure that the shifting and synchronizing processes are as comfortable as is required. The precise design of the parts involves a great deal of know-how, for example in the design of locating faces and in determining functional clearance [9.2, 9.4].

Shifting comfort is especially important when drivers are evaluating a gearshift. In mechanically operated shifting devices the manual force applied by the driver is transmitted by the gearshift lever and transmission element (e.g. shifting linkage) to the gearshift sleeve.

**Table 9.4.** Standard values for acceptable manual effort $F_{H,perm}$ and slipping time $t_{R,perm}$

| Standard value | Gear | Passenger car | Commercial vehicle | | |
|---|---|---|---|---|---|
| | | | Main gearbox | Splitter unit | Range-change unit |
| Permissible manual effort $F_{H,perm}$ | 1 – z | < 120 – 80 N | < 250 – 180 N | Pneumatic | Pneumatic |
| Permissible slipping time $t_{R,perm}$ | 1 – z | < 0.25 – 0.15 s | < 0.4 – 0.25 s | 0.15 s | 0.2 s |

The transmission ratio of this linkage depends on the engineering design, and normally varies between 7:1 to 12:1. The efficiency with which gearshift effort is transmitted must also be taken into account – it is frequently less than 70%. There are standard values for the maximum permissible slipping time $t_{R,perm}$ and the maximum permissible manual effort $F_{H,perm}$ (Table 9.4).

Figure 9.14 gives an example illustrating the relationship between gearshift effort $F$ at the gearshift sleeve and the slipping time $t_R$. The gearshift effort actually applied by the driver depends very much on driving style and traffic conditions. Very low ambient temperatures have a major impact on the manual effort required and slipping time. However, ambient temperature has little effect on the overall shifting comfort, since the gearbox oil warms up relatively quickly. The main problems synchronizers cause for the operator arise from

1/ sticking,
2/ upshift grating and
3/ shifting noise.

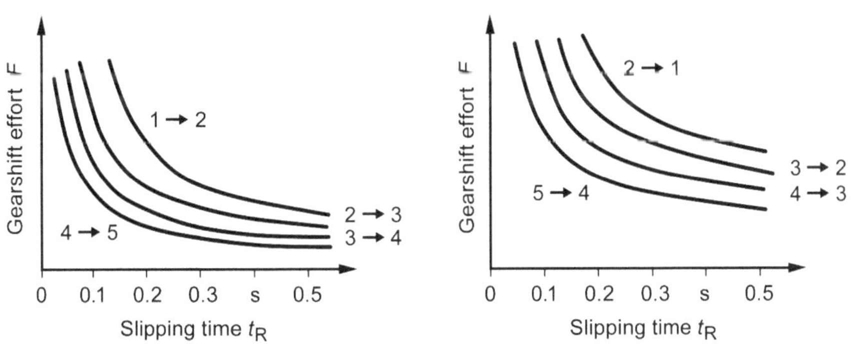


**Fig. 9.14.** Interrelationship between shifting effort $F$ at the gearshift sleeve and the slipping time $t_R$ as a function of the gear selected

### *1/ Sticking*

Once the gearwheel speeds have been synchronized and the synchronizer unit has unlocked, the gearshift effort required from the driver drops noticeably before the gear is engaged. The gearshift sleeve should now twist the synchronizer hub with little effort (in Phase IV, Figure 9.13), and it should be possible to push it easily into the engaged position. Friction is generated between the gearshift sleeve and the thrust pieces (with ball pins) when they slide. If this friction is high or the clearance characteristic is unfavourable, the residual pressure on the synchronizer ring can be so high that a large gearshift effort is required to turn the synchronizer hub relative to the synchronizer ring. The driver perceives the new increase in gearshift effort (referred to in the literature as the "second pressure point") as the gearshift mechanism sticking. The term "clearance characteristic" refers to the synchronizer ring detaching itself from the friction cone of the synchronizer hub.

There are engineering design measures that can be taken to counteract sticking. For example the clearance characteristic can be improved by having an "unwinding" thread in the synchronizer ring friction surface (see also Section 9.2.4 "The Tribological System of Synchronizers"). The bevel angle (opening angle $\beta$) of the internal dog gearing of the gearshift sleeve and the drag torque of the transmission also have a considerable effect. For more information, see Section 9.2.3 Number 4/ "Designing Locking Toothing for Locking Effect".

### *2/ Upshift Grating*

Upshift grating is typical of gearshifting problems that occur at low temperatures (it is often referred to as "cold scraping"). It occurs especially when the gearbox oil is cold and when shifting from first to second gear. It generally no longer occurs at temperatures above 10°C.

During the transition from Phase III to Phase IV, after unlocking, the gearshift sleeve moves along a certain path before the gearshift sleeve and synchronizer hub are positively engaged. During this phase, the ball pin is covered. Only the residual pressure now presses the synchronizer ring against the synchronizer hub. During this period, the idler gear is relatively free. The drag torque can cause a speed difference between the gearshift sleeve and the selected idler gear, which causes the dogs to grate when the gear is engaged.

### *3/ Shifting Noise (Grating)*

If the synchronizer does not function properly, the gearshift sleeve can engage in the selector teeth before the speeds have been synchronized. This then causes grating or scraping noises. Whether or not these noises occur depends largely on the driver's gearshifting action. Grating can occur when the driver forces the gear to engage too quickly. The slipping time is too short and the gearshift sleeve and the synchronizer hub are positively engaged before the speeds have been synchronized.

But grating may also be caused by torsional vibrations, made possible especially by circumferential backlash in the powertrain. The vibration excitation alters the coefficients of friction. This "shaking action" facilitates the movement of the gear-shift sleeve dogs on the bevels of the locking teeth [9.11] (see also Section 9.2.3 Number 4/ "Designing Locking Toothing for Locking Effect").

### 9.2.3 Design of Synchronizers

Synchronizers are subject to high levels of stress. This applies particularly to commercial vehicle synchronizers. Figure 9.15 shows the factors affecting its functioning and service life. A single operator error may permanently damage or destroy the synchronizer. The principal criteria according to which synchronizers are designed are the following:

- *function:*
  - synchronizable masses, shifting comfort,
  - cold shifting behaviour, shifting in new condition ("green shiftability"),
  - locking safety and
  - abuse.
- *service life:*
  - mechanical stress on the selector teeth,
  - mechanical stress on the synchronizer ring,
  - thermal stress on the friction surfaces and
  - nominal service life (see Table 9.3).

#### *1/ Synchronizer Performance Limits*

Figure 9.16 shows a dog gearing of a commercial vehicle synchronizer hub that has been damaged by grating. There is a ring groove in the case-hardened steel (16 MnCr5) of the friction cone as "drainage" to cut through the oil film (see also Section 9.2.4).

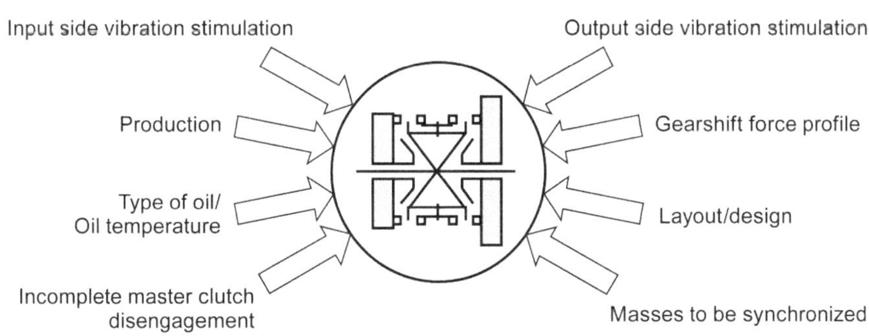


**Fig. 9.15.** External factors affecting the function and service life of synchronizers

Figure 9.17 shows a fracture at the stop boss of a synchronizer ring made of a special brass. Such fractures are the consequence of torsional vibration.

**Fig. 9.16.** Damage to the dogs of the selector teeth caused by grating. Ring groove in the friction cone of the commercial vehicle synchronizer hub

**Fig. 9.17.** Damage to the synchronizer ring caused by torsional vibration, fracture at the stop boss

They occur mostly in engines that run very irregularly (e.g. diesel engines with direct injection). This picture shows clearly the synchronizer ring locking toothing.

**Fig. 9.18.** Scuffing and thermal streaks on the friction cone of a commercial vehicle synchronizer hub

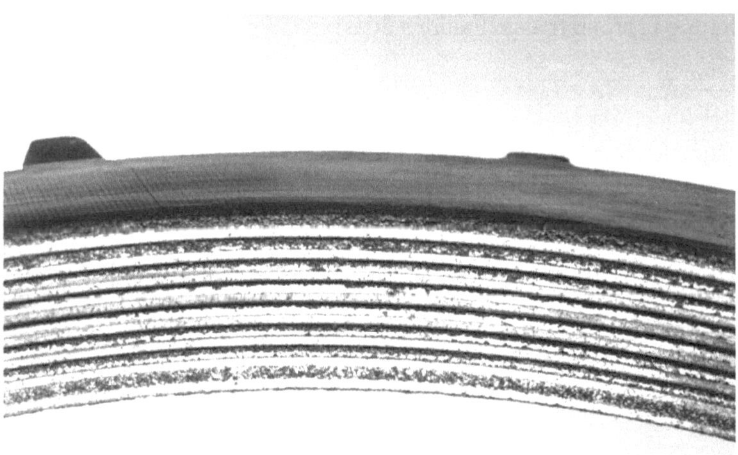

**Fig. 9.19.** Increased wear at the molybdenum friction surface of a commercial vehicle synchronizer ring caused by thermal overload

In a synchronizer with dimensions sufficient for the mechanical stress encountered, the thermal stress determines the performance limits. The surface temperature $\vartheta$ can reach peak levels of up to 1000°C at particular points in less than 0.1 seconds [9.25]. If the thermal stress exceeds the permissible levels, the friction surfaces are damaged.

A distinction is made between transient overload caused by temperature peaks, and continuous overload such as that caused by excessive slipping times. Figure 9.18 shows scuffing and thermal streaks on the friction cone of a commercial vehicle synchronizer hub.

Figure 9.19 shows increased wear on a commercial vehicle synchronizer ring. The synchronizer ring shown has a steel ring with a molybdenum friction surface and ground-in grooving. Even without overload, thermal stress arising during normal service has a detrimental effect on synchronizing action. Acceptable frictional power related to surface, $P_{A,perm}$, is the measure normally used to assess thermal stress.

### 2/ Basis for Design Calculation

As shown in Section 7.4 "Operational Fatigue Strength and Service Life", not all components in a transmission are susceptible to service life calculations (see also Section 17.2.2 "Qualitative Reliability Analysis"). The service life of "B" components in the "A, B, C" analysis cannot be calculated. Synchronizers are "B" components, so design engineers have to rely upon empirical data.

In calculations relating to mechanical synchronizers the general fundamental equations for shiftable friction clutches apply, as described in [9.27]. The torque equilibrium for a synchronizer as in Figure 9.20 is:

$$T_L + \frac{d\omega}{dt} J_{red} + T_V + T_R = 0 \,. \tag{9.4}$$

When the master clutch is fully opened, the load torque $T_L = 0$ throughout the synchronizing process. The torque losses $T_V$ are the result of bearing losses, oil churning losses, oil drag losses and oil squeezing losses. The torque loss figures are specific to each individual transmission. When shifting up, the gearwheel to be shifted is decelerated with the rotating masses reduced to its axis $J_{red}$. Friction torque and torque losses act in the same direction.

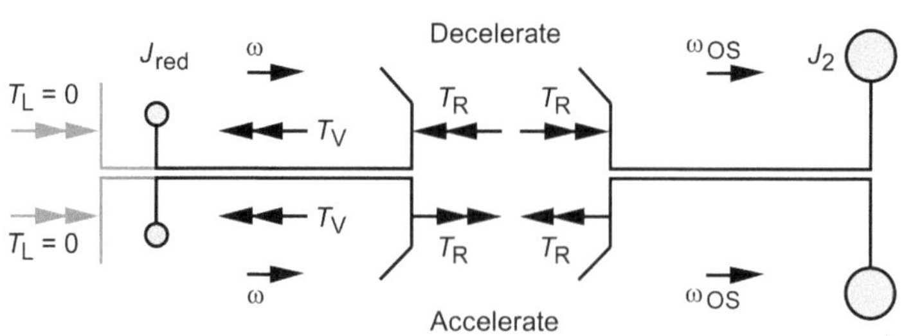


**Fig. 9.20.** Synchronization of two equivalent rotating masses

When shifting down, the gearwheel to be shifted is accelerated with the rotating mass reduced to its axis. Friction torque and torque losses act in opposite directions. The acceleration torque $T_B$ can be calculated as:

$$T_B = \frac{d\omega}{dt} J_{red}. \tag{9.5}$$

Equation 9.4 gives the friction torque $T_R$ as

$$T_R = -\frac{d\omega}{dt} J_{red} - T_V, \quad \left( \frac{d\omega}{dt} < 0 \Rightarrow T_R > 0; \frac{d\omega}{dt} > 0 \Rightarrow T_R < 0 \right). \tag{9.6}$$

The power $P$ transmitted momentarily to the synchronizer is derived from the product of the friction torque $T_R$ and the relative angular velocity $\omega_{rel}$ of the parts to be synchronized

$$P = T_R\ \omega_{rel}. \tag{9.7}$$

From this the frictional work $W$ per gearshift with slipping time $t_R$ may be calculated thus

$$W = \int_0^{t_R} P\, dt. \tag{9.8}$$

### 3/ Practical Design for Acceptable Thermal Stress

This section presents a procedure for designing synchronizers "by hand". Simplifications are needed to make this feasible. For the slipping time $t_R$, if the following assumptions are made

- the gearshift effort $F$ = constant,
- the friction coefficient $\mu$ = constant,
- torque losses $T_V$ = constant,

then

- friction torque $T_R$ = constant,
- change in angular velocity $d\omega/dt$ = constant.

The errors resulting from the simplifying assumptions made are largely offset in the calculation by the acceptable stress values. The acceptable stress values are derived from experience.

### Reduction of Moments of Inertia

As a result of the steps between ratios, the masses involved in the synchronizing process are subject to different angular accelerations. In order to be able to use only one angular velocity for all the masses involved in the calculation, the masses are related to one axis. This is normally the rotation axis of the idler gear to be shifted. In general:

$$J_{\text{red,i}} = J_{\text{i}} + \sum_{\text{k=1}}^{\text{i}} J_{\text{k}} \frac{1}{i_{\text{k}}^2}. \tag{9.9}$$

**Example:** When shifting the transmission shown in Figure 9.21 from second to first gear, the masses are reduced onto the rotation axis of the idler gear *7*. In this case:

$$\begin{aligned} J_{\text{red,7}} &= J_7 + (J_{\text{C}} + J_{\text{IS}} + J_1)\left(\frac{z_7}{z_8}\right)^2\left(\frac{z_2}{z_1}\right)^2 + \\ &(J_{\text{CS}} + J_2 + J_4 + J_6 + J_8 + J_{10} + J_{14})\left(\frac{z_7}{z_8}\right)^2 + \\ &\left[J_3\left(\frac{z_4}{z_3}\right)^2 + J_5\left(\frac{z_6}{z_5}\right)^2 + J_9\left(\frac{z_{10}}{z_9}\right)^2 + J_{11}\left(\frac{z_{10}}{z_{11}}\right)^2 + J_{13}\left(\frac{z_{14}}{z_{13}}\right)^2\right]\left(\frac{z_7}{z_8}\right)^2. \end{aligned} \tag{9.10}$$

On the assumption that the output shaft *OS* and the components connected to it are not subject to any change of angular velocity during synchronization, their moments of inertia may be ignored.

## Coaxial "In-Line" Countershaft Transmission

The term in-line transmission is a general term for a two-stage coaxial countershaft transmission. Sometimes the term is more closely cirmcumscribed and relates only to coaxial countershaft transmissions where all the synchronizers are mounted on the output shaft (main shaft).

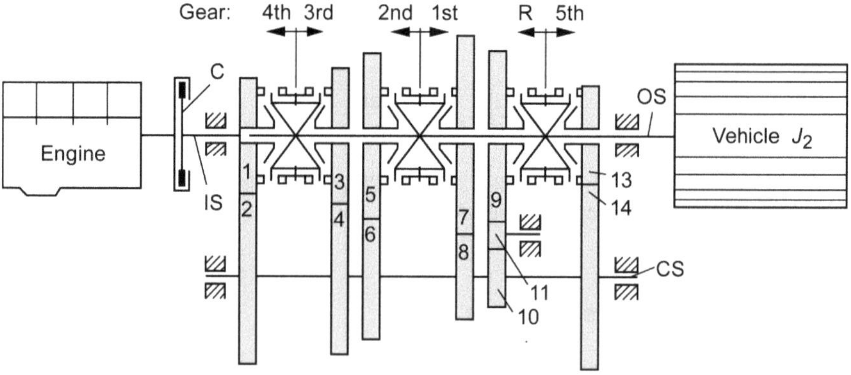


**Fig. 9.21.** Powertrain with coaxial 5-speed countershaft transmission ("in-line")

This closer definition is used in the following. Commercial vehicles with a gross weight of more than 3.5 t generally have in-line gearboxes.

The moments of inertia are reduced onto the input shaft *IS*. This enables all the idler gears involved to be calculated with one and the same moment of inertia. In this case the following reduced moment of inertia applies to the idler gearwheel *i* of the gear *n* to be shifted:

$$J_{\mathrm{red,i}} = J_{\mathrm{IS}}\, i_{\mathrm{n}}^2 \,. \tag{9.11}$$

Table 9.5 gives reference values for moments of inertia $J_{\mathrm{IS}}$ reduced to the input shaft.

### Relative Speed and Friction Speed at the Synchronizer Ring

Speeds before and after synchronization are determined at the synchronizer of a particular gear. The synchronizer is designed to operate at the maximum relative rotational speed.

The friction speed $v$ at the synchronizer ring has a major impact on thermal stress. The temperature at the friction surface rises exponentially with the friction speed. At the maximum angular velocity difference $\Delta\omega_{\mathrm{i}}$ the friction speed is

$$v = \Delta\omega_{\mathrm{i}}\, \frac{d}{2} \,. \tag{9.12}$$

### Torque Losses $T_{\mathrm{V}}$

The torque losses $T_{\mathrm{V}}$ at the synchronizer ring of a given gear are difficult to determine by calculation. Table 9.6 shows the torque losses $T_{\mathrm{V,IS}}$ as reference values measured at the input shaft. $T_{\mathrm{V,IS}}$ is estimated on the basis of empirical data. Together with the permissible stress values, also determined empirically, it allows the design of a synchronizer that is viable in practice.

**Table 9.5.** Moments of inertia $J_{\mathrm{IS}}$ reduced to the input shaft (with clutch plate, without output shaft) for "in-line gearboxes"

| Type of transmission | Maximum gearbox input torque | Overall gear ratio | $J_{\mathrm{IS}}$ |
|---|---|---|---|
| Passenger car 6-speed "in-line gearboxes", upper mid range | 500 Nm | 6 | 0.008 kg m$^2$ |
| Commercial vehicle 6-speed "in-line gearboxes" | 900 Nm | 10 | 0.12 kg m$^2$ |
| Commercial vehicle 9-speed "in-line gearboxes", with rear-mounted planetary range-change unit, 4 x 2 + crawler | 1100 Nm | 13 | 0.17 kg m$^2$ |

**Table 9.6.** Torque losses $T_{V,IS}$ at an oil temperature of 80°C [9.29]

| Empirical values | Passenger car | Commercial vehicle | Commercial vehicle with range unit |
|---|---|---|---|
| Torque losses at the input shaft $T_{V,IS}$ | 2 Nm | 4–8 Nm | 10–14 Nm |
| • "In-line gearboxes": $T_V = T_{V,IS}\, i_n$<br>• For any gearboxes: determine $T_V$ at the idler gear selected, with the values given above from the input shaft (rough estimate) | | | |

### Friction Torque $T_R$ at the Synchronizer Ring

Using Equation 9.6, the following applies for the simplifications made

$$T_R = -J_{red,i} \frac{\Delta\omega_i}{t_R} - T_V . \tag{9.13}$$

When shifting up (deceleration), $\Delta\omega_i < 0$, and when shifting down (acceleration) $\Delta\omega_i > 0$.

### Frictional Work $W$

From a linear plot of angular velocity $\omega$ over the slipping time $t_R$, integration of Equation 9.7 gives frictional work of

$$W = \frac{1}{2}\left(-J_{red,i}\, \Delta\omega_i^2 - T_V\, \Delta\omega_i\, t_R\right). \tag{9.14}$$

The frictional work must be dissipated as heat, and therefore has a negative sign. In practical design calculations the magnitude of the frictional work $|W|$ is used.

### Frictional Power $P_m$

The mean frictional power $P_m$ is given by

$$P_m = \frac{W}{t_R} . \tag{9.15}$$

### Specific Stresses

When designing synchronizers for permissible thermal stress, the stress values calculated are related to the "gross friction surface area" $A_R$. Detail features of the friction surface such as grooves and slots are ignored when considering "gross friction surface area". The proportion of the surface that actually comes into frictional contact during the synchronizing process cannot be precisely calculated anyway. $A_R$ is made up of the sum of the various gross friction surface areas (e.g. as in the case of multi-cone synchronizers).

$$A_R = A_{R,1} + A_{R,2} + \ldots + A_{R,j} = \sum_{i=1}^{j} A_{R,i} \,. \tag{9.16}$$

The error resulting from ignoring part of the contact area is taken into account in the permissible stress values. The levels of stress expected from the calculation are compared with the stress values permitted for the specific material and application. Table 9.7 gives reference values for the permissible stresses of the following popular friction surface combinations: steel friction cone/uncoated special brass synchronizer ring, steel friction cone/steel synchronizer ring with a molybdenum friction coating and steel friction cone/steel synchronizer ring with a sprinkle sinter friction coating (see also Tables 9.8 and 9.9).

These design data have to be considered in the context of the calculation algorithm and its simplifications. Transient peak loads significantly higher than those given may be tolerated. Peak values for specific frictional work $W_A$ in the synchronizer ring friction linings [9.26] are as follows:

- special brass: 1.2 J/mm$^2$,
- molybdenum: 1.5 J/mm$^2$,
- paper: 2.5 J/mm$^2$,
- sprinkle sinter: 4.0 J/mm$^2$.

**Table 9.7.** Design data: standard values for friction pairings [9.29]

| **Friction surface combination** | **Reference values** | | | | |
|---|---|---|---|---|---|
| | Coefficient of friction | Permissible friction speed | Specific frictional work | Specific frictional power | Contact pressure |
| | $\mu$ | $v = \Delta\omega_i \dfrac{d}{2}$ | $W_A = \dfrac{\lvert W \rvert}{A_R}$ | $P_A = \dfrac{\lvert P_m \rvert}{A_R}$ | $p_{R,i} = \dfrac{F_n}{A_{R,i}}$ |
| | | $v_{perm}$ | $W_{A,\,perm}$ | $P_{A,\,perm}$ | $p_{R,\,perm}$ |
| Steel | | (m/s) | (J/mm$^2$) | (W/mm$^2$) | (N/mm$^2$) |
| / special brass | 0.08 – 0.12 | 5 | 0.09 | 0.45 | 3 |
| / molybdenum | 0.08 – 0.12 | 7 | 0.53 | 0.84 | 6 |
| / sprinkle sinter | 0.08 – 0.12 | 9 | 1.0 | 1.5 | 7 |

The specific frictional power $P_A$ is the critical stress in the case of synchronizers subject to high levels of thermal stress. The friction pairing is capable of regenerating itself to a certain degree. A friction lining that has been slightly damaged by violent shifting can regenerate itself by subsequent gentle shifting. Violent shifting can moreover improve frictional characteristics by roughening the smooth lining surface created by gentle shifting. The permissible friction speed $v_{perm}$, defined by the material pairing, restricts the attainable friction surface diameter $d$.

But in about 90% of applications it is not the specific stresses that are the limiting factor in using a synchronizer, but the shifting comfort, which is determined by the variables slipping time $t_R$ and manual effort $F_H$.

### Discussion of the Design Equations

The smaller the slipping time required, the greater the friction torque that must be transmitted – see Equation 9.13. The frictional power $P$ to be transmitted increases as $\Delta\omega^2$. Both the slipping time $t_R$ and the difference in angular velocity $\Delta\omega$ are dependent on operational and design factors. There is little scope for influencing the masses involved in synchronization, expressed by their reduced moment of inertia $J_{red,i}$. The friction speed $v$ at the synchronizer ring increases according to Equation 9.12 with the effective diameter $d$. The main starting points to consider when optimising existing frictional synchronizers and developing new ones are thus:

**Design Measures:**

- enlarge the friction surface $A_R$: – external cone synchronizer,
- increase the number of friction surfaces $j$ and enlarge the friction surface $A_R$: – multi-cone synchronizer, – multi-plate synchronizer,
- transmission of gearshift effort: – cone angle $\alpha$, – lever-assisted synchronizer,
- engageability: – clearance characteristic,
- oil supply: – oil ducts, drip edges, deflectors.

**Material Measures:**

- increase the permissible stress values with "new" friction surface pairings,
- increase the friction coefficient $\mu$.

Most common single-cone synchronizers have reached their performance limits in commercial vehicle transmissions and in the lower gears of passenger car transmissions. In these situations they are replaced by double-cone and triple-cone synchronizers if necessary [9.5, 9.23].

### Calculation Procedure

Figure 9.22 shows an algorithm for the thermal design of synchronizers, based on the simplifications, equations and tables presented above.

The procedure is iterative. First the "simplest" solution is calculated, normally a single-cone synchronizer from the standard production range. If the requirements are not met, then the loops of the algorithm are executed repeatedly with varying design, structure and material parameters, until the synchronizer selected meets the requirements. Economic constraints have to be taken into account as well as the technical parameters.

### 4/ Designing Locking Toothing for Locking Effect

In most common mechanical synchronizers the locking effect depends on the same principle. The friction torque $T_R$ acts to lock the synchronizer, and is opposed by an opening torque $T_Z$ resulting from the decomposition of forces at bevelled surfaces (frequently also referred to as index torque $T_I$). As long as there is a speed difference, the locking friction torque is greater than the opening torque. This is illustrated below by the example of a locking tooth design (Figure 9.23).

The gearing torque $T_Z$ arising at the dog bevels acts as an opening torque and is calculated from the friction coefficient $\mu_D$ between the locking and shifting dogs:

$$T_Z = \frac{F\,d_C}{2}\left(\frac{\cos\frac{\beta}{2} - \mu_D \sin\frac{\beta}{2}}{\sin\frac{\beta}{2} + \mu_D \cos\frac{\beta}{2}}\right) = \frac{F\,d_C}{2}\left(\frac{1-\mu_D \tan\frac{\beta}{2}}{\mu_D + \tan\frac{\beta}{2}}\right). \tag{9.17}$$

For the friction coefficient $\mu_D$: $\mu_D \approx 0.09$. It is hardly possible to give the actual value of $\mu_D$ in operation. The torsional vibration mentioned above causes the dogs of the gearshift sleeve to be "rattled" by the locking toothing. The opening torque may thus be described in a simplified manner, ignoring the coefficient of friction $\mu_D$, as

$$T_Z = \frac{F\,d_C \cot\frac{\beta}{2}}{2}. \tag{9.18}$$

In designing the locking toothing it is assumed that the entire gearshift effort acts on the locking teeth, so that no excess force is conducted to the synchronizer ring via the thrust pieces. The gearshift sleeve is prevented from engaging for as long as the locking condition

$$T_Z < T_R \quad \text{with} \quad T_R = j\,F \frac{d}{2}\frac{\mu}{\sin\alpha} \tag{9.19}$$

is satisfied. Substituting from Equations 9.18 and 9.19 it follows that

$$\frac{F\,d_C \cot\frac{\beta}{2}}{2} < \frac{j\,F\,d\,\mu}{2\sin\alpha}. \tag{9.20}$$

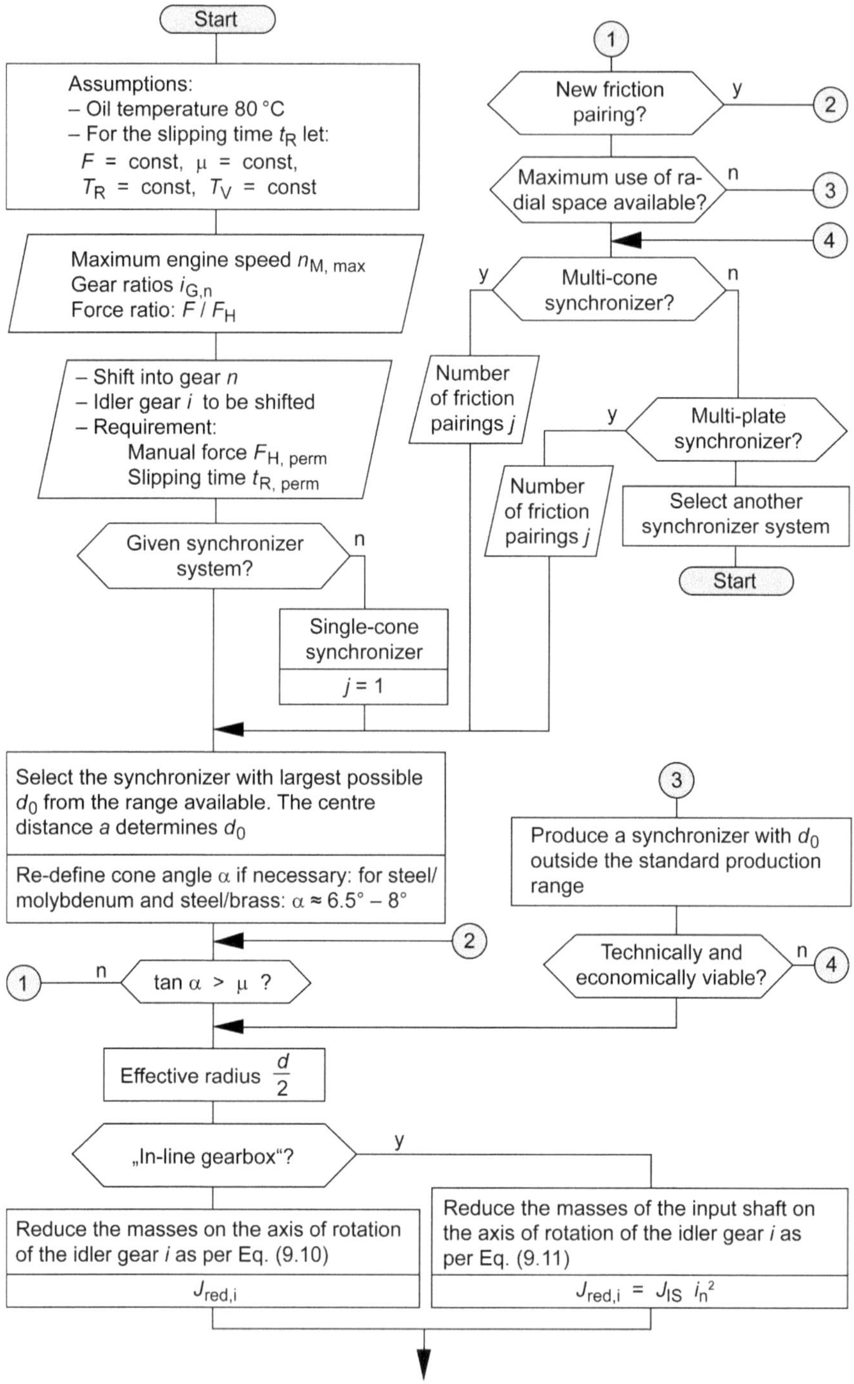


**Fig. 9.22.** Algorithm for the thermal design of synchronizers

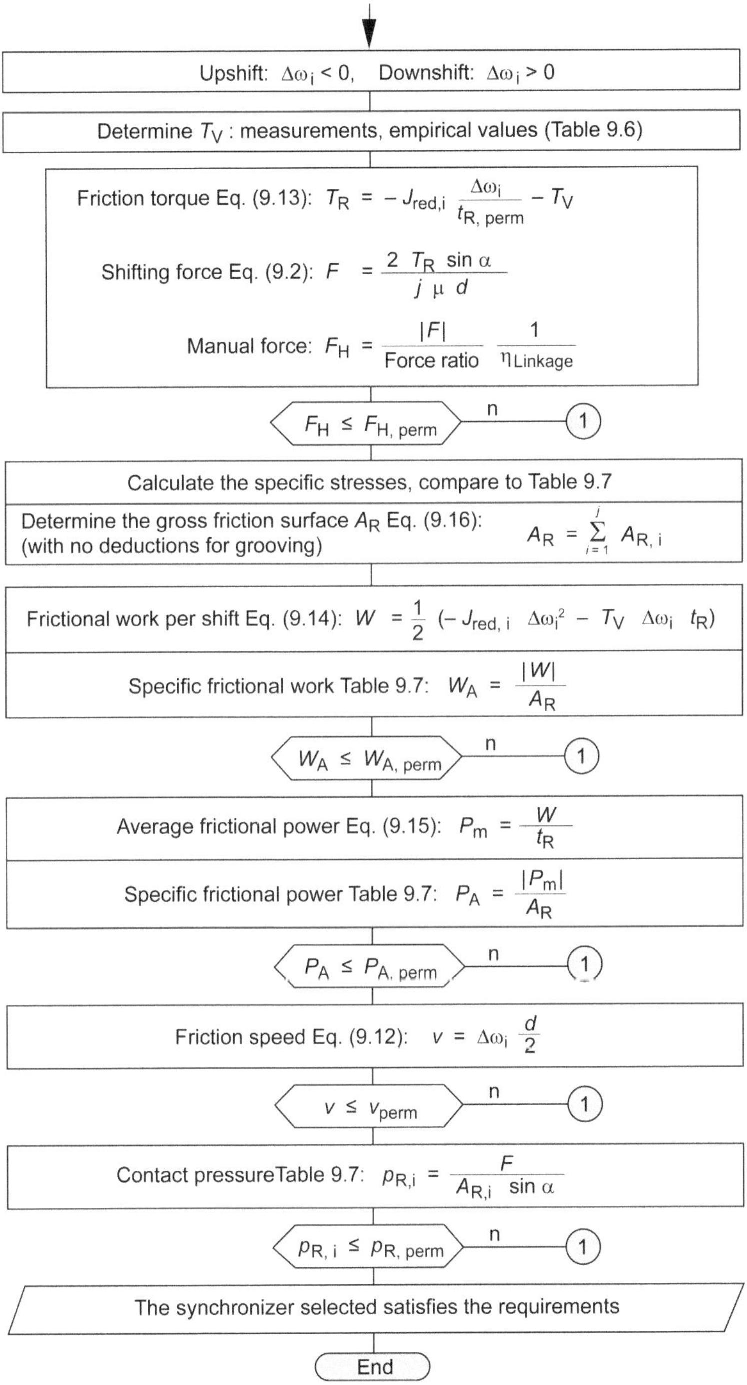


**Fig. 9.22.** *(continued)*

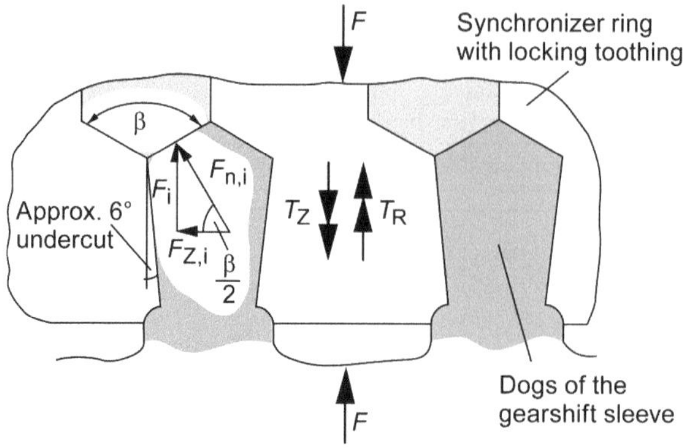


**Fig. 9.23.** Decomposition of forces at a locking toothing. Opening torque $T_Z$

This results in the following design equation for the bevel angle or opening angle $\beta$ of the locking toothing

$$\cot\frac{\beta}{2} < \frac{j\,\mu}{\sin\alpha}\,\frac{d}{d_C}\,\frac{1}{S} \quad \text{with} \quad 105° < \beta < 125° . \tag{9.21}$$

The locking safety factor $S$ is introduced in order to assess the locking effect. If the opening angle $\beta$ falls below the lower limit indicated, then "grating" occurs; if the opening angle exceeds the upper limits, the gearshift effort increases and shifting comfort suffers. In Equation 9.21 the diameter ratio $d/d_C$ appears as an additional influencing variable affecting synchronizer characteristics.

### 9.2.4 The Tribological System of Synchronizers

The synchronizer ring with its friction layer, the friction cone of the synchronizer hub and the lubricant together constitute a tribological system. Both the engineering design characteristics and the tribological characteristics of the synchronizer affect the shifting comfort and the service life of a transmission. The locking toothing geometry and the cone angle must be matched to the friction coefficient of the material pairing used.

To achieve high levels of friction torque $T_R$ combined with the least possible gearshift effort $F$, the dynamic friction coefficient $\mu$ must be as large as possible, in accordance with Equation 9.2. Such a high friction coefficient can only be achieved by boundary friction (see Figure 11.6 "Stribeck curve"). Boundary friction or boundary layer friction is defined as a frictional state in which the normal force $F_n$ is no longer transmitted by hydrodynamic pressure, not even partially (mixed friction) [9.26]. The friction surfaces are only separated by a boundary layer a few nanometres thick, made of chemically formed reaction layers (see also Section 11.2 "Lubrication of Gearboxes, Gearbox Lubricants"). The combined action of the lubricant with the structure and the chemical composition of the friction materials influences the boundary layer and thus the friction coefficient [9.28].

In order to counteract the hydrodynamic formation of a lubricant film, the friction surface of the synchronizer ring is provided with grooves, and in the case of commercial vehicles, that of the synchronizer hub as well. Common types of such "drainage" grooves are:

- threaded grooves in the synchronizer ring (unwinding thread – see p. 320 "Sticking"),
- axial grooves in the synchronizer ring and/or friction cone of the synchronizer hub,
- circular grooves in the friction cone of the synchronizer hub (commercial vehicles).

The function of the grooving in the friction surfaces is to cut through the film of oil and dissipate some of the frictional heat by means of the oil (see also Figures 9.12, 9.16 and 9.19).

Carbon linings tend to have a higher friction coefficient than molybdenum and sprinkle sinter linings. Non-metallic linings also open up more possibilities with respect to overload capability and oil compatibility.

### *Materials*

The pairing of materials affects the service life and reliability of a synchronizer. The wear characteristics of the friction pairing must be matched (Table 9.8).

**Table 9.8.** Friction pairings of cone synchronizers

<table>
<tr><th>Friction surface of the synchronizer hub:<br>highly wear-resistant</th><th colspan="2">Friction surface of the synchronizer ring:<br>scuffing resistant, wearing</th></tr>
<tr><td rowspan="8">Case-hardened steel<br>16 MnCr5,<br>20 MoCr4,<br>with 60 HRC</td><td colspan="2">Commonly used:</td></tr>
<tr><td>Uncoated special brass rings</td><td>Passenger cars, corundum blasted after turning to texture the friction surface</td></tr>
<tr><td>Steel rings with molybdenum thick film approx. 0.5 mm thick</td><td>Commercial vehicles, grooves ground in</td></tr>
<tr><td>Sprinkle sinter friction linings</td><td>Special brass powder with non-metallic constituents</td></tr>
<tr><td>Molybdenum thin film</td><td>Applied to shaped synchronizer ring by flame or plasma spraying</td></tr>
<tr><td>Paper friction linings</td><td>Fibre composites with organic matrix glued to synchronizer ring</td></tr>
<tr><td colspan="2">Trend:</td></tr>
<tr><td>Carbon</td><td>Carbon fabric glued to synchronizer ring</td></tr>
</table>

**Table 9.9.** Comparative evaluation of synchronizer ring friction linings according to [9.26]. Improvement: +++ substantial, ++ noticeable, + little, 0 none; Deterioration: – little, – – noticeable. Basis of comparison: Synchronizer ring made of special brass

| Characteristics | Special brass | Sprinkle sinter | Molyb-denum | Paper | Carbon |
|---|---|---|---|---|---|
| Synchronizer ring wear | 0 | ++ | + | – – | ++ |
| Synchronizer hub wear | 0 | 0 | – | 0 | 0 |
| Specific frictional work $W_{A,perm}$ | 0 | +++ | ++ | ++ | +++ |
| Coefficient of friction | 0 | + | + | ++ | ++ |
| Coefficient of friction constancy | 0 | + | 0 | ++ | ++ |
| Overload capacity | 0 | ++ | + | – | +++ |
| Oil compatibility | 0 | ++ | + | ++ | ++ |

The main requirements of a friction pairing are:

- almost non-wearing with high friction coefficient $\mu$,
- material is easy to machine,
- low material costs,
- almost constant friction coefficient throughout service life,
- resistant to overloading.

Table 9.9 gives a comparative evaluation of friction materials.

### 9.2.5 Engineering Designs

All major synchronizers in use adapt rotational speed by means of slipping friction clutches. In all designs except the Porsche synchronizer, the process of synchronization and the generation of locking torque are achieved in a manner similar to that in the example discussed in Section 9.2.2. This section describes some different types of synchronizer design.

#### *1/ Single-Cone Synchronizer*

See Figures 9.7, 9.12 and 9.13 and the associated variations.

#### *2/ Multi-Cone Synchronizer*

In manual transmissions, the design should aim to make gearshift effort equal for all gears. Multi-cone synchronizers are therefore increasingly being used in the low gears (first and second). The number of friction cones and the friction materials used will depend on the intended use.

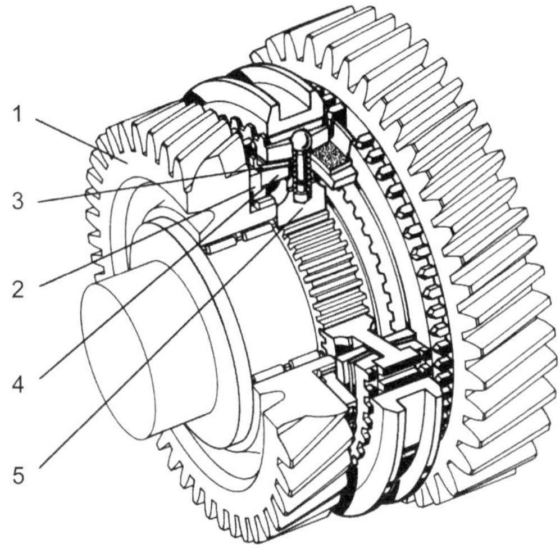


**Fig. 9.24.** Double-cone synchronizer (ZF-D).
*1* Idler gear;
*2* synchronizer hub with dog gearing;
*3* double-cone ring;
*4* counter-cone ring;
*5* synchronizer body

For example, the use of a triple-cone synchronizer may make it possible to use low-cost special brass synchronizer rings [9.5].

Two cone friction surfaces are required to achieve synchronization in the case of double-cone synchronizers (Figure 9.24). The link between the double-cone ring *3* and the synchronizer hub *2* is rotationally fixed using several dogs, but axially flexible. The counter-cone ring *4* is rotationally fixed to the synchronizer body *5*. The gearshift effort is reduced and the torque capacity, i.e. performance, is increased because of the increased number of friction surfaces and the larger friction surface area of the double-cone synchronizer.

The parallel multi-cone design requires closer manufacturing tolerances and therefore entails higher production costs. Multi-cone synchronizers are therefore used only in the lower gears.

### 3/ External-Cone Synchronizer

In the Mercedes-Benz external-cone synchronizer system (Figure 9.25) the synchronizer unit *3* is fixed by an annular spring *2* to the idler gear *1*. The synchronizer ring has three inward-facing locking lugs *6*, which engage in corresponding grooves *7* in the idler gear. It can turn relative to the wheel circumferentially a certain distance and axially once the annular spring has been overcome.

When shifting, the gearshift sleeve is pressed against the synchronizer ring. The friction torque turns the synchronizer ring as far as it will go. Its locking lugs *6* are then so placed before a bevel in the idler gear that the gearshift sleeve *5* and synchronizer ring *3* can move no further as long as the friction torque is not equal to zero. When the speeds are synchronized, the sloping surfaces slide over each other and turn the synchronizer ring back. The noses of the synchronizer ring are pushed into the grooves *7* of the idler gear. Positive engagement can now take place by means of the dogs.

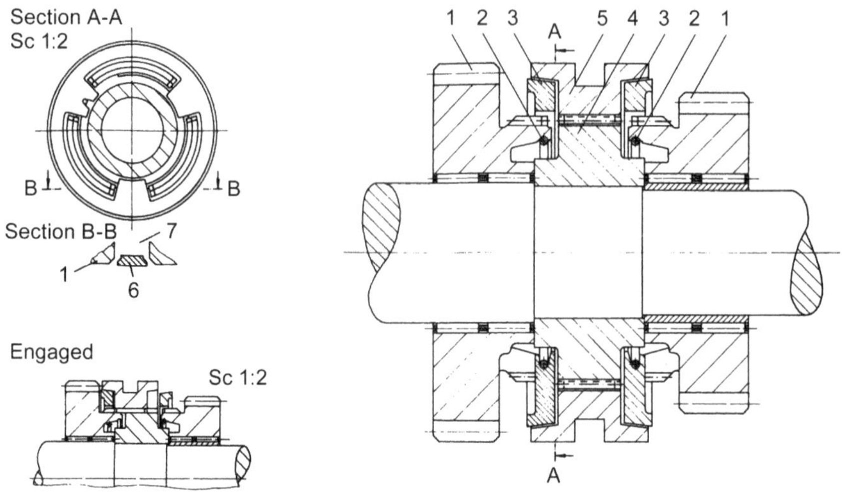


**Fig. 9.25.** Mercedes-Benz external-cone synchronizer system. *1* Idler gear with dog gearing; *2* annular spring; *3* synchronizer ring; *4* synchronizer body; *5* gearshift sleeve; *6* locking lug; *7* groove in idler gear

When the gearshift sleeve is moved, the annular spring *2* is pushed out of its groove, and slides along the cone surface under the selector teeth. The radial tension of the annular ring exerts an axial restoring force on the synchronizer ring, and moves it into its initial position when the gear is released.

The ratio of the effective diameter $d$ to the clutch diameter $d_C$ is greater than 1. According to Equation 9.21, the opening angle $\beta$ can be reduced relative to the Borg-Warner system, with the same safety factor $S$; shifting comfort is improved.

The friction surfaces are located outwards, as compared to synchronizers based on the Borg-Warner system. In accordance with Equation 9.2, this arrangement results in reduced gearshift effort, and in lower specific stresses because of the increased friction surface area $A_R$. Because of the larger effective diameter $d$, the friction speed $v$ increases in accordance with Equation 9.12, and the synchronizable speed difference falls.

### 4/ Locking-Pin Synchronizer

The Spicer or Tompson synchronizer shown in Figure 9.26 is a locking-pin synchronizer. The gearshift sleeve *3* has six drill holes and is rotationally fixed to the transmission shaft, but axially linked by a sliding connection. The locking pins *4* engage in the drill holes parallel to the axis. They are each rigidly connected to a synchronizer ring *2*. The conically countersunk drill holes are larger than the conical part of the locking pin, enabling the synchronizer pin to turn by a certain amount. As long as there is a speed difference, the opening torque at the conical surfaces of the locking pins and the drill holes is smaller than the locking friction torque. The gearshift sleeve does not slide.

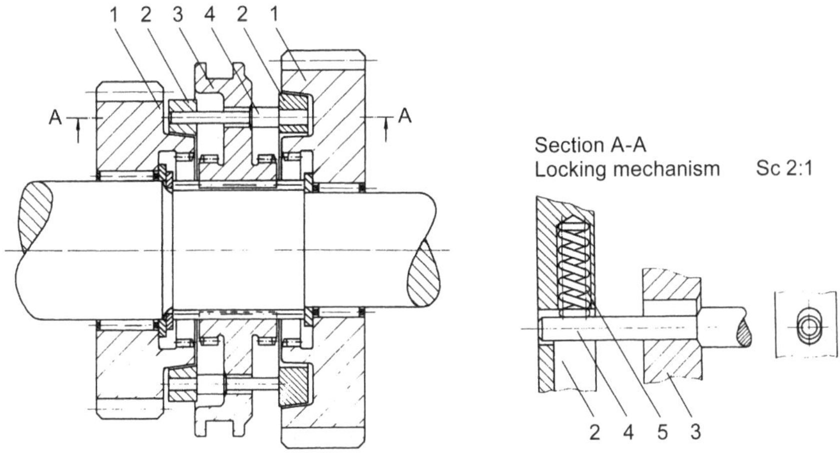


**Fig. 9.26.** Locking-pin synchronizer. *1* Idler gear with dog gearing; *2* synchronizer ring; *3* gearshift sleeve; *4* locking pin; *5* compression spring

When the speeds are synchronized, the circumferential component of the gearshift effort prevailing at the bevelled locating face of the drill holes compresses the compression spring *5*. The gearshift sleeve slides along the locking pins and causes the dogs to engage.

Contrary to synchronizers based on the Borg-Warner system, the shifting dogs are mounted on a smaller diameter and the friction surfaces on a larger diameter. In accordance with Equation 9.2, the same friction torque $T_R$ is achieved with less gearshift effort $F$ at the synchronizer ring because of the larger diameter $d$. The increased friction surface area $A_R$ results in lower specific stresses. The serial configuration of friction surface and shift movement results in a greater overall package length $b_0$ than for synchronizers based on the Borg-Warner system.

### 5/ Eaton LF Synchronizer

The LF Synchronizer (LF = Low Force) is used in multi-range transmissions with two countershafts of the Eaton S Series for simultaneous shifting of idler gears (see Figure 12.38). All synchronizers of this three-range transmission are arranged on the central shaft. An essential aim of this LF synchronizer is reducing the shifting force and increasing the speed of gear change. This is achieved by converting the rotation force resulting from the synchronizing process into an axial force. This axial force strengthens contact between the synchronizer ring and the synchronizer body of the gearwheel to be shifted.

Figure 9.27 shows the structure of this synchronizer. At the beginning of the shifting process, the driver moves the gearshift lever in the direction of the new gear position. In this way, the synchronizer plate *3* brings the synchronizer ring *2* into frictional contact with the synchronizer body *1* of the new gearwheel *6* by means of the sliding sleeve *8* and the axially shiftable pin *10*.

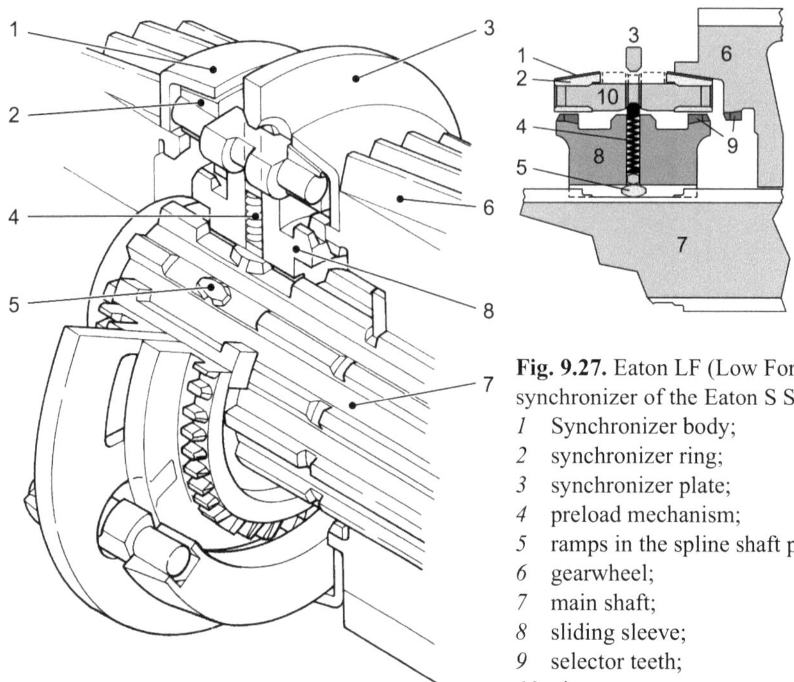


**Fig. 9.27.** Eaton LF (Low Force) synchronizer of the Eaton S Series.
*1* Synchronizer body;
*2* synchronizer ring;
*3* synchronizer plate;
*4* preload mechanism;
*5* ramps in the spline shaft profile;
*6* gearwheel;
*7* main shaft;
*8* sliding sleeve;
*9* selector teeth;
*10* pin

This motion preloads the system by means of the spring-loaded preload-mechanism *4*. Due to this, the synchronizer plate *3* moves against specially shaped ramps *5* in the spline shaft profile of the main shaft *7*.

The ramps, sloping in the direction of shifting, convert the rotation force into an axial force directed against the gearwheel to be shifted *6*. This axial force, which presses the synchronizer plate towards the direction of the gearwheel to be shifted, speeds up the shifting process, shortens the time needed for synchronizing and reduces the force that must be applied by the driver to the gearshift lever. There is thus an increase in force at the friction surface between the synchronizer body *1* and the synchronizer ring *2*. As soon as the speeds of the sliding sleeve *8* and the gearwheel *6* are synchronized, the selector teeth *9* of the sliding sleeve *8* are positively engaged with the selector teeth *9* of the gearwheel *6* [9.6].

### 6/ Multi-Plate Synchronizers

The multi-plate synchronizer in its present form has been developed from the multi-plate clutches used in powershift transmissions (Figure 9.28). Because of its large power transmission surface $A_R$, it is suitable wherever there is a requirement for very high synchronizer performance.

The cone angle $\alpha$ of a multi-plate synchronizer is 90°. To be operated with the same gearshift effort as a single-cone synchronizer with $\alpha = 6.4°$, according to Equation 9.2 a multi-plate synchronizer of the same effective diameter must have $j = 9$ friction surfaces.

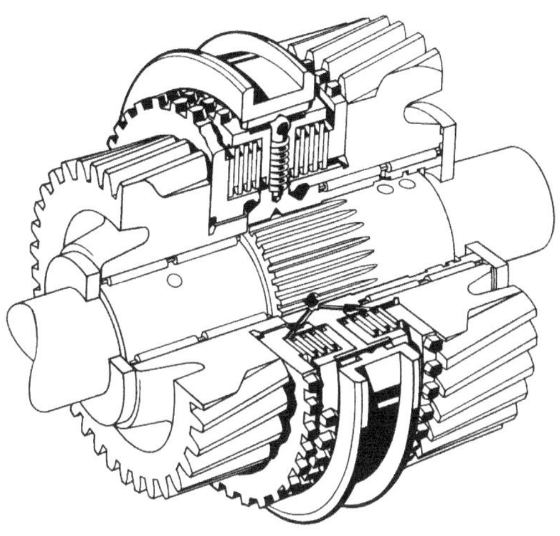

Fig. 9.28. Multi-plate synchronizer

The lengths of the two synchronizers are then roughly equal. Multi-plate synchronizers are complex and costly.

### *7/ Porsche Synchronizer*

The Porsche system locking synchronization (Figure 9.29) has a self-reinforcing locking effect, preventing premature gearshift action before speeds have been synchronized. The Porsche synchronizer requires relatively little gearshift effort, but its high manufacturing cost means it is no longer of practical significance.

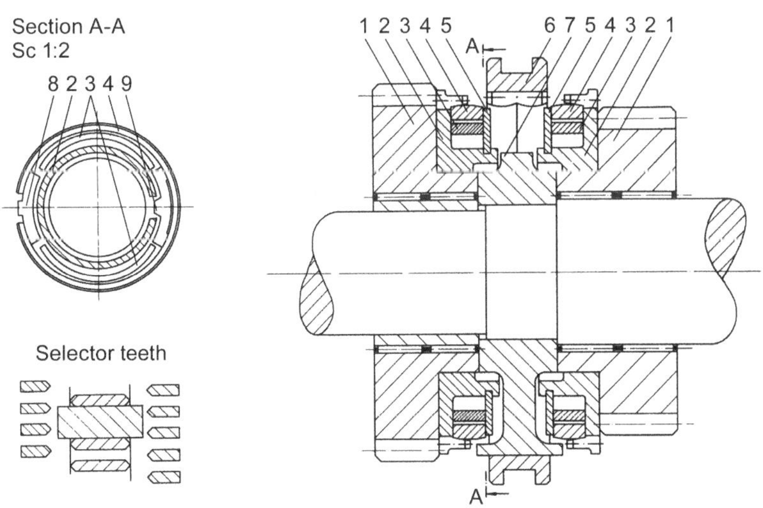


**Fig. 9.29.** Porsche synchronizer. *1* Idler gear; *2* synchronizer hub with dog gear; *3* locking belt; *4* synchronizer ring; *5* circlip; *6* gearshift sleeve; *7* guide sleeve; *8* pad; *9* end stop

The quality of the synchronization process of the Porsche synchronizer is very much subject to variations in the friction coefficient. The synchronization process will be only briefly described.

The slotted synchronizer ring *4* located in front of the selector teeth is crowned. It has to be squeezed together in order to slide into the gearshift sleeve *6*. When there is a speed difference, the synchronizer ring is twisted by the friction torque until it rests at the stop by means of the pad *8* and the locking belt *3*. This gives rise to radial forces which press the locking belts outwards and prevent the synchronizer ring from being pressed together. The greater the axial clamping force of the gearshift sleeve, the more strongly the slotted synchronizer ring is pressed outwards by the locking belts. After synchronization there is no more effective spreading force. The synchronizer can then be pressed together, and the gearshift sleeve can slide over it.

### 9.2.6 Alternative Transmission Synchronizers

As an alternative to having a synchronizing device for each individual gear, a gearbox can also be synchronized in the following ways [9.16]:

1/ central synchronizing device for the whole gearbox,
2/ speed matching by the internal combustion engine.

#### *1/ Central Synchronizer*

In central synchronizers (Figure 9.30), only one synchronizer unit is needed for all upshift and downshift operations. Energy input and output is external. The electrical transmission control determines the relative speeds of the parts that are to be positively engaged, controls the synchronizer unit that carries out the speed synchronization, and initiates the gearshift action.

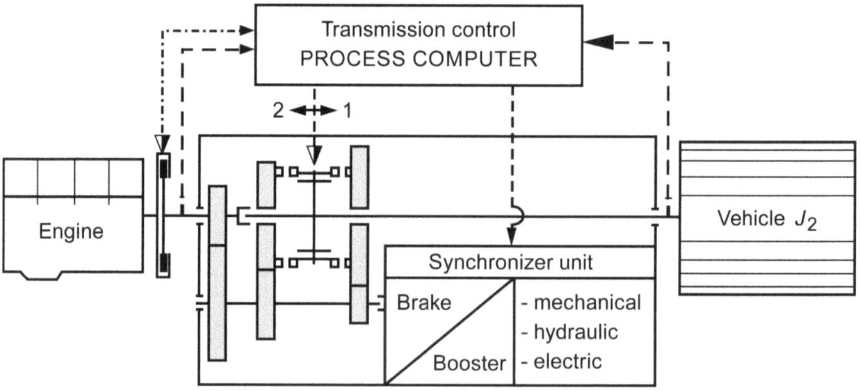


**Fig. 9.30.** Principle of central synchronization

A brake retards the masses to be synchronized when shifting up, and a booster motor accelerates the masses when shifting down. The principle of central synchronization was developed in 1972 and introduced in the SST-10 SA Spicer transmission for use in heavy-duty commercial vehicles.

#### *2/ Speed Matching with the Internal Combustion Engine*

The master clutch is not opened during active speed matching using the internal combustion engine. The speed difference is adjusted by means of brief acceleration or deceleration of the engine.

An electronic device controls the engine, and determines the synchronization point. Transmission synchronizers of this type have the disadvantage that the synchronizing time depends on the engine, and may therefore take too long in certain driving situations. The Faun and Siemens Symo gearshift mechanism introduced in 1954, and further refined in the following years, operates on this principle.

Thanks to advances in electronics and engine technology as well as the use of a gearbox brake, this type of synchronizer has been mass produced in the automated commercial vehicle transmission ZF AS-Tronic since 1998. The engine is responsible for speed matching when shifting down, and the gearbox brake is responsible when shifting up. Section 12.2.2 provides a more detailed explanation of the functioning of such transmissions and gearshifting.

### 9.2.7 Detail Questions

Synchronizer pack production technology is of crucial importance in determining final costs [9.10].

#### *1/ Ensuring Positive Engagement*

The dogs of the gearshift sleeve and the synchronizer hub are undercut by 4° to 6° in order to prevent gear dropout (see Figure 9.23).

#### *2/ Test and Measuring Technology*

When the design calculations for the synchronizer have been completed, tests must be carried out to verify and refine the values. Load profiles are established in complex, expensive vehicle measurements. They provide practical data on gearshift effort, shifting time, oil temperature and frequency of shifting. These data are fed into the simulation and test bench runs.

By varying individual parameters in the bench test, information can be gathered on the friction coefficient, gearshift effort, friction torque, frictional power rating, rotational speed and state of wear.

### *3/ Abuse Test*

Heavy-duty synchronizer tests are used to investigate the performance margin of synchronizers against overloading. This involves shifting gear with high levels of shifting force $F$ and short slipping times $t_R$ under unfavourable conditions (e.g. low gearbox oil temperature). In practice the gearshift profile depends on the individual driver and is therefore random.

It is true of all components that the type and number of operator errors are the major factor determining service life and functioning, and this is particularly true for synchronizers. The General Motors "abuse test" used in the U.S. is well known. In this abuse test it is assumed that the driver presses the synchronizer ring against the friction cone of the synchronizer hub without operating or fully opening the master/gearshifting clutch. This also brings to bear the load moment $T_L$ indicated in Equation 9.4. In practice this represents the "high-performance" driver who changes gear without using the clutch. This test is carried out using a high level of gearshifting force ($F > 2000$ N for passenger cars) and with a high speed difference.

## 9.3 Layout and Design of Multi-Plate Clutches

Shifting clutches are of central importance in the functioning of powershift transmissions. The gear to be shifted is brought into the power flow frictionally and without power interruption. The driver judges the quality of the gearshift according to the level of driving and shifting comfort experienced and thus rates the quality of the transmission. Moving-off and shifting quality are the most frequently mentioned points of complaint for transmissions that are not manually shifted.

Brakes also fall under the category of "shifting clutches". The difference between a clutch and a brake can be described as follows:

- clutches rotate in a closed state,
- brakes are stationary in a closed state and support themselves in the housing.

In the case of automatic transmissions of planetary design, brakes serve to support components of the planetary gear in the housing as needed. Multi-plate brakes and belt brakes are established designs. In belt brakes, a metal belt is looped around a brake drum. The important advantage of the belt brake is:

- small radial installation space requirement.

On the other hand, its limitations include:

- non-harmonic, poorly reproducible torque built-up,
- uneven load share with wear on the belt ends,
- radial forces and
- high sensitivity to adjustment tolerances.

Figure 9.31 shows a multi-plate clutch and belt brake unit from the Mercedes-Benz W5A 030 5-speed automatic transmission (Figure 12.22).

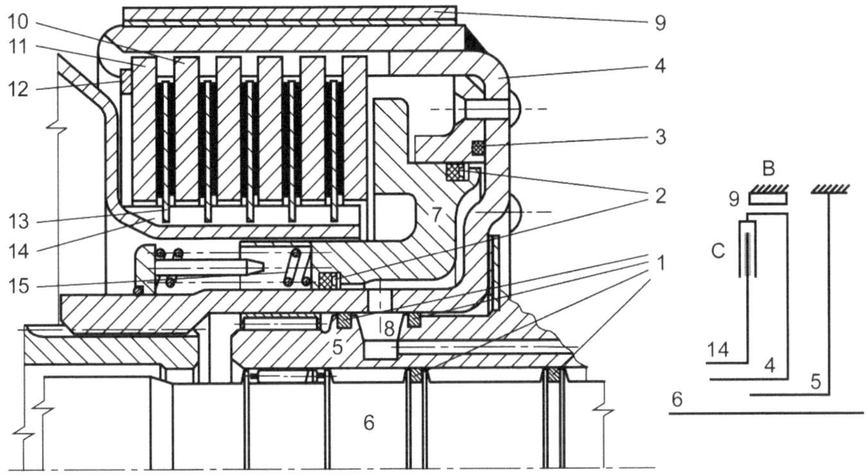


**Fig. 9.31.** Multi-plate clutch and belt brake unit. *1* Rectangular ring (rotating); *2* grooved ring; *3* O-ring (static); *4* outer plate carrier and brake drum; *5* housing; *6* shaft; *7* piston; *8* pressure oil supply; *9* brake belt; *10* steel plate; *11* end plate; *12* snap ring; *13* lined plate; *14* inner plate carrier; *15* piston return spring

In modern transmission designs, belt brakes have been to the greatest extent replaced by multi-plate brakes. Belt brakes will not be discussed in the following, whereas the layout and design of multi-plate clutches and brakes will be explored further.

Wet multi-plate clutches and brakes (i.e. those that are through-flowed with oil) are used in many transmissions. Besides the classic applications as shifting clutches in passenger and commercial vehicle automatic transmissions, they are also used, for example, in reversing gearboxes for ships and construction machines, in connectable power take-offs and axles as well as in differential locks. As master clutches, they can be found in continuously variable and dual clutch transmissions as well as in power take-offs to limit torque ("torque fuse").

Multi-plate clutches will be understood in this section as clutches active in the shifting process, flown through with oil and activated by pressure oil. The multi-plate brake is a special form of the multi-plate clutch.

The following pages will first formulate the requirements of multi-plate clutches and then describe the basics of the shifting process, these considerations providing the basis for presenting the main features of design. A further section is concerned with the tribological system composed of the lined plate, oil film and steel plates as counter-running surfaces (Figure 9.32). Since the torque is transmitted through friction, the friction coefficient $\mu$ between the friction surfaces has a large influence on the system's behaviour. Special oils, ATF oils (Automatic Transmission Fluid), were developed for use in automatic transmission with various gear ratios. The behaviour of friction and thus the transmission of torque can be influenced decisively with the type of friction lining and oil used. The section "Layout and Design of Multi-Plate Clutches" is concluded with design recommendations and information on detailed questions.

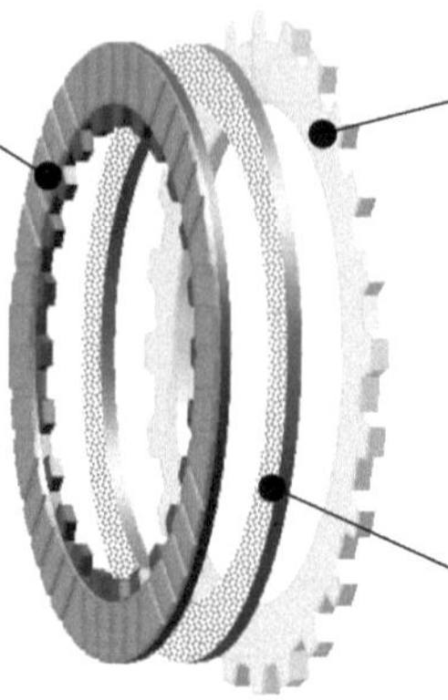


**Fig. 9.32.** Tribological system of a multi-plate clutch: lined plate (shown with internal gearing), oil film and counter-running surface (steel plate, shown with external gearing)

### 9.3.1 Multi-Plate Clutch Requirements

The requirements of multi-plate clutches depend significantly on the main mode of operation. In accordance with the mode of operation, we differentiate between:

- *powershifting*, e.g. shifting clutches in automatic transmissions; characteristics: short slipping times, energy storage and
- *continuous slip*, e.g. master clutches or controlled torque converter clutches; characteristics: long slipping times, thermal equilibrium.

Multi-plate clutches, their friction linings in particular, have the following requirements:

- good shifting behaviour (good dynamic friction coefficient profile),
- high torque capability (high static friction coefficient),
- long service life , i.e. low wear,
- high mechanical strength,
- high thermal load capacity,
- favourable noise behaviour, i.e. no friction vibrations (stick-slip),
- good controllability (friction coefficient profile $d\mu/dv > 0$),
- good friction properties, constant throughout the service life of oil and lining,
- insensitivity to metallic wear/particles in the oil and
- low drag torque.

### 9.3.2 The Shifting Process

Driving and shifting comfort as felt by the driver stands in direct correlation with the characteristic of the transmission output torque and thus with vehicle acceleration during shifting.

**Table 9.10.** Evaluation table for shifting comfort (shifting quality) according to ATZ

| **Grade** | **1** | **2** | **3** | **4** | **5** | **6** | **7** | **8** | **9** | **10** |
|---|---|---|---|---|---|---|---|---|---|---|
| Attribute evaluation (gearshift) | ex-tremely heavy jerk | heavy jerk | jerk | very obvious | obvious | well percep-tible | percep-tible | slightly percep-tible | barely notice-able | not notice-able |
| Customer satisfaction | very dissatisfied | | | | slightly dissat-isfied | pretty satisfied | | very satis-fied | outstandingly satisfied | |

The quality of the gearshift is documented with an ATZ grade (Table 9.10). In modern powershift transmissions, assessments lower than 7 are only acceptable in exceptional situations (e.g. cold start). The appraisal of what makes a good gear-shift is subjective and depends on the desired vehicle image. The target definition ranges from sporty, i.e. gearshifts have to be clearly perceptible, to comfort-oriented, i.e. not noticeable. Criteria for judging the shifting process are:

- shifting comfort (jerk, noise, frequency),
- spontaneity (deadtime, shifting duration, acceleration) and
- careful calibration (i.e. clutch load, see also Section 13.4 "Transmission Calibration").

In the following, both operating modes, powershifting and continuous slip, will be explained in more detail.

### 1/ Powershift Operating Mode

Shifting gears in a powershift transmission involves 5 different shift types:

- upshift with positive torque: *power upshift,*
- upshift with negative torque: *overrun upshift,*
- downshift with positive torque: *power downshift,*
- downshift with negative torque: *overrun downshift,*
- downshift with neutral torque: *rollout shift.*

In powershift systems, the activating element must take over the torque and open the deactivating element exactly at the synchronous speed. Delivery of the torque from one to the other shifting element can be realised in two ways:

- *Freewheel shifting*
  Freewheels are absolutely unbeatable with respect to the temporally precise delivery of the torque (synchronous speed). The activating clutch takes over the torque and overruns the freewheel. The freewheel is removed from the lock by the higher speed and then rotates freely. Freewheels do however have other disadvantages (see Section 9.1).
- *Overlapping shifting*
  The activating and deactivating elements are clutches. This gives more freedom of choice in driving strategy, efficiency improvement and driving comfort.

However, at low torques and speeds, influences from the tolerances of the assembly groups involved demand more elaborate control processes.

Overlapping shiftings are further subdivided into:

- *Positive overlapping*, for superelevation shifting:
  The activating and deactivating clutches are both activated for a short period of overlap with a pressure, which however does not yet cause a blocking of the transmission. Slipping of the clutches in the overlap phase leads temporarily to a somewhat higher power loss. The traction force is preserved to a great extent. The period of overlap is kept as short as possible.
- *Negative overlapping*, for release shifting:
  There is a short period of overlap in which the activating and deactivating clutches are activated so weakly that they release the engine, allowing it to rev up. This is done in such a way that the synchronous speed is reached.

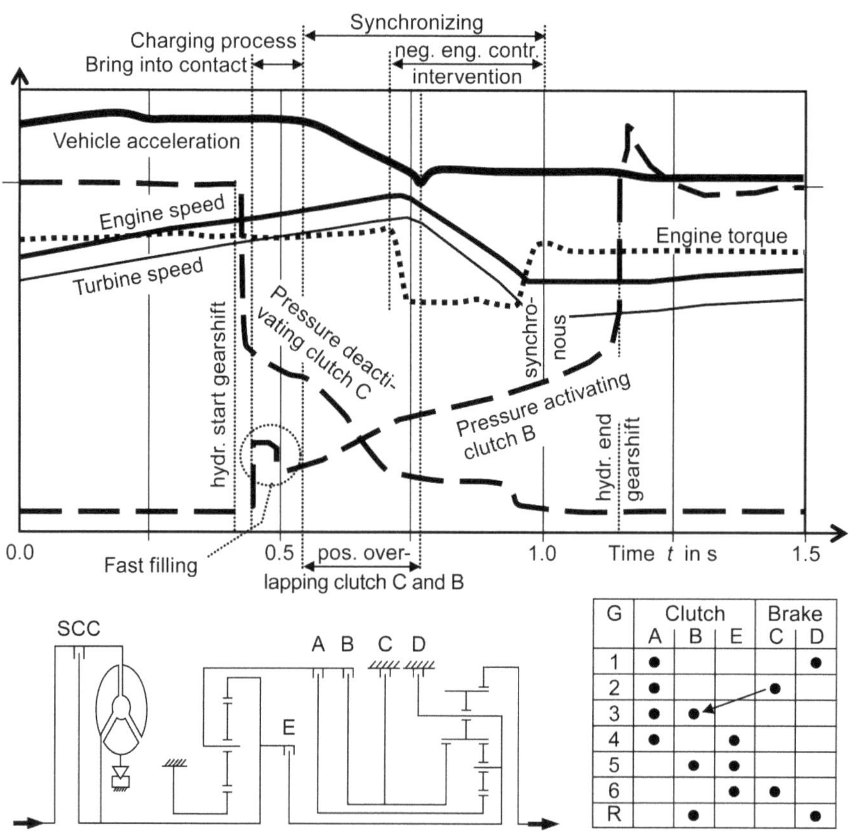


| G | Clutch A | Clutch B | Clutch E | Brake C | Brake D |
|---|---|---|---|---|---|
| 1 | ● | | | | ● |
| 2 | ● | | | ● | |
| 3 | ● | ● | | | |
| 4 | ● | | ● | | |
| 5 | | ● | ● | | |
| 6 | | | ● | ● | |
| R | | ● | | | ● |

**Fig. 9.33.** Positive overlapping (superelevation shifting): power upshift from 2nd to 3rd gear with constant accelerator pedal position using a 6-speed passenger car automatic transmission as example

When referring to the shifting process, we distinguish between superelevation shifting and release shifting. In *superelevation shiftings*, the transmission (i.e. the activating clutch) pulls the engine up to a higher level of speed (overrun downshift) or forces it down to a lower speed (power upshift). Figure 9.33 shows a power upshift from the 2nd to the 3rd gear (see also Figure 13.14 "Calibration").

With the fast filling, the operating clearance of the multi-plate packet is overcome. After a charging equalisation phase, in which charge tolerances are equalised, the shifting pressure is increased as a function of the torque. The clutch slips until the synchronous speed of the new gear has been reached. Afterwards, the pressure is raised according to a closing ramp to normal system pressure.

In *release shifting*, the engine is released during the shifting process and climbs by itself to the higher speed level (power downshift) or falls by itself to the reduced speed level (overrun upshift).

An active engine control intervention can improve the overlapping gearshift process further.

- *Negative engine control intervention*:
  an active reduction of the engine torque requested by the transmission for support in positive overlapping (see Figure 9.33).
- *Positive engine control intervention*:
  an active increase of the engine torque requested by the transmission for support in negative overlapping. This measure requires safeguarding measures so that the possibility of an unwanted vehicle acceleration is excluded in case of an error.

One type of shift that deserves special attention is the rollout shift. These are downshifts at low vehicle speeds that take place when the vehicle is brought to a halt by braking or rolling out. The kinetic energy needed to raise the engine speed when downshifting is removed from the vehicle. This can be experienced as an unpleasant vehicle deceleration. (Note: rollout shifts are easier to master as freewheel shifts.)

### *2/ Continuous Slip Operating Mode (Slip-Controlled Clutch)*

The requirements of slip-controlled clutches are:

- Good controlability,
- no friction vibrations (stick-slip), "shudder",
  - green shudder: during the first continuous slip operations, friction systems (lining, steel, oil) that are not yet "run in" tend to produce torque vibrations that can occur as a result of the frictional behaviour of the slip-controlled clutch.
- favourable continuous slip properties across the entire service life.

See also Section 9.3.4 "Tribological System of Multi-Plate Clutches".

### 9.3.3 Design of Multi-Plate Clutches

Multi-plate clutches are designed for function (i.e. for torque transmission), shifting comfort and service life. Figure 9.34 shows influencing variables on mechanical and thermal stress and thus on multi-plate clutch damage. Besides shifting parameters like shifting work, frictional power, sliding speed, lining pressure, cooling/oil flow and shifting interval, geometric/design parameters such as the geometry of the plates, lining quality, grooving, corrugation, thickness of the steel plates, planarity of the plates as well as the elasticity of the pistons and end plate also play a role.

The driving splines of inner and outer plates are mechanically stressed by pressure and vibration. For multi-plate clutches that have been sufficiently dimensioned mechanically, thermal stress determines its performance limits. Only superelevation shiftings are relevant for thermal stress.

#### *1/ Performance Limits of Multi-Plate Clutches*

In case of damage to the plates, we differentiate between spontaneous and cumulative damage (Table 9.11). The following damages can be seen on the linings of the lined plates:

- *heat discolourations*, carbonisation:
  thermal overstress, e.g. from insufficient lubrication,

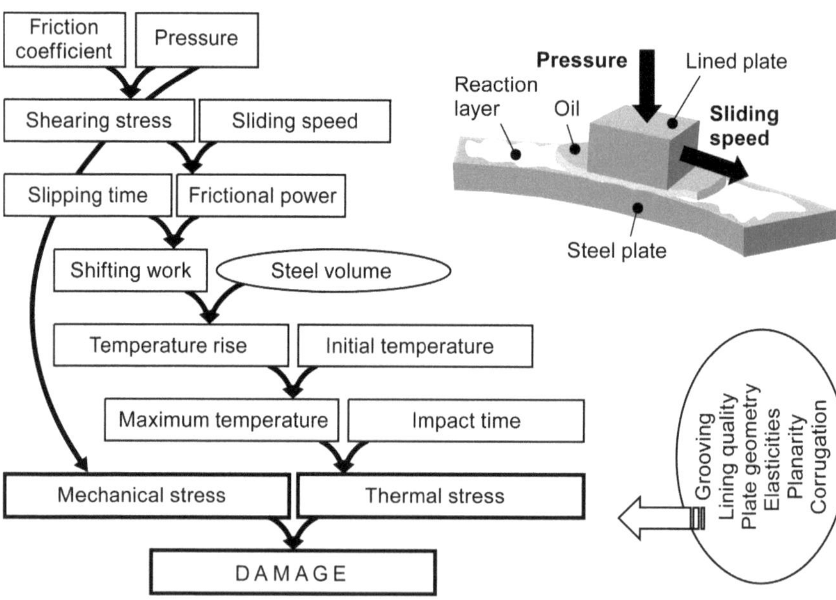


**Fig. 9.34.** Influencing variables on mechanical and thermal stress of multi-plate clutches (source: ZF)

**Table 9.11.** Plate damage (source: ZF)

| Spontaneous damage | Cumulative damage |
|---|---|
| • Local overloading<br>– Scuffing<br>– Hot spots<br>– Lining spalling<br>– Plate wobbling<br>• Global overloading<br>– Deformations<br>– Weldings<br>– Burnings | • Wear of friction parameters<br>• Friction material fatigue<br>• Oil ageing<br>– Thermal<br>– Mechanical<br>• Chemical reactions<br>– Attack or detachment of paper linings<br>– Sulphide formation (on sinter bronze) |

- *hot cracks*:
  high thermal and mechanical stress, usually combined with heat discolouration of the steel plates,
- *pittings, chunking right up to extensive spalling* (Figure 9.35a):
  lining fatigue resulting from high mechanical load,
- *smoothing, glazing* (Figure 9.35b):
  incorporation of oil-carbon residues as a result of high thermal load on the oil,
- *lining separation* (Figure 9.35c):
  due to insufficient lining strength resulting from lack of resin (manufacturing error),
- *lining detachment*:
  poor adhesion or chemical attack, i.e. due to rust under the lining,
- *lining wear*:
  as a result of an unsuitable counter-running surface, abrasive foreign bodies, little operating clearance,
- *frictional martensite on sinter linings* (Figure 9.35d):
  caused by excessive surface temperatures.

The following damage types can be found in steel plates:

- *extensive heat discolouration* (Figure 9.36a):
  high shifting work, long slipping times or not enough operating clearance lead to thermal overload due to excessive friction surface temperature,
- *hot spots* (Figure 9.36b):
  high frictional power with short slipping times, frequently accompanied by unfavourable pressure conditions, leads to local thermal overloading. The hot spots are often distributed randomly. The cause can be inferred from the position of the hot spots. An even distribution indicates corrugated plates, uneven pistons or plate natural vibrations. If the hot spots are on the outer diameter, this suggests plate wobbling.
- *scoring, wear* (Figure 9.36c):
  due to insufficient lubrication or abrasive foreign bodies,

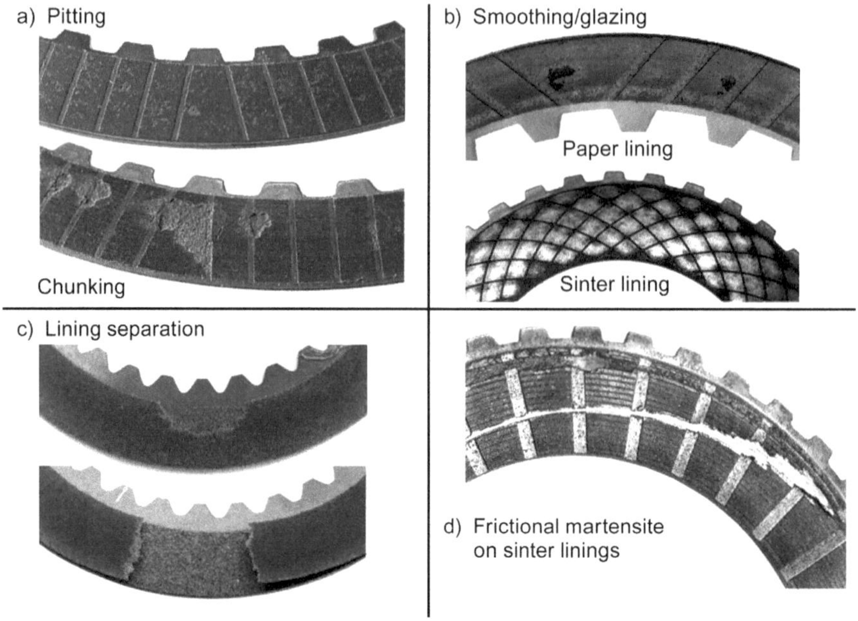


**Fig. 9.35.** Damage to lined plates. *a* Pitting and chunking; *b* smoothing, glazing; *c* lining separation; *d* frictional martensite on sinter linings [9.30]

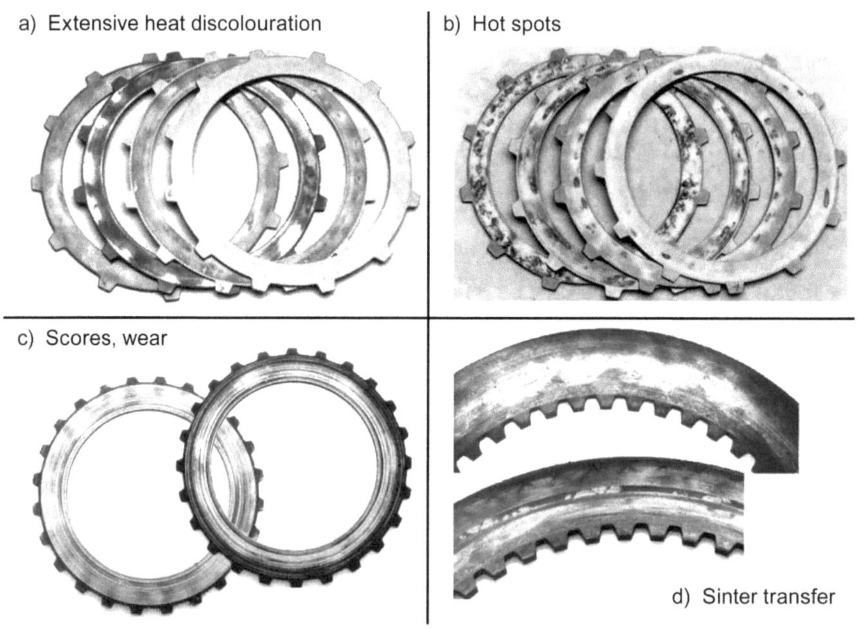


**Fig. 9.36.** Damage to steel plates. *a* Extensive heat discolouration; *b* hot spots; *c* scoring, wear; *d* sinter transfer [9.30]

- *matting*:
  due to additive residues,
- *sinter transfer on sinter linings* (Figure 9.36d):
  thermal overload due to lacking lubrication, insufficient operating clearance or poor design,
- *corrosion*:
  resulting from water in the oil.

### 2/ Basic Principles of Multi-Plate Clutch Design

The static clutch torque is supported/transmitted by the plate gearing in the inner and outer plate carriers. The *pressure on the driving flanks (1)* of the gearing has to be examined. (Note: the number in the parentheses, e.g. *(1)*, refers to the algorithm shown in Figure 9.37.)

The design of multi-plate clutches must guarantee that the torques are transmitted frictionally, i.e. the *static torque capacity (2)* must be ensured.

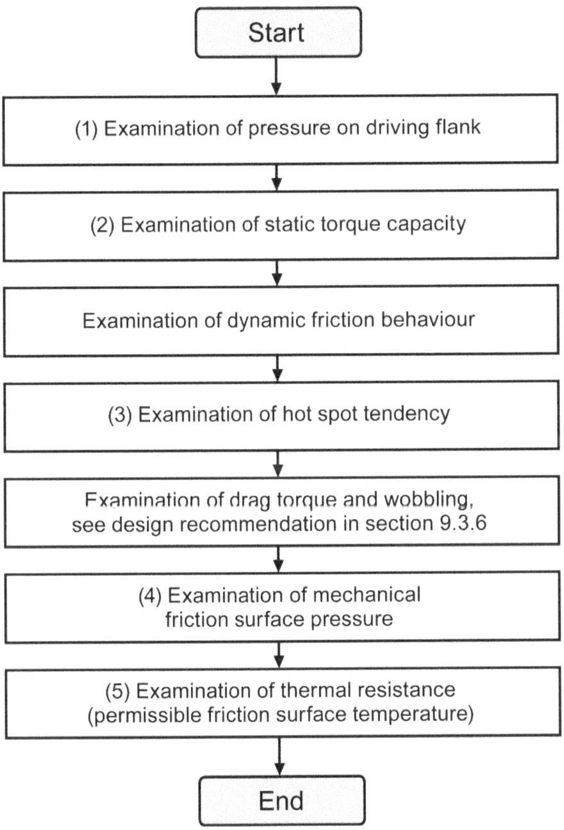


**Fig. 9.37.** Algorithm for designing multi-plate clutches

The hydraulic pressure in the piston chamber is modulated over the slipping time as shown in Figure 9.33. Beyond the shifting process, the hydraulic pressure ensures the static torque capacity of the clutch. In calculating the clutch pressure required, the forces working against closing should be considered. These result from:

- the force of friction of the piston sealing, from pressure as well as surface quality,
- spring plate, frequently between the piston and the first steel plate, to improve the contact behaviour of the clutch (see Figure 9.41, part 5),
- the force of the piston return spring, as a plate spring or a coil spring package,
- the moving force of the plates, friction coefficient between the plates and the plate carriers.

In making design calculations for the piston return spring, these external forces are to be set in the respective direction of action as well as the rotation pressure of the oil in the piston chamber if applicable. On this, see the topic of rotational pressure compensation in Section 9.3.5.

Hot spots play a crucial role when designing for thermal load. They can be seen on steel plates. The *hot spot tendency (3)* of different linings depends strongly on sliding speed and surface pressure. Modern organic linings are less sensitive to hot spots because they are elastic and have a higher pore volume, thus increasing their ability to even out unevenness on the steel counter-surface. The thickness of the friction lining is another influencing parameter on hot spot behaviour. Thicker linings have higher elasticity and exhibit a more homogeneous distribution of pressure. In this way, the entry of heat into the steel plates is also more homogeneous [9.13, 9.22].

The plates are loaded thermally from friction during engagement and mechanically when closed due to *friction surface pressure (4)*. Reference values for load limits for a single shift are provided in Table 9.12. Load profiles are necessary for such designs. The *friction surface temperature (5)* is decisive when investigating the thermal service life of multi-plate clutches. The way this temperature is reached is, however, of secondary significance.

**Table 9.12.** Reference values for load limits during a gearshift

| **Limit values** | **Mean temperature rise** | **Specific shifting work** | **Friction surface pressure** |
|---|---|---|---|
| | $\Delta T$ | $W_A = \frac{\lvert W \rvert}{A_R}$ | $p_{R,i} = \frac{F_n}{A_{R,i}}$ |
| | $\Delta T_{max}$ (°C) | $W_{A,max}$ ($J/mm^2$) | $p_{R,max}$ ($N/mm^2$) |
| Paper/steel | 80 | 1 | 5 |
| Sinter/steel | 120 | 2 | 10 |

In the case of shifting clutches, heat removal during the shifting process (short slipping times) is to be neglected. Heat flows to the steel plates. The heat absorption of the double-sided bonded lined plates is negligible as far as organic friction linings are concerned.

However, in the case of clutches for continuous slip operation (master and modulation clutches), heat must be dissipated by the oil (thermal equilibrium). The formulae and relations for frictional power $P$ and frictional work $W$ per shifting appearing in Section 9.2 "Layout and Design of Synchronizers" and Chapter 10 "Moving-Off Elements" are also valid here (see Equations 9.7 and 9.8).

### 9.3.4 Tribological System of Multi-Plate Clutches

Figure 9.34 shows the tribological system of multi-plate clutches, consisting of a lined plate, oil film and steel plate as counter-running surface. The reaction layer that forms on the steel plate plays an important role (chemical protective film, see Section 11.2.2 "Principles of Lubricating Gearwheel Mechanisms"). The friction coefficient is the decisive quantity for multi-plate clutches [9.9, 9.19]. Essential factors of frictional behaviour include:

- type of lining: paper, sprinkle sinter, carbon (see Table 9.13),
- for organic lining (paper),
  - composition of the lining (fibres, fillers, resin etc.),
  - density of the lining,
- grooving,
- lining surface,
- surface of the steel plate,
- type of oil and oil additive as well as the condition of the oil (new or used oil),
- shifting parameters such as sliding speed, friction surface temperature etc. (see Figure 9.34).

The oil should dissipate the heat built up during the shifting process. To this end, the shifting elements must be adequately supplied with oil. On the size of the *lubricating oil flow*, see Section 9.3.5. On the other hand, the oil is also a decisive partner in the tribosystem. The additives mixed into the base oil determine essential properties (Fig 9.38). The torque peak at the end of the shifting process, known as the "rooster tail", is not desirable because it initiates friction vibrations [9.14, 9.18].

Friction vibrations are a considerable comfort problem. They arise when the friction coefficient $\mu$ goes up with a decreasing sliding speed $v$, i.e. with a decreasing differential rotational speed $\Delta n$. This results in shudder with wet master and converter clutches (Figure 9.39) and in noise with shifting clutches (Figure 9.38).

The control behaviour of slip-controlled clutches, and thus the tendency towards friction vibrations, is determined by the profile of the friction coefficient over the sliding speed. A friction coefficient which falls over sliding speed is bad for control (Figure 9.39).

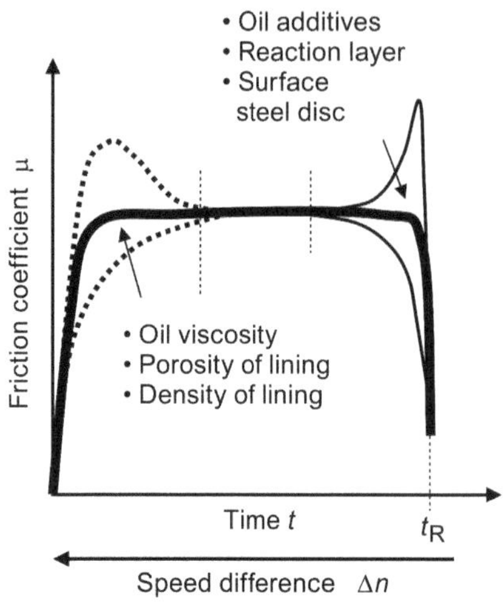


**Fig. 9.38.** Powershift operating mode: tribological system, effect on the friction coefficient profile. $t_R$ Slipping time

Because of the increasing friction coefficient, the system reacts to an increase in pressure in the piston chamber (in order to reduce the sliding speed, differential rotational speed) in a self-reinforcing way and thus tends to produce vibrations and to close abruptly. A friction coefficient profile that increases over sliding speed has a damping effect on the system. Influencing factors for friction vibration include:

- *friction material*,
  - constitution,
  - surface,
  - elasticity and porosity,
  - capacity to absorb oil,
- *tribology*,
  - surface reaction layer,
  - contact angle of the friction surfaces,

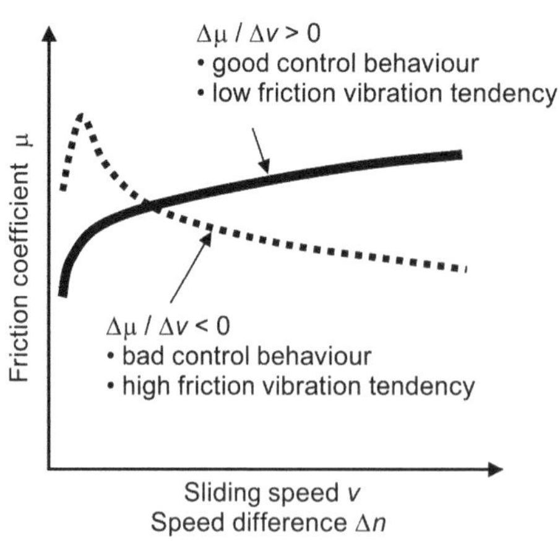


**Fig. 9.39.** Continuous slip operating mode: control behaviour of slip-controlled clutches

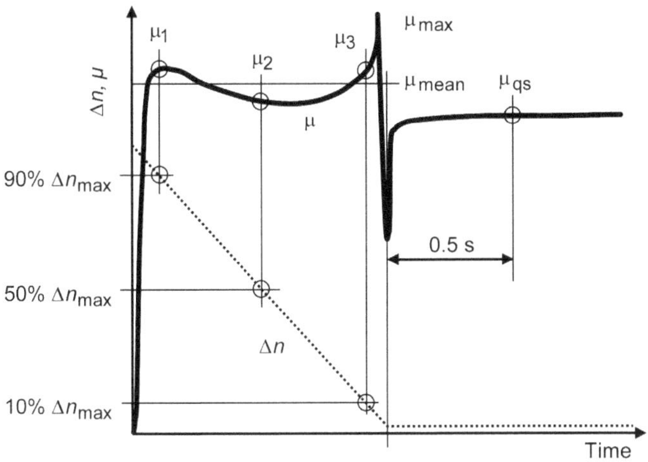


**Fig. 9.40.** Friction coefficient definitions (ZF)

- *oil composition,*
- *operation conditions,*
  - temperature,
  - pressure,
  - differential speed,
- *stiffness of the powertrain components.*

Figure 9.40 shows various friction coefficient definitions depending on the differential rotational speed. Plate test benches are used to determine the parameters. The quasistatic friction coefficient $\mu_{qs}$ is defined as the friction coefficient detected 0.5 s after shifting ends at a differential rotational speed of 10 1/min (hot clutch).

Table 9.13 cites established friction linings as well as their components, properties and fields of application.

**Table 9.13.** Current friction linings, components, properties, layout and field of application

| | Organic ("paper") | Sprinkle sinter | Carbon |
|---|---|---|---|
| **Components** | Fibres, filler material, phenolic resin, 0.4 to 1.2 mm thickness | Copper, tin, iron, graphite and different additives | Carbon fabric impregnated with resin, 100% carbon |
| **Properties** | Elastic, good friction and noise behaviour, high thermal resistance | Inelastic, low shifting comfort, sensitive to noise, high mechanical and thermal resistance | Full carbon, very expensive, high thermal resistance, use in motorsports, no grooves required |
| **Layout** | Glued to carrier plate | Sintered on carrier plate | Glued to carrier plate |
| **Field of application** | E.g. passenger car automatic transmission | E.g. construction vehicles | E.g. motorsports |

The thermal destruction of paper linings commences at a continuous temperature of ca. 200°C. Sprinkle sinter linings are used in conditions of high thermal load because, besides their inherently higher thermal resilience, lined plates absorb additional heat unlike the isolating paper plates.

### 9.3.5 Engineering Designs

In the following, design details of multi-plate clutches will be introduced. Figure 9.41 shows existing designs of multi-plate clutches and brakes using the example of the conventional automatic transmission ZF 6 HP 26 (Figure 6.34 and Figure 12.25).

#### *1/ Grooving*

Friction linings of lined plates are usually provided with grooves (with the exception of full carbon fabric). Figure 9.42 shows some established groove types. Grooving has the following tasks:

- cooling the plates with the oil flow, even when the shifting element is closed,
- cutting the oil film and thus stabilising the friction coefficient,
- promoting the desired friction behaviour during shifting and
- improving open clutch behaviour, reducing drag torque.

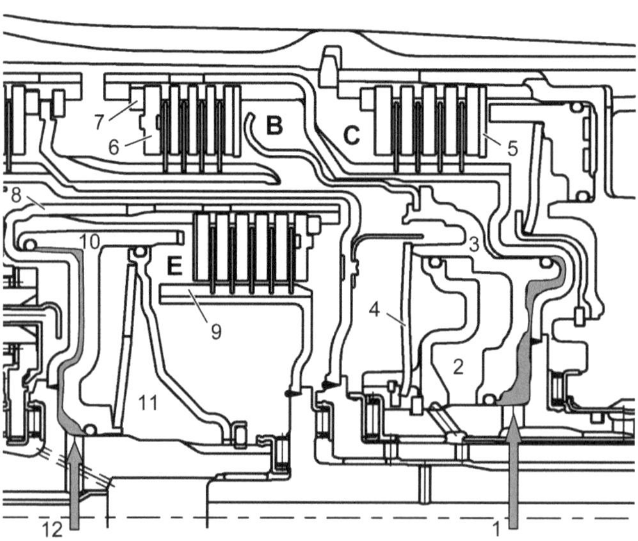


**Fig. 9.41.** Clutches *B* and *E* as well as brake *C* of the transmission ZF 6 HP 26. Gearbox diagram from Figure 9.33. *1* Pressure supply clutch *B*; *2* rotational pressure compensation *B*; *3* piston *B*; *4* return plate spring *B*; *5* spring plate *C*; *6* end plate *B*; *7* snap ring; *8* outer plate carrier *E*; *9* inner plate carrier *E*; *10* piston *E*; *11* rotational pressure compensation *E*; *12* pressure supply clutch *E*

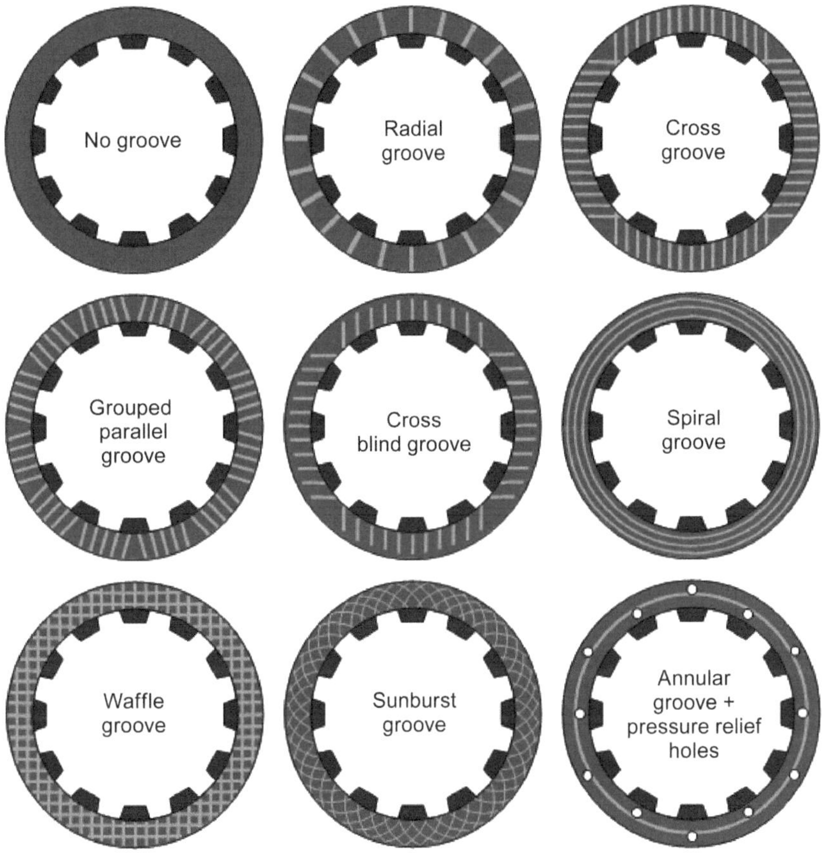


**Fig. 9.42.** Typical grooving of friction linings (Source: ZF)

There is a great amount of variety in groove geometry. Depending on the application, various groove shapes and geometries are used (see Figure 9.42). The groove shape is usually determined empirically and optimised by testing it on the specific case.

In the case of paper linings, the grooving is embedded by embossing. In this case, the depth of the groove is a maximum of 40% of the lining thickness. Alternatively, blanking is possible in case segmented linings are being used. Segmented paper linings are one possibility of manufacturing inexpensive lined plates. In case the groove shape necessitates locking areas ("puzzle links") to connect the individual segments, one must be sure that its mechanical resilience is sufficient.

Sheet metals, C-steels, and microalloyed steels are used as support material for lined plates. They are hardened, heat-treated or nitrided depending on the boundary conditions.

### 2/ Surface Quality of the Counter-Friction Surface (Steel Plate)

The surface quality of the counter-friction surface (steel plate) plays a large role in the friction coefficient profile and therefore on shifting and noise behaviour.

### 3/ Driving Splines

For the purposes of passenger cars, driving splines are usually designed as trapezoidal teeth. In case of higher requirements with respect to resistance against flank wear, involute teeth are also used, as they adapt better to the opposite teeth. The driving spline is designed for permissible surface pressure. Because the plates are mostly blanked parts, not the whole width is bearing the load. The blanking reduction has to be regarded.

Damage to driving splines typically occurs due to vibrations when the clutch is open. Notches appear, induced by the irregularities of the internal combustion engine. The common countermeasure is to increase hardness by means on a corresponding heat treatment.

### 4/ Corrugated Plates, Spring Plates

Corrugated plates not only reduce the drag torque, they also help to improve the contact behaviour. Section 9.3.6 provides design recommendations aimed at reducing drag torque. In order deliberately to influence the contact behaviour of the piston, a spring plate can be introduced additionally, e.g. Figure 9.41, part 5.

### 5/ Cooling/Lubricating Oil Flow

Heat generated during shifting must be dissipated by means of oil. To this end, the shifting elements must be sufficiently supplied with cooling/lubricating oil. In the case of gearshift clutches, the specific cooling/lubricating oil flow is in the order of 0.5 $mm^3/(mm^2s)$, for master clutches it is about 5.0 $mm^3/(mm^2s)$ of oil. The oil has to be evenly distributed on all plates. Turning off or reducing the cooling/lubricating oil flow leads to diminished drag torque (see Section 9.3.6). An open clutch should never run in a dry state. A minimal amount of about 0.1 $mm^3/(mm^2s)$ should constantly be provided.

### 6/ Gearshift Piston

The piston supplies the multi-plate packet with gearshifting force. The end plate serves as a counter bearing. Deformations of the gearshift piston due to shifting forces must be avoided in order to avoid uneven pressure distribution in the clutch packet.

In automatic passenger car transmissions, the gearshift piston is often manufactured from sheet metal, in trucks and for special uses more massively of aluminium. The supporting snap rings must be designed carefully, since they are not only subjected to the axial force but also, due to rotating shafts, to centrifugal forces.

Furthermore, circumferential forces also have an effect, resulting from the closing or opening clutch.

In many passenger cars, plate springs are used for piston resetting, in commercial vehicles and other special applications coil spring packets are also in use. The advantage of plate springs is a force that remains almost completely equal throughout the spring travel. The disadvantages are hysteresis and high tolerances. Coil spring packets exhibit in contrast almost no hysteresis. However, the force required constantly increases with the spring travel.

Piston sealings must be designed carefully (also see Section 11.5 "Gearbox Sealing"). Sealing requirements include:

- low frictional force levels at all relevant temperatures, no significant break away force at low temperatures,
- equal force across the piston stroke,
- minimal lining density influence, insensitivity to manufacturing tolerances,
- unproblematic assembly.

Multi-plate clutches *vary in their response characteristics*. Possible causes could be:

- tolerances in the piston return spring,
- sealing ring friction, twisting of the sealing elements, sealing contact surfaces,
- tilting of the piston,
- air in the system, including the control unit,
- variation in the friction coefficient profile of the lined and steel plates etc.

While some of these causes can be avoided by taking measures in design and production, effects that set in after a certain amount of running time must be countered with adaptations.

### 7/ Rotational Pressure Compensation

In rotating, oil-filled piston chambers, oil centrifugal forces lead to a rotational pressure that exerts an axial force on the piston (Figure 9.43). Beyond a certain speed, the rotational force acting on the piston becomes greater than the restorational force of the spring. In the absence of countermeasures, the piston would close the multi-plate packet. Design measures to compensate the rotational pressure include:

- stronger return spring:
  inexpensive, but reduces the effective shifting pressure,
- ball valve for clutch discharge:
  low effort, but unsuitable for controlled clutches and overlapping shiftings, since the function is influenced by speed and oil temperature,
- spring-controlled clutch discharge valve:
  relatively high effort, unsuitable for controlled clutches and overlapping shiftings, since the function is influenced by speed and oil temperature,

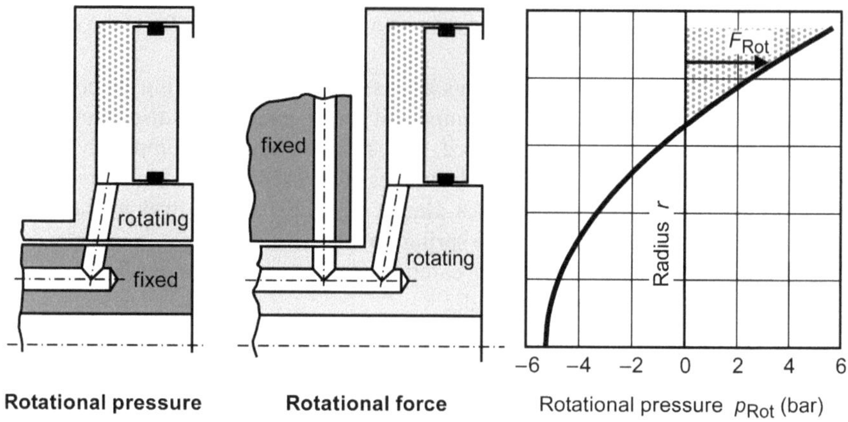


$p_{Rot}(r) = \rho \cdot \omega^2 \cdot \int_r r \cdot dr + C \qquad F_{Rot}(r) = 2 \cdot \pi \cdot \int_r p_{Rot}(r) \cdot r \cdot dr$

**Fig. 9.43.** Rotational pressure in rotating, oil-filled piston chambers

- shift pressure-controlled discharge valve: functionally reliable, but very high effort, also unsuitable for controlled clutches and overlapping gearshifting,
- pressure compensation chamber (Figure 9.41, rotational pressure compensation *B 2* and *E 11)*: requires a lot of installation space, yet always functions and can also be used for controlled clutches and overlapping shiftings.

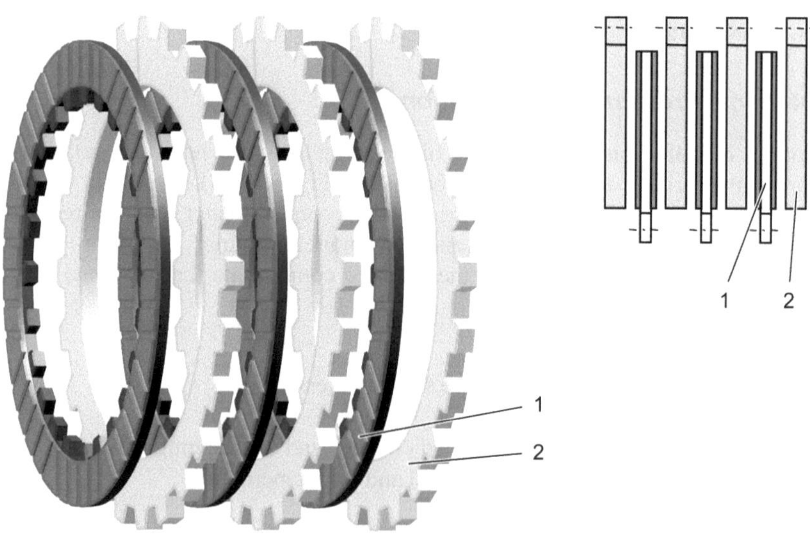


**Fig. 9.44.** Conventional arrangement: double-sided lined plates (DSP = Double-Sided Plates). *1* Inner plate, double-sided lining; *2* outer plate

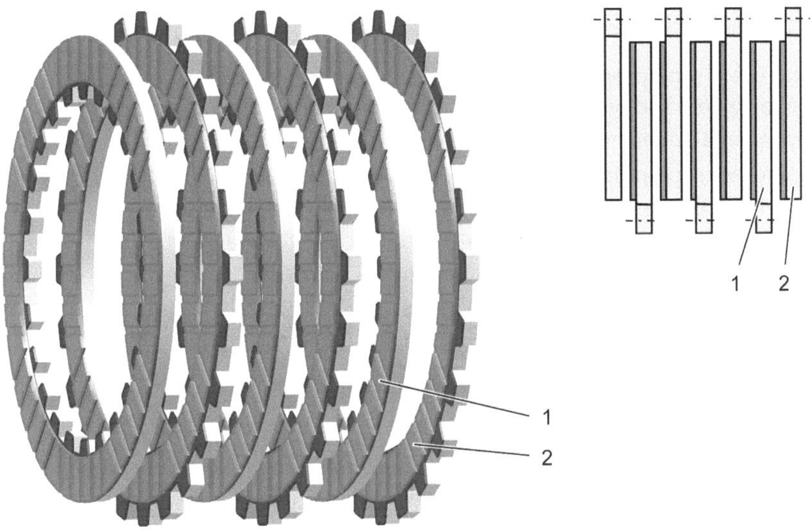


**Fig. 9.45.** Single-sided plates (SSP). *1* Inner plate, one-sided lining; *2* outer plate, one-sided lining

### *8/ Single-Sided Plates*

As opposed to conventional lined plates with double-sided friction lining (DSP = Double-Sided Plates, Figure 9.44), Single-Sided Plates (SSP) have only one friction lining alternating between the inner and outer plates (Figure 9.45).

By increasing the available volume of steel for thermal storage, the load capacity of the clutch can be increased at equal installation space. However, SSPs are more expensive than DSPs, they increase the danger of assembly faults and tend to wobble at high rotational speeds due to their low weight.

## 9.3.6 Detail Questions

In the following, a few questions of detail concerning multi-plate clutches will be addressed.

### *1/ Drag Torque/Drag Loss, Design Recommendations*

Drag torque is defined as the loss torque at open shifting plates through-flowed or dipped in oil. By means of the relative motion between the inner and outer plates, the oil found between them is sheared. The oil shear forces cause the drag torque. Alternatively, “drag torque” is also called “drag loss” [9.15, 9.20–9.21].

In the case of conventional automatic transmissions, drag losses contribute approx. 20% to transmission losses at operating temperatures. The level and profile of the drag torque $T_D$ are contingent on many factors. Some important influencing parameters and resulting design recommendations are:

- $T_D$ increases superproportionally with increasing friction radius (circumferential speed) and proportionally with the number of friction surfaces:
  - better small diameters and more friction surfaces,
- $T_D$ increases superproportionally with a smaller operating clearance:
  - operating clearance should not be smaller than 0.1 mm/friction surface (better 0.2 mm/friction surface),
- $T_D$ increases proportionally with increasing oil viscosity:
  - oil with low viscosity is preferable,
  - $T_D$ increases proportionally with increasing oil flow.

Further positive (drag torque reducing) measures include:

- corrugated lined plates:
  - not at circumferential speeds larger than 60 m/s due to the risk of wobbling,
- design faster rotating plates as lined plates with grooving,
- demand-oriented cooling fluid supply,
- favourable gearbox layout:
  - small amount of open clutches and low relative speeds,
- grooving optimised for the conditions of use.

Demand-oriented cooling fluid supply (i.e. shutting off or reducing the cooling/lubricating oil flow) is an effective way to reduce the drag torque, but it is also relatively elaborate and expensive. Figure 9.46 shows a characteristic drag torque profile over the differential rotational speed. In the ascendant at lower rotational speeds, the Newtonian shear stress equation is approximately valid.

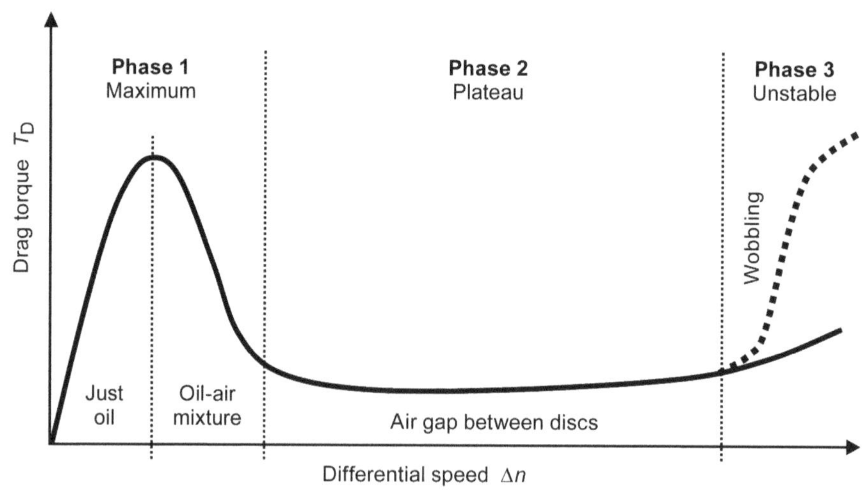


**Fig. 9.46.** Characteristic drag torque profile of open multi-plate clutches

Drag torques are dependent on the gear and are relatively high at speeds smaller than 2000 1/min. The average value over rotational speed and gears are in the order of about 4 Nm (at 40°C) for conventional automatic passenger car transmissions. This corresponds to ca. 1 kW power loss. The oil temperature (viscosity) exerts a strong influence – at 80°C oil temperature, the drag torque is only half of what it is at 40°C.

Wobbling of the plates at high circumferential speeds with reciprocal contact leads to a drastic increase in drag torque. Plate wobbling must therefore be avoided by means of suitable countermeasures.

#### 2/ Test Methods for Friction Linings

On plate test benches, the most varied tests are carried out, both single stage tests and load profile examinations. These are some typical clutch lining tests:

- face tension test to determine tensile strength,
- compressibility test,
- determining the dynamic and static friction coefficient profiles: DKA test (German co-ordination board),
- investigations in friction vibration: green shudder and shudder tests,
- torque-controlled break-off (torque capacity, mainly torque converter clutch),
- control test for controlled clutches,
- bump-plate test to determine the mechanical strength of linings and
- bending test to monitor lining/support plate adherence.

## 9.4 Parking Locks

When the engine is switched off, vehicles with a manual transmission can be kept stationary by engaging a gear with a high ratio in addition or alternatively to applying the parking brake. This option is not available in vehicles with a wet master clutch or a torque converter, since there is no link between the vehicle and the braking power of the engine when the latter is off.

In order to prevent passenger cars and light commercial vehicles with the abovementioned moving-off elements from rolling away unintentionally, they are supplied with parking locks. A parking lock mechanically locks the transmission output shaft against the transmission housing. In the case of commercial vehicles and buses with spring actuator parking brakes, the parking lock is not utilised as an additional element to keep the vehicle stationary. In this case, the transmission selector lever lacks a P position.

The basic principle of parking locks is the radially acting locking pawl. The details of its actual design vary. With respect to the locking elements, we differentiate between:

- systems with a locking roller (Figure 9.47) and
- systems with a locking cone (Figure 9.49).

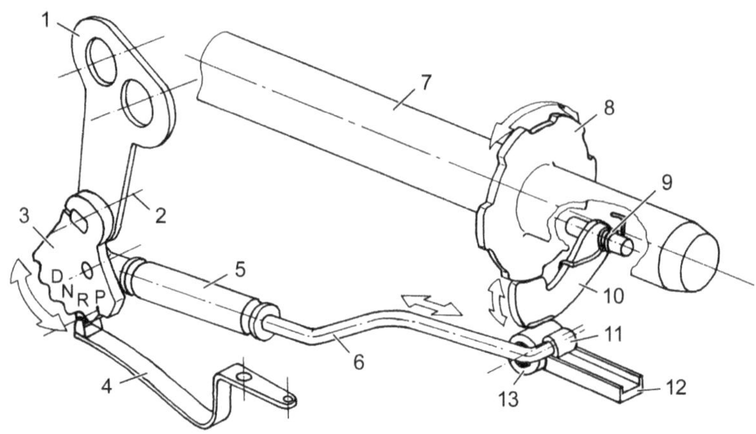


**Fig. 9.47.** Parking lock with radially acting locking pawl (see also Figure 12.22).
*1* Gearshift lever on the transmission; *2* axis selector shaft; *3* detent plate (*P* park position, *R* reverse gear position, *N* neutral position, *D* drive position); *4* detent spring; *5* compression spring; *6* connecting rod; *7* output shaft; *8* parking lock wheel; *9* return spring; *10* pawl; *11* locking roller; *12* guide; *13* roller bearing

Concerning activation we distinguish between:

- mechanical designs (Figure 9.47) and
- electric designs, usually in connection with e-shifting (Figure 9.49).

### 9.4.1 Mechanically Activated Parking Locks

In mechanical designs, the parking lock is activated by means of a Bowden cable connecting the selector lever unit in the vehicle to the transmission, while in electric designs this is done, for example, with button activation.

In the case of mechanical activation, the locking action is initiated by the driver moving the selector lever to engage the park position. The transmission shown in Figure 12.22 will be used as an example to illustrate the design and function of a mechanically activated parking lock of a conventional automatic transmission. The parking lock shown in Figure 9.47 has a radial locking pawl. Moving the selector lever *1* to the park position P has the following effects:

1/ The detent plate *3* rotates about the axis *2* of the selector shaft in the same direction as the gearshift lever, until the roller on the detent spring *4* engages in the park position *P*.
2/ The connecting rod *6* linked to the detent plate, moves the roller *11* (which runs on it) in the guide *12* parallel to the output shaft *7*.
3/ At the end of the guide, the roller runs over a roller bearing *13* fixed in the housing, pressing upwards against the sloping back of the pawl *10*. This

moves upwards against the resistance of its return spring *9* as far as the parking lock wheel *8,* which is rotationally fixed to the output shaft.

4/ When the vehicle is at rest, or moving at a speed below the engagement speed (about 2.5–5 km/h), the pawl engages in a gap in the parking lock wheel, positively locking the driving wheels and preventing the vehicle from moving.

5/ When the vehicle is moving at a speed higher than the engagement speed, the flank angle at the parking lock wheel and the pawl prevents the pawl from engaging. This "grating" continues as long as the vehicle is moving faster than the engagement speed. The compression spring *5* mounted on the connection rod is stretched. As soon as the speed falls below the engagement speed, the compression spring pulls the locking roller under the pawl and moves the latter into a tooth gap, preventing further movement of the vehicle.

6/ The pawl is disengaged when any other driving position is selected. The locking roller *11* moves back into the guide and frees the pawl to move downwards and out of the toothing of the parking lock wheel. This disengagement motion is assisted by the return spring *9.*

The design of the parking lock system must fulfil the following criteria:

- rejection requirement at higher vehicle speeds,
- locking-in requirement both forward and backward on slopes up to about 30%,
- locking reliability on slopes up to about 30%,
- disengagement requirement/actuating force and
- operational safety.

Fulfilment of the *rejection requirement*: unintentional activation of the parking lock at speeds of $v = 2.5$–$5$ km/h is essentially prevented by the tooth gap distance and the flank angle of the parking lock wheel and the pawl. Activating the parking lock during driving (abuse) must be possible for a short time without blocking or destruction of parts.

Fulfilment of the *locking-in requirement*: in the unfavourable "tooth on tooth" position, the edges of the teeth of the locking pawl and the parking lock wheel are exactly on top of each other. The compression spring keeps the system preloaded. If the vehicle on a 30% slope moves due to the downhill slope force, the pawl must lock securely into the next gap. The rotation angle until the next gap of the parking lock wheel must be designed such that the driving speed reached in the interim by the vehicle does not lead to rejection. The engagement speed is defined at the boundary speed at which the locking position (tick point) can just be reached.

Fulfilment of *locking reliability*: the geometry, the deformation under load as well as friction conditions in the system must be chosen such that even dynamic loads on a 30% slope (rocking test) do not lead to automatic deactivation. With a trailer, the requirements must be fulfilled for a 12% slope.

Fulfilment of the *disengagement requirement/actuating force*: for heavy vehicles and corresponding ratio conditions between the tyres and the parking lock wheel, pressure between the locking roller (or locking cone) and the pawl can be very high on a 30% slope. The geometry and choice of material or coating must guarantee sustainable disengagement forces even under these conditions. Related to the axis of the selector shaft (*2* in Figure 9.47), typical values lie between 10 and 20 Nm. The formation of frictional martensite due to excessive pressure between the parking roller (cone) and the pawl should be avoided.

Fulfilment of *operational safety*: the parking lock should be considered from the standpoint of comfort, but even more so from that of safety. The interaction of the geometry of the cams on the detent plate (Roostercomb) and the hardness of the detent spring contributes heavily to shifting feel. The friction in the entire system – from the selector lever unit in the vehicle, through the Bowden cables, to the detent plate and the active roller of the detent spring – must be selected such that only defined conditions are possible. Intermediate positions between P and R can be avoided by a correspondingly shaped cam on the detent plate. This should act as a flip-flop in all conditions (tolerances, temperature, friction etc.). In designs until about 1980, circular paths were also found on the detent plate between P and R. The circular path is less sensitive to frictional influences in the external gearshift system. It must however be set securely such that possible intermediate positions on the circular path remain stable under all conditions and the driver always obtains a clear response from the vehicle about whether P or R is active. Unclear operational conditions ("illusory park") must be safely avoided in all design solutions.

The abovementioned points show how much care is required in the design and approval of a parking lock system. The system has to be robust enough to function safely even in poor tolerance situations, under extreme temperatures and with the deterioration that sets in after extensive mileage. The surface quality of the parking lock wheels and pawls, frequently designed as a blanking part, must fulfil all requirements with respect to surface pressure and friction. Förster provides suggestions for determining the essential geometric conditions of parking lock systems in [9.7].

### 9.4.2 Electrically Activated Parking Locks

The abovementioned criteria for parking lock systems of robustness and functional reliability must also be guaranteed in the case of electric activation. Figure 9.48 shows the assembly situation and Figure 9.49 the working principle of an electrically activated parking lock using the example of the automatic transmission ZF 6 HP 26.

The detent plate in the transmission is omitted, being substituted with a parking plate as well as with a parking lock cylinder with a locking magnet. When leaving the park position, hydraulic pressure is introduced to the parking lock cylinder by means of a solenoid valve. The cylinder pushes back a piston, which extracts the locking cone under the pawl by means of the kinematics shown. An interlock solenoid is turned on and also locks the piston using balls.

When going into the park position, the hydraulic pressure in the cylinder is switched off and the cylinder chamber is deaerated. Mechanical locking of the piston by the balls is lifted and the piston is released.

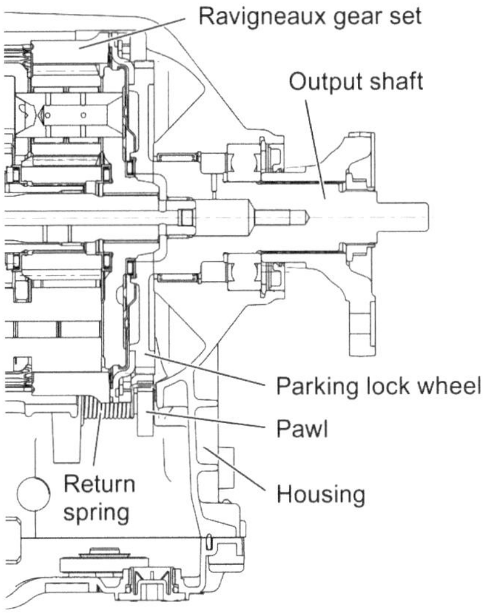


**Fig. 9.48.** Assembly of the parking lock in the transmission ZF 6 HP 26

By means of a preloaded leg spring on the parking plate, the piston is pulled in the park direction and the parking lock is inserted. An additional Bowden cable on the parking plate makes it possible to unlock the parking lock in emergencies and in case of error.

## 9.4.3 Detail questions

The importance of careful testing of parking lock systems prior to release for mass production is now obvious.

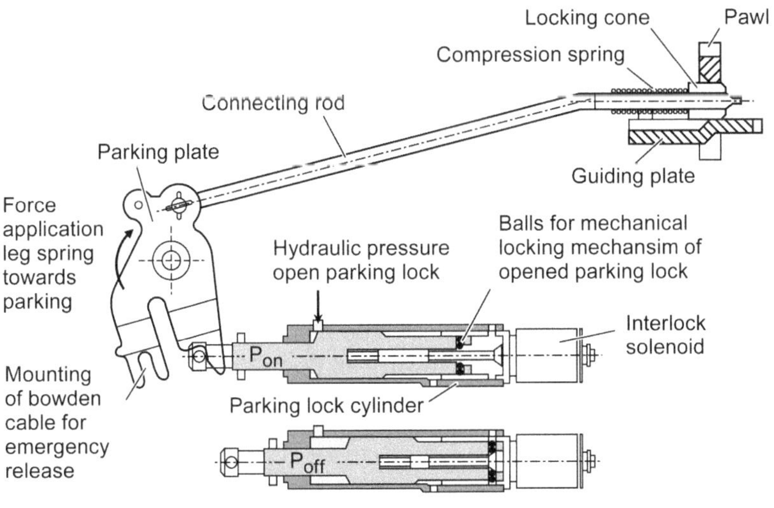


**Fig. 9.49.** Components of the electrically activated parking lock ZF 6 HP 26, Figure 12.25

Typical tests are:

- tests on the parking lock test bench:
  - engagement speed,
  - operational forces/self-locking,
  - load/wear/grating tests,
- vehicle tests forwards/reverse on a 30% slope:
  - locking-in tests: application with the shortest axle, smallest $r_{dyn}$,
  - load tests: application with the longest axle, greatest $r_{dyn}$,
  - trailer tests,
  - rocking and abuse tests (grating, tow-off tests, ...).

Load:

- for front- or rear-wheel drive vehicles, activating the parking lock on a 30% slope near engagement speed presents the greatest load for all involved components,
- for vehicles with permanent all-wheel drive, the greatest load of parking lock components occurs during the tow-off abuse test.

### *Pawl Movement*

The automatic insertion of the pawl into the parking lock wheel when driving on poor-quality roads must be safely avoided. This phenomenon is influenced by the return spring, the dead weight of the pawl, but especially by engine/transmission mounting. The grating caused by such insertion leads ultimately to damage to the parking lock system by intense wear.

### *Disengagement Noise*

Although stipulated in instruction manuals, many drivers of vehicles with automatic transmissions neglect to secure the vehicle with the parking brake. After activating the parking lock and releasing the service brake, vehicle on slopes fall into the pawl. The resulting forces and tensions in the powertrain lead to a disengagement noise (relief knock) when the parking lock is disengaged (by pulling out the locking cone or the locking roller). This is no problem for mechanically activated systems, since the driver activates the process himself "mechanically"; the driver is then familiar with and accepts disengagement noise as a normal response to disengaging the parking lock. In the case of electrically activated parking locks, disengagement noise is occasionally seen as an improper reaction of the transmission to a "touch of a button", thus becoming a comfort problem. Mechanical solutions, partially supported by software, aimed at reducing disengagement noise are well-known.

# 10 Moving-Off Elements

*Moving-off comfortably and dynamically, manoeuvring delicately*

The internal combustion engine has a minimum speed (idle speed). To move-off from vehicle standstill, the speed gap between the lowest engine operating speed and the stationary transmission input shaft must be closed by means of a speed converter (see also Section 4.1 "Powertrain"). Speed converters exhibit the following characteristics (Figure 10.1):

- output torque $T_2$ is equal to the input torque $T_1$: $T_2 = T_1$,
- output speed $n_2$ is smaller than or equal to the input speed $n_1$: $n_2 \leq n_1$ and
- input power $P_1$ is reduced by the power losses $P_V$: $P_2 = P_1 - P_V$.

Figure 10.2 illustrates an idealised clutch engagement during moving-off. Input and output speeds converge during clutch engagement. During continuous slip operation – slipping of the clutch – part of the input power is converted to heat as power loss. The efficiency of the clutch $\eta_C$ is determined by Equations 4.2 and 4.3 as

$$\eta_C = \frac{P_2}{P_1} = \frac{T_2\, 2\,\pi\, n_2}{T_1\, 2\,\pi\, n_1} = \mu_C\, \nu_C\,, \qquad (10.1a)$$

with $T_2 = T_1$, i.e. $\mu = 1$,

$$\eta_C = \frac{n_2}{n_1} = \nu_C\,. \qquad (10.1b)$$

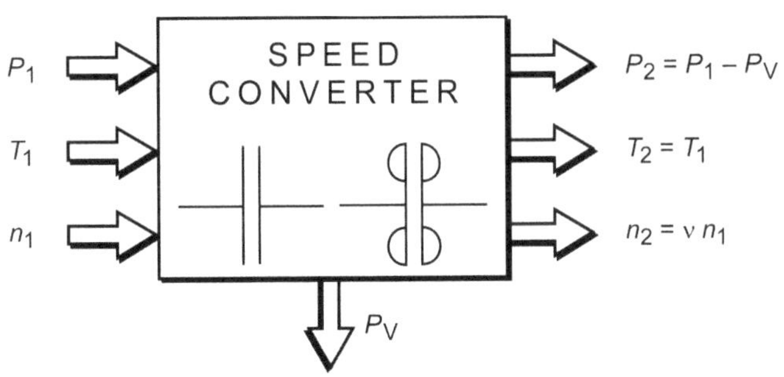


**Fig. 10.1.** Input and output values of a speed converter

H. Naunheimer et al., *Automotive Transmissions*, 2nd ed.,
DOI 10.1007/978-3-642-16214-5_10, © Springer-Verlag Berlin Heidelberg 2011

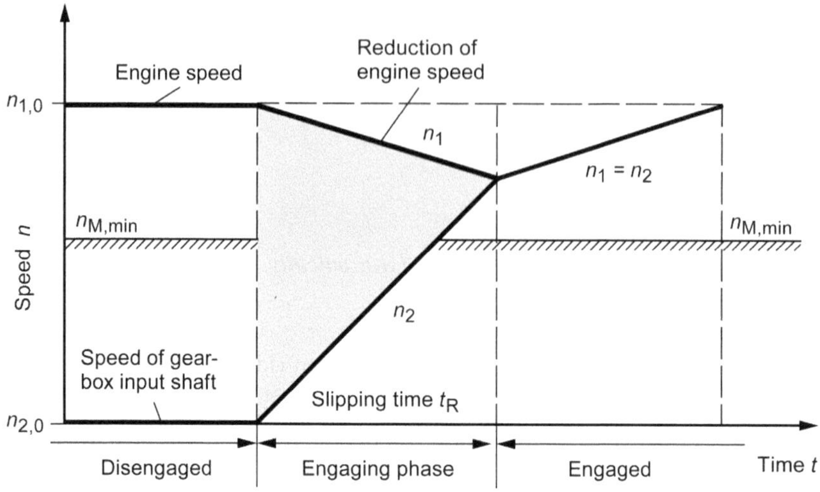


**Fig. 10.2.** Idealised moving-off process for a friction clutch

The slip $S$ is defined as the speed difference based on the input speed between the input and output speeds:

$$S = \frac{n_1 - n_2}{n_1}. \tag{10.2}$$

With Equations 10.1 and 10.2, we obtain the following for efficiency, slip and speed ratio:

$$S = 1 - \eta_C = 1 - \nu_C. \tag{10.3}$$

The moving-off element must be designed such that it transmits the maximum input torque with sufficient reliability on the one hand and resists thermal load on the other, even when moving-off frequently (stop-and-go) [10.33]. From this we can derive the main tasks of moving-off elements:

- transmission of the engine torque to the gearbox,
- gentle and jerk-free moving-off,
- long service life,
- damping of torsional vibrations,
- minimal size and
- overload protection of the internal combustion engine and power transmission components.

Moving-off elements are exclusively force-locking clutches. Figure 10.3 shows a systematic classification of moving-off elements [10.30].

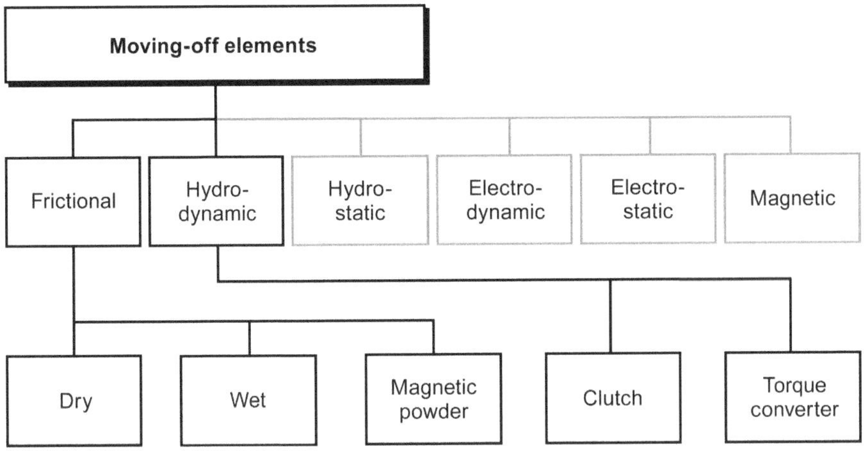


**Fig. 10.3.** Systematic classification of moving-off elements

In the modern vehicle, four basic systems are available as force-locking moving-off elements:

- dry friction clutches with $i_S = 1.0$ as the standard for manual transmissions,
- wet friction clutches for continuously variable and powershift transmissions,
- hydrodynamic torque converters with $i_S \geq 1.0$ as the standard for conventional automatic transmissions and
- hydrodynamic clutches.

Dry clutches are characterised by a high level of efficiency (no drag torque) as well as by a small moment of inertia, whereby the moment of inertia can increase considerably with additional demands regarding damping or automated crawling starts or multiple starts.

The advantages of wet moving-off clutches include their small mass, small moment of inertia, good controllability as well as a high power/weight ratio and large torque capacity. They are suited to vehicles with little installation space and high torque.

Due to their function as hydrodynamic fluid transmissions, torque converters with or without controlled lock-up clutches and torsional vibration dampers provide especially comfortable moving-off, an increase in torque as well as overload safety. The hydrodynamic clutch, the classic speed converter, has a high torque transmission capability. It is for all intents and purposes no longer used in road-going vehicles [10.9].

Frictionally engaged magnetic powder clutches are rare. Here, a magnetisable powder transmits the power frictionally. Table 10.1 shows a functional comparison of the four moving-off elements used in motor vehicles.

**Table 10.1.** A functional comparison of moving-off elements [10.9]

| | | Dry clutch | Wet clutch | Hydrodyn. torque converter | Hydro-dynamic clutch |
|---|---|---|---|---|---|
| Con-trolla-bility | Engagement characteristics | + | ++ | ++ | ++ |
| | Disengagement characteristics | + | + | – – [1)] | – – [1)] |
| Thermal capacity → continuous operation | | – | + | ++ | ++ |
| Thermal capacity → thermal shock | | + | – [2)] | ++ | ++ |
| (Lining) wear | | – | + | ++ | ++ |
| Installation space | | 0 | ++ | 0 | 0 |
| Actuating energy demand | | 0 | 0 | – | – |
| Crankshaft bearing load | | 0 | ++ | + | + |
| Centrifugal force influences | | 0 | ++ | 0 | 0 |

[1)] Disengagement only possible with additional systems, e.g. multi-plate clutches in the gearbox
[2)] Thermal shock can be avoided by means of control

## 10.1 Dry Clutches

Shiftable, externally actuated dry clutches make it possible for motor vehicles driven by internal combustion engines to move-off from a stationary position comfortably. This clutch transmits the engine torque by force-locking by means of friction forces on the transmission input shaft. It allows a rapid and complete separation and then jerk-free closing of the torque flow in both moving-off and shifting engagements. Moreover, dry clutches have a long service life without sacrificing comfort and also damp vibrations.

Dry friction clutches consist of at least one friction system with two friction plates pressed against each other. Depending on the number of clutch plates, we distinguish between single-plate clutches and multi-plate clutches.

### 10.1.1 Structure of Dry Clutches

A single-plate dry clutch of diaphragm spring design for passenger car applications consists of the following assembly groups (see Figure 10.4):

1/ a clutch plate almost exclusively with integrated torsional vibration damper, which is moveable axially on the transmission input shaft,
2/ a pressure plate assembly mounted to the flywheel of the combustion engine,
3/ a clutch actuation with release device, which transfers the release travel from the non-rotating actuation elements to the pressure plate assembly by means of a release bearing and a sliding sleeve

and, as an interface to the engine, the

4/ flywheel.

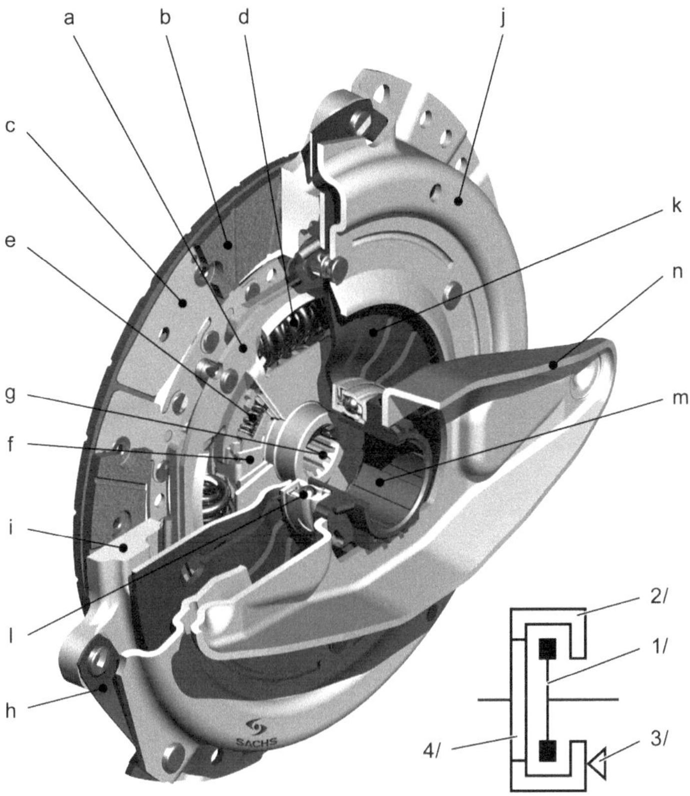


**Fig. 10.4.** Passenger car single-plate dry clutch (ZF Sachs).
*1*/ Clutch plate: *a* driving plate; *b* friction lining; *c* cushion spring; *d* torsion spring (driving operation); *e* torsion spring (idle operation); *f* friction device; *g* hub;
*2*/ pressure plate assembly: *h* return flat spring; *i* pressure plate; *j* pressure plate housing; *k* diaphragm spring;
*3*/ clutch actuation: *l* release bearing; *m* sliding sleeve; *n* release lever;
*4*/ flywheel

In dry clutches, the engine torque is transmitted via the flywheel *4* and the pressure plate assembly *2* by friction to the clutch plate *1* and thus to the transmission input shaft. There is static friction in the engaged state. During engagement, different speeds between the pressure plate assembly and clutch plate cause sliding friction, which generates heat and leads to lining wear.

### *1/ Clutch Plate*

The main components of the clutch plate are the driving plate *a* with friction linings *b*, the cushion spring *c* for comfortable moving-off, a torsional vibration damping system *d*, *e*, *f* for the reduction of torsional vibration induced by the engine as well as the hub *g* with internal spline.

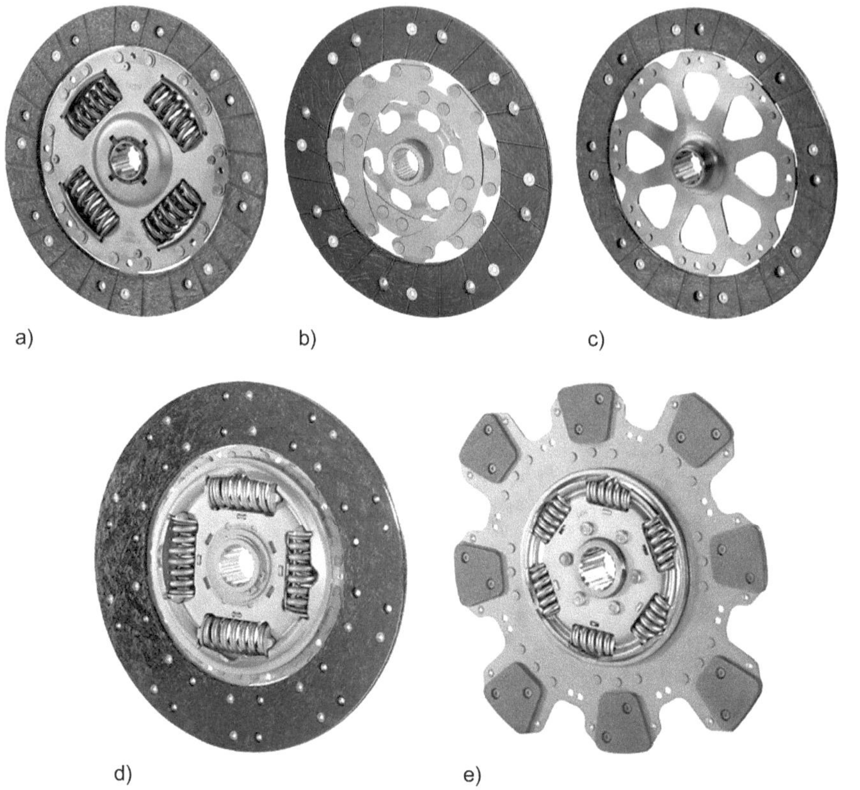


**Fig. 10.5.** Clutch plates (ZF Sachs). *a* Clutch plate with torsion damper (passenger car); *b* flexible clutch plate (passenger car); *c* rigid clutch plate (passenger car); *d* clutch plate with torsion dampers (commercial vehicle); *e* clutch plate with cerametallic lining pads (commercial vehicle)

Figure 10.5 shows various clutch plate designs for passenger cars (Figure 10.5a–c) and commercial vehicles (Figure 10.5d–e).

### 1.1/ Friction Linings

The flywheel *4* and the pressure plate *i* (see Figure 10.4) are the counter-friction surfaces to the friction linings on both sides of the clutch plate. The most important criteria in evaluating clutch friction linings are the friction coefficient (especially its minimal value under extreme load), the burst speed, the wear rate of the linings and the counter-friction surfaces, distortion due to thermal load, specific weight as input parameter for moment of inertia, controllability of the torque build-up and the tendency to self-excited friction vibrations (judder) and friction noises (squeal) [10.5].

Three types of clutch linings have established themselves for dry-running plate clutches in motor vehicles:

- *Organic linings*: these are, for comfort reasons, mostly used in the automobile sector. They consist of yarns (primarily of glass, aramid and cellulose fibres with brass or copper wire), that are embedded in friction cement made of resins, rubbers and fillers like carbon black, graphite and kaolin [10.5]. Organic clutch linings are always riveted in the shape of a ring on both sides of the clutch plate, in exceptional cases also (additionally) bonded.
- *Metallic (ceramic) sinter linings*: these materials are utilised for heavy loads, especially for tractors or excavation machines. The lining material is either sintered directly on the clutch plate or riveted by means of support plates, usually trapezoidal in shape. Due to their tendency to judder, they are of only limited suitability for road vehicles.
- *Carbon linings*: these are a subgroup of organic linings. The carbon fibres are inserted into the linings in order to increase the stiffness and temperature resistance of the linings. Due to their high costs and reduced shifting comfort, the most common area of application is in racing vehicles.

Radial grooves are used almost exclusively for dry clutches. They run straight from the inner radius of the friction lining radially to the outer diameter. Radial grooves promote removal of the abrasion generated, improve cooling air circulation and avoid adhesion on the running surfaces of the clutch linings.

### 1.2/ Cushion Deflection

Cushion deflection is a defined axial elasticity between both clutch lining rings. Its task is a gentle engagement for jerk-free moving-off. The pressure plate *i* of the clutch must first, against the spring force of the cushion springs *c,* press the clutch plate *1* against the flywheel *4* (see Figure 10.4). This slows down the build-up of pressure and thus makes the engagement process smoother.

Deflection is realised by means of corrugated steel plate segments and usually has an axial spring travel of 0.5 to 1.0 mm. Further advantages are favourable wear characteristics as well as an improved contact pattern leading to a more homogeneous distribution of heat. Basically, there are four types of cushion deflec tion:

- *Simple segment deflection*: riveting of the linings on thin, curved segments, which are riveted in turn on the driving plate.
- *Double segment deflection*: riveting the linings on two contrarily acting segments lying on each other.
- *Multi-plate deflection*: riveting of the linings on cushion springs that are slotted and corrugated on the outer fringe.
- *Intermediate plate deflection*: used in commercial vehicles. Riveting of the linings on curved segment-like spring plates that are connected on one side with an intermediate plate.

### *1.3/ Torsion Damping Systems*

Together with the damper system, the torsion spring system integrated into the clutch plate serves to reduce torsional vibration, a result of ignition-related rotational irregularities of the internal combustion engine or load change (load reversal) due to rapid accelerator pedal/clutch activation. This torsional vibration damping system consists of a friction device *f* as well as one torsion spring set each for driving *d* and idle operation *e* (see Figure 10.4). The torsion springs provides the working travel with which the damping device can become effective. The decisive criteria for torsion springs are the amount of angular deflection, the selectable progressive spring characteristic and end stops to protect the spring device.

The torsion spring system is mostly designed with coil compression springs arranged tangentially in the clutch plate in corresponding windows. By means of tangential enlargement of individual windows and the use of variously stiff springs, multi-level torsional vibration dampers can be created, leading to a broken progressive damper characteristic curve profile.

The function of a friction device is to eliminate part of the torsional vibration energy, thereby reducing the amplitude. The principle of mechanical friction is the sole operating principle. It is produced by pressing flat surfaces together axially, usually friction rings. The required friction torque can be adjusted by varying the amount of axial pressure, the material of the friction rings and the size of the friction diameter.

Figure 10.6 shows a characteristic curve for a torsional vibration damper with load and overrun capacity and end stops that absorb torque peaks and prevent the coil springs from blocking.

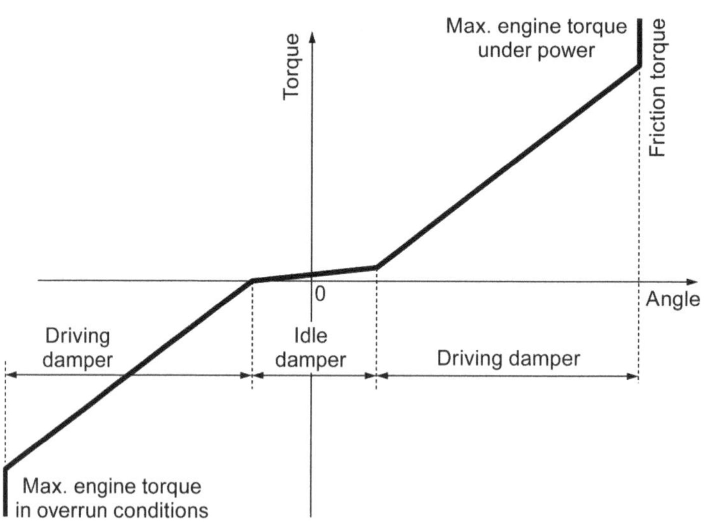


**Fig. 10.6.** Torsion damper characteristic curve [10.36]

The driving damper is adjusted such that, at lowest possible spring stiffness, the friction hysteresis is set which results in the best compromise between resonance damping and degradation of supercritical decoupling. This can help to reduce, for example, rattling noises in the transmission. In order to minimise rattling noises in the transmission when no gear is engaged, the idle damper of the torsional vibration damper is adjusted for idle engine running, the spring stiffness of which is about 1% of the driving damper.

### 2/ Pressure Plate Assembly

The pressure plate assembly transfers the engine torque via the clutch plate to the transmission input shaft. The task of the pressure plate assembly is to produce an axial pressure force that is as homogeneously distributed as possible over a large surface. In the motor vehicle clutch sector, the mechanical solution using spring force has established itself, as it is simply constructed, inexpensive, maintenance-free and temperature-resistant. In the case of mechanical clutch pressure plate assemblies therefore, the axial force required for friction coupling is produced solely by means of preloaded steel compression springs. These compression springs can be distinguished according to their design as

- diaphragm spring pressure plate assemblies (slotted plate springs),
- plate spring pressure plate assemblies and
- coil spring pressure plate assemblies.

Depending on whether or not the release force acts as a pressure or pulling force on the compression springs, we can differentiate between:

- pressure plate assemblies actuated by pushing and
- pressure plate assemblies actuated by pulling.

The essential components of a diaphragm pressure plate assembly mounted on a flywheel are the diaphragm spring clamped between the housing and the pressure plate, the pressure plate and the clutch housing.

### 2.1/ Diaphragm Spring

In the case of a diaphragm spring pressure plate assembly, the axial pressure force is produced by a preloaded plate spring characterised by several slots starting from the inner diameter. The clutch characteristic curve of a diaphragm spring pressure plate assembly is shown in Figure 10.7.

The structure of a diaphragm spring characteristic curve is of particular importance. In the second half of the pedal travel, it affects the pedal force directly. In the first half of the pedal travel, the pedal force is also influenced by the cushion deflection. The characteristic curve depends on the one hand on the inner and outer diameter of the diaphragm spring and on the other hand on the selection of diaphragm spring thickness, on the alignment angle and on material hardening. The mathematical relation is described in detail in German standard DIN 2092 [10.4].

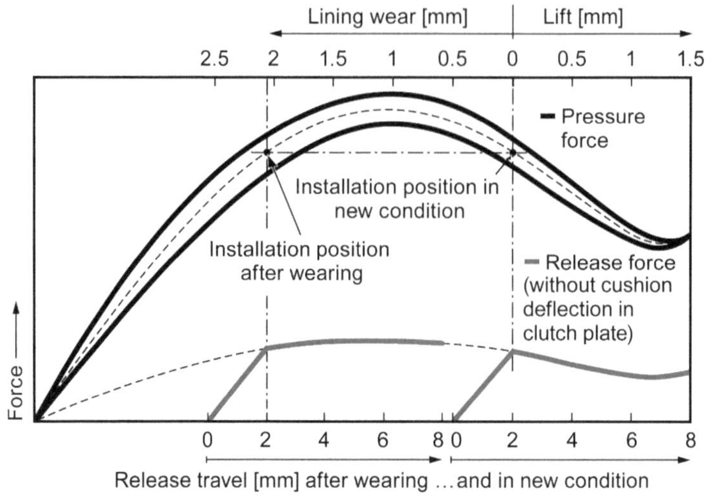


**Fig. 10.7.** Characteristic clutch curve of a diaphragm spring pressure plate assembly [10.5]

The design pressure force (arithmetically necessary pressure force) required for clutch actuation is valid for the installation position of the diaphragm spring in a new condition. When released, the diaphragm spring characteristic curve is passed from the installation position in the new condition towards the right in the lifting direction of the pressure plate (see Figure 10.7). When lining wear appears, the installation position of the diaphragm spring, and thus the starting point of the lifting movement on the characteristic curve, moves to the left. The clutch is designed such that the pressure force is again just as large as in the new condition even after lining wear sets in as anticipated by the design. After this, the pressure force drops steeply, so the driver determines the necessity of a clutch change when the clutch starts to slip.

The distance between both of the continuous pressure force curves in Figure 10.7 corresponds to the friction losses of the diaphragm spring bearing. The release force characteristic curve corresponds to the pressure force characteristic curve under consideration of the leverage ratio of the diaphragm spring [10.5].

### 2.2/ Pressure Plate

When designing the pressure plate *i* (Figure 10.4), which must dissipate a considerable amount of the heat originated, it is a compromise between a large heat storage capacity – and thus more weight – and the reduction of mass in order to unburden the return flat springs *h* (Figure 10.4) of the mounting. One sensible measure to optimise heat evacuation is expanding the surface by ribbing the reverse side of the pressure plate and by applying cool air by means of holes in the surrounding pressure plate housing *j* [10.33].

The pressure plate is mounted backlash-free and frictionlessly by means of return flat springs. These are responsible for centring the pressure plate, transferring the torque and lifting the pressure plate. During disengagement, the flat springs act as return springs against the force of the diaphragm springs.

### 3/ Clutch Actuation

Clutches are classified according to the type of actuation of the diaphraghm spring *6* of the pressure plate assembly as (Figure 10.8):

- *push-type clutches:*
  stepping on the clutch pedal pushes the release bearing in the direction of the clutch onto the diaphragm spring tips, and the clutch is separated.

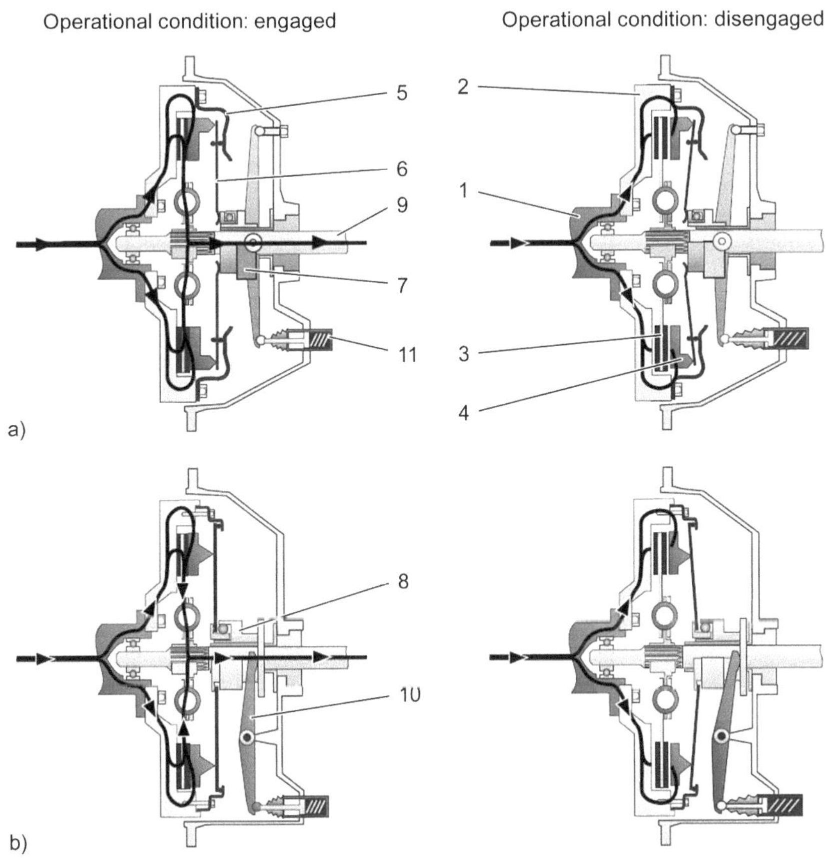


**Fig. 10.8.** Schematic diagram of push-type (*a*) and pull-type (*b)* clutch actuation with operational conditions and power flows. *1* Engine crankshaft; *2* flywheel; *3* clutch plate; *4* pressure plate with support for *6*; *5* pressure plate housing with support for *6;* *6* diaphragm spring with release tongues; *7* push-type release device; *8* pull-type release device; *9* transmission input shaft; *10* release lever; *11* slave cylinder

- *pull-type clutches:*
  stepping on the clutch pedal pulls the release bearing on the tips of the diaphragm springs, the pressure plate is lifted, and the clutch is separated.

Figure 10.8 shows a schematic representation of a push-type and a pull-type clutch. In the passenger car market, pushed actuation is used almost exclusively as it is less expensive and allows for a relatively simple fastening mechanism for the release bearing, simplifying assembly and disassembly.

Pulled actuation on the other hand are used in all applications in which large torque must be transmitted and/or only limited installation space is available, such as in commercial vehicles. Their advantages are:

- improved pressure and release force characteristic,
- higher pressure forces with the same installation space,
- lower release forces due to the larger leverage ratio of the diaphragm spring,
- fewer housing deflection losses and
- fewer parts, lighter weight.

### 3.1/ Release Device

The release device consists of the release bearing as the transmitter between the rotating clutch and the stationary actuation system as well as a sliding sleeve that slides on the guide tube attached to the transmission and engages the release levers on both of its lateral cams. Release bearings are generally designed as thrust ball bearings. The function of the release device is transmitting the release force from the stationary release mechanism to the pressure plate assembly rotating with engine speed. Figure 10.9 shows both of the current release systems, central release device and lever release.

Transferring the actuation force from the clutch pedal to the release device can be done mechanically as well as hydraulically, whereby manual transmissions use almost exclusively a hydraulic and automated manual transmissions use electrohydraulic, electropneumatic or electromechanical actuation system.

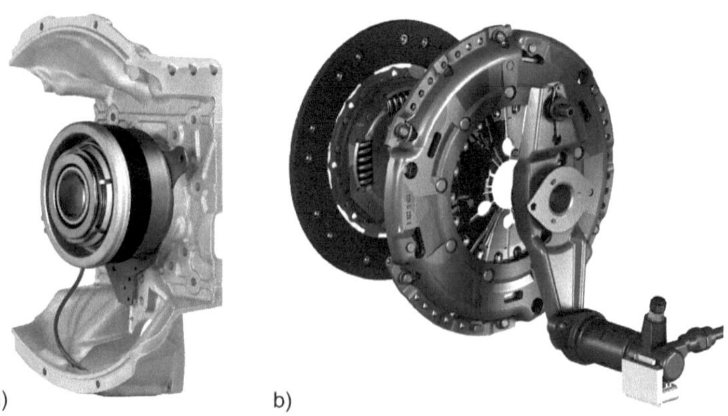

a) b)

**Fig. 10.9.** *a* Central release device and *b* lever release device (ZF Sachs)

Both hydraulic and pneumatically supported hydraulic actuation systems consist of a master cylinder, a pressure line and a slave cylinder. Varying the dimensions of the cylinder diameter can also lead to hydraulic/pneumatic force multiplication. In the case of electromechanical actuation systems, the release force is supplied by an electric motor and transferred mechanically to the release bearing.

### 3.2/ Wear Compensation

Dry-running friction clutches are subject to wear. The largest amount of wear occurs on the friction linings of the clutch plate. The friction partners flywheel and pressure plate are almost wear-free. In order to guarantee a constant actuation force – the release force – throughout the service life, today adjustment mechanisms are integrated into the pressure plate that make automatic adjustments after every actuation of the pressure plate. In this way, lining wear is disconnected from the movement of the diaphragm spring. The wear mechanism registers lining reduction and compensates for the distance by turning the adjustment ring.

These automatic wear compensation mechanisms are known, for example, by the product names "SAC" (Self-Adjusting Clutch) by LuK, "XTend" by ZF Sachs or "SAT" (Self-Adjusting Technology) by Valeo. Figure 10.10 shows a representative self-adjusting clutch by the LuK company.

In the clutch shown, the mounting of the degressive diaphragm spring *4* is supported by a sensor spring *5* on the pressure plate housing, possessing a sufficiently long travel with almost constant force.

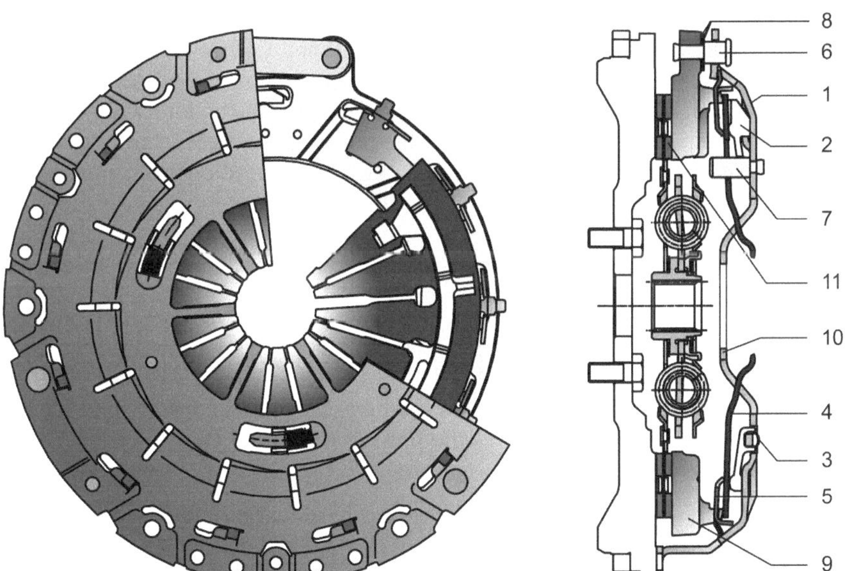


**Fig. 10.10.** The self-adjusting clutch SAC by LuK. *1* Pressure plate housing; *2* adjustment ring (ramp ring); *3* compression spring; *4* diaphragm spring; *5* sensor spring; *6*/*7* bolts; *8* flat spring; *9* pressure plate; *10* end stop; *11* clutch plate

If the release force is smaller than the retention force of the sensor spring, the pivotal support of the diaphragm spring stays in the same place when the clutch is released. Since the release force is increased when the clutch linings become worn, the counterforce of the sensor spring is exceeded and the pivotal mount swerves in the direction of the flywheel until the release force is back down to the level of the sensor force. When the sensor spring drifts, a gap appears between the pivotal mount of the diaphragm spring and the pressure plate housing that fills the space between the diaphragm spring bearing and the pressure plate housing by turning the ramp ring *2*. The ramp ring itself moves on opposing ramps in the pressure plate housing and is preloaded by small compression springs *3* in the direction of the circumference.

## 4/ Flywheel

The function of the flywheel is to store and pass on the kinetic energy provided by the internal combustion engine. The flywheel is utilised primarily to smooth out temporary deviations in load and power, to achieve higher power peaks and to bridge power interruptions. This is necessary because of the working principle of the piston engine, which delivers uneven torque and speed.

Heavier flywheels lead to improved running smoothness of the powertrain but it also increases the rotating masses, which leads to a deterioration of the powertrain's response characteristics and an increase in fuel consumption. For this reason, heavy flywheels are only preferable for commercial vehicles with large diesel engines. For passenger cars, particularly in the sports car market, the goal is to reduce the weight of the flywheel as much as possible.

The two basic flywheel designs are designated as pot type flywheel and flat flywheel, which are defined with recommended size gradations in the SAE standards J618 [10.20] for single-plate dry clutches and J619 [10.21] for two-plate dry clutches. In comparison, the advantage of the pot type flywheel is its larger moment of inertia at lighter weight as well as a safeguard against possible bursting of the pressure plate or the friction linings. The flat flywheel is less expensive to produce.

### 4.1/ Dual Mass Flywheel (DMF)

Since the end of the 1980s, dual mass flywheels (DMF) have been used increasingly in mid-range and luxury vehicles. DMFs are also being used more often in commercial vehicles. The function of the dual mass flywheel consists in reducing rotational irregularities originating in the powertrain. In the case of a dual mass flywheel, a conventional flywheel is separated into two plates:

- the primary flywheel with starter ring gear assigned to the engine is attached in a rotationally fixed fashion to the engine crankshaft and
- the secondary flywheel on the transmission end is fastened to the pressure plate assembly.

The secondary flywheel, comprising one of the two counter-friction surfaces to the clutch plate is supported, rotatable to the primary side, by a separated axial and radial plain bearing. Both decoupled masses are connected to each other by a grease-lubricated spring/damping system.

The main advantage of the dual mass flywheel is the damping of torsional vibrations across the entire speed range, which in turn increases driving comfort (reduced noise, improved shifting comfort etc.). The disadvantages are, besides the high cost, the deteriorated response characteristics of the engine due to the higher moment of inertia as well as the required installation space, which makes assembly difficult, especially in front-transverse engine designs, and necessitates integrative solutions. Depending on the design, we can differentiate between the following DMFs:

- bow spring DMFs,
- radial DMFs and
- special designs.

### 10.1.2 Design of Dry Clutches

When determining clutch size, both the ability to transfer the maximum amount of input torque with sufficient reliability and the anticipated thermal load caused by heating during moving-off and shifting are crucial factors.

#### *1/ Design Calculations*

The clutch must transmit the maximum engine torque including excessive dynamic increases. The clutch torque to be transferred $T_C$ is dependent on the pressure force of the pressure plate, the friction coefficient of the friction lining, the average friction radius and the number of friction surfaces:

$$T_C = F_{ax}\, r_m\, \mu\, z\,. \tag{10.4}$$

| | | | |
|---|---|---|---|
| $F_{ax}$ | Pressure force of the pressure plate (N), | | |
| $r_m$ | average friction radius (m) (see also Equation 10.7), | | |
| $\mu$ | friction coefficient, | | |
| $z$ | number of friction surfaces: | single-plate clutch: | $z = 2$, |
| | | two-plate clutch: | $z = 4$. |

The friction coefficient for static friction (fully engaged clutch) and for dynamic friction (during slip) are nearly identical and are between 0.3 and 0.45 for organic friction linings. In the case of extreme thermal overloading – "fading" – the coefficient can drop to less than 0.2.

To cover friction coefficient deterioration during extreme fading as well as pressure force dropping caused by relaxation of the diaphragm spring, return spring force, the centrifugal force of the diaphragm spring tongues, friction and preload in the actuation system, a safety factor $S$ is introduced.

The safety factor $S$ is the ratio of the clutch torque to be transmitted to the maximum engine torque $T_{M,max}$. It lies between $1.2 \leq S \leq 1.4$:

$$S = \frac{T_C}{T_{M,max}}. \tag{10.5}$$

Figure 10.11 shows the bandwidths of practical designs for passenger cars and commercial vehicles as a function of the clutch plate diameter. Exact calculation of shiftable, externally actuated friction clutches follows VDI guideline 2241, page 1 [10.31]. The following literature is also helpful in the calculation process: [10.1, 10.3, 10.6, 10.8, 10.10–10.11, 10.13–10.15, 10.24–10.27, 10.33].

### *2/ Thermal Balance*

The service life of friction linings is crucial for the service life of a motor vehicle's clutch. The service life of friction linings depends on the temperature, load duration, the quality of the linings and the amount of wear. Moreover, the operation of the dry clutch also has a lot of influence on the service life of the friction linings, which can be seen as the weakest link in the wear chain.

In the case of thermal loading of the dry clutch, a consideration of thermal balance of the entire clutch system is mandatory. Here, the added heat is related to the heat capacity of the clutch (pressure plate, flywheel) and the dissipated heat. The temperature of the pressure plate is a decisive criterion, since it generally has less heat capacity than the flywheel and its cooling is impaired by the surrounding pressure plate housing.

### *3/ Load Limits*

As long as the clutch of a motor vehicle is sufficiently dimensioned mechanically and designed with an eye to wear, temperature will be the critical load limit.

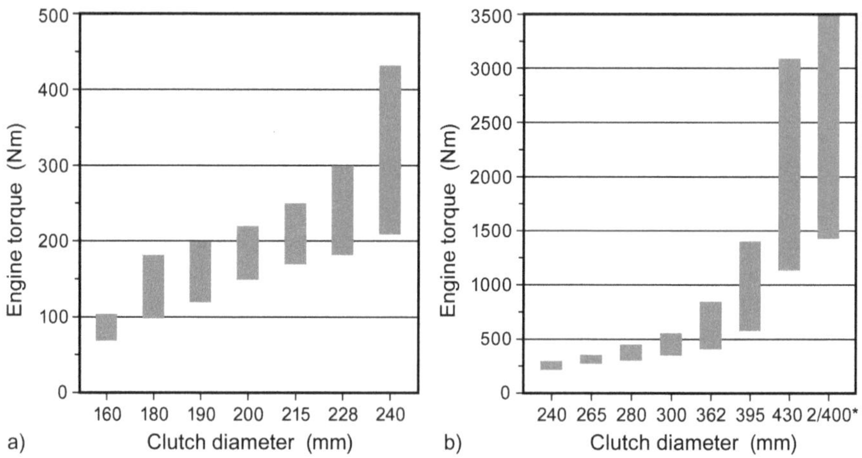


**Fig. 10.11.** Bandwidths of practical clutch designs for *a* passenger cars/vans and *b* commercial vehicles [10.37]. * two-plate clutch

**Table 10.2.** Maximum allowed shift energy relative to the friction surfaces when using organic clutch linings [10.33]

| Type of vehicle | Moving-off on the level ($J/mm^2$) | 15% gradient ($J/mm^2$) |
|---|---|---|
| Passenger cars | 1.2 | 2.2 |
| Light-duty commercial vehicles | 1.0 | 2.0 |
| Heavy-duty commercial vehicles | 0.3–0.6 | 1.5 |

To assess load, the value of surface-specific shift energy is consulted. Table 10.2 shows empirically established figures for the maximum allowed shift energy relative to the friction surface.

The permissible temperature of the pressure plate is limited by the reduction in pressure force induced by the relaxation of the diaphragm spring. However, it is generally higher than the maximum allowed temperature of the clutch plate. This is decisive for the service life of the tribological system respectively the friction linings. The limit temperature is defined here as the highest temperature at which the clutch is still operative. However, this also implies the fact that, when the limit temperature of the friction lining is reached, the surface is already irreversibly damaged. Only its core remains undamaged.

For organic linings, this temperature, measured in the pressure plate about 0.5 mm under the friction surface, is about 280°C. The air temperature in the clutch bell housing is then somewhere between 180 and 200°C. When inorganic metallic linings are used, the limit is determined by relaxation of the diaphragm spring to about 450°C. The linings themselves can withstand still higher temperatures for a short time.

With respect to loading of the release device, the temperature of the grease is very important. The temperature limit in continuous operation is about 130°C measured on the outer diameter of the outer ring of the release bearing.

### 10.1.3 Dry Multi-Plate Clutches

Highly motorised vehicles with high torques (sports cars and commercial vehicles) are too demanding on dry single-plate clutches. In these cases, dry multi-plate clutches are usually used in the two-plate design. Figure 10.12a shows a two-plate clutch for a passenger car with a clutch plate diameter of 228 mm and an engine torque of 800 Nm. By multiplying the number of friction surfaces, a higher torque can be transmitted at increased heat capacity.

Figure 10.12b shows a commercial vehicle two-plate clutch with both clutch plates as well as 6 torsion springs per plate. One great advantage of this clutch is its enhanced heat capacity as a result of the intermediate plate. By means of intermediate plate control, exactly half of the lift of the pressure plate is regulated in order to guarantee an exact, proportional torque transmission to both clutch plates [10.7].

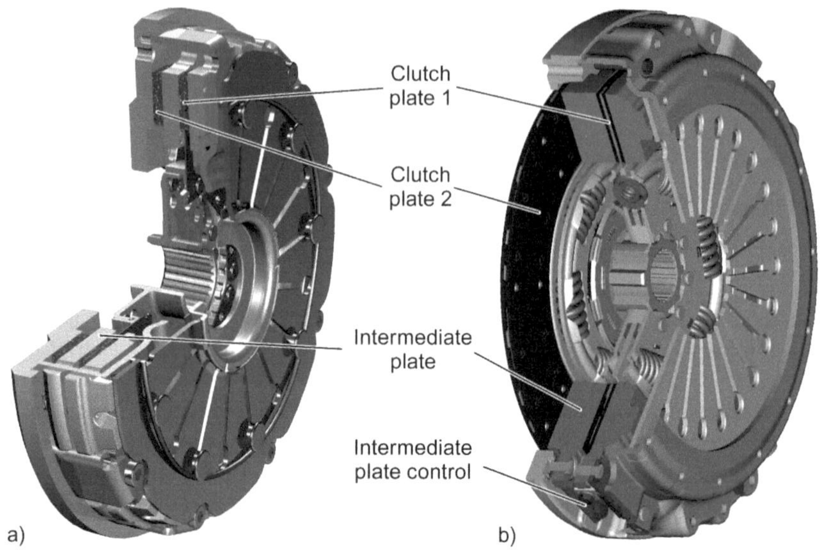


**Fig. 10.12.** *a* Two-plate clutch for highly motorised passenger cars (ZF Sachs MF2/228), *b* commercial vehicle two-plate clutch with intermediate plate control (ZF Sachs) [10.7]

## 10.2 Wet Clutches

In contrast to dry clutches, wet master clutches are of a technically higher order because of the oil supply required. An example for this type of master clutch is the continuously variable passenger car transmission Audi Multitronic, in series production since 1999 (see Figure 12.30).

It is worth mentioning that wet master clutches have been effective for decades, in motorcycles as well. In the latter, the clutch and the gear set are built in a common housing with splash and spray lubrication. Such motorcycle clutches will not be discussed further in the following.

Wet master clutches are generally built in a multi-plate design, immersed by transmission oil in an oil-tight housing. The valid relations are given in Section 9.3 "Layout and Design of Multi-Plate Clutches" and 11.3.1 "Oil Supply". The flow of cooling oil removes the energy introduced into the clutch by frictional work. The advantages of wet master clutches include, among other things:

- small moment of inertia, depending on the type of damping system,
- small installation space requirement,
- high power/weight ratio,
- large torque capacity,
- high heat capacity as a result of higher thermal energy inputs,
- high durability at low wear and
- good controllability.

A variety of parameters are important for the durability of the clutch (both dry and wet) such as lining pressure, specific frictional power, type of friction lining, lining thickness, lining grooving, type and amount of oil, shifting frequencies, moving-off load profiles and temperature. As opposed to hydrodynamic torque converters, the moving-off strategy of master clutches is selectable. This means that the moving-off speed of the engine can be freely adjusted.

Due to this clutch type's mode of operation in an oil bath, drag torques are generated. Besides the internal design features of the clutch, the drag torque depends heavily on the system temperature, oil viscosity and on system resistances, e.g. as a result of a preselected gear in DCTs [10.9]. Moreover, it should be noted that the clutch linings do not get worn, yet the oil is worn and oil changes may be necessary to avoid clutch judder [10.19].

As in dry clutches, torsional vibration damping is necessary for wet clutches as well. Such torsional vibration damping can be achieved by means of a torsional vibration damper, controlled slip or by combination of a torsional vibration damper and continuous slip operation.

The transmissible torque $T_C$ of a wet clutch in multi-plate design can be calculated with the axial force $F_{ax}$, the friction coefficient $\mu$ and the number of friction surfaces $z$ to

$$T_C = F_{ax}\, r_m\, \mu\, z \tag{10.6}$$

$$\text{with} \quad r_m = \frac{2}{3}\frac{(r_o^3 - r_i^3)}{(r_o^2 - r_i^2)} \tag{10.7}$$

as the mean friction radius $r_m$ from the outer and inner friction surface radius $r_o$ and $r_i$.

Figure 10.13 shows a wet master clutch in which the friction in the torsional vibration damper *5* can be altered additionally by the clutch pressure, making it especially suitable for engines with high rotational irregularities. Damping is high in idle operation as well as at low torques and low speeds, whereas damping is low at high torques and high speeds. Depending on the combustion engine being used, one can dispense with the damper, making the unit more compact in design. The controllability of wet master clutches makes it possible to pass through the low speed range, where torsional vibration is critical, with light, controlled slip.

Crawling or stopping on slopes is a possibility with wet master clutches. Accruing power losses can be removed by the cooling oil, comparably to a torque converter. However, relatively high volumetric flow rates are required for this. For master clutches with oil-tight housings (as shown in Figure 10.13), different concepts are possible:

- the *2-line principle*: one channel for the clutch pressure oil, from which the cooling oil is channelled off through suitable throttles, as well as one channel for cooling oil recirculation,
- the *3-line principle*: one channel each for the clutch pressure oil, cooling oil and cooling oil recirculation.

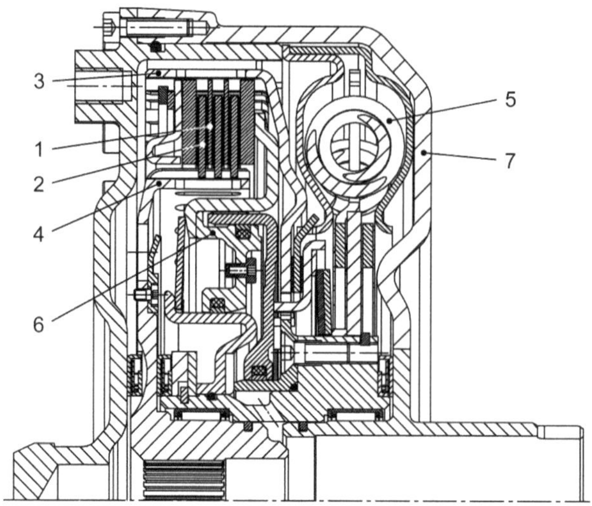


**Fig. 10.13.** Wet master clutch with variable torsional vibration damper (BorgWarner).
*1* Outer plates;
*2* inner plates;
*3* outer plate carrier;
*4* inner plate carrier;
*5* torsional vibration damper;
*6* piston;
*7* oil-tight housing

## 10.3 Dual Clutches

Dual clutches are two clutches actuated independently of each other that each serve one of two independent sub-gearboxes of a dual clutch transmission, making powershifts possible. The principles are given in Sections 10.1 and 10.2 – see also Section 6.6.3 and 12.1.3 "Dual Clutch Passenger Car Transmissions (DCT)". The gears are preselected in the respective load-free sub-gearbox. Gears are changed under full load – without power interruption – by means of controlled torque delivery from the first sub-gearbox clutch to the second clutch of the second sub-gearbox. We distinguish dual clutches according to their design as

- dry dual clutches and
- wet dual clutches.

These clutches can be designed as single-plate or multi-plate units, whereby the single-plate design is primarily used for dry clutches and the multi-plate design for wet clutches. Dry dual clutches, because of their smaller torque capacity, are used more in small and compact cars, while wet multi-plate dual clutches are generally utilised in vehicles with higher torques.

Dual clutches can be actuated mechanically (electromechanically or electrohydraulically) or directly hydraulically. With respect to actuation, there are two basic types of dual clutches:

- clutches that are closed in the absence of actuation energy ("normally closed"): actuation acts against the spring force and
- clutches that are open in the absence of actuation energy ("normally open").

Clutches that are closed without actuation energy do theoretically offer the potential of somewhat reduced fuel consumption, but the "normally open" type of clutch is usually preferred for safety reasons. In the case of clutches that close without actuation energy, blocking of the transmission in case of error has to be avoided at considerable effort (see Section 6.6.3).

### 1/ Dry Dual Clutches

Dry dual clutches are usually used for lower engine torques to about 250 Nm in a single-plate design. The torque is transmitted by means of the pressure plate assembly and clutch plate to the corresponding sub-gearbox input shaft.

### 2/ Wet Dual Clutches

Wet dual clutches allow for much higher energy inputs compared to dry clutches and are used in vehicles with higher torques of about 250 Nm. They are mounted either directly on the shafts or on a clutch carrier connected to the gearbox housing. With respect to the assembly of the multi-plate packets we distinguish between

- radially arranged (also called "concentric") and
- axially arranged (also called "parallel") design.

Concentric dual clutches are especially suitable for short installation spaces, while parallel dual clutches lend themselves mostly to small diameters.

Figure 10.14 outlines an excerpt of a variety of possible arrangements, whereby the clutch is mounted here on a stationary clutch carrier and the clutches are actuated directly hydraulically.

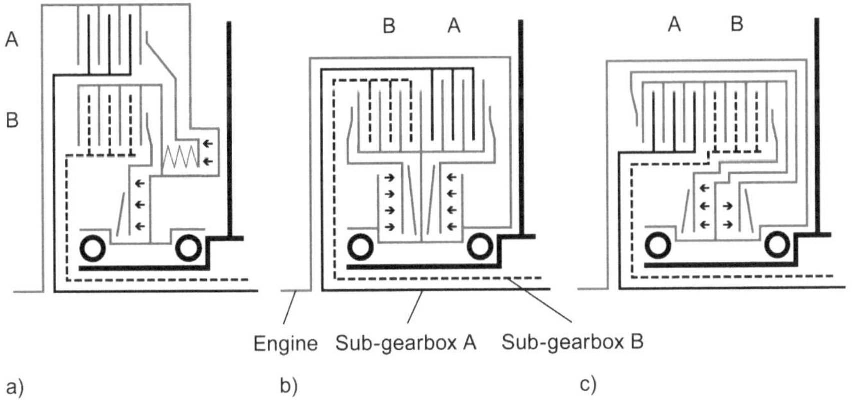


**Fig. 10.14.** Schematic multi-plate arrangement. *a* Radially arranged (concentric); *b* and *c* axially arranged (parallel)

Since we generally try to keep the masses to be synchronized on the gear shaft small, arrangement c) is more advantageous than arrangement b) due to the smaller plate carrier. The outer clutch "A" of the concentric arrangement a) in Figure 10.14 is preferred as a master clutch because of its higher thermal capacity and thus for uneven gears. Arrangements b) and c) make it possible to set the 1st gear optionally on the inner or outer shaft. In the concentric arrangement, the cooling oil first goes through the inner and then through the outer clutch. In the case of the parallel arrangement, the cooling oil can be applied to both clutches separately.

Wet dual clutches can be actuated electromechanically as well as electro-hydraulically. The hydraulic unit required to actuate and cool the clutch consists of a pump to deliver the cooling oil flow and to actuate the clutch as well as a valve block for control.

The centrifugal force effect of the hydraulic fluid is compensated at the piston in order to minimise speed influences in control (see Section 9.3.5 Number 7/ "Rotational Pressure Compensation"). These compensation cylinder chambers are not shown in Figure 10.14 for the sake of simplicity.

In order to reduce torsional vibration, usually a torsional vibration damper is necessary. This is ideally integrated with the dual clutch in the wet chamber. The torsional vibration damper can however also be arranged in the dry space between the engine and the dual clutch.

Figure 10.15 shows the wet dual clutch of the dual clutch transmission DSG by Volkswagen in a concentric multi-plate design. The outer clutch (clutch *C1* in Figure 12.17) is dimensioned for thermal reasons as a moving-off element for the 1st gear as well as for the R gear. This clutch will be used as shifting clutch in gears 3 and 5. The inner multi-plate clutch (*C2*) serves as a shift element for even shifting in gears 2, 4 and 6.

In the case of this transmission, the engine torque of 350 Nm is conducted via a dual mass flywheel and a spline shaft at a control pressure of 10 bar into input hub *7* of the clutch unit. An intermediate plate separates the dry from the wet clutch chamber. From there, the torque arrives via the driving plate *10* in the clutch housing respectively the outer plate carrier *2* of clutch *C1* and then further in the main hub *8* as well as in the outer plate carrier *4* of clutch *C2*.

The main hub *8* of the clutch is mounted with two needle roller bearings on the outer transmission input shaft almost frictionlessly. The torque flows from the steel plates arranged on the engine-side to the friction plates allocated to the inner plate carriers *1*, *3* and further to the inner or outer transmission input shaft [10.23].

The actuation pistons of both clutches have rotational pressure compensations and work against friction-optimised return springs. The pressure oil for clutch actuation is fed to the piston chambers from ring channels via a rotary feedthrough sleeve with axially running channels. The dual clutch is supplied intensively with a continuously adjustable oil flow by means of channels running axially in the main hub. A sensor in the clutch chamber monitors the temperature of the sling oil emerging from the clutch and serves to control the amount of cooling oil that is optimal for clutch functionality. By means of the available cooling oil flow of up to 20 *l*/min in combination with a high dual clutch heat storage capacity, frictional powers of up to 70 kW can be briefly induced [10.23].

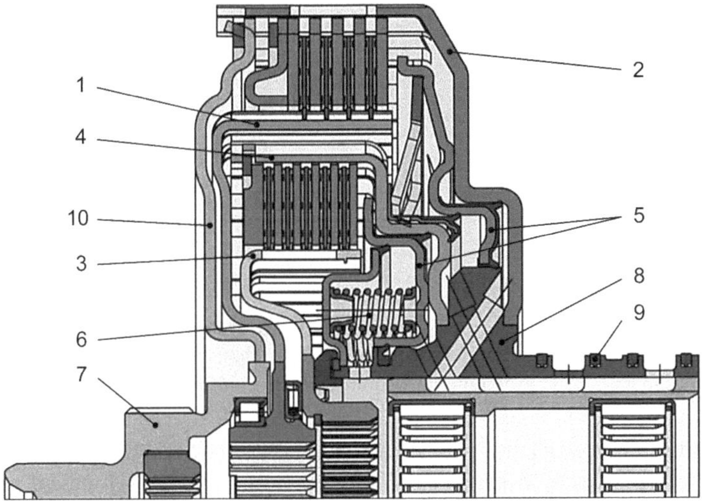


**Fig. 10.15.** Wet dual clutch of the dual clutch transmission DSG® by VW (BorgWarner). *1* Inner plate carrier of the outer clutch *C1*; *2* outer plate carrier of the outer clutch *C1*; *3* inner plate carrier of the inner clutch *C2*; *4* outer plate carrier of the inner clutch *C2*; *5* piston; *6* compression spring; *7* input hub; *8* main hub; *9* sealing ring; *10* driving plate

## 10.4 Hydrodynamic Clutches and Torque Converters

Internal combustion engines have a minimum engine speed. To move the vehicle from rest, the speed difference between the lowest engine speed and the stationary transmission input shaft has to be overcome. The torque converter is the standard moving-off mechanism in automatic transmissions. It converts not only rotational speed (as a clutch), but both speed and torque (as a transmission). This chapter is devoted to the torque converter. The hydrodynamic clutch and the hydrodynamic retarder are "reduced" converters, governed by the same theory.

In contrast to hydrostatic transmissions, which operate on the principle of displacement and pressure transmission, hydrodynamic transmissions use the inertia of a fluid flow. The individual components of such a transmission are fluid flow devices forming a closed fluid flow circuit. A rotary pump performs the function of machine, and the turbine that of prime mover. The mechanical energy applied through the drive shaft is converted in the pump into the hydraulic energy of the fluid, and then back into mechanical energy in the turbine, which is available (less losses arising) at the output shaft (Figure 10.16a). Friction losses in the pipelines and outlet losses make the efficiency achievable with such an arrangement very low.

The crucial development was the idea of engineer Hermann Föttinger who largely avoided fluid flow losses by combining the impeller pump, turbine wheel and a reactor to absorb the reaction torque together in one housing. This also reduced its weight and size (Figure 10.16b).

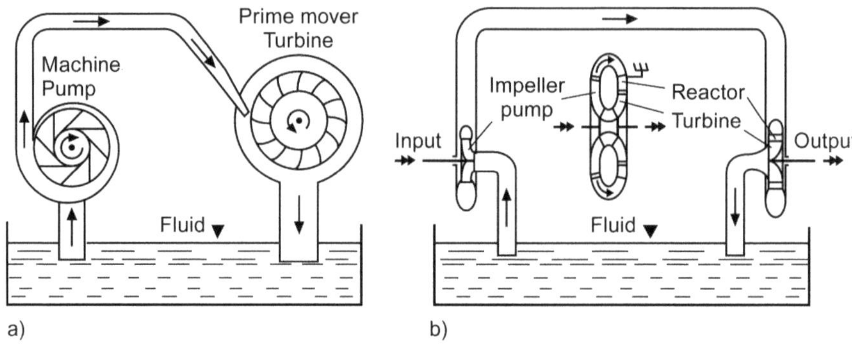


**Fig. 10.16.** *a* Schematic view of hydrodynamic power transmission; *b* H. Föttinger's seminal invention of machine and prime mover combined in one unit [10.28, 10.32]

This Föttinger transmission, named after its inventor, is the last real basic innovation in the vehicle transmission sector (see also Section 1.2.5 "Development of Torque Converters and Clutches"). The advantages of the hydrodynamic transmission are as follows:

- *Load-dependent, continuously variable ratio changing:* Adapting the ratio to the load on the output shaft.
- *Virtually non-wearing:* No abrasion.
- *Elastic connection between engine and powertrain:* Vibration and torque shock loads are damped since input and output are not positively engaged.
- *Reaction effect can be eliminated:* No stalling of the engine.

However, they have the following disadvantages:

- *Low efficiency over broad operating ranges:* Requires a rear-mounted gearbox.
- *Complexity of the rear-mounted gearbox:* The gearbox must be power-shiftable (conventional automatic transmission, CVT) or have an additional gearshifting clutch (torque converter clutch transmission).

### 10.4.1 Principles

A hydrodynamic clutch with the two main components impeller and turbine wheel permits no torque conversion, since no torque can act against the housing. A torque converter thus must have in addition at least one reactor to provide reaction forces (Figure 10.17).

The system uses ATF (Automatic Transmission Fluid). The fluid flows through the pump, then the turbine, then the reactor, following the particular blade contour, assuming the blades are as close together as required. Figure 10.18 shows the speeds on entering and leaving the blades. The torque converter operation is shown at the optimum point $M$ (Figure 10.20); i.e. the fluid suffers no impact losses in this operating mode since it always encounters the blades tangentially.

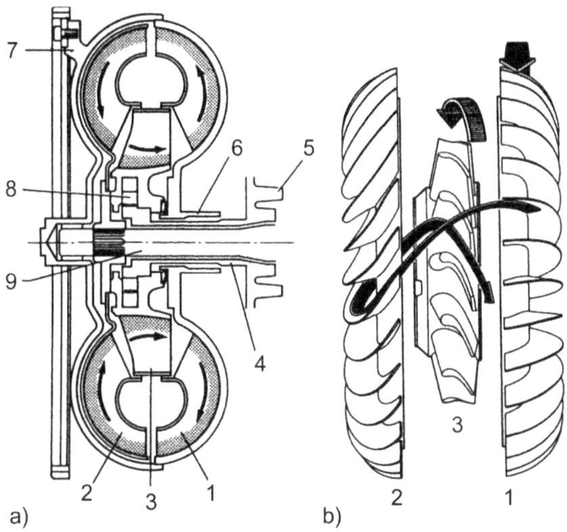


**Fig. 10.17.** *a* Components of a hydrodynamic torque converter [10.8].
*1* Impeller;
*2* turbine;
*3* reactor;
*4* hollow shaft bracing reactor to housing;
*5* housing, fixed;
*6* impeller hollow shaft for transmission fluid pump drive;
*7* converter cover, linked to the impeller;
*8* reactor freewheel;
*9* turbine shaft (transmission input);
*b* flow pattern

From Equation 2.7 the following torque equilibrium applies for the torque converter as a whole

$$T_P + T_T + T_R = 0\,. \tag{10.8}$$

The individual torque values can be determined using Euler's turbine equation

$$T = Q\,\rho\,\Delta(r\,c_u)\,. \tag{10.9}$$

They therefore depend on the flow rate $Q$, the fluid density $\rho$ and the angular momentum difference $\Delta(r\,c_u)$ between the blade input and output.

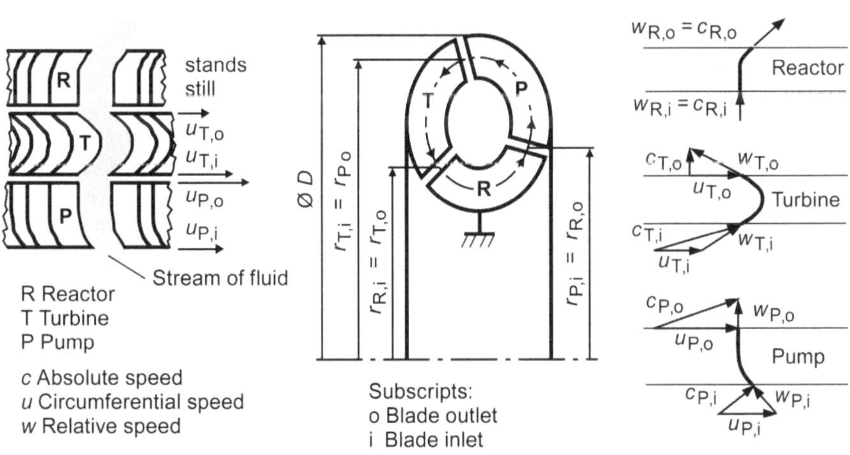


**Fig. 10.18.** Flow cycle in the converter with flow speeds for flow without impact losses [10.32]

The angular momentum is the product of the radius $r$ and the circumferential component $c_u$ of the absolute speed $c$

$$\Delta(r\,c_u) = r_o\,c_{u,o} - r_i\,c_{u,i}\,. \tag{10.10}$$

Since it is a closed system in which the fluid flow passes through all the wheels in sequence, and there is thus the same mass flow everywhere, the angular momentum balance $\Sigma\,\Delta(r\,c_u) = 0$ results in addition to the torque equilibrium. If the power of one wheel $P = T\,\omega$, it follows, given that the reactor remains fixed, that the power equilibrium is

$$\sum P = P_P + P_T + \sum P_V = 0\,. \tag{10.11}$$

The power losses $P_V$ are made up of friction and impact losses, windage and gap leakage. The efficiency $\eta_{TC}$ of a torque converter, from Equation 10.1a, is

$$\eta_{TC} = \frac{P_T}{P_P} = \frac{T_T\,\omega_T}{T_P\,\omega_P} = \mu\,\nu\,. \tag{10.12}$$

using the torque ratio from Equation 4.3 $\mu = T_T / T_P$ and the speed ratio from Equation 4.2 $\nu = \omega_T / \omega_P$ .

Torque converters are designed according to hydraulic model laws using characteristic values derived by experiment. The following two preconditions must be fulfilled for models to be considered:

- *Geometrical similarity:*
  The same linear scale $m$ for all parts relevant to hydraulic design, in this case given as the ratio of the profile diameter of the model, $D_M$, to that of the original, $D$

$$m = \frac{D}{D_M}\,. \tag{10.13}$$

- *Kinematic similarity:*
  Corresponding speeds of the original and the model must be in the same ratio, i.e. the speed triangles must be similar. Using the designations in Figure 10.18, the following is true

$$\frac{c}{c_M} = \frac{w}{w_M} = \frac{u}{u_M}\,. \tag{10.14}$$

Substituting the circumferential speed $u = \omega\,D/2$, the speed scale $m_v$ is given by

$$m_v = \frac{u}{u_M} = \frac{\omega\,D}{\omega_M\,D_M} = m\frac{\omega}{\omega_M}\,. \tag{10.15}$$

From (10.15), the scale of the angular momentum $m_{AM}$ is

$$m_{AM} = \frac{\Delta(r\, c_u)}{\Delta(r\, c_u)_M} = m^2\, \frac{\omega}{\omega_M}\,. \tag{10.16}$$

The flow rate $Q$, equal to the product of speed and area according to the continuity equation, is proportional to the product of the speed scale (Equation 10.15) and the area scale (Equation 10.13)

$$m_Q = \frac{Q}{Q_M} = m_v\, m^2 = \frac{\omega\, D^3}{\omega_M\, D_M^3}\,. \tag{10.17}$$

Using Equation 10.9, the power is given by

$$P = T\,\omega = Q\,\rho\,\Delta(r\, c_u)\,\omega\,. \tag{10.18}$$

By substituting Equations 10.15, 10.16 and 10.17 into 10.18, it follows that

$$\frac{P}{P_M} = \frac{\rho\,\omega^3\, D^5}{\rho_M\,\omega_M^3\, D_M^5} \quad \text{or simply} \quad P \sim \rho\,\omega^3\, D^5\,. \tag{10.19}$$

Since $T = P\,/\,\omega$ it follows that for the torque

$$\frac{T}{T_M} = \frac{\rho\,\omega^2\, D^5}{\rho_M\,\omega_M^2\, D_M^5} \quad \text{and correspondingly} \quad T \sim \rho\,\omega^2\, D^5\,. \tag{10.20}$$

Adding the proportionality factor $\lambda$ gives the law of similarity

$$T_P = \lambda\,\rho\,\omega_P^2\, D^5\,, \tag{10.21}$$

where $\lambda$ is a function of the speed ratio $v$ and is designated as performance coefficient. It can be used to compare various torque converters. The density of automatic transmission fluids $\rho = 800–900\ \text{kg/m}^3$.

### 10.4.2 Hydrodynamic Clutches and their Characteristic Curves

Hydrodynamic clutches contain only a turbine and impeller; the fixed reactor to provide reaction force is omitted. Torque cannot be converted, since this configuration does not allow reaction torque to be absorbed. Only speed is converted.

Torque can only be transmitted where there is a speed difference between the impeller and the turbine. The pressure difference arising from differing centrifugal forces circulates the fluid, enabling momentum exchange between the two wheels. The speed difference relative to the speed of the pump is referred to as slip $S$ (see also Equation 10.3)

$$S = \frac{\omega_P - \omega_T}{\omega_P} = 1 - \frac{\omega_T}{\omega_P} = 1 - \nu\,. \tag{10.22}$$

The external air friction can no longer be ignored relative to the transmitted torque when the slip is very small, and thus the transmitted torque tends to zero. This has an impact on the efficiency profile, which drops rapidly towards zero when $S$ is very small, i.e. for $\nu$ approaching 1. This range is not reached in normal operation where the residual slip is of the order of $S=2$–$4\%$ as is normal for hydrodynamic clutches (and converters). Substituting Equation 10.22 into Equation 10.12

$$\eta = \frac{T_T\,\omega_T}{T_P\,\omega_P} = \frac{T_T\,\omega_T}{(T_T + T_{fric})\,\omega_P} = \frac{T_T}{(T_T + T_{fric})}(1 - S)\,. \tag{10.23}$$

Figure 10.19 shows the characteristic curves of a hydrodynamic clutch for a constant test pump speed $n_{PV}$. The non-dimensional representation on the right shows the comparison of different clutches more clearly.

The profile of the performance coefficient $\lambda$ as a function of the slip $S = 1 - \nu$ can be influenced by the design of the blade geometry and fluid flow path, and by altering the fluid fill level. The aim is usually to limit the torque transmitted with stationary output, and thereby avoid stalling the engine when idling and moving-off.

The hydrodynamic retarder is a special version of the hydrodynamic clutch. The turbine wheel is in this case usually a part of the housing and remains stationary, so the clutch is only operated at the stall point $S$ (see Figure 10.20). The torque transmitted, and thus the braking effect, is very much dependent on rotational speed, and can be controlled by the fluid level and additional orifices (see Section 11.6 "Vehicle Continuous Service Brakes").

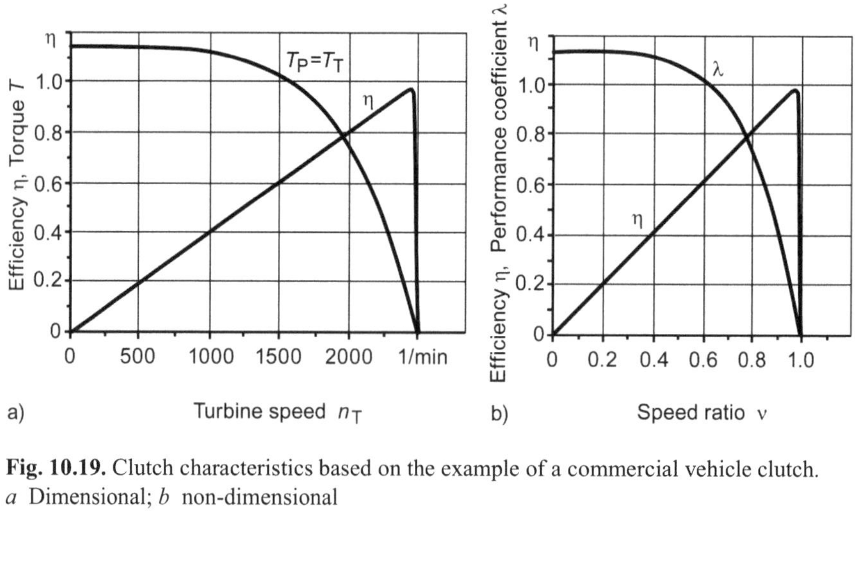


**Fig. 10.19.** Clutch characteristics based on the example of a commercial vehicle clutch. *a* Dimensional; *b* non-dimensional

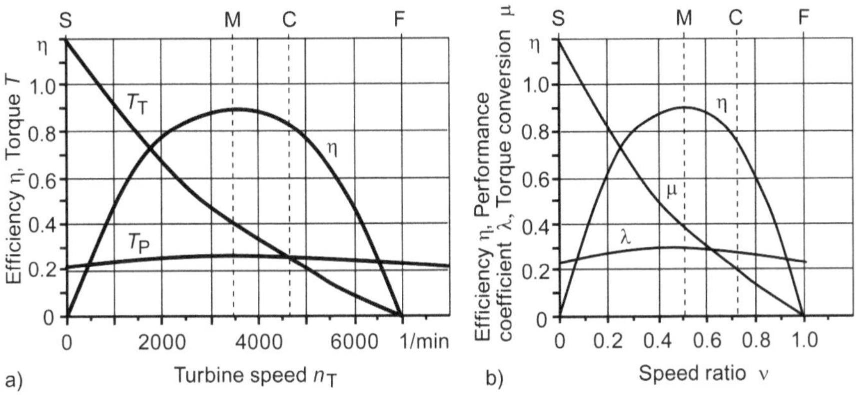


**Fig. 10.20.** Characteristic curves of a torque converter. *a* Dimensional; *b* non-dimensional

## 10.4.3 Torque Converters and their Characteristic Curves

The converter can absorb a moment of reaction by means of its fixed reactor, and is thus able to convert the input torque. Therefore its efficiency is better than that of clutches at speed ratios below $\nu = 0.7$–$0.8$ (depending on torque converter type). Plotted against speed ratio, the turbine torque drops, at first approximation, from the stall torque with the ratio $\mu_{\text{stall}}$ linearly to $T_\text{T} = 0$ at speeds in the region of $\nu = 1$. With constant input power this gives rise to a parabolic output power curve and thus a parabolic efficiency curve (Figure 10.20). Figure 10.20 shows the following key operating points:

*S* Stall point, the turbine is at rest, the stall torque ratio is $\mu_{\text{stall}} = T_{\text{T,stall}} / T_{\text{P,stall}}$,
*M* Optimum point (design point) with maximum efficiency,
C Lock-up point, $T_\text{P} = T_\text{T}$, $T_\text{R} = 0$,
*F* Free-flow point, no turbine load, $T_\text{T} = 0$.

At the design point *M*, the point of optimum efficiency, the fluid flows without impact losses from one wheel to the next.

The torque converter performance coefficient curve $\lambda(\nu)$ can be influenced by the configuration and design of the wheels. In the simplest configuration, with the reactor located before the pump and a single-unit turbine design, $\lambda$ remains approximately constant so that the engine is evenly loaded regardless of the output speed (Figure 10.21a).

For motor vehicle use it can be more advantageous if $\lambda$ falls as turbine speed increases. The engine speed is depressed by increased torque at low turbine speeds, so that the engine contributes to speed conversion. This speed depression also gives the driver more feel for the acceleration process, since the pump and engine speed increase as road speed (turbine speed) increases (see also Figure 10.23). This falling $\lambda$ characteristic curve can be achieved by locating a turbine immediately before the pump in the direction of flow.

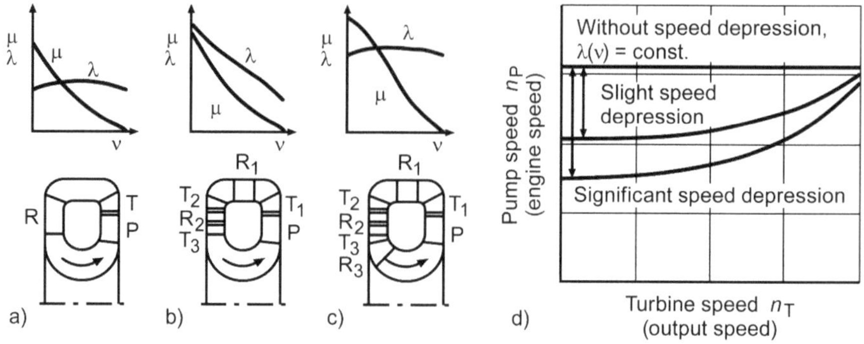


**Fig. 10.21.** Modifying the torque converter characteristics. *a* Single-stage; *b* three-stage with speed depression; *c* three-stage without speed depression; *d* effect of speed depression [10.32]

In order to still utilise the high pump output velocity and resultant high efficiency, the turbine may be of multi-stage design. This greatly increases the stall torque ratio in particular (Figure 10.21b). If a reactor is again fitted in front of the pump in a multi-stage turbine design, the value of $\lambda$ can be increased, but the curve will remain largely constant over the entire range (Figure 10.21c) as in Figure 10.21a.

### *Trilok Converters*

The advantages of the hydrodynamic clutch and the torque converter can be combined to avoid the falling section of the torque converter efficiency parabola.

In the first phase up to the lock-up point *C*, in which the reaction torque $T_R$ becomes zero, the torque converter operates. In the second phase, the reactor is released from the housing by means of a freewheel. Since the reactor now revolves freely it no longer absorbs any reaction torque. This results in the straight efficiency line typical of clutches (Figure 10.22).

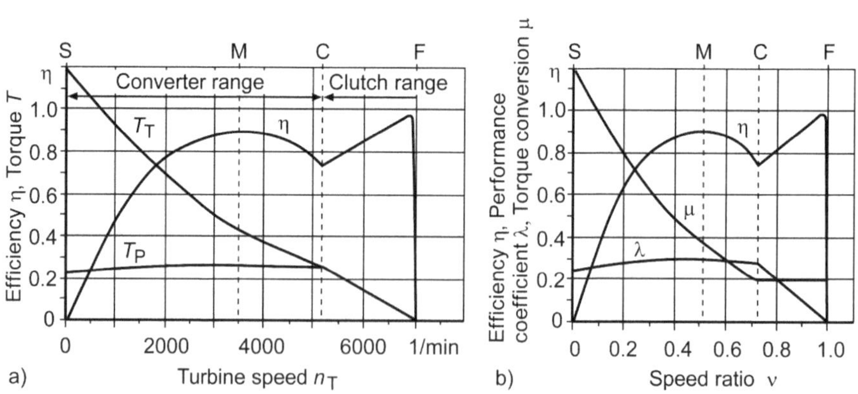


**Fig. 10.22.** Characteristic curves of a Trilok converter. *a* Dimensional; *b* non-dimensional

This type of single-stage two-phase torque converter is called a Trilok converter, after the TRILOK research consortium that developed it. Its high level of efficiency and simple construction make it particularly suitable for vehicle transmissions, so that Trilok converters with centripetal flow through the turbine are the only type used in passenger cars.

### 10.4.4 Engine and Torque Converter Working Together

Since the torque absorption of the impeller in a torque converter without speed depression is independent of the turbine speed, there is only one parabola in the engine performance map derived from Equation 10.21, with $\lambda = \text{const}$ as the operating curve.

Three different single-phase torque converters are shown in Figure 10.23. The diameter of torque converter 1 is designed so that its operating curve intersects the full load characteristic curve of the engine at the point of rated power. The diameter of torque converter 2 was selected to keep maximum engine torque available. The two torque converters are assumed to be geometrically similar.

$$D_1 = \sqrt[5]{\frac{T_\mathrm{n}}{\lambda\, \rho\, \omega_\mathrm{n}^2}}\,, \quad D_2 = \sqrt[5]{\frac{T_\mathrm{max}}{\lambda\, \rho\, \omega_\mathrm{T,max}^2}}\,, \quad D_1 < D_2\,. \tag{10.24}$$

The third torque converter is characterised by a falling performance coefficient $\lambda(v)$ curve, a speed depression type. The operating curve is therefore expanded into an operating map in the engine performance map, extending from the left-hand operating curve when $v = 0$ to the right-hand line when $v = 1$.

There are no points of intersection with the rated power of the engine in the case of torque converters 2 and 3. This is illustrated in Figure 10.23b, where power input is plotted against turbine speed.

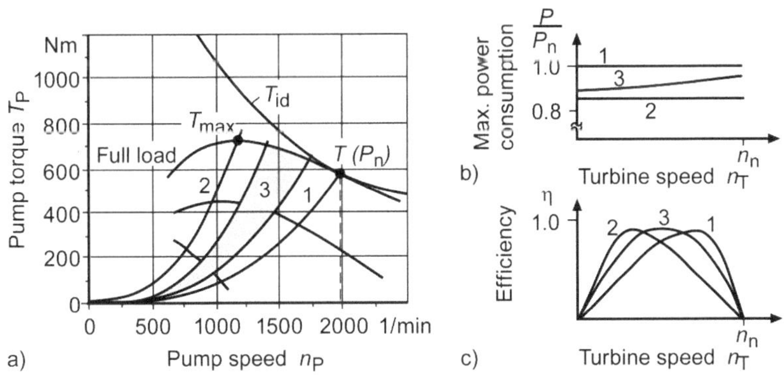


**Fig. 10.23.** Three torque converters with different $\lambda$ characteristic curve. Converter 1, 2 with constant performance coefficient $\lambda = \text{const} \Rightarrow$ operating curve; converter 3 with speed depression $\lambda \neq \text{const} \Rightarrow$ operating map. *a* Engine performance map; *b* maximum power consumption; *c* efficiency

Torque converter 3 still approaches 95% of the rated power by means of speed depression as the speed increases, but torque converter 2 can take up only a maximum 85% of the rated power.

This has no effect on the torque converter's maximum efficiency. Whilst the point of maximum efficiency of the first torque converter is at a speed ratio of approximately $\nu = 0.75$, in the case of torque converter 2 it moves towards smaller speed ratios. Torque converter 3 lies between the two, with a broad range of high efficiency. At high speed ratios, efficiency can be improved by using a two-phase Trilok converter.

Figure 10.24a again shows torque converter 3 as above with speed depression, but now in the form of a two-phase torque converter. At the lock-up point ($\nu_C = 0.75$), the reactor becomes free to move by means of a freewheel, and the torque converter acts as a clutch. The expanded operating map applies to this range. The maximum power input for these three Trilok type torque converters is given in Figure 10.24b.

The efficiency curve of these three Trilok converters is shown in Figure 10.24c. The third torque converter shows the advantages of a high starting torque and a broad range of high efficiency at intermediate values of $\nu$. Since operation at low speed ratios occurs almost only when moving-off, its effect on fuel consumption is of minor significance.

Figure 10.25a shows the three Trilok converters on the turbine map. They approximate closely to the maximum power hyperbola through torque conversion. A rear-mounted gearbox nevertheless remains indispensable. Figure 10.25b again shows torque converter 3 in the turbine map, to establish a relation to fuel consumption. The lines of constant normalised specific fuel consumption are also shown.

Figure 10.26a shows three different versions of a torque converter in the engine performance map of a 55 kW passenger car, to illustrate the effect of "torque converter characteristic – hardness" on fuel consumption.

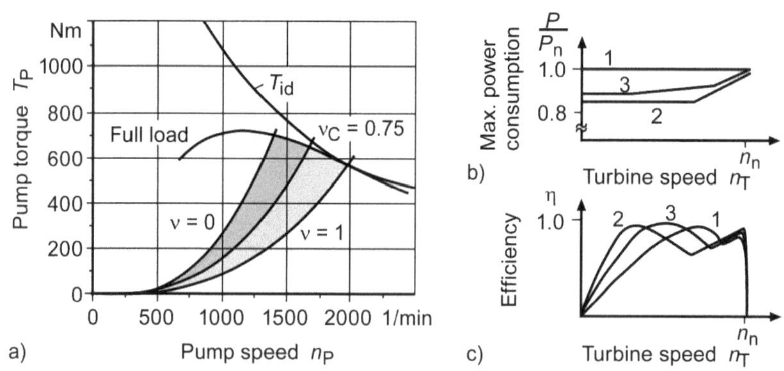


**Fig. 10.24.** Trilok versions of the converters in Figure 10.23. *a* Engine operating map with torque converter 3; *b* maximum power consumption; *c* efficiency

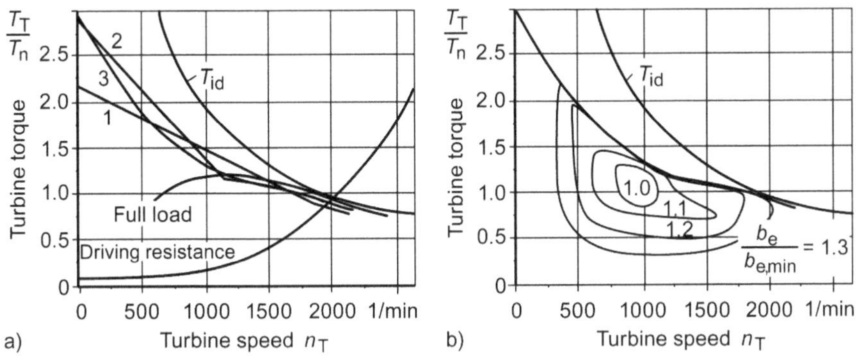


**Fig. 10.25.** The three Trilok converters from Figure 10.24. *a* In the turbine map; *b* Trilok converter 3 in the consumption map

A "soft" torque converter was created by reducing the diameter of a standard torque converter by 8.5%, and takes up only 64% of the pump torque of the standard torque converter at the same engine speed (Equation 10.21). The operating map between $v = 0$ and $v = 0.96$ is flatter in the primary map.

A "hard" torque converter can arise from enlarging the diameter of the initial torque converter. Thus for example a torque converter with a diameter 8.5% greater than the standard torque converter will take up 1.5 times the torque of the standard torque converter at the same pump speed: $T_P \sim D^5$ (Equation 10.21). Since this relates to a Trilok torque converter with speed depression, there are three resultant operating ranges, of which only two pump parabolas each are shown in Figure 10.26a, $T_P(v)$ for $v = 0$ and $v = 0.96$. The engine operating points are thus displaced towards lower engine speeds in hard torque converters, and towards higher engine speeds in soft torque converters.

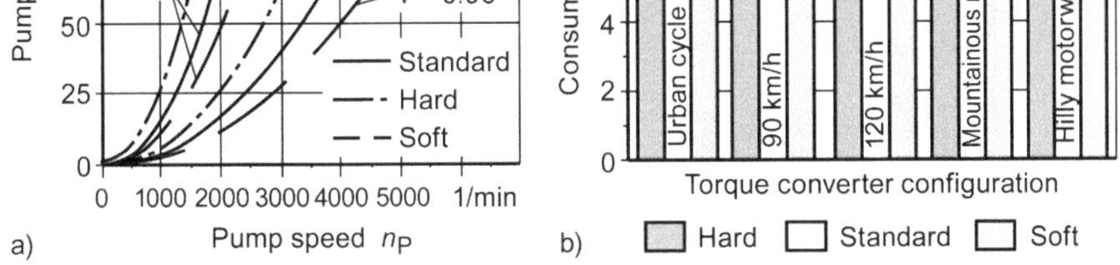


**Fig. 10.26.** Effect of converter diameter. *a* Primary map; *b* fuel consumption

Figure 10.26b shows consumption under various conditions based on a simulation calculation for the three torque converters in a mid-size passenger car with a conventional automatic transmission and without torque converter lock-up [10.12]. Under full load, the interaction of the engine and soft torque converter results in operating points with greater engine power for the same torque converter slip, provided that the torque converter parabolas intersect the engine full-load curve before the maximum speed. This also makes more tractive power available to the vehicle, which is reflected in the acceleration figures. But if limited engine speed means that the torque converter parabolas no longer intersect the engine full-load curve at high $v$ values, then the relationships are reversed. The harder torque converter displaces the engine operating points at low engine speeds, making it possible to reduce fuel consumption.

### *Torque Converter Test Diagram, Interaction of Engine and Trilok Converter*

The torque converter test diagram, (Figure 10.27) is the basis for calculating the engine operating points and the available traction of a powertrain with a hydrodynamic clutch or torque converter. See also Section 5.1.4 "Geared Transmission with Torque Converter" and Figures 5.7 and 5.8.

In Section 5.1.4 the example of a passenger car torque converter was used; in this case a commercial vehicle torque converter with a profile diameter of 370 mm is used (Figure 10.27a). The torque converter is to be used in a commercial vehicle with a 150 kW diesel engine. Characteristic values of the torque converter, the torque converter test diagram, were captured in bench tests at a constant pump test speed of $n_{PV}$ = 1600 1/min and a varying speed ratio $v$ (Figure 10.27b). For some operating points the resultant values for pump torque $T_P$ and torque conversion $\mu$ are listed in Table 10.3. To simplify the calculation, in place of the performance coefficient $\lambda(v)$, the factor $k(v)$ is used, which includes the density of the fluid and the torque converter diameter.

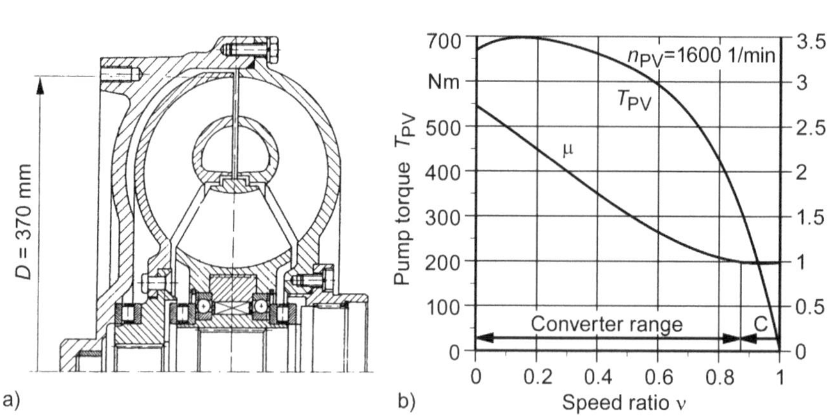


**Fig. 10.27.** *a* Torque converter with $D$ = 370 mm; *b* converter test diagram where $n_{PV}$ = 1600 1/min; $v_C$ = 0.88

**Table 10.3.** Pump side data from the converter test diagram and conversion for full load to the turbine side

| Measurement series No. | | 1 | 2 | 3 | 4 | 5 | 6 | 7 |
|---|---|---|---|---|---|---|---|---|
| Test data pump side | $n_{PV}$ (1/min) | 1600 | 1600 | 1600 | 1600 | 1600 | 1600 | 1600 |
| | $v$ | 0.00 | 0.10 | 0.40 | 0.60 | 0.80 | 0.88 | 0.94 |
| | $\mu$ | 2.73 | 2.49 | 1.74 | 1.35 | 1.023 | 0.997 | 0.997 |
| | $T_{PV}$ (Nm) | 672 | 690 | 669 | 585 | 427 | 301 | 185 |
| Calculate converter parabolas | $k(v)$ ($10^{-6}$ Nm min$^2$) | 262.5 | 269.5 | 261.3 | 228.5 | 166.8 | 117.6 | 72.3 |
| Converter parabolas – full load line intersection | $T_P$ (Nm) | 710 | 715 | 710 | 705 | 680 | 625 | 550 |
| | $n_P$ (1/min) | 1640 | 1630 | 1650 | 1770 | 2020 | 2300 | 2760 |
| Conversion to turbine side | $T_T = \mu\, T_P$ (Nm) | 1938 | 1780 | 1235 | 952 | 696 | 623 | 548 |
| | $n_T = v\, n_P$ (1/min) | 0 | 163 | 660 | 1062 | 1616 | 2024 | 2594 |

Since

$$k(v) = \frac{T_{PV}}{n_{PV}^2} \tag{10.25}$$

the pump parabolas (converter parabolas) in the primary map are given by

$$T_P = k(v)\, n_P^2 . \tag{10.26}$$

The values of the factor $k(v)$ calculated using Equation 10.25 show that the converter in this example is one with speed depression. The pump parabolas can now be calculated from this information using Equation 10.26, and recorded on the pump map for various speed ratios. This is combined with the full load characteristic curve of the engine to give the primary map (Figure 10.28a).

The points at which the engine characteristic intersects the pump parabolas represent possible full load operating points. The pump torques and speeds associated with these points are read off and entered in Table 10.3. In order to derive the secondary map, the turbine speeds and the associated engine torque values also have to be calculated from the torque ratio $\mu$ and the converter speed ratio $v$ using Equations 4.2 and 4.3. Figure 10.28b shows the turbine map derived in this way (see also Section 5.1.4).

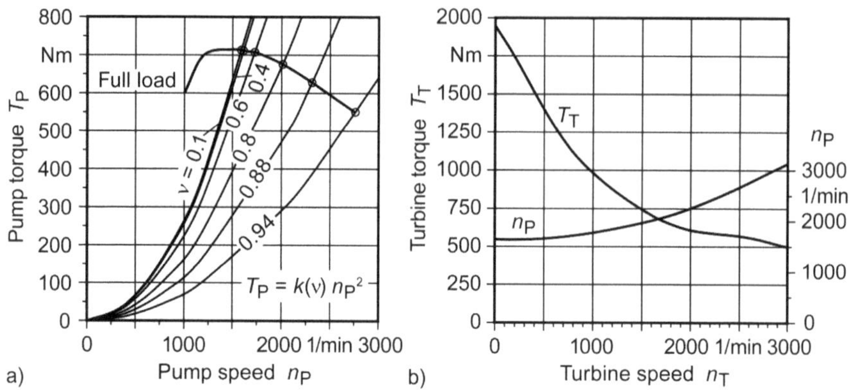


**Fig. 10.28.** *a* Primary map; *b* turbine map (secondary map)

### 10.4.5 Practical Design of Torque Converters

Figure 10.29 shows a manual calculation algorithm illustrating the rough design of a torque converter based on the preceding example.

### 10.4.6 Engineering Designs

The various requirements made on torque converters lead to conflicts of goals (Table 10.4).

Figure 10.30 shows an existing design of a passenger car Trilok torque converter of sheet metal design with a profile diameter of $D = 280$ mm and designates the most important components. The torque converter is equipped with a lock-up clutch with two friction surfaces. Section 10.4.7 "Design Principles for Increasing Efficiency" goes into lock-up clutches in more detail.

The Trilok converter has straight pump blades and bent turbine blades. The blade numbers of the individual wheels often vary in order to suppress resonance. The reactor is made by casting. The sheet metal design makes it less expensive to produce with low moments of inertia. With respect to soldering the blades and welding the shells, production does require a considerable amount of manufacturing know-how. For transmissions for special purposes or heavy vehicles, there are also torque converters with cast impellers and turbines.

The oil in the torque converter is applied with a filling pressure in order to prevent the formation of air pockets and cavitations. Excess pressure is kept at a minimum level of 1–2 bar with a valve. At higher speeds, the pressure produced by centrifugal forces also plays a role, which raises the pressure in passenger car converters to about 6 bar. This pressure increases with pump speed and is also contingent on the turbine speed and thus on load. It reaches its highest level when the pump and turbine are at the same speed, i.e. when the lock-up clutch is closed.

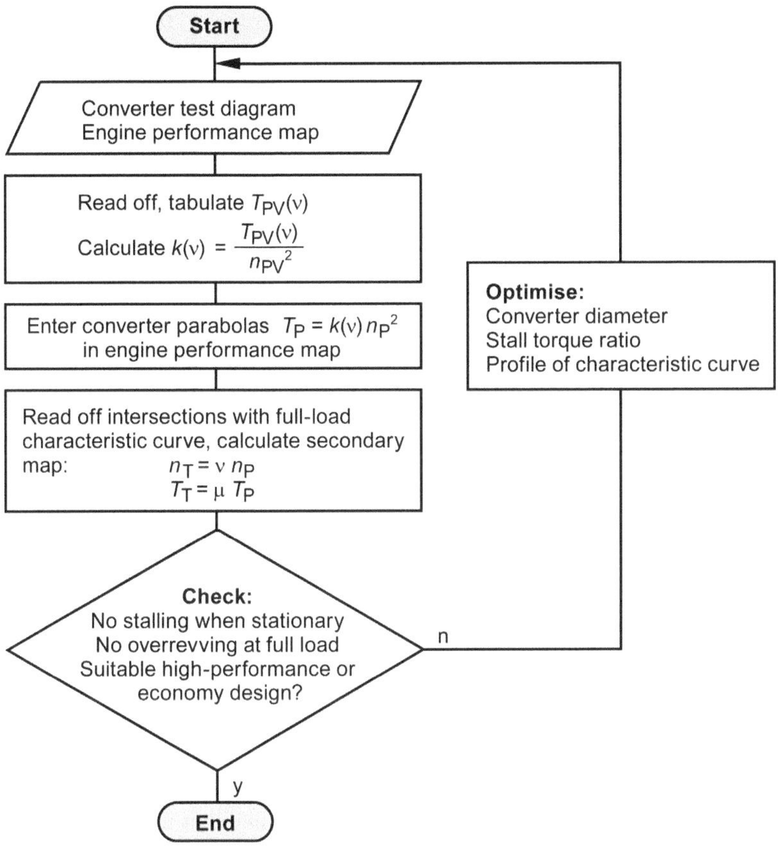


**Fig. 10.29.** Manual calculation algorithm for converter calculation

**Table 10.4.** Standard torque converter requirements [10.35]

| Optimisation | Performance | Fuel consumption | Comfort |
|---|---|---|---|
| Requirement | • Raise traction force<br>• Reduce moment of inertia<br>• No slip at $v_{max}$ | • Non-slip operation<br>• High efficiency in hydraulic operation | • Optimal vibration damping<br>• Reduced engine speed |
| Solution approach | • Soft converter characteristic<br>• High stall torque ratio<br>• Converter lock-up clutch closed at $v_{max}$ | • Hard converter characteristic<br>• High driving share with closed converter lock-up clutch | • Hard converter characteristic<br>• High driving share with open converter lock-up clutch |

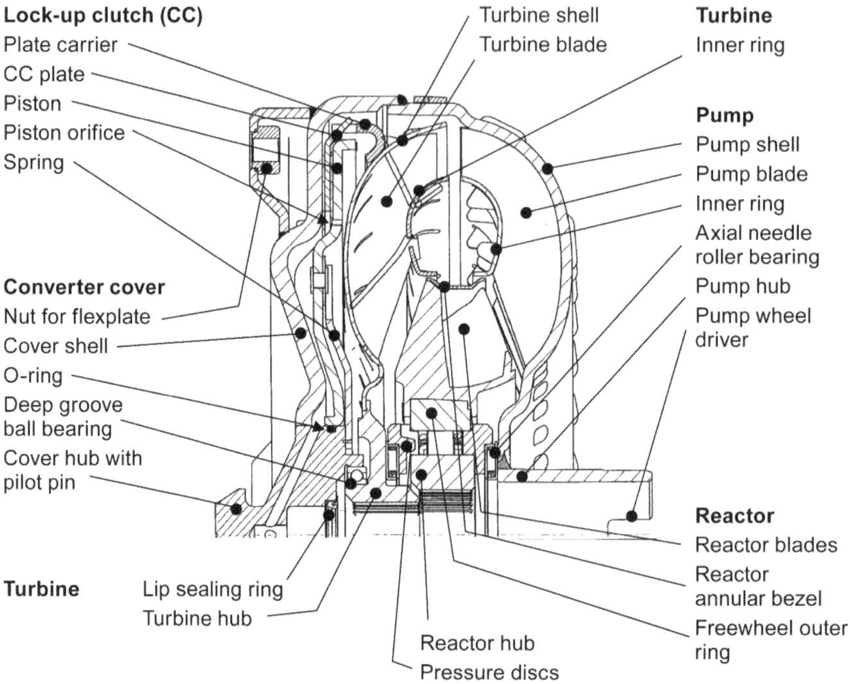


**Fig. 10.30.** Components of a passenger car Trilok torque converter of sheet metal design (ZF Sachs)

The design should model the torque converter interfaces such that its functions are not impaired even at maximum converter expansion. The pump hub in Figure 10.30 propels the oil pump and serves as the running surface of a rotary shaft seal.

In most designs, the impeller is joined to a shell-shaped cover that encloses the other converter blade wheels, thus comprising a revolving housing. In this way, axial forces on the shafts caused by internal pressure can be compensated to a great extent (see Figures 10.30 and 10.31).

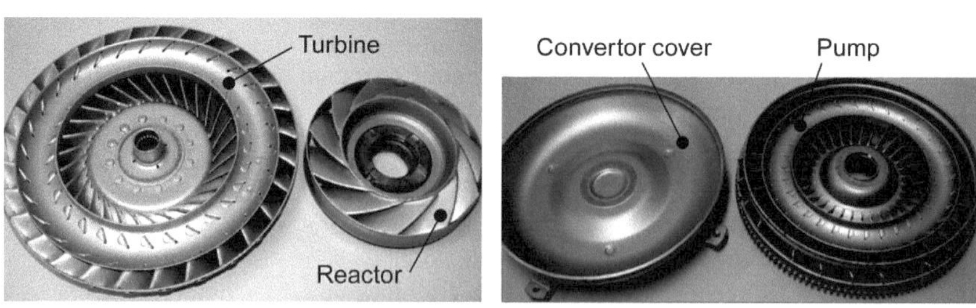


**Fig. 10.31.** Trilok torque converter of sheet metal design

The parts that form the torque converter housing (especially the pump) can be furnished with cooling plates to improve heat removal. These cooling plates also increase stiffness and counteract against expansion due to internal pressure. The flexplate is screwed onto the torque converter cover as an axially soft connection between the engine and the gearbox. Often, the starter ring gear and trigger wheels are also attached in order to determine the crankshaft angle. Prior to the mounting of the gearbox to the engine, the converter is in an overhung position. Therefore a retaining bracket is required to fix the converter during transport of the gearbox.

The commercial vehicle torque converter shown in Figure 10.32, with a profile diameter of $D$ = 400 mm, is also of Trilok-type design. In addition to the lock-up clutch, it is also equipped with a coasting freewheel *F1*, which allows for the exploitation of the engine braking torque during coasting. With the coasting freewheel, the pump speed and with it the coasting engine speed are equal to the turbine speed.

The torque converter is again of sheet metal design. The shells, inner rings and blades are made of deep-drawn sheet steel. Slots and beads in shells and inner rings determine the position of the blades fitted with lobes.

When the parts are joined, the lobes are bent over and the joint welded oil-tight with an electron beam. The cover and reactor are made of die cast light metal. The reactor runs on two ball bearings and is supported by a roller freewheel *F2* on a hub. The pump shell is supported on a hub on an angular contact ball bearing on the output side. There is also external toothing supported by the hub to provide a power take-off for engine-driven auxiliary units (Figure 6.78) as well as the oil pump.

While the torque converter shown in Figure 10.30 is an example of a so-called 2-line converter, a 3-line converter is shown in Figure 10.32. The torque converter lock-up clutch is designed here as an independent multi-plate clutch supplied with pressure oil. The principles of 2-line and 3-line converters are discussed in Section 10.4.7.

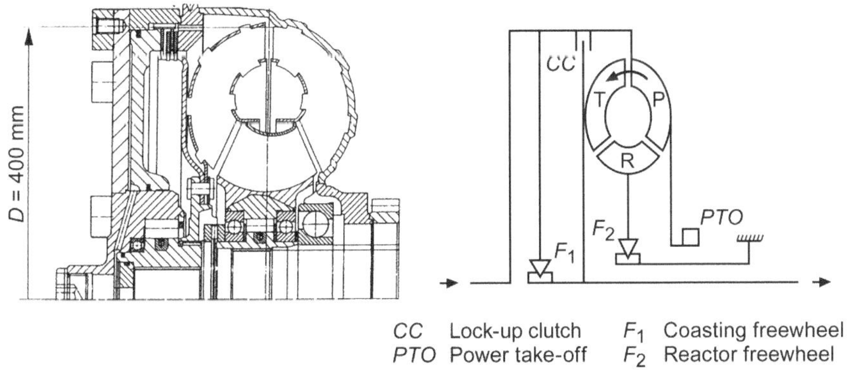


**Fig. 10.32.** Commercial vehicle Trilok torque converter with lock-up clutch and coasting freewheel

| Meridian profile **nearly round**: | Meridian profile **flat**: | Meridian profile **round-high**: |
|---|---|---|
| • Good hydrodynamic characteristics<br>• Big installation space necessary<br>• High moment of inertia | • Adequate hydrodynamic characteristics<br>• Small installation space<br>• High moment of inertia | • Good hydrodynamic characteristics<br>• Small installation space<br>• Low moment of inertia |

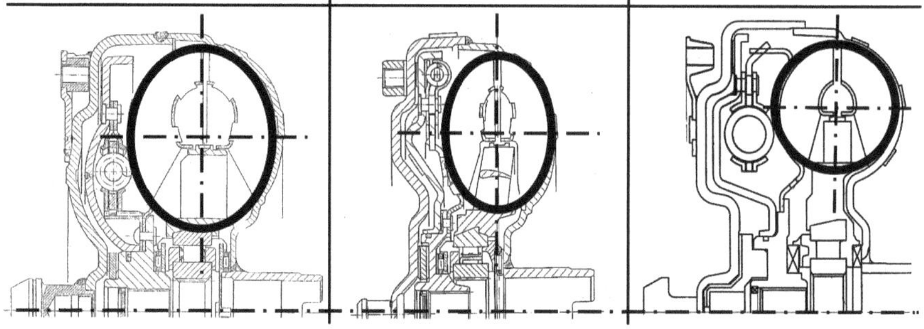

**Fig. 10.33.** Influence of the meridian profile on hydrodynamic characteristics, installation space and moment of inertia [10.35]

When developing torque converters, simulations, models and tests are utilised to produce the optimal hydrodynamic circulation design. Besides blade geometry, the meridian profile also exerts an influence on the characteristics of a torque converter (Figure 10.33). Efficiency in the relevant area of operation is of foremost importance. Boundary conditions of the installation situation and demands for a low moment of inertia, i.e. driving dynamics, must be taken into consideration. The question of whether additional functions such as torsional vibration dampers or starter generators must be integrated should also be considered.

The hydraulic efficiency of the torque converter at the lock-up point is relevant despite lock-up clutches, since the converter is opened and the resulting torque increase is used to improve driving dynamics in the full-load range. Purposeful adaptation of the speed depression of the converter to dynamic engine characteristics of modern petrol and diesel engines leads to advantageous linkage of the engine to the powertrain [10.16, 10.22].

### 10.4.7 Design Principles for Increasing Efficiency

There are two ways of increasing the efficiency of transmissions with torque converters:

- increasing the efficiency of the torque converter (hydraulic properties) and
- partially or completely bypassing the torque converter in certain operating ranges.

## 1/ Torque Converter Lock-Up Clutch

One possibility of reducing torque converter losses consists in bypassing the converter from a particular speed ratio by means of a clutch.

### 1.1/ Design Configuration

Figure 10.30 shows the components of a lock-up clutch (see also Section 5.1.4). There are manifold requirements concerning fuel consumption, shifting quality, vibration decoupling and driving dynamics. Modern systems with slip-controlled torque converter lock-up clutches (SCC) are already engaged in first gear and are operated in all gears slip-controlled (Figure 10.34). This requires sophisticated actuation and control concepts [10.18]. An efficient controlled torque converter lock-up clutch thus also includes a temperature model in the software. Design measures must be taken with respect to thermal loading and cooling as well as decoupling vibrations.

The demand to provide for opening and closing of the passenger car controlled torque converter lock-up clutch even at higher differential speeds (with up to 4 kW frictional power during the engagement control process) requires a high performance lock-up clutch (see also Section 9.3 "Layout and Design of Multi-Plate Clutches"). One approach for raising the performance is to increase the number of friction surfaces. In this way, surface pressure and the specific frictional work are reduced. Furthermore, the multi-plate clutch must be supplied with enough cooling fluid flow in order to lower peak temperatures. To this end, grooving in the friction lining and in the opposing surface and additionally, if applicable, piston orifices that supply the surrounding area of the friction surfaces with cooling fluid, are helpful (Figure 10.30 [10.2]).

### 1.2/ Oil Supply

The commercial vehicle torque converter with lock-up clutch shown in Figure 10.32 works in accordance with the 3-line principle. According to this principle, the torque converter lock-up clutch is designed as an independent multi-plate clutch (Figure 10.35).

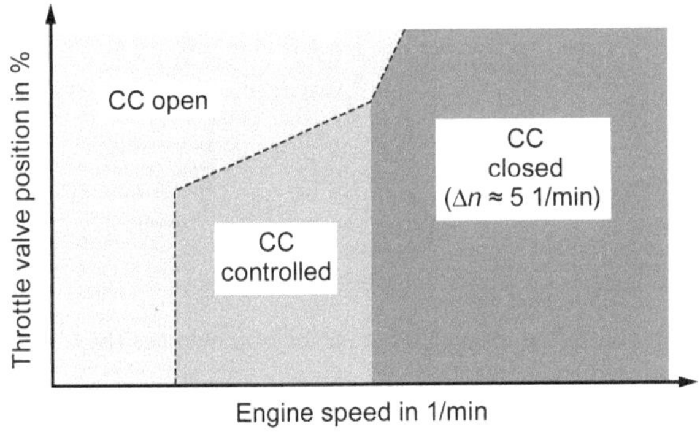


**Fig. 10.34.** Operating ranges of a controlled torque converter lock-up clutch

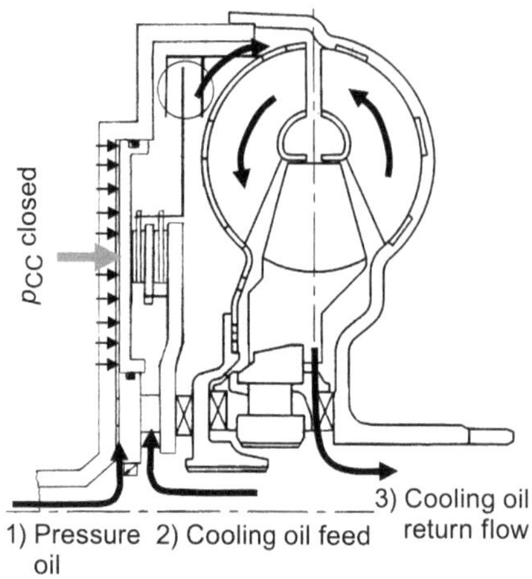


**Fig. 10.35.** Working principle of a 3-line torque converter. Lock-up multi-plate clutch supplied with its own pressure oil [10.34]

A line serves to supply the lock-up clutch piston with pressure oil, while a second line leads oil into the converter and a third serves the purpose of oil recirculation. This design tends to have better control properties than the 2-line principle shown in Figure 10.30, but it is less advantageous with respect to design complexity and cost.

One design approach that allows for compact and inexpensive solutions of torque converters is the 2-line principle (Figure 10.36). In this case, two lines take over feed and return of the converter cooling fluid as well as supplying the lock-up clutch piston.

The torque converter lock-up clutch has two friction surfaces in the design shown. The lined plate has a paper lining on both sides of the externally toothed carrier plate and is attached to the turbine with a driving spline.

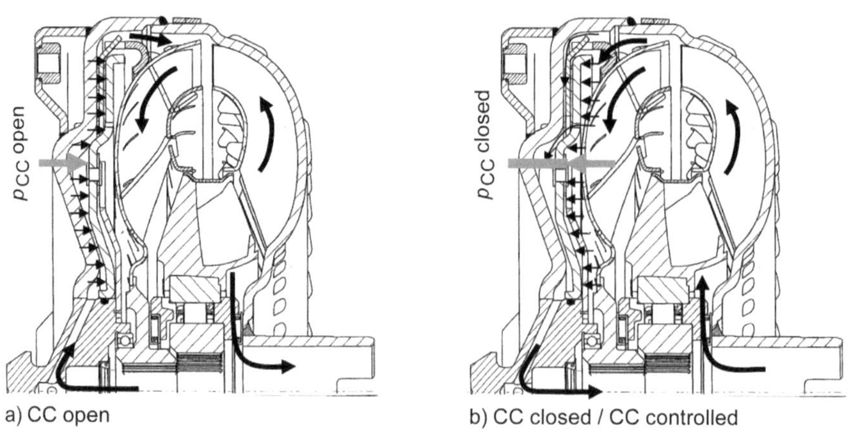


**Fig. 10.36.** Working principle of slip-controlled torque converter lock-up clutches (SCC) with a 2-line converter. *a* Open; *b* closed and controlled

The lock-up clutch piston is connected to the pump housing with flat springs. They create a light preload of the piston against the lined plate and the pump housing. An O-ring seals the inner diameter of the piston towards the housing.

When the torque converter lock-up clutch is open (Figure 10.36a), oil is supplied through the central channel in the input shaft. After the converter, the oil is led to the cooler. In order to close the lock-up clutch (Figure 10.36b), the flow direction is reversed in the lines. The central channel is now unpressurised and drains into the sump. The pressure difference at the piston determines the torque transmitted by the torque converter lock-up clutch.

Making the simplification that the pressure percentages arising from rotation are equal in front of and behind the piston, only the input pressure determines the torque $T_{CC}$ transmissible with the torque converter lock-up clutch.

The oil flow through the converter and the multi-plate clutch required for cooling is based on the amount of heat to be removed. Section 9.3.5 shows reference values for multi-plate clutches. In automatic passenger car transmissions for upper mid-range cars, the torque converter is flowed through with about 5 to 10 *l*/min of oil for cooling while the lock-up clutch is open. When the torque converter lock-up clutch is closed or controlled, both converter and clutch are immersed in about half of this amount of cooling oil. The flow rate depends on various parameters.

### 1.3/ Vibration Decoupling

One basic customer expectation of automatic transmission with torque converters is comfort. As shown in Table 10.4, vibration decoupling belongs to this. The hydrodynamic circulation ideally decouples engine irregularity, but it is lossy due to relatively high slip. Slip-controlled actuation of the torque converter lock-up clutch should prevent humming noises, engine vibrations and load change backlashes when the converter is bypassed. A slipping lock-up clutch is clearly less lossy than an open converter, but it has a significantly reduced decoupling capacity. A torque converter can only be bypassed when the excitation frequency lies above the eigenfrequency. Depending on the type and response characteristics of the engine, an additional mechanical torsional vibration damper (TD) may be necessary. For cost reasons and with respect to the moment of inertia, it is more desirable to get along without mechanical damping systems.

Mechanical damping is tuned by means of the masses of the torque converter, friction in the torsional vibration damper, stiffness of the springs and the arrangement of damper elements. A low eigenfrequency $\omega_0$ can be reached by lowering the spring rate $c$ of the TD springs or by increasing the primary and secondary inertia masses (Figure 10.37). Yet both methods have their limits. Low spring rates lead to high rotation angles and thus to high load change sensitivity (load reversal sensitivity). Furthermore, large inertia masses are only possible at the cost of driving dynamics and installation space [10.17, 10.22, 10.29].

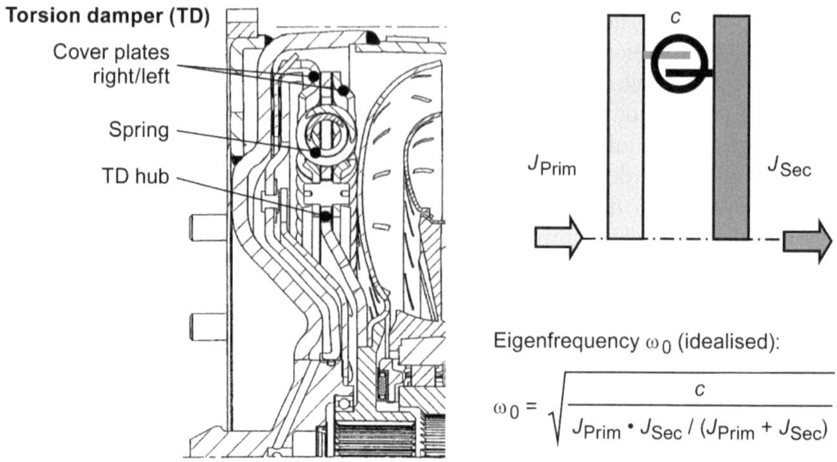


**Fig. 10.37.** Mechanical torsional vibration damper in a torque converter

Figure 10.37 shows a conventional torsional vibration damper. There are currently designs with single-stage and multi-stage torsional vibration dampers depending on the spring sets used. The spring sets decouple the primary and secondary sides of the torque converter, comparably to a dual mass flywheel.

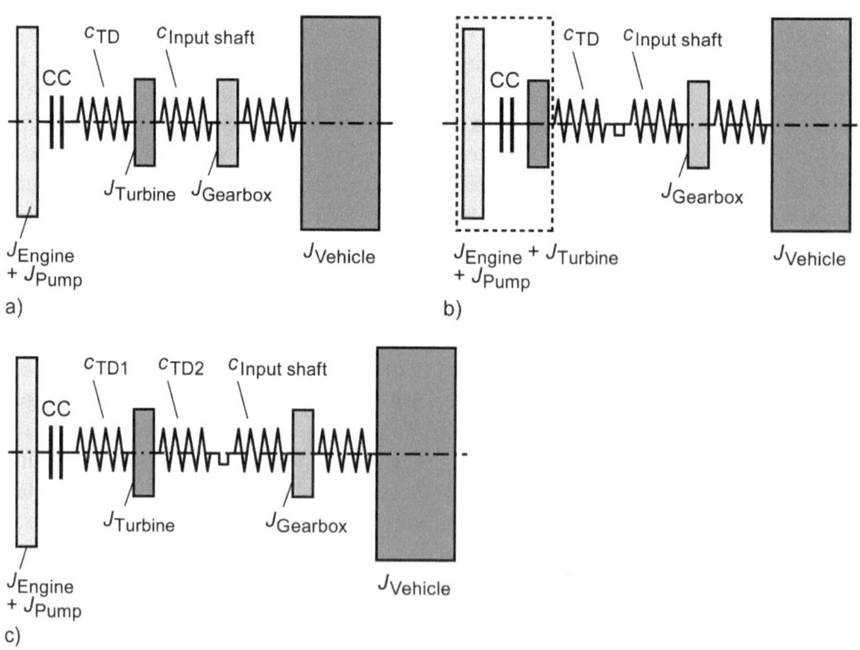


**Fig. 10.38.** Mechanical damper concepts. *a* Conventional torsional vibration damper (TD); *b* turbine torsional vibration damper (TTD); *c* two-damper torque converter (TDTC)

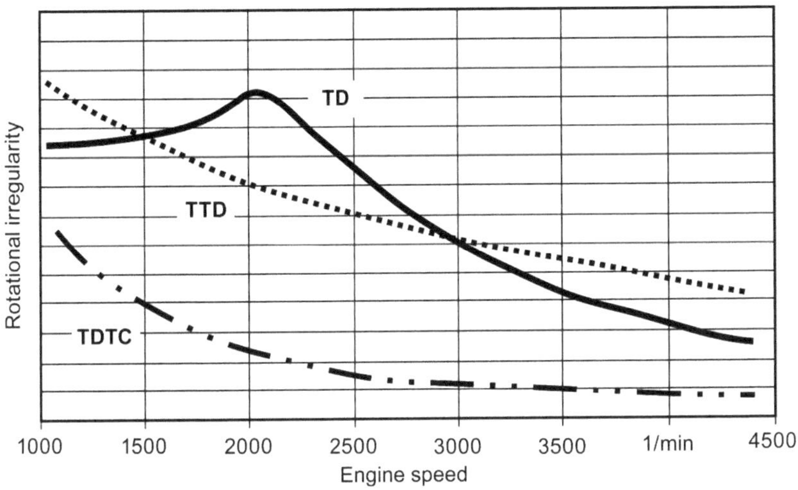


**Fig. 10.39.** Excitation characteristics of powertrain systems when using various damper concepts. Damping concepts according to Figure 10.38 [10.22]

Figure 10.38 shows the layout of individual torsional vibration dampers for various mechanical damper designs. Associated excitation characteristics of these damper concepts are simplified in Figure 10.39.

A comparison of the excitation of conventional torsional vibration dampers (TD) with turbine torsional vibration dampers (TTD) reveals advantages in both systems depending on the operating range, whereby TTDs tends to respond more advantageously. TTDs make the transmission input shaft softer, increases the moment of inertia, exhibits a smaller degree of freedom and also functions with an open torque converter lock-up clutch. The two-damper torque converter (TDTC) unites the advantages of both systems and further lowers rotational irregularities behind the converter [10.22]. With this approach, the lock-up clutch can be closed even earlier.

### *2/ Power-Split Transmission*

The efficiency of a transmission can be increased without foregoing the advantages of a torque converter by applying the principle of power-splitting. We distinguish between external and internal power-splits (Figure 10.40). Power-splits for enhancing torque converter efficiency and for increasing transmission conversion have not caught on to a great extent. There are however probably typical representatives.

### *2.1/ External Power-Split Transmission*

Only part of the input power flows through the torque converter of an external power-split transmission; the other part is transmitted purely mechanically.

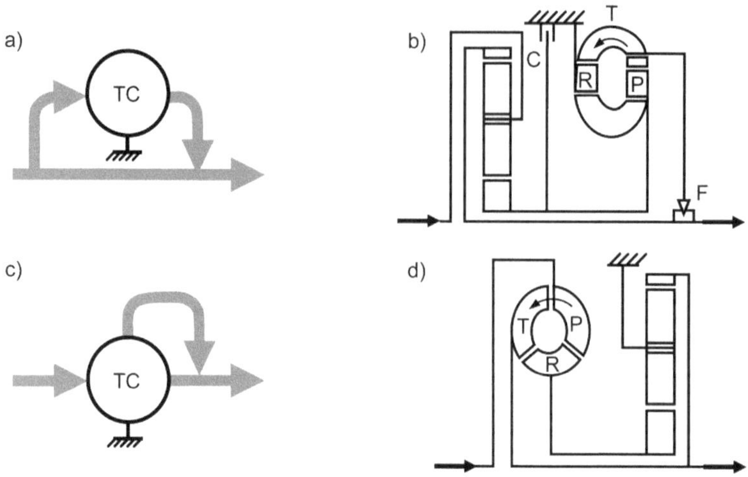


**Fig. 10.40.** *a* External power-split transmission; *b* example of external power-split transmission; *c* internal power-split transmission; *d* example of internal power-split transmission

There are two ways the power can be split in this case. The two power paths can be given a fixed torque ratio by means of a planetary differential gear unit, either on the input side (distributor gear) or on the output side (summarising gear). The power paths are then joined together again, either directly or by a gearbox, so that speed ratio is constant.

Figure 10.41 shows a schematic view of both versions. To illustrate the effect of the design on operating performance, the torque converter input power $P_{TC}$ is also plotted related to total input power $P$ over the speed ratio between input and output.

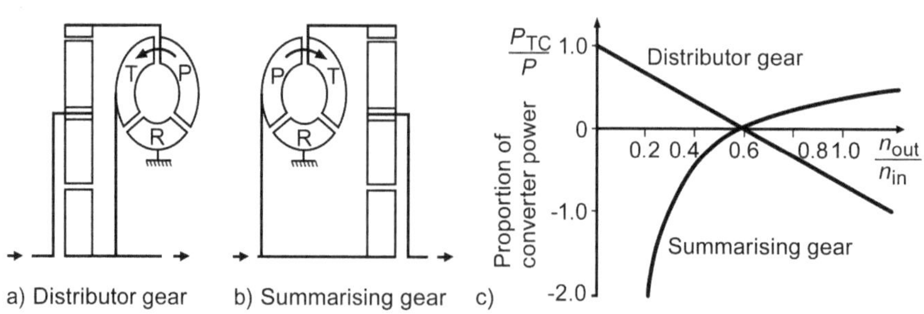


**Fig. 10.41.** External power-splitting. *a* Distributor gear; *b* summarising gear; *c* power transmitted through the converter as a proportion of the total input power, as a function of the gearbox ratio. The speed ratio belonging to the zero crossover of the converter power is a function of the ratio of the summarising gear or distributor gear

Since all the power is transmitted through the torque converter at the stall point in the case of power distribution with a distributor gear, this arrangement is particularly suitable for moving-off, and is therefore used mostly for low gears. Where summarising gears are used, high reactive power is generated at the stall point, which can amount to several times the input power. The gear mechanisms therefore have to be correspondingly larger and heavier. Summarising gears are therefore used primarily for high gears that always run above a particular speed ratio.

It is true for both types of power-splitting that efficiency is greatest when the torque converter input power is zero. The operating range thus extends in the case of a distributor gear from the stall point to around the torque converter input power crossover, and in the case of a summarising gear from the zero-crossing to higher speed ratios.

Figure 10.40b shows a diagram of the first two gears of a truck and bus transmission with power distribution using a distributor gear. The impeller of the torque converter is linked to a sun gear of a planetary gear differential; the unit can be stalled by a clutch on the housing. The drive passes along the spider, the mechanical branch flows through the ring gear of the differential. The hydraulic and the mechanical branch are combined at the same speed on the output side of the transmission.

When moving-off, the ring gear remains fixed and all the drive power is transmitted through the torque converter. As the road speed increases, the proportion of mechanical power transmitted increases with the torque converter speed ratio. When the torque converter maximum efficiency is exceeded, the transmission shifts to direct drive and the clutch *C* is engaged. The pump is now at rest, and the turbine is disengaged from the output shaft by a freewheel.

Figure 10.42 shows a diagram of external power-splitting with a summarising gear. All the power flows through the torque converter in first and second gear only. In third gear, the power is split between two paths at the same speed by means of a torsionally damped clutch.

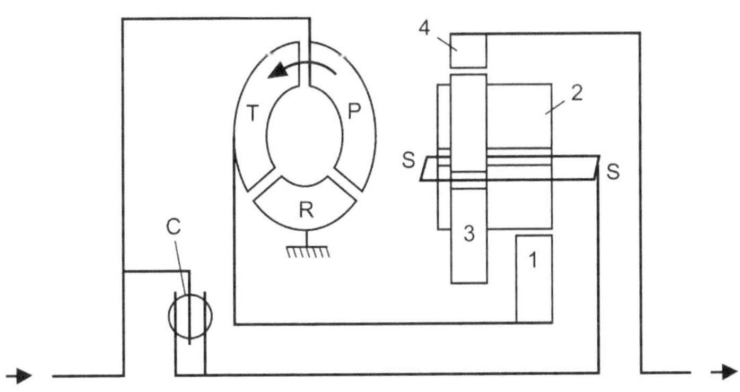


**Fig. 10.42.** Power-split with summarising gear (see Figure 6.30): Power flow in 3rd gear of a 4-speed passenger car automatic transmission. *C* Torsionally damped clutch; *P* impeller; *T* turbine; *1* sun gear; *2* 1st planetary gear set; *3* 2nd planetary gear set; *4* ring gear (output); *S* shared spider

The hydraulic and mechanical paths are then summarised again by the Ravigneaux set acting as a summarising gear. The output drive is through the ring gear. The Trilok torque converter used operates largely in the clutch range, and its slip has little effect on the transmission's speed ratio and efficiency because of the power-split. The transmission shown in Figure 10.42 is described in detail in Section 6.6.4. See also Figures 6.30 and 6.31.

### 2.2/ Internal Power-Split Transmission

In an internal power-split transmission, the torque converter turbine is of multipart design, or the stationary reactor also serves as a turbine at certain speed ratios (Figure 10.40d). This results in two hydraulic paths that are summarised on the output side by a summarising gear. In this way the torque conversion characteristic curve can be further modified.

The following presents the example of the single-phase Renk/SRM torque converter, with which high stall torque ratios can be reached (Figure 10.43). This 1960s design was used in 3- and 4-speed bus transmissions. The torque converter has two turbines. The reactor *R* lying between them can be operated with the brake *BCR* such that it is shifted by the oil flow in counter-rotation to the impeller and turbines ("double rotation").

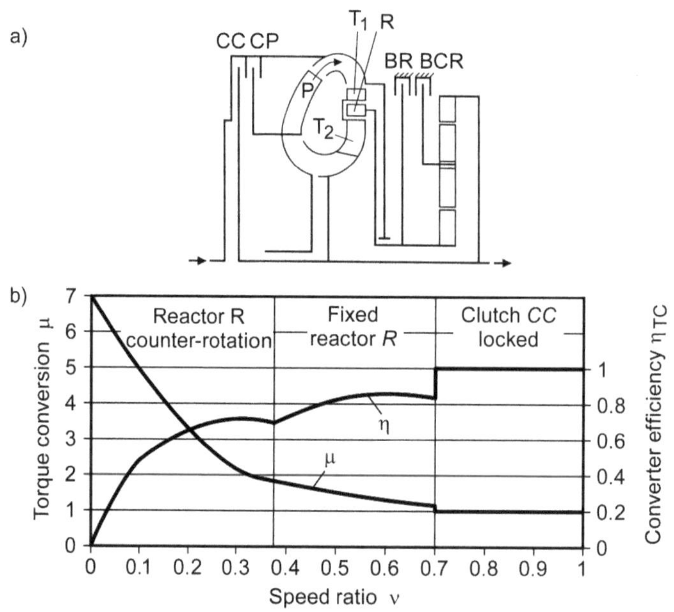


**Fig. 10.43.** Torque converter with internal power-split (SRM principle Doromat, Renk).
*a* Diagram: *CC* Torque converter lock-up clutch; *CP* impeller clutch; *P* impeller; *T1* first turbine stage; *T2* second turbine stage; *R* reactor; *BR* reactor brake; *BCR* counter-rotation brake; *b* characteristic curves

This internal power-split only works in the first operating range and results in a high torque conversion (Figure 10.43b). In the “reactor fixed” range, the brake *BR* is engaged.

Since the single-phase torque converter has no clutch range, the torque converter is locked-up with the clutch *CC* when small differential rotational speeds are reached between the impeller and the turbine. The impeller can be disengaged with the clutch *CP* – in neutral when the vehicle is at rest, for example.

# 11 Design and Configuration of Further Design Elements

*Let us proceed from what we know*
*/Aristotle/*

This chapter deals with the theory, design and configuration of vehicle transmission *bearings, lubrication, oil supply, oil pumps, housings, seals* and *continuous service brakes*. The aim is to give guidance for tackling these design elements. Sophisticated modern quantitative techniques, such as the Finite Element Method (FEM) for calculating housings or designing bearings for operational fatigue strength, are not examined in fine detail. Further literature references are given where appropriate.

## 11.1 Bearings

The function of a bearing is to support or guide components that move relative to each other, to absorb the forces arising and transmit them to the housing. A distinction is made between plain bearings and rolling bearings depending on the type of movement involved. The bearings most commonly used in vehicle transmissions are rolling bearings (see Figure 11.1).

"Bearings" as machine elements become effective only when they are positioned between a supporting housing and the shaft to be supported (see also Chapter 8 "Specification and Design of Shafts"). A distinction is made between *locating/non-locating bearing arrangements* and *semi-locating bearing arrangements* depending on the engineering design and arrangement of the bearings. Semi-locating bearing arrangements can be further subdivided into adjusted and floating bearing arrangements. Both are the same in terms of their layout. While the *adjusted bearing* has zero clearance or is even preloaded, in the case of *floating bearings* some axial clearance is deliberately left. This amounts to approximately 0.3 to 0.5 mm, depending on the size of the bearing.

Shafts normally have two bearings. Multiple bearings are also used in the case of long shafts with large deflections (see Figure 11.1). Shafts with multiple bearings are statically indeterminate and require a high degree of calculation. This currently no longer presents a problem (see also Section 8.4 "Calculating Deformation" of shafts). Vehicle transmissions have a high power/weight ratio. This creates high demands on the rolling bearings.

H. Naunheimer et al., *Automotive Transmissions*, 2nd ed.,
DOI 10.1007/978-3-642-16214-5_11, © Springer-Verlag Berlin Heidelberg 2011

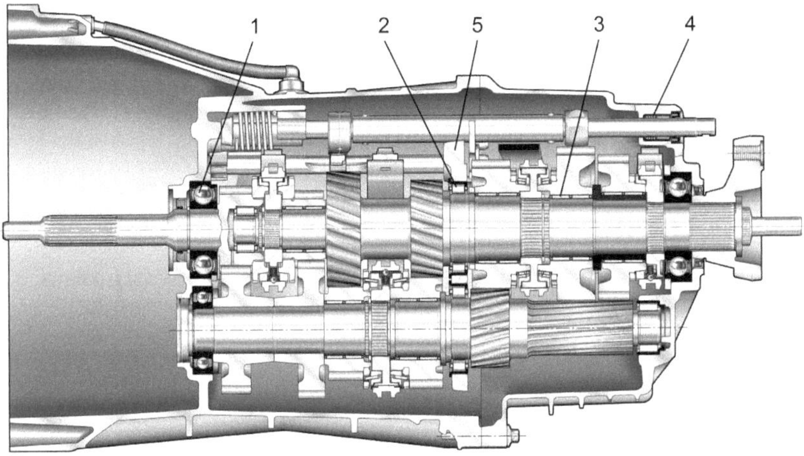


**Fig. 11.1.** Transmission bearings using the manual transmission Getrag Ford MT82 as example. *1* Deep groove ball bearing; *2* drawn cup roller bearing with open ends; *3* needle roller and cage assembly; *4* drawn cup ball bearing with open ends; *5* intermediate plate

Bearings are bought-in parts, so close co-operation with the bearing manufacturer is necessary when selecting and calculating bearings. Typical requirements for rolling bearings for use in vehicle transmissions are [11.19]:

- guaranteeing bearing load capacity, even when skewed as a result of shaft deflection,
- compensation for major heat expansion with light alloy housings,
- resistance of bearings to high operating temperatures, and consequent low oil viscosity,
- high radial and axial rigidity during tooth contact and
- resistance to dirt particles.

### 11.1.1 Selecting Rolling Bearings

When selecting and arranging bearings, it is necessary to consider their loading, ease of assembly and disassembly and type of lubrication or lubricant. There are further requirements relating to maximum rotational speed, operating temperature, bearing internal clearance and tolerances, depending on the operating conditions.

The preferred bearing types for shafts in vehicle transmissions are *deep groove ball bearings, angular contact ball bearings, four-point contact bearings, cylindrical roller bearings, drawn cup roller bearings with open/closed ends* and *tapered roller bearings*. Deep groove ball bearings are suitable for many applications because of their high running accuracy, minimal space requirements and low price. Space and price considerations make the use of special bearings with non-standard dimensions usual. Table 11.1 shows various types of bearing, their advantages and disadvantages, and their uses in motor vehicles.

**Table 11.1.** Advantages and disadvantages of some rolling bearings

| Type of bearing | Advantages (+) / Disadvantages (–) | Applications |
|---|---|---|
| Deep groove ball bearings 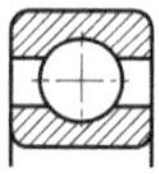 | +: Radial and axial load capacity;<br>Insensitive to angular deviation;<br>Simple installation, no adjustment;<br>Simple design, cost-effective<br>–: Low load capacity; High loads require large external diameter;<br>Sensitive to dirt | Manual gearboxes,<br>Differential gears,<br>Axle gearboxes<br>Passenger cars,<br>e.g. Figure 12.2<br>Commercial vehicles,<br>e.g. Figure 12.34 |
| Angular contact ball bearings 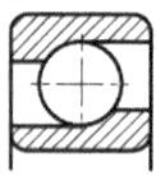 | +: High radial and axial load capacity;<br>Tight guidance in axial and radial direction<br>–: Radial load generates axial reaction forces | Manual gearboxes,<br>Pinion shafts<br>Passenger cars,<br>e.g. Figure 12.5<br>Commercial vehicles,<br>e.g. Figure 12.45 |
| Four-point contact bearings 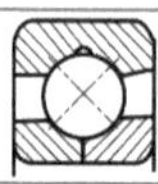 | +: High radial and axial load capacity;<br>Small overall width<br>–: Low axial clearance required | Mostly used as purely thrust bearing |
| Cylindrical roller bearings 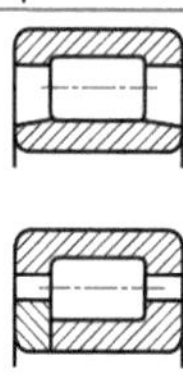 | +: High radial load capacity;<br>Can be used without inner ring;<br>Suitable types have axial load capacity;<br>Easy to mount and disassemble;<br>High resistance to dirt<br>–: Sensitive to angular deviation;<br>Expensive | At highly stressed bearing points,<br>Manual gearboxes,<br>Transfer boxes<br>Passenger cars,<br>e.g. Figure 12.10<br>Commercial vehicles,<br>e.g. Figure 12.41 |
| Tapered roller bearings 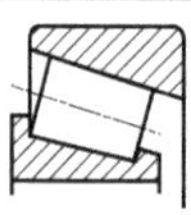 | +: High radial and axial load capacity;<br>Inner ring with roller set and outer ring can be mounted seperately;<br>Simple to fix to shaft and in housing;<br>cost-effective<br>–: Bearing clearance adjustment when mounting; Sensitive to skew postion;<br>(may be improved by suitable profiling of rollers and/or races);<br>Mutual interaction;<br>Different thermal expansion coefficients of shaft/housing affects clearance | In pairs in manual gearboxes,<br>Final drives,<br>Steering gears<br><br>Passenger cars,<br>e.g. Figure 12.10<br>Commercial vehicles,<br>e.g. Figure 12.40 |
| Spherical roller bearings 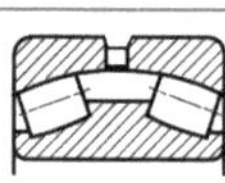 | +: High radial and axial load capacity;<br>Compensate for angular misalignment and shaft displacement<br>–: Expensive | Transfer boxes |
| Clean bearings (Sealed deep groove ball bear.) 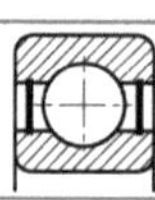 | See open deep groove ball bearings<br>+: Insensitive to dirt | See open deep groove ball bearings |
| Drawn cup roller bearings with open/closed ends 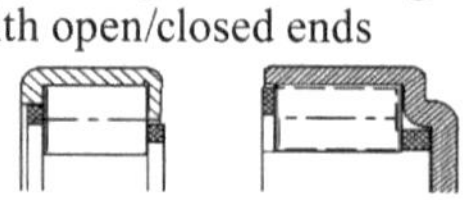 | +/–: Like roller bearings, but lower load capacity and considerably more cost-effective<br>–: "Rigid" support of drawn cup in housing necessary | For support of shafts in passenger car gearboxes<br>Passenger cars,<br>e.g. Figure 12.15 |

Idler respectively shift gearwheels are usually supported on single-row or double-row *needle roller and cage assemblies* on the transmission shafts. In the case of coaxial countershaft transmissions, the main shaft runs in the input shaft (Figure 8.1a). This type of bearing is normally fitted with a *roller and cage* or *needle roller and cage assembly*, and is known as a *stub shaft bearing* or *pilot bearing*.

### 11.1.2 Rolling Bearing Design

Vehicle transmission bearings are designed for operational fatigue strength. This means that they are designed to be reliable for a particular period, subject to typical operational stresses. Bearings are classified as so called "A" components in the "A, B, C" analysis. "A" components (e.g. bearings, shafts, gearwheels) can be designed by service life calculation or rather service life estimation (see also Section 17.2.2 "Qualitative Reliability Analysis" and 7.4 "Operational Fatigue Strength and Service Life").

Bearing forces in transmissions result chiefly from the gear-tooth forces of the gearwheels mounted on the shafts and preload forces with adjusted bearing arrangements. To determine the necessary load profiles, see Section 7.4.2 "Load Profile and Counting Procedure".

The dynamic capacity of roller bearings is calculated using a fatigue calculation to German standard DIN ISO 281. The cause of failure is attributed to pitting. The commonly used service life formula for rolling bearings is [11.10]

$$L_{10} = L = \left(\frac{C}{P}\right)^{p} \ (10^6 \text{ revolutions}). \tag{11.1}$$

$L_{10}$ Nominal service life in millions of revolutions, reaching at least 90% of a large number of like bearings,
$p$ service life exponent ($p = 3$ for ball bearings, $p = 10/3$ for roller bearings),
$C$ basic dynamic load rating in N (given in the bearing catalogues),
$P$ equivalent bearing load in N, Equation 11.2.

The service life formula for roller bearings is based on the equation of the fatigue zone of Wöhler curve (Equation 7.21). The relations are shown again in Figure 11.2. As regards the load $\sigma_i$ and the bearing load $P$, normally shown in the Wöhler diagram, this relation applies:

$$\sigma_i \sim \sqrt{P}\,.$$

The dynamic equivalent load $P$ for combined loading is derived from

$$P = X\,F_r + Y\,F_a\,. \tag{11.2}$$

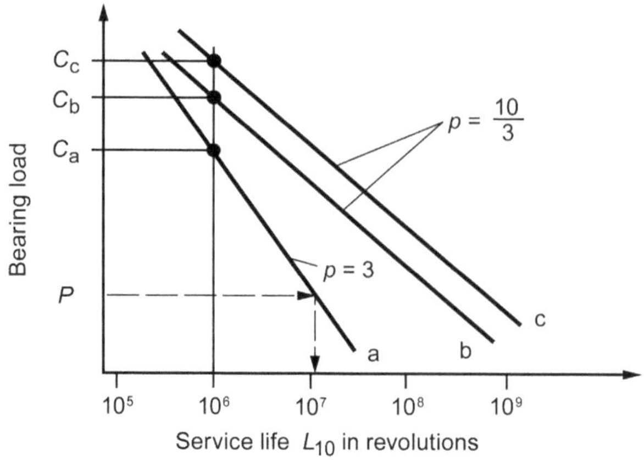


**Fig. 11.2.** Nominal service life $L_{10}$ – various types of bearing, 10%-probability of failure.
*a* Deep groove ball bearing;
*b* cylindrical roller bearing;
*c* tapered roller bearing

$F_r$ Constant radial bearing load in N,
$X$ radial factor,
$F_a$ constant axial bearing load in N,
$Y$ axial factor.

The special features of calculating non-locating and locating bearings, and the individual calculation factors for the various types of bearing, are given in the applicable rolling bearing catalogue.

A rolling bearing is subjected to differing loads and speeds over time in a motor vehicle (load profile). The total service life is calculated with the help of the equivalent load and the equivalent speed. The equivalent speed is calculated taking into account the portions $q_i$ of the duration in % as follows

$$n = \frac{q_1 n_1 + q_2 n_2 + q_3 n_3 + \ldots + q_n n_n}{100\%} \tag{11.3}$$

and the equivalent load as

$$P = \sqrt[p]{\frac{q_1 n_1 P_1^p + q_2 n_2 P_2^p + q_3 n_3 P_3^p + \ldots + q_n n_n P_n^p}{q_1 n_1 + q_2 n_2 + q_3 n_3 + \ldots + q_n n_n}}\,. \tag{11.4}$$

The total service life is then obtained from the equivalent load and equivalent speed:

$$L_h = \frac{16666}{n}\left(\frac{C}{P}\right)^p, \tag{11.5}$$

with the service life $L_h$ in h and speed $n$ in 1/min. The effective service life can differ from the calculated bearing service life if operational loads and speeds are not exactly known, if there are variations in loads or other influences such as inadequate lubrication, installation and assembly errors, or if dirt impairs the bearings environment. Table 11.2 gives the usual service life for vehicles in hours. The method of calculation shown and the service life figures are based on extensive experience with many transmission that have proved themselves in practice. Modern, more refined calculation methods [11.11] make it possible to consider

- axial clearance or preload in the bearing, and therefore changing load zones,
- Hertzian contact stresses, pressure ellipses and edge stresses,
- edge freedoms in the contact geometry,
- shaft deflections and
- statically indeterminate systems.

Greater certainty in the design for operational fatigue strength of rolling bearings in vehicle transmissions can be achieved by using refined service life calculation methods. This makes it possible to use smaller rolling bearings, thus saving energy and weight [11.37].

Usually a particular oil is used depending on the transmission design. In the case of manual transmissions, mineral oils with wear and high pressure additives are used. For automatic transmissions, special automatic transmission fluids (ATF) are employed. Manual gearboxes and rear axle drives with hypoid toothing require a hypoid oil. In the case of modern vehicle transmission, lifetime lubrication is provided, i.e. oil changes are no longer necessary (see Section 11.2 "Lubrication of Gearboxes, Gearbox Lubricants").

The metallic, mineral and organic impurities contained in the oil have a major influence on bearing service life. Rolling contact with these impurities produces impressions in the raceways, depending on the type of particle. Each subsequent rolling contact produces increased stress in the area of the impression, leading to early fatigue of the raceway. Especially in the case of small ball bearings, this leads to substantial reductions in service life. This can be countered by choosing suitable heat treatment processes and materials, and by installing dirt-protected bearings [11.3]. Dirt-protected or clean bearings are bearings sealed on both sides and packed with grease. Their service life is longer than bearings with gearbox oil flowing directly through them (see also the 5-speed transmission in Figure 12.2 and the associated versions).

**Table 11.2.** Guidelines for dimensioning motor vehicle transmission bearings at 100% engine torque. * using dirt-protected bearings

| Type of vehicle | Intended $L_{10}$ service life (h) |
|---|---|
| Motorcycles | 300 – 600 |
| Light passenger cars | 500 – 1100 (200 – 500)* |
| Heavy passenger cars, SUV | 400 – 1000 (200 – 500)* |
| Light-duty trucks | 1400 – 4000 |
| Heavy-duty trucks | 4000 – 6000 |
| Buses | 1400 – 4000 |

Typical bearing damage resulting from installation and operating errors is [11.45]:

- *Scoring:*
  In the case of cylinder roller bearings, by tilting the ring without flanges, or because of insufficient internal bearing clearance,
- *Depressing:*
  In ball bearings when the static load of the bearing is exceeded,
- *Grooving:*
  In different types of rolling bearings as a result of small swivelling movements, or because of shocks when at rest.

### Axial Load Capacity of Radial Cylindrical Roller Bearings

Certain cylindrical roller bearing designs can handle large radial forces as well as high axial forces. They can thus be used as locating bearings or as semi-locating bearings, always assuming that $F_a \leq 0.4\, F_r / i$ (where $i$ = number of rows of rollers). The axial load capacity depends on the size and bearing capacity of the contact surfaces between the roller faces and the bearing flanges. It is however also affected by sliding speed, load duration and by lubrication (lubricant, quantity and viscosity). Calculation of axial bearing capacity is based on the frictional power arising at the sliding contact faces. The axial continuous load of a cylindrical roller bearing can be determined using the following equation [11.23]:

$$F_{\mathrm{a,perm}} = k_{\mathrm{S}}\, k_{\mathrm{B}}\, d_{\mathrm{M}}^{1.5}\, n^{-0.6} \;\leq\; F_{\mathrm{a,max}} = 0.075\, k_{\mathrm{B}}\, d_{\mathrm{M}}^{2.1}\,. \tag{11.6}$$

$F_{\mathrm{a,perm}}$ Permissible axial load in N,
$F_{\mathrm{a,max}}$ axial load limit in N,
$k_{\mathrm{S}}$ coefficient dependent on lubrication method, given in rolling bearing catalogue ($k_{\mathrm{S}} = 7.5$–$15$),
$k_{\mathrm{B}}$ coefficient dependent on bearing model, given in rolling bearing catalogue ($k_{\mathrm{B}} = 4.5$–$30$),
$d_{\mathrm{M}}$ average bearing diameter $(D + d) / 2$ in mm,
$n$ operating speed in 1/min.

When using radial cylindrical roller bearings for axial load, it is essential to ensure precise production and installation of the bearing points, so that the bearing flanges under load are supported over the whole locating face as far as possible.

Should cylinder roller bearings or drawn cup roller bearings be used under higher axial loads, close collaboration with a rolling bearing manufacturer is necessary.

## 11.1.3 Design of Rolling Bearings

Please refer to the design examples in Chapter 12 for bearing design.

### *Selector Gearbox*

The bearings used in selector gearboxes are mostly deep groove ball bearings, drawn cup roller bearings, cylindrical roller bearings, tapered roller bearings and four-point contact bearings. For production reasons, the deep groove ball bearing type with a groove in the outer ring is preferred. This means that there can be through bores in the housing. The axial locating is then by means of a circlip in the groove.

If there are high axial forces in the transmission, the radial and axial forces must be separately absorbed in the locating bearing. Then a cylindrical roller bearing is often fitted to absorb the radial force, in combination with a deep groove ball bearing or four-point contact bearing to absorb the axial force.

Countershafts often have floating bearing arrangements, i.e. both bearings are capable of absorbing axial and radial forces. When cylindrical roller bearings or drawn cup roller bearings with open or closed ends are used, their axial bearing capacity has to be borne in mind.

With shaft mountings, the inner ring is subject to circumferential load, and must therefore have the tighter fit. But the outer ring should also not have a loose seating because of alternating load.

Often needle roller and cage assemblies and roller and cage assemblies are fitted instead of complete rolling bearings because of space constraints. The machining tolerances of the shaft seats or housing drill holes are related to the radial internal clearance required. The contact surfaces must be hardened at all events. Table 11.3 gives reference values for shaft and housing tolerances.

Idler gear bearings require special attention. Because of the gearwheel design, for example on account of the synchronizer hub, there is often eccentric application of force to the gearwheel. The resultant tilting movement stresses the needle rollers and cages of the idler gears (see also Figure 8.10).

**Table 11.3.** Reference values for transmission bearing fits

| Type of bearing | Shaft | Housing |
|---|---|---|
| Deep groove ball bearings | k6 | J6 – K6 (steel or cast steel)<br>M6 – N6 (light alloy) |
| Four-point contact bearings<br>(for absorbing pure axial load) | k5 | E8 |
| Cylindrical roller bearings | k6 – m6 | K6 – M6 (steel or cast steel)<br>N6 – P6 (light alloy) |
| Needle roller and cage assemblies,<br>roller and cage assemblies | h5 – h6 | G6 |
| Tapered roller bearings:<br>Adjustment of the inner ring<br>Adjustment of the outer ring | <br>h6 – j6<br>k6 | <br>M6 – N7<br>J6 |

Until the 1960s, idler gears were predominantly fitted with plain bearings. They were later fitted almost exclusively with rolling bearings. For idler gears in passenger car transmissions running on rolling bearings, single and double-row needle roller and cage assemblies are mostly used. Because of the larger widths of the gearwheels in the case of commercial vehicle gearboxes, double-row needle roller and cage assemblies are normally used. Double-row needle assemblies with a one-part cage are very common. Double-row needle assemblies with two separate cages provide superior bearing properties, but are more expensive. In coaxial countershaft transmissions, the main shaft runs in the input shaft with a *stub shaft bearing (pilot bearing)*. These bearings normally use roller and cage assemblies.

In the case of adjusted bearing arrangements, it is important to ensure the correct degree of adjustment, since heat expansion has to be compensated for by clearance or preload, depending on the bearing arrangement. For example large fluctuations in bearing clearance can occur in bearing arrangements with tapered roller bearings installed in light-alloy housings, because of the difference in rates of heat expansion of the shaft and the housing [11.38]. A recommended value for the X arrangement is axial clearance of 0.05 mm per 100 mm distance between bearings, and for the O arrangement a preload which gives a bearing friction torque of 1 to 2 Nm, depending on the size of bearing. This does not apply to tapered roller bearings arranged in pairs to form a locating bearing; in this case there should be as little clearance as possible.

#### *Final Drives*

With front-wheel drive transverse-mounted engines, the final drive consists of a helical spur gear pair, and the above engineering design guidelines apply to its bearing arrangement likewise.

Two pinion shaft configurations are possible for longitudinal-mounted engines and bevel gear final drive. In one case, the pinion is overhung, and in the other case, mounted between the bearing points. With an overhung pinion arrangement, the gap between the two bearings should be at least 2.5 times the distance between the pinion and the first bearing. It is normal to use an adjusted bearing arrangement or a locating/non-locating arrangement in the overhung configuration. Where the pinion is supported in the middle, only the locating/non-locating arrangement is used. The crown wheel bearing is generally in the form of an adjusted bearing arrangement.

The same dimensions as shown in Table 11.3 apply as fitting guidelines for the individual bearings. For the double-row angular contact ball bearings also used with axle drives, the tolerance of the shaft should be k6 and that of the housing drill-hole J6. Final drives are systematically reviewed in Section 6.8. Section 12.3 sets out some design examples.

### 11.1.4 Plain Bearings – Bearing Bushes and Thrust Washers

Plain bearings, thanks to their compact design and advantageous acoustics, can always be found where shafts rotating relative to each other are used in gearboxes.

This is especially true for transmissions with planetary gear sets. The gearbox ZF 6 HP 26 (Figure 12.25), for example has 9 radial plain bearings (bearing bushes) and three axial plain bearings (thrust washers). Figure 11.3 illustrates some of them.

The layout and calculation of plain bearings is discussed thoroughly in the literature and have been standardised in guidelines. General recommendations on the classification, calculation and function of plain bearings are given in German standards DIN 31651 [11.7] and 31652 [11.8] as well as in the guideline VDI 2204 [11.52]. DIN ISO 12128 [11.14] provides recommendations for designing lubrication holes, lubrication grooves and lubrication pockets. DIN ISO 3547 [11.13] contains information about plain bearings for transmissions. Similarly to rolling bearings, manufacturers of plain bearings support transmission designers by means of design calculation programs.

Bearings are specified by the transmission designer according to outer and inner diameter, bearing width, bearing clearance, speed range, journal material, housing material, oil viscosity as well as the dominant oil temperatures of – 40 to +150°C.

Properties demanded of plain bearing materials are complex and partially contradictory (Figure 11.4). Plain bearings must have emergency running properties in case of insufficient lubrication. Towing is a critical case of application here (see Section 11.3.3 "Detail Questions") on oil supply and oil pumps.

Lead has good emergency running properties, but it is problematic as a bearing material with respect to environmental pollution and recycling. Current developments are pointing to lead-free plain bearing materials.

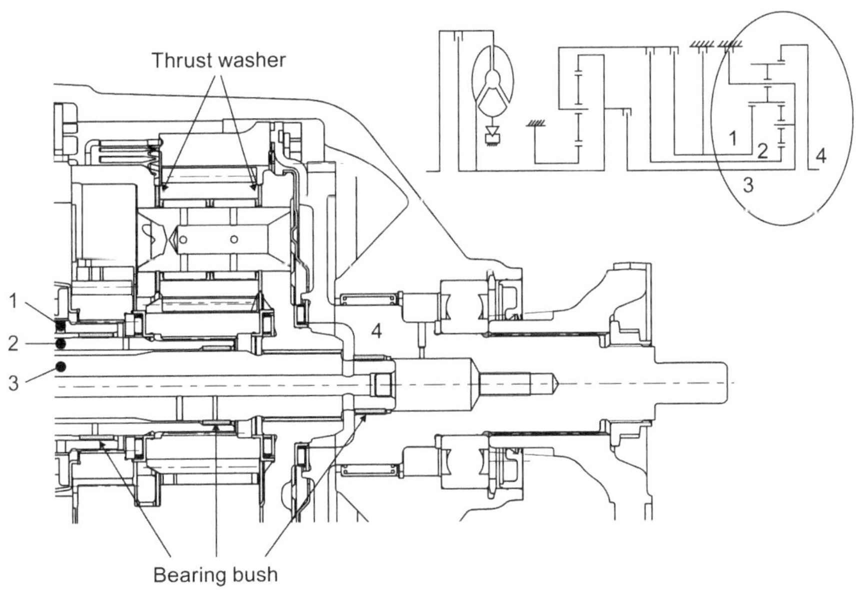


**Fig. 11.3.** Bearing bushes and thrust washers in a conventional automatic transmission, ZF 6 HP 26. *1* Sun gear shaft II; *2* sun gear shaft III; *3* intermediate shaft; *4* output shaft

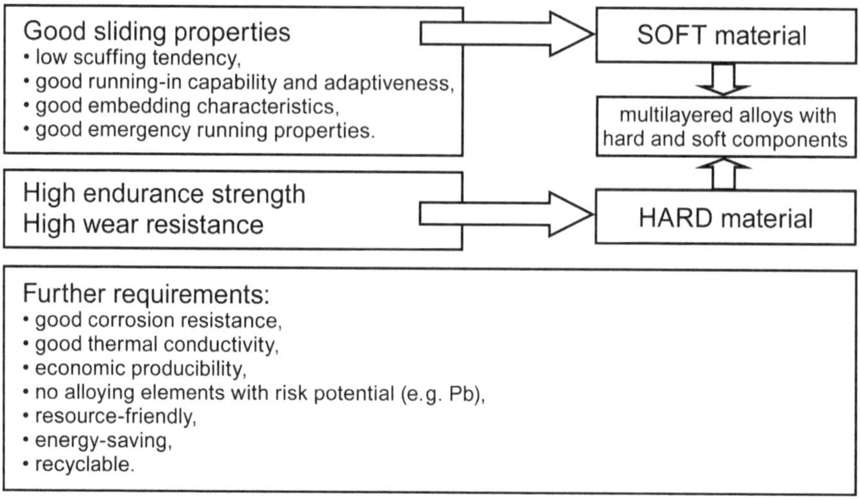


**Fig. 11.4.** Requirements for plain bearing materials [11.17]

In vehicle transmissions, rolled bearing bushes made of coated sheet metal are used primarily. Figure 11.5 shows the layout of a bearing bush. Bearing bushes are manufactured from strip material. The bearing metal is sintered, poured or rolled on the support plate. After cutting the individual plates to length, grooves, holes, and the latch are embedded by broaching, embossing or blanking. Then it is shaped by rolling. In case the bearing bush also has a sealing function or the outer diameter still needs to be ground, the butt joint is latch closed. Finishing can take place only after pressing into the housing bore, or a previously machined plain bearing can be pressed in. Typical wall strengths of plain bearings lie between 1.5 and 2 mm.

Hollowly supporting bearing bushes should be avoided. The contact share must be greater than 60% in order to guarantee a firm seat in the housing bore and therefore to prevent turning of the bush in the bore. It is also necessary to ensure heat removal.

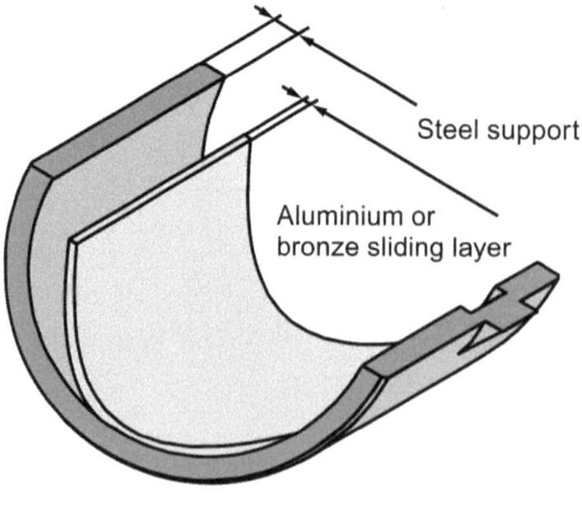


**Fig. 11.5.** Layers of bearing bushes [11.17]

The contact share is difficult to determine. One can induce it indirectly by means of the press-out force necessary to remove a pressed-in bearing bush again.

Bearing bushes are secured axially in order to avoid axial migration during operation. This is especially the case when they take on the function of sealing pressure oil chambers against each other. They can be secured axially, for example, by caulking or by rolling. In the case of rolling, part of the shell of the bush is plastically pressed into a groove of the locating bore in an additional operation.

## 11.2 Lubrication of Gearboxes, Gearbox Lubricants

A tribological system consists of the following three components: Base body (e.g. rolling element), mating body (e.g. bearing shell), and an intermediary (e.g. lubricating oil). There is relative movement between the base body and the mating body. Lubricants are design elements whose function is to keep the base bodies and the mating bodies apart under all loads [11.15]. Regardless of system conditions, dry friction, mixed friction and hydrodynamics can arise (Figure 11.6).

As part of the move towards reducing vehicle weight, attempts are being made to reduce the weight of gearboxes, and thus the quantity of lubricant. An additional requirement is that it should not be necessary to change the lubricant throughout the vehicle's service life (lifetime lubrication).

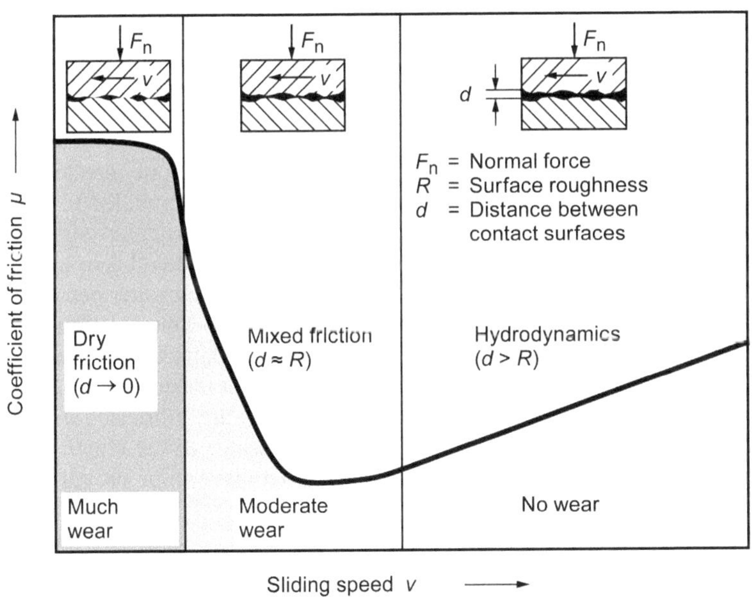


**Fig. 11.6.** Relation between coefficient of friction and sliding speed [11.15] ("Stribeck curve"; *Stribeck*: Basic research into plain bearings)

In summary, the lubricant must fulfil the following functions:

- reduce friction and wear (saving energy),
- prevent possible damage, or prevent or delay further damage where mechanisms are already damaged,
- dissipate heat,
- create hydrodynamic lubricating wedges,
- form barrier layers in the mixed friction zone,
- protect materials used in the gearbox against corrosion,
- non-aggressive to seals and paintwork,
- good dirt removal/cleaning,
- good dirt absorption capability,
- water separation,
- stability at high and low temperatures,
- resistant to ageing and
- low cost.

In contrast to slow-running, grease-lubricated industrial gear units, vehicle transmissions are oil-lubricated. Fluid media are better suited for creating the necessary hydrodynamic bearing film than solid materials (greases or pastes). The constant oil circulation also provides better heat dissipation from the stressed components. It is easier for the design engineer to ensure all the various lubrication points in the gearbox are washed with lubricant. It is also easier to remove impurities in the gearbox through appropriate filter systems with oil than with pasty materials.

### 11.2.1 Bearing Lubrication

Bearings in selector gearboxes are normally lubricated by the oil spray created by the gearwheels in the housing. Traps and feed ducts have to be provided to feed bearing points in unfavourable locations. Good lubrication is particularly important with cylindrical roller bearings subject to axial load, since the oil also has to dissipate the heat created by friction. The same applies to tapered roller bearings, and in this case it also has to be borne in mind that the oil is pumped from the smaller taper opening to the larger one. Gearwheel debris impairs the service life of roller bearings, so an oil circulation system with an oil filter is beneficial.

The bearings and gearwheels of bevel gear final drives are lubricated exclusively with a pressure resistant hypoid oil. Whilst the bearings of the crown gear shaft are well lubricated by the oil spray, flow ducts often have to be provided to take the oil to and from the pinion shaft.

### 11.2.2 Principles of Lubricating Gearwheel Mechanisms

When gear teeth mesh, two types of movement take place: rolling and sliding movement. The sliding speed is at its maximum at the beginning $A$ or the end $E$ of tooth contact, i.e. at the root or the tip of the tooth; at the pitch point *C* it is zero (Figure 11.7). The wear increases as the proportion of sliding increases.

At the tooth flanks, the most favourable lubrication conditions arise around the pitch point, whilst conditions are less favourable at the tooth tips, because of the meshing impact and high temperatures resulting from higher sliding speed. Figure 11.7 shows typical friction zones on the tooth flanks:

- Boundary friction:
  There is dry friction. The tooth flanks are only separated by a boundary layer of chemical reaction products a few nanometres thick, intended to prevent metal-to-metal contact (boundary lubrication).
- Mixed friction:
  The tooth flanks are only partially separated by a film of lubricant. There is liquid friction and dry friction at the same time. Where the surfaces touch, there is boundary lubrication.
- Fluid friction (hydrodynamics):
  The tooth flanks are completely separated by a film of lubricant. There is elasto-hydrodynamic lubrication.

The lubricant thus has a two-fold effect in reducing friction and wear at the tooth flanks [11.1, 11.32, 11.35, 11.42]:

1/ *elasto-hydrodynamic lubricant film*: EHD lubricant film,
2/ *chemical protective film*: frequently also referred to as boundary layer or reaction layer.

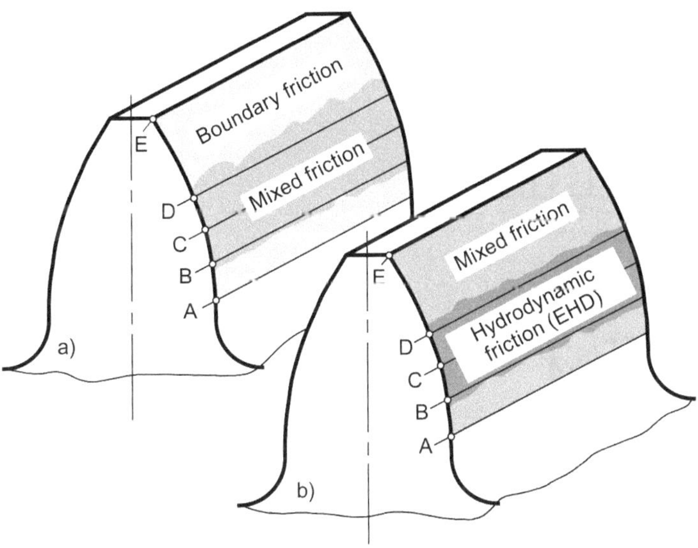


**Fig. 11.7.** Typical friction zones on tooth flanks at high contact pressures [11.34–11.35]. *a* Low circumferential speed (up to 5 m/s); *b* high circumferential speed. *A* first point of contact; *B* internal single contact point; *C* pitch point; *D* external single contact point; *E* last point of contact

### *1/ Elasto-Hydrodynamic Lubricant Film*

The lubrication process is discontinuous since the bearing film has to be re-established for each meshing action. Hydrodynamic lubrication theory of plain bearings is not applicable because of the high contact pressures for the toothing. Elasto-hydrodynamic lubrication theory has to be applied. This theory takes into account the pressure viscosity of the oil and the elasticity of the tooth flanks. Elasto-hydrodynamic lubrication is characterised by two fundamental features:

- The viscosity of the oil film increases erratically because of the high surface stress.
- Because of the high surface stress, there is elastic deformation at the tooth flank contact points. The crowned engaged tooth flanks flatten under load.

The tooth flanks are kept out of direct contact by the increase in contact surface, and the load capacity of the lubricant film related to its viscosity. The formation of the lubrication gap and the pressure distribution in the contact zone is shown in Figure 11.8. A pressure peak is formed before the end of the lubrication gap, and the end of the gap at the lubricant outlet is contracted.

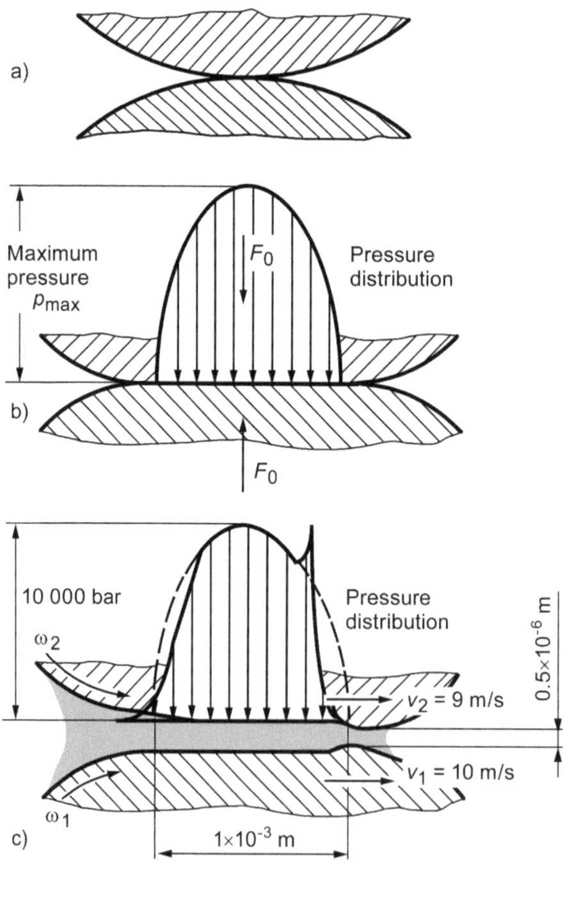


**Fig. 11.8.** Formation of the lubrication gap with elasto-hydrodynamic lubrication [11.16].
*a* Without load;
*b* under load without movement;
*c* under load and rotating with film of lubricant

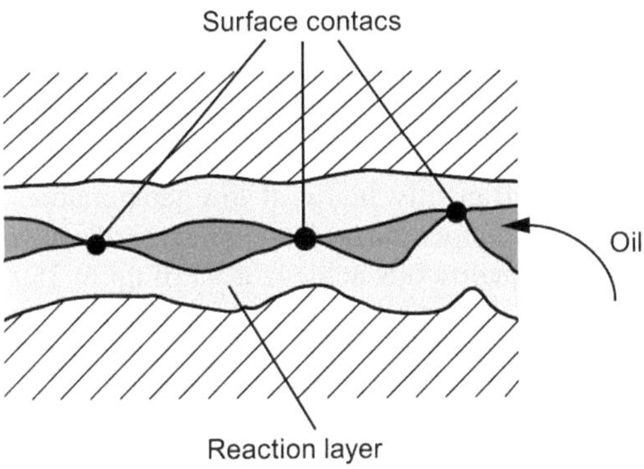


**Fig. 11.9.** Mixed friction, surface reaction layer – protecting the metal layer beneath

The thickness of the film of lubricant depends chiefly on the toothing geometry, the viscosity of the lubricating oil, the circumferential speed, the contact pressure, the tooth flank temperature and the surface roughness.

### *2/ Chemical Protective Film*

If the surfaces touch with mixed friction or boundary friction, the wear-reducing additives in the oil come into effect, forming a chemical protective film on the tooth flanks [11.27, 11.50].

The wear-reducing additives are called EP (Extreme Pressure) additives. The alternative term AW (Anti-Wear) has gained acceptance, referring directly to the function of these materials.

Put simply, mild EP/AW additives are first physically absorbed, and only then in the second stage (under load) are chemical reaction layers formed. They prevent the contact surfaces bonding by forming surface reaction layers with lower shear strength than in the case of pure materials [11.49] (Figure 11.9).

Highly reactive EP/AW additives lead to measurable reaction layers even before the trigger temperature is reached. They then quickly regenerate the abraded reaction layer at very high stresses. If the tooth flanks heat up under friction, this boundary layer will be destroyed if a characteristic temperature for the lubricant is exceeded.

The composition of the reaction layers depends very much on the mechanical conditions, materials, temperature, the lubricant base fluid and the additives. Investigations show that the chemical reaction between additive and tooth flank is the crucial factor in the scuffing load. The scuffing load speed curves are crucially dependent on the additive [11.4].

## 11.2.3 Selecting the Lubricant

In normal operation, the oil temperature in the sump of passenger car and commercial vehicle gearboxes is approximately 60–90°C. Under extreme conditions, for example on mountain roads with a trailer, the oil sump temperature can reach approximately 110°C. Oil temperatures may reach 130 to 160°C locally.

Modern lubricants are fundamentally made up of several constituents. They consist of a base oil and appropriate additives. Mainly mineral oils are used as base oil for manufacturing gearbox oils. The various mineral oils are distinguished by their viscosity index VI [11.12]. The viscosity index describes the high/low temperature properties of the base oil, i.e. remaining sufficiently liquid at low temperatures, without becoming excessively liquid at high temperatures. Good mineral oils have a VI of approximately 95 to 105, and high-quality oils achieve a VI of up to 150. Where extremes of temperature are anticipated (below –20°C and above 140°C), synthetic oils are used.

The characteristics of gearbox oils are affected to a significant degree by the additives and packages. The term "package" refers to a gearbox oil constituent made up of several additives comprising approximately 2–10% of the total volume. The most common additives for gearbox lubricants in accordance with [11.31] are:

- EP (Extreme Pressure) additives for improving high-pressure characteristics,
- corrosion inhibitors for preventing rust, verdigris and similar harmful products of oxidation,
- D/D (Detergent/Dispergent) additives for removing dirt,
- friction modifiers for reducing friction and wear,
- VI improver to improve high and low temperature performance.

This enables lubricants with different characteristics to be created by appropriate choice and composition of additives. The action of the various additives should be complementary.

### 11.2.4 Selecting Lubricant Characteristics

#### *1/ Viscosity*

Probably the most important characteristic of gearbox oils is their flowability, or *viscosity*. Viscosity describes the internal friction of a fluid. A distinction is made between dynamic viscosity $\eta$ and kinematic viscosity $\nu$; it is almost always the kinematic viscosity that is quoted. It is calculated as the quotient of dynamic viscosity and density of the oil $\rho$

$$\nu = \frac{\eta}{\rho}. \tag{11.7}$$

Lubricating oils are classified into viscosity groups. Both automotive engine oils and gearbox oils are usually given SAE grades (Figure 11.10). The appropriate viscosity for a gear mechanism can be selected in accordance with German standard DIN 51509 [11.9]. The kinematic nominal viscosity is thereby determined as a function of the surface stress and the sliding speed.

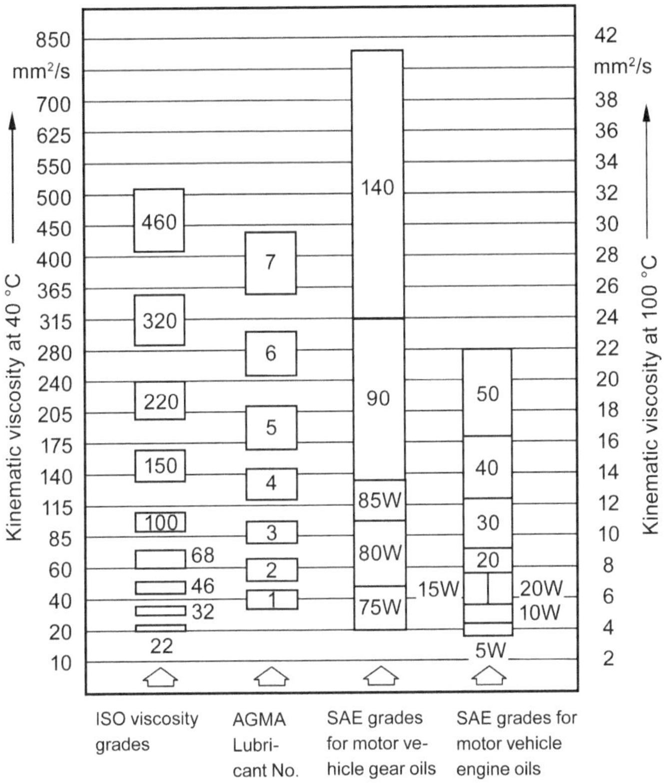


**Fig. 11.10.** Comparison of various systems of viscosity classification (basis: VI=100) [11.16]

A fluid bearing film can be created according to the EHD theory, even in friction combinations with unfavourable lubrication relations and large contact pressures, such as those encountered in meshing. The thickness of the film of lubricant can be determined by applying this theory, as a function of the stress, the circumferential speed, the effective temperature and the viscosity of the lubricant [11.44]. This film thickness can be regarded as adequate when it is larger than the average surface roughness of the tooth flank surface. Alternatively, the lubricant viscosity required to provide a bearing film of sufficient thickness for a given gear/tooth system can be determined when the operating conditions are known.

The gears resistance to scuffing and pitting also improves as viscosity increases. Damping capacity increases and load losses decrease. If the viscosity of the lubricating oil is too great, there are also negative effects. For example friction losses and thus temperatures can become very large [11.48]. Lower viscosity improves low temperature fluidity, air release behaviour and cooling capacity. Idling losses are reduced.

Selecting the viscosity is always a compromise. It is frequently determined by other components in the lubrication system, such as the torque converter.

### 2/ Viscosity/Thermal Behaviour

The viscosity of lubricating oils decreases exponentially as temperature increases. The viscosity/temperature profile of mineral oil based lubricating oils (*a*) is a straight line in the log-log Ubbelohde diagram (Figure 11.11). The change in viscosity depends on the base oil.

Synthetic lubricating oils based on poly-α-oleofins (*b*) also give straight lines on an Ubbelohde diagram, whereas polyglycol-based oils (*c*) appear as curves [11.23]. The empirical "Ubbelohde-Walther formula" (German standard DIN 51563) is derived as

$$m = \frac{W1 - W2}{\log T_2 - \log T_1} . \tag{11.8}$$

The variables in Equation 11.8 are as follows:

- $m$ slope; normal values are $2 < m < 5$,
- $W$ log-log $(v + 0.8)$,
- $v$ kinematic viscosity,
- $T$ test temperature in K.

The gradient of the straight line, the slope $m$, is a measure of the temperature dependence of the lubricating oil. The characteristic value of the viscosity/thermal behaviour is given by the viscosity index VI (German standard DIN ISO 2909) [11.12].

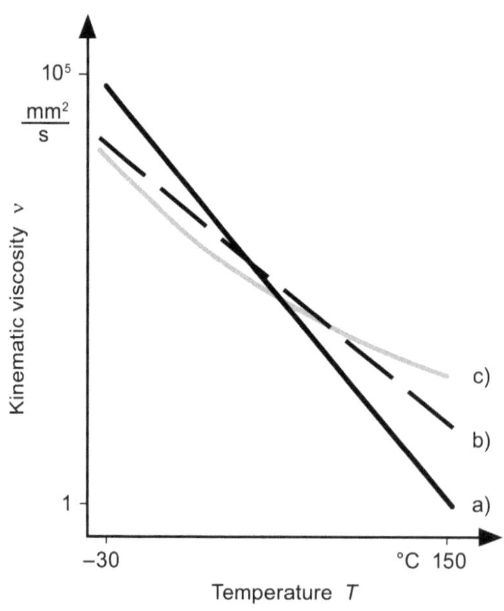


**Fig. 11.11.** Ubbelohde diagram (log-log scale). Basic viscosity/temperature profiles of various gearbox oils [11.23].
*a* Mineral gearbox oil;
*b* poly-α-oleofin based gearbox oil;
*c* polyglycol based gearbox oil

### 3/ Viscosity/Pressure Behaviour

Very high Hertzian stresses arise briefly in vehicle transmissions during meshing (in excess of 2000 $N/mm^2$), and the dynamic viscosity $\eta_p$ of the oil rises

$$\eta_p = \eta_0 \, e^{\alpha p} . \tag{11.9}$$

The variables in Equation 11.9 are as follows:

- $\eta_p$ dynamic viscosity at working pressure in Pa s,
- $\eta_0$ dynamic viscosity at 1 bar,
- $\alpha$ viscosity-pressure coefficient of the oil in $Pa^{-1}$,
- $p$ working pressure in Pa.

The pressure coefficients of common types of oil are in the range

$$0.7 \leq \alpha_{25°C} \leq 8 \, Pa^{-1} .$$

### 4/ Pour Point and Flash Point

The pour point describes the flow properties at low temperatures. The pour point must be at least 5 K below the lowest operating temperature.

The flash point can be ignored in all but a few critical high-temperature applications that do not apply in automotive engineering.

### 5/ Foaming Tendency and Air Release Behaviour

A distinction has to be made between surface foam and bubble foam. Surface foam can be prevented by design measures (baffles, settling chambers). Bubble foam is prevented by adding a small amount of silicone oil, which, however, impairs the air release characteristics.

### 6/ Water Demulsibility

In order to prevent foam formation, any water entering the oil should not emulsify with it.

### 7/ Ageing Characteristics, Oxidation Characteristics

Oil ageing is a chemical change that takes place under the influence of high temperatures and the catalytic effect of metals. Ageing is caused mainly by oxidation of oil molecules. The high flank temperatures arising in gear mechanisms, the high turbulence of the oil, and the impurities present in the oil are major factors in ageing. The resistance of gearbox oils to oxidation can be improved by oxidation inhibiting additives.

#### *8/ Corrosion Protection*

The corrosion protection characteristics of the oil can be improved by additives that create a protective film on the metal surfaces in the gearbox, and/or neutralise the corrosive decomposition products formed in the process of oil ageing.

#### *9/ Seal Compatibility*

Seals must not change in terms of their material characteristics under the influence of gearbox oils. For example, they must not shrink, swell or become brittle, nor should any material components be deposited which could lead to impairment of the oil's characteristics and thus of the gearbox's function.

### 11.2.5 Lifetime Lubrication in Vehicle Gearboxes

Lifetime lubrication of vehicle gearboxes is of particular significance in the light of the discussion about more environment-friendly and cost-effective products. Lifetime lubrication offers the following advantages:

- reduced lubricant consumption,
- reduced lubricant costs related to gearbox service life and
- low maintenance costs (down time).

Lifetime lubrication means that oil change intervals extend to the service life of the vehicle. But an oil change after a particular running-in period can maybe not be avoided. Lifetime lubrication has already been introduced for manual passenger car transmissions, but lifetime lubrication for more highly stressed commercial vehicle transmissions and automatic transmissions will only be introduced with new generations of transmission [11.26].

During the service life of a transmission, the lubricant changes because of oxidation, decomposition, additive degradation, changing viscosity, and absorption of moisture and particles. This also affects the service life of the gearwheels, bearings and shifting elements. The increased requirements on lubricants for lifetime lubrication are already largely satisfied by synthetic oils, given appropriate selection of base oil and additives.

### 11.2.6 Testing the Scuffing Resistance of Gearbox Lubricants

Two categories of testing machine types are mostly used for experimental investigation of scuffing resistance:

- gearwheel and roller type testing machine and
- two, four and five ball tester.

In the case of the development of new and existing oils and their quality control, costly and time-consuming bench tests are necessary before the application. With the increasing demands on lubricants and friction materials due to light-weight design and higher power transmission, the scuffing resistance requirements for the friction pairings also increase. Scuffing resistance is finally evaluated on gearwheel test stands. The Integral Temperature Method now provides a satisfactory method for calculating scuffing resistance of production transmissions from gear bench test programs [11.39]. Investigations have shown that the Integral Temperature can also be determined without using a gearwheel test stand, using the test results of the four ball tester.

Gearbox scuffing resistance is evaluated on the basis of a mean, weighted contact temperature (Integral Temperature). The criterion of constancy of the Integral Temperature is regarded as the best method for transferring gear test bench findings to practical performance [11.56]. With this method, the permissible Integral Temperature $T_{int,perm}$ is determined with a gearwheel torque test bench (FZG test bench, FZG: "Forschungsstelle für Zahnräder und Getriebebau") for a given lubricant/material combination:

$$T_{int,perm} = T_{oil} + a\,T_{fl}\,. \tag{11.10}$$

The variables in Equation 11.10 are:

| | |
|---|---|
| $T_{int,perm}$ | permissible Integral Temperature, |
| $T_{oil}$ | oil bath temperature, |
| $a$ | constant $a = 1.2$, |
| $T_{fl}$ | flash (friction) temperature. |

This temperature has a constant value for a lubricant/material combination, regardless of the operating conditions. If this temperature is exceeded, there is scuffing damage to the gear pair (20% of the active tooth flank height of the pinion shows signs of wear).

If this calculation is carried out with the associated operating parameters for a production gearbox analogously to the calculations of permissible Integral Temperature for the gearwheel torque test bench, then the resultant Integral Temperature value $T_{int}$ must not exceed the permissible temperature value $T_{int,perm}$ determined with the test bench for scuffing-resistant operation. The following applies for scuffing resistance:

$$S_s = \frac{T_{int,perm}}{T_{int}} \geq 1.2\,. \tag{11.11}$$

The variables in Equation 11.11 are as follows:

| | |
|---|---|
| $S_s$ | scuffing resistance, |
| $T_{int,perm}$ | permissible Integral Temperature, |
| $T_{int}$ | Integral Temperature. |

The test method conforms to German standard DIN 51354 (FZG gearwheel torque test bench), i.e. at a constant circumferential speed, the stress is increased at intervals until scuffing occurs. Investigations [11.43] have however also shown that as circumferential speed increases, the tooth loading transmitted without wear decreases up to a certain speed. Yet a further increase in speed results in another increase in scuffing wear resistance. This phenomenon is probably explained by the familiar elasto-hydrodynamics of the lubricant layer formed. The primary requirements for a gearbox lubricant according to [11.48] are

- good FZG-performance, minimum power level 9 and
- good four ball tester welding load value, starts at 3000 N welding load (steel/steel material pairing).

The four ball tester welding load is determined in accordance with German standard DIN 51350. The load-bearing capacity is conditional on effective viscosity and additive action, and increases as four ball tester welding load increases and wear scar size decreases.

A suitable Extreme Pressure additive has a major impact on scuffing resistance, as can be demonstrated with the testing methods described. Scuffing resistance can in some cases be increased fivefold by the use of suitable additives [11.40].

## 11.3 Oil Supply and Oil Pumps

In automatic transmissions with various gear ratios and mechanical continuously variable transmissions (according to the definition given in Figure 1.2), the gearbox oil has several functions. It transmits power hydrodynamically in the torque converter or retarder, affects the friction coefficient profiles of the clutches, dissipates heat and lubricates the gears and bearings. It provides information and pressure energy for the actuation of valves and shifting elements.

### 11.3.1 Oil Supply

These varied tasks are mirrored in the various oil circuits. Figure 11.12 shows a simplified representation of oil supply for a conventional automatic transmission. The oil pump sucks oil from the oil sump and creates oil pressure. An oil filter purifies the oil of dirt particles. The main pressure has the highest level in the system. It is directly behind the pump and amounts to up to 20 bar in conventional automatic transmissions. All other pressures are derived from this main pressure.

Pressure control and distribution takes place in the hydraulic control unit. Pressure control and control solenoid valves reduce pressure to the extent necessary for different functions in the gearbox. Pressure distribution is the task of control valves and control solenoid valves. The hydraulic control unit is located in the oil sump and is controlled by means of electronic transmission control signals (TCU). Transmission signals, like rotational speeds or the sump temperature, are captured by sensors in the gearbox. Signal communication with the vehicle and engine occurs by means of an interface.

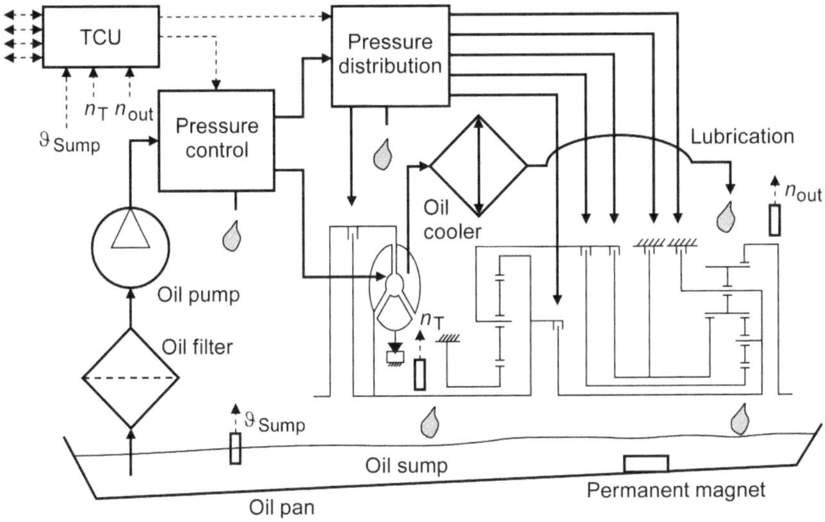


**Fig. 11.12.** Oil supply using a conventional automatic transmission as an example

During gear change, there is a variable shift pressure of 6–12 bar at the shifting element. After the shifting engagement, the shift pressure is replaced by the main pressure (pressure outside the gearshift system). The main pressure is regulated as a function of engine torque. Pressure in the torque converter lock-up clutch is also set according to the engine torque to be transmitted.

The torque converter is filled in a separate circuit. The latter also supplies the lubrication points, and usually the oil cooler is located within it. The lubricating oil must be distributed to all clutches, dynamic clutch pressure compensation chambers, bearings and gearwheels. The lubrication pressure is 3–6 bar and is available at the lubrication points. The oil flows back into the sump from the lubrication points of the gear sets and shifting elements as well as from the reflow points of the valves. There should be a sufficient amount of drain openings provided for this. The operating temperature of the oil measured in the oil sump is between –30°C and +130°C. The permanent temperature of the oil sump should be under 100°C.

### 1/ Oil Volume

The correct oil filling volume is important for the functioning and efficiency of the transmission. The determination of the oil filling volume (oil setting) is thus given a lot of attention in the development phase. In the case of conventional passenger car automatic transmission, the volume of oil is about 10 litres, while it is approximately 30 litres for commercial vehicle automatic transmissions.

The oil level must have sufficient distance to the rotating components for all relevant vehicle tilts and oil temperatures. Oil emission through the transmission ventilation system, foaming due to churning gearwheels and suction noises are all impermissible. The transmission should also be able to sustain a certain amount of over-filling (up to 0.5 litres for passenger car automatic transmissions) without such negative effects. Countermeasures could include baffle plates or oil compensation reservoirs. The oil setting is of particular importance for transmissions with integrated axle differentials (e.g. front-transverse transmissions).

### 2/ Oil Pan

Ground clearance is a significant requirement in designing the oil pan. Furthermore, the designer must also pay heed to noise sensitivity due to its diaphragm effect. Plastic oil pans are more advantageous than sheet metal oil pans for this reason and because of their lighter weight. The oil pan is sealed to the transmission housing by means of a surface seal (moulded seal or paper seal, see Section 11.5). A favourable distribution of pressure (pressure pattern) must be ensured. The suction point for the oil must be designed such that enough oil is taken in during acceleration, cornering and vehicle tilts.

Dry sump lubrication can also be encountered in case of transmissions built for special purposes, e.g. for military or motorsports, where a lot of ground clearance or an especially low-lying crankshaft is required. Here, the oil volume is not stored in the oil sump but in a separate oil reservoir under light pressure (about 1 bar). A suction pump conveys the returning oil from the flat oil sump to the reservoir.

### 3/ Oil Purification, Filters

Particles of various sizes and materials are introduced into the gearbox oil not only during manufacturing of the transmission but also from wear during its service life. Both automatic transmissions and mechanical continuously variable transmissions contain components with narrow, yet functionally essential clearances (valves, pumps) and are therefore sensitive to oil impurities. An oil filter purifies the oil of dirt particles that may cause wear. Depending on the type of filter circuit, oil filters are designated as main flow or partial flow filters. The entire quantity of oil flows through a main flow filter. In addition to that, a partial flow filter can be utilised to filter out finer particles. Only about 30% of the oil flows through it. We also differentiate between suction and pressure filters. Suction filters are located before and pressure filters after the oil pump.

Suction filters represent a compromise between the amount of separation and sufficient flow rate (pressure loss), especially in the case of cold starting. The fineness of a filter is often given in terms of mesh size or pore size. The suction filter of the conventional passenger car automatic transmission ZF 6 HP 26 (Figure 12.25) has a fleece filter with a pore size that captures particles larger than 60 μm. The statistical retention value is of ultimate relevance. It takes into consideration both irregular pore sizes as well as irregular dirt particle shapes. The $\beta$-value defines this degree of retention. $\beta_{60} = 100$ means that, of 100 particles of 60 μm size, only one particle is allowed through.

In the case of the transmission shown in Figure 12.25, the oil filter and suction tube are integrated into the plastic oil pan. Permanent magnets are placed in the oil pan to attract metallic particles. They are found at a location where there is no oil flow so that they are not washed away again.

#### *4/ Oil Cooling*

In case of transmissions with torque converters (AT, CVT), slipping wet master clutches (DCT, CVT) or integrated retarders in commercial vehicles, the surface of the housing is not sufficient for heat dissipation. The gearbox oil is led to an oil cooler, usually an oil-water cooler. This can be fastened on the vehicle side or directly on the transmission. Oil-air coolers are seldom used.

### 11.3.2 Oil Pumps

Oil pumps are used in various types of automotive transmissions. They always have the function of supplying oil in defined quantities and with a defined pressure as lubricating or control oil. Ideally, transmission designs dispense with additional oil pumps whenever possible. We classify oil pumps according to use and pressure range as:

- pure lubricating oil pumps: up to about 6 bar:
  - commercial vehicle manual transmissions with injection lubrication,
  - transfer boxes etc.
- control oil pumps:
  - also take on the task of lubricating oil supply,
  - all-wheel drive locking systems: up to about 20 bar to supply actuators and clutches,
  - automated manual transmissions (AMT): up to about 20 bar to supply transmission and clutch actuators,
  - automatic transmissions with various gear ratios (AT, DCT): up to about 20 bar,
  - mechanical continuously variable transmissions: up to about 70 bar and
  - hydrostatic continuously variable transmissions: up to about 450 bar.

In automatic transmissions with various gear ratios and mechanical continuously variable transmissions, the oil pump is a central assembly unit. It must fulfil the following requirements [11.21] that are also applicable by analogy to other transmission types:

- unrestricted functionality from –30 to +150°C oil temperature (no damage should occur at – 40°C oil temperature),
- failure safety in case of oil contamination equivalent to the defined filter fineness,
- reliable functioning across the entire engine speed range,

- cavitation safety, even at high engine speeds,
- ensure the oil peak demand during shifting (reference value for conventional passenger car automatic transmissions: approximately 50 *l*/min) and
- critical operating points with respect to undersupply and therefore high temperature also becomes relevant to the design:
  - gear engagement at idling speed and hot oil (reference value for conventional passenger car automatic transmissions: about 10 *l*/min),
  - rollout shifts at low speeds and hot oil and
  - torque converter lock-up clutch control at high torques, low speeds and hot oil.

We distinguish pumps according to the type of volume displacement as rotating pumps or stroke pumps. Depending on the type of displacer element, we can further classify them into gear pumps, vane pumps and piston pumps (Figure 11.13). In the case of fixed displacement pumps, the volumetric flow increases linearly with speed, while the flow rate is variably controlled in the case of variable displacement pumps. The displacement volume $V$ indicates the flow rate per rotation ($cm^3$/R) respectively, in the case of piston pumps, piston displacement.

### *External Gear Pump*

External gear pumps consist of only a few components. Basically, these are the input shaft, both gearwheels, the pump housing and the pump cover (Figure 11.14).

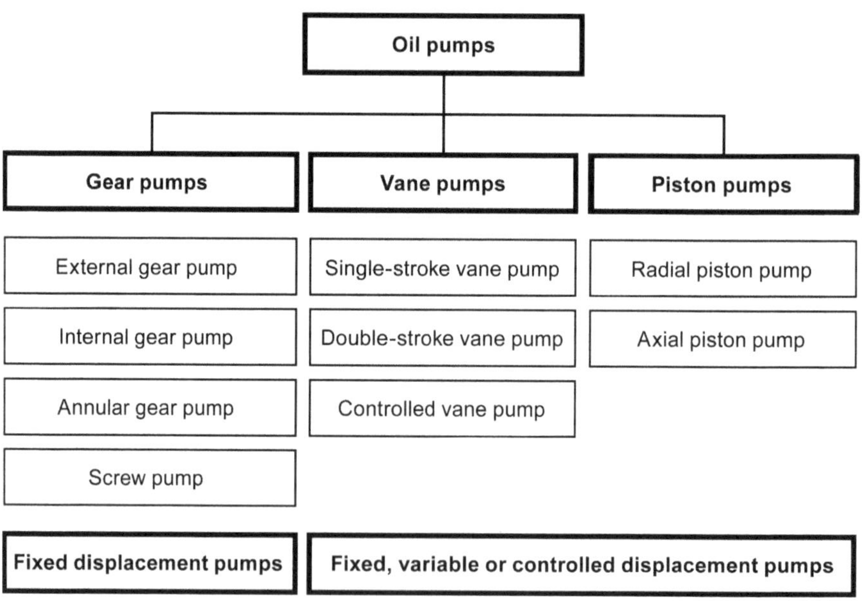


**Fig. 11.13.** Classification of oil pumps for automotive transmissions

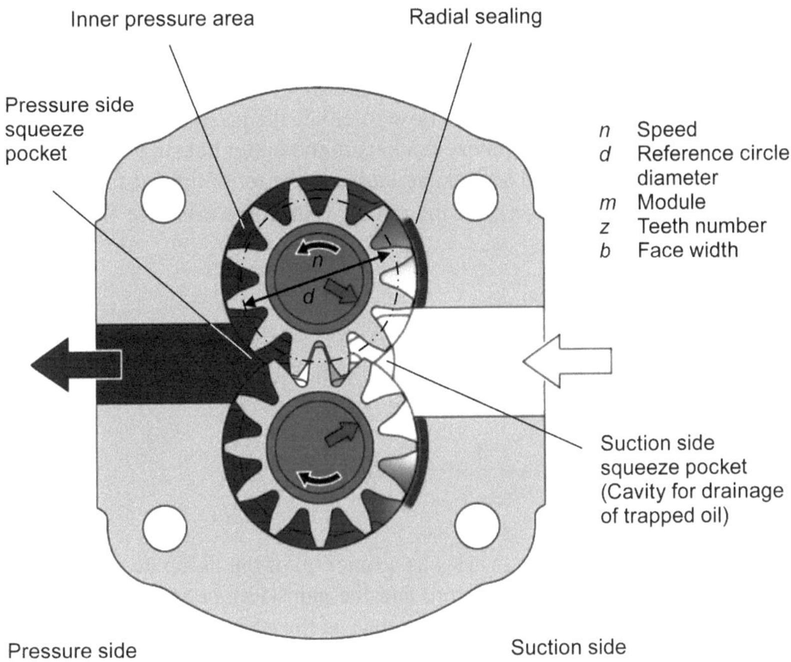


**Fig. 11.14.** Structure of an external gear pump

The tooth gaps are filled on the suction side with oil conveyed along the housing towards the pressure side.

In gear pumps, the displacement volume corresponds to the sum of the tooth gap volumes. With the face width $b$ and the module $m$ we can approximate the tooth gap volume of an external gear pump using the circular ring surface:

$$V \approx \pi\, d\, m\, b\,. \tag{11.12}$$

With two equally large gearwheels and $m = d / z$ as well as a speed $n$ we obtain the theoretical flow rate as

$$\dot{V}_{\text{th}} = 2\, V\, n = \frac{2\, \pi\, d^2\, b\, n}{z} \tag{11.13}$$

or as a unit equation

$$\dot{V}_{\text{th}}\ (l/\text{min}) = \frac{2\, \pi\, d^2\ (\text{mm}^2)\, b\, (\text{mm})\, n\, (1/\text{min})}{z\, 10^6}\,. \tag{11.14}$$

External gear pumps have three sealing positions. The faces of the gearwheels seal axially by means of the gap appearing on face end, the tooth tips seal radially against the housing. The engaged tooth flanks make the third sealing position. The oil is enclosed in the area of the meshing teeth and squeezed by the cell volume, which narrows down to the size of the tooth backlash.

The quantity of the squeezed oil can flow to either the pressure or suction side by means of specially designed squeeze pockets in the pump housing and/or pump cover. Squeeze oil losses as well as leakage losses at the tooth tip and face gap are designated as volumetric losses $\eta_{\mathrm{vol}}$ of the pump. With it the effective flow rate is calculated as follows:

$$\dot{V} = \dot{V}_{\mathrm{th}} - \dot{V}_{\mathrm{Leakage}} = \dot{V}_{\mathrm{th}}\ \eta_{\mathrm{vol}}\,. \tag{11.15}$$

There are also hydraulic/mechanical oil shear and friction losses $\eta_{\mathrm{h,m}}$. The required input power $P$ of the pump is given by

$$P = \dot{V}_{\mathrm{th}}\ \Delta p \frac{1}{\eta_{\mathrm{h,m}}} = \dot{V}\ \Delta p \frac{1}{\eta_{\mathrm{h,m}}\ \eta_{\mathrm{vol}}} = \dot{V}\ \Delta p \frac{1}{\eta_{\mathrm{tot}}}\,. \tag{11.16}$$

The total efficiency $\eta_{\mathrm{tot}} = \eta_{\mathrm{h,m}}\ \eta_{\mathrm{vol}}$ varies as a function of the pump design, the dynamic viscosity of the oil, oil temperature and the gap width between 0.6 and 0.9. To keep leakage losses small, the axial clearance of the gearwheels should be kept as small as possible. Gap compensation is defined as design measures (e.g. spring pressurisation) intended to keep the gap size as small as possible and independent of pressure deformations.

Current oil pump designs for automatic transmissions with various gear ratios and continuously variable transmissions include:

1/ internal gear pump (crescent pump),
2/ annular gear pump (gerotor),
3/ double-stroke vane pump,
4/ controllable vane pump and
5/ suction-throttled radial piston pump.

### *1/ Internal Gear Pump (Crescent Pump)*

Crescent pumps are the standard design for conventional automatic transmissions as they are short, allow for a coaxial assembly around the torque converter impeller hub and are suited to high rotational speeds (Figure 11.15, see also the assembly example in Figure 12.25). The suction $S$ and pressure $P$ sides are sealed against each other by a crescent (radial gap tooth tips), the gearwheel faces (axial gap) and the meshing teeth. Oil is delivered and removed axially via the pump housing or the pump cover. The displacement principle is comparable to that of the external gear pump. The rotating tooth gaps transport the oil from the suction side to the pressure side. Here too, corresponding squeeze pockets must be provided. The pump should be designed such that the pump gears are as balanced as possible with respect to axial pressurisation and are thus free of axial force.

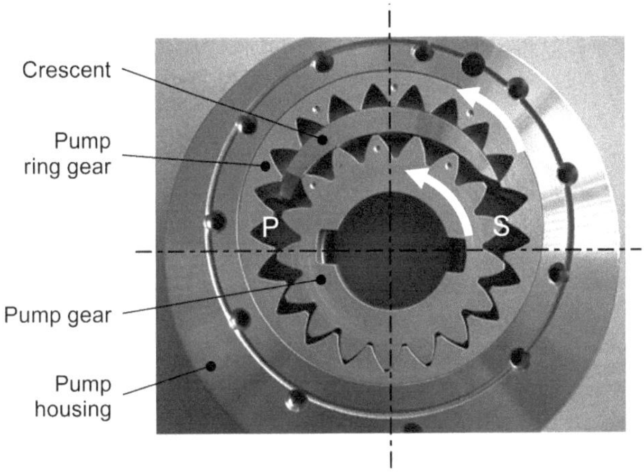


**Fig. 11.15.** Crescent pump. Pump gear as the driving and pump ring gear as the driven gearwheel

The displacement volume of crescent pumps can be approximated to

$$V \approx \frac{\pi}{4} b \left(d_{\mathrm{a1}}^2 - d_1'^2\right). \tag{11.17}$$

Here, $b$ is the face width, $d_{\mathrm{a1}}$ the tip circle diameter of the pump gear and $d_1'$ the reference circle diameter of the pump gear minus twice the addendum of the pump ring gear [11.36]. With this we obtain the theoretical flow rate:

$$\dot{V}_{\mathrm{th}}\ (l/\mathrm{min}) \approx \frac{\pi}{4} b\,(\mathrm{mm}) \left(d_{\mathrm{a1}}^2 - d_1'^2\right)(\mathrm{mm}^2)\, n\,(1/\mathrm{min}) / 10^6 . \tag{11.18}$$

The crescent pump of the passenger car automatic transmission ZF 6 HP 26 (Figure 12.25) supplies approximately 20 $\mathrm{cm}^3$/R. Excess pressure oil at higher speeds is recycled via a flow control valve directly to the suction side of the pump.

### 2/ Annular Gear Pump (Gerotor)

The annular gear pump (often called Eaton pump or gerotor) is also a common pump type in automatic transmissions with various gear ratios (Figure 11.16). It can also be attached very space-efficiently to the converter impeller hub. The gerotor is easier to manufacture than the crescent pump but is not as efficient. However, gerotors are not as susceptible to cavitation at higher speeds. The displacement volume of the annular gear pump can be approximated as

$$V \approx \frac{\pi}{4} b \left(d_{\mathrm{a1}}^2 - d_{\mathrm{f1}}^2\right), \tag{11.19}$$

where $b$ is the rotor width, $d_{\mathrm{a1}}$ the tip circle diameter of the inner rotor and $d_{\mathrm{f1}}$ the root circle diameter of the inner rotor.

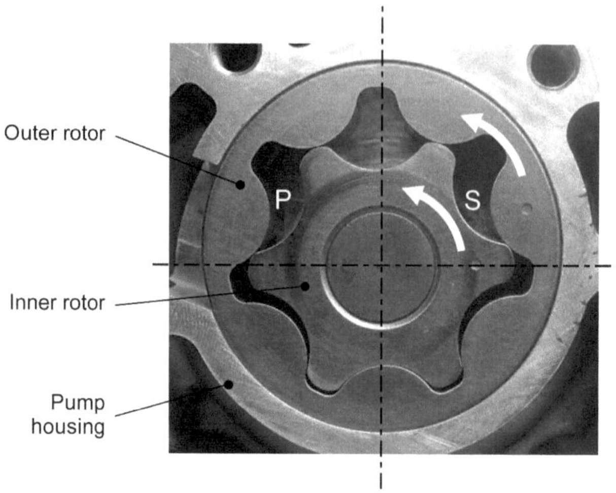


**Fig. 11.16.** Gerotor. Inner rotor as the driving and outer rotor as the driven gearwheel

From this we can obtain the theoretical flow rate:

$$\dot{V}_{\mathrm{th}}\ (l/\mathrm{min}) \approx \frac{\pi}{4}\, b\ (\mathrm{mm}) \left(d_{\mathrm{a1}}{}^{2} - d_{\mathrm{f1}}{}^{2}\right) (\mathrm{mm}^{2})\ n\ (1/\mathrm{min}) / 10^{6}\,. \tag{11.20}$$

### *3/ Double-Stroke Vane Pump*

Rather than a coaxial arrangement on an input shaft, vane pumps lend themselves to an “off axis” assembly as a compact pump unit. They are acoustically more sensitive than crescent pumps, but they are more efficient. Their essential advantage it that they have two strokes, i.e. suction and pressure occurs twice for every rotation (Figure 11.17). This pump type is used, for example, in continuously variable transmissions like the Audi Multitronic (Figure 12.30) [11.46].

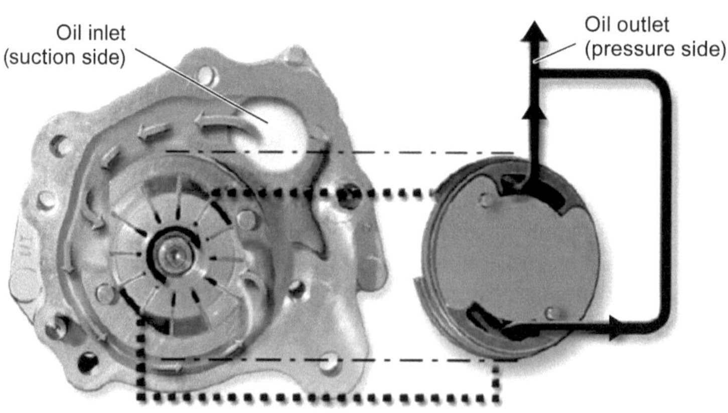


**Fig. 11.17.** Double-stroke vane pump [11.46]

The two pump halves are, hydraulically, two separate pumps. One pump half can therefore be switched off at higher engine speeds, reducing the flow rate. For example the double-stroke vane pump for the continuously variable transmission Mercedes-Benz Autotronic (Figure 12.31) is designed to be switchable.

#### 4/ Controlled Vane Pump

In the case of controlled vane pumps, the flow rate can be changed by varying rotor eccentricity [11.33]. The cam ring is turned around a pivot by means of a control pressure. In this way, rotor eccentricity is adjusted, and with it, the flow rate.

Transition from the comparatively simple fixed displacement pump to switchable or controlled pumps should be carefully considered with respect to cost and use. This is especially true of automatic transmissions with various gear ratios (AT and DCT) with their comparatively low levels of pressure. Furthermore, with large overall gear ratios in real vehicle operation, the time intervals of higher pump speeds are relatively slight. Hydraulic control systems of mechanical continuously variable transmissions have a higher level of pressure. To limit pump input power, switchable or controlled pumps can there be found in actual designs.

#### 5/ Suction-Throttled Radial Piston Pump

In the radial piston pump, pistons arranged in a cylinder block and impelled by an eccentric tappet perform reciprocating motions. In the suction-throttled radial piston pump, the suction windows are closed by the movement of the pistons according to geometrical conditions. From a boundary speed (i.e. piston speed), the filling time is no longer sufficient to fill the entire piston chamber. Volumetric flow then remains almost constant at a further increasing rotational speed.

The pump of the continuously variable transmission ZF CFT 30 (Figure 12.32) is designed for pressures of up to 70 bar. The required flow rate is proportional to the engine speed up to an engine speed of 2000 l/min. At higher speeds, it is held at a constant level of 28 *l*/min [11.54].

Suction-throttled radial piston pumps are demanding in the passenger car market with respect to mastering the noise emissions. In the design shown in Figure 12.32, the pistons convey oil into special damping chambers that are surrounded by a steel ring that has different rigidities in radial and axial directions.

### 11.3.3 Detail Questions

#### 1/ Efficiency

To obtain high oil supply efficiency, the following points should be borne in mind:

- minimal leakages, especially in the hydraulic control unit,
- minimal oil viscosity,

- main pressure as small as possible,
  - torque-contingent main pressure,
  - if necessary, separation into high and low pressure circuit,
- examine alternatives to fixed displacement pumps,
- pump concepts with high efficiency $\eta_{tot} = \eta_{vol}\, \eta_{mech}$,
  - internal leakage should be reduced to a minimum,
- pumps should be designed for the smallest possible flow rate.

### 2/ Noise Behaviour, Pulsation and Cavitation

When designing pump gears, load plays a subordinate role. Besides efficiency, acoustics are also a main criterion. The causes of gear pump noise can be distinguished as mechanical or hydraulic excitation mechanisms [11.47].

Mechanical causes include gear and bearing excitation. For this reasons, oil pumps often have plain bearings. Hydraulic causes include volumetric flow pulsation and cavitation. When fine-tuning the gears, the degree of irregularity (volumetric flow pulsation) $\delta$ should be taken into account. This is defined as

$$\delta = \frac{\dot{V}_{max} - \dot{V}_{min}}{\dot{V}_{max}} . \tag{11.21}$$

Typical values for the degree of irregularity are:

- external gear pump approx. 16% ,
- internal gear pump approx. 3% .

As the rotational speed increases, so does the tendency towards cavitation. High expansion speeds on the suction side of the pump causes gas bubbles to form due to reduced pressure. The gas bubbles then implode on the pressure side, leading to noises, pressure pulsation and wear. One countermeasure against cavitation is suction flow charging. In this case, compression energy from too much supplied oil is recycled to the suction side. In terms of the design, the cavitation tendency has to be reduced with minimal flow speeds in the suction channel. Target values are below 2 m/s. Large channel cross-sections, soft transitions and straight channel guides contribute to this end.

### 3/ Pump Damage

Due to their functionally necessitated small clearances, oil pumps are especially sensitive to abrasive particles. During manufacturing and assembly, provisions must therefore be made against the entry of dirt into the oil.

Pump systems should be rendered more resilient against dirt particles by design and choice of materials. To this end, the following can be helpful:

- friction partners in the tribosystem with a large hard/soft distance and
- separating layers with poor heat transfer prevent local bonding and thus scuffing damage.

#### 4/ Ancillary Pumps

In vehicles with engine start/stop technology, the engine is automatically shut off every time the vehicle stops in order to save fuel. When the engine is not running, the oil pump is not impelled by the automatic transmission. Therefore, there is no oil pressure to activate the shift elements. The transmissions of such vehicles can be equipped with electrically operated ancillary pumps so that, after the engine is started again, there is no down time before the oil pressure builds up again and the shift elements are activated. These ancillary pumps take over pressure oil supply when the internal combustion engine is off. The necessity of an electrically run *start/stop pump* depends on the required oil quantity and the acceptable down time. The driver's threshold of perception of down time is about 0.3 s.

One approach towards consumption reduction is to use an electrically operated *booster pump*. Here, the mechanically impelled oil pump is designed for optimal efficiency in normal operation. The switchable booster pump covers oil peak demands. This approach can be combined with the aforementioned start/stop pump.

If a vehicle, the transmission of which includes wet clutches (AT, DCT), must also be able to be tow-started, a pump that is impelled by the output is required to close the clutches. These are designated as *secondary* or *tow pump*. This requirement is customary in military vehicles.

For normal towing of vehicles with automatic transmissions with various gear ratios and mechanical continuously variable transmissions, no tow pump is necessary at low speeds in position N. The oil spray is sufficient to supply the bearings and gears with lubricating oil.

## 11.4 Gearbox Housing

The gearbox housing houses all the components of a gearbox. The following requirements have to be taken into account when designing the housing:

- absorb the acting operational forces and moments,
- guarantee the exact position of the shafts and gearwheels relative to each other in the various operating states,
- ensure good heat conduction and radiation,
- insulating and damping the gearbox noises,
- easy to fit and remove and
- rigid layout and good strength characteristics, combined with low weight.

### 11.4.1 Gearbox Housing Design

The housing can be designed as a classical *trough housing* or as a split housing. Split housings are divided into *end-loaded* and *top-loaded* depending on the position of the parting plane or of the shaft. The types of housing are listed in Table 11.4, with their advantages and disadvantages.

**Table 11.4.** Types of housing. Advantages and disadvantages

| Type of housing | Advantages | Disadvantages |
|---|---|---|
| **Trough housing** 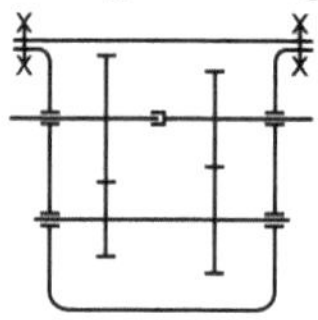 | – Bearing bores easy to produce<br>– Precise production in one clamping operation | – Unfavourable fitting conditions<br>– Fitting cannot be automated |
| **End-loaded housing** 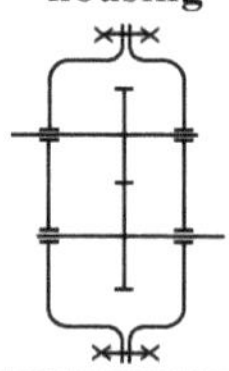 | – Rigid housing<br>– Easy to fit<br>– Fitting can be automated | – Expensive production equipment<br>– Critical bores in two clamping operations |
| **Top-loaded housing** 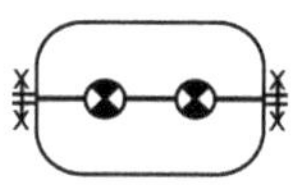 | – Precise production in one clamping operation<br>– Very easy to fit<br>– Fitting can be automated | – Reference surface expensive to machine<br>– Not very rigid |

Vehicle gearboxes are predominantly of end-loaded design. The clutch bell housing can be flange-mounted on the gearbox (3-part housing) or integral to the housing. The latter configuration results in a 2-part housing, which involves lower manufacturing costs.

Due to the ever increasing number of gear ratios, it may be necessary to provide gearbox shafts with a third bearing point to help reduce deflection. This can be designed as additional intermediate housing or as bearing plate that is screwed separately onto a housing component (see Figure 11.1).

As a result of the demands to improve performance and at the same time reduce weight, motor vehicle gearbox housings are now primarily made of die cast light alloy. Die cast aluminium is a good compromise between cost and weight. In order to avoid having thicker walls than cast iron for the same strength and rigidity, the housing has ribs. These ribs increase rigidity and strength, and at the same time reduce sound emission from the gearbox housing.

Besides known aluminium alloys (e.g. AlSi9Cu3: 2.75 kg/dm$^3$), special magnesium alloys are also used. These are optimised with respect to creeping and contact corrosion. Due to their lower density (about 1.8 kg/dm$^3$), the weight of the housing can be reduced. This advantage is partially lost because of necessary additional stiffening measures (wall strength, ribbing).

Section 14.1 concerns FEM-based calculation methods for optimising housings with respect to stiffness, weight, noise and structure. [11.30] deals with knowledge gained concerning gearbox housing designing with respect to sealings. Information on machining technology of housings is given in Section 16.2 "Process Chains for Cast Part Processing". Vehicle transmission housing design was systematically investigated [11.55] using the Finite Element Method, and design guidelines for die cast light alloy housings derived:

- housing ribs should always run in the direction of the principal normal stress, thus reducing the tensile stresses which are dangerous for casting by enlarging the supporting cross-section,
- ribs on bearing walls should be arranged in a star shape from the bearing bores, dimensions of the ribs relating to the wall thickness $t_W$:
  Height $h = (3–4)\ t_W$,
  Width $b = (1–2)\ t_W$,
- reverse gear bearing should be strengthened by high ribs ($h = (3–5)\ t_W$) mounted at an angle of 0° or 90°,
- strengthen longitudinal walls with wide ribs ($b = (1–2)\ t_W$) with large radius of curvature $r = 1.2\ t_W$, mount ribs at an angle of 45° to the longitudinal axis of the gearbox,
- heavy ribs with spacing ($s = (5–15)\ t_W$) achieves good acoustic transmission characteristics.

Figure 11.18 shows a commercial vehicle end-loaded gearbox housing made of die cast aluminium with flange-mounted power take-off.

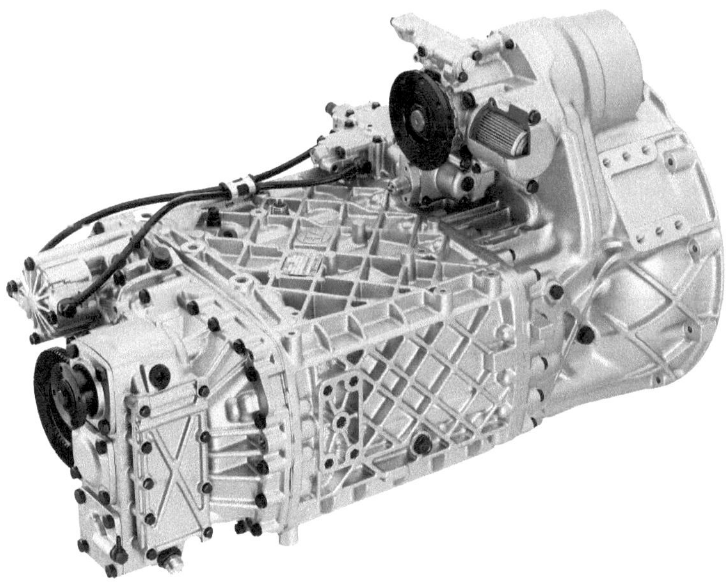

**Fig. 11.18.** Gearbox housing of the ZF Ecosplit 16 S 151 commercial vehicle gear unit; end-loaded housing

### 11.4.2 Venting Gearboxes

Vehicle transmissions are oil-lubricated. The seals around the input and output shafts and the gearbox housing are in most applications rotary shaft seals. These sealing rings are designed to seal unpressurised fluids. The amount of air contained in the gearbox means that temperature fluctuations in a completely enclosed gearbox would cause pressure fluctuations. Because of increasing sealing edge contact pressure of the rotary shaft seal, overpressure in the housing can lead to increased heat build-up, wear of the sealing edge, and thus to leakage. As the negative pressure adjusts, it can cause air, water and dirt to be sucked in at the seals because of the falling sealing edge contact pressure.

To prevent the above problems arising at the seals, gearboxes are fitted with vents to enable the pressure to be equalised by air flowing in and out. For reasons of operational reliability and environmental pollution, there must be no emission of oil, oil foam, oil vapour or oil mist from the vents. Water, dust and dirt must be prevented from entering, and the vents must be kept clear.

#### *1/ Operational Modes of a Breather System*

The laws of thermodynamics dictate two different operational modes for a breather system, because of the temperature fluctuations (Figure 11.19):

- Venting:
  When there is excess pressure, air flows out of the housing through the breather system into the environment, because of the rising temperature in the gearbox housing. The flow of air can discharge oil, oil foam, oil mist and oil vapour.
- Aerating:
  When there is negative pressure, air flows from the environment through the breather system into the housing, because of the falling temperature in the gearbox housing. The flow of air can draw in water, moisture, dust and dirt.

When non-contact seals are used, e.g. labyrinth seals at the gearbox shaft exits, air propulsion of the labyrinth can lead to much higher air flows through the breather system.

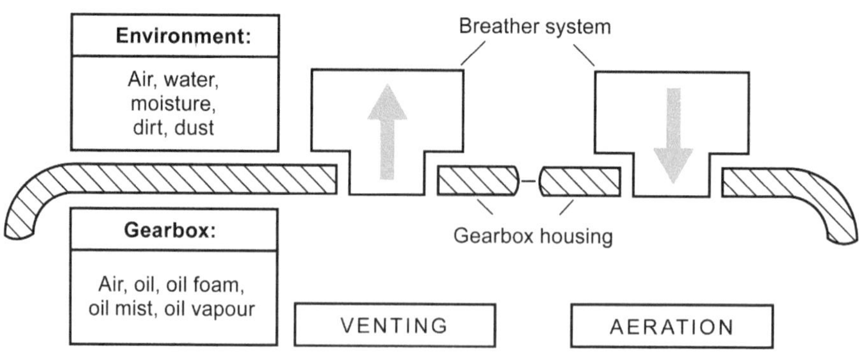


**Fig. 11.19.** Operational modes of a breather system: venting and aeration [11.24]

Non-contact seals should be designed so that they do not accelerate the air flow. Provided this requirement is met, the air flow rate through the breather is restricted to that caused by temperature fluctuations.

### 2/ Designing Breather Systems

Figure 11.20 shows a typical breather system design. The design of the interior is of particular importance for deflecting oil spray. Many breather systems therefore have a spray guard on the inside. The flow of air is guided through tube extensions, and can be diverted and restricted in several places. Breather systems with valve inserts can also be used. The filters used are made of various materials or forms of material (e.g. flat wire mesh, foam, sintered bronze, filter fabric). Externally, there is usually a cap to protect against water splashing. There may also be an externally fitted filter for extreme operating conditions.

Most breather systems are made of metal. The various parts of the system are made by machining and forming. For complex designs, it can be more suitable to use plastic injection moulding. Considerations of weight and cost also play a major role when selecting the material. Breather systems can generally be fitted externally. There are some special designs in which the breather is combined with other functions.

### 3/ Structural Elements of Breather Systems

The structural elements listed below are used in current breather systems. The breather system as a whole is made up by mounting these individual elements in series.

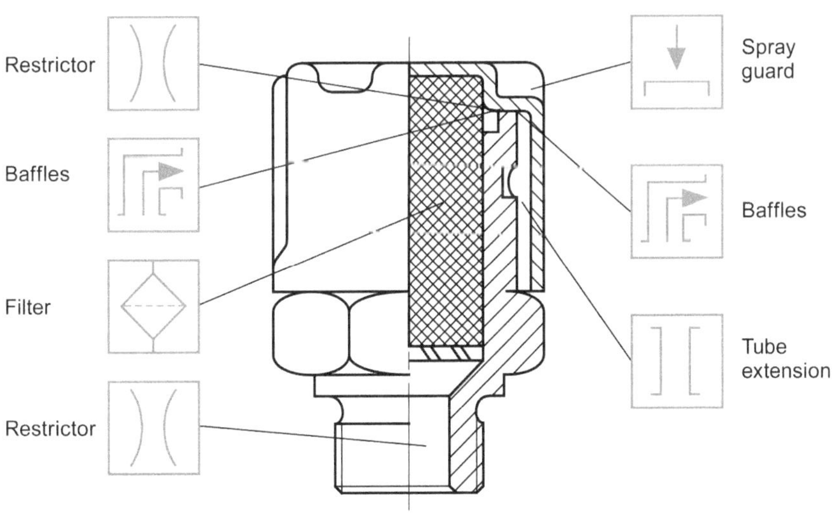


**Fig. 11.20.** Structural elements of a breather system based on the *Hunger 1019 R1/8", shape A*

*External spray guard:*
Caps, covers and external deflector surfaces are used to prevent water spray and dirt directly entering the breather system.

*Filter:*
The function of a filter in the breather system is to retain the dirt in incoming air. The oil mist and oil foam carried by the exhaust air is also deposited on the filter surface. A contaminated filter can lead to an increase in pressure.

*Vent valve:*
Air is allowed to escape from the gearbox only at a certain pressure, controlled by spring tension or by the weight of a cover. No air can flow in through the vent valve.

*Aerating valve:*
A spring arrangement with a sealing plate is used to ensure that air can enter the gearbox from outside only when a given negative pressure is reached. Air cannot escape through the aerating valve.

*Baffles:*
The baffles in the flow path prevent oil spray from being directly emitted, and water spray from directly entering.

*Labyrinth:*
A labyrinth is made up of various air flow baffles and channels. This has the desired effect of lengthening the flow path in the breather system, and thus increasing the accretion of oil on the internal surface of the labyrinth. This oil can then be returned to the gearbox.

*Tube extension:*
A tube extension is often necessary to connect the various structural elements. Not only round cross-sections, but also rectangular, square or circular cross-sections can be used. For example, the quantity of oil spray reaching the top end of a vertical tube can be decreased using this tube.

*Restrictor:*
A restrictor in the flow restricts the flow cross-section and thus increases the speed of flow. It is often difficult to distinguish between a restrictor and a tube extension.

*Internal spray guard:*
Spray guards are fitted to the inside of the breather system to prevent oil spray directly entering the breather system.

### 4/ Design Guidelines for Breather Systems

Comprehensive bench tests in [11.24] have been carried out to establish the ideal design for various operating conditions (presence of oil spray, oil foam, water and dust). Figure 11.21 shows a possible overall structure of a breather system for these operating conditions.

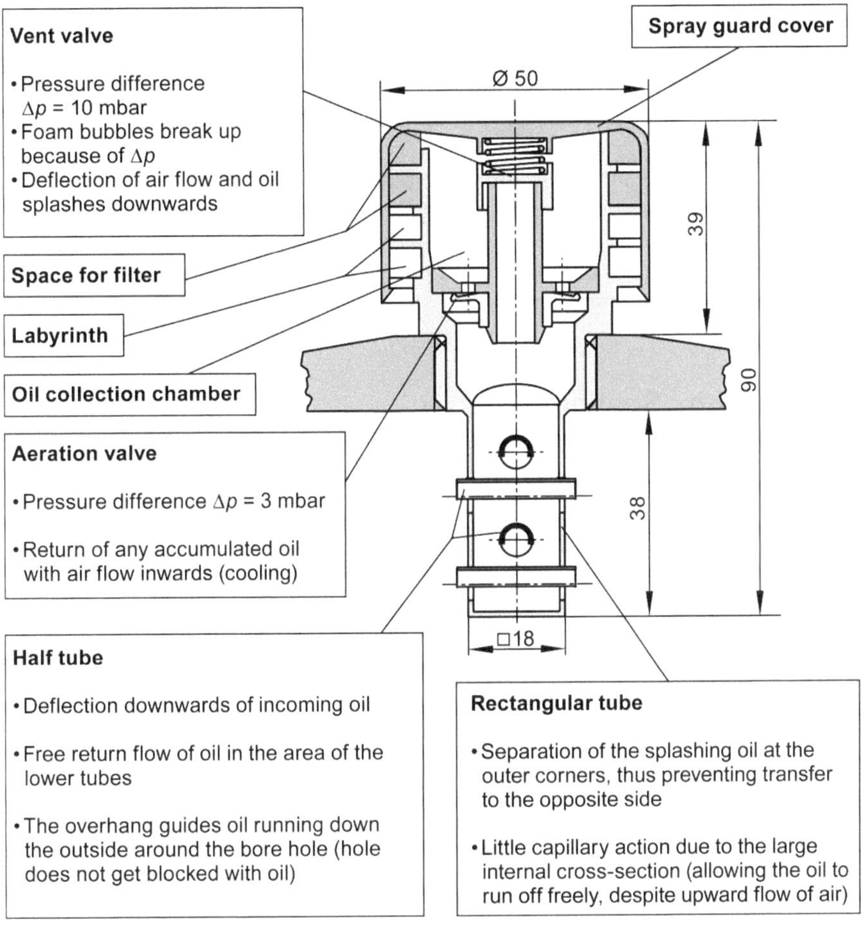


**Fig. 11.21.** Breather system recommendation for different operating conditions [11.24]

If oil splashing against the inner side of the breather cannot be avoided, then the oil spray should be separated from the exhaust air at the inlet on the inner side of the breather. The type of breather system proposed, comprised of an intruding rectangular tube with several half tubes, offered the best characteristics for preventing oil escaping when oil splashes against the inner side of the breather.

When there is oil foam on the inner side of the breather combined with an outward flow of air, there is no simple means of preventing an oil leak. In the double valve type of breather, the oil foam emitted through the vent valve, which has dispersed at the vent valve, is held in the breather. When air is drawn in, the oil held in the breather is fed back inside through the aerating valve, thus acting to reduce the oil leakage.

Where the outside of the breather system is subject to external water splashing, the inlet point must not have any small orifices. Especially when there is water running down the outside of the breather system, there is a danger of the orifices being blocked by the retained water. Any incoming flow of air will draw large quantities of water into the gearbox in these circumstances. Special filter units are a suitable means of preventing dust from entering the gearbox. The filter elements should be fitted in the external upward air intake, but not directly at the point of entry.

The decisive factor for the performance of the breather system is its location in the gearbox. When assessing or designing a breather system, the gearbox system and its environment as a whole must be taken into account. There is no panacea. But it is possible to define the preferred parameters for a breather system on the basis of the information provided.

## 11.5 Gearbox Sealing

The demands of service life and environmental considerations have made reliable, durable gearbox sealing a major factor. If a seal fails, the repair costs amount to many times the cost of the seal.

There are numerous points to be sealed in vehicle transmissions, such as shaft input and output seals, housing joints, selector shaft output, and speedometer drive. There are three types of seal for these purposes:

1/ seals for static components (e.g. flat gaskets),
2/ seals for rotating components (e.g. rotary shaft seals) and
3/ seals for reciprocating round components (e.g. grooved rings).

### 11.5.1 Seals for Static Components

Radial seals for static components are normally O-rings, radially compressed (Figure 11.22a). Static radial seals (flange seals) can be in the form of O-rings or surface seals. O-ring type flange seals are axially compressed (Figure 11.22b). O-rings are usually made of nitrile butadiene rubber (NBR).

#### *Guidelines for Installing O-Rings:*

- compress as necessary (10 to 20% of the cord thickness),
- use largest possible cord diameter, to compensate for relaxation and production variances,
- avoid twisting radially fitted O-rings when installing,
- the surface roughness at A, B and C (Figure 11.22) depend on the operating conditions; further details will be given in the manufacturers catalogues,
- preferably fit O-rings in the front face to facilitate disassembly (Figure 11.12b),
- use lead-in chamfers to avoid damage when installing radially fitted O-rings and
- use back-up rings to prevent gap extrusion at high pressures.

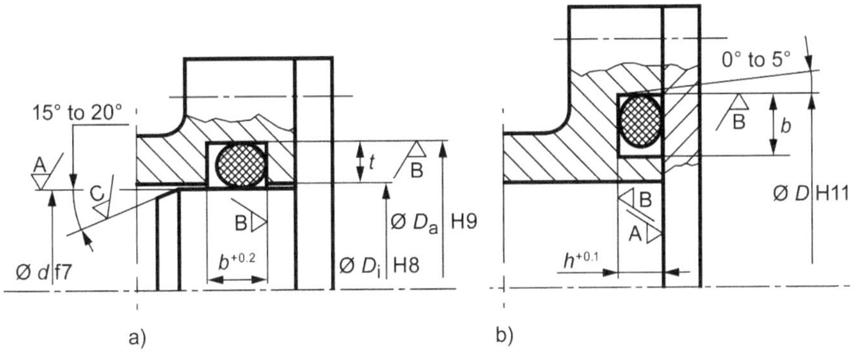


**Fig. 11.22.** Static components sealed with O-rings. *a* Radial static seal (groove in outer part), O-ring radially compressed; *b* axial static seal (flange seal), O-ring axially compressed

Surface seals are used to seal the parting planes of gearbox housings and covers that are bolted together (flange seals) (Figure 11.23). Together with the flanges and the bolted connection, the surface seal forms a *sealing joint*.

Surface seals can be pre-formed seals (*flat gaskets, metal bead gaskets*) or unformed seals (*sealing compounds*). Sealing compounds assume the required shape only when applied to the sealing faces. Flat gaskets have to change shape when fitted to the sealing faces. This ensures microscopic matching of the seal surface to the sealing faces, and closes the pores in the sealing material. Usual materials are:

- *flat gaskets* (soft gaskets):
  - cellulose fibre seals (e.g. paper seals, sealing paperboard),
  - fibre reinforced seals (e.g. aramide fibres),
- *sealing compounds* (liquid seals):
  - chemical curing (e.g. anaerobic sealing compounds, silicones),
  - non-curing (e.g. glycol compounds).

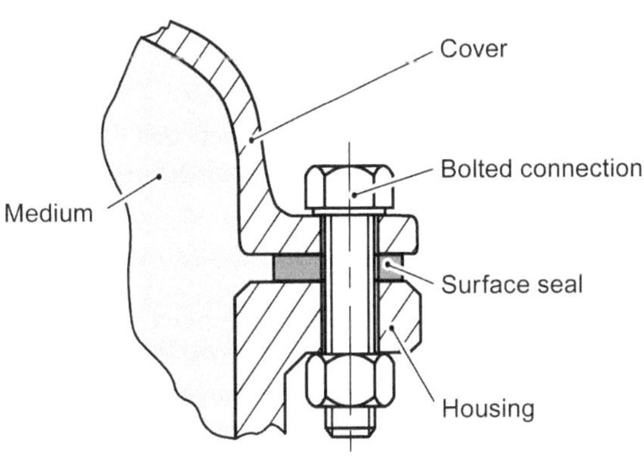


**Fig. 11.23.** Surface sealed joint

High demands are made on surface sealing joints of light-weight housings. There are considerable stresses on sealing joints in vehicle transmissions because of the high power density. As well as providing a seal, the surface sealing joint must provide a suitable mechanical bond between the parts of the housing so that force and torque can be transmitted. Together with the housing, the sealing joint must be made as stiff as necessary. There must be no excessive deformation in operation. Deformation generates microscopic relative movements in the sealing faces, which impair the function of the gearbox if they are excessive.

The sealing joint must be capable of withstanding the mechanical stresses arising throughout the life of the gearbox. There must be no breakdown due to loss of pre-tension in the bolted connections (relaxation in service), or break-up of the sealing material, or surface wear to the sealing flange [11.22, 11.25, 11.28–11.29].

### *Guidelines for Using:*

### *1/ Flat Gaskets:*

- keep within the recommended contact pressure range, which is normally approximately 2 to 50 $N/mm^2$; differential pressure across the seal usually approximately 1 bar (up to 50 bar in the case of AT or AMT control unit housings),
- ensure adequate contact pressure,
- ensure contact pressure between the tensed up parts is as even as possible,
- select the flat gasket to suit the macroscopic characteristics of the flange (e.g. undulations); the surface coating of the gasket adapts itself to the microscopic characteristics of the flange (flange roughness),
- sealing joint design with operational fatigue strength; avoid excessive thermal and dynamic stressing, to prevent the seal flowing (relaxation in service).

### *2/ Sealing Compounds:*

- check surfaces before fitting (ensure the surfaces to be joined are clean),
- ensure gaps are bridged as far as possible,
- provide lifting screws for removal,
- allow time for de-aeration when fitting and for curing after fitting,
- there must be a minimum contact pressure in the flange joint between the bolts to prevent permanently plastic (non-curing) sealing material being flushed out.

### *Determining the Contact Pressure in the Sealing Joint*

It is very important for the designer to know the distribution of contact pressure in the sealing joint of a housing seal, since it is an important indicator of the quality of design and the selection of sealing material. There must be a minimum contact pressure on a closed track inside the sealing face to achieve a seal. Contact pressure should ideally be as evenly distributed as possible.

| Pressure pattern | Characteristic feature |
|---|---|
|  | Very **even pressure pattern** with high values in the middle between the bolts. "Very good". |
| 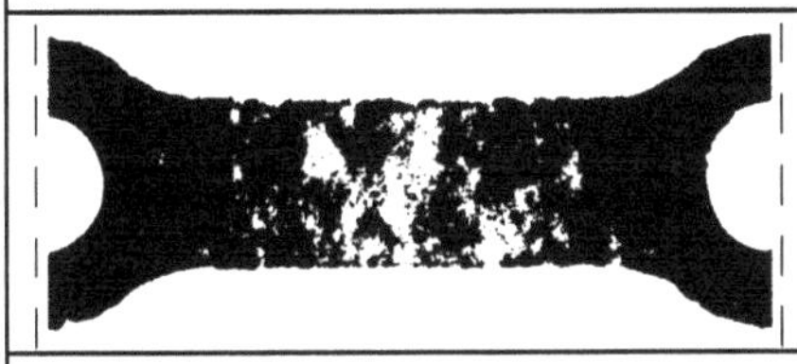 | Marked **inhomogeneities** due to the structure of the sealing material. |
| 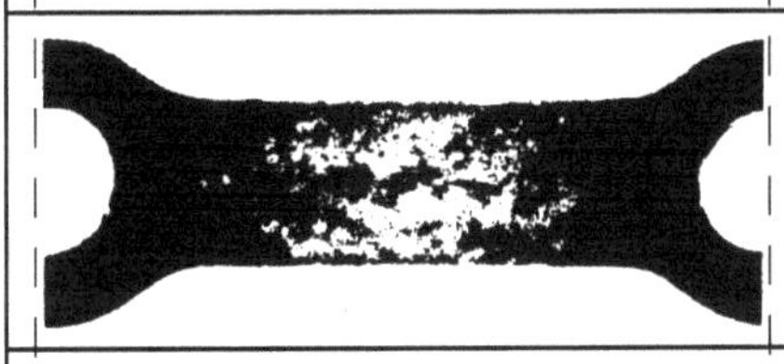 | High contact pressure in the **peripheral zones of the seal,** caused by deformation during production of the seal (blanking, cutting). |
| 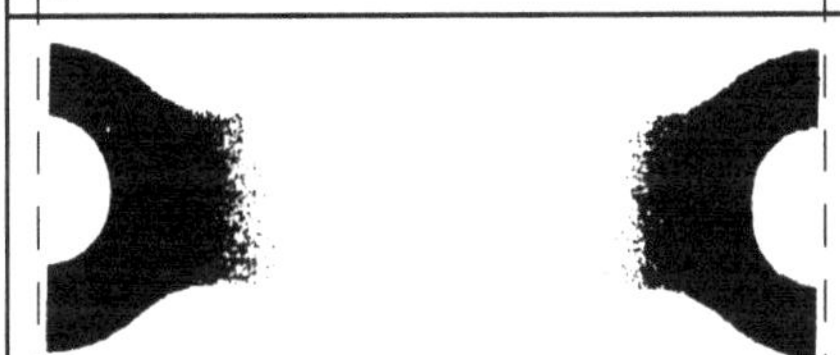 | **Uneven distribution**, no contact pressure in the middle between the bolts. |
| 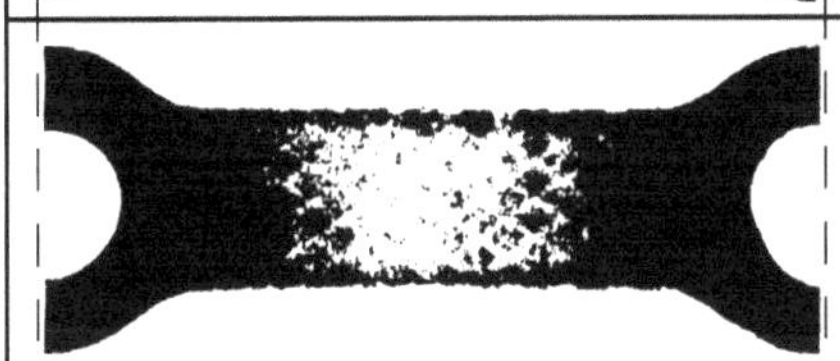 | Effect of the **sheet steel carrier.** |

**Fig. 11.24.** Characteristic pressure patterns recorded using Fuji Film Prescale pressure-sensitive film (low pressure range)

One suitable method of determining the distribution of contact pressure in the sealing joint of the test housing is to use the pressure-sensitive Fuji Film Prescale film. Figure 11.24 shows some typical segments of the pressure patterns of a housing. The following characteristic features emerge:

- very *even pressure pattern* with high contact pressure evident in the middle between the bolts,
- marked *inhomogeneities,* due to the structure of the sealing material,
- increased contact pressure in the *peripheral zones of the seal,* caused by deformation during production of the seal (blanking, cutting),

- *uneven distribution of contact pressure*; no contact pressure in the middle between the bolts and
- effect of *sheet steel carrier* (structure of the sheet steel carrier is visible).

Beside the qualitative appearance described above, the contact pressure in the middle between the bolts is the decisive criterion when assessing the pressure pattern, and thus the leak tightness of the joint.

### 11.5.2 Seals for Rotating Components

Elastomer *rotary shaft seals* are used in vehicle transmissions wherever rotating shaft through holes are subject to unpressurised fluid or have to be sealed against lubrication oil spray [11.41]. The basic types of rotary shaft seal are shown in Figure 11.25. Figure 11.26 shows the main dimensions and terms. Rotary shaft seals provide a seal in one direction only (oil side). Inside the gearbox, usually small rectangular rings are used to separate chambers of varying pressure (see Figure 11.27).

#### *Guidelines for Using:*

#### *1/ Rotary Shaft Seals:*

The damage caused by leaking rotary shaft seals is many times greater than the price of a new seal. Guidelines for installing, operating, testing and designing seals are comprehensively set out in German standards DIN 3760 [11.5] and 3761 [11.6], to eliminate seal failures as far as possible.

- Contact surface of the shaft, diameter *d*:
  - tolerance zone h11,
  - surface roughness $R_z = 1$ to 4 μm; the surface must not be too smooth, to ensure the wear due to running in necessary for proper functioning,
  - surface free of scrolling, i.e. grind or smooth without feed, trim grinding wheel scroll-free beforehand,
  - surface hardness: $\geq 45$ HRC at a circumferential speed of $v_u \leq 4$ m/s, $\geq 55$ HRC at a circumferential speed of $v_u > 4$ m/s ,

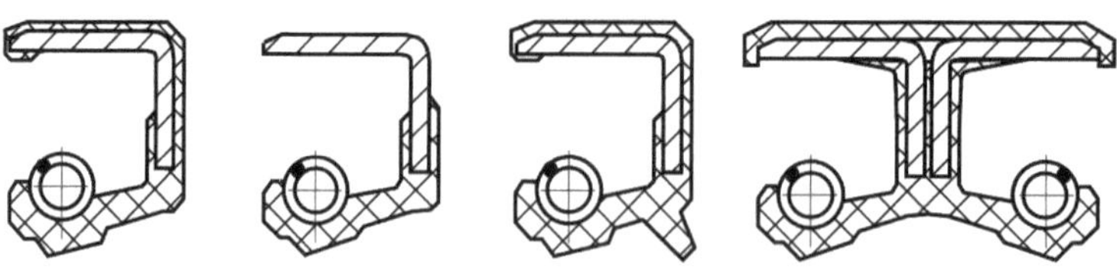

a) b) c) d)

**Fig. 11.25.** Rotary shaft seals, basic types. *a* Rubber elastic external shell; *b* metallic housing; *c* with additional dust lip; *d* double seal for fluid separation

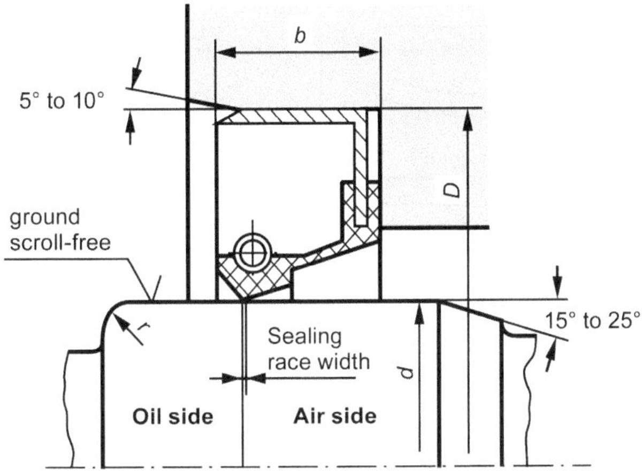


**Fig. 11.26.** Dimensions and terms for rotary shaft seals

  - depth of hardness at least 0.3 mm,
  - the grey film must be smoothed after nitriding,
  - soft bushes made of stainless steel have however also proved effective in practice.
- Locating bore, diameter *D*:
  - tolerance zone H8,
  - surface roughness $R_z \leq 16$ µm.

### 2/ Rectangular Rings (Piston Rings)

In the case of radial pressure oil feeding in rotating systems (e.g. pressure oil supply into automatic transmission piston chambers), rectangular rings are used for sealing. Cost and scant installation space often allow for no alternative sealings (see Figure 11.30 and Section 9.3 "Layout and Design of Multi-Plate Clutches").

The ring sits with clearance in the groove of the transmission shaft and is pressed against the housing bore and groove flank with oil pressure. Frequently, there is also a direction-changing pressurisation. When correctly designed, the ring geometry determines friction coupling in the bore and relative movements at the ring flank. Surface pressure and the sliding speed of the ring flank ("*p*-*v* value") are important design quantities and should be taken from the manufacturer's specifications. Wear is automatically compensated and should be considered during design (axial wear during lifetime). In order to keep wear within allowable limits, great care should be taken to abide by geometrical guidelines during groove manufacture and to meet surface requirements.

The ring is slotted to enable assembly (see Figure 11.27). The design of the joint is decisive for the assembly and leakage. Minimal internal leakage rates are important with respect to the flow rate of the pump to be installed and thus for transmission efficiency [11.46] (see also Section 11.3.3 "Detail Questions" on oil supply and oil pumps).

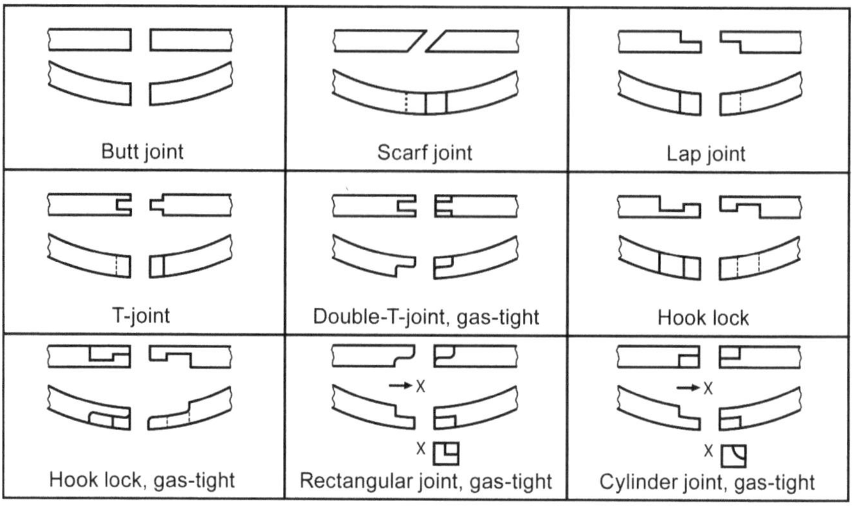


**Fig. 11.27.** Types of joints and locks of rectangular rings

Due to the high tribological loading, grey cast iron or high-performance plastics such as PTFE compounds, PEEK, PI and PAI are used as materials.

### 11.5.3 Seals for Reciprocating Round Components

Seals for reciprocating round components are required at the selector bar through hole for example, and in the case of automatic transmissions, at the pistons of oil-pressured clutches and brakes as well as for taper disc adjustment in the case of CVT transmissions. The seals used are *O-rings*, *X-rings* and *grooved rings* (Figure 11.28).

#### *Guidelines for Using:*

#### *1/ O-Rings and X-Rings*

O-rings and X-rings are used where sliding speeds are low, and a little leakage (necessary for lubrication and cooling) is acceptable. X-rings have the advantage over O-rings that they do not twist. The following points should be noted:

- adequate lubrication is essential (no dry running),
- no long periods out of service, since the ring can "stick" to the contact surface,
- select materials with Shore A hardness of 70 to 80,
- provide back-up rings where there is high pressure, to prevent gap extrusion,
- provide lead-in chamfers and rounded edges to facilitate fitting,
- permissible initial compression 15%,
- surface roughness must comply with manufacturers specification and
- use rectangular rings or X-rings where there is a danger of twisting.

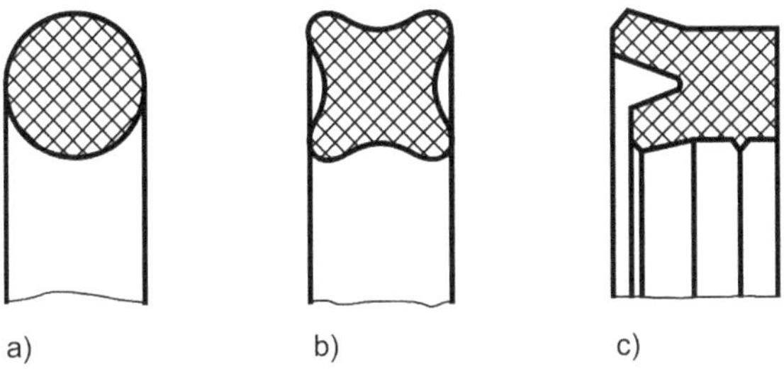


**Fig. 11.28.** Types of seal. *a* O-ring; *b* X-ring; *c* grooved ring

### 2/ Grooved Rings

O-rings and X-rings can only be used where radial and tilting movements are small. If the design requires considerable freedom of movement between piston and cylinder, grooved rings are used. They seal in one direction only, and are available with a great variety of profiles. Further details are given in the manufacturers catalogues.

### 3/ Special Designs

If especially low and above all consistent friction is required, as in slip-controlled clutches or CVT transmissions, special designs are employed. There are characterised by relatively hard materials in the sliding zone, crowned contact faces and little contact pressure.

## 11.5.4 Practical Examples

The designs discussed in Chapter 12 offer a large number of practical examples of seals. Figure 11.29 shows the sealing of a gearbox output shaft.

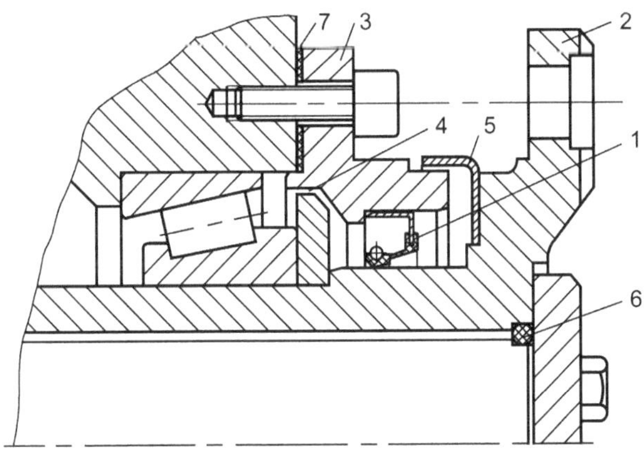


**Fig. 11.29.** Sealing a gearbox output shaft.
*1* Rotary shaft seal;
*2* output flange;
*3* housing cover;
*4* gap seal;
*5* splash guard;
*6* O-ring;
*7* flat gasket

The rotary shaft seal *1* prevents oil spray escaping between the rotating output flange *2* and the housing cover *3*, whereas some of the oil is retained by means of the gap seal *4* between rolling bearing and housing. To prevent the rotary shaft seal being directly splashed with dirt and water from outside, a splash guard *5* is mounted on the output flange. The spline profile of the gearbox output shaft is sealed with an O-ring *6*.

Figure 11.30 is a half-section view of a multi-plate clutch of a passenger car automatic transmission. The seal between the rotating clutch cage *4* and the housing *5* or between the rotating shaft *6* and the housing *5* is formed by rectangular rings *1*, mostly butt jointed. The rings are tensioned outwards, and run in the groove of the respective inner part. The piston *7* moves linearly, and is sealed by the groove rings *2*.

### 11.5.5 Final Inspection for Detecting Leakage

As shown at the beginning of the section, repair costs far outweigh the price of a sealing element in case of leakage. External leakage is visible from outside. Examples of this include leaky oil pans or rotary shaft seals.

Internal leakage can appear wherever pressure oil or different oil chambers in the transmission have to be sealed. Before delivery, transmission manufacturers generally test every transmission for leakage [11.51]. In addition to complete transmission tests on final inspection test benches at the end of the assembly line (EoL = end of line), it can be sensible for more complex assembly groups to carry out sealing tests on the assembly group level as well so as to recognise faults early and to avoid expensive reassemblies. The degree of automation depends on the method and the number of units to be tested.

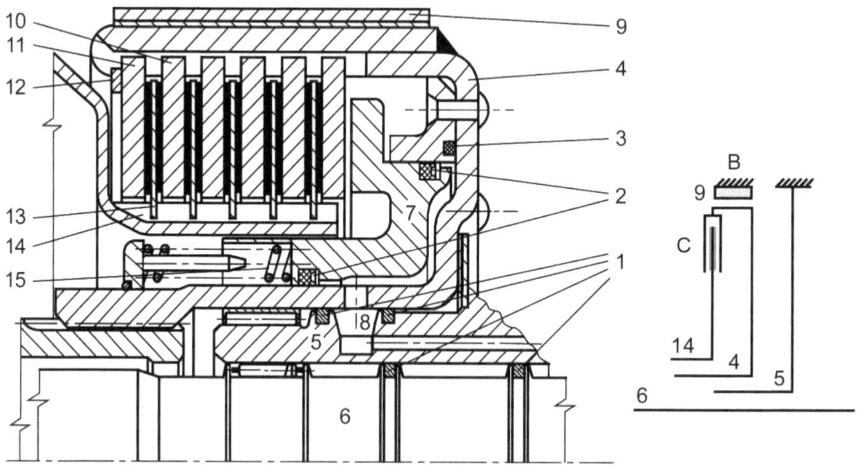


**Fig. 11.30.** Sealing a multi-plate clutch (see also Figure 9.31). *1* Rectangular ring (rotating); *2* grooved ring (linear); *3* O-ring (static); *4* outer plate carrier; *5* housing; *6* shaft; *7* piston; *8* pressure oil supply; *9* brake belt; *10* steel plate; *11* end plate; *12* snap ring; *13* lined plate; *14* inner plate carrier; *15* piston return spring

In the following, a few methods used in practice for testing transmission seals will be introduced:

- *differential pressure measurement* (test medium: air, testing pressure loss in the housing),
- *water bath* (immersion test with a small amount of excess pressure in the housing),
- *helium leak test* (test medium: helium, "sniffing test" with detectors) and
- *mass flow measurement* to determine leakage rates.

Thus discovered leakages are the result of error. These can arise due to:

- damage to the sealing elements during assembly (twisting, shearing off, ...),
- sealing surfaces that are damaged or do not comply with guidelines,
- mix-ups or wrong installation,
- turning down the rotary shaft seal sealing lip (especially in the case of double-seals) during assembly and
- faulty positioning or tilting of the rotary shaft seal during assembly.

Early and conscientiously executed process analyses (P-FMEA) as well as permanent quality control must be used to counter problem points on the design and on the process side. This includes, for example, optimising auxiliary assembly tools (e.g. lead-in sleeves), measuring equipment, lead-in chamfers etc. Beyond assembly and mix-up safety, sealing elements and their surroundings should be designed such that errors made in manufacture and assembly can be recognised reliably in the final inspection.

## 11.6 Vehicle Continuous Service Brakes

*You can drive faster with good brakes*

Engine power has increased significantly relative to overall vehicle weight in recent years. Vehicles are thus capable of travelling at higher average speeds. Maintaining high speeds even when travelling downhill, requires steady-state braking, sometimes with high energy content, especially in the case of vehicles with a high gross vehicle weight (see also Section 5.1.2 "Engine Braking Force").

Heavy-duty commercial vehicles normally have compressed air operated service brakes. These are designed to be capable of safe deceleration braking. They are designed for continuous operation only to a limited extent. Friction service brakes can be subject to thermal overload on long downhill runs with permanent steady-state braking which leads to impairment of the braking effect. This effect is called "fading".

Equation 5.7 demonstrates that a commercial vehicle with a gross vehicle weight of 40 t needs some 360 kW of braking power to negotiate a downhill gradient of 7% at a constant speed of 60 km/h. This braking power must be continuously dissipated in the form of heat. The service brake can effectively carry out this function only for brief periods, because of poor heat dissipation.

Continuous service brake systems increase the economic efficiency of commercial vehicles by considerably reducing brake lining wear, and they allow for higher average speeds, especially on long downhill runs. But above all they contribute to improving active safety by reducing thermal load on the service brake.

There are two types of continuous service brakes in general use, namely engine brakes and retarders, which in turn have different basic concepts (Figure 11.31).

### 11.6.1 Definitions

A *continuous service brake* is a supplementary braking system capable of producing and maintaining braking force over a long period without any noticeable wear. The continuous service brake must therefore function reliably regardless of the condition and effectiveness of the other braking system. The requirements a motor vehicle braking system has to fulfil in the EU are set out in the ECE regulation 13 and in EU directives [11.2, 11.53].

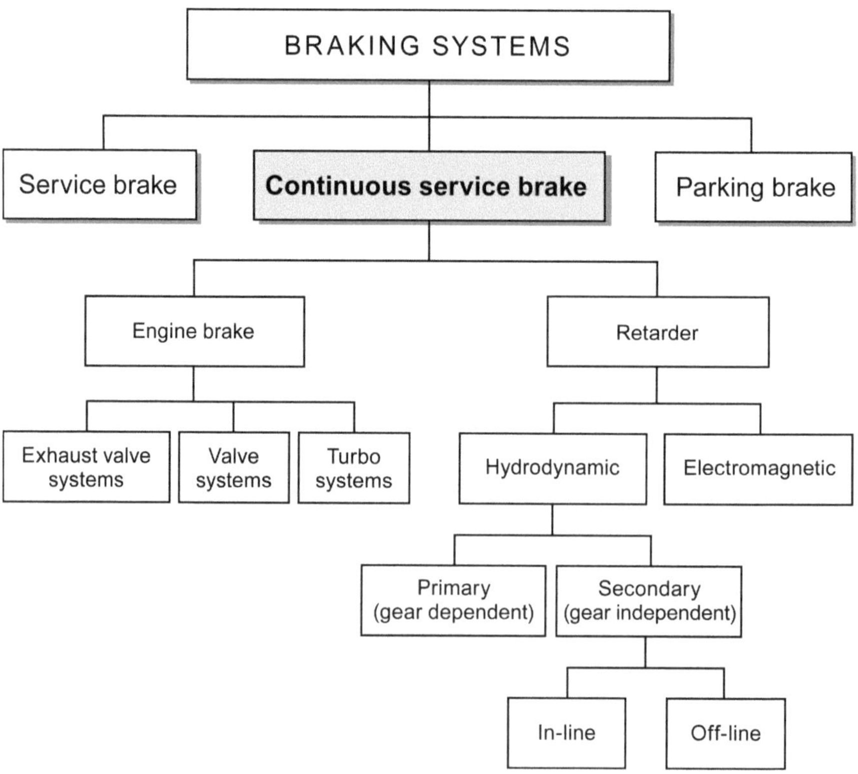


**Fig. 11.31.** Overview of braking systems

Whereas the service brake acts on all wheels, the continuous service brake brakes only on the vehicle's wheels connected to it (drive wheels). There are also continuous service brake systems for trailers and semi-trailers.

ECE regulation 13 requires in its Type II Test that continuous service brakes be fitted to all commercial vehicles with a gross weight rating of more than 12 t and to all buses of more than 5 t. The continuous service brake fitted must enable the vehicle to sustain a steady-state speed of 30 km/h on a 6% downhill gradient for a distance of 6 km.

### 11.6.2 Engine Braking Systems

Combustion engines generate braking torque in overrun conditions (Figure 3.14). Engine braking torque arises principally from pumping work. The engine braking force $F_B$ depends on the gear (Figure 5.6). Additional engine brakes are mounted directly in the exhaust section or in the cylinder head.

In the case of engine brakes with *exhaust valve*, a butterfly valve installed in the exhaust section closes, creating back pressure in the exhaust system (Figure 11.32). The back pressure increases the engine drag torque by inhibiting the gas exchange, thus braking the vehicle. Exhaust valves are not used in coaches because of the valve noise. The exhaust valve is also known as "exhaust throttle valve" or simply "throttle valve".

The *constant throttle valve*, which belongs to the valve system group (Figure 11.31), is an additional outlet valve in the cylinder head of the engine. Opening the constant throttle valve when braking dissipates the unwanted expansion energy during the second cycle in the combustion chamber, amplifying the engine braking effect.

The braking effect of the constant throttle valve is much better than that of the exhaust valve in the lower engine speed and velocity range, which in turn provides better braking performance in the upper speed range. Figure 11.32 shows a combined solution comprising both exhaust valve and constant throttle valve [11.20].

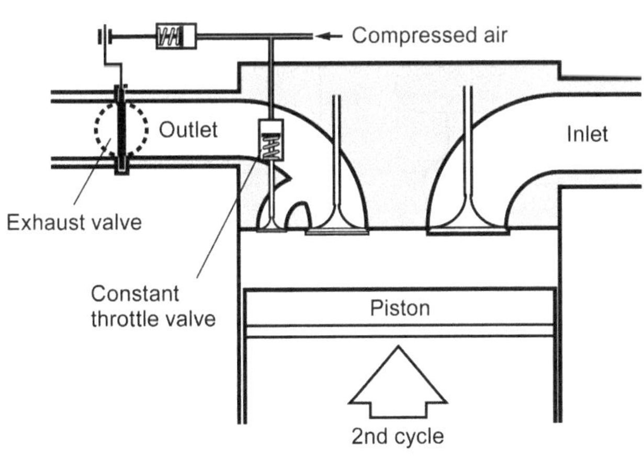


**Fig. 11.32.** Engine brake with exhaust valve and constant throttle valve

As a representative of turbo systems, the *turbo brake* is a further development of the engine brake of naturally aspirated engines with exhaust valves. In this case, the exhaust driven turbocharger is activated in the engine braking phase and thus the filling amount of the cylinder increased in order to increase brake performance. This is achieved by means of a baffle arranged in an axially adjustable fashion in the nozzle channel of the turbocharger. In the engine braking phase, this guide vane is pushed over the turbine wheel, which leads to an increased inflow velocity and an increased air flow-rate in the engine.

The continuous shiftability of the guide vane allows for fine dosability of the turbo brake. An additional blow-off system permits the targeted modification of brake performance and prevents overloading of the engine during engine brake operation.

### 11.6.3 Retarders

Retarders are virtually non-wearing continuous service brakes. They are capable of converting and dissipating a vehicle's kinetic and potential energy arising over long periods into thermal energy. They are used in commercial vehicles and buses. Retarders used in practice vary principally in the way they convert energy. In the case of hydrodynamic retarders, brake torque is built up by the principle of fluid angular momentum change and the energy is converted by fluid friction, whereas electromagnetic retarders use a magnetic field.

#### *1/ Hydrodynamic Retarder*

In hydrodynamic retarders, the hydraulic energy of a fluid is used to brake the vehicle. The physical principle of operation corresponds to that of a hydrodynamic clutch with a fixed turbine (see Section 10.4 "Hydrodynamic Clutches and Torque Converters"). The rotor $R$ (impeller) is located in the power flow. The stator $S$ is fixed to the retarder housing (Figure 11.33).

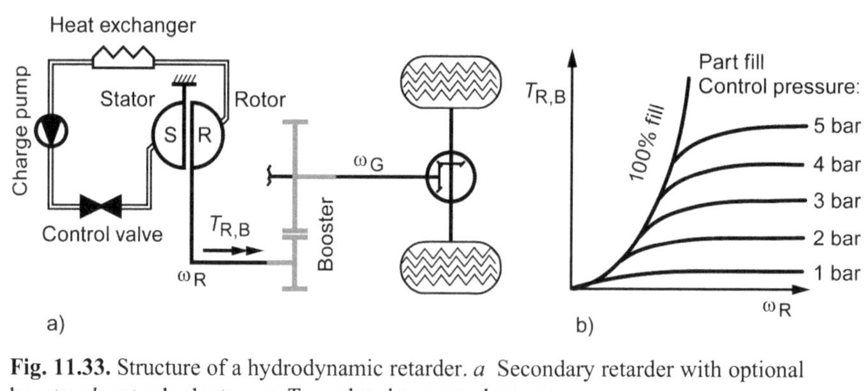


**Fig. 11.33.** Structure of a hydrodynamic retarder. *a* Secondary retarder with optional booster; *b* rotor brake torque $T_{R,B}$ related to control pressure

When the hydrodynamic retarder is activated, a quantity of oil proportionate to the brake position is fed into the blade chamber. The braking torque is controlled by the retarder fill level.

The rotating rotor carries the oil that reacts on the stator, thus producing a braking effect on the rotor shaft. Hydrodynamic retarders can produce no braking torque when the vehicle is at rest, and very little when the rotor is rotating slowly. Hydrodynamic retarders are therefore also just as unsuitable as electromagnetic retarders as service brakes.

The relationship of retarder torque or rotor brake torque $T_{R,B}$ to the rotor speed is given by the theory of hydrodynamic machines (Equation 10.21), thus

$$T_{R,B} = \lambda\, \rho\, \omega_R^2\, D^5 \,. \tag{11.22}$$

The performance coefficient $\lambda$ is a function of the speed ratio $v = n_S / n_R$ (Equation 4.2). In the retarder the stator $S$ stands still, making the speed ratio $v$ of stator to rotor zero. This corresponds to the stall point of a hydrodynamic clutch.

The performance coefficient of the 100% fill parabola is influenced by the structure of the hydrodynamic system and fluid viscosity. The fill level of the retarder, and thus the rotor braking torque, is a function of the control pressure. The oil seeking to escape because of the centrifugal force in the retarder is retained in the retarder by means of the control pressure (also known as back pressure), depending on the braking torque required. This gives a brake torque control range $T_{R,B}$ (Figure 11.33b).

The symbol for oil density is $\rho$, and the symbol for rotor diameter is $D$. The angular velocity of the rotor shaft $\omega_R$ is either equal to the angular velocity of the gearbox input shaft or in a fixed ratio to the angular velocity of the propeller shaft $\omega_G$ (Figure 11.34), depending on where the hydrodynamic retarder is installed.

Primary retarders are located on the engine side and secondary retarders on the gearbox side in the powertrain (Figure 11.34). Secondary retarders are further subdivided into in-line and off-line retarders. In the case of in-line retarders, there are both variants that are flanged onto the transmission as well as variants that are mounted between the propeller shafts. Off-line retarders are flange-mounted onto the transmission and connected to it by means of a booster gear stage.

There is gear-dependent braking torque at the wheels for the primary retarder, which increases in proportion to the transmission ratio as the gear becomes lower. Primary retarders are therefore effective even at low road speeds. They generate relatively high braking torque levels at the drive wheels, which fall substantially with increasing angular velocity of the propeller shaft $\omega_G$, i.e. with increasing road speed (Figure 11.35).

The effectiveness of primary retarders is impaired by the interruption of the power flow and thus the braking effect when changing gear in manual transmissions. In commercial vehicle transmissions with torque converters, primary retarders are often integrated into the transmission.

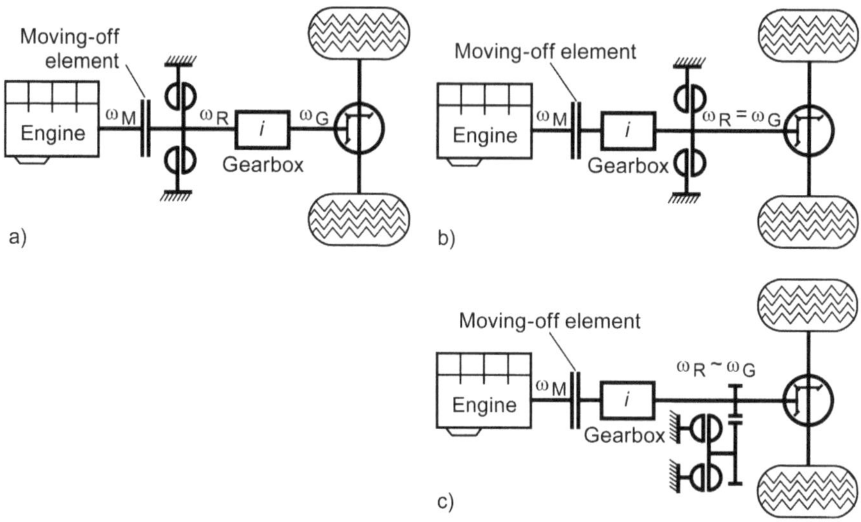


**Fig. 11.34.** Alternative positions for mounting hydrodynamic retarders. *a* Primary retarder; *b* secondary retarder; in-line variant; *c* secondary retarder, off-line variant

Please refer in this connection to the production examples in Figure 12.45 "16-speed semi-automated manual commercial vehicle gearbox (torque converter clutch transmission)", and Figure 12.48 "6-speed automatic gearbox".

The braking torque of the secondary retarder is only dependent on its characteristic. The secondary retarder is advantageous for dimensioning the transmission, since its braking torque – besides the booster step common in secondary retarders – is not an additional load on the transmission.

A high level of braking torque is virtually constantly available over a wide velocity range (Figure 11.35). This design is particularly suitable for commercial vehicles with a high gross vehicle weight and high speed. A booster gear stage can also be fitted before the rotor shaft (Figure 11.33a). The booster is one way of increasing the braking torque of secondary retarders at low road speeds, and fitting twin rotors is another. The secondary retarder with booster is flange-mounted to the side of the gearbox at the rear, and driven by a spur gear stage – a ratio of approximately 1 :2 is usual.

The achievable braking power is limited primarily by the capacity of the cooling circuit rather than by the continuous brake itself. The permanent cooling capacity is limited to about 300 kW. With a shared oil circuit, the additional heat exchanger of the retarder can be used to cool the transmission when the retarder is switched off. This increases the service life of toothing and bearings, and slows down oil ageing. When the retarder is switched on, the oil circuit is separated from the transmission. The transmission oil temperature is thus not affected during braking.

If there is a retarder in the powertrain, then there will be power loss under power when the retarder is unfilled, due to air re-circulation (fan losses).

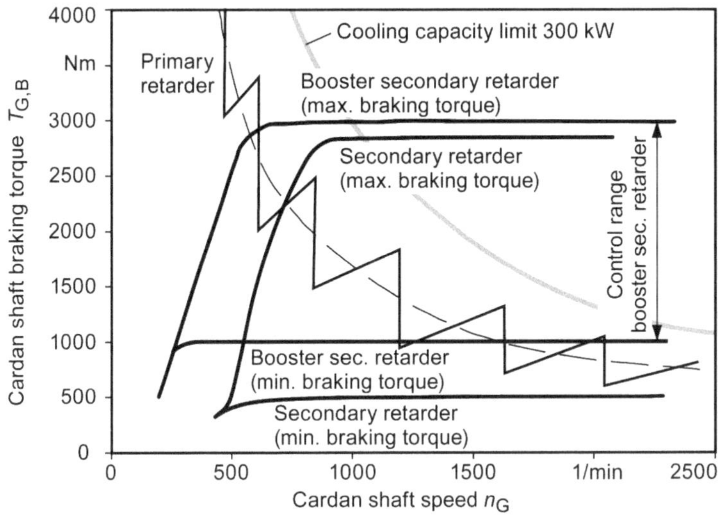


**Fig. 11.35.** Braking torque levels of hydrodynamic retarders as a function of the propeller shaft speed, i.e. the road speed

The losses of the unfilled retarder can be minimised by reducing the air circulation by means of annular slide valves and restrictor slide valves pivoting between rotor and stator [11.18] (see also Figure 12.48 and associated discussion). Another established possibility is automatic rotor displacement, in which the rotor withdraws from the stator by a small distance when the retarder is switched off.

### *2/ Hydrodynamic Water Retarder*

The Voith Aquatarder is a hydrodynamic retarder consisting of a housing, rotor, stator and valves. It is attached to the engine front end of a commercial vehicle diesel engine and connected directly with the crankshaft of the engine (Figure 11.36). As such, it belongs to the class of primary braking systems.

The distinctive feature of the Voith Aquatarder is that the operating medium is not oil but the engine coolant – a water-glycol mixture. The Aquatarder is directly and hermetically integrated into the cooling system. During the working circuit, the operating medium is accelerated by the rotor impelled by the crankshaft of the engine and decelerated again in the stator. Braking torque is built up in proportion with the fill level between the rotor and the stator. The braking torque affects the powertrain of the vehicle via the rotor and crankshaft, causing it to brake. The kinetic energy arising in the process is converted completely into thermal energy in the hydrodynamic circuit of the retarder and directly absorbed by the coolant. The heated coolant is cooled by the vehicle's cooling system.

In traction mode, the volumetric flow created by the water pump is guided directly to the engine without flowing through the Aquatarder system.

**Fig. 11.36.** Voith Aquatarder® WR190/D20 with MAN common rail engine D20

In braking operations, the switching valve is adjusted such that the volumetric flow of the water pump is guided up to 100% into the retarder circuit (Figure 11.37). From there the retarder boosts the flow, as it itself acts as a strong pump. To achieve the desired braking power, the retarder has to work against an outlet resistance. This restrictor is a pneumatically controlled outlet control valve attached to the Aquatarder outlet, which performs a continuous control of the braking torque. When the retarder is switched off, the switching valve allows the coolant to flow by the retarder again while the latter drains itself via the outlet control valve.

The wear-free and maintainence-free commercial vehicle braking system Aquatarder is dependent on engine speed and has a maximum braking torque of up to 1450 Nm and a maximum braking power of about 300 kW at a weight of about 30 kg. The Aquatarder, together with the controlled engine brake EVBec of the MAN engines D20 and D26, makes the MAN PriTarder Braking System.

Due to its continuous controllability, the PriTarder Braking System is integrated into the cruise control and controlled by the vehicle management computer. In addition, the retarder system can be optionally operated with the activation lever or the braking pedal.

### *3/ Electromagnetic Retarder*

Electromagnetic retarders – often also referred to as eddy current retarders – are of much simpler design than hydrodynamic retarders. They are generally designed as separately mounted secondary retarders. The braking effect is based on the physical principle of the action of force in electromagnetic fields. The stator, in the form of a disc, is fitted with several field coils, and fixed to the gearbox housing.

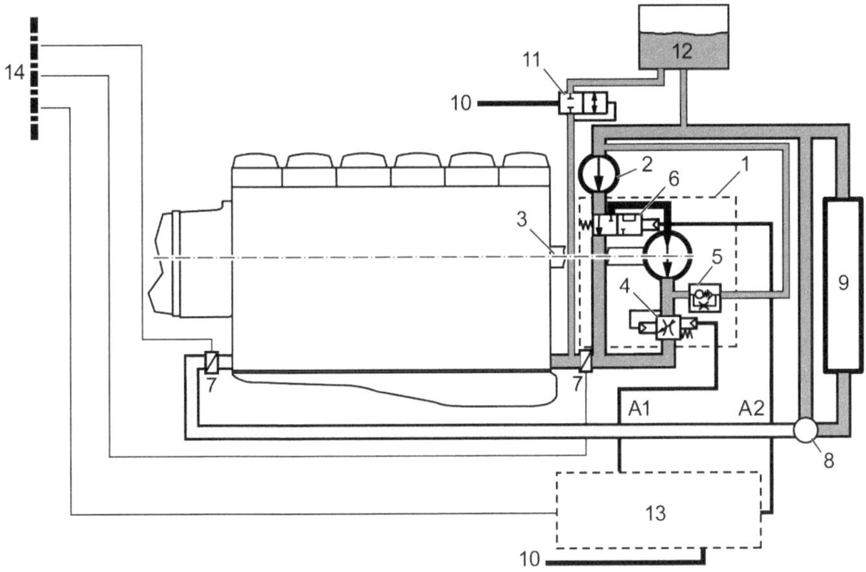


**Fig. 11.37.** Control diagram of the Voith Aquatarder in braking operation. *1* Aquatarder; *2* water pump; *3* crankshaft; *4* control valve; *5* non-return valve; *6* switching valve; *7* temperature sensors; *8* thermostat; *9* vehicle radiator; *10* connection supply air; *11* 2-port/2-way valve; *12* compensation reservoir; *13* valve block; *14* vehicle electronics; *A1* pneumatic line to the control valve; *A2* pneumatic line to the switching valve

On the transmission and rear axle side there are rotors linked to the propeller shaft (Figure 11.38). The rotors are air-cooled, and have fins to facilitate heat dissipation to the environment.

The field coils are fed with current from the battery or the generator during braking. Eddy currents are induced in the rotors when they pass through the magnetic field, which impede the rotation of the rotors. The level of braking torque depends on the excitation of the stator coils and of the air gap between the rotor and the stator. The braking power can be activated in several stages by passing current through the field coils in pairs.

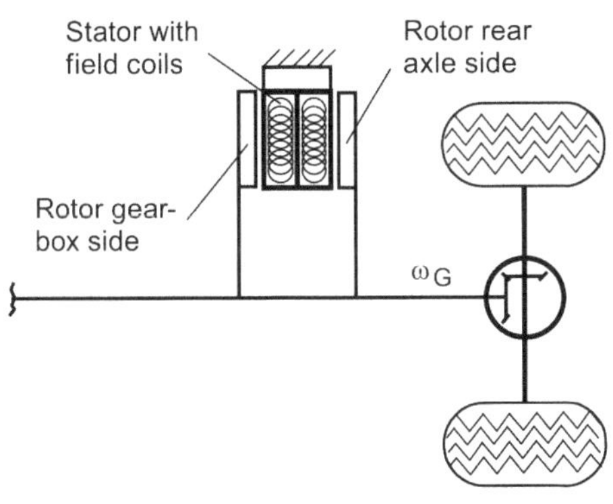


**Fig. 11.38.** Basic structure of an electromagnetic retarder (eddy current retarder)

The features of electromagnetic retarders are their simplicity of design, and their rapid response; on the other hand, they are relatively heavy and require an adequate electrical power supply to function properly.

In contrast to hydrodynamic retarders, they have relatively high braking torque even at low rotor speeds. However, due to declining braking torque characteristics, they too cannot be used completely until the vehicle is at rest.

But as the rotor speed increases, the temperature of the electromagnetic retarder increases too, causing the braking torque to go down. The number of active coils, or the power supply to the coils, is accordingly limited in order to protect against thermal damage.

### 11.6.4 Actuation and Use

The aim of modern retarder systems is to relieve the driver. When the driver activates the braking pedal, the individual braking systems are controlled collectively by the vehicle management computer via a data bus. When the brake pedal is operated, first the retarder can be engaged for example. If the braking force is not sufficient, the engine brake is automatically activated, providing additional braking power. The service brake only cuts in finally if needed.

In older systems, the engine brake (exhaust valve, constant throttle valve) is activated by means of foot switches. The retarder is usually operated either with the brake pedal or a hand lever with which different braking levels can be selected.

Electronic engine brake and retarder control combined with automatic speed control can keep road speed constant on downhill gradients. Speed control is activated via the cruise control or the brake pedal. The driver only needs to activate the brake pedal until the vehicle slows down to the desired speed. When the pedal is raised, the current speed is held. The braking system is working until the accelerator pedal is activated.

Whereas primary retarders and engine brakes have advantages in the lower speed range and on steep downhill stretches, secondary retarders are suitable for higher speeds. There is normally a hydrodynamic primary retarder integrated in commercial vehicle automatic transmissions, since various essential peripherals such as the charge pump have to be fitted anyway.

Both primary and secondary retarders brake only through the drive wheels. In modern systems, if the drive wheels are only partly loaded, or the coefficient of friction of the wheels to the road surface is reduced, the retarding effect is limited by load sensors in order to avoid excessive slip of the driving wheel. If the ABS system recognises blocking of the drive wheels, the retarder is switched off.

# 12 Typical Designs of Vehicle Transmissions

*The process of human thought proceeds from the concrete to the abstract, then back to the concrete /J. Dewey, 1910/*

This chapter examines some particular transmission designs, and considers their structural design. With regard to the gearwheel configurations in the transmissions examined, you may refer to the gearbox diagrams in Sections 6.6 "Passenger Car Transmissions", 6.7 "Commercial Vehicle Transmissions", 6.8 "Final Drives" and 6.9 "Power Take-Offs".

It is not the intention of this chapter to present the newest and most current developments, nor to provide an exhaustively detailed account of the subject. For this purpose, please refer to the relevant technical literature. Instead of this, the following chapter will outline the basic, general principles. Moreover, the following will also discuss gearboxes no longer in production.

## 12.1 Passenger Car Transmissions

Table 12.1 shows in summarised form the passenger car transmissions treated in systematic sections 6.6.1 to 6.6.6 and designs sections 12.1.1 to 12.1.6. To aid orientation, consecutive numbers are assigned in this section to the designs discussed (e.g. 1/ Single-Stage 5-Speed Manual Passenger Car Gearbox; VW MQ).

**Table 12.1.** In Sections 6.6.1–6.6.6 and 12.1.1–12.1.6 introduced automotive gearboxes. *FT* front-transverse drive; *S* standard drive; *FL* front-longitudinal drive; *FLA* front-longitudinal all-wheel drive; *RL* rear-motor longitudinal drive; *RT* rear-motor transverse drive; *CC* converter lock-up clutch; *TCC* torque converter clutch

| No. | Design Fig.-No. | Speeds | Characteristics | Configuration | Manufacturer | Name | Diagram Fig.-No. |
|---|---|---|---|---|---|---|---|
| — | — | 4 | MT, 1-stage | FT | VW | MQ | 6.18a |
| 1/ | 12.1 | 5 | MT, 1-stage | FT | VW | MQ | 6.18b |
| — | — | 4 | MT, 2-stage | S | Getrag | 4-speed | 6.19a |
| 2/ | 12.2–12.4 | 5 | MT, 2-stage | S | ZF | S 5-31 | 6.19b |
| 3/ | 12.5 | 6 | MT, 2-stage | S | Getrag | 286 | 6.20a |

H. Naunheimer et al., *Automotive Transmissions*, 2nd ed.,
DOI 10.1007/978-3-642-16214-5_12, © Springer-Verlag Berlin Heidelberg 2011

**Table 12.1.** *(continued)*

| | | | | | | | |
|---|---|---|---|---|---|---|---|
| 4/ | 12.6 | 6 | MT, 1-stage | FT | Opel | F28-6 | 6.20b |
| 5/ | 12.7–12.8 | 6 | MT, 1-stage, 3-shaft | FT | Getrag | 285 | 6.21a |
| 6/ | 12.9–10 | 6 | MT, 1-stage, 3-shaft | FT | MB | FSG 300-6 | 6.21b |
| 7/ | 12.11 | 6 | MT, 2-stage, petrol engine | S | Getrag | 217 | 6.22a/b |
| 8/ | 12.12 | 6 | MT, 1-stage | FL | Audi | ML350-6F | 6.23a |
| | 12.64 | 6 | MT, 1-stage | FLA | Audi | ML450-6Q | 6.23b |
| — | — | 3 | AMT, 1-stage, TCC | RL | VW | Built 1967 | 6.24 |
| 9/ | 12.13–14 | 6 | AMT, 1-stage, multi-range | RT | Getrag | 431 | 6.25a |
| 10/ | 12.15–12.16 | 7 | AMT, 2-stage | S | Getrag | 247 | 6.25b |
| 11/ | 12.17–12.20 | 6 | DCT | FT | VW | DSG | 6.26 |
| 12/ | 12.21 | 7 | DCT | S | ZF | 7 DCT 50 | 6.27 |
| — | — | 4 | AT, w/o CC | FT | ZF | 4 HP 14 | 6.30–6.31 |
| 13/ | 12.22 | 5 | AT, w/o CC | S | MB | W5A 030 | — |
| 14/ | 12.23 | 5 | AT | S | MB | W5A 580 | 6.32 |
| 15/ | 12.24 | 7 | AT | S | MB | W7A 700 | 6.33 |
| 16/ | 12.25 | 6 | AT | S | ZF | 6 HP 26 | 6.34 |
| 17/ | 12.26 | 6 | AT | FT | AISIN | TF 80-SC | 6.35 |
| 18/ | 12.27 | 5 | Countershaft-type automatic transm. | FT | MB | W5A 180 | 6.36 |
| 19/ | 12.28 | 6 | Hybrid, parallel | S | BMW/ ZF | Active transmission | 6.37–6.38 |
| 20/ | 12.29 | ∞ | Hybrid, power-split | FT | Toyota/ Lexus | P310 | 6.39 |
| — | — | ∞ | CVT, toroid principle | — | — | — | 6.41 |
| 21/ | 12.30 | ∞ | CVT, chain variator | FL | Audi | Multitronic | — |
| 22/ | 12.31 | ∞ | CVT, chain variator | FT | MB | Autotronic | — |
| 23/ | 12.32 | ∞ | CVT, chain variator | FT | ZF | CFT 30 | 6.43 |
| — | — | ∞ | CVT, geared neutral | — | — | — | 6.44 |

### 12.1.1 Manual Passenger Car Transmissions (MT)

4-speed manual gearboxes were standard for passenger cars in Europe until the early 1980s. As engine power and vehicle weight increased and $c_W$ ratings improved, larger overall gear ratios became necessary. Large overall gear ratios facilitate moving-off, provide good acceleration, and also reduce engine speed and hence fuel consumption at high speeds. Manual transmissions therefore now usually have either five or six speeds.

#### *1/ Single-Stage 5-Speed Manual Passenger Car Gearbox; VW MQ*

A large proportion of vehicles are fitted with front-wheel drive and transverse-mounted gearbox. Single-stage countershaft transmissions with integral final drive are always used in this drive configuration. This format is compact, and reduces the space needed to accommodate the powertrain, importantly by eliminating the propeller shaft. It has the major advantage over the standard drive configuration that the complete powertrain including the engine can be preassembled as a "package" and fitted to the car body. The 5-speed gearbox in Figure 12.1 is a typical example of the front-transverse format *(the basic concept is explained in Section 6.6.1 and Figure 6.18b).*

The gearwheels of first gear to fifth gear are mounted on the input shaft in sequence, starting from the clutch side. A striking feature is the fifth gear located outside the cast gearbox housing, which is encased against the environment by a separate sheet metal pan. The reason for this feature is that this gearbox is developed from a 4-speed gearbox, with the fifth gear added on. The distance between bearings is kept small by the resultant location of a main bearing between fourth and fifth gear. This has a beneficial effect on shaft deflection under load, although the fifth gear is overhung.

The tapered roller bearings of the input shaft have a collar at the bearing outer ring to counteract the axial forces against the housing. These special bearings allow the wall of the housing to be thinner, since no collars are required on the housing, and no grooves for circlips.

The toothing of the reverse gear is located on the first and second gear sliding sleeve. All forward gears have a single-cone synchronizer. The synchronizers are activated by swing forks, whose plain bearings are clearly visible over the input shaft. The swing forks are in turn activated by means of a central selector shaft (mounted vertically in the middle of the gearbox).

The gearbox housing is open to both sides and has an integrally cast half of the axle drive housing. The clutch bell housing, which carries the other half of the axle drive housing, is bolted to the gearbox. The other side of the gearbox housing is closed off with a metal cover, as already mentioned. The bevel gear differential is lubricated by the gearbox oil circulation. The worm gear of the speedometer drive can be seen at the drive cage of the differential. The clutch release bearing with its operating lever is located in the clutch bell housing, sitting on the gearbox input shaft. The lever is designed as a formed sheet metal part.

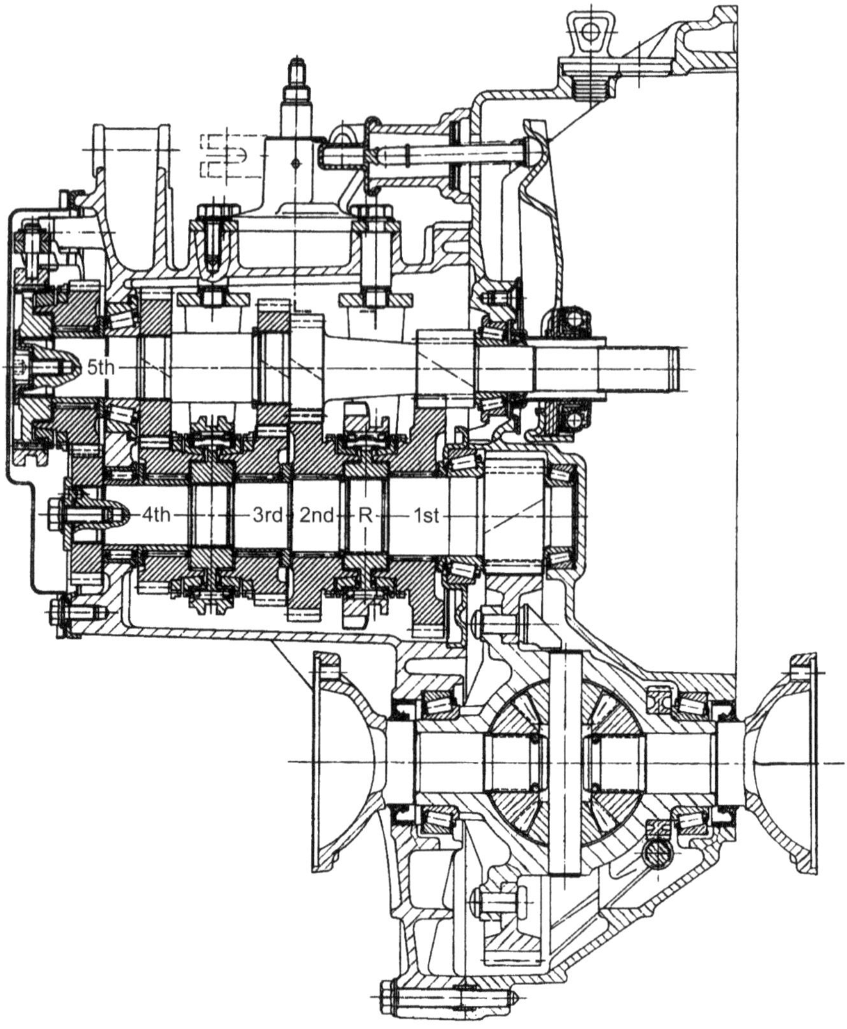


**Fig. 12.1.** 5-speed manual passenger car gearbox VW MQ, *gearbox diagram Figure 6.18b*

### 2/ Two-Stage 5-Speed Manual Passenger Car Gearbox; ZF S 5-31

Figure 12.2 shows a two-stage coaxial 5-speed passenger car countershaft-type manual gearbox with direct drive in fifth gear *(the basic concept is explained in Section 6.6.1 and Figure 6.19b)*. In this design, first and second gear are roughly in the middle of the main shaft. This contravenes the principle whereby gears with higher torque conversion should be located as close as possible to a main bearing (Section 8.2 "General Design Guidelines" for Shafts). But the resultant shaft deflection can be controlled by appropriate gearing geometry.

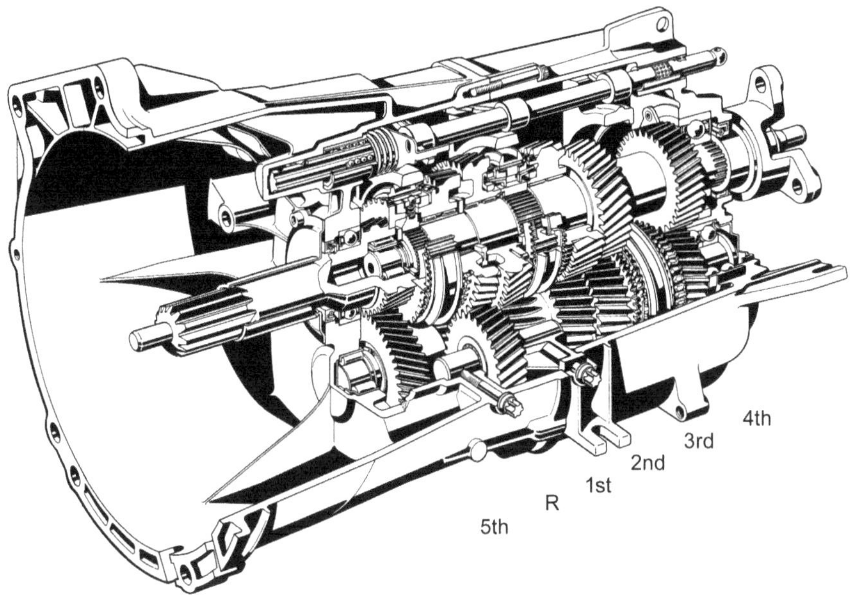


**Fig. 12.2.** 5-speed manual passenger car gearbox ZF S 5-31, *gearbox diagram Figure 6.19b*

The advantage of this structural design is that the more frequently used gears of third and fourth gear are near a bearing point, making them run more quietly.

In contrast to *in-line gearboxes*, where all synchronizers are mounted on the gearbox main shaft, in this gearbox the synchronizers for third and fourth gear are moved to the countershaft. This arrangement means the idler respectively shift gears for third and fourth gear are no longer on the countershaft itself, but linked to the output side. Their speeds thus no longer have to be matched to the output speed during synchronization, which reduces the frictional work and shifting force required to change gear. It is not possible to move any more idler gears to the countershaft, since the differential rotational speeds are too great. The synchronizers for first and second gear are of double-cone design, significantly reducing shifting force. All other gears, including reverse gear, have single-cone synchronizers.

Since the idler gears in third and fourth do not rotate when the vehicle is stationary in neutral, they do not produce as much rattle as idler gears mounted on the main shaft (see also Section 7.5).

The kinematics of the shift system is shown in Figure 12.3. Instead of separate selector bars for each individual shift fork, a central selector shaft *1* is used with swing forks *5–7*. This has weight advantages, and is more cost-effective. The central selector shaft runs in linear ball bearings *4*, which reduces gearshift effort. The swing forks have a pivot *9* supported in the housing, around which they pivot on the lever principle. This enables gearshifting effort to be reduced by selecting a suitable lever ratio. In this type of design, the swing forks (or gates) are changed by the shaft turning.

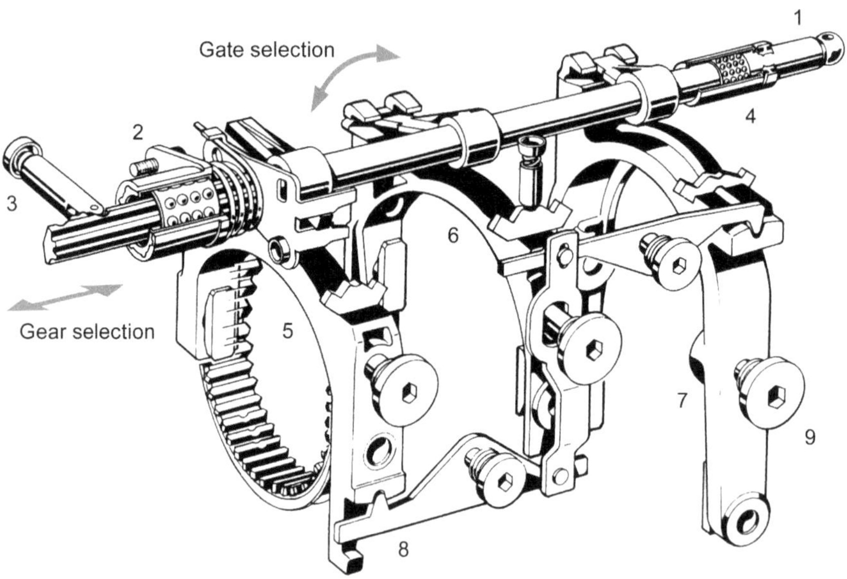


**Fig. 12.3.** Operation of the shifting elements in the gearbox in Figure 12.1. *1* Central selector shaft; *2* swing fork; *3* detent pin; *4* ball sleeve; *5–7* swing forks; *8* interlock; *9* pivot of the swing fork

**Fig. 12.4.** Individual parts of the 5-speed manual passenger car gearbox ZF S 5-31

The countershaft runs in a cylindrical roller bearing on the input side and in a double-row angular contact ball bearing on the output side, instead of the more usual tapered roller bearings or single-row deep groove ball bearings, making shifting more precise and the gearbox quieter.

No tapered roller bearings were used anywhere in the gearbox, since they entail a number of disadvantages. These include changing the internal bearing clearance as a result of temperature-related changes in length in the gearbox, and increased clearance caused by relaxation of the housing. Tapered roller bearings also make it impossible to use "clean bearings" that are sealed against dirt by a sealing collar. Using them makes it possible to extend the oil change intervals. The gearbox housing itself is of end-loaded design with an integrally cast clutch bell housing (Figure 12.4), which ensures the housing is very rigid [12.54].

### *3/ Two-Stage 6-Speed Manual Passenger Car Gearbox; Getrag 286*

Changes in technological general conditions in automotive engineering have lead to the development of manual gearboxes with six speeds. These changes were improved $c_W$ ratings enabling higher top speeds, increased vehicle weight due to more comprehensive equipment, and the desire for improved elasticity figures [12.54].

An example of 6-speed manual gearboxes for passenger cars is shown in Figure 12.5 *(the basic concept is explained in Section 6.6.1 and Figure 6.20a)*. This is a two-stage countershaft gearbox with constant gear mounted on the input side.

The transmission shafts are longer than the gearbox shown in Figure 12.2. This is firstly because of the additional gear pair, and secondly because of the larger face widths needed for vehicles with powerful engines. In order to minimise the resultant shaft deflection under load, the gear pairs for first and second gear are located near the output side main bearing.

The gear pairs in third and fourth gear are mounted in the centre of the gearbox. The gear pattern resulting from the configuration of the gear pairs, with a gate each for first and second gear, third and fourth gear, and fifth and sixth gear, represents a logical extension of the familiar gear pattern of 5 speed gearboxes.

To reduce rattling noise, the third and fourth gear idler gears were moved onto the countershaft. To reduce the gearshift effort, the first and second gear synchronizers are triple-cone synchronizers, and the third and fourth gear synchronizers are double-cone synchronizers. The fifth and sixth gear and reverse gear synchronizers have a single cone.

Figure 12.5 shows the central selector shaft running in ball bearings. This view does not show the four additional selector bars for the respective gates. Shifting between these selector bars is controlled by turning of the central selector shaft, acting through lever mechanisms. The gear selection detent is located on the input side end of the central selector shaft. The torsion spring mounted on the right of the selector shaft defines the initial position for the gate selection rotary movement.

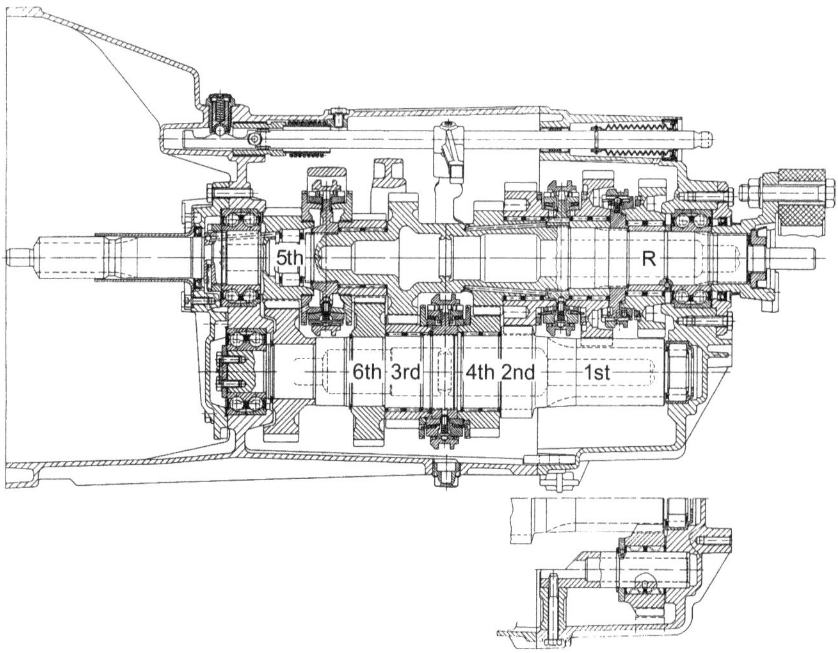


**Fig. 12.5.** 6-speed manual passenger car gearbox Getrag 286, *gearbox diagram Figure 6.20a*

The countershaft runs in a cylindrical roller bearing on the output side and in a double-row angular contact ball bearing on the input side, which makes the gearbox run more quietly. Bearings sealed to exclude dirt (clean bearings) increase bearing service life, and can enable smaller and therefore lighter bearings to be used.

The main shafts and countershafts of the gearbox are hollow drilled to contour. This is achieved using two-piece shafts, which are friction welded after the internal contour has been bored. This confers weight advantages. The gearbox housing itself is of end-loaded design with integrally cast clutch bell housing, and is thus very rigid [12.54].

### 4/ Single-Stage 6-Speed Manual Passenger Car Gearbox; Opel F28-6

More and more powerful models of passenger cars with front-wheel drive are being produced. This has led to the development of 6-speed gearboxes as single-stage passenger car gearboxes as well [12.3] *(the basic concept is explained in Section 6.6.1 and Figure 6.20b)*.

The gearbox in Figure 12.6 is an example of this type. The gearwheels of the individual gears are mounted on the input shaft as follows, starting from the clutch side. The gearwheels of first and second gear are located at the bearing, and their idler gears and synchronizers are located on the countershaft.

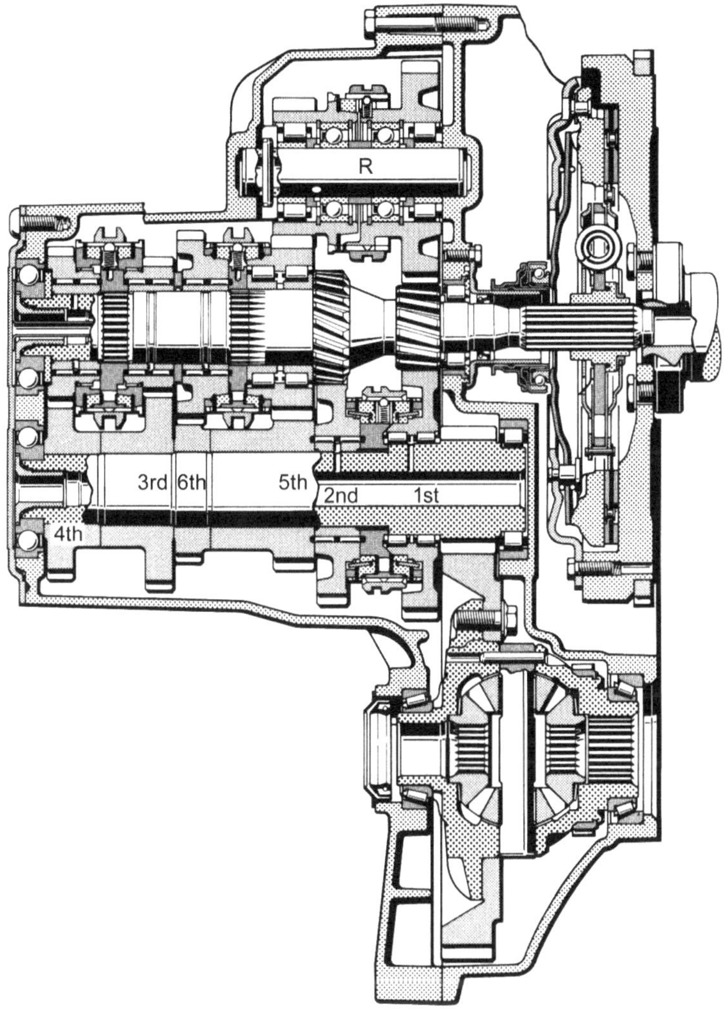


**Fig. 12.6.** 6 speed manual passenger car gearbox Opel F28-6, *gearbox diagram Figure 6.20b*

Then come gearwheels of fifth and sixth gear, and finally those of third and fourth gear. Their idler gears are mounted on the input shaft. The design solution adopted for reverse gear is of interest, which dispenses with additional gearing on the input shaft. It is shifted on its own countershaft. This is not located in one plane with the other shafts, as shown in the Figure, but is spatially displaced [12.5].

The power flow in reverse gear is from the fixed gear of first gear to its idler gear, from there to the countershaft of the reverse gear, and from there on to the fixed gear of fifth gear via its idler gear. This design makes it possible for the overall length of the gearbox to be very short. In this case, it was actually possible to reduce the width compared to its predecessor with five gears.

All gears are synchronized. The first and second gears are fitted with a double-cone synchronizer. The operating elements are not shown in the figure; they have rolling bearings at all bearing points. The structure of the gearbox housing is similar to the gearbox presented in Figure 12.1, so the comments in that case apply here too.

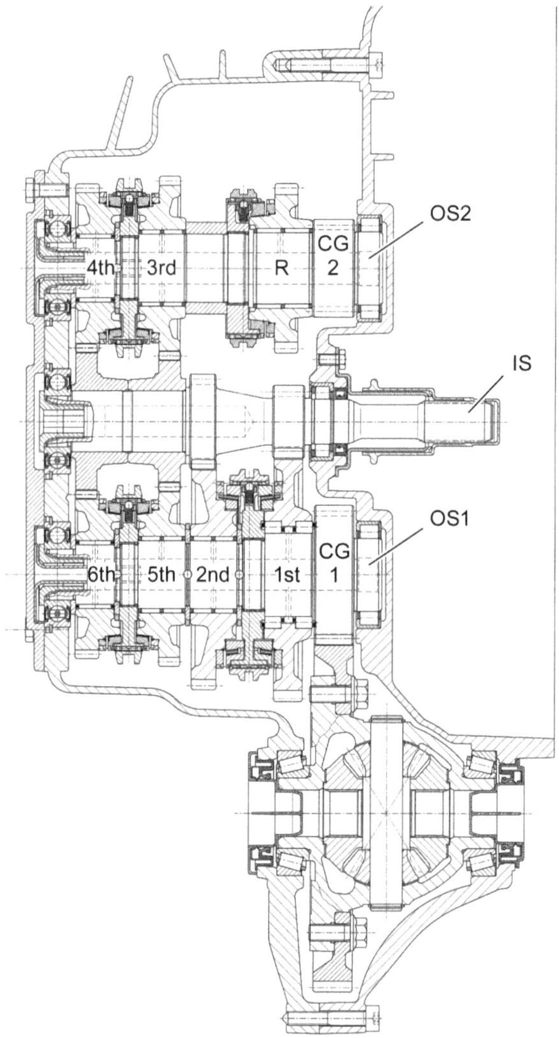


**Fig. 12.7.** 6-speed manual passenger car gearbox Getrag 285, *gearbox diagram Figure 6.21a*

### 5/ Single-Stage 6-Speed Manual Passenger Car Gearbox; Getrag 285

Figure 12.7 shows a three-shaft gearbox developed to have the shortest possible overall length. All gears are synchronized, including the reverse gear. The first and second gears have double-cone synchronizers, and the remaining gears have single-cone synchronizers [12.27] *(the basic concept is explained in Section 6.6.1 and Figure 6.21a)*.

The individual gears are arranged on the gearbox input shaft *IS* as follows (starting from the clutch side). Located directly next to the bearing of the input shaft are the first and reverse gears. The shift gear for first gear is on the output shaft *OS1*, the shift gear for reverse gear on the output shaft *OS2*. The reverse gear uses the shift gear of the first gear as a reverse idler gear. Running on two cylindrical roller bearings, the mounting is therefore more elaborate than for the other shift gears, which run on simple needle roller bearings. Following the fixed gear of the second gear are the two fixed gears for the third and fifth gears and for the fourth and sixth gears. Due to the fact that the third and fifth gears and the fourth and sixth gears share their respective fixed gear (double-uses), the input shaft *IS* only has four fixed gears, which allows it to remain short. The power flow is through the output shaft *OS1* in fifth and sixth gear as well as in first and second gear. In third and fourth gear as well as reverse gear, the power flow is through the output shaft *OS2*. All three shafts are mounted via a classic locating/non-locating bearing arrangement, using ball bearings as locating bearings and drawn cup roller bearings as non-locating bearings.

Figure 12.8 shows the side view. Because the synchronizing units on the two output shafts are far apart, a long selector shaft with two selector fingers is required.

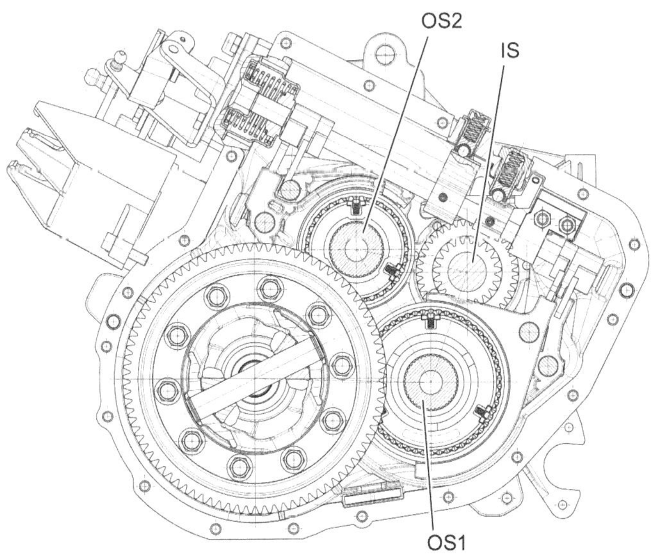


**Fig. 12.8.** 6-speed manual passenger car gearbox Getrag 285

There are two synchronizing units on each output shaft activated by identical selector fingers. The cam profiles necessary for locking are between the selector fingers. The selection forces holding the gearshift lever in the third/fourth gear gate are created by means of coil springs in the gearshift cover. Selecting in other gates yields corresponding restoring forces.

### 6/ Single-Stage 6-Speed Manual Passenger Car Gearbox; Mercedes-Benz FSG 300-6

The gearbox shown in Figures 12.9 and 12.10 is similar in layout to that shown in Figure 12.7. However, only one fixed gear with double-use is used in this case. While this increases the freedom of ratio selection, it also increases the gearbox length *(the basic concept is explained in Section 6.6.1 and Figure 6.21b).*

Located directly next to the main bearing is the first gear, followed by the second gear. The sixth and third gears, with separate fixed gears, are located at the centre, followed by the fourth and fifth gears, which share a common fixed gear.

The power flow of first to fourth gear is through the upper output shaft *OS1*, while that of fifth, sixth and reverse gear through the lower output shaft *OS2*. With this arrangement, the output shaft, with its smaller number and geometrically smaller gears, dips into the oil sump, which results in smaller drag torques [12.16].

The reverse gear has a separate reverse idler gear for reversing the rotational direction. This idler gear is also stepped so as to achieve the necessary ratio. In the gearshifting pattern, reverse gear is located to the immediate left of second gear.

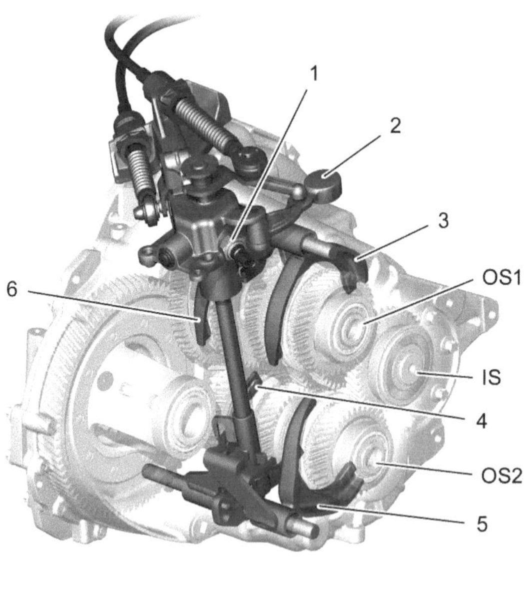


**Fig. 12.9.** Shifting mechanism of the 6-speed manual passenger car gearbox Mercedes-Benz FSG 300-6.
*1* Reverse light switch;
*2* shifting weight;
*3* shift fork, 3rd/4th gear;
*4* shift fork, reverse gear;
*5* shift fork, 5th/6th gear;
*6* shift fork, 1st/2nd gear

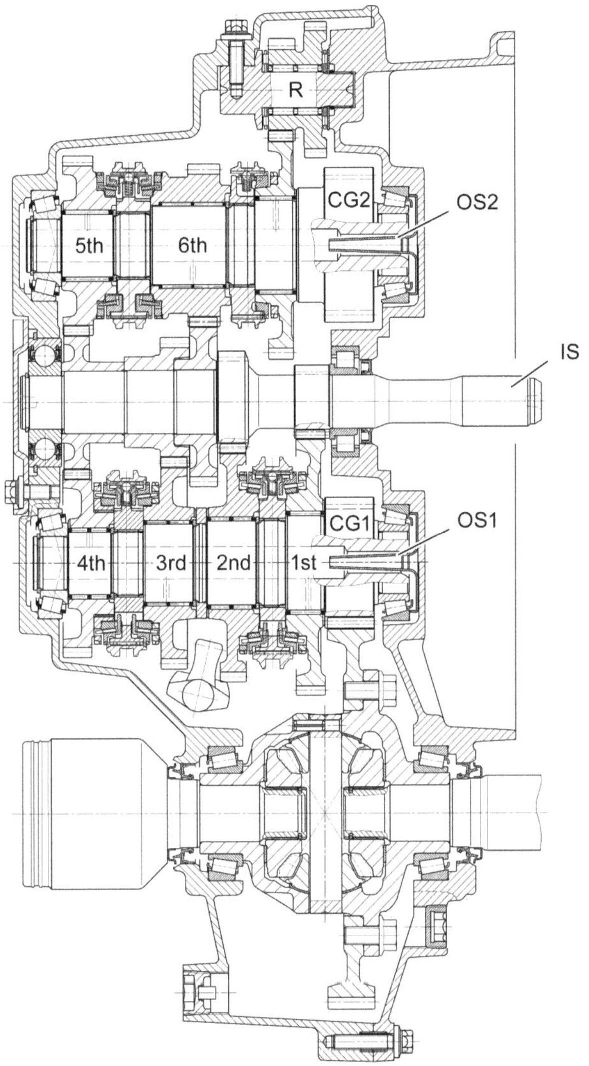


**Fig. 12.10.** 6-speed manual passenger car gearbox Mercedes-Benz FSG 300-6, *gearbox diagram Figure 6.21b*

The selector shaft with two selector fingers is also similar to that of the Getrag 285 gearbox, but instead of the selector shaft lying diagonally over the input shaft, it lies in this case between the gear set and the differential, which yields a highly compact design (see Figure 12.9).

### *7/ Two-Stage 6-Speed Manual Passenger Car Gearbox; Getrag 217*

Another example of 6-speed manual transmission is provided by the Getrag 217 shown in Figure 12.11 *(the basic concept is explained in Section 6.6.1 and Figure 6.22a)*. The gear configuration in the variant for petrol engines shown here corresponds to that of the transmission shown in Figure 12.5. Since this gearbox is designed for smaller engines, the design is more compact and cost-effective that the one in Figure 12.5. Thus, simple ball bearings are used as main bearings instead of double angular contact ball bearings. The shafts are not friction welded, but rather deep drilled. Although the gearbox is, relatively speaking, somewhat heavier, this is compensated for by its cost-effectiveness. Because the masses needing to be synchronized are smaller, smaller synchronizers can be used; in first and second gear, double-cone synchronizers are used instead of triple-cone synchronizers.

This gearbox is designed for petrol and diesel engines. Since a greater overall gear ratio is required for diesel engines because of the smaller rotational speeds, the fourth gear is used as direct gear instead of fifth gear, as in the petrol variant.

The configuration of the third and fourth gears is thus exchanged for that of the fifth and sixth gears and the direct gear is an even gear, not an odd one. This must be compensated for by modifications to the internal gearshift system in such a way that the usual gearshifting pattern is realised on the gearshift lever. By changing few components, shifting can be achieved for both variants, while simultaneously fulfilling increasing standards of shifting comfort. Since for one of the variants direct gear is an even gear and an odd gear for the other one, the shifting direction must be reversed for one of the variants.

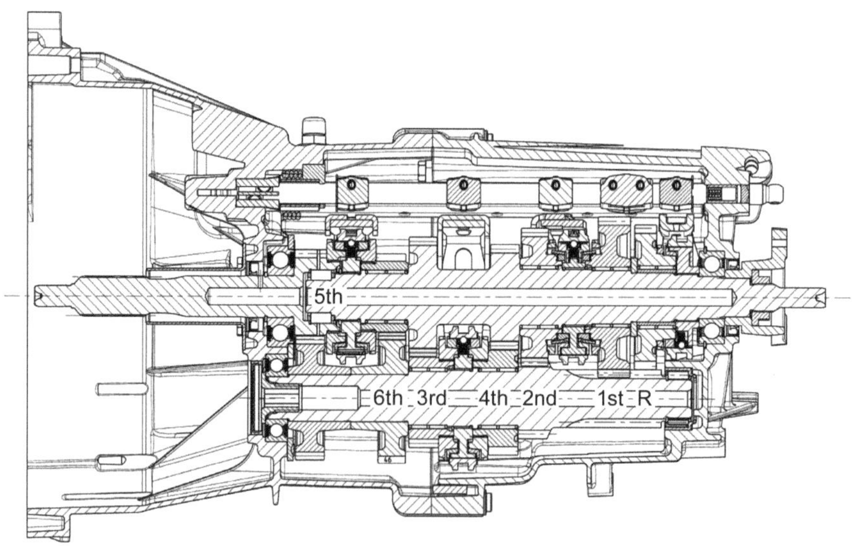


**Fig. 12.11.** 6-speed manual passenger car gearbox Getrag 217, spark ignition engine design; *gearbox diagram Figure 6.22a*

This is achieved by using swing forks or shift forks. In the case of swing forks, the motion at the gear sliding sleeve is opposed to that of the selector shaft. The interchanging of gear pairs 3, 4 and 5, 6 is compensated for through different positions of the selector fingers on the central selector shaft and through modified engaging elements on the forks. The interlock mechanism preventing the simultaneous selection of multiple gears is the same for both variants.

### 8/ Single-Stage 6-Speed Manual Passenger Car Gearbox; Getrag 466 (Audi ML350/450)

The Audi ML350-6F single-stage 6-speed gearbox shown in Figure 12.12 is designed for front-longitudinal applications *(the basic concept is explained in Section 6.6.1 and Figure 6.23a)*. This gearbox thus contains a final drive with differential *1*, as with front-transverse drives. In order to achieve a compact design with small centre distance, while being able to transmit high torques, the gearbox has a triple bearing system, i.e. the main shafts run at the centre on additional roller bearings.

Since the bearings are thus statically overdeterminate, the components involved must be correspondingly closely tolerated. Also, the roller bearing internal clearance is increased to prevent distortions. The output shaft runs on the pinion side in a double-row angular contact ball bearing which can absorb the high radial and axial forces of the powertrain. The input shaft is very long as a result of the differential between the clutch and the gear set.

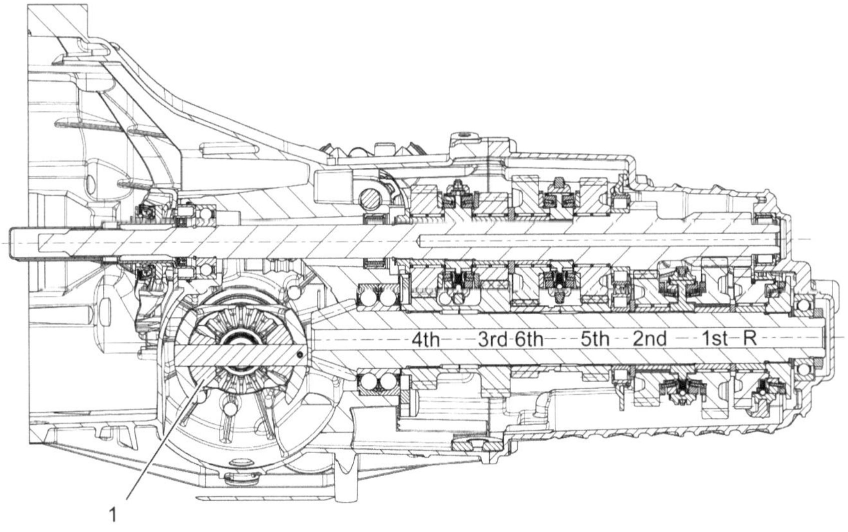


**Fig. 12.12.** 6-speed manual passenger car gearbox for front-longitudinal drive Audi ML350-6F; *gearbox diagram Figure 6.23a. 1* Front axle differential

The locating bearing arrangement is thus divided into two bearings: a roller bearing for taking up the radial forces near the gear set and a four-point contact bearing which principally absorbs the axial force, but also minimises deflection in the clutch area [12.14].

The front-wheel gearbox can be converted with relatively little effort into a gearbox for all-wheel drive (see Figure 12.64). The output shaft with pinion head is thereby replaced by a hollow shaft and a pinion shaft mounted inside it. A centre differential – in this case a TORSEN differential – is integrated at the end of the gearbox. The centre differential distributes the torque at a ratio of 50 : 50 to the front and rear axles.

### 12.1.2 Automated Manual Passenger Car Transmissions (AMT)

#### *9/ Single-Stage 6-Speed Passenger Car AMT; Getrag 431*

The first gearbox in mass production developed purely as an AMT is the Getrag 431, implemented in Smart cars (Figure 12.13). Since this gearbox was developed as an AMT, no account had to be taken of the gearshift pattern in designing the gear set and gearshift system *(the basic concept is explained in Section 6.6.2 and Figure 6.25a).*

This is a single-stage 6-speed gearbox with integrated actuator technology. The six gears are created via a 3-speed main gearbox with reverse gear and the output constant gears $CG_{\mathrm{H}}$ and $CG_{\mathrm{L}}$ of the rear-mounted range-change unit. The gears on the input shaft are arranged as follows. The third/sixth gear is on the clutch side. Contrary to the usual "design rule" dictating that gears with high forces be arranged near bearing planes, the gear pairs for the first and reverse gears are in the shaft centre. Due to the relatively small torques, other criteria were more highly weighted, such as package and costs. The gearing for the second/fifth gear is at the end of the shaft.

This concept allows for cost- and space-effective 6-speed transmissions designs. There are limitations with respect to ratio selection and gear steps resulting from the multi-range design. As opposed to conventional 6-speed gearboxes, however, this gearbox cuts down on one synchronizing unit plus gearshift mechanism. As with all AMTs, gear set components such as gears, shafts, and synchronizers are identical to those of manual transmissions. All main bearings are deep groove ball bearings. Because of the small torque to be transmitted, the differential can also run on ball bearings instead of the usual tapered roller bearings.

The advantages of this design are an improved level of efficiency and a more simple assembly. The differential cage consists of two aluminium die cast shells bolted together with the two output gear rings.

With this gearbox, the internal gearshift system is not designed as with manual transmissions. The gear sliding sleeves are operated by an electrically activated gearshifting drum *2* (Figure 12.14). The shift forks *3* which engage in the grooves are moved axially by the rotation of the gearshifting drum; they are simultaneously locked against each other.

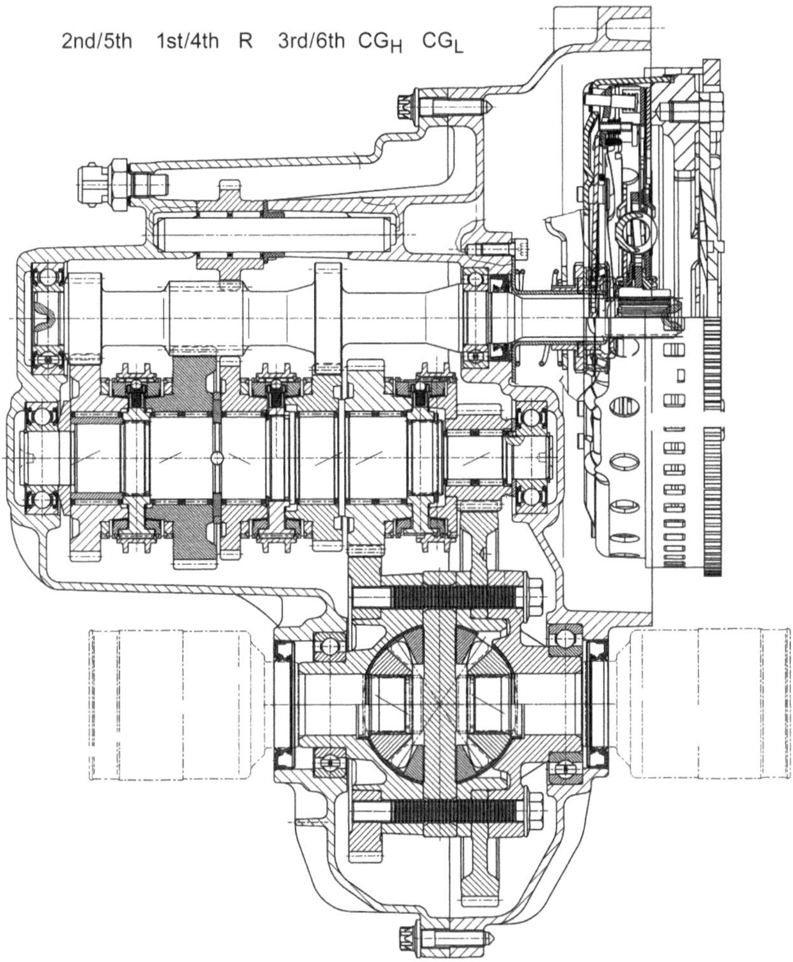


**Fig. 12.13.** 6-speed passenger car AMT Getrag 431 in multi-range design, *gearbox diagram Figure 6.25a*

This prevents the simultaneous, unintentional shifting of two gears. The groove profile determines the shifting characteristic. When shifting from third into fourth gear, two gear sliding sleeves are engaged via the gearshifting drum at the same time. Both the gear stage and the axle drive are shifted.

Thus only automated transmissions can be built with this gear set arrangement. The electromechanical actuator technology is an integral part of the transmission design. Due to the gearshifting drum, only serial shifting is necessary, i.e. skipping gears is not possible, thus all the gears must be shifted in sequence. Newer designs with two independent gearshifting drums offer advantages here in terms of multiple downshifts.

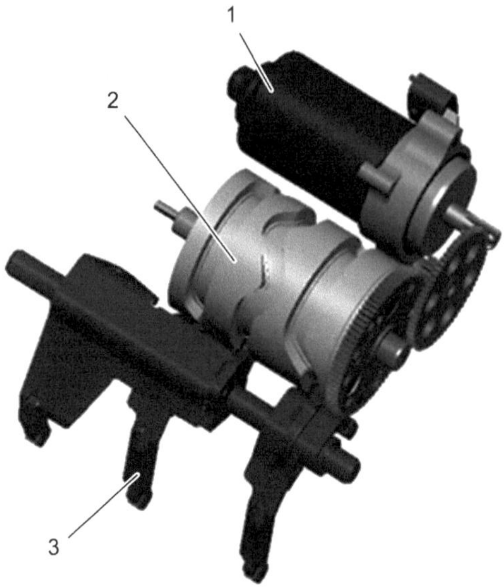


**Fig. 12.14.** Electromechanical activation of the sliding sleeves of the gearbox Getrag 431.
*1* Electric actuator;
*2* gearshifting drum;
*3* shift fork

### *10/ Two-Stage 7-Speed Passenger Car AMT; Getrag 247*

Figure 12.15 shows a two-stage 7-speed AMT corresponding to the gearbox which went into series production in the BMW M5 (E60) in 2004 *(the basic concept is explained in Section 6.6.2 and Figure 6.25b).*

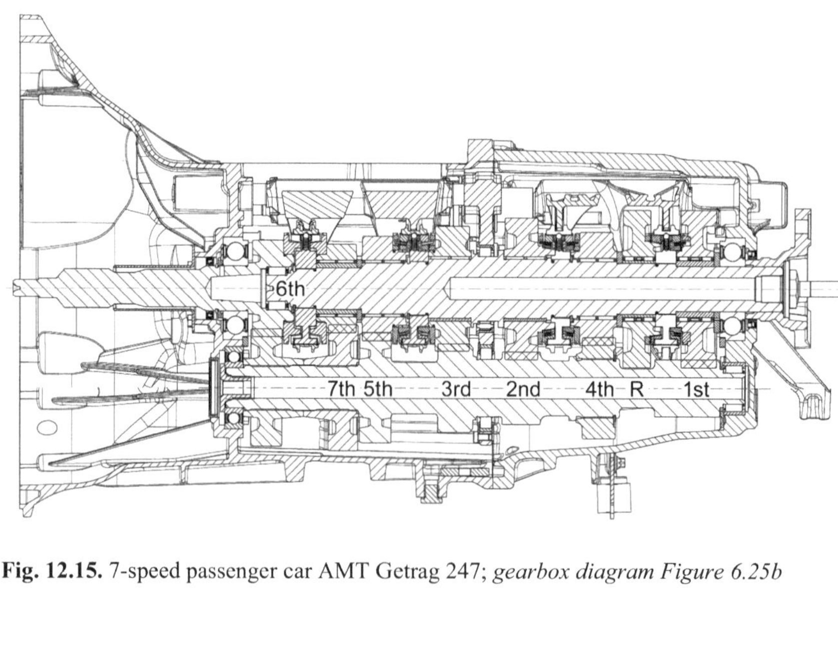


**Fig. 12.15.** 7-speed passenger car AMT Getrag 247; *gearbox diagram Figure 6.25b*

This gearbox was developed as an automated gearbox with integrated actuator technology [12.40]. This enables an AMT-specific gear set arrangement to be realised in which, unlike MTs, consecutive gears – with the exception of sixth and seventh gear – are located on different synchronizing units. Independent activation of the current and target gears when shifting allows for overlapping and thus shorter shifting processes. This minimises the duration of power interruptions. When disengaging the current gear, the target gear can already be synchronized. Due to the positive engagement, however, real overlapping gearshifts, such as with powershift transmissions with friction clutches, must be absolutely avoided. A simultaneous positive engagement of two gears would cause the blocking of the gearbox. To ensure the synchronizing units stand up to the high thermal stresses resulting from the short shifting times and high differential speeds, the friction linings are made of highly resilient carbon coatings.

The sixth gear is direct gear, i.e. the input and output shafts are connected. There are no gearwheels in the power flow. In order to minimise shaft deflection and to transmit torques which are high relative to the centre distance, the gearbox has an intermediate bearing plane. The bearings are thus overdeterminate, and the precision of the components must be correspondingly high in order to prevent bearing stress due to the design. However, all highly stressed gears can be arranged on a bearing plane and thus input torques can be transmitted which are high relative to the centre distance.

The gearshift mechanism functions hydraulically by means of single-bar actuators, producing extremely rapid gearshifts. The hydraulic actuators engage directly in the shift forks, which are made of die cast aluminium. Figure 12.16a shows the hydraulic unit mounted in a completely filled state on the gearbox and Figure 12.16b shows the shift fork set.

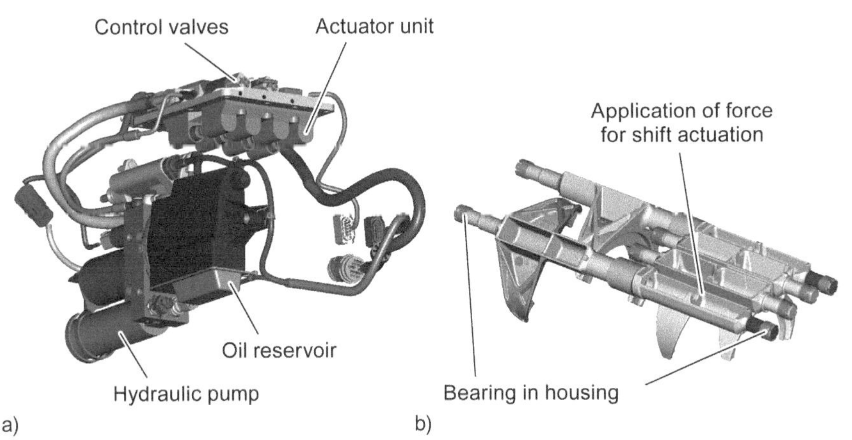


**Fig. 12.16.** *a* Hydraulic unit for 7-speed passenger car AMT Getrag 247; *b* shift fork set

### 12.1.3 Dual Clutch Passenger Car Transmissions (DCT)

#### *11/ Single-Stage 6-Speed Passenger Car DCT; VW DSG®*

Figure 12.17 shows the VW 6-speed front-transverse DCT which went into production in 2003 *(the basic concept is explained in Section 6.6.3 and Figure 6.26).* The two sub-gearboxes are nested in each other. The main section bears similarity to a manual transmission with a three-shaft design [12.62].

The engine drives the outer plate carriers of the clutches *C1* and *C2* via a dual mass flywheel. The inner plate carriers of the two clutches are connected via stub shaft spline to the input shafts of the two sub-gearboxes: *sub-gearbox 1* with the internal throughout input shaft *IS1* (first, third, fifth and reverse gear) and *sub-gearbox 2* with the shorter input shaft *IS2* designed as a hollow shaft (second, fourth and sixth gear).

The synchronizers are mounted on the output shafts. First and third gear, second and fourth gear, and sixth and reverse gear are combined respectively into synchronizer units. The lower gears are designed as multi-cone synchronizers in order to satisfy the high demands with respect to shifting speed.

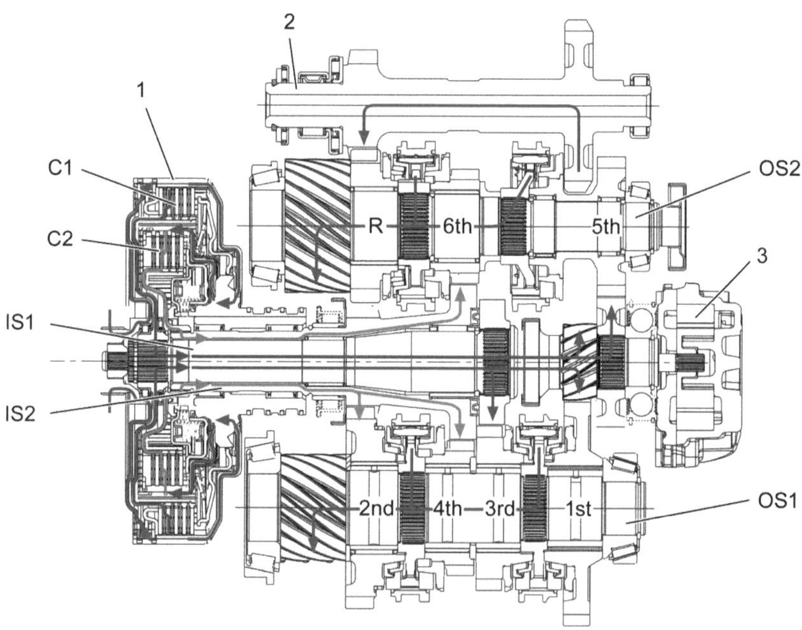


**Fig. 12.17.** 6-speed passenger car DCT VW DSG®; *basic diagram Figure 6.26.*
Main section with power flow. *1* Wet-running dual clutch with *C1* and *C2*; *2* reverse idler shaft; *3* oil pump; *IS1* input shaft of sub-gearbox 1 (1st/3rd/5th/R); *IS2* input shaft of sub-gearbox 2 (2nd/4th/6th); *OS1* output shaft 1 with output constant pinion; *OS2* output shaft 2 with output constant pinion

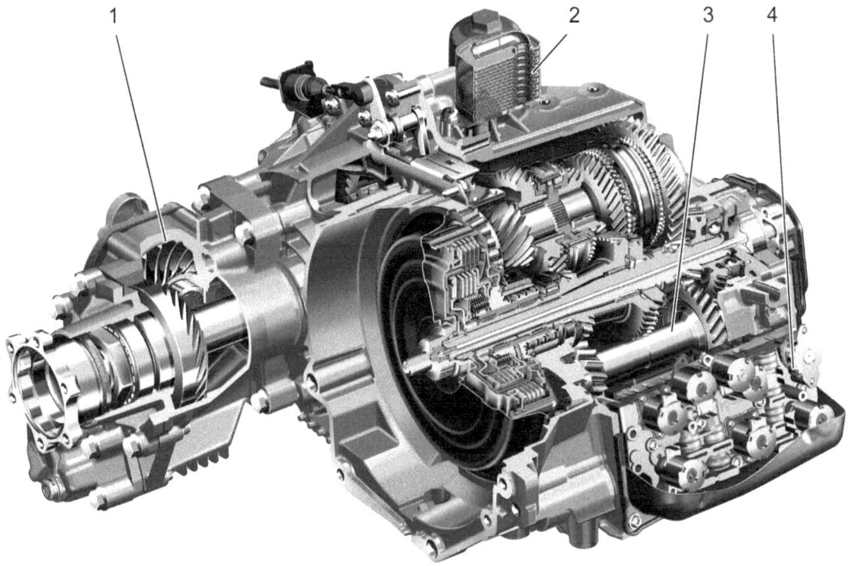


**Fig. 12.18.** 6-speed passenger car DCT VW DSG®. *1* Transfer gearbox for all-wheel drive; *2* oil cooler; *3* reverse idler shaft; *4* mechatronic module

The two output constant gears engage into the final drive gear not illustrated here, which drives a bevel gear differential. A transfer gearbox for all-wheel drive can also be flanged to the end of the gearbox power flow. This and the spatial arrangement of the gearbox components are shown in Figure 12.18.

The wet-running dual clutch is one of the essential elements of the gearbox (see Figure 10.15 in Section 10.3 "Dual Clutches"). Power is transmitted from the engine via the external stub shaft spline of the dual clutch, which is positively engaged with the dual mass flywheel. Power is transmitted to the input shafts *IS1* and *IS2* via the plate carriers of the respective clutches *C1* and *C2*. In the process, the torque is transmitted via the plate set from the external gearing to the internal gearing. The two input shafts are each positively engaged with the internal gearings of the dual clutch. The clutch is designed for 350 Nm.

The mechatronic module is mounted as a local controller on the front of the gearbox (see Figure 12.19). It combines the hydraulic module *1* with the clutch modulators *2* (pressure control valve), the shift valves and the corresponding hydraulic valves, as well as an electronic module *3* consisting of the gearbox computer and a sensor cluster. This actual control unit of the gearbox is connected directly to the vehicle wiring harness by means of a central connector *4*.

In addition to input and output speeds, the sensor system detects the respective positions of the four gear actuators and diverse temperatures. The essential components of the actuator technology are the two clutch modulators, the cooling oil modulator, as well as the shift valves of the gear actuator unit.

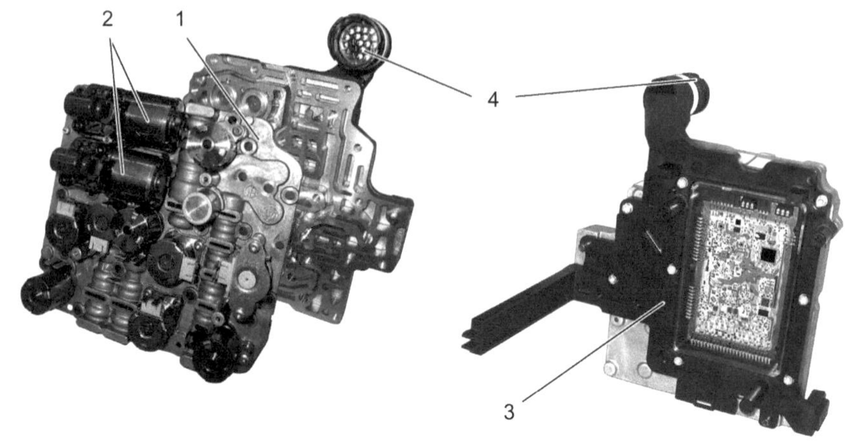


**Fig. 12.19.** Mechatronics of the VW DSG®. *1* Hydraulic module; *2* clutch modulators; *3* electronic module; *4* vehicle-side connector

In order to achieve high shifting speeds and resultantly high shifting comfort, especially high importance is attached to the quality with respect to control quality and to the speed of the clutch modulators.

Analogously to manual transmissions, the individual gears are shifted via gear actuators. In the case of the VW DSG, however, these are hydraulically controlled and run on both sides in low-friction ball sleeves *1* (Figure 12.20). The location of the respective shift fork can be determined by means of magnets *2* mounted on the fork and a hall sensor on the sensor system module. It is possible to detect an arbitrary number of intermediate positions between defined centre and end positions, which makes it possible to control the movements of gear actuators.

Because of the manifold demands which had never presented themselves before in this combination, a special gearbox oil was developed for use in the VW DSG.

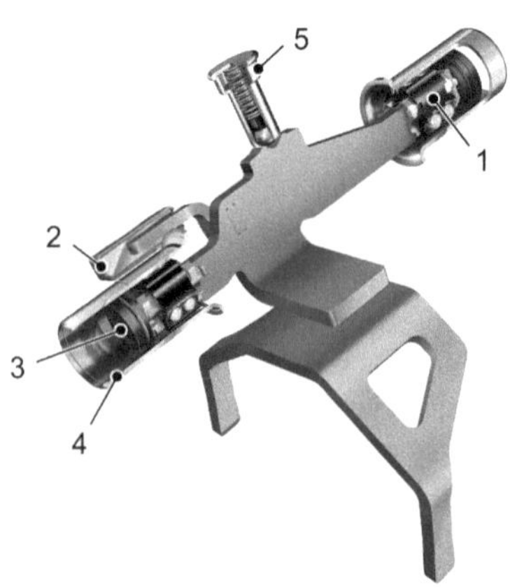


**Fig. 12.20.** Shift fork with ball sleeves for 1st/3rd gear. *1* Ball sleeve; *2* magnet with position sensor; *3* gear actuator piston ; *4* gear actuator cylinder; *5* detent unit

The essential criteria were the demands of the tribological system clutch in terms of the friction coefficient profile and stability as well as the protection of the gearing from wear. Further considerations are suitability in the synchronizing area, power loss in the bearing system, the general compatibility of materials and the long-term behaviour of the gearbox.

The oil pump (*3* in Figure 12.17) for supplying the oil circulation with pressure oil is located on the side of the gearbox away from the engine and is driven directly with the engine speed via a central shaft running within the input shaft *IS1*. The pump is designed as a crescent pump with a cavern design and trochoidal gearing. Since the total pump power absorbed represents a major difference to manual gearboxes in terms of overall power loss and a significant single source of power loss, optimising the pump efficiency and power absorption became particularly important. See also Section 11.3 "Oil Supply and Oil Pumps".

Within the oil circulation, the pump first sucks the gearbox oil through a suction filter. In addition, excess pressure oil, especially with higher speeds, is conducted directly by the mechatronics into the suction side of the pump, which limits the volume flow at the suction filter. The pressure oil is available to the mechatronics for controlling the clutches and gear actuators. The clutch coolant is also fed directly from the pressure oil. A partial flow is conducted via an oil cooler (*2* in Figure 12.18) which, because of the small gearbox power loss, is directly integrated into the water circulation of the engine. Then this oil flow is guided via a pressure filter into the oil injection lubrication system.

### 12/ Two-Stage 7-Speed Passenger Car DCT; ZF 7 DCT 50

The 7 DCT 50 dual clutch transmission introduced in 2005 by ZF is a 7-speed dual clutch gearbox for standard drive [12.37]. Its coaxial output distinguishes it structurally from the VW dual clutch gearbox for front-transverse applications treated above. While the latter gearbox has two output constant gears, the ZF design has two input constant gears *(the basic concept is explained in Section 6.6.3 and Figure 6.27)*.

The engine torque is induced via the dual mass flywheel (not shown in Figure 12.21) mounted in the dry area. The input hub transmits the torque to the outer plate carriers of the two wet-running clutches of the dual clutch module *1*. Short shifting times constitute a major goal in designing DCTs. This requires small moments of inertia to be synchronized. A contributing factor is that the input shafts are connected by means of the inner plate carriers having lined plates which are more advantageous with respect to the moment of inertia. See also the gearbox diagram in Section 6.6.3, Figure 6.27. If one continues to follow the power flow in the gearbox, one reaches the countershaft via the two input constant gears connected to the two clutches via a solid and a hollow shaft.

To achieve high gearbox efficiency and to be able to realise engine speeds of up to 9000 l/min, an axially parallel pump *2* and injection lubrication are included as further features of this design. The oil pump is designed as an internal gear pump driven via a spur gear stage. The driving gear is directly mounted on the outer plate carrier of the dual clutch rotating with the engine speed.

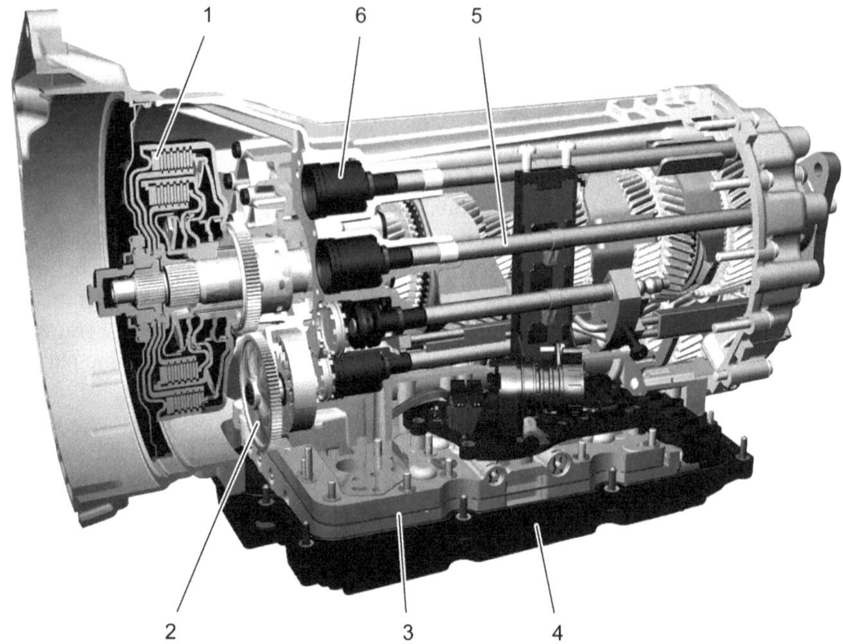


**Fig. 12.21.** 7-speed passenger car DCT for standard drive ZF 7 DCT 50; *gearbox diagram Figure 6.27. 1* Dual clutch; *2* axially parallel oil pump; *3* hydraulic/mechatronic module; *4* oil pan; *5* selector bars; *6* double-acting shift cylinder

Drive via spur gear stage has the advantage that the pump can be adjusted over different ratio steps in terms of its pump characteristics or its maximum speed, depending on the application. A further advantage is that, because of the resulting variable spatial design, the ratio of pump width to pump diameter can be optimised for the overall level of pump efficiency. Through the injection lubrication, combined in this gearbox with a dry sump lubrication system, oil can be directly injected into the gearing for improved heat removal. Moreover, there are no perceptible losses caused by gearings splashing in the oil sump.

Gears are preselected through the double-acting cylinder *6* activated via an electrohydraulic control unit *3* with pilot-operated valves and pressure regulators.

### *Manually Shiftable Dual Clutch Transmissions*

According to their design principle, dual clutch transmissions offer a high flexibility with respect to possible ratios; however, the gears are arranged differently than in manual transmissions. This means that DCTs cannot be shifted by hand. Although the ratios of dual clutch and manual transmissions can principally be identical, the gear sets of the respective systems differ in the configuration of their gear pairs, i.e. more specifically in terms of shaft designs, synchronizer connections and toothing correction. Component sharing strategies for MT/DCT require design measures at the gearshift system.

There are approaches for kinematically decoupling the gearshift lever from the internal gearshift system, thus facilitating a flexible arrangement of gear planes independent of the usual H gearshift pattern. Not only can a manually shiftable dual clutch gearbox be derived this way on the basis of a dual clutch gear set; a gearbox kit can also be designed which has a common gear set for dual clutch gearboxes, automated manual gearboxes and manual gearboxes [12.11].

### 12.1.4 Automatic Passenger Car Transmissions (AT)

The first 5-speed conventional automatic transmissions for passenger cars went on the market in 1989. They were initially designed for standard drive. Twenty years later, 4-speed automatic transmissions are still being offered for front-transverse applications, depending on the vehicle and the region.

#### *13/ 5-Speed Conventional Automatic Passenger Car Gearbox; Mercedes-Benz W5A 030*

The design of the 5-speed automatic passenger car gearbox W5A 030 from Mercedes-Benz from the year 1989 stands at the transition between a purely hydraulic control unit to an electronic-hydraulic control unit. It also exhibits other interesting characteristics which will be discussed in the following (Figure 12.22). The gearbox was derived from the W4A 040 4-speed automatic transmission from 1979 by adding a planetary gear [12.61]. The 4-speed part still has a purely hydraulic control unit; only the 5th gear attached onto it is electrohydraulically controlled.

The gearbox housing with integrally cast clutch bell housing is divided into a front and a rear chamber by a bulkhead at the level of the breather unit *1*. The front chamber contains basically the same assemblies as in the 4-speed gearbox, and the fully hydraulic control of the 4 speed part is also taken from the 4-speed gearbox. The planetary gear set *2* in the rear chamber rotates as a block in the first four gears, providing a speed increasing ratio in fifth gear, with the 4-speed part rotating as a block. The 4-speed part comprises a Ravigneaux gear set *3* with a rear-mounted simple planetary gear set *4*.

The front chamber is separated from the torque converter by a cover *5*, which also accommodates the primary oil pump *6*. This cover has channels for the oil pump, and also serves as the cylinder for the actuating piston *7* of the multi-plate brake *8*. Some of the brakes are multi-plate and some belt brakes *9*. The belt brakes in particular are very compact, using the outer surface of a clutch plate carrier *10* as brake surface. No lock-up clutch was included in the "hard" torque converter *11* (see also Section 10.4.4).

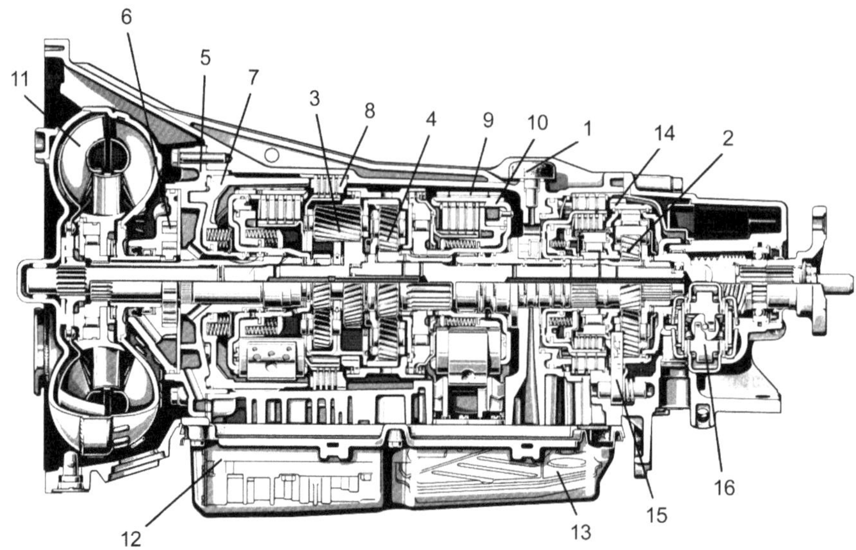


**Fig. 12.22.** 5-speed automatic gearbox Mercedes-Benz W5A 030. *1* Breather unit; *2/4* planetary set; *3* Ravigneaux gear set; *5* cover; *6* primary oil pump; *7* actuation piston; *8* multi-plate brake; *9* belt brake; *10* clutch plate carrier; *11* converter; *12* control plate; *13* oil pan; *14* parking lock wheel; *15* parking lock pawl; *16* centrifugal governor

The control plate *12* of the hydraulic transmission control unit has valves directing the flow of oil into the various channels to control the transmission, and is located on the bottom of the transmission. The control plate is enclosed by the sheet metal oil pan *13* containing the oil sump.

The parking lock wheel *14* is located in the rear chamber, and locks the gearbox output shaft by means of the parking lock pawl *15* (see Section 9.4). The gearing of the parking lock wheel is fitted externally on the ring gear of the planetary gear set, saving axial space. To enable towing of the vehicle, the gearbox has a secondary oil pump mounted on the shaft of the centrifugal governor *16*, which provides the control pressure in the oil when the vehicle is moving and the engine is at rest. The centrifugal governor actuates a pressure control valve, and depends on the velocity of the vehicle, which supplies a speed-dependent control pressure.

### *14/ 5-Speed Conventional Automatic Passenger Car Gearbox; Mercedes-Benz W5A 580 (330, 900)*

The 5-speed passenger car automatic gearboxes W5A 330, 580 and 900 by Mercedes-Benz were introduced in 1995 with the W5A 580. *The basic function of this design is described in detail in Section 6.6.4 and Figure 6.32.* The three gearbox variants differ in their torque capacity – 330, 580 and 900 Nm, respectively. Each having three planetary gear sets and six powershifting elements, the gearboxes are all based on a simple and structurally very short system. Depending on the design, the weight ranges from 75 to 80 kg. Figure 12.23 show a middle section of the W5A 580.

The gearbox has an internally toothed crescent oil pump *3* (see Section 11.3.2) driven by the converter *1* [12.56]. This is a 3-line converter (see Section 10.4.6). A special feature of this gearbox is the two-piece housing *5* and *15* of die cast aluminium. By disposing of a partition wall, it is both lighter and adaptable to different engines. All planetary gears are mounted on needle roller and cage assemblies, with two planetary gear carriers (*17* and *19*) being made of die cast aluminium. In addition to weight and production costs, another advantage is the integration of inner plate carriers. The parking lock is designed as a cone system (see Section 9.4).

The electrohydraulic control unit *18* consists of two die cast aluminium components, an intermediate sheet and the electric kit, with the transmission control unit mounted externally in the vehicle. Gearshifting is controlled by electrohydraulically regulated overlapping control (see Section 9.3.2). The engaging gearshift element is controlled with the shift pressure. By means of the overlapping, pressure reduction takes place at the disengaging element.

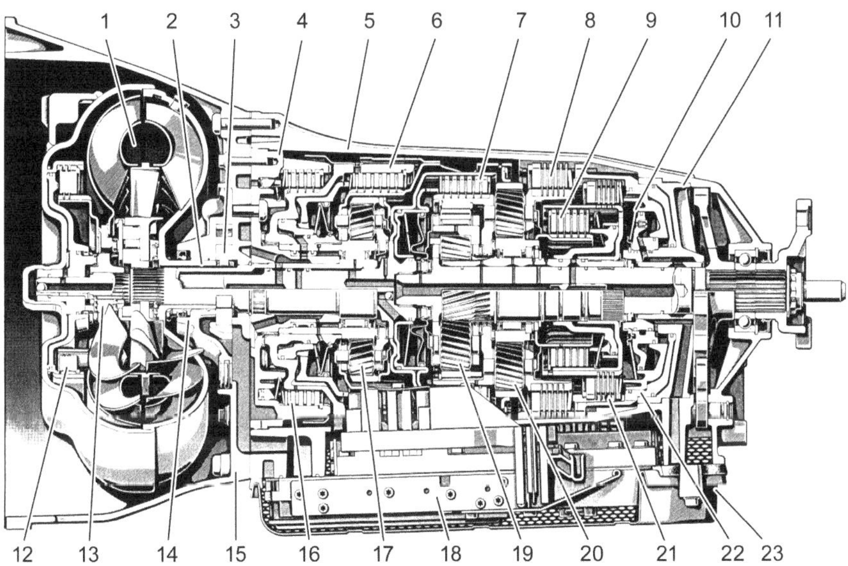


**Fig. 12.23.** 5-speed automatic passenger car gearbox Mercedes-Benz W5A 580; *gearbox diagram Figure 6.32. 1* 3-line converter; *2* oil pump drive; *3* oil pump, *4* outer plate carrier; *5* gearbox housing; *6* outer plate carrier; *7* multi-plate clutch *C2*; *8* multi-plate brake *BR*; *9* multi-plate clutch *C3*; *10* return spring *B2/BR*; *11* cylinder guidance B2/BR; *12* torque converter lock-up clutch; *13* turbine wheel flange; *14* oil pump housing; *15* converter housing; *16* multi-plate brake *B1*; *17* front planetary gear set; *18* electrohydraulic control plate; *19* rear planetary gear set; *20* middle planetary gear set; *21* multi-plate brake *B2*; *22* piston *B2*; *23* oil pan

The electric kit consists of a plug connector, two magnet control valves, three switching valves, a pulse width modulation valve, two speed sensors, a temperature sensor and a starter interlock contact. The entire unit is held together by a plastic injection moulding part which accommodates all the electrical components.

## 15/ 7-Speed Conventional Automatic Passenger Car Gearbox; Mercedes-Benz W7A 700

In 2003, Mercedes-Benz introduced the first conventional automatic gearbox for passenger cars with seven forward and two reverse gears. The W7A 700, designed for a maximum torque of 700 Nm, is distinguished by a consistently light and compact design. Developing further on the W5A 580 5-speed conventional automatic transmission discussed above, the W7A 700 differs in its concept through the replacement of the front single planetary gear set with an inverse Ravigneaux gear set and the integration of an additional brake [12.21–12.22] *(the basic concept is explained in Section 6.6.4 and Figure 6.33).*

In order to realise the seven forward gears and two reverse gears, the W7A 700 requires three clutches and four multi-plate brakes. It weighs approximately 82 kg and is 621 mm long with an overall gear ratio of 6.02. The main housing is made of the magnesium die cast alloy AS 31 HP, and the flange-mounted converter bell is made out of aluminium. The housing thus weighs 2.5 kg less than if it were made entirely out of aluminium. Figure 12.24 shows a longitudinal section of the gearbox.

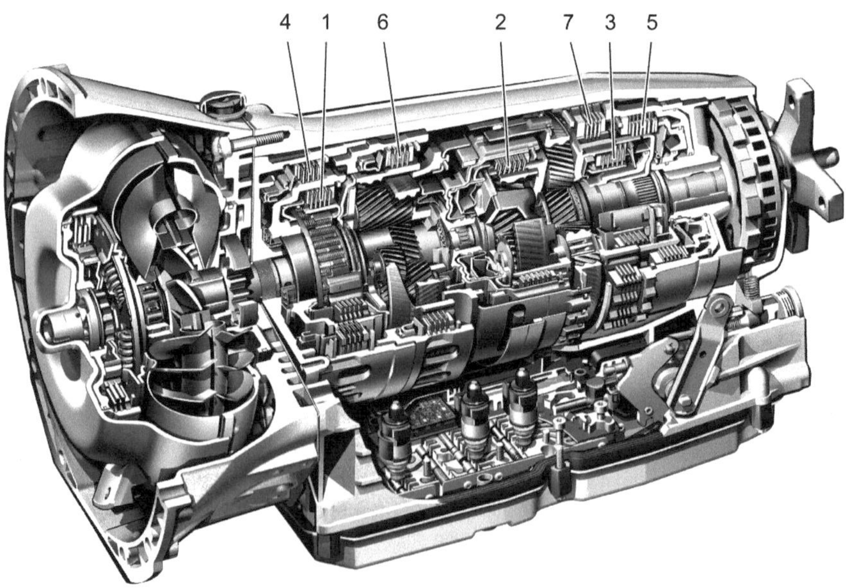


**Fig. 12.24.** 7-speed automatic passenger car gearbox Mercedes-Benz W7A 700; *gearbox diagram Figure 6.33. 1* Clutch *C1*; *2* clutch *C2*; *3* clutch *C3*; *4* brake *B1*; *5* brake *B2*; *6* brake *B3*; *7* brake *BR*

The torque converter, including the lock-up clutch with integrated turbine torsion damper, and the complete parking lock system were adopted from the W5A 580 gearbox design. The integration of the gearbox control unit into an electrohydraulic control unit (mechatronics) directly in the gearbox is also new to the W5A 580 gearbox. The sub-components of the electrohydraulic control unit integrated in the gearbox are the hydraulic control plate, an electrics kit with control unit, speed sensor technology and the eight magnet control valves. The external electrical connection is only made via five pins. The control plate comprises a valve housing in die cast aluminium, an intermediate sheet and slide valves. Multiple downshifts are made possible by the principle of direct control. This facilitates highly dynamic double or multiple downshifts of up to four gears, either direct or embedded in each other, which is perceived by the driver as one gearshift.

### 16/ 6-Speed Conventional Automatic Passenger Car Gearbox; ZF 6 HP 19, 6 HP 26 and 6 HP 32

The six-speed gearbox family ZF 6 HP 19, 6 HP 26 and 6 HP 32 for engine torque of up to approximately 350, 600 and 750 Nm is based on a gear set principle of Lepelletier and has been in use since 2001. Its overall gear ratio of 6.04 places it in an area which up to that time, as far as passenger cars are concerned, had been reserved for continuously variable transmissions (CVT) *(the basic concept is explained in Section 6.6.4 and Figure 6.34).*

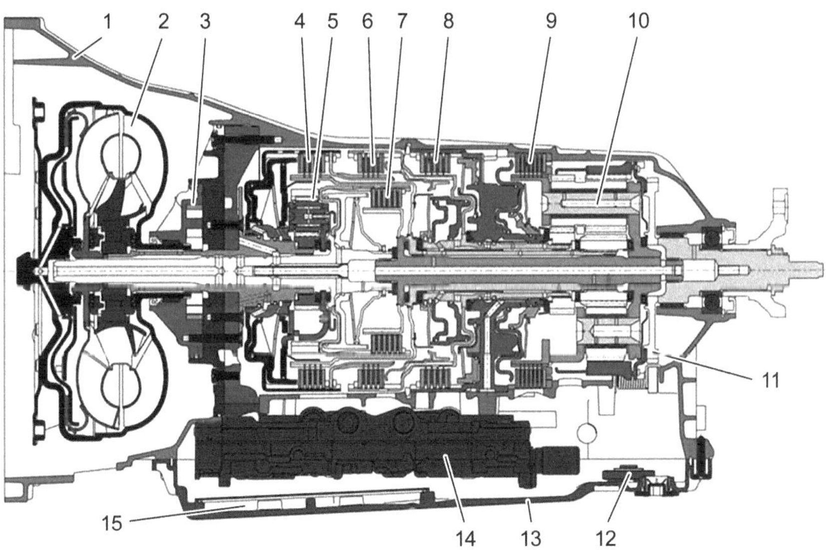


**Fig. 12.25.** 6-speed automatic passenger car gearbox ZF 6 HP 26; *gearbox diagram Figure 6.34. 1* One-piece gearbox housing; *2* 2-line converter; *3* oil pump; *4* clutch *A*; *5* gear set 1; *6* clutch *B*; *7* clutch *E*; *8* brake *C*; *9* brake *D*; *10* Ravigneaux gear set; *11* parking lock wheel; *12* magnets; *13* plastic oil pan; *14* mechatronic module; *15* oil filter

With the gear set design with single planetary gear set *5* and rear-mounted Ravigneaux gear set *10*, six forward gears and one reverse gear can be realised while using only five shifting elements (Figure 12.25). The gearbox has three multi-plate clutches (*4*, *6*, *7*) and two multi-plate brakes (*8*, *9*). The shifting elements are all built with the same diameter except for the clutch *E 7*, which is smaller in diameter. This gives the gearbox design a compact structure [12.23, 12.65].

The gearbox housing *1* with integrated converter bell is a one-piece construction made of die cast aluminium. The converter is a two-line converter (see Section 10.4.6). The front cover contains the oil supply. This comprises two die cast aluminium parts (intermediate and centring plates), as well as a pump *3*. The pump is designed as an internal gear crescent pump.

To reduce weight and costs, many parts of the gearbox are manufactured from cut or formed sheet steel (see also Section 16.4 "Process Chains for Sheet Metal Machining"). The inner and outer plate carriers are thus primarily manufactured from sheet metal. Even the Ravigneaux gear set is a rigid sheet steel design. Its planetary gears have a wide design, so as to minimise tilting and to reduce noise. To minimise weight, the oil pan is manufactured as a plastic composite part *13* with an integrated filter *15* and seal. Cast-in steel cores are used to bolt it to the gearbox housing.

These 6 HP conventional automatic gearboxes appear as examples in this book at several places. Further conceptional, design-related and manufacturing-related details regarding the 6 HP gearboxes can be found in:

- external gearshift system of an automatic transmission, *Figure 9.5*,
- positive overlapping shifting, *Figure 9.33*,
- engineering designs of multi-plate clutches, *Figure 9.41*,
- assembly of the parking lock in the gearbox, *Figure 9.48*,
- electrically activated parking lock, *Figure 9.49*,
- components of a passenger car Trilok torque converter of sheet metal design, *Figure 10.30*,
- oil supply, *Figure 11.12*,
- parallel hybrid, 6-speed automatic gearbox, BMW Active Transmission, *Figure 12.28*,
- automatic gearbox with integrated all-wheel drive, Audi Quattro AL420-6Q, *Figure 12.65*,
- manufacture of shaft, housing, cast and sheet metal parts, *Figures 16.2, 16.3* and *16.9*.

The 6 HP 26 represents the first integrated shift-by-wire system for passenger cars to go into series production. This allows for more design freedom with respect to its operating concept (see also Section 9.1.3 "Shift-by-wire"). The mechatronic module *14* in the oil sump comprises a hydraulic control unit, actuators, sensors and gearbox electronics (TCU).

The control unit comprises, among other things, six electronic pressure control valves, which control the shifting elements and the torque converter lock-up clutch, two speed sensors and an oil temperature sensor. The control plate is composed of two die cast aluminium parts with an intermediate aluminium sheet.

### 17/ 6-Speed Conventional Automatic Passenger Car Gearbox for Front-Transverse Drive; AISIN AW TF 80-SC

The AISIN AW TF 80-SC was the first 6-speed conventional automatic transmission for front-transverse drive allowing an engine torque of 450 Nm. With a length of 357 mm, the gearbox in Figure 12.26 weighs approximately 95 kg. Like the gearbox discussed in the last section, this gearbox is based on the Lepelletier gear set concept *(the basic concept is explained in Section 6.6.4 and Figure 6.35)*. A special feature of this design is the brake *C*, which has a belt brake design. The belt brake *C* is active in second and sixth gear [12.31-12.32].

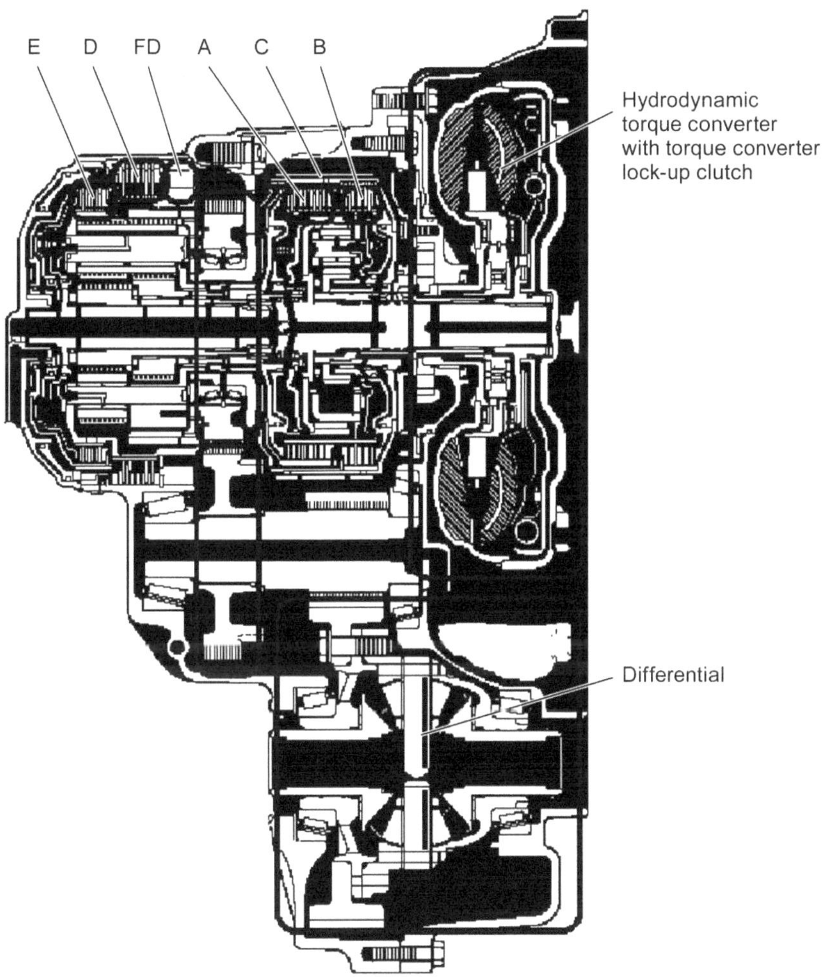


**Fig. 12.26.** 6-speed automatic passenger car gearbox for front-transverse drive AISIN AW TF 80-SC; *gearbox diagram Figure 6.35*

The gearbox has a two-piece die cast aluminium housing. The torque converter with a hydraulic diameter of 260 mm has a flat meridian profile. The oil pump is designed as a gerotor. The electronic-hydraulic control unit functions with six pressure regulators and two magnet valves and has two speed sensors and a temperature sensor. The transmission control unit is externally mounted in the vehicle.

### *18/ 5-Speed Countershaft-Type Automatic Passenger Car Gearbox for Front-Transverse Drive; Mercedes-Benz W5A 180*

Figure 12.27 shows the 5-speed countershaft-type automatic gearbox Mercedes-Benz W5A 180 for front-transverse A-class installation from 1997 [12.19]. With a length of 305 mm and a weight of 69 kg, it is designed for engine torques up to 180 Nm. The slanted outer surface of the housing is a continuation of the crash glide surface of the engine along the vehicle width.

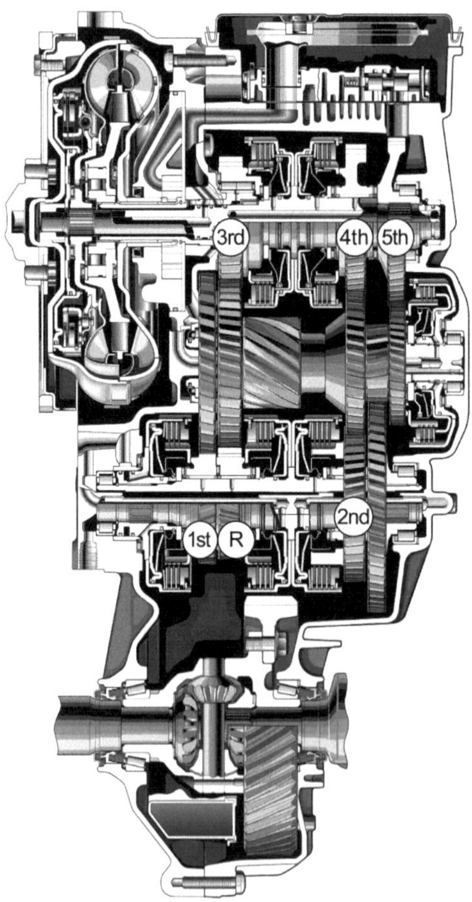


**Fig. 12.27.** 5-speed automatic passenger car countershaft-type gearbox for front-transverse drive Mercedes-Benz W5A 180; *gearbox diagram Figure 6.36*

The basis of the gearbox is a three-shaft system with six multi-plate clutches *(the basic concept is explained in Section 6.6.4 and Figure 6.36)*.

A converter with lock-up clutch is mounted between the engine and the gearbox. All the plate carriers and pressure pistons have a cost-effective thin-sheet design. The two halves of the housing are manufactured from die cast aluminium. An innovation of this design is the direct hydraulic control of the shifting elements and the exclusive use of pulse width modulation valves. The hydraulic control elements and the selector slide valve are accommodated in the valve housing manufactured from die cast aluminium. The actuators contained in the electrics kit and the electronic transmission control unit are arranged on the rear side of the valve housing. The direct control of the shifting elements is achieved via pressure reducer valves controlled via pulse width modulation valves.

### 12.1.5 Passenger Car Hybrid Drives

#### *19/ 6-Speed Hybrid Automatic Passenger Car Gearbox as Parallel Hybrid; BMW Active Transmission*

*As explained in Section 6.6.5 and Figure 6.37*, the 6-speed automatic gearbox 6 HP 26 (as shown in Figure 12.15) is the basis for the parallel hybrid known as the "BMW Active Transmission" (Figure 12.28).

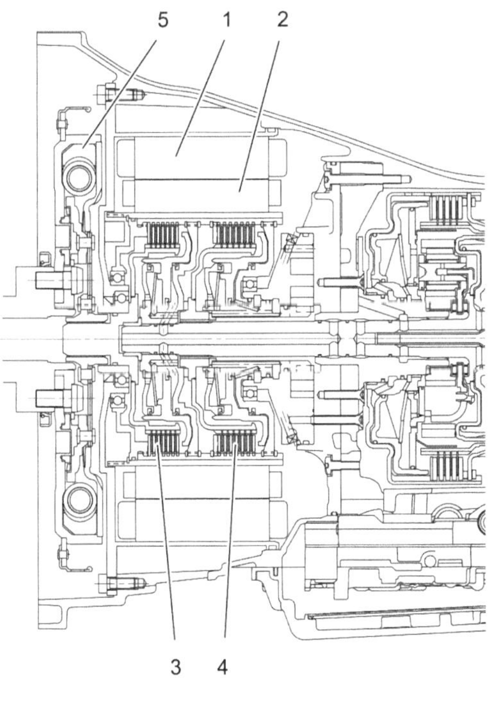


**Fig. 12.28.** 6-speed hybrid automatic passenger car gearbox. Parallel hybrid with one electric machine, BMW/ZF/Continental; *gearbox diagram Figure 6.37, main gearbox Figure 12.25.*
*1* EM stator;
*2* EM rotor;
*3* multi-plate clutch *C1*;
*4* multi-plate clutch *C2*;
*5* dual mass flywheel

In comparison to conventional torque converter transmissions, here the torque converter is replaced in its function as moving-off element by the permanent-magnet three-phase synchronous motor (PSM), comprising a stator *1* and a rotor *2*. It is integrated together with the two wet-running multi-plate clutches *C1 3* and *C2 4* as a hybrid module into the bell-shaped housing, requiring no additional installation space. The separating clutch *C1* and the master clutch *C2* are engaged and oil-cooled by means of the hydraulic transmission control unit. A dual mass flywheel *5* decouples rotational irregularities of the engine in the direction of the gearbox input. In the highly integrated "Active Transmission" expansion stage, the power electronics for controlling the electric machine can be mounted within the bell-shaped housing directly at the periphery of the stator, which means the space required for the power electronics no longer has to be reserved in the vehicle structure [12.17]. This mechatronic design also does away with the necessary wiring running between the externally arranged power electronics and the electric machine.

### *20/ Power-Split Hybrid Transmission; Toyota/Lexus P310*

Figure 12.29 shows the design of the P310 power-split hybrid transmission with two electric machines from Toyota/Lexus *described in Section 6.6.4 and Figure 6.39*. This hybrid transmission with integrated front axle differential is used in SUVs with transverse engines and front-wheel drive.

Both the electric motor (*1* and *2*) and the generator (*5* and *6*) are permanent-magnet three-phase synchronous motors (PSM) operated inside the gearbox, i.e. in the oil area. The two electric machines are cooled both via the ATF oil slung past them, which conducts their waste heat away through the gearbox housing and into the surrounding area, and via a cooling water jacket around the stators of the electric machines, which is coupled with the cooling water circulation of the vehicle.

A central assembly of this hybrid transmission is the first planetary gearbox, the summarising gear *4*, which couples the engine crankshaft (spider shaft) with the output (connected ring gears of the summarising and reduction gears) and the generator (sun gear). In this way, it is possible to exploit the engine power both mechanically to drive the gears – supported when necessary by the electric motor – and electrically to drive the generator. The torque ratios and power flows for different operational states of the P310 hybrid gearbox are explained in more detail in Section 6.6.5.

The second planetary gear, the reduction gear, has the task of lowering the speed of the electric motor in the direction of the output and increasing torque. This allows for more compact dimensions for the high-performance electric motor. Moreover, the electric motor is connected to the sun gear, while the ring gear, which is coupled to the ring gear of the summarising gear, represents the output in the direction of the front axle differential. The spider shaft is connected firmly to the gearbox housing.

The hybrid gearbox has a length of 417 mm and weighs 125 kg, including the oil filling.

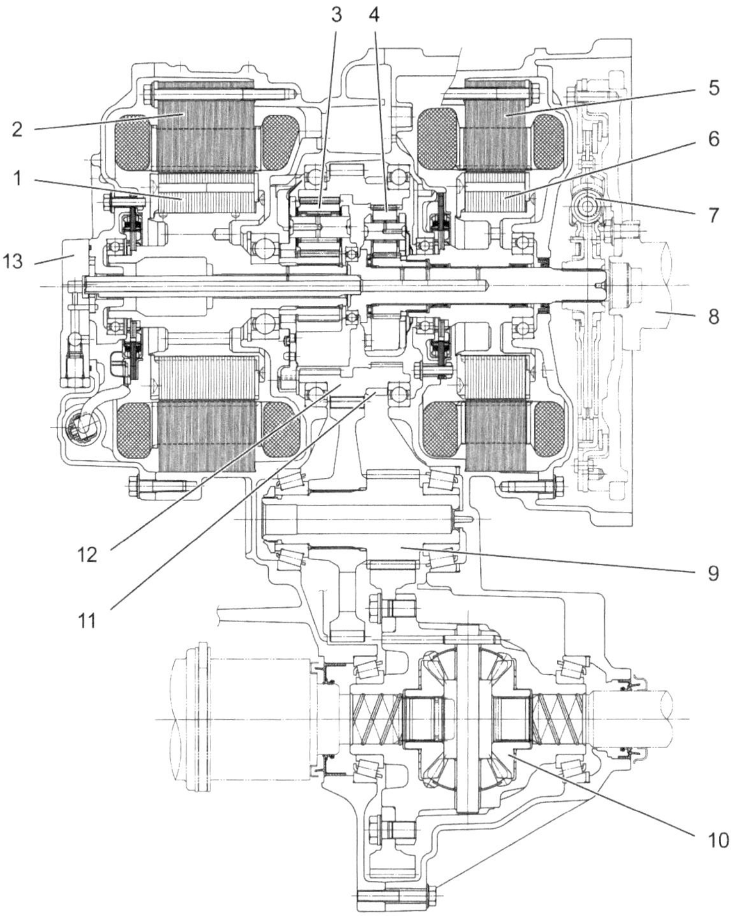


**Fig. 12.29.** Power-split hybrid gearbox P310 by Toyota/Lexus, *gearbox diagram Figure 6.39*. *1* Electric motor rotor; *2* electric motor stator; *3* reduction gear (planetary gear) for electric motor; *4* summarising gear (planetary gear); *5* stator of generator; *6* rotor of generator; *7* torsion damper with flywheel mass; *8* crankshaft flange; *9* output spur gear stage; *10* front axle differential; *11* parking lock wheel; *12* coupled ring gears of the summarising and reduction gears; *13* oil pump

The powertrain can take on certain properties of an all-wheel drive if the rear axle of the vehicle is electrified by means of a combination of the rear axle differential with an additional third electric machine. This electric machine can be energised from the electric energy accumulator or, when the engine is running, from the electric power path of the generator. The electrified rear axle differential will not be treated here any further [12.26, 12.35].

### 12.1.6 Continuously Variable Passenger Car Transmissions (CVT)

The desire to equip a vehicle with a continuously variable automatic transmission is almost as old as the automobile itself. After numerous developments throughout the history of automotive engineering, none of which proved viable, continuously variable transmissions were again developed by various manufacturers in the 1980s to the point where they could go into production. Please refer to Section 6.6.6 for a discussion of their principle of operation. All mechanical CV transmissions transmit torque by means of friction. There will therefore inevitably be a goal conflict at all operating points in the transmission between ensuring sufficient contact pressure for the friction elements, and keeping the excessive contact pressure as small as possible. This is one of the transmission's two main control functions, the other being controlling the transmission ratio, which affects the dynamic behaviour (driveability) and fuel consumption of the vehicle [12.12].

#### *21/ Continuously Variable Passenger Car Gearbox; Audi Multitronic®*

Figure 12.30 shows the continuously variable Audi Multitronic gearbox used in passenger cars with petrol and diesel engines with up to 330 Nm and 188 kW [12.58].

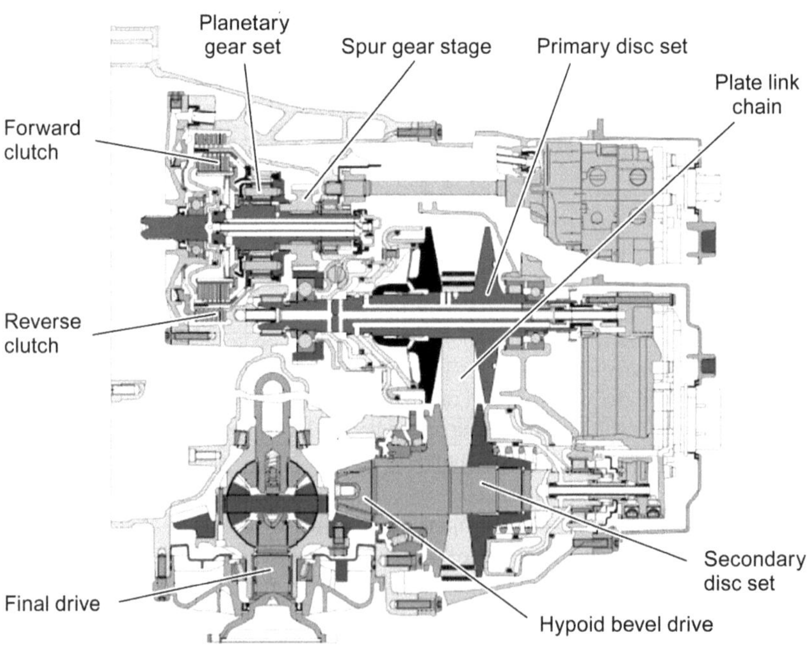


**Fig. 12.30.** The continuously variable automatic passenger car gearbox Audi Multitronic

For vibration isolation, the engine torque is guided via a flywheel damping unit into the transmission. Two multi-plate clutches have the appropriate dimensions to serve as moving-off elements for moving-off in forward and reverse, respectively. In order to ensure a consistent quality of moving-off processes and to guarantee the durability of the clutches for the entire service life of the vehicle, each clutch is cooled in a targeted way during the moving-off process with a defined oil flow.

When driving in reverse, the rotational direction is reversed by means of a planetary gear set. During forward driving, the planetary gear set rotates as a block, thus achieving optimal efficiency. Via a countershaft, the engine torque is further guided to the variator. Along with the front axle, this gear stage enables the adaptation of the total ratio to the most variable of vehicle and engine combinations.

The drive power is transmitted from the primary to the secondary disc set through a tensional link chain (plate link chain). This consists of link plates of varying lengths which build the strand as well as of the rocker pins of the joints and the retaining pins.

The pressure of the plate link chain in the variator is secured with a high degree of accuracy and dynamics by means of the hydromechanical torque sensor placed directly in the power flow. It sets a constant contact pressure proportional to the torque, thus contributing to the gearbox efficiency by avoiding excessive pressure, especially during operation at part load.

The contact pressure acts directly in the pressure cylinder of the primary and secondary disc set. The double-piston principle separates the functions of applying pressure and displacement, thus facilitating, in conjunction with the torque sensor, highly dynamic displacement processes with a contact pressure and pump dimensions favourable to the gearbox efficiency. The pressure force available for the transmission of the respective torque results ultimately from the sum of the displacement and contact pressures. The torque is transmitted from the secondary disc set via the hypoid bevel drive to the axle drive and, from there, to the front wheels.

The electronic-hydraulic control unit is integrated in the gearbox housing. The hydraulics provides pressure oil for applying contact pressure and displacing the chain, as well as for applying pressure to the moving-off clutches. It provides cooling oil for the variator and the multi-plate clutches as well as both cooling and lubricating oil for diverse bearing points and gear meshes. The pressure oil is created by means of a small, efficiency-optimised pump with a flow rate of 10 litres per minute for every 1000 l/min (see also Figure 11.17). The pump is driven by the input shaft via a gear stage and the pump shaft. The compact dimensions of the pump allow it to be arranged directly on the hydraulic control unit.

To ensure sufficient cooling of the two starting clutches, especially for moving-off with heavy loads, quantities of oil are required which the oil pump alone cannot provide. For this reason, an ejector pump is engaged during moving-off processes which provides additional cooling oil at a low pressure level. The amount of oil used for cooling the clutch can thus be roughly doubled when necessary without requiring additional pump power.

Information is exchanged between the transmission control unit mounted directly on the hydraulic control unit and the other control devices in the vehicle via the CAN bus.

## *22/ Continuously Variable Passenger Car Gearbox; Mercedes-Benz Autotronic®*

The continuously variable gearbox Autotronic by Mercedes-Benz was developed especially for the A-Class powertrain (from 2004). The transmission has an overall gear ratio of 6.4 and is designed for engine torques of up to 280 Nm [12.34].

Figure 12.31 shows the middle section of the Autotronic and the power flow through the gearbox. The main assemblies are the torque converter with a slip-controlled lock-up clutch, the oil pump, the taper disc variator with thrust link belt and the output-side clutch with reversing gear set (downstream drive/neutral/ reverse set).

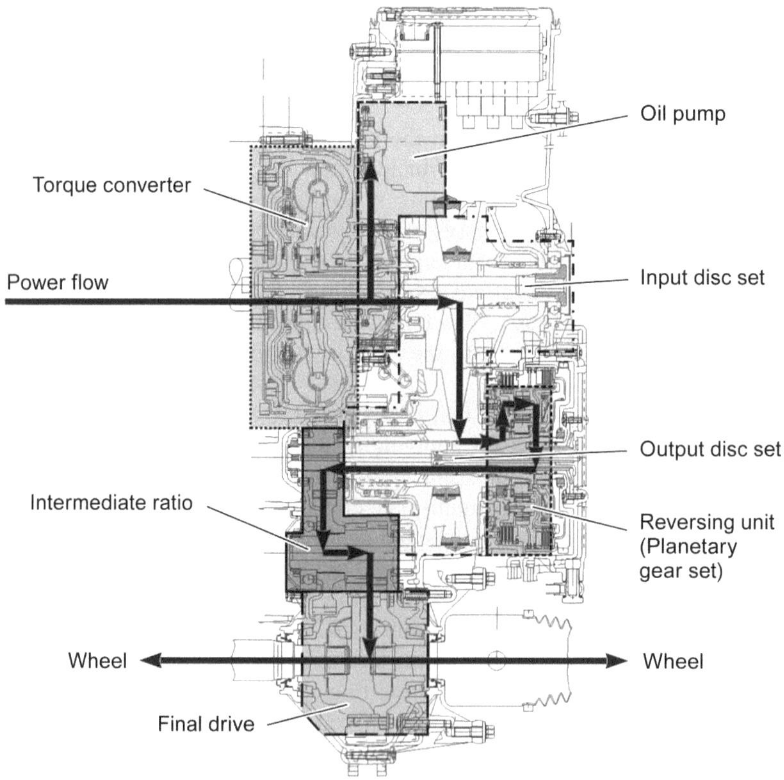


**Fig. 12.31.** The continuously variable automatic passenger car gearbox Mercedes-Benz Autotronic

Adjustment to the different engine variants is achieved via an intermediate ratio before the differential.

A hydrodynamic torque converter is used as moving-off element in the Autotronic. The combination of package requirements and high engine torques made a new design necessary. The 3-line design of the converter otherwise typical of Mercedes-Benz was replaced by a more compact 2-line design.

The Autotronic is designed with an output-side reversing gear set. The reasons for and advantages of this are:

- the special package situation involving reduced space on the input side,
- the possibility of reducing the pressure safety at the variator or the thrust link belt and of protecting from output-side torque impulses (torque-fuse function) and
- the adjustability of the variator during vehicle standstill.

This design leads to the characteristic features of the variator:

- The connection of the gearbox input shaft to the input-side fixed disc by means of driving spline.
- One-piece travel disc with forged-on cylinders for taking up the axial sealing ring race.
- The design of the output-side fixed disc of the disc set as a hollow shaft with the output shaft inserted in it, through which the torque is transmitted further after the reversing gear set to the intermediate shaft.
- Basic pressure applied to the thrust link belt in a pressure-less state via a coaxially arranged compression spring in the output disc set (engine start).
- Centrifugal oil pressure compensation at the output disc set through two sheet metal centrifugal oil covers.

A 30-mm-wide thrust link belt with twelve rings is used as a pulley element. In designing the variator and the belt, the greatest value was placed on maximising the overall gear ratio. The variator disc sets have a centre distance of 180 mm.

The reversing gear set arranged on the output side basically comprises a planetary gear and two hydraulically activated multi plate clutches. During forward driving, the clutch forward (*CF*) is closed and the brake reverse (*BR*) is open. This way, the planetary gear rotates as a block.

If the rotational direction is to be reversed, the forward clutch is opened and the reverse brake is closed so that the planetary gear carrier is completely braked. The power flows in this case via the sun gear and the planetary gears into the ring gear and thus into the output shaft. The rotational direction is reversed by the stationary gear ratio of the planetary gear set.

A double-flow vane pump is used as an oil pump in the Autotronic (see also Figure 11.17). It consists essentially of a rotor with 12 vanes and a housing with two separate chambers, with one chamber in full-time operation and the second according to necessity and thus only pressurised on a part-time basis. An advantage of this is that the power absorption of the pump can be reduced almost by half, especially at higher speeds. This has a beneficial effect on the driving performance and fuel consumption of the vehicle.

The electrohydraulic control unit comprises a hydraulic control unit with 12 hydraulic slide valves, four electronically controlled magnet control valves and the electrics kit, which includes a sensor system and fully integrated control unit. In conjunction with the hydrodynamic torque converter, the electrohydraulic control unit allows for purely hydraulic driving as emergency operation function.

A hybrid design was selected for the electronic control unit in order to increase quality and reliability. This goal is mainly achieved by avoiding wiring and vehicle-side power supply connectors between the control unit and the gearbox. Moreover, through the blocking of the control unit and the gearbox, the entire mechanics-hydraulics-control system can be checked and, with special test runs, adapted already in the transmission plant, increasing the product quality.

The control unit itself is designed for a temperature range of –40°C to +140°C. It has two CAN controller modules and four current-controlled output stages for the magnet control valves as well as three speed sensors, one temperature sensor, the selection range sensor (comprising four hall sensors) and one pressure sensor. The control unit is connected to the vehicle power supply and the vehicle CAN via a 6-pin gearbox connector. The gearbox control unit is connected to other vehicle control units (e.g. the engine control unit, the ESP control unit etc.) via the CAN data bus.

### *23/ Continuously Variable Passenger Car Gearbox; ZF Ecotronic CFT 30*

Figure 12.32 shows the continuously variable transmission CFT 30 used, for example, in the Ford Five Hundred or the Ford Freestyle. It represents the most powerful variant of the Ecotronic transmission family developed by ZF *(the basic concept is explained in Section 6.6.6 and Figure 6.43)*. The gearbox has a torque capacity of 310 Nm (maximum engine power: 172 kW) [12.46, 12.66].

A torque converter with converter lock-up clutch serves as moving-off element. It has a small, circular torus located on a big diameter. This allows for an optimal hydraulic circulation with a slim design. This converter was especially adapted to the continuously variable gearbox with respect to its design and to its characteristic features. The radial piston pump, the planetary gear set with the two clutches for forward and reverse gears and the input taper disc sets are arranged on the primary axis behind the converter. A 33-mm-wide tensional link chain (plate link chain) is used for transmitting power in the variator.

Mounted on the secondary axis is the output-side taper disc set with integrated parking lock wheel as well as the pinion for the first stage of the final drive ratio. A shaft carries the intermediate gears of the final ratio. Mounted on the axle shaft are the output gear of the final drive ratio and the differential. The final drive ratio can be varied in six stages over a range of 4.11 to 5.95 through different intermediate shaft gears and differential gears.

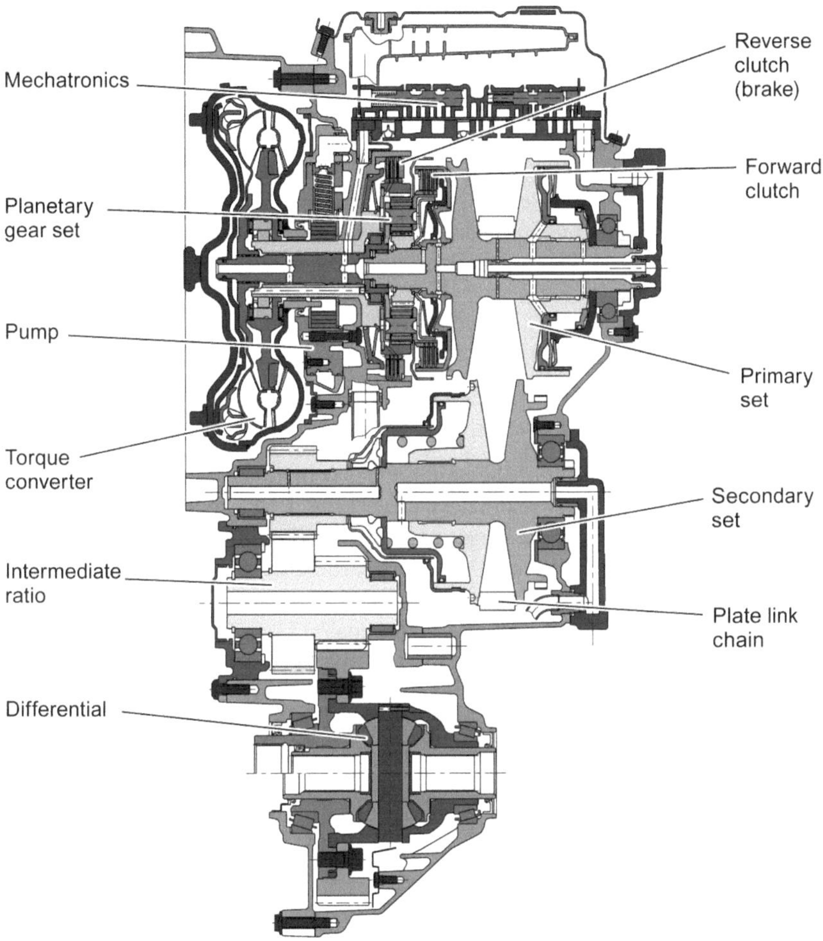


**Fig. 12.32.** The continuously variable passenger car gearbox ZF Ecotronic CFT 30; *gearbox diagram Figure 6.43*

The transmission control unit is arranged under the mechanical gearbox part and is in the oil sump. The hydraulic transmission control unit, the electronic transmission control unit, the electrohydraulic valves, the sensors, the wiring and the connectors all form a functional unit referred to as the mechatronics. The lack of any mechanical sensors on the discs for controlling the variator is clearly recognizable. This design requires neither a mechanical torque sensor nor a mechanical feedback of the variator disc position.

The variator control unit functions on a fully electronic basis and can individually set the primary and secondary pressure in the variator. In this way, a suitable pressure setting can be achieved via the electronic transmission control unit for each operating point. A detailed description of the contact pressure system for the Ecotronic family is found in [12.46].

A special feature of the oil supply system in the CFT 30 is the radial piston pump developed specifically for CVT applications. This is a suction-throttled pump designed for pressure values of up to 70 bar. The supplied flow rate is proportional to the engine speed up to a speed of 2000 1/min. Given higher speeds, it is regulated down to a constant level of approximately 28 *l*/min. Through this constant flow rate, the pump requires less drive power than a classical gear pump does. This has a positive effect on fuel consumption in customer-oriented driving operation.

## 12.2 Commercial Vehicle Transmissions

Table 12.2 shows an overview of the commercial vehicle transmissions treated in systematic sections 6.7.1 to 6.7.6 and designs sections 12.2.1 to 12.2.6. To aid orientation, consecutive numbers are assigned in this section to the designs discussed (e.g. 1/ 6-Speed Manual Commercial Vehicle Gearbox; ZF S 6-66).

**Table 12.2.** Automotive gearboxes introduced in Sections 6.7.1–6.7.6 and 12.2.1–12.2.6. *CS* countershaft; *TCCT* torque converter clutch transmission

| No. | Design Figure No. | Speeds | Characteristics | Manu-facturer | Name | Diagram Figure No. |
|---|---|---|---|---|---|---|
| — | — | 4 | MT, 1-range gearbox | — | — | 6.45a |
| 1/ | 12.33 | 6 | MT, 1-range gearbox | ZF | S 6-66 | 6.45b |
| — | — | 5 | MT, direct drive/ overdrive gearbox | — | — | 6.46 |
| — | — | 5 | MT, output constant gear | — | — | 6.47 |
| — | — | — | MT, multi-range gearbox | — | — | 6.48–6.50 |
| — | — | — | MT, multi-stage gearbox | — | — | 6.51–6.52 |
| 2/ | 12.34 | 9 | MT, 2-range gearbox, direct | ZF | 9 S 109 | 6.53 |
| 3/ | 12.35–12.36 | 16 | MT, 3-range gearbox | ZF | 16 S 221 | 6.54 |

**Table 12.2.** *(continued)*

| | | | | | | |
|---|---|---|---|---|---|---|
| 4/ | 12.37 | 12 | MT, 2-range gearbox 2 CS | Eaton | TSO-11612 | 6.55 |
| 5/ | 12.38 | 16 | MT, 3-range gearbox 2 CS | Eaton | RTSO-17316A | 6.56 |
| 6/ | 12.39 | 6 | AMT, 1-range gearbox | ZF | eTronic 6 AS 380 VO | 6.58 |
| 7/ | 12.40 | 16 | AMT, 3-range gearbox | MB | PowerShift G241-16K | 6.59 |
| 8/ | 12.41–12.44 | 16 | AMT, 3-range gearbox, 2 CS | ZF | AS-Tronic 16 AS 2230 TD | 6.60 |
| 9/ | 12.45 | 16 | TCCT, 3-range gearbox | ZF | WSK 400 + 16 S 221 | 6.61 |
| 10/ | 12.46 | 12 | TCCT, 3-range gearbox, 2 CS | ZF | TC-Tronic 12 TC 2740 TO | 6.62 |
| 11/ | 12.47 | 5 | AT | Allison | 2000 | 6.63 |
| 12/ | 12.48 | 6 | AT, retarder | ZF | 6 HP 602 C | 6.64 |
| — | — | — | Serial hybrid, variants | — | — | 6.65 |
| 13/ | 12.49 | — | Serial hybrid | ZF | EE Drive 1 | 6.66 |
| — | — | — | CVT, hydrostatic unit, power-split | — | — | 6.67 |
| 14/ | 12.50 | — | CVT, hydrostatic unit, power-split | ZF | Eccom | 6.68 |

### 12.2.1 Manual Commercial Vehicle Transmissions (MT)

As discussed in Section 6.7.1, most commercial vehicle gearboxes with more than six speeds are generally not designed as single-range, but rather as multi-range gearboxes. If the gearbox is so designed that the range units are only flange-mounted at defined interfaces, it is possible to produce a gearbox kit with a uniform basic gearbox.

### 1/ 6-Speed Manual Commercial Vehicle Gearbox; ZF S 6-66

The two-stage 6-speed manual commercial vehicle gearbox ZF S 6-66 was designed in the 1960s (Figure 12.33). This gearbox was used for input torques of up to 660 Nm for mid-range trucks (110–210 kW).

The special characteristic of this gearbox is its capacity to be expanded into a 12-speed gearbox by means of a front-mounted splitter unit with second constant gear. The gear sequence of the gearbox is thereby "compressed".

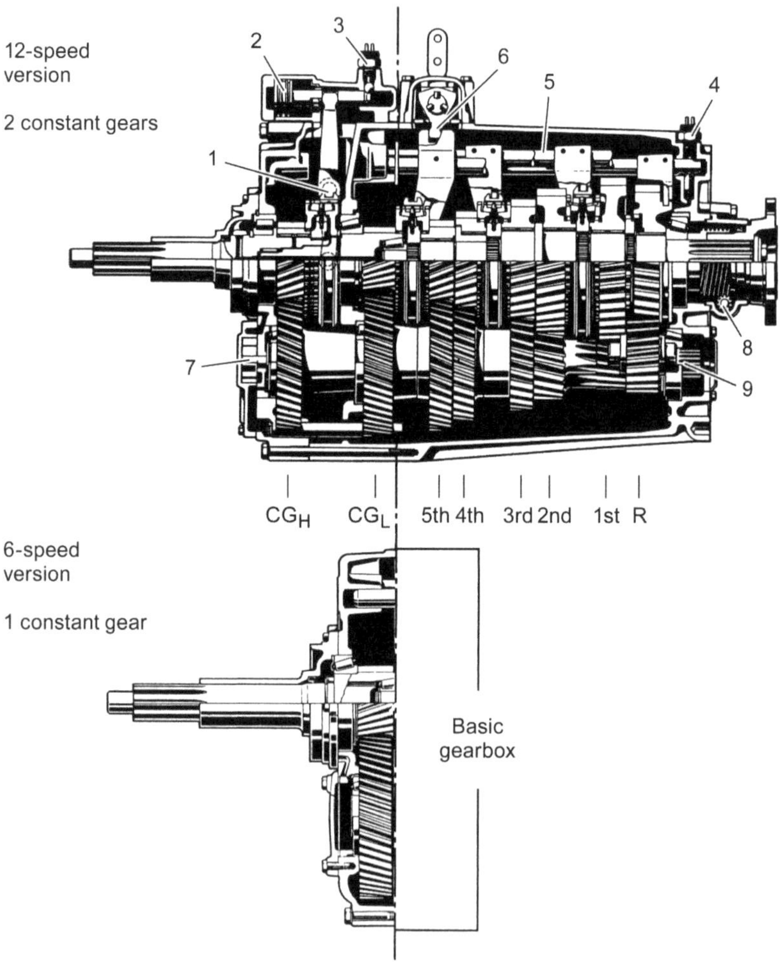


**Fig. 12.33.** 6(12)-speed manual commercial vehicle gearbox ZF S 6-66 with ZF splitter unit GV 66; *gearbox diagram Figure 6.45b. 1* Swing fork; *2* compressed-air cylinder; *3/4* switches; *5* selector bar; *6* selector finger; *7* oil pump; *8* speedometer drive; *9* power take-off shaft; $CG_H$ constant gear high (high speed); $CG_L$ constant gear low (low speed)

There is no design change to the basic gearbox. *Please refer to Figure 6.45b to compare the gearbox diagrams.*

In the 6-speed basic gearbox, the reverse and first gears are located directly at the output side bearings to minimise the shaft deflection caused by the large tooth forces arising from the high transmission ratios (see Section 8.2). The gear pairs for second gear to fifth gear are mounted in sequence from right to left. The highest (sixth) gear of the basic gearbox is direct drive (direct drive gearbox). Two different constant ratio gear pairs are available in the optional splitter unit. The front-mounted splitter unit makes the 6-speed direct drive gearbox into a 12-speed overdrive gearbox ($CG_H$ constant gear high = high speed). The transmission shafts run in tapered roller bearings and cylindrical roller bearings.

The synchronizer for first and second gear is a double-cone synchronizer, reducing the shifting force required. Reverse gear is unsynchronized, and shifted by means of a dog clutch. All the other gears have single-cone synchronizers. The splitter unit is shifted by means of a swing fork *1*, in contrast to the shift forks in the basic gearbox. The shifting force is provided by the compressed-air cylinder *2* located above the swing fork. This is operated by the driver using a pilot valve mounted on the gearshift lever.

The electric switch *3* mounted on the right-hand end of the splitter unit selector bar gathers information on the splitter stage selected. The selector bars *5* with their shift forks and the selector finger *6* moving the shift forks are shown in section in the top part of the gearbox. Depending on the gate selected, the selector finger grips one of the selector bars with an axial movement along its axis of rotation. In the top right-hand part of the gearbox the electric switch *4* can be seen, which gives the signal for the reversing lights. The locking pins that define the end of the selector bar travel are located on the opposite side of the shaft.

The gearbox housing is of rigid end-loaded design. The worm gearing *8* of the speedometer drive is located on the output flange. The serration gear *9* located on the output end of the countershaft enables a clutch-controlled power take-off to be attached (Section 6.9). This also requires no design change to the basic gearbox, as for connecting the splitter unit. An oil pump *7* can optionally be mounted at the input end of the countershaft, which supplies the input side bearings with lubricating oil, and pumps the gearbox oil through a separate oil cooler if required.

### *2/ 9-Speed Two-Range Manual Commercial Vehicle Gearbox; ZF 9 S 109*

The 9-speed commercial vehicle two-range gearbox in Figure 12.34 consists of a 4-speed basic gearbox with a crawler gear (high traction starting gear) and reverse gear as well as a flange-mounted range-change unit (cf. also the *gearbox diagram in Figure 6.53*). The gearbox with a maximum input torque of 1100 Nm is used in upper mid-range trucks (180–240 kW).

In the basic gearbox, the reverse gear and crawler are located close to a bearing, in order to minimise shaft deflection. The crawler's large ratio facilitates moving-off on mountainous roads and is also advantageous for manoeuvring. Then come the first, second and third gear pairs in order.

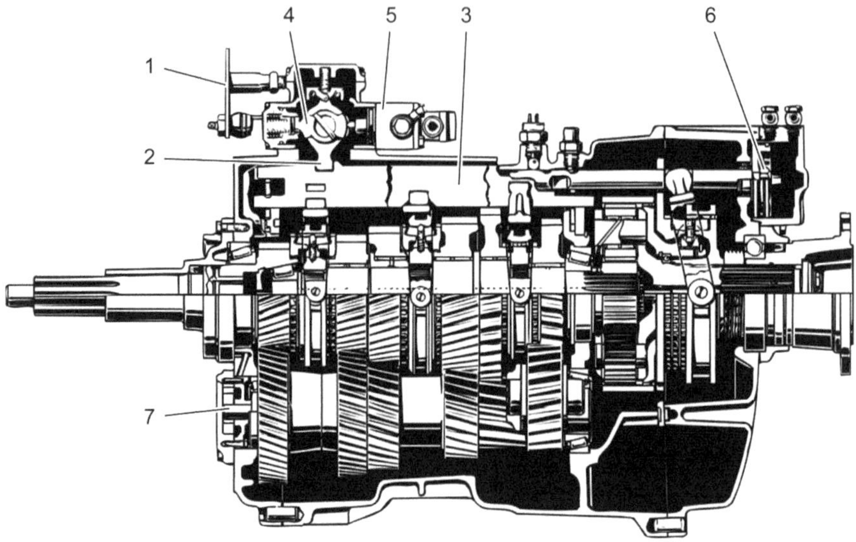


**Fig. 12.34.** 9-speed (4 x 2 + crawler) commercial vehicle gearbox ZF 9 S 109 with direct drive design; *gearbox diagram and power flows Figure 6.53. 1* Connection for turning shaft remote control; *2* selector finger; *3* selector bars; *4* cam plate; *5* shift valve for the range-change unit; *6* shift cylinder for the range-change unit; *7* lubricating oil pump; *CG* constant gear; *C* crawler; *R/D* range-change unit: *R* range; *D* direct

The fourth gear of the basic gearbox is direct drive. The rear-mounted range-change unit is of planetary design, and has two gears: one direct gear in which the planetary gear set revolves as a block (toothed clutch), and a second gear with a ratio larger than the overall gear ratio of the basic gearbox multiplied by the gear step of the main gearbox. This serves to double the gear sequence from 4 to a total of 8 selectable gears. The constant gear and the first, second and third gear pairs are used to double the gear number. The crawler ratio is only used when the range-change unit is engaged.

The gearbox shafts run in tapered roller bearings and in deep groove ball bearing. The central shafts of the planetary gear (sun, ring gear, spider) are interlinked by means of a single-cone synchronizer, located on the output side beside the gearwheels of the planetary gear. The output shaft of the gearbox runs on a locating ball bearing on the one side, and on the toothing of the planetary gears on the other side. Since no major axial forces arise in the planetary gear, the ball bearing of the output shaft is primarily required to counteract external forces from the propeller shaft in the gearbox housing.

All the gears except reverse have single-cone synchronizers. The gearshift sleeves are shifted by swing forks. The connection *1* for the turning shaft remote control can be seen in the upper gearbox housing. This moves the selector finger *2*, which in turn operates the selector bars *3*. A cam plate *4* can be seen on the rotational axis of the selector fingers, which activates the shift valve *5* located on the right when the selector finger is moved axially (changing from gate 3/4 to gate 5/6), causing the shift cylinder *6* mounted above the output flange to automatically change the range-change unit.

At the input end of the countershaft there is a lubricating oil pump *7* that supplies the entire gearbox (gearwheels, bearings etc.) with lubricating oil, and pumps the gearbox oil through a separate oil cooler when necessary, since cooling the oil makes the gearwheels more resistant to pitting, thus increasing their service life. The gearbox housing is of three-part, end-loaded design.

The gearshift mechanism is mechanical and functions via a double-H gearshift gate or with superimposed H (for details, see also Figure 12.36), with each gear being allotted its own position in the gearshift pattern [12.68].

### 3/ 16-Speed Three-Range Manual Commercial Vehicle Gearbox; ZF 16 S 221

This 16-speed three-range gearbox 16 S 221 from the ZF Ecosplit 3 Series comprises a 4-speed main gearbox with reverse gear, a front-mounted 2-speed splitter unit and a rear-mounted 2-speed range-change unit (see also the *gearbox diagram in Figure 6.54*). Figure 12.35 shows the gearbox with direct-drive design. With a maximum input torque of 2200 Nm, it is used in heavy-duty trucks with an engine power of up to 420 kW.

In the Ecosplit main gearbox, due to the transmission of high torques, the reverse and first gears are located near a bearing in order to minimise shaft deflection. The gear pairs of the second and third gears follow. Fourth gear is the direct gear.

In the front-mounted splitter unit, the gear sequence is "compressed" by means of two constant ratio gear pairs, leading to an increase of selectable gears up to eight (see Figure 6.49a). The rear-mounted planetary range-change unit has a direct gear in which the planetary gear set rotates as a block and a second ratio greater than the overall gear ratio of the main gearbox, multiplied with the gear step in the main gearbox (see Figure 6.49b). This doubles the number of selectable gears, increasing them from eight to sixteen (see Figure 6.49c).

The planetary range-change unit is shifted by means of a single-cone synchronizer. A double-cone synchronizer is used for first and second gear in the main gearbox to reduce gearshift effort. All other gears have a single-cone synchronizer optimised for shifting effort and shifting travel.

The gearbox is activated by means of the shift turret unit *1*. The selector bars *3* with respective gearshift sleeves are moved via the selector finger *2*. The shift valve *5* of the range-change unit is activated with the turning shaft *4* in the rotational axis of the selector finger *2*. The range-change unit is shifted by means of a pneumatic shift cylinder *6*. The used 5-port/2-way valve (shift valve *5*) is positioned in the shift turret unit [12.69]. The splitter unit is also shifted pneumatically via a relay valve and an integrated shift cylinder *7* (not shown in Figure 12.35).

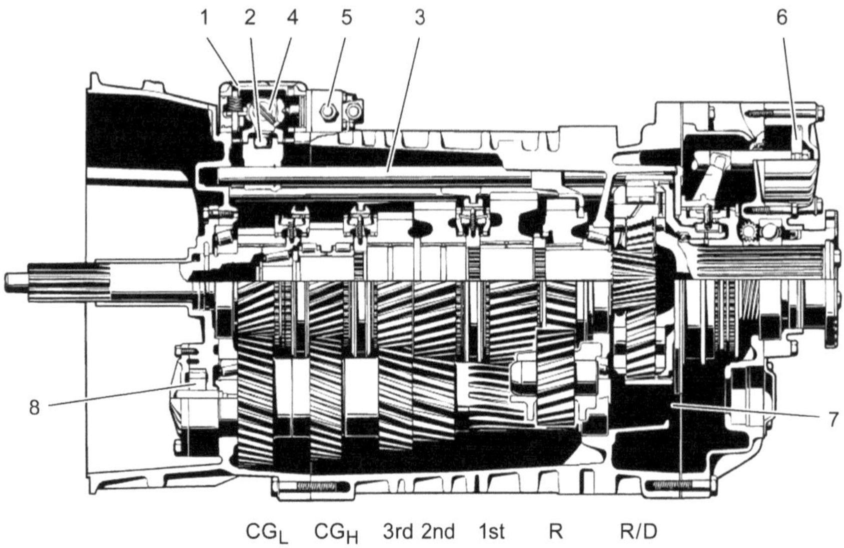


**Fig. 12.35.** 16-speed (2 x 4 x 2) commercial vehicle gearbox ZF 16 S 221; *gearbox diagram and power flows Figure 6.54. 1* Shift turret unit; *2* selector finger; *3* selector bars; *4* turning shaft; *5* shift valve for the range-change unit; *6* shift cylinder for the range-change unit; *7* relay valve and shift cylinder for the splitter unit (not shown); *8* lubricating oil pump; $CG_H$ constant gear high; $CG_L$ constant gear low; *R/D* range-change unit: *R* range; *D* direct [12.69]

The lubricating oil pump *8* located at the countershaft supplies the entire gearbox (gearwheels, bearings, sliding surfaces etc.) with lubricating oil. The gearbox housing with integrated clutch bell has a die cast aluminium design.

The planetary gear set is helical-cut to reduce noise. Thrust cone technology, known from fast-running, stationary spur gear drives, is used to absorb the large axial forces. A thrust cone pairing consists of two bevelled contact faces on the front sides of the two intermeshing gearwheels. The conical form of the contact faces on which the two surfaces make contact on a cone envelope line causes a lubricating gap in which a hydrodynamic lubricating film can form. Via these contact faces, the two gearwheels are braced axially against each other in such a way that the axial forces from the toothing are not guided to the shafts and thus do not put a load on the bearings. This prevents the gears and shafts from tilting [12.7].

The gearbox may also be equipped with a secondary retarder (see Section 11.6.3). The retarder is integrated in the gearbox, along with the heat exchanger belonging to it. Also, the gearbox can be combined with clutch-, drive- and engine-controlled power take-offs (see Section 12.2.2 Number 8/, Power Take-Offs).

Two gearshift mechanisms are available for actuating the gearbox:

- double-H gearshift pattern (Figure 12.36a) and
- gearshifting with superimposed H (Figure 12.36c).

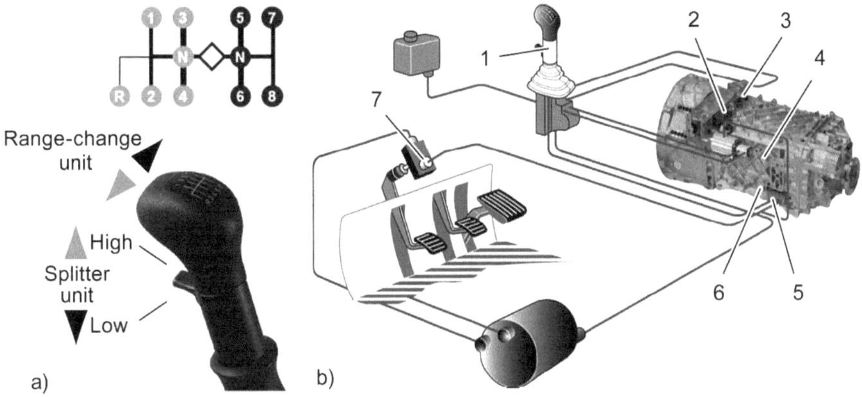


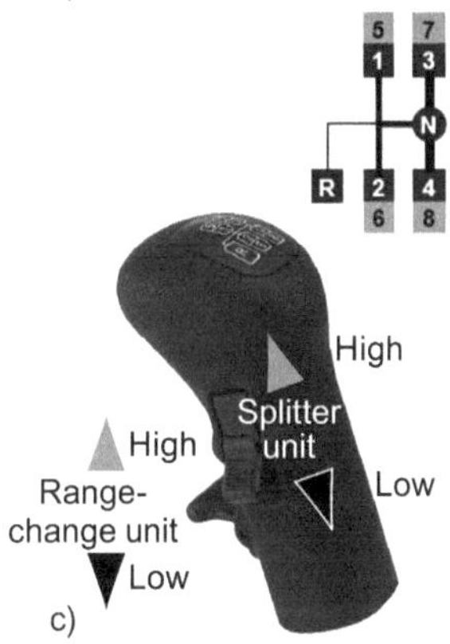


**Fig. 12.36.** *a* Gearshift lever for double-H gearshift pattern with integrated rocker switch for splitter unit selection;
*b* connection diagram for pneumatic control device with double-H gearshift pattern of ZF commercial vehicle transmission series Ecosplit 3.
*Splitter unit control device: 1* pilot valve; *5* relay valve; *7* triggering valve;
*Range-change unit control device: 2* shift valve;
*Display switch for: 3* neutral position; *4* range-change unit; *6* splitter unit;
*c* gearshifting with superimposed H [12.69]

In the case of the double-H gearshift pattern, each of the eight forward gears and the reverse gear receive their own position in the gearshift pattern (Figure 12.36a). By means of the pilot valve *1*, the splitter position Low or High can be preselected via the relay valve *5* and shifted via the triggering valve *7* (Figure 12.36b). The range-change unit shifts automatically when changing from the third/fourth into the fifth/sixth gear gate and vice versa. In the neutral position arranged between the third/fourth and the fifth/sixth gear gates, the shift cylinder of the range-change unit is applied with compressed air via the shift valve *2*.

In the case of gearshifting with superimposed H, gears 5 to 8 superimpose gears 1 to 4 (Figure 12.36c). The neutral position is located in gate 3(7)/4(8). Switching the range-change unit is achieved by activating the rocker switch on the front side of the gearshift lever while in neutral. Gears 1–4 and reverse are assigned to the rocker position Low and gears 5–8 to High. The corresponding High and Low splitter units are preselected by means of the pilot valve *1* on the side of the gearshift lever and switched by activating the clutch.

Both gearshift mechanisms can be equipped with a servo system in order to reduce shifting effort and to enhance shifting comfort. This particularly improves shifting comfort at low temperatures.

### 4/ 12-Speed Two-Range Manual Commercial Vehicle Gearbox; Eaton Twin Splitter TSO-11612

The gearbox in Figure 12.37 is an interesting type of countershaft gearbox developed in the 1970s (compare also with *Figure 6.55*). This is a 12-speed gearbox with a 4-speed basic gearbox and a 3-speed rear-mounted splitter unit fitted with two countershafts. The gearbox is used in heavy-duty trucks for input torques of up to 1600 Nm and an engine power of 280 kW.

The special feature of the rear-mounted splitter unit is that it has a low-speed and a high-speed splitter stage in addition to the direct drive. Splitter mode is preselected by the driver at the gearshift lever, and implemented by means of pretensioned shifting elements when power is interrupted (clutch operation or lifting the accelerator pedal).

The basic idea behind both the basic gearbox and the rear-mounted splitter unit is to reduce the overall length of the gearbox by reducing the face widths required. To achieve this, the rolling contact power is transmitted by two opposed countershafts. These two countershafts are located together with the gearbox main shaft in one plane for reasons of symmetrical loading. This distributes the load in one gearwheel stage to two meshings, thus halving the theoretical contact stress arising. In practice, face widths could be reduced by 40%. This serves to achieve the desired reduction in overall length, at the expense of making the whole gearbox wider. The load on the main shaft gearwheels is balanced by precise "float" on the main shaft in a radial direction. They are centred between their respective countershaft gearwheels under load.

**Fig. 12.37.** 12-speed (4 x 3) manual commercial vehicle gearbox Eaton Twin Splitter TSO-11612. Main gearbox and rear-mounted splitter unit in two-countershaft design; *gearbox diagram and power flows Figure 6.55*

Since the main shaft gearwheels are only guided and do not run in bearings, they are not capable of transmitting substantial axial forces. Helical gearing is therefore not appropriate for them. The noise level of straight-cut gearing is improved by high contact ratios and high contact gearing.

All gears except the friction-synchronized splitter stages are constant mesh. The splitter stages are shifted pneumatically. The gearbox has a gearbox brake connected to the gearbox input shaft. It provides for quick upshifts. For this, the rotational speed of the gearbox input shaft is slowed down considerably in order to achieve a quick engagement of the splitter gear stage. The gearbox brake is pneumatically powered when the clutch pedal is fully pressed down [12.9].

The gearbox housing is of end-loaded design, and has an internal bulkhead for the bearings of the main shafts and countershafts. The clutch bell housing is bolted-on.

### *5/ 16-Speed Three-Range Manual Commercial Vehicle Gearbox; Eaton S Series Gearbox RTSO-17316A*

The Eaton S transmission series represents a further development of the Eaton twin splitter gearbox (Figure 12.38). An essential feature of this three-range transmission is that all three range units are equipped with two countershafts. See also the *gearbox diagram in Figure 6.56.*

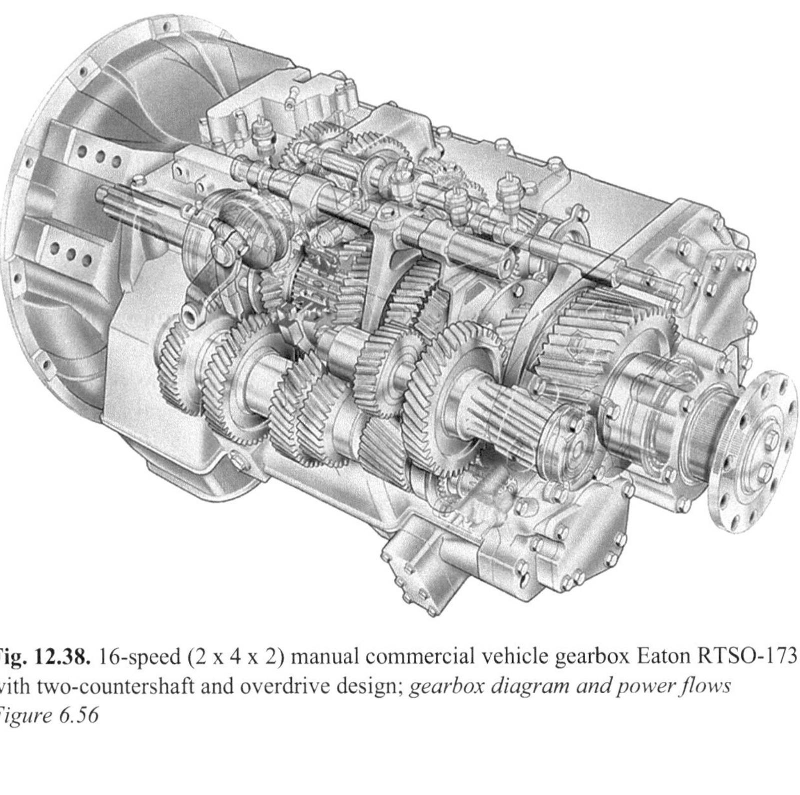

**Fig. 12.38.** 16-speed (2 x 4 x 2) manual commercial vehicle gearbox Eaton RTSO-17316A with two-countershaft and overdrive design; *gearbox diagram and power flows Figure 6.56*

The gearbox is used in heavy-duty trucks with input torques of up to 2400 Nm and a total weight of up to 44 t. All range units of the gearbox are fully synchronized. All synchronizers are arranged on the central shaft.

Each synchronizer switches two countershaft gearwheels into the power flow. For this, Eaton LF synchronizers are used (LF = Low Force, see Section 9.2.5 "Engineering Designs" of Synchronizers).

The gears are shifted by means of a double-H gearshift gate. The gearbox has an aluminium alloy clutch housing. The use of helical gearing contributes to maintaining a low noise level [12.10].

### 12.2.2 Automated Manual Commercial Vehicle Transmissions (AMT)

In the case of commercial vehicle applications, there is no clear transition from manual to partially and fully automated transmissions. In principle, all transmissions introduced in Section 12.2.1 can be automated by means of appropriate moving-off and actuation elements.

All manufacturers of commercial vehicles or commercial vehicle transmissions offer automated manual transmissions. They have been introduced into the market under names such as "AS Tronic", "eTronic" (ZF), "Telligent EAS"/"PowerShift", "Sprintshift" (Mercedes-Benz), "I-Shift/Geartronic" (Volvo), "Opticruise" (Scania) and "SAMT B" (Eaton) [12.18, 12.57].

#### *6/ 6-Speed AMT for Light-Duty Commercial Vehicles and Vans; ZF eTronic 6 AS 380 VO*

An example of a fully automated manual gearbox for vans and light commercial vehicles with electromechanical actuators is the eTronic 6 AS 380 VO (Figure 12.39). See also the *gearbox diagram in Figure 6.58*. The basis of this two-stage single-range gearbox is the fully synchronized manual gearbox ZF 6 S 380 VO. The gearbox is used for input torques of up to 380 Nm and engine powers of up to 120 kW. The overall gear ratio of the overdrive gearbox amounts to 6.79. The components required for automation – clutch actuator and transmission actuator, as well as the electronic transmission control unit including gearbox wiring harness – are mounted externally as add-on components on the gearbox housing. The advantage of this is that the same parts are used for both the manual and the automated gearbox designs. By means of the fully automated, self-adjusting dry clutch, a two-pedal solution can be realised in the vehicle (accelerator and brake pedals).

Opening and closing the clutch is achieved by means of an electromechanical clutch actuator *2*. The main feature of this actuator is the compensation of the clutch force by means of a compression spring which mainly takes over releasee work.

Because of the release ratio (clutch actuator – clutch release fork – clutch release bearing), the clutch can be activated with a small-sized electric motor, which only compensates for the friction losses of the clutch release system.

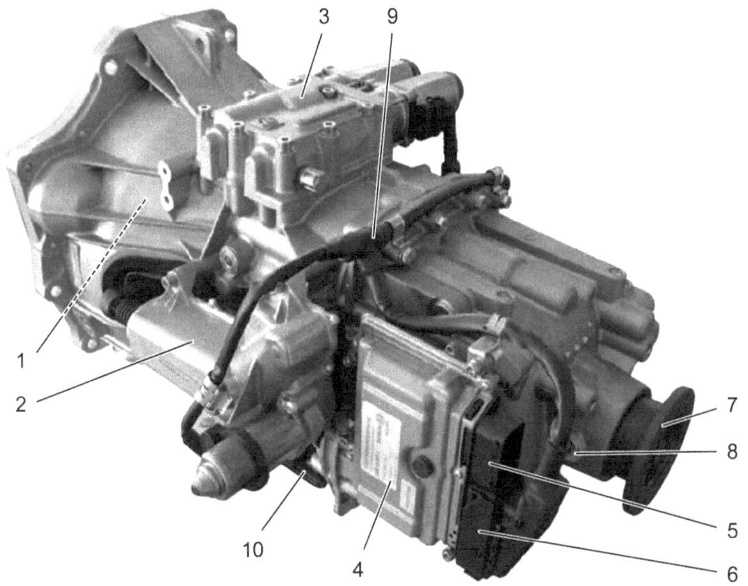


**Fig. 12.39.** 6-speed AMT ZF eTronic 6 AS 380 VO with electromechanical gearshift mechanism; *gearbox diagram and power flows Figure 6.58. 1* Gearbox input shaft (hidden in clutch bell); *2* clutch actuator; *3* transmission actuator; *4* electronic transmission control unit (TCU); *5* electric connection for vehicle connector; *6* electric connection for gearbox connector; *7* output flange; *8* output speed sensor; *9* gearbox wiring harness; *10* power take-off connection [12.71]

For determining the release travel, the electric motor of the clutch actuator has an incremental sensor. The shifting process is accomplished by means of an electromechanical transmission actuator *3*.

The shifting principle thus realized requires a transmission actuator with two electric motors: one for selecting and one for shifting. Both electric motors have a contactless incremental sensor for determining the angle/travel. The rotational motion of the shaft of the electric motor is transmitted via a recirculating ball screw/recirculating ball nut to the selector/gearshift lever. The two levers are arranged in turn in a central driver mounted on the central selector and shifting shaft (CSS). This way, the CSS can be rotated accordingly during the selection process and moved axially during the shifting process.

Communication between the three electric motors of the two actuators and the electronic transmission control unit (TCU) *4* takes place via the gearbox wiring harness *9*. The TCU has two electric connections, one for the gearbox wiring harness connector *6* and one for the vehicle wiring harness connector *5*. The CAN communication with the vehicle occurs via the vehicle wiring harness connector. The gearbox output speed sensor *8* determines the current speed of the vehicle and makes it available to the TCU. The gearbox can also be equipped with a clutch-controlled power take-off *10* driven with the rotational speed of the countershaft.

The driving strategy software calculates the optimal gears and shifting points from the data regarding the current driving condition. As with most automated transmission systems, both an automatic and a manual operation mode are possible here too. The drive selector (console switch) is used to send control instructions (automatic or manual operation as well as gear selection, where appropriate) to the gearbox via "shift-by-wire". Driving condition, operational mode, current gear and, where necessary, diagnostic information are displayed on the display connected to the CAN bus [12.71].

### *7/ 16-Speed Three-Range Automated Gearbox for Heavy-Duty Trucks; Mercedes-Benz PowerShift G241-16K*

The PowerShift gearbox by Mercedes-Benz is designed as a fully-automated constant-mesh gearbox with 16 forward gears and 2 reverse gears, as well as an electropneumatic gearshift system. The direct drive gearbox with an overall gear ratio of 17.03 is used for input torques of up to 2400 Nm, an engine power of up to 420 kW and in commercial vehicles with total weight of truck and trailer of 45 t. The moving-off process, the activation of the gearshifting clutch and gear selection are fully automated, though a manual operational mode is also possible. The two-pedal system of this model helps to ease use and thus reduces driver stress.

The three-range (2 x 4 x 2) 16-speed gearbox is illustrated in Figure 12.40. The corresponding *gearbox diagram is shown in Figure 6.59*. The splitter unit *I* and the range-change unit *III* are shifted via synchronizers, while the gears of the main gearbox *II* are engaged via dog clutches.

When upshifting, speed is adjusted by means of an electropneumatically activated multi-plate brake *4* at the forward end of the countershaft *3*. When downshifting – after the previous gear has been disengaged – the synchronizing process in the main gearbox *II* is realised via an increase in engine speed with the vehicle clutch closed and an engaged splitter unit.

The oil pump *6* is connected via a shaft *7* directly to the rear end of the countershaft *3*, so that it rotates at a speed proportional to the engine speed and the splitter unit ratio. It supplies all the required locations in the gearbox (bearings, gearing) with lubricating oil on the one hand; on the other, it is dimensioned in such a way that it can feed a gearbox oil cooler, provided such a cooler is used, with the necessary volume of oil at an appropriate pressure level.

The gearbox is activated with a so-called PowerShift shifting – an automated electropneumatic gearshift system. This gearshift system is the successor to the Telligent EPS. The splitter unit *I* and the range-change unit *III* are each activated by a separate actuator cylinder. The range-change unit *III* is controlled by a pneumatic actuator cylinder *8*. Here, the corresponding piston rod *9* activates the sliding sleeve *11* of the planetary gear set *5* via a swing fork *10*. The three sliding sleeves *12*, *13*, *14* of the main unit are activated via the corresponding swing forks *15*, *16*, *17* with the aid of a selector shaft *18* equipped with selector fingers. The selector shaft can be swivelled for the shifting process or axially displaced for the selection process by means of a redirection mechanism not shown here using two pneumatic cylinders. Thus the three sliding sleeves *12*, *13*, *14* of the main gearbox can be activated using only two pneumatic cylinders.

The gears of the main gearbox are locked against each other mechanically in order to avoid a simultaneous engagement of two gears. A further actuator cylinder pneumatically activates the clutch (not shown in Figure 12.40) via a clutch release lever.

The electronic control unit of the PowerShift gearbox is fastened directly onto the gearbox along with the gearshift actuators (shift cylinder of the main gearbox, shift valves). The "intelligence" required for the process of gearshifting (interplay of vehicle and clutch, activation of the sliding sleeves, countershaft brakes etc.) is stored in the electronic transmission control unit (TCU). The actual shift command, i.e. based on driver behaviour (accelerator pedal condition) and the current driving situation, is sent by the superordinate vehicle control to the TCU. During gearshifting, the speeds of the countershaft (indirectly via the idler gear of the second gear on the main shaft) and of the output shaft are determined by means of contactlessly functioning sensors in order to monitor synchronization.

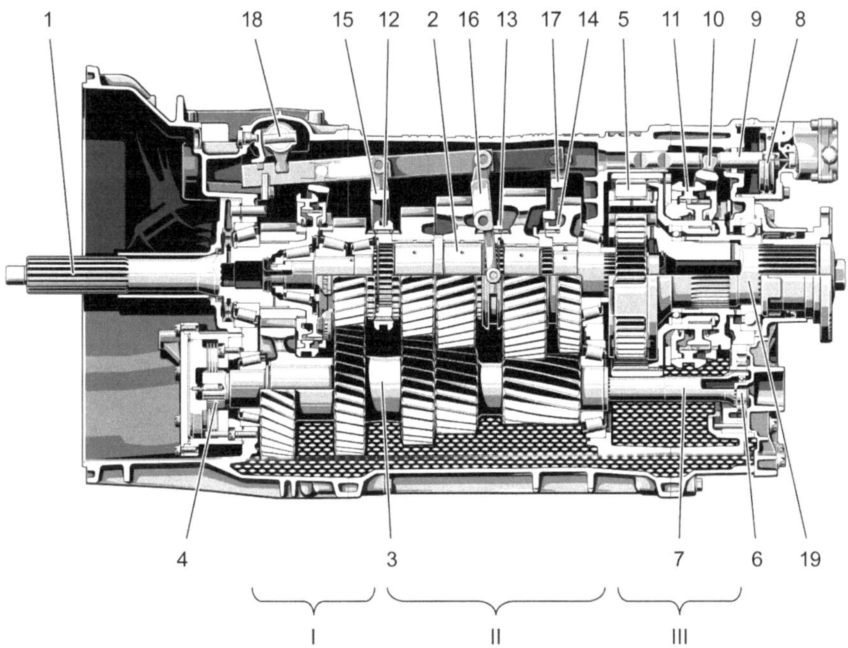


**Fig. 12.40.** 16-speed three-range automated gearbox Mercedes-Benz PowerShift G241-16K; *gearbox diagram and power flows in the gears Figure 6.59. I* Splitter unit; *II* main gearbox; *III* range-change unit; *1* input shaft; *2* main shaft; *3* countershaft; *4* countershaft brake; *5* planetary gear set (range-change unit); *6* oil pump; *7* oil pump shaft; *8* range-change unit shift cylinder; *9* piston rod; *10* range-change unit swing fork; *11* range-change unit sliding sleeve; *12*/*13*/*14* main gearbox sliding sleeve; *15*/*16*/*17* main gearbox swing fork; *18* selector shaft with selector fingers; *19* output shaft

Also, the positions of the four pneumatic actuators are also monitored contactlessly and sent on to the TCU, which means that the current shifting state of the gearbox is always known. Optionally, a hydrodynamic retarder or an eddy current brake can be built on. Power take-offs are also possible.

### *8/ 16-Speed Three-Range Automated Gearbox for Heavy-Duty Commercial Vehicles; ZF AS-Tronic 16 AS 2230 TD*

Another example of an automated gearbox for heavy-duty commercial vehicles is the two-countershaft ZF AS-Tronic, developed as a purely automated gearbox. Gearboxes of the AS-Tronic Series are designed as three-range gearboxes with a front-mounted splitter unit, a three/four speed main gearbox and a rear-mounted planetary range-change unit. This allows for gearbox systems with 10, 12 and 16 gears. This gearbox is a combination of an electropneumatically shifted constant-mesh and synchronized gearbox with an automated clutch. It allows for a two-pedal solution in the vehicle. The gearbox system is used in commercial vehicles (trucks and coaches) with an input torque of up to 3100 Nm, an engine power of up to 500 kW and a total vehicle weight of up to 44 t.

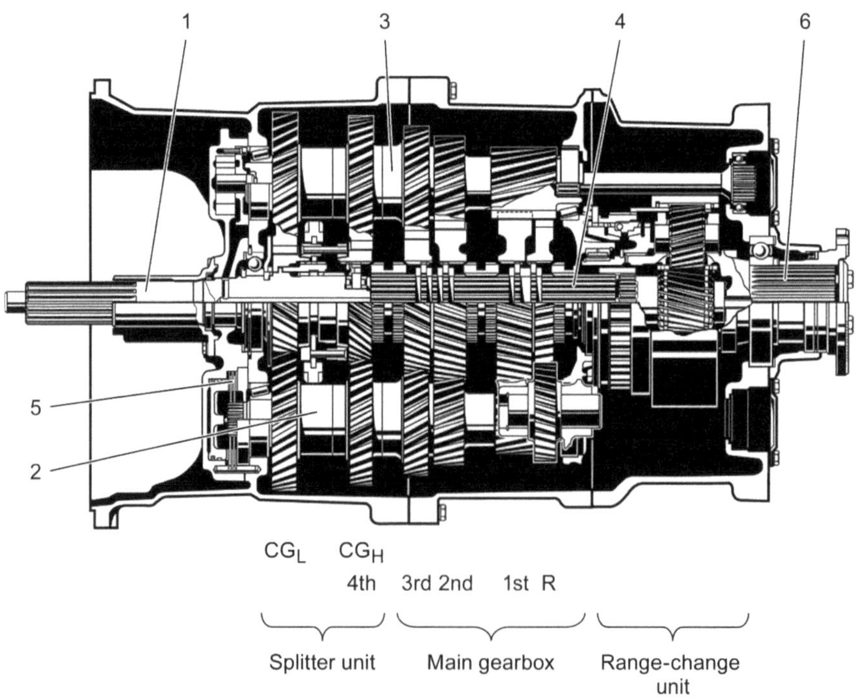


**Fig. 12.41.** 16-speed three-range automated gearbox ZF AS-Tronic 16 AS 2230 TD with direct drive design; *gearbox diagram and power flows Figure 6.60. 1* Gearbox input shaft; *2* countershaft I; *3* countershaft II; *4* main shaft; *5* gearbox brake; *6* output shaft

Figure 12.41 shows the 16-speed gearbox 16 AS 2230 TD in direct drive design as an example of the ZF AS-Tronic Series. The corresponding *gearbox diagram is shown in Figure 6.60*. For this gearbox, the allowable gearbox input torque amounts to 2200 Nm; the overall gear ratio is 17.03.

Both the front-mounted splitter unit and the main gearbox have a two-countershaft design. In this way, the power flow from the gearbox input shaft *1* is split among the two countershafts *2* and *3* via the constant gears $CG_L$ and $CG_H$. The power-split is redirected to the main shaft *4* in the main gearbox in the respectively engaged gear. Then the power is conducted via the rear-mounted planetary range-change unit to the gearbox output shaft *6*. Because of the power-split to two countershafts, the gearbox has a short design.

The main gearbox is unsynchronized and is shifted with dog clutches. Synchronization during downshifting is realised by means of controlled speed increases of the internal combustion engine. Speed is adjusted during upshifting via the multi-plate gearbox brake *5* mounted on the countershaft *3*. The splitter and range-change units are fully synchronized [12.53]. The gearbox housing made of aluminium die cast alloy has three parts and an integrated clutch bell.

Figure 12.42 shows the design of the internal gearshift system. This comprises a selector cylinder *1* for the main gearbox, the three shift cylinders *2*, *3* and *4* for every range unit and the corresponding selector bars *5* and shift forks. The shift cylinder *2* activates the splitter unit, the shift cylinder *3* shifts the respective gears in the main gearbox and the shift cylinder *4* activates the range-change unit.

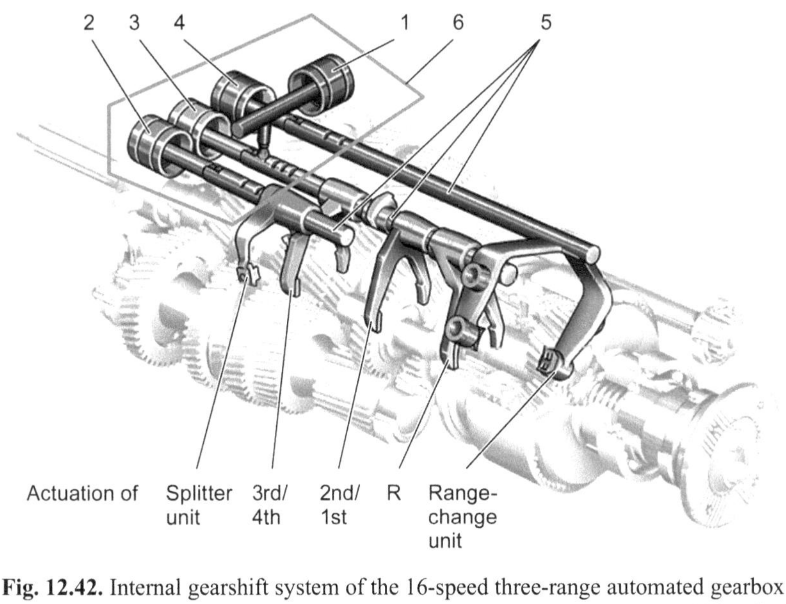


**Fig. 12.42.** Internal gearshift system of the 16-speed three-range automated gearbox ZF AS-Tronic 16 AS 2230 TD. *1* Selector cylinder for main gearbox; *2* shift cylinder for splitter unit; *3* shift cylinder for main gearbox; *4* shift cylinder for range-change unit; *5* selector bars and shift forks; *6* transmission actuator module

Together, the pneumatically activated shift cylinders with their valves, sensors and external connections form a transmission actuator module *6*.

The dry clutch is activated by means of an electropneumatic clutch actuator module integrated in the gearbox housing (Figure 12.43a). The actuating piston *1* pneumatically activates the dry clutch via a clutch release bearing and a clutch release lever. Four synchronized magnet valves *2* are responsible for the sensitive motion control. Two valves operate together to open and close the clutch (see also Section 13.3 "Control Systems").

The displacement of the actuating piston and thus of the clutch is determined by means of a contactless travel sensor *3*. The sensor signal is supplied by the evaluating electronics *4* of the TCU *7*. Depending on the operational mode and situation, the TCU controls the clutch movement according to the selected moving-off and shift processes.

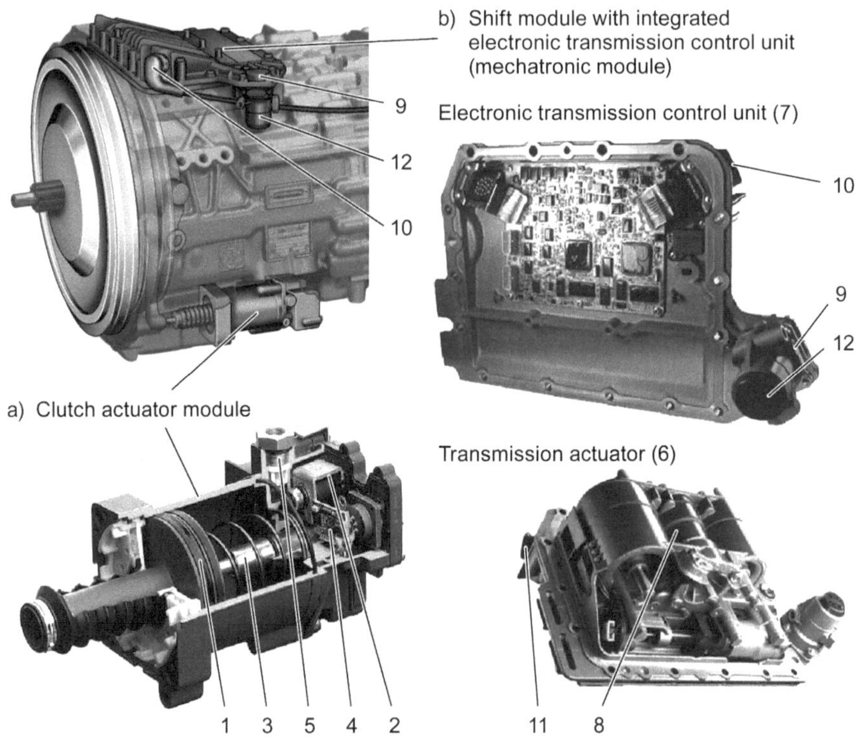


**Fig. 12.43.** Position and structure of the clutch and shift module of the 16-speed three-range automated gearbox ZF AS-Tronic 16 AS 2230 TD. *a* Clutch actuator module: *1* actuating piston; *2* magnet valves for controlling clutch actuators; *3* travel sensor; *4* evaluating electronics; *5* compressed air connection with non-return valve; *b* transmission actuator with integrated electronic TCU (shift module): *6* transmission actuator; *7* electronic TCU; *8* magnet valves, shift cylinders and sensors; *9* pressure limiting valve; *10* gearbox connector; *11* vehicle connector; *12* compressed air connection

With every moving-off process, the current clutch engagement point is newly taught in and stored in the electronics for future use (clutch control). In this way, the control software has constant access to the current data regarding the state of wear of the clutch lining.

The shift module built into the upper side of the gearbox contains, in two housing parts, the transmission actuator *6* and the integrated electronic transmission control unit (TCU) *7* (Figure 12.43b). The transmission actuator contains the valves for selecting, shifting and activating the gearbox brake as well as the shift cylinder *8* for shifting the respective gearbox part (see Figure 12.42). The sensors for determining the current position of all shift cylinders, the gearbox input speed, the operating temperature and the adjacent air pressure are also integrated. Furthermore, a pressure limiting valve *9*, which ensures a constant working pressure, the gearbox connector *10* and the vehicle connector *11* are housed here as well [12.13, 12.25, 12.33, 12.53]. The vehicle-side connector is the interface for supplying power and, via the vehicle CAN, to the display, drive selector and diagnostic devices. The actuation energy is supplied via the compressed air connection *12*.

The software of the TCU contains the necessary functions for controlling the transmission system and clutch, for engine control, as well as for the total driving strategy. The TCU is networked with the control units for the engine, ABS/ASR, EBS, retarders and power take-offs by means of the CAN data bus.

The gearbox is operated by means of a "shift-by-wire" system with a console drive selector (Figure 12.44) or a steering column switch. The operating mode is set with a rotary switch for the operating modes *neutral*, *drive* and *reverse*. Another switch is used to preselect the operating mode (automatic/manual). Tipping the switch once or more in manual mode causes an up- or downshift. Information such as operating mode, the current gear, an imminent clutch overload and other information regarding system malfunctions are shown in the display [12.13, 12.24–12.25, 12.33, 12.53, 12.67].

The gearbox has a secondary retarder integrated on the output side of the gearbox and thus directly connected with the wheels.

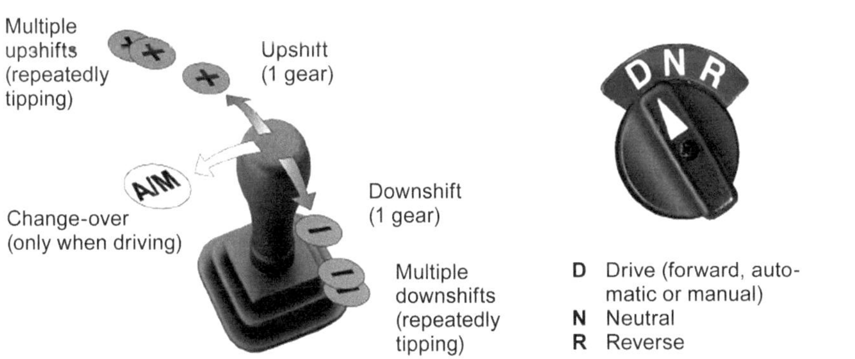


**Fig. 12.44.** Operation of the 16-speed three-range automated gearbox ZF AS-Tronic 16 AS 2230 TD with a console drive selector

The advantage of a secondary retarder is that the braking force is independent of the engine speed and that there is no braking force interruption during clutch disengagement or shifting. The maximum braking torque is 3200 Nm with the brake power limited to 500 kW (see also Section 11.6 "Vehicle Continuous Service Brakes").

#### *Power Take-Offs*

Engine-, clutch- and drive-controlled power take-offs are possible for AS-Tronic gearboxes (see also Section 6.9 "Power Take-Offs").

*Engine-controlled power take-offs* are designed for high performance levels during continuous operation with maximum engine torque. They can be switched on and off either while driving or when stationary, even under load. They are arranged between the engine and the transmission and are driven directly by the engine via a special clutch. Engine-controlled power take-offs are used, for example, in vehicles with high-pressure pumps, such as fire-fighting vehicles, sewer cleaning vehicles and suction vehicles, as well as in ground-boring devices and concrete mixers and pumps.

*Clutch-controlled power take-offs* are used, according to the type, for short-term or continuous operation. They are used during driving or when stationary (in neutral gear) and are mounted on the output side or on the lateral side of the gearbox housing. Clutch-controlled power take-offs are only used when the engine is running and the vehicle clutch is closed. They are used, for example, in water, sludge and hydraulic pumps, compressors, cable winches, fire engine ladders and working platforms.

*Drive-controlled power take-offs* supply the hydraulics of two-circuit systems with working pressure. This means that vehicles remain steerable even when the engine fails. These power take-offs are connected to the output shaft of the gearbox and are thus active as soon as the drive wheels rotate.

### 12.2.3 Commercial Vehicle Torque Converter Clutch Transmissions (TCCT)

While torque converter clutch transmissions fall under the category of automated manual transmissions, they receive special treatment here because of their special design. This transmission type is particular in that both manual gearboxes and automated gearboxes can be used for heavy goods vehicles, while the difficult moving-off processes characteristic of such vehicles are made considerably easier by means of a torque converter.

#### *9/ 16-Speed Semi-Automated Manual Commercial Vehicle Gearbox; ZF-Transmatic WSK 400 + 16 S 221*

The torque converter clutch gearbox ZF Transmatic WSK 400 + 16 S 221 developed in the 1960s is a combination of a torque converter clutch (TCC) and a 16-speed synchromesh gearbox from the Ecosplit Series (Figure 12.45; cf. also the *gearbox diagram in Figure 6.61*). With this semi-automatic gearbox, the driver engages the desired gear by activating a gearshift lever. Moving-off is automatic and is achieved solely by means of the torque converter. The dry clutch is activated by foot and only when shifting for the purpose of power interruption. The 16-speed manual gearbox ZF 16 S 221 mounted behind the torque converter clutch is a fully synchronized three-range transmission with splitter unit and range-change unit with a 2 x 4 x 2 design.

The design and arrangement of the gearwheels of the manual gearbox 16 S 221 are shown in Figure 6.54 and Figure 12.35. Please refer to the explanations provided there. The gearbox as a whole is designed as a kit of series production parts and can therefore be customised simply to certain requirements. The Transmatic allows for input torques of up to 2300 Nm with a ratio of 16.47 to 1 with direct drive design or a ratio of 13.86 to 0.84 with overdrive design.

The foot-activated clutch pedal acts hydrostatically on the clutch release lever *2*. Since a lot of force is required to disengage the clutch, the action is pneumatically assisted by a servo unit. Beside the selector finger *3*, the bellows *4* is to be seen, which protects the piston rod of the compressed-air cylinder. The cylinder acts through the clutch release lever and the clutch release bearing *5* by opening the plate spring to disengage the gearshifting clutch *6*. Gears are changed manually.

The Trilok torque converter *7* is used for moving-off. This increases the starting torque to 2 to 2.5 times the nominal torque. The torque converter makes the whole powertrain largely free of shocks and jolts.

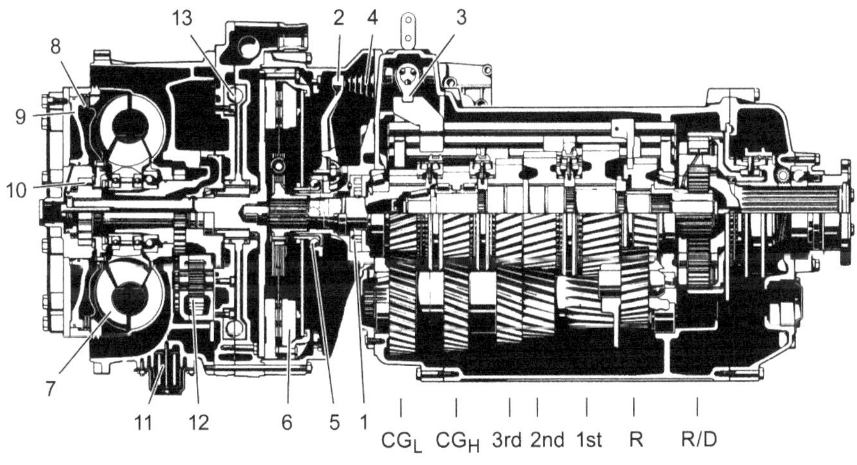


**Fig. 12.45.** 16-speed semi-automated manual commercial vehicle gearbox ZF-Transmatic (ZF WSK 400 + ZF 16 S 221); *gearbox diagram Figure 6.61*. *1* Oil pump for manual gearbox; *2* clutch release lever; *3* selector finger; *4* bellows; *5* clutch release bearing; *6* gearshifting clutch; *7* torque converter; *8* lock-up clutch; *9* hydraulic piston; *10* coasting freewheel; *11* oil filter; *12* oil pump for torque converter clutch; *13* primary retarder; $CG_H$ constant gear high; $CG_L$ constant gear low; range-change unit: *R* range; *D* direct

To prevent moving-off with the gearshifting clutch slipping and the torque converter not locked up, this state is eliminated electronically or by an accelerator interlock.

The torque converter has a lock-up clutch *8*, which is engaged by means of an annular hydraulic piston *9*. The cylindrical rollers of the coasting freewheel *10* can be seen below the torque converter lock-up clutch. This freewheel is located between the impeller and the turbine wheel of the torque converter, and grips when the vehicle is in overrun conditions. This harnesses the engine braking torque, and enables the vehicle to be tow-started. The oil filter *11* of the torque converter clutch oil circuit is located below the torque converter. Above the oil filter is the oil pump *12*. The vacant space beside the torque converter is occupied by an engine-controlled power take-off, if fitted.

An integral hydrodynamic primary retarder *13* is fitted between the torque converter and the gearshifting clutch. The retarder braking torque, which is dependent on the input speed, is 1700 Nm, and the brake power limited to 320 kW. The function of the retarder is to relieve the service brake in heavy vehicles. The torque converter clutch does not have to be flange-mounted directly on the selector gearbox as in Figure 12.45. They can both be mounted separately from each other in the vehicle, and linked by a cardan shaft. The same applies to the link between the engine and the torque converter clutch.

### *10/ 12-Speed Automated Commercial Vehicle Torque Converter Clutch Transmission; ZF TC-Tronic 12 TC 2740 TO*

An example of a younger generation of torque converter clutch transmissions is the fully automated 12-speed TC-Tronic 12 TC 2740 TO by ZF. The automated ZF AS-Tronic gearbox 12 AS 2740 TO is mounted behind the torque converter clutch unit (Figure 12.46, *the basic concept is explained in Section 6.7.3 and Figure 6.62*).

This gearbox allows for input torques of up to 2700 Nm with engine power values up to 500 kW. It is used in heavy-duty commercial vehicles, such as heavy cranes, heavy-duty towing machines and special off-road vehicles with a total weight of up to 250 t. The ratio of the overdrive transmission system in the first gear is 12.29, that of the highest gear is 0.78.

The torque converter clutch unit TC2 (Torque Converter 2: basic model WSK 400) has a torque converter *1*, a lock-up clutch *2* with torsional vibration damper *3*, an integrated primary retarder *4* mounted between the converter and the gearbox, an oil pump for the converter oil circuit *5* and an electronic control unit which is generally mounted in the interior of the vehicle. The converter is only used for moving-off and manoeuvring. A stall torque ratio of up to 1.6 times the input torque can be reached. With a roughly 80% speed synchronization between the turbine wheel and the impeller, the torque is no longer increased and the converter is locked up by the lock-up clutch. In this condition, a mechanical connection between the engine and the gearbox is established.

The torque converter lock-up clutch is opened and closed automatically at a defined turbine speed and a predetermined turbine/engine speed ratio.

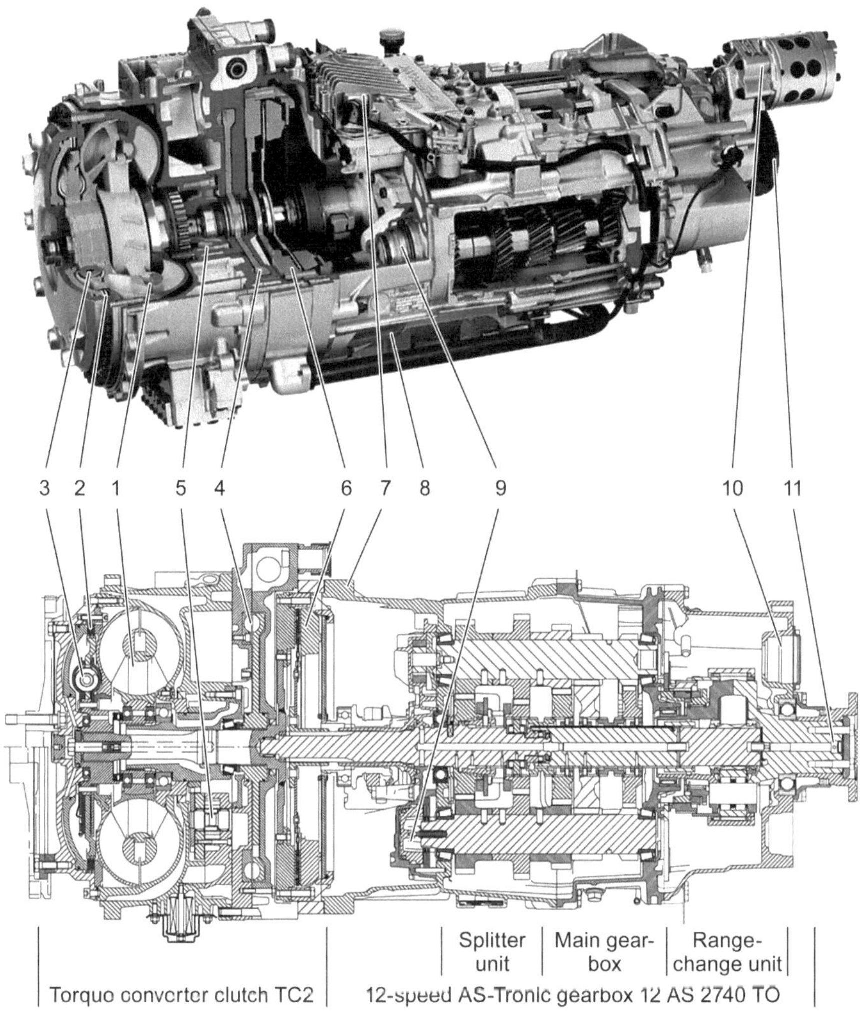


**Fig. 12.46.** Automated 12-speed torque converter clutch gearbox ZF TC-Tronic 12 TC 2740 TO (12-speed gearbox shown rotated 90°); *gearbox diagram and power flows Figure 6.62*. *1* Torque converter; *2* torque converter lock-up clutch; *3* torsional vibration damper; *4* primary retarder; *5* oil pump for oil circuit TC2; *6* gearshifting clutch; *7* transmission actuator with integrated electronics; *8* clutch actuator; *9* gearbox brake; *10* power take-off connection; *11* output shaft with output flange

In addition, the control of the torque converter lock-up clutch can be influenced via the kick-down function at the accelerator pedal, thus broadening the application range of the converter. The braking effect of the primary retarder is directly dependent on which gear is engaged. Thus, a maximum braking torque of up to 1700 Nm is available at the gearbox input shaft in the lower gears. The brake power of the primary retarder is limited to 320 kW. Since operation of both the converter and the retarder is dependent on oil temperature, the latter is continuously monitored. If it exceeds defined limiting values, information is shown on the display and the retarder brake power is reduced. An oil-water and/or oil-air heat exchanger is used to cool the oil.

The gearbox mounted behind the torque converter clutch unit has 12 forward and two reverse gears. The gearbox has a three-range (2 x 3 x 2) design. The three-speed main gearbox is constant-mesh, the splitter and range-change units are synchronized. Both the main gearbox and the splitter unit have two countershafts. The range-change unit has a planetary design. The torque converter clutch unit and the AS-Tronic gearbox have their own oil circuits. Further features of and explanatory material regarding the ZF AS-Tronic gearbox can be found under 8/ and in Figures 12.41-12.44. With the assistance of a drive selector, the fully automated variant of the gearbox unit allows for both a manual and an automatic operating mode [12.72].

### 12.2.4 Automatic Commercial Vehicle Transmissions (AT)

The basic design principles of automatic commercial vehicle transmissions are explained in Section 6.7.4. The load profiles for automatic commercial vehicle transmissions differ greatly from those for passenger cars. As a result, there are different design criteria, and therefore different designs. The use of automatic passenger car transmissions in light-duty commercial vehicles must be evaluated on an individual basis. The permissible gearbox input torque is not the only factor involved. According to the individual case, automatic commercial vehicle transmissions may also have retarders and power take-offs.

#### *11/ 5-Speed Automatic Commercial Vehicle Gearbox for Light-Duty Trucks and School Buses; Allison 2000*

Figure 12.47 shows the 5-speed automatic gearbox for light-duty trucks, delivery vehicles, mobile homes and school buses from Allison Transmission's 1000/2000/2400 Series. This gearbox is used in commercial vehicles with a weight up to 13.6 t, an engine power up to 250 kW and a gearbox input torque up to 750 Nm.

The gearbox structure is similar to that of a planetary passenger car AT. The input power of the engine is transmitted to the impeller of the torque converter *4*. The 2-line converter has a lock-up clutch *2* with torsion damper *3*. The torque converter lock-up clutch is activated by means of a magnetic pulse-width modulation valve (PWM). The oil pump *6* and the control plate *7* comprise the oil supply unit of the gearbox. The pump pinion of the internal gear pump is driven via a positive engagement with the pump side of the converter and thus directly with the engine speed.

*Figure 6.63 shows the gearbox diagram and power flow in the individual gears*. The planetary gearbox comprises three coupled, helical-cut single planetary gear sets. The planetary gear carriers *13*, *15* of planetary gear sets I and II are each connected with the ring gears of the next planetary gear set. The individual gears are shifted by activation of the two clutches *10* and *11* as well as the three brakes

*12*, *16* and *18*. The inner clutch *A* (*10*) connects the gearbox input shaft *1* with the intermediate shaft *14* of the gearbox. The outer clutch *B* (*11*) serves to connect the gearbox input shaft *1* with the planetary carrier of the second planetary gear set *15*. The three brakes *D* (*12*), *E* (*16*) and *F* (*18*) are supported by the housing and are connected with the respective ring gear of the planetary gear set. The clutches and brakes have a multi-plate design and are actuated hydraulically. Two proportional valves and three magnet valves contained in the hydraulic control device *22* are used to control gear changes. The output is through the planetary carrier of the third planetary gear set *17* and the output shaft *20*.

The externally mounted pressure oil filter *24* exclusively serves to protect the pressure regulators and valves. It supports the suction filter in the oil sump in cleaning the oil. The oil is cooled by means of a separately mounted oil cooler.

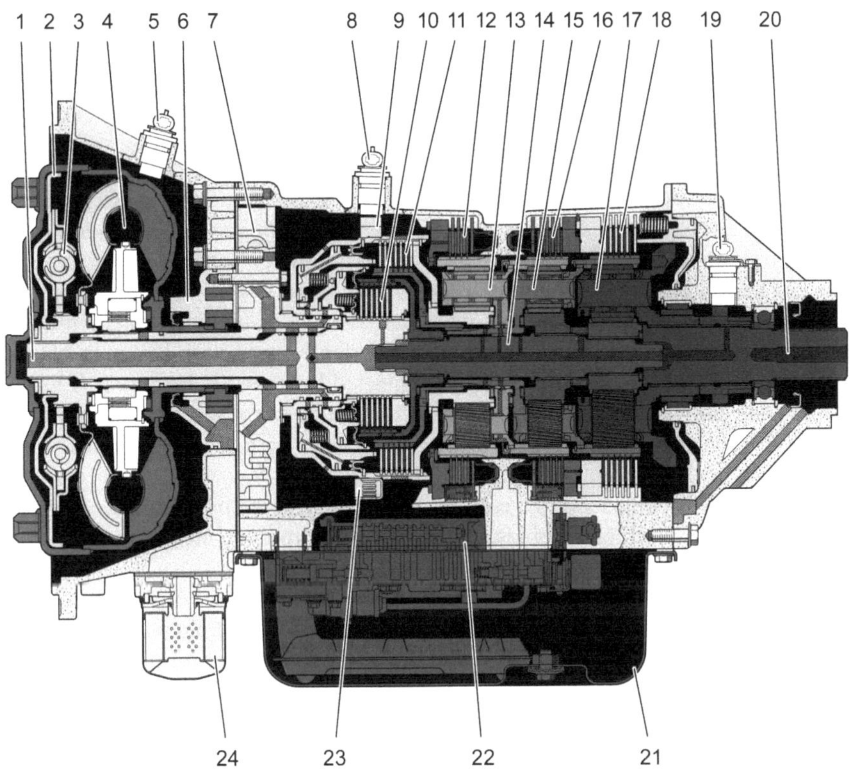


**Fig. 12.47.** 5-speed automatic gearbox Allison 2000 for light-duty commercial vehicles; *gearbox diagram und power flows, Figure 6.63*. *1* Gearbox input shaft; *2* lock-up clutch; *3* torsion damper; *4* torque converter; *5* input speed sensor; *6* oil pump; *7* control plate; *8* turbine speed sensor; *9* PTO gearwheel; *10* clutch *A*; *11* clutch *B*; *12* brake *D*; *13* planetary gear carrier I; *14* intermediate shaft; *15* planetary gear carrier II; *16* brake *E*; *17* planetary gear carrier III; *18* brake *F*; *19* output speed sensor; *20* output shaft; *21* oil sump; *22* transmission control; *23* speed sensor wheel; *24* pressure oil filter

Section 11.3.1 "Oil Supply" explains the oil supply of an automatic transmission and Section 9.1.2 discusses external gearshift mechanisms of automatic transmissions.

The individual gears and operational states are selected with a selector lever which transmits the position via cable control to the hydraulic transmission control unit and to the position switch mounted on the gearbox housing. This transfers the information regarding the current shift position to the transmission control unit in the form of an electric signal. The hydraulic selector shift valve coupled mechanically to the selector lever releases the control lines of the shifting elements for forward and reverse as well as the neutral and park positions.

The gearbox can be optionally equipped with a parking lock and a primary retarder (not shown in Figure 12.47). Also, this transmission series can also be equipped with one or two power take-offs (PTOs). PTOs are connected to the PTO gearwheel *9* of the automatic gearbox mounted on the outer clutch housing of clutch *B* (*11*) [12.1–12.2, 12.29, 12.36, 12.55].

### *12/ 6-Speed Automatic Gearbox for Trucks and Buses; ZF 6 HP 602 C*

The ZF 6 HP 602 C conventional automatic transmission for trucks, buses and special-purpose vehicles shown here (Figure 12.48) is one of the Ecomat Series (Ecomat 2). Please *refer to Figure 6.64 for the corresponding gearbox diagram.* There are 4- to 7-speed versions of this transmission for various power classes from 125 to 280 kW and with input torques of 500 to 1600 Nm.

The torque converter *3* has a lock-up clutch *2*, actuated by a separate hydraulic piston (3-line converter). After the torque converter housing with its pump blades is the engine-driven primary oil pump *4* connected by a gear interstage (off-axis). Between the torque converter and the selector gearbox is the integrated primary retarder *5*. Because it is mounted on the engine side, high braking torques up to near vehicle standstill can be achieved, especially with low vehicle speeds. The primary retarder is controlled in a continuously variable way by an electric proportional magnet valve which receives switching pulses from the electronic transmission control unit.

The selector gearbox itself can again be subdivided into two sections, the space for the clutches, and the space accommodating the brakes and planetary gear sets. The three gearshifting clutches *6*, *7* and *8* are grouped together on one clutch carrier and have a multi-plate design. (The sequence of shifting actions in automatic transmissions is described in Section 9.3.2 "The Shifting Process").

The inner clutch *A* (*6*), characterized by its large number of plates, is engaged by the annular piston located beside it on the input side. The two outer clutches *B* (*7*) and *C* (*8*) share the same multi-plate external gearing. The supporting plate required to engage the clutches is located between the two multi-plate sets of the clutches. The left-hand one of the two clutches *7* is activated directly by the annular piston beside it acting on it. The piston of the right-hand clutch *8* is located at the left-hand outermost diameter of the clutch carrier, and acts on the multi-plate set on the opposite side through a formed sheet metal part enclosing the entire clutch carrier.

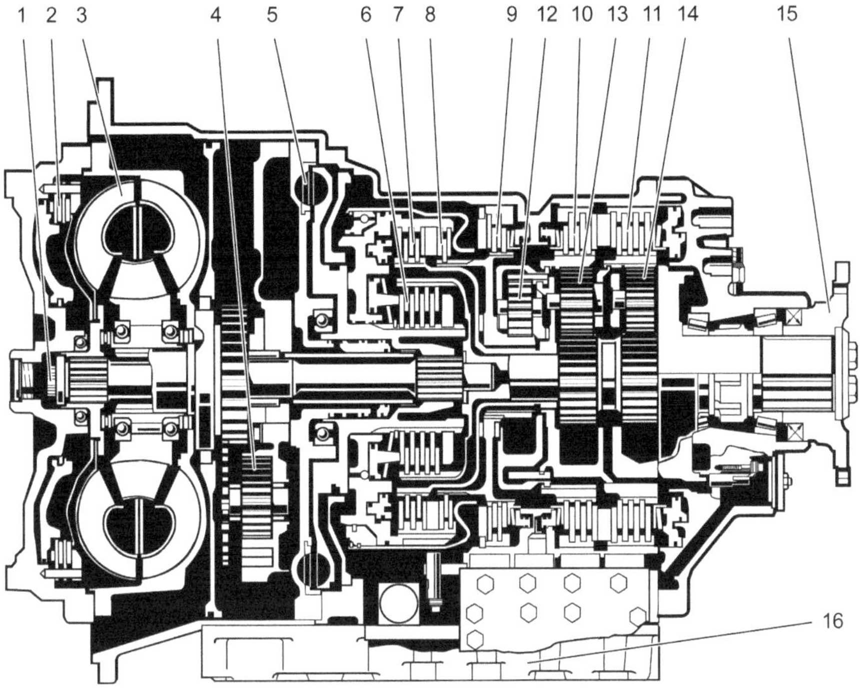


**Fig. 12.48.** 6-speed automatic gearbox ZF 6 HP 602 C; *gearbox diagram and power flows Figure 6.64*. *1* Gearbox input shaft; *2* torque converter lock-up clutch; *3* 3-line converter; *4* oil pump; *5* retarder; *6* clutch *A*; *7* clutch *B*; *8* clutch *C*; *9* brake *D*; *10* brake *E*; *11* brake *F*; *12* planetary gear set I; *13* planetary gear set II; *14* planetary gear set III; *15* output shaft with output flange; *16* transmission control

The three brakes *9*, *10* and *11* and the three single planetary gear sets *12*, *13* and *14* are located on the output side. The multi-plate sets of the brakes with their actuating pistons and counter bearings are mounted on the inner wall of the housing. The contact pressure of the hydraulic pistons and of the clutches must be adapted to the torque to be shifted for the particular change of ratio. This is carried out by an electronic shifting pressure control, and by means of stepped pistons with varyingly effective piston areas. The speed sensor for the output speed or the velocity signal can be seen on the output side, and that of the turbine speed signal in the lower central area.

Since this transmission has one single oil circuit for the torque converter, retarder, shift control, lubrication and for heat dissipation, a heat exchanger is flange-mounted on the transmission output (not shown).

The Ecomat 2 automatic gearbox has an electrohydraulic transmission control *16*. The necessary signals come from the electronic transmission control unit (TCU). The TCU is in constant communication with other electronic control units in the vehicle. Data exchange with the engine control unit makes for optimal gear-shifting by engine control. Even the control of the retarder is integrated into the vehicle's engine and brake management. This makes it possible to adjust the driving strategy to the individual application profile of the vehicle. Moreover, the use of CAN technology allows for a simple diagnosis of the transmission system [12.15, 12.70].

### 12.2.5 Commercial Vehicle Hybrid Drives

The following will consider a serial hybrid powertrain for commercial vehicles. With serial hybrids, there is no mechanical coupling of the engine to the wheels (see Sections 3.2.4 and 6.7.5).

#### *13/ Electric Drive Axle for Trucks and Buses, ZF EE Drive 1*

Section 6.7.5 describes the electric drive axle EE Drive 1 *(see Figure 6.66)*, which is used in serial hybrid drives for commercial vehicles. Figure 12.49 shows a low-profile axle with axle support, on which the two torque-controlled traction motors are each flanged to a wheel. The torque-controlled traction motors are designed as air-cooled asynchronous machines (ASM). Their maximal torque is 350 Nm with a maximum power of 75 kW. At maximum speed, the traction motors rotate at a speed up to 11 000 1/min. The two-stage planetary gearbox mounted on both sides concentrically to the axis of the wheel and the drive motor has a total ratio of $i_{\text{tot}} = 25.5$ and is helical-cut to reduce noise. Due to the high speeds arising within the wheel hub drive and to the relatively high temperature level, a temperature-stable synthetic oil must be used. Figure 12.49 also shows the wheel brakes with ventilated brake discs whose design is adjusted to the electric motors in comparison to the standard variant.

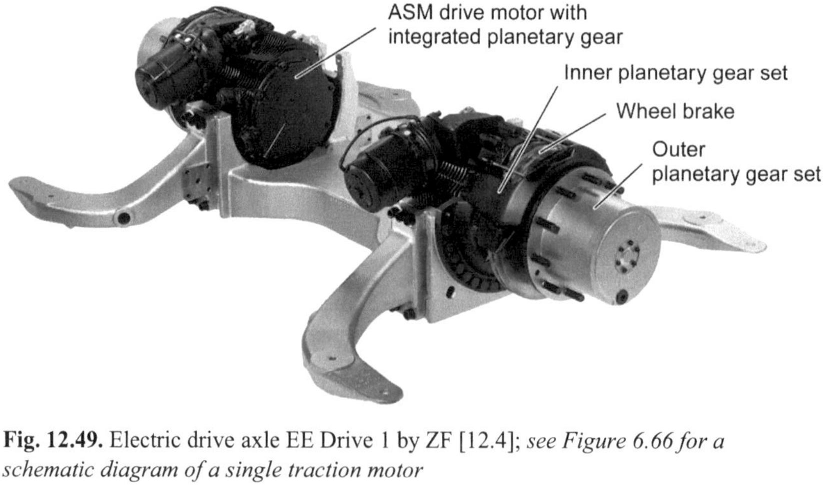


**Fig. 12.49.** Electric drive axle EE Drive 1 by ZF [12.4]; *see Figure 6.66 for a schematic diagram of a single traction motor*

By keeping the axle coupling points, the electric drive axle EE Drive 1 or, alternatively, a conventionally driven axle can be used in the same vehicle platform. When using the electric drive axle, the differential gear can be omitted due to the torque-controlled single-wheel drive. Moreover, this drive configuration offers a more flexible distribution of the axle loads and can thus help combat the frequent problem of high rear axle loads.

### 12.2.6 Continuously Variable Commercial Vehicle Transmissions (CVT)

Continuously variable hydrostatic transmissions in combination with gear transmissions represent a common form of drive in commercial vehicles and particularly in mobile working machines.

#### *14/ Continuously Variable Hydrostatic-Mechanical Power-Split Gearbox for Tractors, ZF ECCOM*

In agricultural tractors, there is a distinct need to adjust the speed as finely as possible. The demand for very low driving speeds (partly up to 0.1 km/h) on the one hand and transport speeds of 50 km/h on the other means a very large overall gear ratio. While in the past multi-stage transmissions shiftable under load were used, continuously variable transmissions working on the basis of a hydrostatic power-split have been seen more and more in tractors (see Section 6.7.6).

The design of this type of CVT is determined by the necessary overall gear ratio, the necessary tractions and efficiency requirements on the one hand and the allowable amount of stress placed on the transmission elements on the other hand, particularly the maximum allowable hydrostatic pressures.

Hydrostatic power-split transmissions as implemented in tractors thus have either two or four operating ranges. These operating ranges are either preselected or generated via a set of coupled planetary gears with automated range selection.

The gearbox system shown in Figure 12.50 is a continuously adjustable, hydrostatic-mechanical power-split gearbox. It consists of the following components: planetary coupled gear, hydrostatic unit, reversing gear unit, electrohydraulic control, sensor system and electronic control unit. The total ratio of the gearbox can be continuously adjusted within a range of $\infty$ to $+/-0.58$. The hydrostatic unit comprises axial piston machines in the form of a variable displacement pump and a fixed displacement motor with a swash plate and a back-to-back configuration having the same displacement and which are run in a closed circuit.

The electronic system of the gearbox controls all transmission functions, such as continuously adjusting the gear ratio or changing the direction of travel in line with the driver's demands. The transmission's continuously variable operation allows for ultra low driving speeds of up to 10 m/h and even force-locking standstill (gearbox output speed = 0), frequently referred to as geared neutral operation.

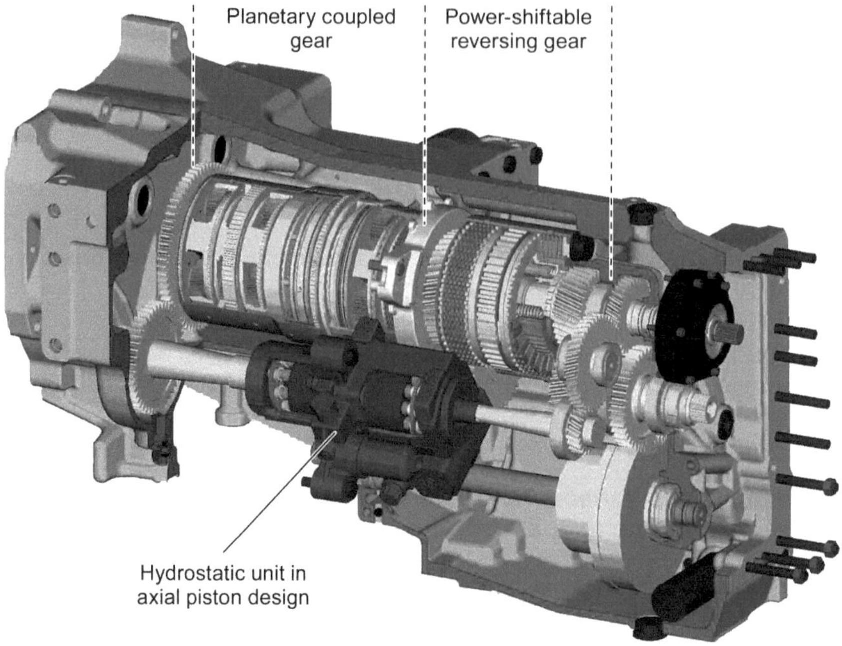


**Fig. 12.50.** Continuously variable hydrostatic-mechanical power-split gearbox ZF ECCOM for use in tractors; *gearbox diagram and power flows, Figure 6.68.* Hydrostatic power fraction as function of driving speed and gearshift pattern respectively power-transmitting components in the continuously variable gearbox part

The essential advantage of continuously variable transmissions in such working machines as tractors, however, is that the engine speed optimal for the working process and the desired driving speed can be continuously adjusted independently of each other and under load. In the CAN-bus connection with the engine and vehicle control units, an arbitrary number of superordinate control strategies for employing the tractor is possible, such as power-optimised vs. fuel-efficient driving, cruise control etc. [12.20, 12.52].

## 12.3 Final Drives

Section 6.8 systematically treats final drives of vehicle powertrains and introduces basic designs. The follow will now discuss design examples with the aim of gaining a better understanding of the principles by looking at how they are implemented. While current developments are highly centred on making improvements [12.44, 12.49], the focus of this section will be placed on basic topics. For that reason, historical designs will also be considered. For the sake of completeness, gearbox diagrams are given for the examples used in Section 6.8.

### 12.3.1 Axle Drives for Passenger Cars

Examples of integrated axle drives for passenger cars are given in Sections 12.1.1–12.1.6. From the multiplicity of production designs, the following final drives are considered:

- spur gear axle drive, Figure 12.51,
- bevel gear axle drive, independent assembly with hypoid drive, Figure 12.52,
- bevel gear axle drive, flange-mounted to gearbox housing, Figure 12.53 and
- worm gear axle drive, Figure 12.54.

The axle drive of the Opel Kadett D (1984 model) is shown in Figure 12.51 as an example of a spur gear axle drive. The torque is transmitted by the output shaft of the selector gearbox *1* by means of a helical-cut spur gear stage *2* to the cage of the differential gear *3*. The spur gear and the differential gear cage are bolted together.

The Mercedes Benz mid-range W 124 axle drive (1988 model) is shown in Figure 12.52 as a representative example of a vehicle with bevel gear axle drive. The input torque is transmitted to the axle drive from the cardan shaft via a flexible clutch. The drive pinion shaft and the differential cage run in tapered roller bearings.

Differential bevel gears and the flanges for connecting the drive shafts to the rear wheels run in plain bearings. Output is via drive shafts with constant velocity joints. The ratio in this example is $i_{E,A} = 3.07$. The crown gear has a diameter of $d_m = 185$ mm.

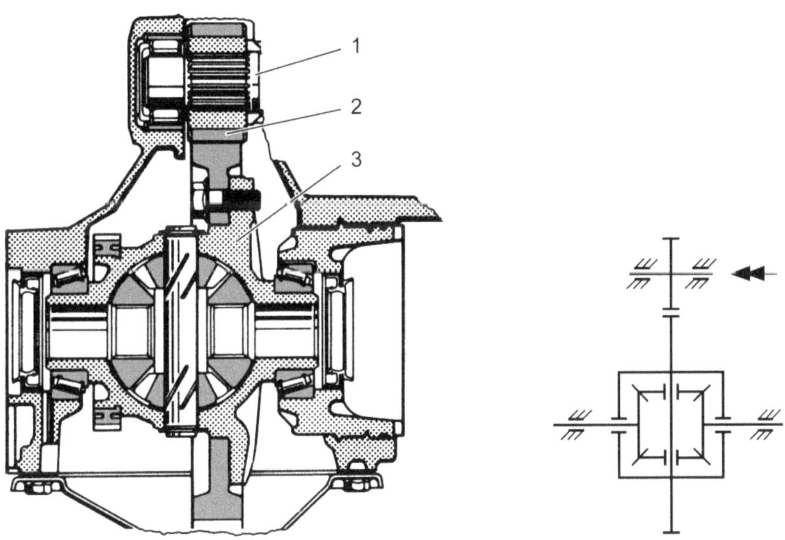


**Fig. 12.51.** Spur gear axle drive for vehicles with transverse-mounted engine, Opel Kadett, 1984 model. *1* Output shaft; *2* spur gear stage; *3* differential cage

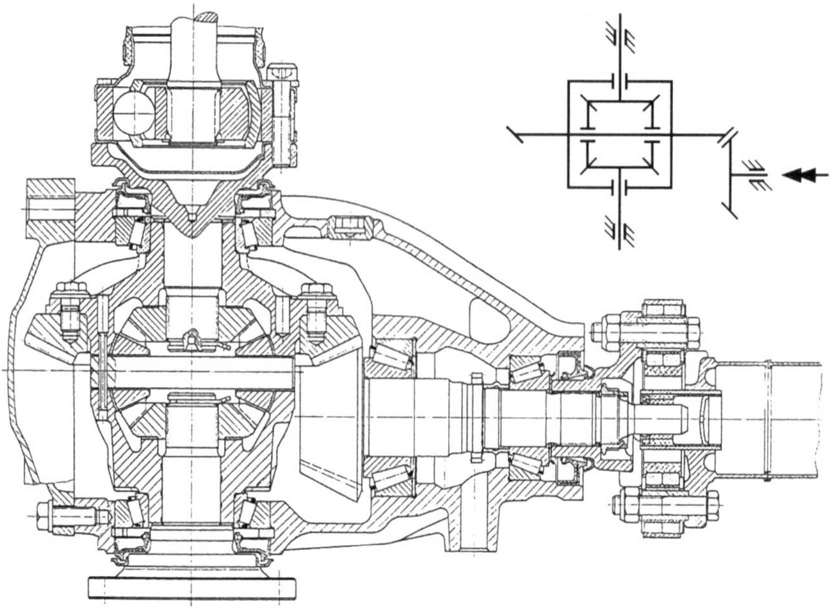

**Fig. 12.52.** Example of an axle drive as independent assembly for standard or all-wheel drive, Mercedes-Benz mid-range W 124, 1988 model

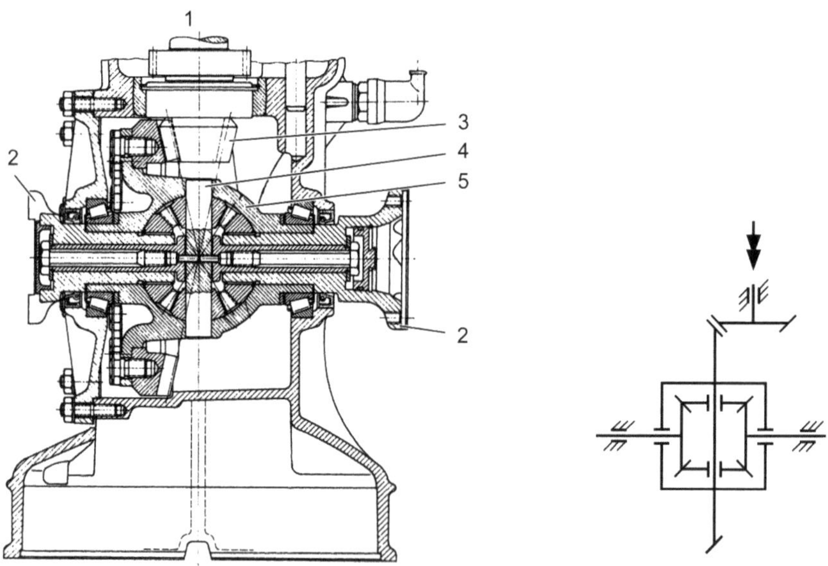


**Fig. 12.53.** Bevel gear axle drive directly flanged onto the gearbox with spiral gearing (Porsche 911). *1* Gearbox output shaft; *2* flange for drive shafts; *3* spiral bevel drive; *4* differential shaft; *5* differential cage

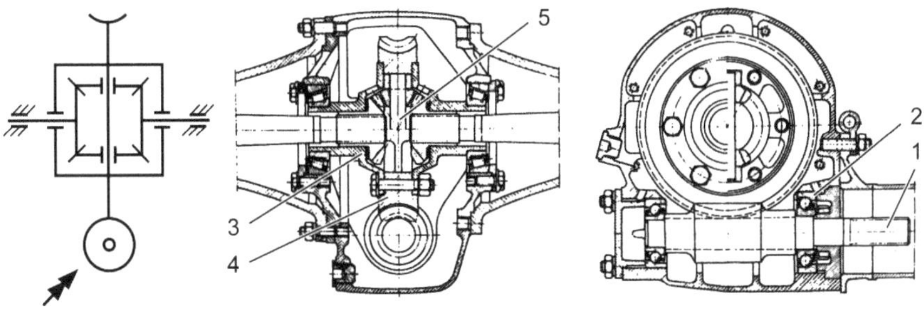


**Fig. 12.54.** Axle drive of the Peugeot 404 with worm gear drive. *1* Worm shaft; *2* angular contact ball bearing; *3* differential cage; *4* worm gear; *5* bevel gear differential

Figure 12.53 shows the axle drive of the Porsche 911. It is attached to the selector gearbox, and has a spiral bevel drive.

A typical example of the worm gear drive is the Peugeot 404. Figure 12.54 shows its axle drive. In this case, the worm shaft *1* runs in angular contact ball bearings *2*. The differential cage *3* is of two-piece design to accommodate the worm gear *4*. Speed compensation between the wheels is provided by the bevel gear differential *5* in the differential cage. Differential bevel gears and axle bevel gears run in plain bearings.

### 12.3.2 Axle and Hub Drives for Commercial Vehicles

Numerous possible axle designs can be composed from the axle drives and hub drives described in Sections 6.8.2 and 6.8.4 [12.30, 12.39, 12.41]. For commercial vehicles, the term "centre gearbox" is occasionally used instead of "axle drive" when a hub drive is present. Four commonly used drive axles in commercial vehicles are:

1/ axle with single ratio, Figure 12.55,
2/ two-speed axle, Figure 12.56,
3/ pinion axle (gantry axle), Figure 12.57,
4/ outer planetary axle, Figure 12.58.

These axle designs are described below.

#### *1/ Axle with Single Ratio*

These axles are generally used in light to medium-duty commercial vehicles, and offer ratios of up to approximately $i_E = 7.0$. Increasing the ratio or the power rating would involve enlarging the crown gear. This would further reduce the ground clearance underneath the axle drive.

Where high levels of power and torque are to be transmitted, a high ratio in the axle drive means that all the following parts have to be adapted to the increased torque, which militates against light-weight design and efforts to minimise unsprung weight. Figure 12.55 shows the Mercedes Benz HL 2/11 single-ratio axle.

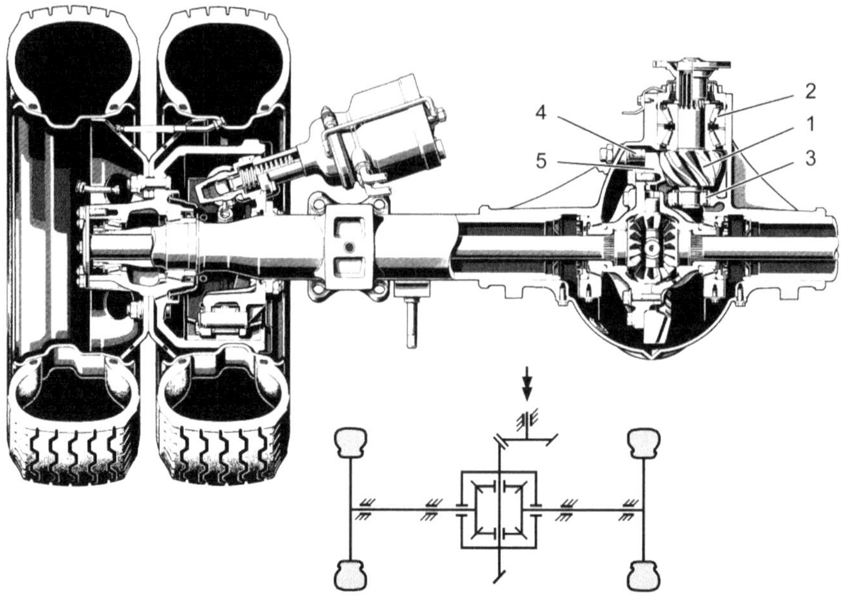


**Fig. 12.55.** Axle with single ratio, Mercedes-Benz HL 2/11, axle drive shown rotated 90°. *1* Pinion; *2* tapered roller bearing; *3* cylindrical roller bearing; *4* fulcrum pad; *5* crown gear

Because of the large axial and radial forces of the hypoid bevel gears, the pinion *1* has bearings on both sides. Tapered roller bearings *2* are used in an O arrangement with the largest possible angle of contingence. A cylindrical roller bearing *3* is used as supporting bearing. The adjustable thrust piece with fulcrum pad *4* prevents the crown gear *5* being deformed excessively under load. This enables the full capacity of the bevel gears to be exploited.

### 2/ Two-Speed Axle

The type of axle shown in Figure 12.56 is encountered in commercial vehicles with one driven axle and a requirement for large final ratios, especially buses. The axle is broadly similar to the single-ratio version. For the larger ratio, a spur gear stage or a planetary gear set in the crown gear is engaged as well. This enables the final ratio to be increased to approximately $i_{E,A} = 9.0$. Some of the disadvantages of the single-ratio axle can be offset with this design. The high inertia torque of the powertrain creates shiftability problems where ratio changes are too great.

### 3/ Pinion Axle (Gantry Axle)

In contrast to the axle designs described above, the driving torque is increased not only in the axle drive but also in the hub drives (Figure 12.57).

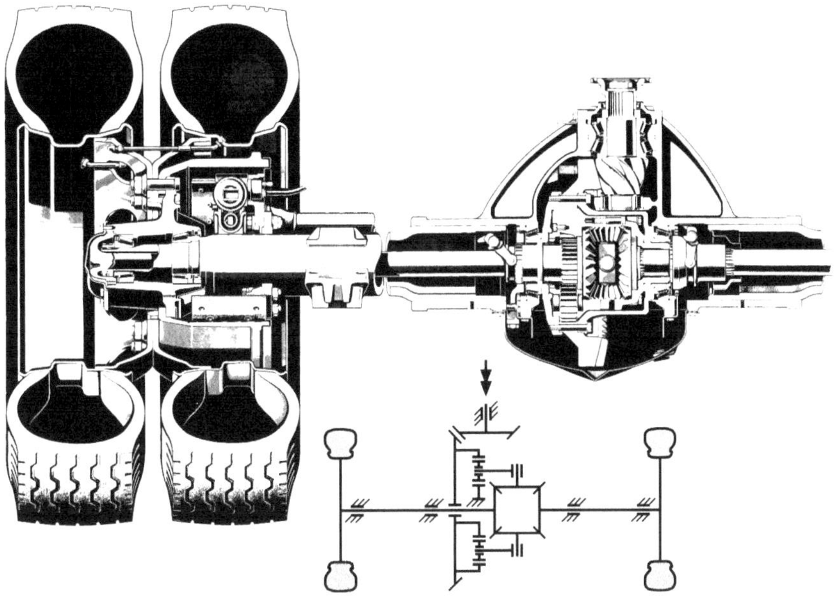

**Fig. 12.56.** Two-speed axle, Mercedes-Benz HL5/2Z-10. Top schematic diagram: gear ratio engaged. Bottom schematic diagram: gear ratio not engaged. Axle drive shown rotated 90°

The axle drive and the axle shafts can be smaller if the ratio is divided. This gives adequate ground clearance even with high outputs. The spur gear drive is in most cases located inside the brakes. This means the brake drums have to be very large.

If the output rating were further increased, it would become difficult to transmit the relatively large forces with the single meshing of the spur gear drive. The gearwheel bearings would also need to be correspondingly sophisticated. Figure 12.57 shows the pinion axle (gantry axle) of the Mercedes-Benz Unimog as a typical example of this axle design. The helical-cut wheel hub drive limits the maximum ratio possible in the wheel hub $i_{E,N,max} \approx 2.5$. Internal gearing provides improved transmission of large forces, with the disadvantage of less ground clearance. See also Figure 6.75c.

#### 4/ Outer Planetary Axle

In the outer planetary axle, the high torque is also only created at the wheel, where it is converted directly into traction. This has a number of advantages. The axle drive can be kept relatively small, as with the pinion axle (gantry axle), allowing good ground clearance. The planetary gear is located outside the wheel brake, so there are no problems when designing the brake. The power to be transmitted is distributed in the planetary gear to several gear meshes (up to 5 planetary gears). This is what makes this type of gear unit compact.

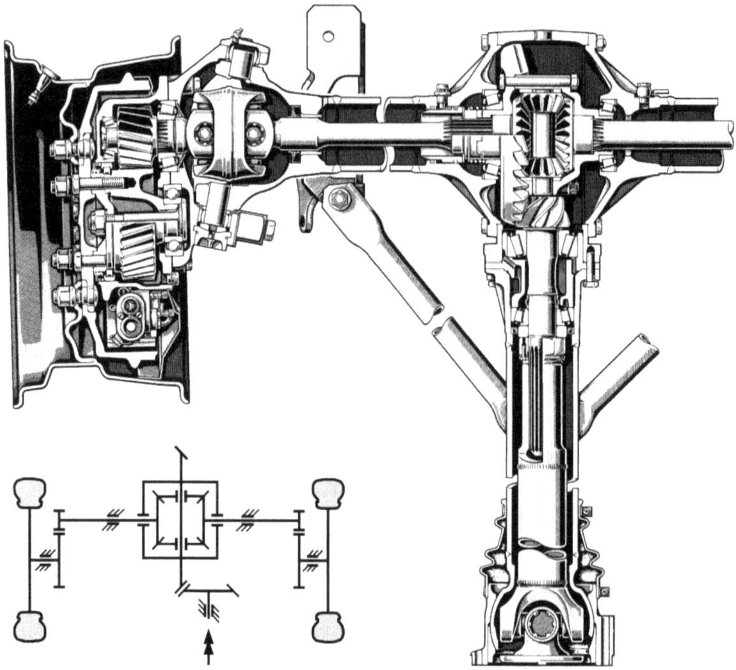

Fig. 12.57. Pinion axle (gantry axle) AU 2/2S-2.6 of the Unimog 407 as driven steering axle. Axle drive shown rotated 90°

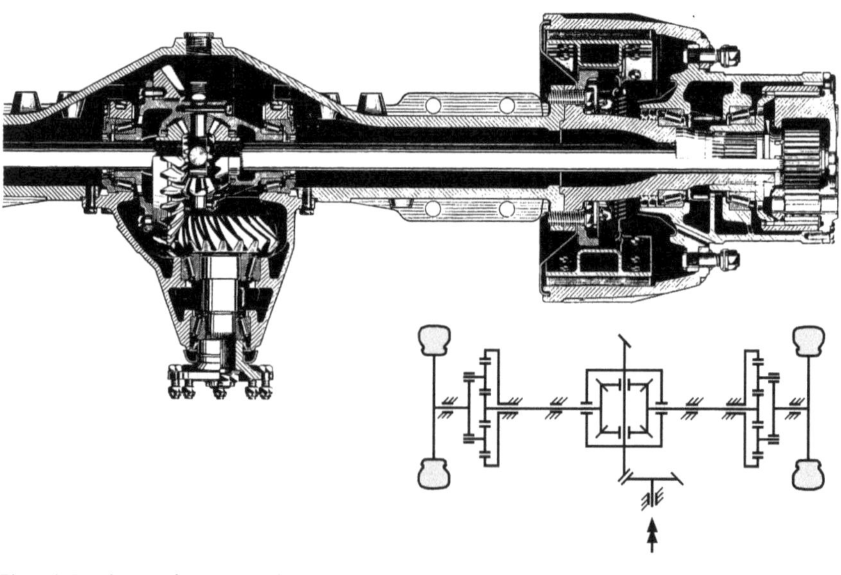

Fig. 12.58. Outer planetary axle with spur gear planetary drive, ZF. Axle drive shown rotated 90°

Low rolling and sliding speeds at the tooth flanks lead to a high level of efficiency. A floating bearing arrangement is possible due to the balance of static forces within the planetary gear. The outer planetary axle can easily provide nearly any desired ratio. Figure 12.58 shows a ZF outer planetary axle.

### 12.3.3 Differential Gears and Locking Differentials

As discussed in Section 6.8.3, differential mechanisms can be divided into transfer boxes (longitudinal split of speed and torque) and differential gear units (transverse split). Some examples of differential gears and locking alternatives are examined below, selected from numerous production designs. The following differential gears and locking differentials are examined below:

1/ bevel gear differential, Figure 12.59,
2/ self-locking differential with multi-plate clutches, Figure 12.60,
3/ self-locking differential with worm gears (TORSEN), Figure 12.61,
4/ self-locking differential with fluid clutch, Figure 12.62.

#### *1/ Bevel Gear Differential*

Rear axle drives, such as the one shown in Figure 12.59, are used in practically all vehicles with longitudinal engine and rear-wheel drive (standard drive).

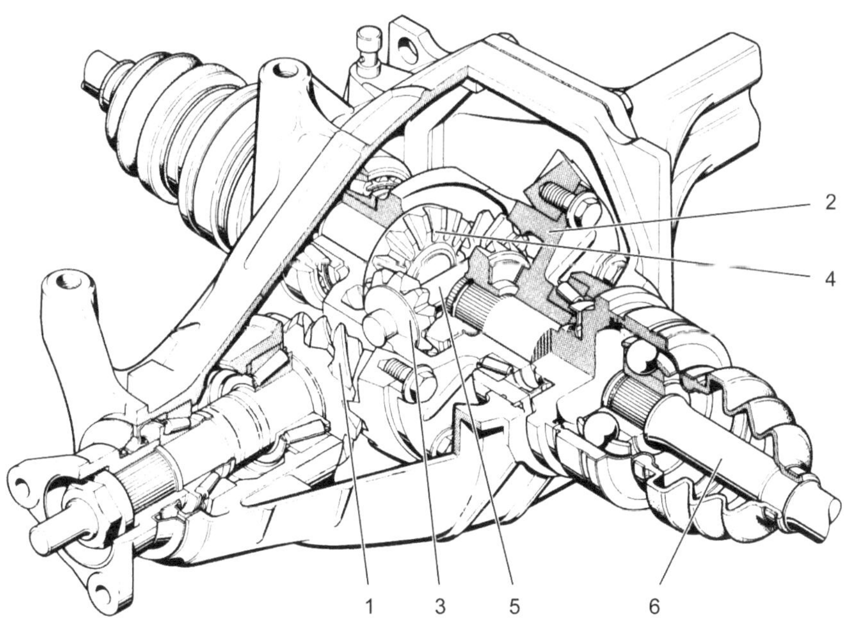


**Fig. 12.59.** Non-locking Opel rear axle differential. *1* Bevel gear axle drive; *2* differential cage; *3* differential bevel gears; *4* axle bevel gears; *5* differential shaft; *6* axle shafts

The torque applied through the bevel gear axle drive *1* is transmitted through the differential cage *2* and the differential shaft *5* to the differential bevel gears *3* and from there to the axle bevel gears *4*, which are torsionally locked to the axle shafts *6*. When driving in a straight line, the differential cage *2*, the axle bevel gears *4*, the axle shafts *6* and the differential bevel gears *3* rotate inside the cage as a block. There is no relative movement between the differential shaft *5* and the differential bevel gears mounted on it. One axle shaft has to turn faster than the other when cornering. Axle bevel gears and differential bevel gears roll on each other. This enables speed compensation between the wheels.

### 2/ Self-Locking Differential with Multi-Plate Clutches

The locking effect of a self-locking differential with multi-plate clutch relies on the torque-dependent internal friction generated in two multi-plate clutches mounted symmetrically in the differential cage. The self-locking action results from a combination of the load dependency and spring loading of the multi-plate clutches. The load-dependent locking effect (Figure 12.60, top) relies on the input torque $T_1$ applied to the differential cage *1* being transmitted via the differential shaft *2* to two pressure rings *3* in the differential cage *1* that are torsionally locked but slide axially. Under load, locking forces arise automatically at the surfaces of the prism-shaped recesses *8* in the pressure rings (see detail in 12.60), pressing the clutch plates together.

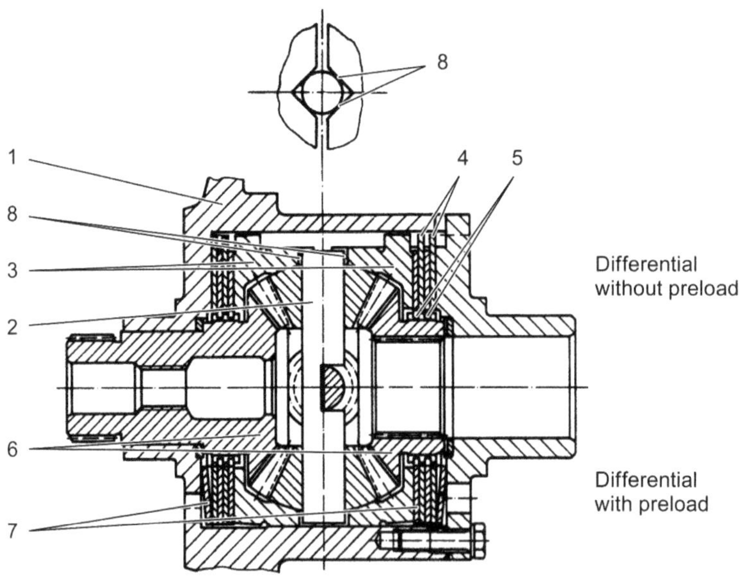


**Fig. 12.60.** Locking differential with preloaded multi-plate clutches, Lok-O-Matic. Top half-section: differential without preload. Bottom half-section: differential with preload. *1* Differential cage; *2* differential shaft; *3* pressure rings; *4* outer plates; *5* inner plates; *6* axle bevel gears; *7* plate springs; *8* recesses

The outer plates *4* are torsionally locked to the differential cage *1*, and the inner plates *5* are torsionally locked to the axle bevel gears *6*.

The frictional contact between the plates thus opposes the different axle shaft speeds (for example when a wheel spins) with a precisely defined force. This effect increases as the input torque increases. Since the locking forces are proportional to the torque transmitted, the locking effect adapts to the changing engine torque and to the torque increase in the various gears, but the interlock value does not (see also Section 6.8.3 Number 3/ "The Interlock Value").

The plate springs *7* (shown in the bottom half of Figure 12.60) that can be fitted to preload the multi-plate clutch create a constant initial locking effect that is independent of the torque transmitted, but sometimes makes noticeable creaking noises. This makes the system capable of locking even on extremely unfavourable surfaces, for example one wheel on black ice. There is nevertheless still the disadvantage that a differential of this type always has a basic locking torque. This can be undesirable when parking and when cornering without slip.

The torque-dependent contact pressure can also be created by applying pressure just through the gear spreading forces of the bevel gear differential. These contact pressures are less than those achievable with pressure rings by about a factor of 3. A further drawback that should be borne in mind is that during the self-locking or compensating process, the tooth geometry of the bevel gears changes adversely, because the friction clutches that have to be applied must not be free of clearance. Unsymmetrical self-locking differentials of this type are also used in transfer boxes with just one multi-plate clutch.

### *3/ Self-Locking Differential with Worm Gears (TORSEN)*

The design and function of the self-locking differential with worm gears, known as the TORSEN transfer differential, is described with reference to Figure 12.61. TORSEN stands for "torque sensing".

The driving crown gear *1* is linked to the differential housing *2*. The six worm gears *3* run tangentially on pins in the housing, moving freely. They are supported by the housing at the face. Spur gear toothing *4* are torsionally fixed on each front face of the worm gears. The spur gear toothing of neighbouring worm gears are engaged, and three worm gearwheels of one half of the mechanism mesh with one of two worms *5*. The worms are radially centred by the symmetrical arrangement of the worm gearwheels relative to them. The worms have axial support available in the housing or at their common locating face. The worms are hollow, and carry gearing inside, into which the output shafts *6* are inserted.

When travelling straight ahead and with good grip on both wheels, the differential rotates as a block. The input torque reacts against the spur gear toothing of the worm gears. When cornering or when there is a high tendency for a wheel to spin, rolling compensation can be provided through the spur gearing. But since all the components are rotating at different relative speeds, and are also under load, there are losses at all friction points.

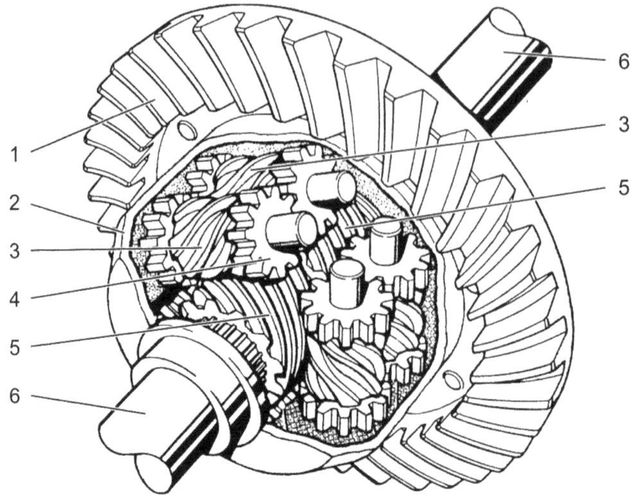


**Fig. 12.61.** Self-locking differential with worm gears, TORSEN. *1* Driving crown gear; *2* differential housing; *3* worm gears; *4* spur gear toothing; *5* worms; *6* output shafts

The losses in the worm gear pairs arise from the effect whereby the friction in the worm gearing increases the worm gear circumferential forces on one worm, and decreases them on the other.

This creates an uneven torque distribution to the two output sides, which is further amplified by the friction in the contact faces of the worms. The total of all torque losses is equal to the locking torque. Rolling compensation is therefore not possible without load or without simultaneously building up the design-dependent basic interlock value.

The basic interlock value of this mechanism is 20%. In practice the interlock value is higher than 20% due to the friction between various components and the differential housing. The teeth of the worms and worm gearwheels may also be additionally deliberately mismatched to achieve interlock values greater than 60%.

To build up the locking effect, this mechanism needs both a speed differential between the output shafts and also a positive input torque. The locking effect is suspended when the load is removed. Vehicles with this differential are therefore fully suitable for use with antilock braking systems.

### 4/ Self-Locking Differential with Fluid Clutch

A visco self-locking differential comprises of the components of a conventional differential, and also those of a fluid friction clutch. This is a completely sealed, annular cylindrical component, and has inner and outer plates and the working fluid, a high-viscosity silicone oil.

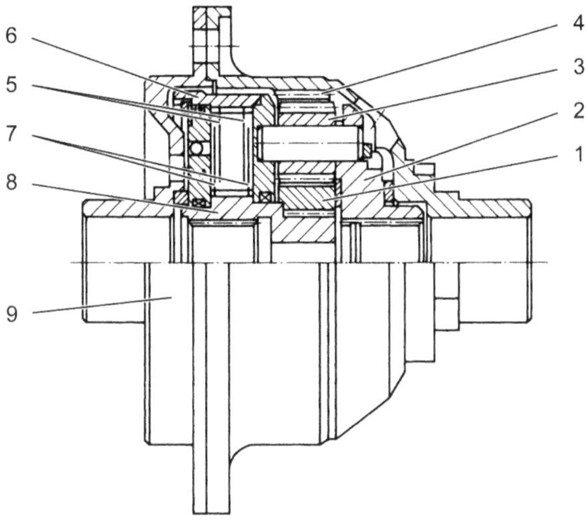


**Fig. 12.62.** Self-locking spur gear differential with visco-clutch. *1* Sun gear; *2* planetary gear carrier; *3* planetary gears; *4* ring gear; *5* outer plates; *6* outer plate carrier; *7* inner plates; *8* inner plate carrier; *9* housing

In the example shown in Figure 12.62 with spur gear differential, the outer plates *5* are torsionally fixed in the outer plate carrier *6*, which is linked to the planetary gear carrier *2* or the right-hand axle shaft inserted in it. The planetary gears *3* rotate on pins in the planetary gear carrier. The inner plates *7* run torsionally fixed in the inner plate carrier *8*, which is connected to the sun gear *1* or to the left-hand axle shaft inserted in it. Drive is through the housing *9* with integral ring gear *4*. Other production versions also have the clutch between the input shaft and an output shaft.

With a slipping drive wheel or when cornering, relative movement arises between the two axle shafts, and so between the sun gear shaft and the planetary gear carrier. The planetary gears *3* roll on each other, in the ring gear and on the sun gear. Since this relative movement also arises between the inner and outer plates, the working fluid of the clutch is subject to shearing stress, giving rise to viscous resistance to the cause – the speed difference. The level of this resistance torque depends on the speed difference. It acts to brake the wheel that is rotating faster, thus increasing the torque on the wheel with the better grip.

The viscous braking torque generated depends exclusively on the speed difference between the output shafts. The heat generated under protracted stress raises the pressure inside the clutch. This pressure presses contiguous inner and outer clutch plates together with greater force, displacing the working fluid from the space between them. The resultant friction contact transforms the clutch into a rigid component, and the increased interlock value of 100% persists even when the speed difference declines. This process is called the "hump effect" because it represents a hump in the speed/braking torque curve of a visco-clutch.

A differential of this type is controlled entirely by the differential speed. As a consequence, undesirable distortions can arise in the powertrain on tight bends, even without load, depending on the clutch design. On the other hand, differential brakes of this type bite very gently, yet they reach very high interlock values of up to 100% when required.

## 12.4 All-Wheel Drives, Transfer Gearboxes

From the broad range of passenger car and commercial vehicle all-wheel powertrains, the following will zero in on passenger car all-wheel drives. The focus will not be on the most current designs, but rather on the basic design concepts. A great number of highly varying all-wheel drives exist on the market. Section 6.1.3 "All-Wheel Drive Passenger Cars" presents the systematics and basic features of all-wheel drives.

In the following, selected all-wheel drive parts will be shown as examples (Table 12.3).

**Table 12.3.** All-wheel drives and transfer gearboxes presented in Section 12.4

| No. | Characteristics | Gearbox name | Figure No. |
|---|---|---|---|
| 1/ | Manual gearbox with power take-off | Fiat Panda 4x4<br>Power take-off: Getrag PTU735 | 12.63 |
| 2/ | Manual gearbox with transfer box and front axle drive | Audi Quattro manual gearbox ML450-6Q (Getrag 466) | 12.64 |
| 3/ | Automatic gearbox with transfer box and front axle drive | Audi Quattro automatic gearbox AL420-6Q (ZF 6 HP 19A) | 12.65 |
| 4/ | Clutch-controlled all-wheel drive | VW Golf R32 4Motion<br>manual gearbox MQ 350-6 with power take-off VAA 350<br>rear axle unit HAA 350 | 12.66<br><br>12.67 |
| 5/ | Transfer gearbox with reduction gear stage | Mercedes-Benz G-Model<br>transfer gearbox MB VG 150 | 12.68 |
| 6/ | Transfer gearbox with spur gear countershaft | BMW 3 Series/5 Series 4x4<br>transfer gearbox MP-3010G-ATC | 12.69 |
| 7/ | Transfer gearbox with chain drive | Landrover Discovery<br>transfer gearbox MP-2624-ITC | 12.70 |
| 8/ | All-wheel drive with variable torque vectoring | Honda SH-AWD™ | 12.71 |

### 1/ All-Wheel Drive Gearbox Fiat Panda 4x4 / Getrag PTU735

The gearbox shown in Figure 12.63 is a 5-speed manual gearbox for the Fiat Panda 4x4 with an all-wheel drive design. This single-stage gearbox with a front-transverse design is identical to that of the front-wheel drive variant, with the power take-off (PTO) Getrag PTU735 flange-mounted on the gearbox to drive the rear axle.

As opposed to the 4Motion system by VW (Figure 12.66), this PTO has a countershaft *1* which is driven by a gearwheel *3* additionally mounted on the differential cage *2* and which conducts the drive torque to the rear axle via a bevel gear stage *4* with hypoid gearing. A visco-clutch mounted on the rear axle drive distributes the torque among the front and rear axles.

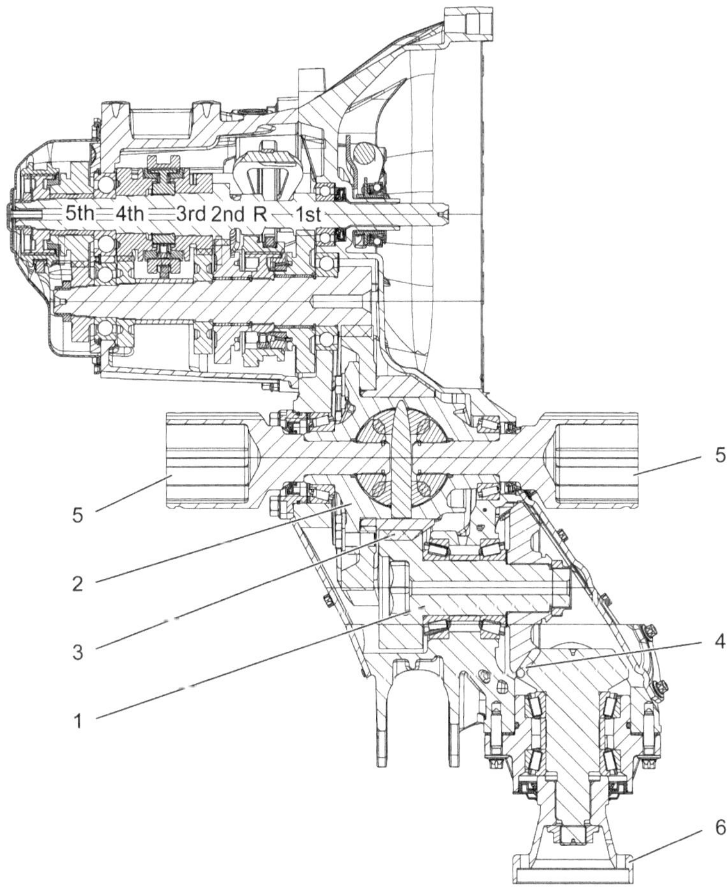


**Fig. 12.63.** Fiat Panda 4x4 manual gearbox with power take-off. *1* Countershaft; *2* differential cage; *3* PTO drive gear; *4* PTO bevel gear; *5* axle flange of front axle; *6* rear axle output flange

With the aid of the centre distance of the countershaft and the hypoid drive, this design allows for a space-efficient placement in the vehicle engine compartment and is often used when the distance between the differential axle and the engine block is small. This dimension is often critical when designing conventional bevel drives, since this dimension limits the diameter of the crown gear.

### *2/ All-Wheel Drive Gearbox with TORSEN Differential; Audi Quattro Manual Gearbox ML450-6Q*

The 6-speed manual gearbox Audi ML450-6Q shown in Figure 12.64 is used in permanent all-wheel drive Audi models A4, S4 and A6 (model year 2006) with engine torques upwards of 350 Nm [12.14]. This gearbox represents an extension of the front-longitudinal gearbox Audi ML350-6F (Getrag 466) shown in Figure 12.12, which was expanded by a TORSEN transfer gearbox.

As opposed to the gearbox variant for front-wheel drive, the output shaft *1* in this case is designed as a hollow shaft. The front axle is driven through the sun gear *3* of the TORSEN differential which is in the front in relation to the vehicle and which is connected firmly to the drive shaft *5*. The sun gear is driven from the input shaft via the hollow shaft to the worm gears *4* housed in the differential housing *2* which transmit the drive torque to the front sun gear. The output to the rear axle is analogous via the rear sun gear *3* of the TORSEN differential.

The locking function in the longitudinal transfer differential is realised by the TORSEN transfer gearbox.

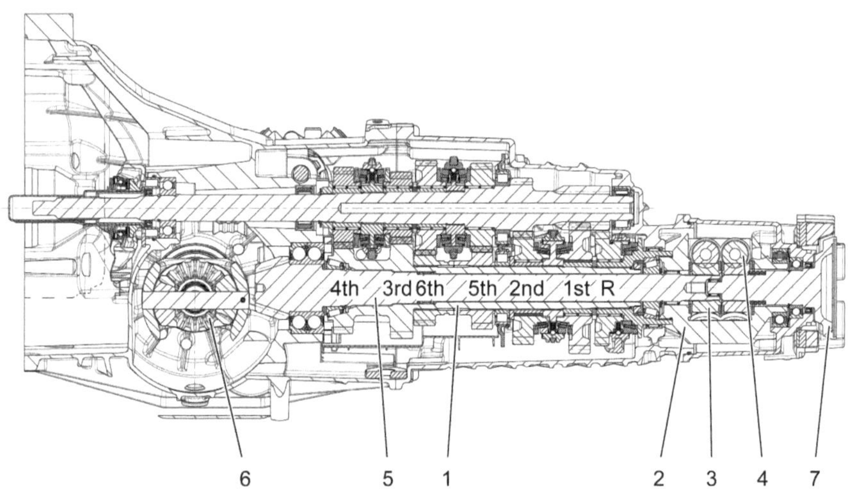


**Fig. 12.64.** Audi Quattro manual gearbox ML450-6Q with TORSEN centre differential; *gearbox diagram Figure 6.23b. 1* Hollow shaft; TORSEN differential: *2* differential housing, *3* sun gear, *4* worm gear; *5* front axle drive shaft; *6* front axle differential; *7* rear axle output flange

Through the dependency of the locking effect on the induced torque, the drive torque can be distributed between the front and rear axles under load. In a load-free state, there is no locking effect. This guarantees complete ABS suitability.

The nominal torque distribution lies at 50 : 50 and is adjusted between 25 : 75 and 75 : 25 in dynamic vehicle operation. The functioning of the TORSEN transfer differential is described in detail in Section 6.8.3 "Differential Gears and Locking Differentials" and in Section 12.3.3 "Differential Gears and Locking Differentials".

### 3/ All-Wheel Drive Gearbox with Side Shaft; Audi Quattro Automatic Gearbox AL420-6Q

This automatic gearbox for Audi Quattro all-wheel drive vehicles with longitudinal engine is an automatic transmission with various gear ratios with a hydrodynamic torque converter. The 6-speed gearbox Audi AL420-6Q (ZF 6 HP 19A) shown in Figure 12.65 is designed for engine torques up to 420 Nm for petrol engines and 450 Nm for diesel engines and is used in Audi models A4, A6 and A8 (model year 2006).

The torque of the automatic gearbox is transmitted via a spur gear stage *2* into the TORSEN transfer gearbox *1*. As with manual Audi Quattro gearboxes (see Figure 12.64), the sun gear of the transfer gearbox, which is in the front, conducts the torque to the front axle and the rear sun gear to the rear axle.

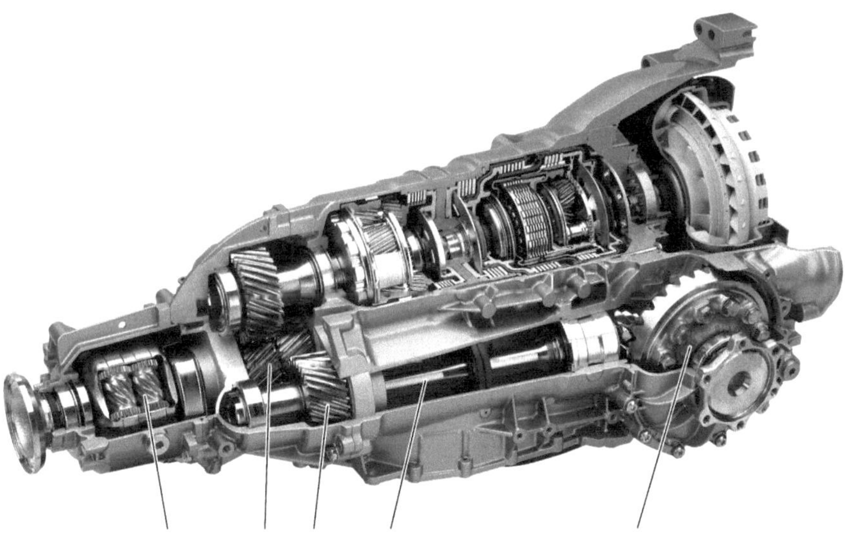


**Fig. 12.65.** Audi Quattro automatic gearbox AL420-6Q with transfer gearbox and front axle drive. *1* TORSEN differential; *2* spur gear drive; *3* spur gear stage with beveloid gearing; *4* side shaft; *5* front axle drive

The drive of the front axle drive integrated in the gearbox housing is through a spur gear stage with beveloid gearing *3* and a side shaft *4*. The beveloid angle in the gearbox shown in the figure is approximately 6°.

For a description of the basic gearbox ZF 6 HP 19 of the automatic gearbox shown here, please refer to Section 12.1.4 "Automatic Passenger Car Transmissions (AT)" Number 16/. The gearbox diagram is identical to the one shown in Figure 6.34.

### *4/ Clutch-Controlled All-Wheel Drive; VW 4Motion*

The following will look at the Volkswagen "4Motion" system as an example of a clutch-controlled all-wheel drive. The basis of this all-wheel drive is a vehicle with front-wheel drive and front-transverse engine with manual, automatic or dual clutch gearbox flange-mounted on the left side [12.48].

Figure 12.66 shows the three-shaft manual gearbox MQ350-6 with the modular power take-off VAA350 with bevel gear *1* and cardan shaft connection *2* built in the Volkswagen Golf R32 (model year 2006) with standard 4Motion all-wheel drive. The front bevel gear is flange-mounted on the right of the manual gearbox. The drive is directly from the differential cage of the manual gearbox through an additional spline.

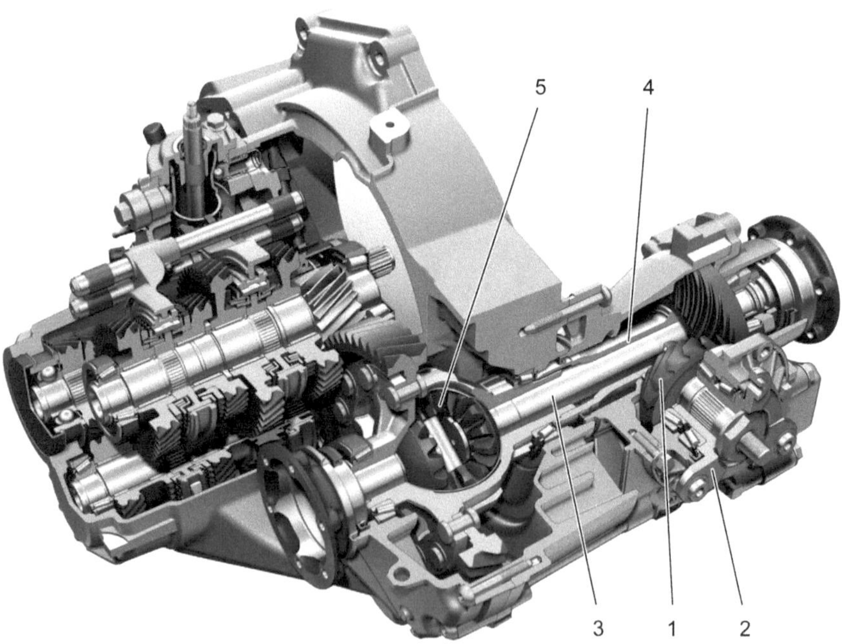


**Fig. 12.66.** Manual gearbox MQ350-6 with power take-off VAA350 of the Volkswagen Golf R32. *1* Bevel gear; *2* cardan shaft connection for driving the rear axle; *3* drive shaft of the right front wheel; *4* hollow shaft; *5* front axle differential

The output to the right front wheel *3* as seen from the direction of travel is conducted through the hollow shaft *4* on which the bevel drive pinion is fixed. The bevel gear with hypoid gearing has a ratio of $i_{\text{front}} = 0.63$ ($z_1 = 27$ to $z_2 = 17$).

Figure 12.67 shows the rear axle drive HAA350 used in the Volkswagen Golf R32 with bevel gear *3*, differential *4* and the heart of the 4Motion all-wheel drive: the Haldex clutch *2*. The bevel drive of the rear axle drive is similarly designed to the one in the front. The ratio is selected in such a way that the powertrain receives a ratio of $i = 1.0$.

The Haldex clutch integrated in the rear axle drive is a multi-plate clutch running in oil which, depending on the speed difference between the input and output side, slip-dependently controls the torque distribution between the front and rear axles. As opposed to the self-locking visco-clutch, the Haldex clutch is activated by a control unit.

Due to the possible speed compensation between the front and rear axles when the all-wheel clutch is open, this all-wheel drive is fully compatible with all brake control systems. Clutch pressure is created through two parallel, but phase-shifted axial piston pumps. This means that the hydraulics only require a minimal slip for its functioning.

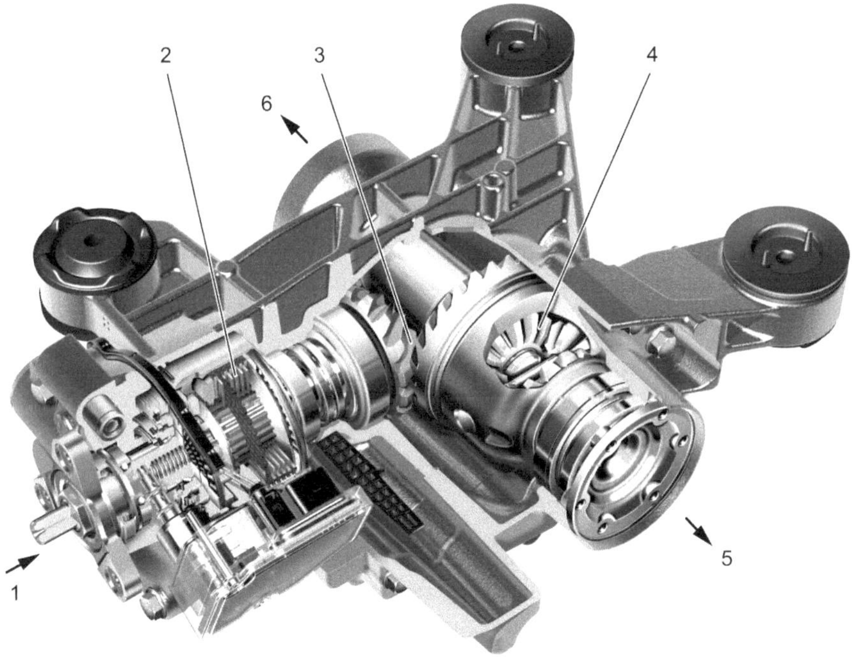


**Fig. 12.67.** Rear axle drive HAA350 with Haldex clutch of the Volkswagen Golf R32. *1* Input; *2* Haldex clutch; *3* bevel gear; *4* differential; *5* left rear wheel output; 6 right rear wheel output

The design of the electrohydraulic system includes a flow proportional valve and also functions along with the integrated pressure sensor in a torque-controlled way, since the maximum transmittable torque is proportional to the pressure.

Acting together with the vehicle sensors, the Haldex clutch recognises at all times whether an increased drive to the rear axle is advantageous and shifts more torque to the rear axle in a fraction of a second. The magnitude of the drive torque at the rear axle is dependent, for example, on the traction requirements of the driver, the state of the road surface and the vehicle operating situation and stability.

When moving-off on steep gradients or with front wheels with little traction, up to 100% of the drive torque, for example, can be conducted to the rear axle. When travelling on the motorway at 140 km/h, however, only about 15% of this drive torque reaches the rear axle. This is sufficient for driving stability, thus reducing the fuel consumption of the vehicle.

The control strategy of the 4Motion all-wheel drive is additionally influenced by comfort and safety demands. In manoeuvring operation, for example, the front and rear axles are only minimally coupled so that there are no disturbing influences caused by distortions in the powertrain. Also sacrifices to comfort typical for all-wheel drives in terms of noise in both stationary operation and when accelerating under full load in high gear are eliminated.

To ensure fully the driving stability and ABS/ESP suitability of the vehicle when braking, the clutch is opened during ABS/ESP use, thus decoupling the rear axle wheels when braking. Following the ABS/ESP event, it is closed again automatically with the next speed difference and thus always ready for full all-wheel drive operation. To this end, the control unit of the Haldex clutch is fully networked with the high-speed drive CAN of the vehicle. All the necessary information from the engine and ABS control units are evaluated in the Haldex clutch control unit, with all the necessary information regarding the operating state of the Haldex clutch made available in turn. This fully integrates the all-wheel drive into the safety and driving dynamics of the vehicle.

The 4Motion all-wheel drive is one which, depending on the situation, controls the distribution of the drive torque between the front and rear axles. In the process, it fulfils the demands on traction, comfort and safety.

### *5/ Transfer Gearbox with Reduction Gear Stage; Mercedes-Benz G-Model*

The following will discuss the transfer gearbox of the Mercedes-Benz G-Model as an example of an all-terrain vehicle with permanent all-wheel drive and reduction gear stage (cross-country gear).

The manual or automatic gearbox has no specific all-wheel drive features in this vehicle. In this powertrain, the engine torque is transmitted via the manual/automatic gearbox and a short input shaft to the transfer box, which has a double-offset design and is independently anchored to the ladder frame. From there, the torque is distributed equally to the front axle and the rear axle by propeller shafts.

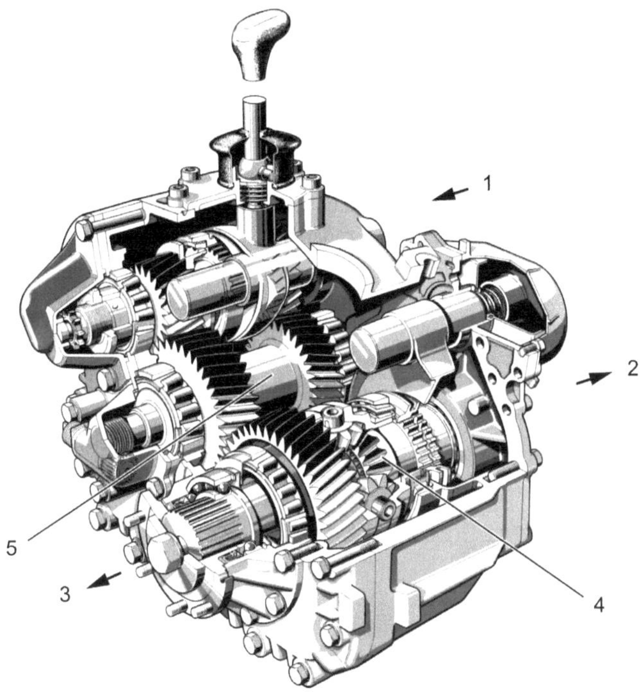


**Fig. 12.68.** Transfer gearbox VG 150 Mercedes-Benz G Model (permanent all-wheel drive). *1* Gearbox input; *2* front axle output; *3* rear axle output; *4* bevel gear differential; *5* secondary shaft

The VG 150 transfer box (Magna Powertrain: MP-2223G) shown in Figure 12.68 is a three-shaft gearbox with a bevel gear differential *4* for interaxle compensation. As a result of the double-offset design, the two drives for the front *2* and rear axle *3* are powered by the secondary shaft *5*.

Torque distribution between front and rear axle is fixed at 50% to 50% in the unlocked state. With the positive differential lock, the interaxle differential can be locked 100%. The all-synchromesh transfer gearbox houses both the differential and the switchover from the on-road ratio ($i = 1.05$) to the off-road ratio ($i = 2.16$). The front and rear axle interwheel differentials are likewise equipped with positive locks. When all three locks are engaged, off-road traction is optimised.

### 6/ Transfer Gearbox with Spur Gear Countershaft; BMW 3 Series/5 Series

The transfer gearbox MP-3010G-ATC (Magna Powertrain) for all-wheel drive variants of the BMW 3 and 5 Series is a typical example of all-wheel technology for passenger cars and SUVs without centre differential and cross-country gear.

On the basis of a standard drive, the BMW 3 and 5 Series vehicles were designed with dynamic all-wheel drive. For this purpose the transfer gearbox shown in Figure 12.69 is flanged onto the manual/automatic gearbox. This transfer gearbox contains an electronically controlled and electromechanically actuated multi-plate clutch *2* and a spur gear countershaft *3* for driving the front axle *4*.

The ToD technology (Torque-on-Demand) guarantees a drive torque distribution between the front and rear axles *5* which is meeting the demands. With ToD, the transmission torque is freely selectable by means of a processor-controlled multi-plate clutch.

The torque distribution is varied electronically, with high dynamics and precision of adjustment, from zero drive torque at the front axle to fixed direct drive automatically according to the driving situation (correction of over- and understeering, maximum traction). The logic for the control is integrated into the driving dynamics control system.

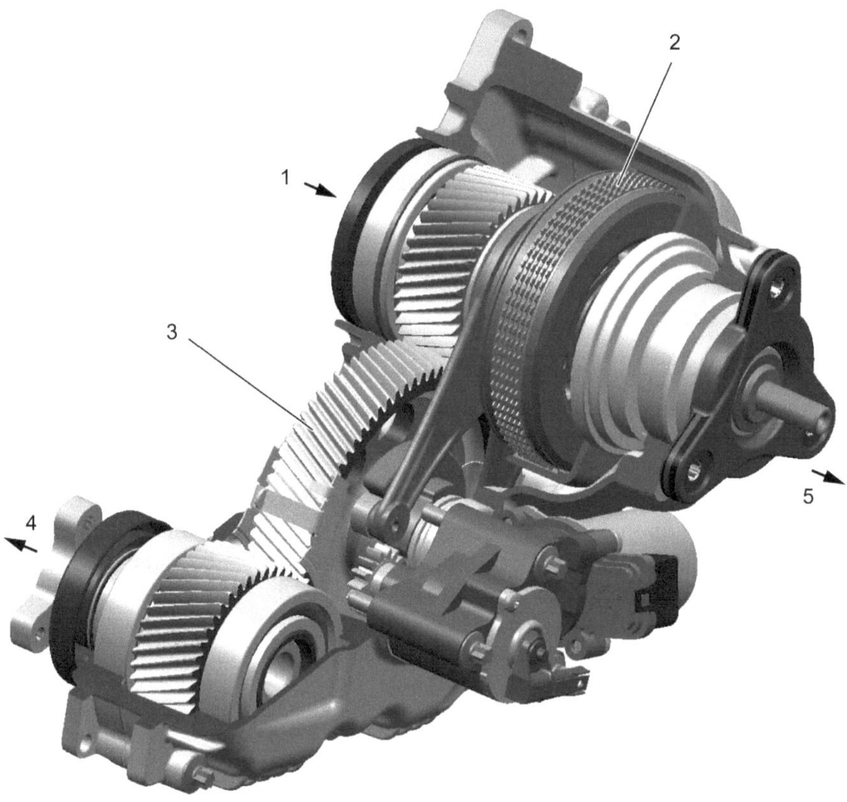


**Fig. 12.69.** Transfer gearbox with spur gear countershaft MP-3010G-ATC of the BMW 3 and 5 Series all-wheel drive models. *1* Gearbox input; *2* electromechanically actuated multi-plate clutch; *3* spur gear countershaft; *4* front axle output; *5* rear axle output [12.42]

By means of a possible speed compensation between the front and rear axles with open all-wheel drive clutch, this all-wheel drive is fully compatible with all brake control systems.

The front and rear axle drives are driven by the transfer gearbox via propeller shafts. The front axle drive lies directly left of the engine oil pan, to which it is firmly bolted. The output shaft to the right front wheel leads through the engine oil pan and is supported again at the right by a supporting bearing. Thus it is possible to realise equally long drive shafts without asymmetrical self-steering properties.

### *7/ Transfer Gearbox with Reduction Stage in Chain Drive Design; Land Rover Discovery*

The transfer gearbox of the Land Rover Discovery MP-2624-ITC made by Magna Powertrain is one which exhibits all the features typical for an all-terrain vehicle, such as a centre differential with lock and cross-country gear.

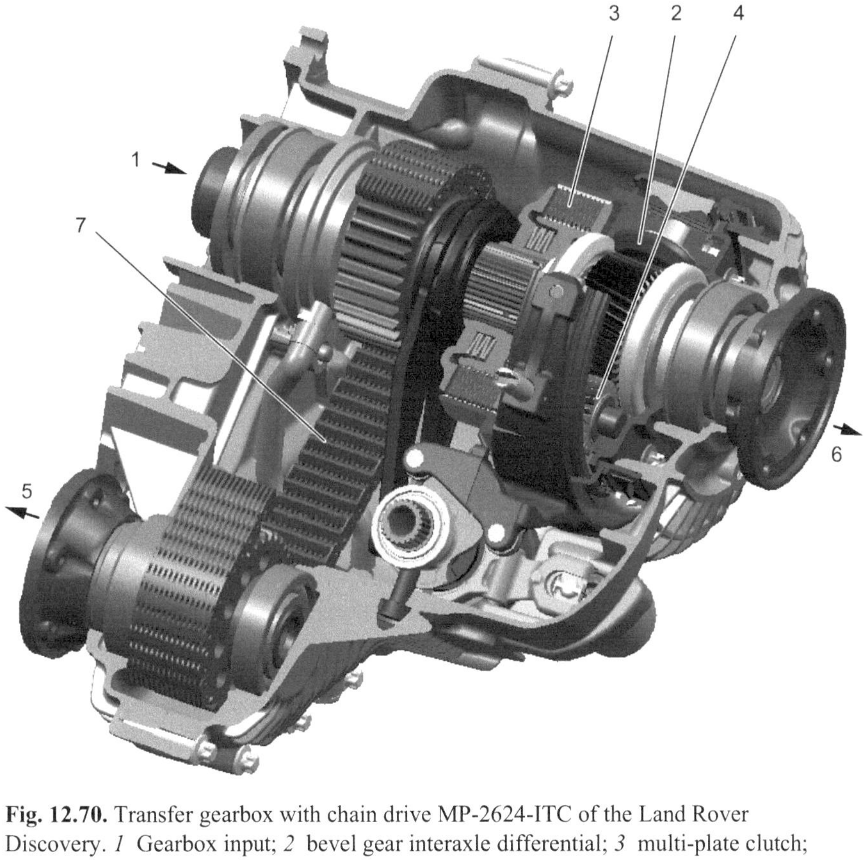


**Fig. 12.70.** Transfer gearbox with chain drive MP-2624-ITC of the Land Rover Discovery. *1* Gearbox input; *2* bevel gear interaxle differential; *3* multi-plate clutch; *4* planetary gear set for cross-country gear; *5* front axle output; *6* rear axle output; *7* chain drive [12.42]

The drive torque is conducted via a manual or automatic gearbox to the transfer gearbox *1* (Figure 12.70). The transfer gearbox contains a bevel gear interaxle differential *2*, a multi-plate clutch *3* for locking the interaxle differential and a planetary gear set *4* for shifting the cross-country gear ($i = 2.93$). With the bevel gear interaxle differential, 50% of the torque is conducted to the front axle and 50% to the rear axle. The locking toque of the multi-plate lock is modulated in a processor-controlled way up to fixed all-wheel drive, with a torque distribution corresponding to the axle load distribution. If needed, the cross-country gear can be shifted during operation by means of a synchronization.

By means of the propeller shaft, the drive of the rear axle *6* is directly from the transfer gearbox. The drive to the front axle *5* is also through a propeller shaft from the transfer gearbox and a chain drive *7* in the transfer gearbox (single-offset design). The axle drive and the differential are mounted in their own auxiliary frame in the chassis subframe. The drive shaft to the right front wheel runs under the engine oil pan. An electronically-controlled differential lock is optionally integrated into the rear axle.

A special feature of the transfer gearbox shown is the integrated design of the cross-country gear and the centre differential. This prevents the rolling motion of the cross-country gear during road operation which is typical of other transfer gearboxes, thus increasing the level of efficiency. Moreover, this layout leads to a compact and light design, which in turn positively influences the natural resonance of the powertrain.

The Land Rover Discovery includes the "Terrain Response™" drive system, which allows the settings for the engine, the transmission, the chassis and the traction to be selected by the driver by means of a rotary switch. Full ABS/ESP suitability is achieved through the disengagement of all locks when braking.

### *8/ Rear Axle Drive with Multi-Plate Clutches; Honda SH-AWD*

The Honda SH-AWD™ is the first all-wheel drive system in series production belonging to the 5th generation in Table 6.4 "Generations of passenger car all-wheel drives". This system has a permanent all-wheel drive based on the principle of "torque vectoring". Not only is the torque distribution between the front and rear axle freely selectable, but also that between the left and right wheel of the rear axle.

When travelling straight or cornering moderately with part load and with a throttle valve position below 50%, up to 70% of the torque is conducted to the front axle. When accelerating under full load in a straight line, up to 40% of the drive power is transmitted to the rear axle. When cornering strongly, up to 70% of the available input torque is conducted to the rear axle in order to improve the vehicle handling. Depending on the driving situation, up to 100% of the torque available to the rear axle can be conducted to the outer wheel in order to directly influence the yaw moment and thus the steering of the vehicle and to minimise understeering.

The Honda SH-AWD system comprises a 5-speed countershaft-type automatic gearbox and a rear axle drive (Figure 12.71).

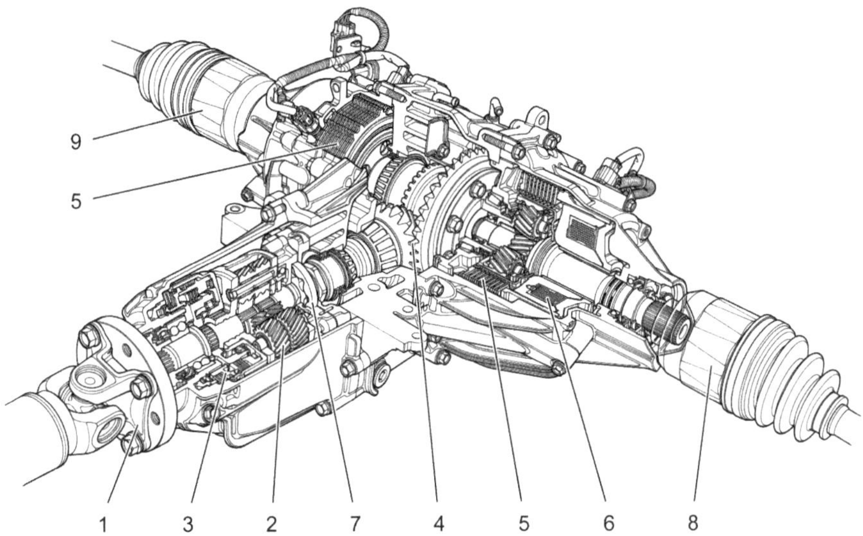


**Fig. 12.71.** Rear axle drive of the Honda SH-AWD. *1* Input; *2* acceleration mechanism; *3* hydraulically activated multi-plate clutch; *4* bevel gear stage; *5* electromagnetically activated multi-plate clutch; *6* electromagnet; *7* oil pump; *8* left rear wheel output; *9* right rear wheel output

A power take-off is connected with the manual gearbox via a spur gear stage. Its driving gearwheel is bolted to the differential cage of the gearbox. The rear axle drive is connected with the power take-off via a propeller shaft made of a carbon-fibre reinforced composite.

The rear axle drive comprises three planetary gear sets with clutches. At the input side of the rear axle drive *1*, the torque of the input shaft passes through the first planetary set, designated as the acceleration mechanism *2*. The output torque of the acceleration mechanism is diverted via a bevel gear stage *4* with hypoid gearing to the rear axle shafts and from there, issued via two separate, electromagnetically controlled clutches *5* to the wheels. The left and right clutch systems can either be controlled identically in order to change the torque distribution between the front and rear axle, or they can be controlled independently in order to conduct up to 100% of the input torque at the rear axle to the outer wheel.

The transmission ratios vary according to the shifting states of the acceleration mechanism. When driving in a straight line, the input shaft is connected with the planetary gear carrier/spider, and there is no speed conversion. When cornering, the spider is connected via a hydraulically-controlled multi-plate clutch *3* to the housing, which leads to a slight increase of the output speed due to the transmission ratio of the planetary gear. This speed increase relative to the average speed of the front wheels improves the handling and driving stability of the vehicle. Also, a speed sensor at the bevel gear stage monitors the output speed of the acceleration mechanism and relays the speed signals on to the SH-AWD control unit.

There is an electromagnetic clutch system on each side of the hypoid gear which controls the magnitude of the drive torque of the two rear wheels and allows for the function of a partially-locking differential. Electric coils control the contact pressure of the clutches, which brake the sun gears of the planetary gears and allot the torque to the wheels. Depending on the driving situation, the torque at the individual rear wheels can be varied continuously between 0 and 100%. A coil sensor makes it possible for the electronic control unit to determine the heat-dependent friction coefficient of the multi-plate clutches and to compensate it via an adjustment of the coil voltage of the electromagnetically controlled clutch. In order to guarantee that the torque transmission remains the same during the entire service life of the gearbox, a control system is made available to the control unit via a coil to compensate for possible clutch wear via the coil voltage.

The control of the SH-AWD is integrated in the engine control unit and the driving stability system of the vehicle. The engine control device indicates the engine speed, the intake manifold pressure and the transmission data of the manual gearbox and the control unit of the driving stability system provides data regarding the longitudinal acceleration, yaw rates, wheel speeds and the steering angle of the steering wheel. The control unit of the SH-AWD monitors the status of the acceleration mechanism and the torque of the left and right electromagnetic clutches. The traction is calculated by means of the data of the engine control unit and the torque distribution to the rear wheels is determined on the basis of the acceleration situation, the longitudinal acceleration and the steering angle. This data is simultaneously used to control the acceleration mechanism.

### Other All-Wheel Drives

Other all-wheel drives of interest are to be found in the following models: BMW X3/X5 [12.63], Chrysler Grand Voyager, Hyundai Santa Fe [12.6], Mercedes-Benz 4matic [12.38], Mitsubishi Evo 8 [12.45], Opel Vectra 4x4 [12.43], Porsche Carrera 4 [12.51, 12.60], Porsche Cayenne [12.28], Renault Scenic 4x4 [12.6, 12.64], VW Golf Syncro [12.59] and VW Transporter Syncro [12.8]. The Getrag Twinster PTO [12.47] and the Magna hybrid design with shiftable, electrically-driven front axle [12.50] are current developments in the area of all-wheel drive technology.

# 13 Electronic Transmission Control

*Information networking*

Innovations in the area of transmissions have been primarily made by means of the integration of electrics, electronics, hydraulics, actuators and sensors. A great amount of transmission functionality is realised by software. Electronic and software functions do not only complement mechanics, they also open up further possibilities. The aim is to create superordinate functions in the vehicle by means of information networking. In this way, functions can be obtained of which one system alone could never have been capable [13.2].

In this chapter, we explore transmission control units (TCU) as both hardware and software, explaining their elements and showing the interaction thereof (regarding hydraulics, see Section 11.3.1 "Oil Supply", and regarding powershifting, see Section 9.3.2 "The Shifting Process").

## 13.1 Networked Systems

While vehicle systems were formerly mostly autarchic in their functions, networking by means of bus systems offers potentials with high customer benefit (Figure 13.1). Signal exchange of systems among each other alone leads to a clear reduction of sensors, thereby reducing cost and complexity while increasing reliability.

The use of immanently present system redundancies helps to increase the safety and availability of vehicle systems. Function networking can help create "virtual functions", of which one system alone would not at all be capable. So called management functions like

- powertrain management,
- driving dynamics control and
- vertical dynamics management

are thereby important whereby vehicle behaviour as a whole is optimally controlled and each sub-system makes its own specific contribution.

H. Naunheimer et al., *Automotive Transmissions*, 2nd ed.,
DOI 10.1007/978-3-642-16214-5_13, © Springer-Verlag Berlin Heidelberg 2011

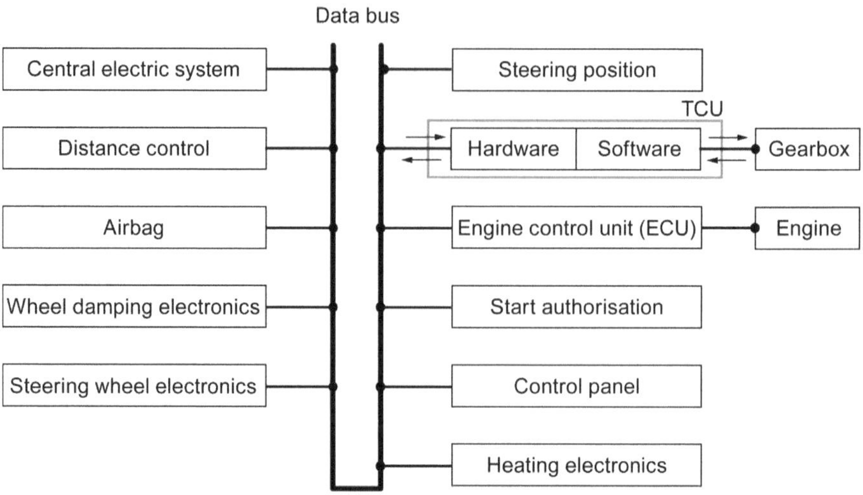


**Fig. 13.1.** Information networking. Communication of the transmission control unit (TCU) with electronic engine control and other systems in the vehicle

## 13.2 Electronic Transmission Control Unit (TCU)

The electronic control unit (electronic hardware) is the "computer centre" of every transmission with controlled functions [13.1]. It receives electric signals from all sensors as well as signal information from other electronic control systems, converts them, evaluates them, provides control values for the actuators and sends control information to other vehicle systems (Figure 13.2).

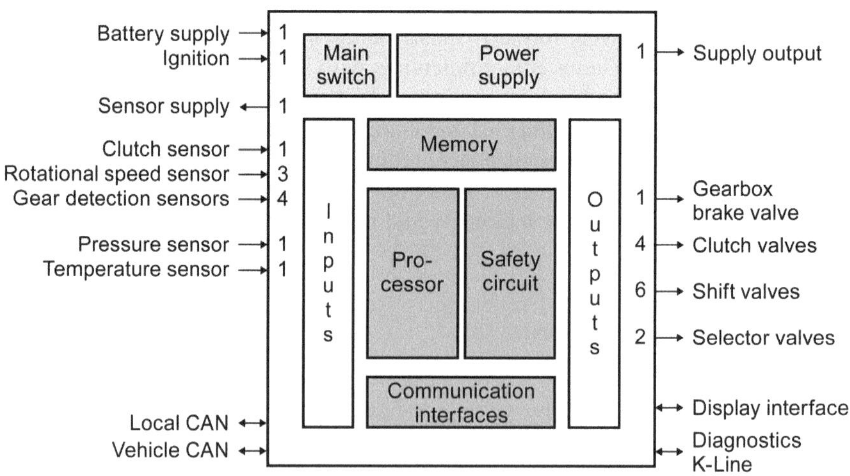


**Fig. 13.2.** Block diagram of a transmission control unit (TCU)

To evaluate signals, compute control algorithms and control the actuators exclusively microcomputer systems are used. Together with the stored program (software), these microcomputer systems control all functions of a transmission system. Referring to Figure 13.2, the structure and operating conditions of a transmission control unit will be explained in the following.

### 13.2.1 TCU Structure

#### *1/ Power Supply*

The power supply protects the unit from extreme increases in voltage, from the coupling of interference voltages and from polarity reversal damage. It creates the voltages required by the internal components. Frequently, the power supply also has a central power cut-off switch used for protection and safety functions.

#### *2/ Signal Processing/Inputs*

In this block we find signal adjustment to various sensor systems, which for their part record the most varied conditions in the transmission or the vehicle. Several signal types with different electric interfaces are captured, converted, interference suppressed (filtered) and also already pre-processed to some extent. The recognition of error conditions (signal diagnostics) is an essential function contained in signal adjustment. Besides guaranteeing system safety, signal diagnostics is an important means of efficiently finding errors when the vehicle is in service.

#### *3/ Processor/Memory*

Everything associated with the microcomputer is found in this block. On the one hand, there is the processor itself with all the circuits (e.g. the clock) necessary for its operation. The read-only memory, RAM and (if necessary) EE-PROM can be housed either in the processor chip itself or outside the chip. The development of highly integrated circuits makes it possible to integrate the maximum amount of processor components on the chip, which also has advantages with respect to operating speed.

#### *4/ Safety Circuit*

Since transmission control units (or the functions they are responsible for) have a certain amount of relevance to safety, monitoring devices and corresponding shutdown paths must be provided. Normally, the transmission control unit includes a complex watchdog – sometimes dual processor systems are also used. The safety circuit constantly monitors whether or not the computer is working properly. Usually, also system-specific plausibility checks must be executed. In accordance with the safety concept, emergency and substitute functions are activated by the control unit, culminating in the case of a critical error where the central power supply is turned off, thus restoring the system to a secure condition.

The internationally standardised safety integrity level (SIL) is defined relative to the functions of the transmission control. SIL ratings help to evaluate the transmission control unit with respect to the reliability of its safety functions. Safety-oriented design principles for hardware and software that must be followed in order to minimise the risk of malfunction are derived from the given SIL rating.

#### *5/ Data Interface (Communication Interfaces)*

Control units include external communication interfaces for inspection, communication with external diagnostic systems and communication with other vehicle systems. The number of such interfaces can vary. Separate diagnostic interfaces are being increasingly replaced by communication interfaces. One or two communication interfaces (the CAN bus and in the future, the FlexRay bus) represent the connection with the outside. Since communication data also has relevance with respect to safety and availability, vehicle communication interfaces are equipped with extensive integrated data protection mechanisms.

#### *6/ Output Stages/Outputs*

In the output stages, there are mostly power amplifiers and power switches, bridge circuits or specific current control outputs in order to run electric motors, magnetic valves, pressure control valves, positioning magnets, relays and so forth. The integrated diagnostic capability and (if applicable) integrated switch-off function is important for guaranteeing system safety.

### 13.2.2 Operating Conditions and Construction Technologies

As opposed to home units, vehicle systems are subject to extreme conditions. Nevertheless, they must still guarantee a minimal failure rate.

#### *1/ Operating Conditions*

Temperature ranges extend, depending on the location, from –40°C to +140°C. Vibration stress ranges from 2 g to about 30 g. Special demands also include those on tightness, resilience against aggressive media, resilience against humidity etc.

Electromagnetic compatibility (EMC) describes the reciprocal influence of electronic devices by the electromagnetic fields produced by their operation. Field strength is given in V/m. Interference immunity requirements of transmission control units range from 100 V/m to about 200 V/m.

#### *2/ Construction Technology*

Varying technologies are used depending on the assembly location and operating conditions. The typical electronic technologies are:

- standard circuit board devices up to approx. 100°C,
- high temperature circuit board devices up to approx. 120°C and
- hybrid microelectronic modules up to approx. 140°C,

with very high vibration and shock strength.

As is customary everywhere in electronics, surface mounted devices (SMD) are normally used for circuit board devices. On hybrid boards, chips are also bonded directly, and in addition, components can be imprinted.

Control unit housings are usually made of metal. The thermal connection of the components to the housing is optimised in order to guarantee good heat dissipation from the unit. Should the electronics be built into the transmission and integrated together with other components (such as sensors, actuators and connection technology), we then speak of mechatronic modules (see e.g. Figure 12.18 or 12.25).

## 13.3 Control Systems

The following control systems will now be explained with the help of system diagrams:

- automated manual transmission, AMT (Figure 13.4),
- automatic transmission, AT (Figure 13.11),
- dual clutch transmission, DCT (Figure 13.12) and
- continuously variable transmission, CVT (Figure 13.13).

The structure and basic functioning of transmission control systems will be shown using the representative example of the automated commercial vehicle three-range gearbox ZF AS-Tronic shown in Figure 13.3. The following can also be applied to AT, DCT and CVT transmission control systems.

The transmission is explained in detail in Section 12.2.2 “Automated Manual Commercial Vehicle Transmissions (AMT)”. Transmission design is shown in Figure 12.41, while 12.42 shows the internal gearshift system. Pneumatically activated shift cylinders, themselves controlled by the electronics, are arranged in front of the selector bars.

Splitter and range-change units are each activated with a shift cylinder with two positions, and the main gearbox with reverse gear with a shift cylinder with three positions. All selector bars are equipped with travel sensors in order to feed back the important positions or in order to make targeted control processes possible.

The gearbox brake consists of a multi-plate packet which can also be actuated with a pneumatic cylinder. The system is completed with a pneumatic clutch actuator attached below on the gearbox which acts directly on the dry-running master/gearshifting clutch by means of a lever.

Speed sensors and a pressure/temperature sensor top off the total sensor design. This arrangement results in a control layout such as is shown in Figure 13.4.

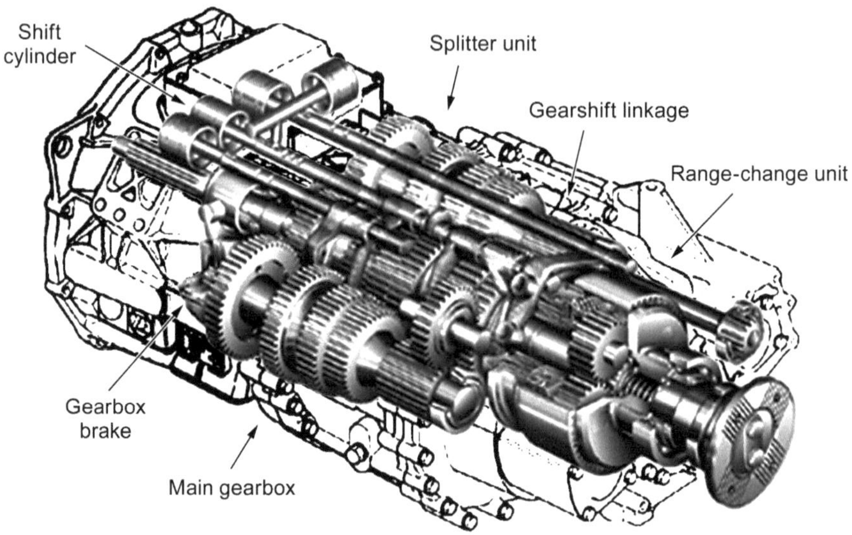


**Fig. 13.3.** Automated 16-speed commercial vehicle three-range gearbox ZF AS-Tronic; design and internal gearshift system *(see also Figures 12.41 and 12.42)*

The central element is the electronics, a computer system that captures the necessary signals and co-ordinates all the actions in the system.

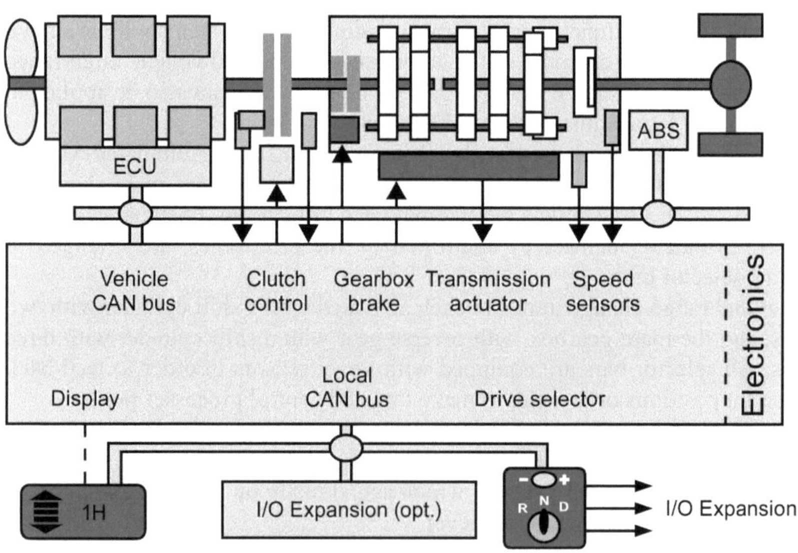


**Fig. 13.4.** System diagram for an automated manual transmission (AMT) using the example of a 16-speed commercial vehicle transmission

Besides the direct connection of the transmission periphery, various other mutually interacting systems are connected via a CAN bus. The most important system is the engine control unit (ECU), with the help of which speed synchronization can be controlled during downshifts. Positive engine control intervention is an important requirement in order to shift the constant-mesh transmission at all. In the case of upshifting, the integrated gearbox brake takes care of speed synchronization.

Further systems, such as the secondary retarder, can be functionally networked with the transmission and also exchange necessary signals via a data bus. The vehicular periphery, such as the drive selector, display or additional input and output channels, are connected by means of a second local CAN bus.

Optional standard interfaces, such as a power take-off control, can also be connected to the drive selector. Special, seldom-used interfaces are connected by an additional I/O expander. In this way, the system remains open and ready for many different cases of operation. Connection of the drive selector with the CAN bus and the signal distribution in the driver zone associated with this conserves essential wiring between the gearbox and the driver seat, making it an essential cost factor and an advantage in reliability.

### 13.3.1 Transmission Actuator

In the case of the AS-Tronic gearbox, the complete transmission actuation unit (electronics, pressure limiting unit, valve block, pneumatic cylinders and associated sensors) is designed as a compact mechatronic assembly module (see also Figure 12.43). This approach follows the principle of assembly modularisation common in the automobile industry. The goal is to realise both cost advantages and high system reliability by means of cost-effective electrical connection technology (Figure 13.5).

The shift module can be produced and pre-tested completely separately from transmission manufacture. When it is built into the transmission, it is automatically connected with the gearbox at a small number of predefined connection points. When being serviced, this unit can also be completely replaced. Flexible circuit boards are used to connect the electric modules. Larger-scale control modules can also be made with this technology.

### 13.3.2 Clutch Actuator

For automatic control of dry clutches, a clutch actuator module is used that is equipped with a pneumatic shift cylinder and four control valves as well as its own travel sensor electronics (see also Figure 12.43a). Control of the clutch valves to control the clutch position is achieved directly via the transmission actuator unit. The design of the clutch actuator guarantees sufficient emergency system operation, even in case of valve or travel sensor failure.

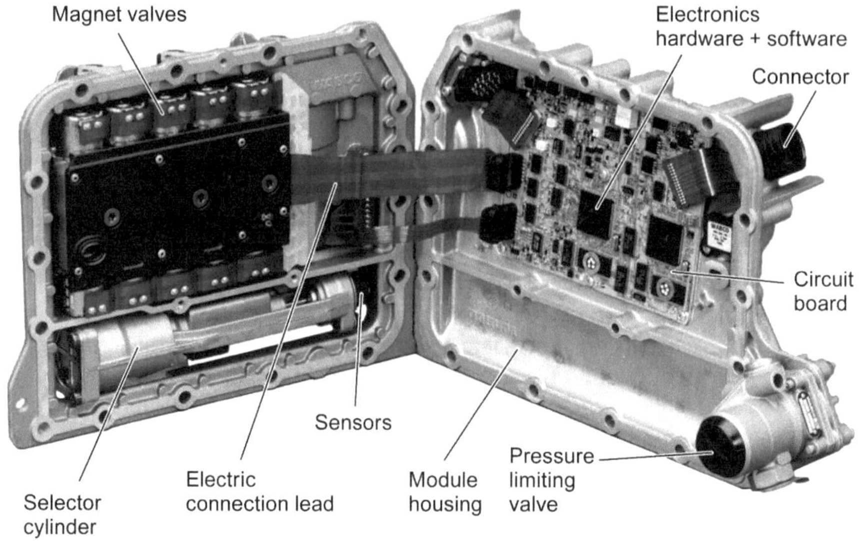


**Fig. 13.5.** Mechatronic shift module of an automated 16-speed commercial vehicle gearbox; transmission actuator with integrated electronic transmission control unit *(see also Figure 12.43)*

### 13.3.3 Transmission Control Functions

The functions of transmission control fall into three functional groups:

- vehicle functions,
- basic functions and
- hardware-related functions.

Function and software development go hand in hand. For this reason, see also Section 13.3.4 "Software" including the process model for function and software development (Figure 13.10).

The functional groups are mirrored on the software level with about the same size (see Figure 13.6). The layer of hardware-related functions includes everything necessary to operate the control itself, comparable to the operating system of a PC. Basic functions encompass everything required to operate the transmission. Vehicle functions are understood as those that are important with respect to the behaviour of the vehicle as a whole. We will examine a few basic functions more closely in the following.

#### *1/ Clutch Control*

In the case of automated manual transmissions, clutch control controls moving-off as well as gearshifts during driving. Manoeuvring, which is essential for trucks, is a special form of moving-off operation.

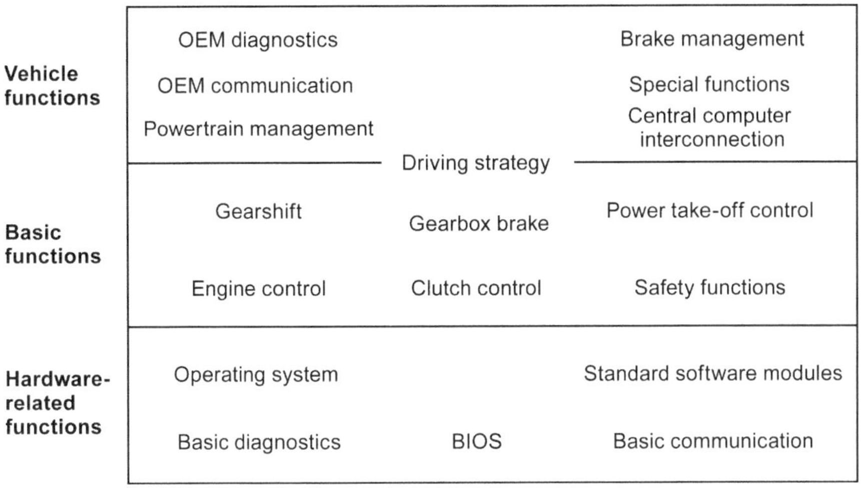


**Fig. 13.6.** Functional overview: e.g. an AMT system

Analogously to manual transmissions, the simultaneously correct actuation of the accelerator pedal is also necessary in order to achieve a certain level of moving-off quality. In the case of automated manual transmissions, clutch control takes over interaction with the accelerator pedal.

When the vehicle moves-off, generally a predefined speed profile between the engine and transmission speed is applied until the clutch, proceeding from the maximum speed difference, reaches speed equality (Figure 13.7a). Then the clutch is shifted to the end position. Free travels at the clutch are passed through very quickly in a controlled way in order to increase the actuation dynamics. In the case of manoeuvring, which is automatically recognised or alternatively can also be selected at the control panel, the clutch is controlled to a defined degree of engagement instead of speed. In this way, the driver can manoeuvre sensitively with the accelerator pedal.

In the case of downshifting, the dry clutch is opened and the shift process started as soon as a load change is introduced at the engine by means of accelerator reduction (Figure 13.7b). In the case of constant-mesh transmissions, if the old gear is disengaged and the new gear already selected, the transmission speed is synchronized with the help of the clutch and a positive engine control intervention (intermediate throttle application, double-declutching), and the new gear is engaged. At the end of the shift, engine control is transferred back from transmission control to the driver and the clutch is engaged by control.

### *2/ Shift Process Control*

Shift process control is responsible for the correct actuation sequence for engaging or changing gears. This process is, as described above, intimately associated with clutch control and engine control intervention.

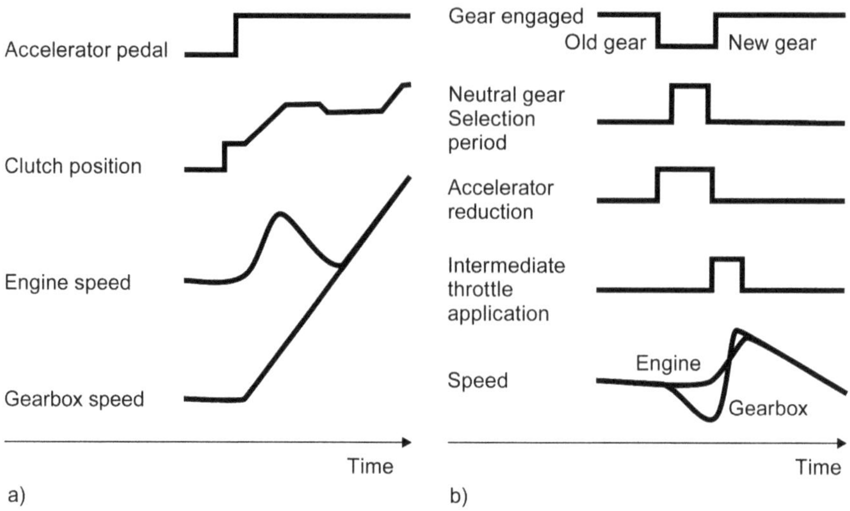


**Fig. 13.7.** *a* Moving-off process; *b* shift process of an automated commercial vehicle transmission

If the powertrain is load-free and the dry clutch is open, the existing gear is first disengaged. This takes place by actuating the shift cylinder in the direction of the neutral gearbox position. If the gearbox is in neutral, the gate of the new gear is selected and then the shift cylinder is activated again in order to engage the new gear.

This process is valid for a relatively simple gearbox. In the case of multi-range transmissions, one or two further shift processes for splitter and range-change units occur simultaneously for certain shift changes. This results in a complex process with nesting and sequence control parts. The shift process is monitored the entire time with sensors so that, on the one hand, the processes take place optimally and, on the other, faults can be recognised early and removed with countermeasures. Such a fault can occur, for example, if the shifting gears come into contact unfavourably ("tooth-on-tooth position") and, as a countermeasure, the shift process must be finished by means of an additional load reduction of the powertrain.

### *3/ Driving Strategy*

Besides the shift process itself, selection of the correct gearshift point, the correct gear and the number of gears to be shifted is an especially crucial function. The aim is to reach optimal levels with respect to fuel consumption, emissions and performance. Furthermore, driver expectations must also be met. This is not an easy task, which requires not only a representation of physical data in the form of control algorithms but also an empirical component for the specific adjustment of *vehicle calibration* (see also Section 13.4).

An essential component of driving strategy is a module for recognising the actual driving states and conditions. Conditions such as standstill, acceleration, constant speed and so forth are detected. In addition, signals concerning speed, acceleration, consumption and excess power are continuously being captured. Moreover, state observers for the most varied of state variables, like load, slope and driving resistance, are applied. All this information is analysed and evaluated in a complex controller for the current gear and for the gears potentially to be engaged. A gearshift recommendation finally ensues from this. It is also possible to incorporate additional and anticipatory road information into the driving strategy, e.g. GPS road topology or traffic signs, and such possibilities have been utilised to some extent.

For certain situations in which the driver decides to intervene, manual intervention into the gear selection is possible. This intervention possibility is important, such as when the traffic situation or the road topology demand correction of the gearshift recommendation (Figure 13.8). As mentioned, there are also subjective adjustment criteria which vary among different vehicle manufacturers, which must then be capable of targeted adjustment by means of calibration drives.

### *4/ Other Functions*

Besides the previously mentioned basic functions, there are many other functions within a transmission control. For example, there are separate processes for activating and for controlling power take-offs, display functions for driver communication, functions for monitoring control panels and functions that are based on data exchange with other vehicle systems, e.g. hill hold control.

Diagnostic and emergency functions make at least 50% of the total extent of functions and software. These should, on the one hand, help to find potential component faults quickly and reliably. On the other hand, they should guarantee a high level of system availability so long as no especially critical components have failed.

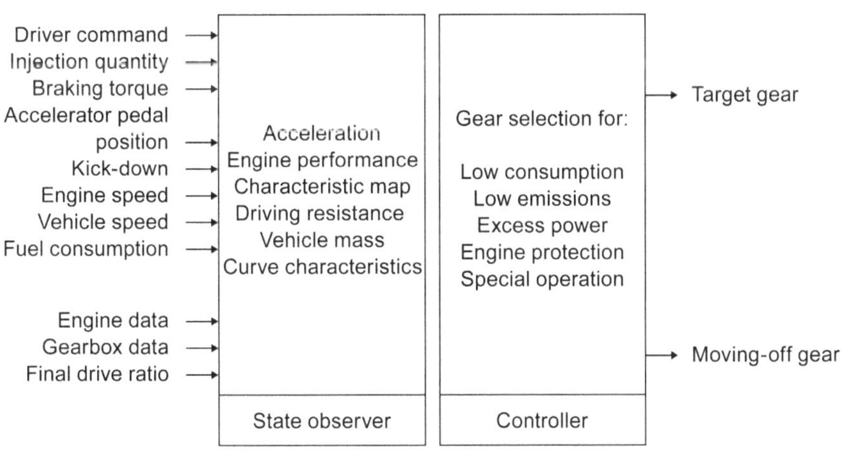


**Fig. 13.8.** Driving strategy: e.g. an AMT system

### 13.3.4 Software

Software for transmission controls are designed modularly based on the principles of object orientation. Function modules are encapsulated as much as possible to give them a high degree of reusability (Figure 13.9).

The development of functions and software follows the V model (Figure 13.10). For all sub-processes from the model, there are detailed process descriptions which must be continually developed further and measured. Linked auxiliary tools (software tools) ensure that the processes with their complex requirements can be carried out in daily operation.

In the left branch of the model, special attention is paid to the consistent analysis of requirements and on simulation already on the level of specifications. In this way, errors can be recognised and removed already in the early stages so that they do not spread into very cost-intensive sub-processes of the right branch [13.4].

Various simulation and testing tools are employed for tests within the sub-processes. SIL (software in the loop) system simulation allows the integration of the most varied of program modules on one operational level, making it possible to represent complete systems. For example, road models, control codes, monitor functions, adjustment functions and debugger functions can be integrated together in this way. This opens up the possibility for testing complete systems without any special hardware.

Testing possibilities range from single steps to complete simulation in quasi real-time. Script files and previously captured road data allow for fully automatic test sequences, the results of which are evaluated partially automatically.

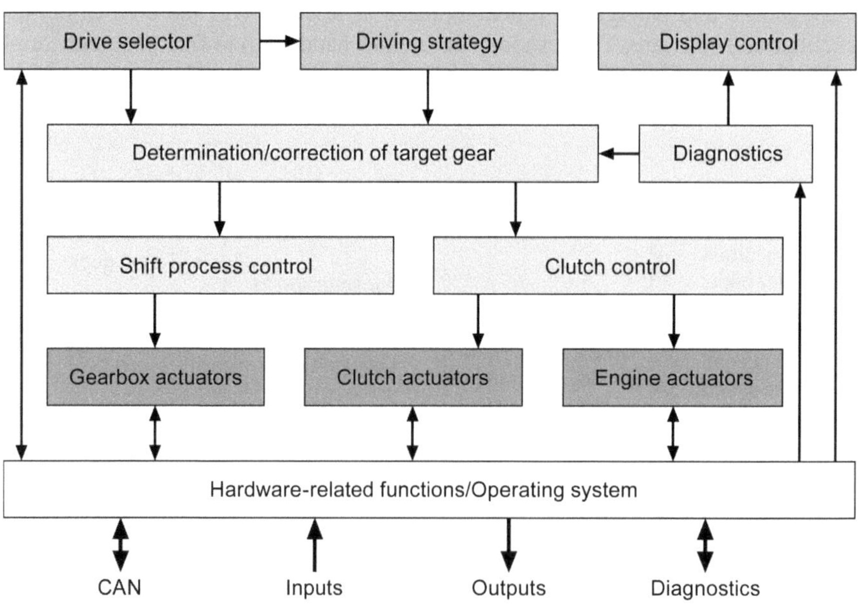


**Fig. 13.9.** Software structure of an automated commercial vehicle manual transmission

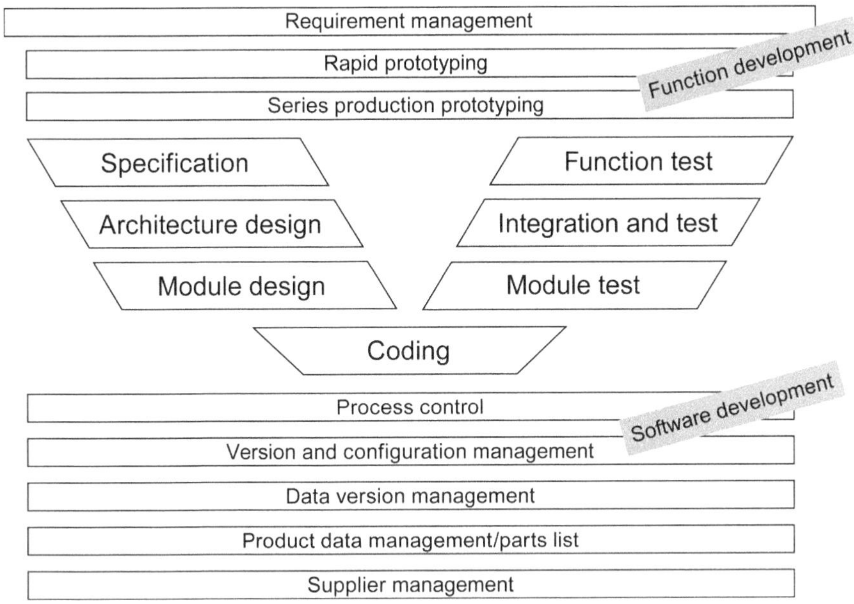


**Fig. 13.10.** Process model for function and software development

Because there are often no functioning vehicles yet during the function/software development phase and there is a high and reproducible transparency given by simulation, SIL simulation is of great importance in the area of analysis and verification [13.3].

For simulations and system tests with the final components, a HIL (hardware in the loop) test system is utilised. Similarly to the SIL simulation, roads and monitor functions are recreated in a complex computer system. Due to the high real-time demands and additional emulation of the electronic periphery, an efficient signal processing system is required in this case. Application or measurement systems can also be connected to the HIL test system.

Software quality is of the greatest importance. Process indexes help to ensure and keep track of the quality of the software. These indexes are thoroughly examined at regular intervals by process auditing (e.g. following the SPICE model "Software Process Improvement and Capability Determination" based on ISO/IEC 15504). This guarantees that the required level of quality is reached in software development.

### 13.3.5 Further Examples of Transmission Control Systems

The typical design and basic functioning of transmission control system were described in the previous pages (Sections 13.3.1 to 13.3.4). These descriptions can be applied to the following AT, DCT and CVT system diagrams.

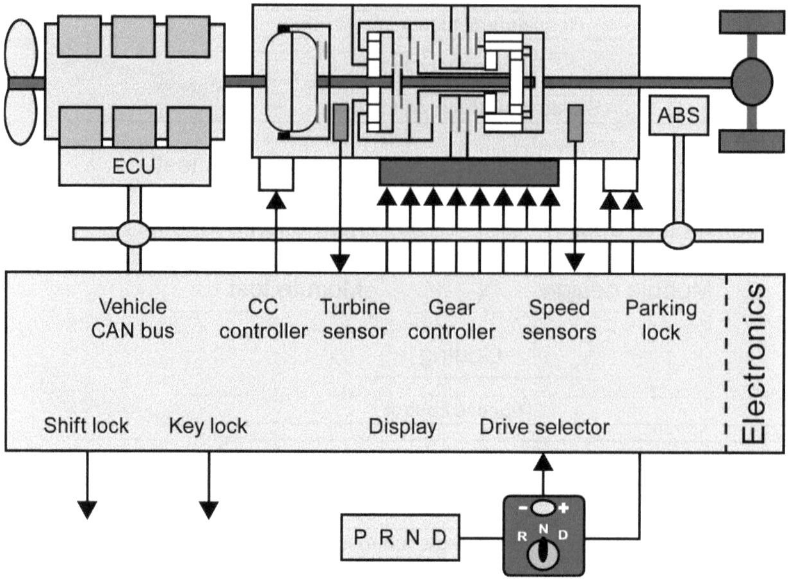


**Fig. 13.11.** System diagram for an automatic transmission (AT)

### *1/ Automatic Transmission (AT)*

The basic design of control systems for automatic transmissions (AT) is similar to that of the AMT transmission described above. The interface between the electronics and the transmission is basically determined by the number of electrohydraulic control valves (Figure 13.11).

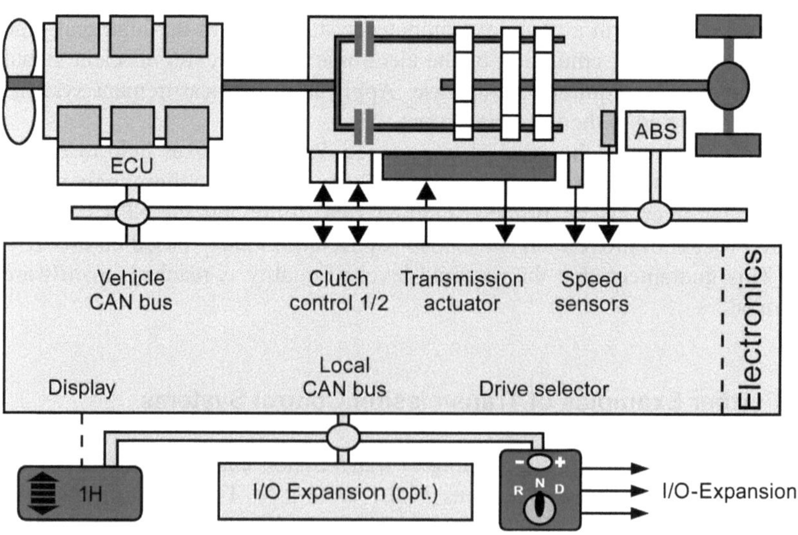


**Fig. 13.12.** System diagram for a dual clutch transmission (DCT)

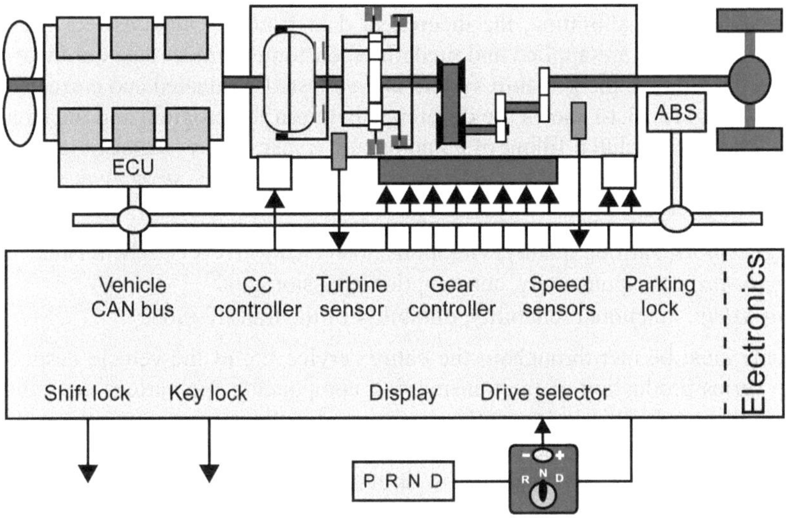


**Fig. 13.13.** System diagram for a continuously variable transmission (CVT)

For passenger car uses, gearboxes are additionally equipped with an integrated parking lock that can secure the vehicle. Depending on the design, the parking lock can be operated mechanically by cable control or electrically by means of actuators (see also Section 9.4.2 "Electrically Activated Parking Locks").

### *2/ Dual Clutch Transmission (DCT)*

Dual clutch transmissions are comprised of two automated transmissions, one built inside the other, in which every transmission branch is operated by a separate clutch (Figure 13.12). In this way, the associated transmission control also represents a kind of combination of an AMT control (activation of shifting elements such as synchronizers) and an AT control (clutches with powershift-algorithm).

### *3/ Continuously Variable Transmission (CVT)*

In the case of continuously variable transmissions, based on pulley transmissions, electronic transmission control basically controls the variator (Figure 13.13). Besides ratio control, control of the contact pressure in the variator is decisive, as this determines the efficiency of the entire transmission.

## 13.4 Transmission Calibration with Vehicle-Specific Software Data Input

Calibration is defined as the adaption of the transmission properties to the dynamics and behaviour of the entire vehicle by inputting data into the transmission software. Working out the characteristics typical of specific brands also belongs to this process. During calibration, the theoretical data from calculations (e.g. permissible clutch loads) are applied and predefined customer project data are incorporated. In the vehicle, the gearshift system is then tested, evaluated and corrected if necessary. Figure 13.14 shows the difference between the program and the data using the example of clutch filling of an automatic transmission (see also Section 9.3.2 "The Shifting Process"). The *shifting process* and *shifting strategy* are calibrated with respect to:

- *driving comfort*: shifting quality, vibrations, load change (reversal) behaviour,
- *driving behaviour*: spontaneity, consumption/emissions and
- *driving safety*: functional reliability, durability of the transmission.

These goals must be met throughout the entire service life of the vehicle despite spread in series production of the transmission components and various environmental conditions. To this end, adaptive (learning) functions are put into the software, into which data is entered during calibration. The correct functioning of the adaptive functions has to be validated by limit sample examination. In this case, transmissions are tested utilising the tolerance limits. This can include:

- clutch clearances,
- spring force tolerances,
- O-ring groutings,
- pressure controller tolerances,
- etc.

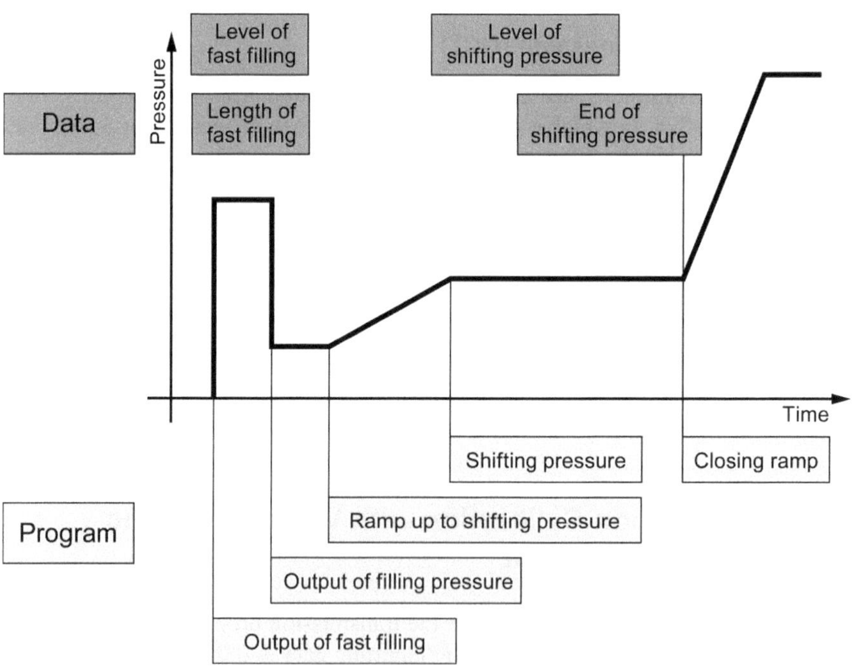


**Fig. 13.14.** Difference between the program (software) and data (calibration)

# 14 Computer-Aided Transmission Development

*Simulation means representing*
*real-world phenomena on a computer*

Increasing time and cost pressure in automotive transmission development demands the systematic use of CAE methods (Computer-Aided Engineering, Figure 14.1). Early computer testing and optimising of as many properties of the individual components and of the entire transmission system as possible do indeed raise expenditures in the early phase of development. However, conserving developmental loops results in the end in obvious advantages in time, cost and quality.

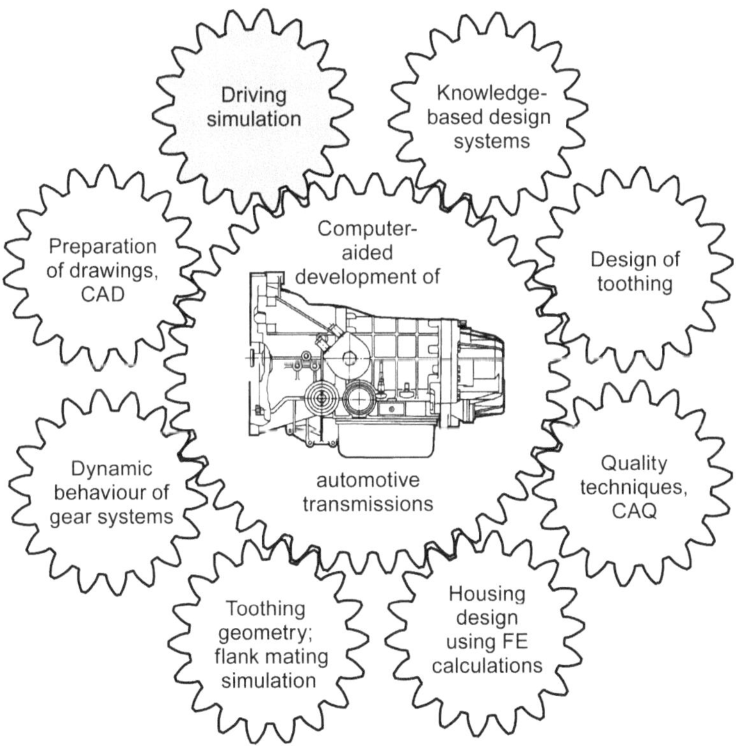


**Fig. 14.1.** Computer-aided development of automotive transmissions

H. Naunheimer et al., *Automotive Transmissions*, 2nd ed.,
DOI 10.1007/978-3-642-16214-5_14, © Springer-Verlag Berlin Heidelberg 2011

Ideally, calculations and simulations are carried out concurrently during development, beginning with the evaluation of the first drafts of the gearbox diagram right up to optimising the manufacturing process [14.5]. On this, see also the description of the development process given in Section 15.4.

CAE methods can help in the decision-making process already in evaluating different transmission concepts, for example by means of comparative performance and consumption simulations or by analysing and synthesising transmission systems. If the transmission concept and the gearbox diagram have been determined, first the gear sets are dimensioned. Therefore, the design-relevant load cases and corresponding load profiles have to be determined from the later operating conditions. Commercial software firms offer numerous calculation programs for the initial rough dimensioning of machine elements like tooth systems, shafts, bearings, clutches or screws.

With increasing level of detail in the design, individual components can be investigated more closely with more refined calculation methods. However, the interaction of the gearbox with other components of the entire vehicle system must now also be taken into consideration. Several tasks can be directly solved within the CAD environment such as tolerance analyses, kinematic investigations for the gearshift system or assembly simulation. Other areas, such as manufacturing simulation or flow calculations require special tools. For complex functional analyses such as shifting comfort simulation, mechanical, hydraulic, pneumatic and electronic models must to some extent be coupled [14.4].

## 14.1 Principles and Tools

Even though individual calculation methods differ greatly, the basic sequence of calculations can still be generally be resolved into the sub-steps shown in Figure 14.2.

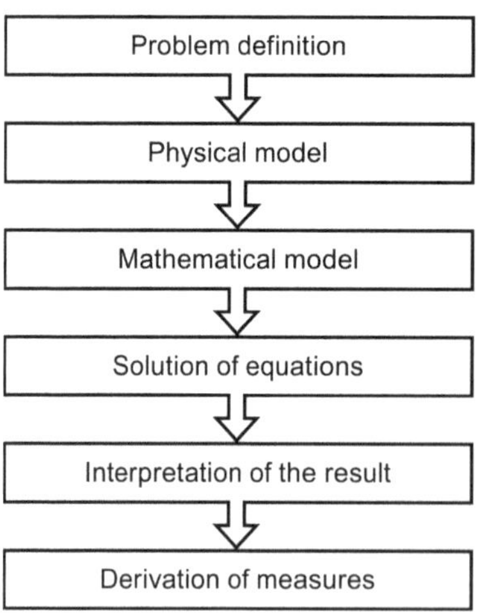


**Fig. 14.2.** Basic calculation sequence, sub-steps

Before starting the calculations, a clear problem definition with explicit goals must absolutely be formulated. On this basis, one must then first create a physical model from the technical design. Depending on the task, suitable system limits, realistic boundary conditions and an idealisation suitable to the problem must be chosen for this. There are always assumptions and simplifications to be made, which should be in keeping with the problem under investigation. With a linear-elastic FE model for example, no stresses should be calculated above the yield strength, and a pure torsional vibration model cannot make any predictions regarding bending natural frequencies. Since not only the precision, but also the complexity increases with increasing degrees of discretisation, the models should be kept as simple as possible and as complex as necessary.

In the next step, a mathematical model is derived from the physical model, which contains, for example, the equations of motion or the formulae for calculating stresses. The actual calculation, i.e. the solution of the mathematical equations, is usually completed by suitable software without a problem. In the case of numerical methods, considerations of convergence with various depths of discretisation are helpful in limiting the effects of discretisation on the result. The analysis and interpretation of the calculation results are of particular importance. Analyses with numerous decimal places suggest high precision, which often does not exist during critical examinations because the assumptions and simplifications made initially lead to a certain degree of uncertainty in the result. The exactness of the calculation results is always directly contingent on the quality of the input data. Plausibility checks, extreme value examinations and comparative calculations with known designs help to judge the calculation results correctly. Sensitivity investigations and parameter studies, which can be carried out quickly by the computer for the model at hand, are helpful for deriving suitable measures in case the goals are not met. The effectiveness of such derived design measures should always be tested by renewed calculations before they are realised [14.18].

### *1/ Finite Elements Method (FEM)*

The Finite Elements Method (FEM) is widely used for detailed investigations, particularly of complex components. Close linking with CAD data and the potentials of automatic meshing have considerably lowered the expense required to create meshes. FEM calculation of deformations and stresses in simple components in static loading conditions can be partially executed by the designer himself using CAD with proper training. For more sophisticated tasks, specially trained calculation engineers must be called upon. Improper application can lead the layperson to completely misleading results. However, when properly used, FEM is a powerful tool with many possibilities [14.1, 14.22].

For example, FEM can help determine the stresses on transversely bored shafts or complex cast housings, the notch geometry of which is not covered by the nominal stress approach, thus being inaccessible to analytical computation. With subsequent operational fatigue strength programs, damage and the expected service life can be derived from the computed stresses with the local approach. Another application example for FEM is the calculation of thermal stresses in dry clutches. Here, we can consider the relation of heat input to component deformation during clutch engagement. From this we can compute the time-dependent distribution of temperature, and thermal stresses can be superimposed over mechanical stresses.

FEM-based structure-optimisers are being used with great success. The user selects the objective functions and defines restrictions – for example, stiffness should be maximised while the weight may not exceed a given limit. The computer then suggests optimal geometries that the designer often does not think of intuitively. We differentiate between topology, shape and parameter optimisation. Topology optimisation helps to find the basic form of a component under optimal usage of a pregiven installation space. Shape or form optimisation improves that form by means of local modification of the surface, exploiting the material in an even fashion. With parameter optimisation, we can optimise individual parameters such as the distribution of wall strength. Modern optimisers can compute several load cases simultaneously and take manufacturing restrictions into consideration, such as the pull-out direction of deep drawing tools or casting moulds.

### 2/ Housing Design with Topology Optimisation

Special consideration should be paid in computer-aided transmission development to the transmission housing with its usually highly complex geometry. In vehicle transmissions, housings are usually made as die cast components in light metal. Since designing the housing and constructing die cast moulds are very time and cost intensive operations, a reduction of the development loops has a lot of potential for savings.

In new product developments, it is advisable to begin with topology optimisation. On the basis of the available installation space and the loads acting on the housing, the optimisation software suggests wall structures and rib arrangements that are an optimal compromise between lightweight design and stiffness. These suggestions are conveyed to the CAD system by surface transfer, where further designing takes place. The fully designed housing geometry is then in the next step investigated thoroughly with FEM with respect to deformations, stresses and natural frequencies, whereby optimisation methods can be again utilised. Only when the computation results show that the demands have been met with high probability is it advisable to approve it for the production of a prototype for conformational tests.

### 3/ Vibration and Acoustic Simulation

Due to the increasingly lightweight design and comfort demands of customers, simulations of dynamic and acoustic behaviour have become much more important. With special torsional vibration programs, the complete powertrain is modelled with rigid mass elements and massless spring elements, whereby only the degree of torsional freedom is taken into consideration. One typical application for linear models is in the design of torsional vibration dampers or dual mass flywheels for vibration decoupling of the gearbox from the engine. With nonlinear models, more complex problems like rattling noises or friction vibrations can also be investigated.

For advanced problems, such as bending vibrations of shafts, Multibody System (MBS) simulations and FEM are employed. In MBS simulation, rigid bodies are flexibly connected to each other, whereby reduced FE models can also be incorporated. Usually, the Boundary Elements Method (BEM) is used to calculate sound emission from the gearbox housing. Comparison with experimental methods is essential for obtaining reliable calculation results in the area of acoustic simulation, e.g. to determine damping values [14.7–14.8, 14.11–14.12, 14.20, 14.25].

### 4/ Efficiency and Thermodynamics Simulation

Calculation principles for the power losses of all transmission components are derived from physical relations and empirical approaches. From this, the efficiency characteristic map of the transmission can be calculated as a function of speed, torque and temperature (see also Figure 3.5).

The high power density in vehicle transmissions and the increasing tendency to encapsulate the engine compartment require a careful consideration of the thermodynamic behaviour of transmissions. The computational equations can be supplemented with parameters for heat transfers, heat capacities and boundary conditions and expanded into thermodynamic network models. With such models, the fluid flows of the lubricating and cooling oils in the gearbox can be calculated and local temperature peaks in critical operating points discovered. With optimisation algorithms, the oil lines can then be designed such that the temperature level in the gearbox is as even as possible.

The model must be extended to the entire vehicle cooling system for heat exchanger design for transmissions with oil coolers. Realistic boundary conditions for air circulation in the engine compartment can be determined by CFD calculations (Computational Fluid Dynamics).

By coupling the thermodynamic model with driving simulation, the time-dependent behaviour of transmission oil temperature can be computed in certain driving cycles or on virtual routes. If the electronic shift program is incorporated into these calculations, temperature-dependent control strategies for retarders in commercial vehicle automatic transmissions can be researched [14.9, 14.27].

### 5/ Consumption and Performance Simulation

See Section 14.2 “Driving Simulation”.

### 6/ Calculation and Testing

The co-ordination of calculation and testing has shown that simulations and experiments are equally important methods for problem solving, each with its own pros and cons. The best results are obtained when both methods are combined using the respective advantages of each method.

Experimental verification is always advisable when introducing new calculation methods and when applying tried methods to new areas of applications. But experiments are also very important in the case of determination of unclear input data for calculations, such as damping values. In the case of complex transmission systems, measurements have the further advantage that all nonlinear effects and influences from the interaction of components are automatically contained in the results.

Calculations on the other hand have the advantage of reproducibility, the possibility of isolated variation of single parameters and in the investigation of components that are difficult to access. Calculations yield first information already before prototypes are available, and variant investigations can be quickly and cheaply executed by computer. In many cases, measurements can determine more easily whether or not there is a problem and how serious it is, while calculations are more suitable for finding causes.

Experimental verification of computational models requires equal boundary conditions and identical design levels. The consequence of the assembly effects and the unavoidable diversification of material properties and geometries is that certain differences between the computational and experimental results always remain.

#### 7/ Computation Quality and Effectiveness

In order to utilise the advantages of CAE methods effectively, associated processes must all be optimised. Hardware costs are increasingly losing relevance to software costs. Reusability of data, reduction of frictional losses at the interfaces to neighbouring processes, standardisation of calculation sequences and automatisation of standard calculations are important points. With the growing importance of CAE methods in the development process, the responsibility of computational engineers for the quality of the results is increasing as well [14.19, 14.24, 14.30].

## 14.2 Driving Simulation

Computer-aided driving simulation is discussed in greater detail below, as an important application of computers in the development of vehicle transmissions. Driving simulation methods for automotive and traffic engineering questions can be divided into three categories, as shown in Figure 14.3 [14.15].

### Submicroscopic Driving Simulation (Driving Simulation)

Submicroscopic driving simulation is the classic application of driving simulation with regard to optimising powertrain components. The first applications arose in the 1970s. The vehicle is represented in exploded form in the computer, i.e. all of the vehicle's or the powertrain's significant components are modelled. The depth of modelling depends upon the requirements.

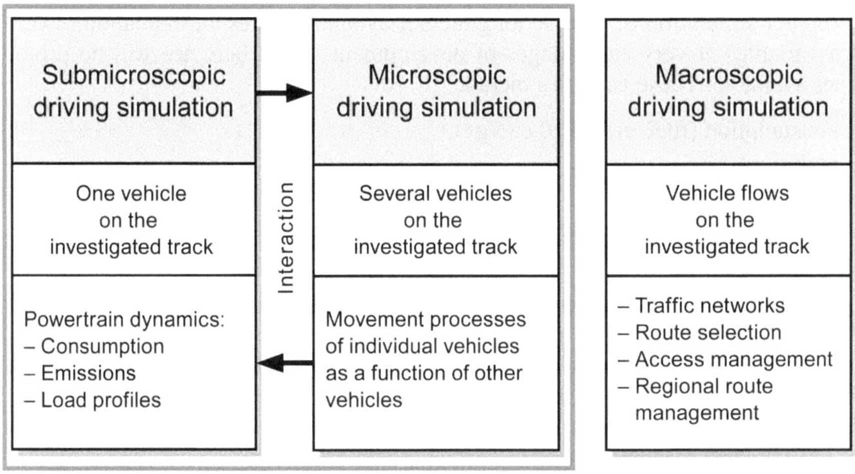


**Fig. 14.3.** Rough classification of driving simulation methods. Submicroscopic and microscopic driving simulation are also used in combination

### *Microscopic Driving Simulation (Traffic Simulation)*

The driver rarely encounters unconstrained traffic conditions leaving him free to choose his speed and style of driving. Drivers increasingly encounter partly constrained or even constrained traffic situations. Nose-to-tail traffic with no means of overtaking is an example of constrained driving conditions.

In microscopic driving simulation, the vehicle is considered in its traffic context [14.26]. Attention is focussed on the movements of the vehicle under investigation related to the prevailing traffic situation. These movements in turn act on the components of the individual vehicle. Submicroscopic and microscopic driving simulation are therefore increasingly combined (Figure 14.3).

### *Macroscopic Driving Simulation (Traffic Flow Simulation)*

In the case of macroscopic driving simulation, whole vehicle flows are considered using continuum mechanics. The technique is generally used to study traffic engineering questions [14.28].

## 14.2.1 Simulation of Vehicle Longitudinal Dynamics

The term "driving simulation" discussed in the next section refers to submicroscopic driving simulation at the component level, viewing the powertrain as a set of components. Simulation methods for investigating driving stability (so-called double-track models) are not considered here.

Computer simulation of vehicle longitudinal dynamics gives an indication of certain variables at very early stages of development when there are still no prototypes available. These variables include:

- consumption (fuel, electrical energy),
- emissions,
- performance,
- load profiles for predicting service life [14.6, 14.13] and
- driveability.

Considerable requirements are placed on the quality of simulation results. Above all, the consumption values achievable with a new gearbox are decisive. However, performance is also investigated in detail. The values are already determined in the concept phase for a defined engine, gearbox and axle configuration and fixed in the specifications.

The complex system comprising driver, vehicle and road can be represented in abstract form so that the effect of individual changes on the system as a whole can be investigated. Driving condition data derived from simulations are also used to specify reference values for component and powertrain test benches ("hardware and software in the loop"). The key advantages of driving simulation compared to road tests are [14.16]:

- reproducible conditions,
- good time/cost ratio,
- calculations can be carried out throughout the whole development process and
- reduced development time with parallel product development.

The core of driving simulation is the driver-vehicle-route control loop (Figure 14.4). The driver acts as controller on the vehicle as controlled system with a view to adapting the actual speed of the vehicle as closely as possible to the desired-speed reference variable. The speed desired is the speed at which the vehicle would move if there were no environmental disturbance variables. Environmental disturbance variables influence both the driver as controller and the vehicle as controlled system. This includes other road users, traffic regulations, weather conditions, and the three-dimensional route profile.

The conditions in which the vehicle is used have a major impact on the vehicle as a system. The route and driving style depend on the driver, and therefore vary considerably. The design of the transmission has to take account of representative operating conditions and also extreme operating conditions. It is therefore not sufficient for the driving simulation to take account only of standard cycles (e.g. NEDC2000, Table 5.4), but real routes have to be specified. The more realistic the route specification data are, the better different operating conditions can be simulated on the computer.

The operating conditions can be recorded in the course of a *pilot run*. During the pilot run, the three-dimensional route profile is recorded as well as road space information and driven speed. This speed profile is called the pilot speed profile. The pilot speed takes account of the environmental disturbance variables in effect at the time of the pilot run on the road.

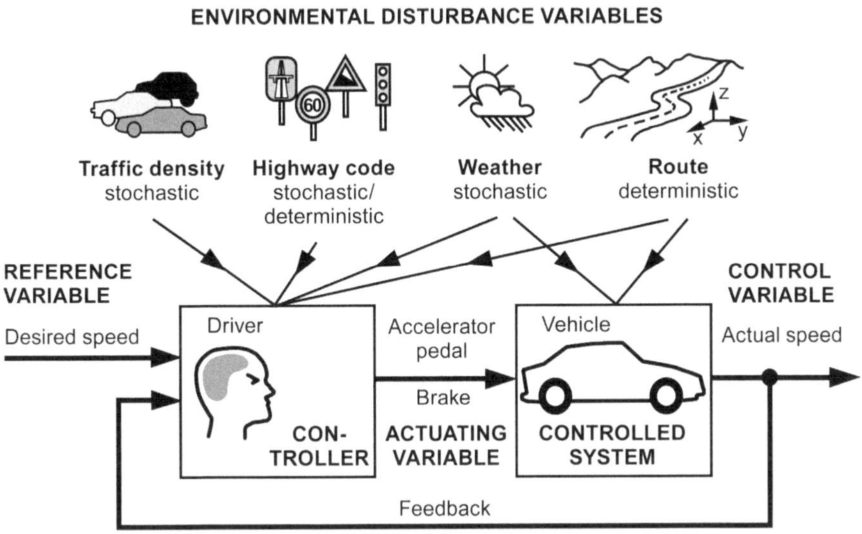


**Fig. 14.4.** Driver-vehicle-route control loop. Acting environmental disturbance variables

Using the pilot speed as the specified speed for the driving simulation calculation is a proven procedure. Alternatively, the specified speed can be obtained at using traffic simulation (microscopic driving simulation) [14.21]. In traffic simulation, the speed of the considered vehicle is calculated from the desired speed and the environmental disturbance variables.

It is advisable to link submicroscopic driving simulation, which considers the individual vehicle and its components, to microscopic driving simulation. But too many specified or setting parameters and excessively complex traffic models can also be a hindrance to achieving clear results from the driving simulation calculations.

The simulation methods applied are analytical methods. They are based on the fundamental equations of the dynamics of vehicle movement, Equations 3.18–3.22. They are also referred to in the literature as geodetic or dynamic driving simulation. They have a direct time/distance relation. The equations of motion are integrated along individual section intervals in accordance with the reference speed. The real operating conditions are determined by pilot runs.

In order to be able to model arbitrary powertrains vividly and clearly, a modular simulation program is necessary (Figure 14.5). Well-known, module-oriented programs such as Simulink [14.33] are less appropriate for driving simulations, because they are based on signal coupling. Preferred are tools with which submodels can be connected with a power coupling [14.3, 14.23, 14.29, 14.31–14.32].

In the case of power coupling, the driver controller reacts to deviations of the actual speed from the desired speed by altering the position of the accelerator. The program calculates from the "software driver" via the accelerator pedal position and the engine performance map to the drive wheels.

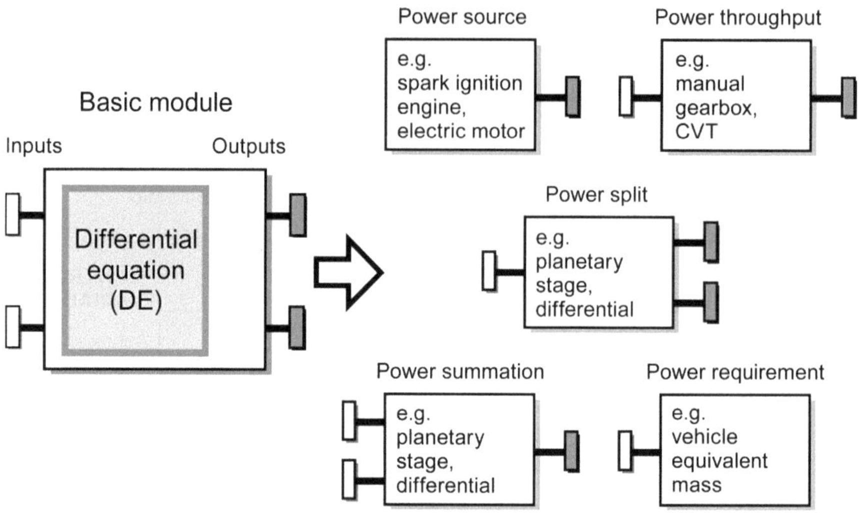


**Fig. 14.5.** Basic module and examples of submodules with power coupling

Simulation programs operating on the reverse principle are easier in terms of driver control, much faster in terms of calculation speed, but also less versatile. Calculation proceeds in this case from required power at the wheels resulting from the driving resistance, to the operating point in the engine performance map.

Powertrain simulation as thus presented represents a modular problem. The individual components of the powertrain can be regarded as modules linked by precisely defined interfaces. Object-oriented programming languages are therefore a suitable tool for computational implementation.

Figure 14.6 gives the example of a parallel hybrid powertrain comprising the powertrain modules described. Losses arising from pumps, auxiliary units etc. and the efficiency of the various powertrain elements must be taken into account. All information on engine speed, torque and losses can be read at the module interfaces.

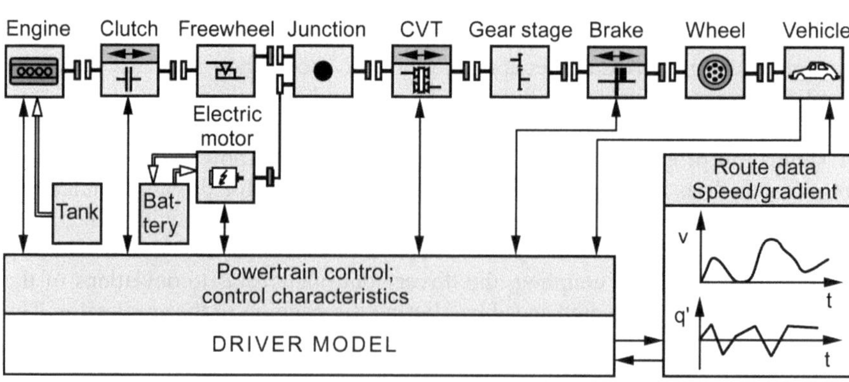


**Fig. 14.6.** Hybrid powertrain made up of the powertrain modules [14.23]

Depending on the object of the investigation, the powertrain modules can be shown in finer resolution, or summarised in larger units. The driver model requires great care, since the driver has a major impact on energy consumption, emissions, performance and load profiles.

### 14.2.2 Route Data Set, Route Data Acquisition

It is necessary to acquire the vertical route profile for the driving simulation in order to simulate the gradient resistance. The height or

- gradient profile

of a route must therefore be known in the greatest possible detail.

- Horizontal route profile and
- road space information such as built-up areas, speed limits, carriageway widths etc.

can be used as additional decision criteria for the driver controller, or as parameters for traffic models and traffic simulation calculations derived from them. If the recorded pilot speed profile is also used as the

- specified speed for the simulation,

the type and engine of the pilot vehicle must be similar to that of the vehicle to be simulated. The route data recording device must therefore be mobile, and usable in any vehicle without major expenditure of resources. Some methods of recording route data are listed below:

- *Distance and speed measurement with MDS:*
  Microwave Doppler Sensors (MDS) register the distance and speed of the vehicle. The MDS analyses the frequency shift of a radar emission to the signal reflected by the road surface and then received [14.13–14.14].
- *Barometric height measurement:*
  With height profile measuring equipment based on the measurement of barometric air pressure, the vertical route profile is recorded over the route. The analysis software enables the air pressure to be converted into altitude in metres, applying various corrections [14.10].
- *Gyroscope systems:*
  Gyroscope systems are also used for three-dimensional mapping of topographic route data [14.2, 14.13]. Such inertial navigation systems provide precise results.
- *Satellite-based data acquisition using GPS:*
  The satellite based Global Positioning System (GPS) enables the three-dimensional position and the inherent speed of a vehicle in normal traffic conditions to be determined and recorded [14.17].
- *Coupled positioning systems:*
  Combination of two or more methods [14.13–14.14].

# 15 The Automotive Transmission Development Process

*Engineering design is creativity with discipline*

The purpose of this book is to present the development process for automotive transmissions in its totality. A product is only successful if people buy it!

A healthy product range (Section 15.1) requires product planning oriented to strategic goals (Section 15.2). The product development process (Figure 15.1) starts with product planning, taking into account the main boundary conditions of the product environment. Next follows the conceptual design phase, which is based on formulating the requirements list (specification); in this phase, solution variants are proposed and evaluated and the most suitable solution selected. This solution concept is then realised by general and detailed form design of all partial solutions. The new product finally is released for series production when all the documentation has been completed and tests have been successfully concluded.

All the functions of the value chain are involved in the Product Development Process (PDP) of a new transmission (Figure 15.2). A suitable project organisation ties all the interdisciplinary sub-processes together.

In order to meet the developmental goals and to control the progress of the project, the entire PDP is subdivided into clear stages, the release stages. Section 15.3 shows this procedure.

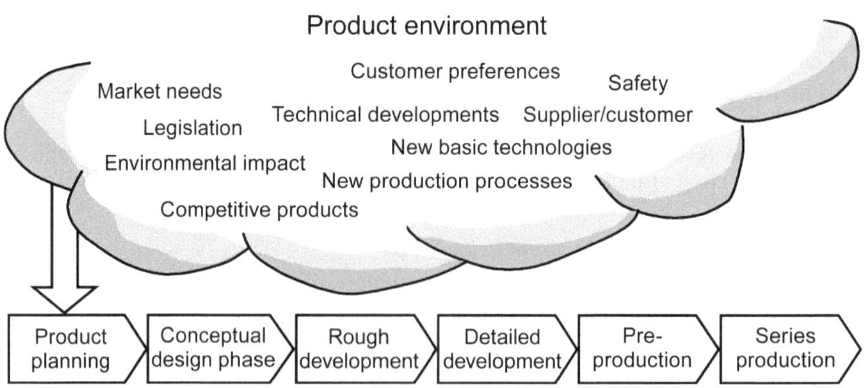


**Fig. 15.1.** Product environment and the development process

H. Naunheimer et al., *Automotive Transmissions*, 2nd ed.,
DOI 10.1007/978-3-642-16214-5_15, © Springer-Verlag Berlin Heidelberg 2011

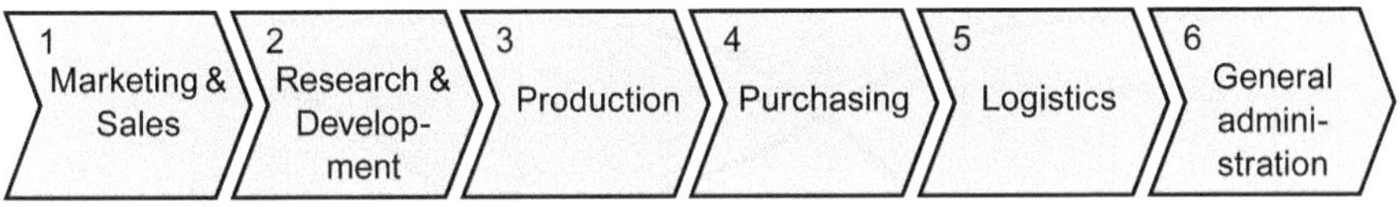


**Fig. 15.2.** Value chain

Finally, Section 15.4 explores the design process and systematic design.

## 15.1 Product Life Cycle

All products have a limited life. Every product is replaced sooner or later by a new one. There are several reasons for this:

- new technical developments offering improved functionality,
- more efficient methods of producing new products,
- fluctuating demand and fashion trends,
- consumer attitudes,
- legal and economic requirements and
- inadequate or inappropriate market policy.

All products pass through various life cycle phases during its production time, which can have different weightings (Figure 15.3). Companies have to be aware of this life cycle – i.e. they have to know where each product is in its life cycle, so that action is taken in good time to develop new products.

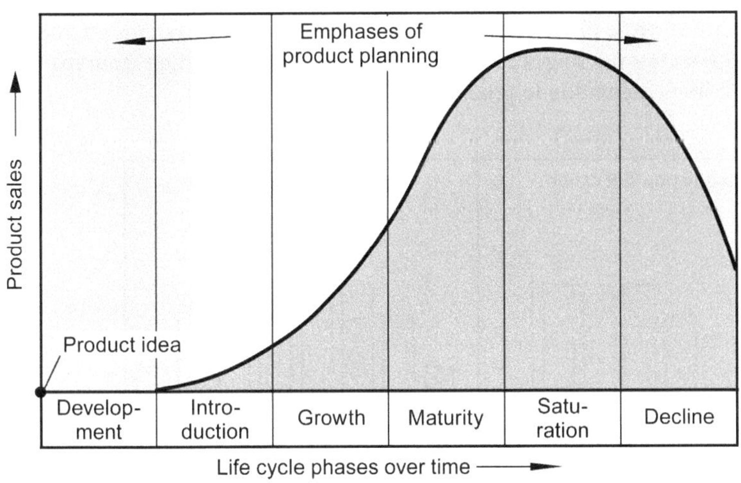


**Fig. 15.3.** Life cycle of a product

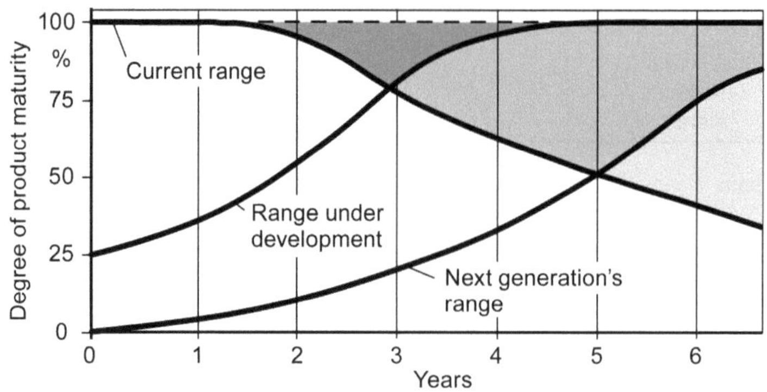


**Fig. 15.4.** Maintaining sales volume by continuously developing new products

The decline in sales of a product as it approaches the end of its life cycle has to be offset by timely development of new products (Figure 15.4).

Most companies produce a variety of products to achieve a balanced mix of sales volume and profit levels. The age structure of this product range has to be balanced, and it must not be allowed to become outdated. A company is only healthy if products no older than three to five years account for 50% of its turnover. Figure 15.5 shows the age structure of various product ranges, illustrating a healthy product range and an outdated product range.

The lifespan of various products can differ greatly, and depends very much on how well attuned the product is to consumer needs. With rapid advances in technological development, the pace of product innovation is continuing to quicken. Companies need to provide a balanced product range to satisfy market demands. They have to be in a position to quickly replace an outdated product with a new one. Product flexibility is essential for a company to remain competitive.

The product life cycle phases can be defined with the aid of systems engineering. Figure 15.6 shows the stages of the product life cycle proceeding from market needs, through implementation to processing of the waste product.

**Fig. 15.5.** Age structure of product ranges

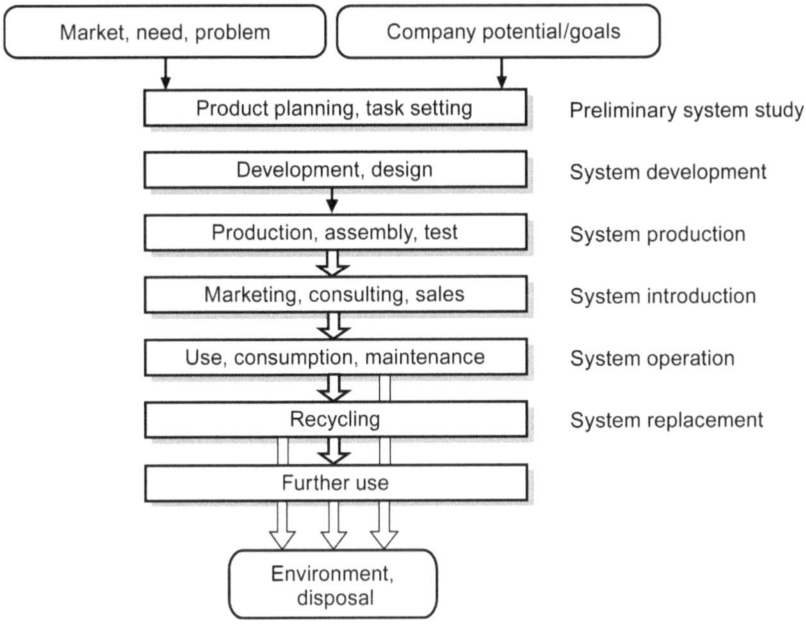


**Fig. 15.6.** Life cycle phases of a product (VDI guideline 2221) [15.18]

## 15.2 Product Strategy, Product Planning

The correct assessment of market, technical and technological trends is of the greatest importance. In order to avoid unsuccessful product developments, planning of new products must be systematic and permanent (take place in defined cycles). Future demands must be recognised, evaluated and new strategies and products derived from this at an early stage (see also Section 2.5 “Trends in Transmission Design”). For topics regarding strategy management please refer to the relevant literature.

Mistakes arise typically from inadequate research into what is required of the product, and inadequate formulation of the task. Product planning comprises “systematic integration, co-ordination and evaluation of all the product-relevant factors from the market, science, technology and industry which are aimed at optimal product development” [15.3].

*Product planning*

- seeks and promotes new ideas,
- investigates the environment of new products (market, legislation),
- assesses the feasibility of development projects and directs them and
- plans and monitors the development process.

Products can be developed for traditional markets or for new ones, using traditional technologies or new ones. Figure 15.7 shows a modified Ansoff diagram.

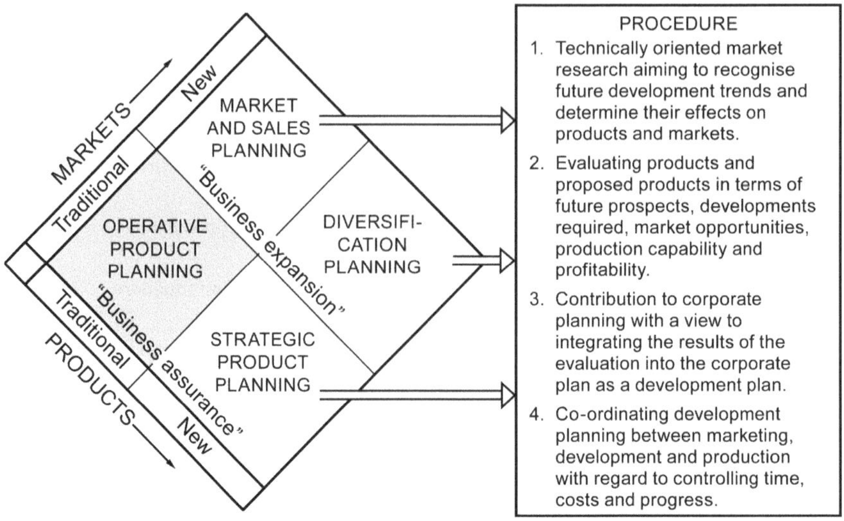


**Fig. 15.7.** Tasks of product planning (Ansoff diagram)

The most risky are new technology products (new products) destined for new markets.

## 15.3 Release Stages in the Product Development Process

In order to meet the developmental goals regarding functionality, performance data, quality, cost, deadlines etc., the product development process (PDP) must be based on an assurance system. To this end, the entire PDP is subdivided into manageable stage goals, called release stages or "quality gates". These are checkpoints at which previously stipulated requirements are evaluated with the help of measured values (definition of degree of maturity) with respect to quality and completeness. VDA Volume 4, Part 3 [15.17] describes what such a process for quality assurance can look like.

Many companies have followed this suggestion or have defined comparable sequences for their PDP. Based on VDA Volume 4, Part 3, Figure 15.8 shows the individual process steps from the product idea to series production (SOP = start of production). The associated quality gates with their essential contents are shown as well as some of the activities involved in the interdisciplinary sub-processes.

Measured values and minimum requirements that must be met to successfully pass through a gate can be represented in the form of a checklist. Estimation of the degree to which the requirements have been fulfilled and a grading system assist in degree of maturity evaluation in the committee of experts. A corresponding reporting system (release reports) accompanies the PDP. A standardised procedure, mandatory responsibilities and permanent control of project progress guarantee transparency and that countermeasures are taken in a timely way in case of goal deviation [15.16].

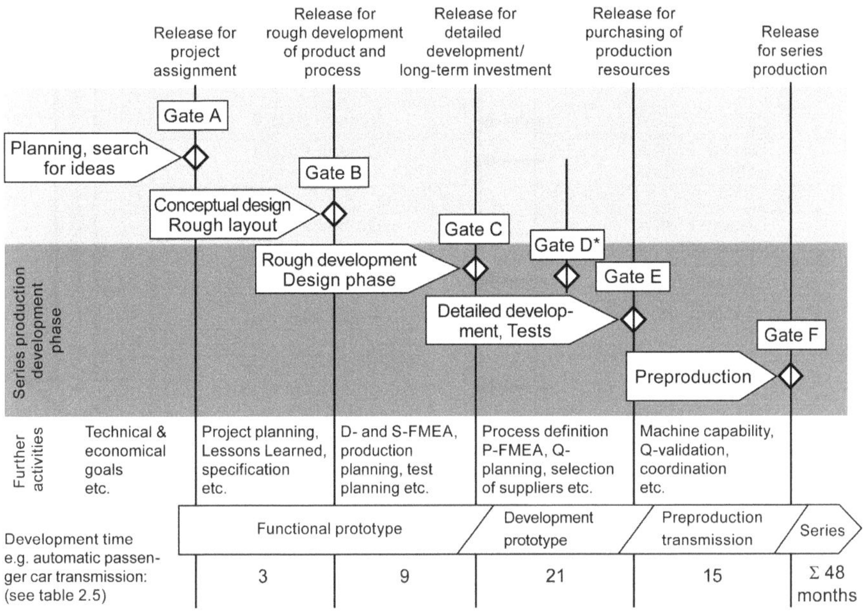


**Fig. 15.8.** Release system in the product development process, quality gates. From VDA Volume 4, Part 3 "Quality Assurance prior to Series Application". *Abbreviations: D-(Design-)FMEA, S-(System-)FMEA, P-(Process-)FMEA.* **) Gate D: Release for detailed planning of the production process*

It is advisable to distinguish between adaptive and new developments, and in case of an adaptive development, to apply a reduced release method.

The PDP is accompanied by experimental and simulative assurance steps (Figure 15.8). Functional prototypes help to test basic functions and interfaces in the early phases. Development prototypes are used to test endurance strength, for function assurance as well as for vehicle tests. Preproduction transmissions are used to carry out vehicle tests, to calibrate gearshift programs and to validate product and manufacturing (process capability verification). These transmissions are classified either as series prototypes, which are manufactured according to series specifications but not yet completely on series machines, and those that already are running completely on series equipment (standard production parts). Initial sample release, with which the customer releases the product for series production, takes place with standard production transmissions.

An essential part for reaching the required quality goals is the knowledge transfer ("Lessons Learned") from the transmissions in series production to the new development. Figure 15.9 shows how such a process might look and its incorporation into the PDP and names a few of the tools used.

Parallel to project processing, activities aimed at continuous improvement in the PDP must be carried out. The Continuous Improvement Process (CIP) is defined as the permanent and continuous testing and optimisation of working systems, sequences and results. Its goal is the elimination of deviations.

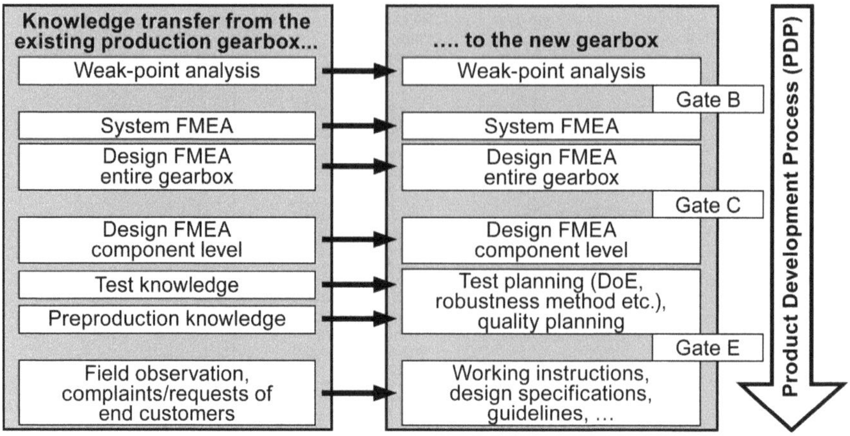


**Fig. 15.9.** Lessons Learned. Knowledge transfer from series production to the new development

Identified improvement potentials with respect to the product or product management are immediately implemented.

It is self-evident that the quality gate cycle of the vehicle manufacturer and the transmission supplier must be calibrated. This affects deadlines and measured values for the gates in particular. Standardisation and interlocking of assurance measures in PDP between OEM, the transmission supplier and sub-suppliers avoids duplication of work. This is especially true for costly experimental activities.

With respect to the large group of themes, development methods for productivity and quality improvement with the subtopics of variant management (complexity costs) and engineering change management, please refer to the relevant literature.

## 15.4 The Design Process and Systematic Design

The terms of the development process: "research", "development", "design", "form design" and "testing" are defined in Figure 15.10. Design is generally divided into four activities of varying degrees of complexity (Table 15.1). The creative/intuitive elements of the activity recede as the design process advances, whilst the deterministic activities increase (Figure 15.11).

**Table 15.1.** Main activities involved in the design process

| | New development | Further development | Product adaption | Product modification |
|---|---|---|---|---|
| Frequency | 15% | 20% | 30% | 35% |
| Difficulty | Very high | High | Moderate | Low |

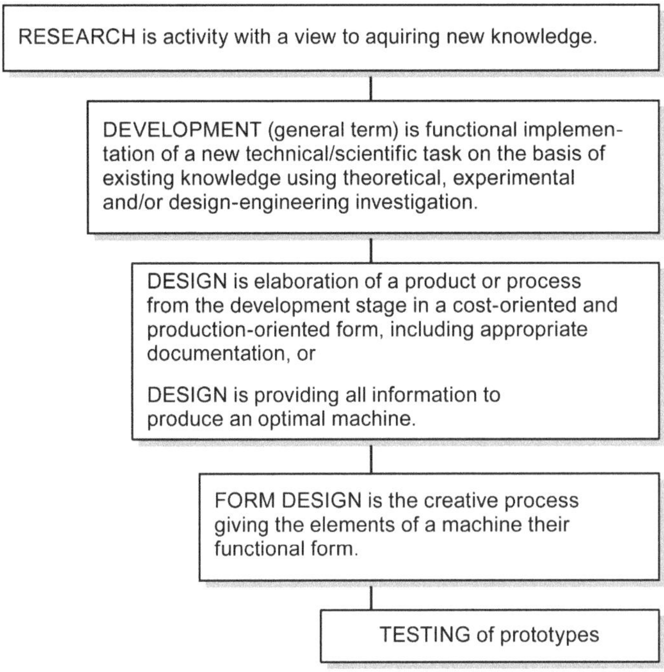


**Fig. 15.10.** Definition and hierarchy: Research, development, design, form design and tests

But the design process as a component of the development process includes not only developing of new products but also maintaining their value. Changes have to be made to the products throughout their life to optimise their function, adapt their performance, prevent failures and adapt to market trends (styling, design). Figure 15.12 shows the various activities involved as a proportion of the overall activities.

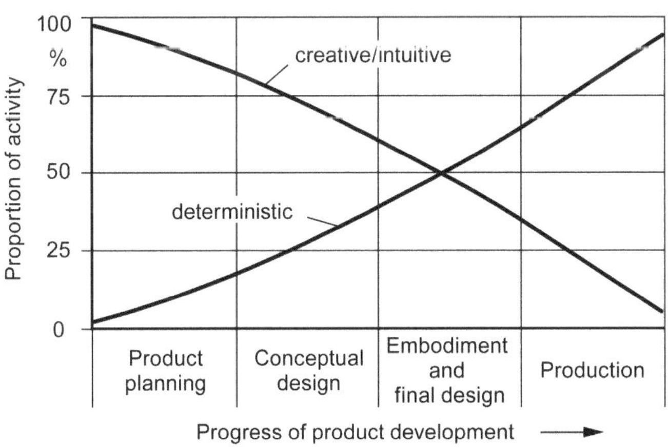


**Fig. 15.11.** Changing activity profile as product development proceeds

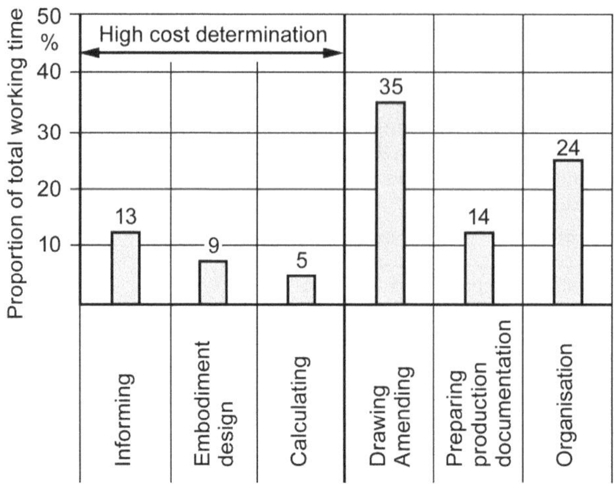


**Fig. 15.12.** Time required for typical development activities

The design engineer has the main responsibility for the product. In developing the product, he has to take into account its functional design, its operating, strength and wear behaviour, reliability and maintenance aspects, technological questions and cost considerations. He is also responsible for modifications to the product during service life as a result of the use of the product and market influences. He is responsible for the environmental impact and recycling of the product.

The necessary creative design is supported by *"Systematic design"*. *Systematic design cannot replace creativity!* Systematic design has been promoted especially in Germany since 1965 [15.4–15.5, 15.7–15.10, 15.11–15.15, 15.20]. In the Anglo-Saxon countries, the main emphasis is placed on creative design [15.1, 15.6].

Systematic design is of particular relevance for new and further development. For product adaptations and modifications, an abbreviated process is appropriate. The main phases in systematic engineering as defined in VDI guideline 2222 [15.19] (Figure 15.13) are: *planning, conceptual design, embodiment design and final design.*

These phases are separated by product evaluations, and decisions on the future of the project. In practice, this systematic development process is often adapted to the circumstances of the particular company.

Beitz [15.2] offers some new insights into design methodology. He suggests breaking down the self-contained, strictly systematic sequence into a number of individual "design modules". These individual modules can be combined individually for the specific development project. The individual design tasks identified by Beitz (Figure 15.14a-h) are as follows:

*a/* formulation and definition of the task,
*b/* setting up function structures,
*c/* searching for solution principles,
*d/* combining partial solutions to achieve overall solutions,

*e/* selecting solutions,
*f/* general form design,
*g/* detailed form design and
*h/* detailing and final design.

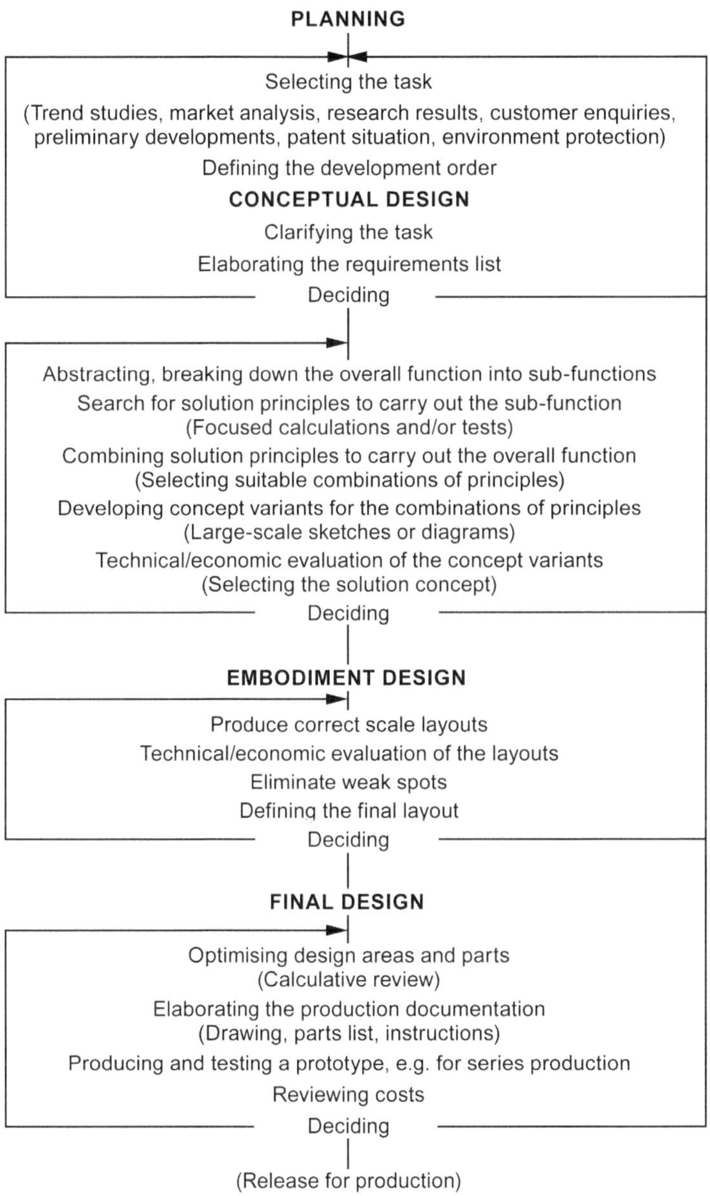


**Fig. 15.13.** Procedure for product development [15.19]

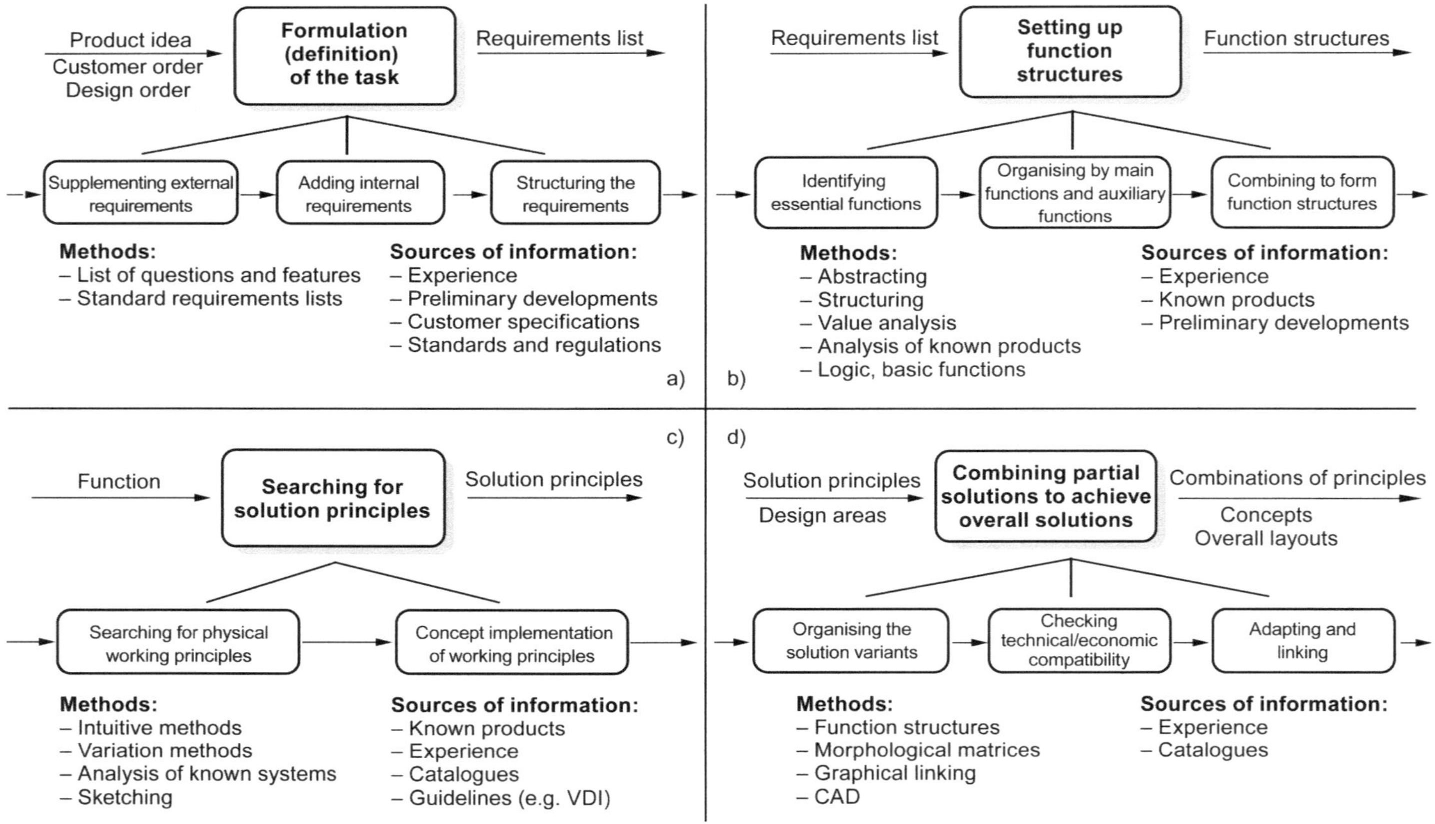


**Figure 15.14.** The modules of the systematic design process [15.2]

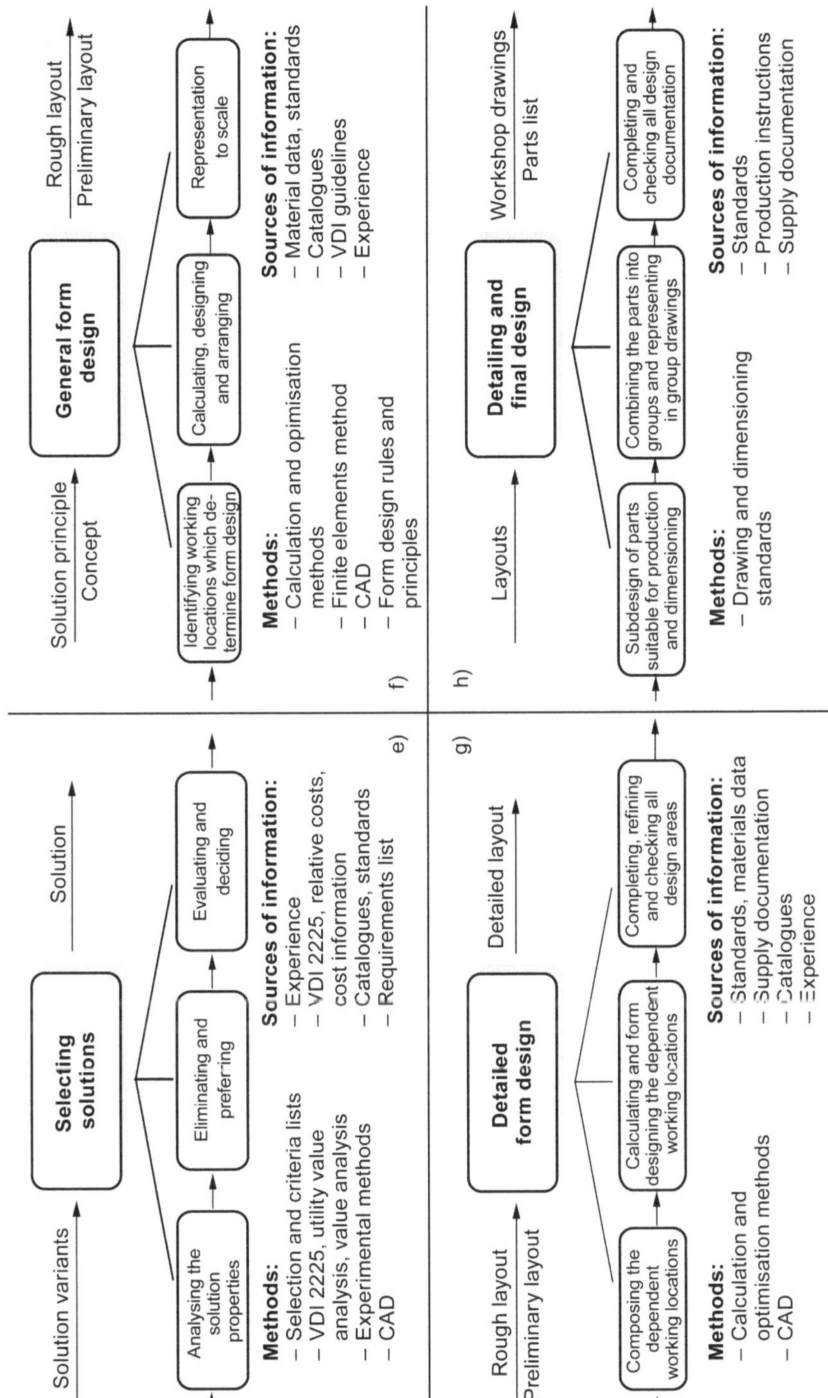


**Figure. 15.14.** *(continued)*

As the design process advances, the requirements are gradually actualised, which requires optimisation at each actualisation stage. The decision-making processes involved are a key factor affecting the efficiency of the design process. The optimal solution selected at each stage defines the parameters for the stage that follows. The golden rules are:

- If a solution seems too good to be true, it probably is!
- A successful new product that performs well comprises around 70% proven parts and only 30% newly developed parts!

### a/ Formulation of the Task

Clarifying the task helps to gather information on the requirements the solutions has to satisfy, and generates a requirements list. This specification includes both *functional* and *operational* requirements. Functional requirements are broken down into main functions and auxiliary functions; operational requirements include safety of operation, cost-effectiveness and human aspects.

The external form of a requirements list depends on the in-house circumstances, but should always include the various requirements the product has to meet (indicating whether they are demands or wishes), the department responsible and a description of the changes.

It is advisable to follow the *"Must – Want – Nice"* rule, so as not to overburden the specification with too many requirements. *"Nice"* features and even *"Wants"* may have to be sacrificed if the project is to remain feasible.

### b/ Setting up Function Structures

The design structure can be abstracted from the design drawing; the function and working structure of a technical system can then be derived by further abstraction.

The *design structure* is a schematic representation of the production version of the design. Unlike the function structure, it is dimensioned. The *function structure* is the simplification of a design to a function that can be described in mathematical terms, and can be regarded as a kind of "circuit diagram" of an assembly. Pahl and Beitz [15.11] propose that the *function* is the general relation between the input and the output of a system, for the purpose of performing a task (Figure 15.15). The main function of the vehicle transmission as a system is to convert torque and speed.

An *overall function* can usually be broken down into distinct *sub-functions.* (The sub-functions of a transmission are to enable moving-off, transmit power and control torque/speed conversion.) Certain functional constraints apply to combining the sub-functions to perform the overall function; often certain sub-functions must be concluded before other sub-functions can be performed. A functional analysis has to be carried out to find solutions for the sub-functions, and is embedded in the conceptual design phase of the development process.

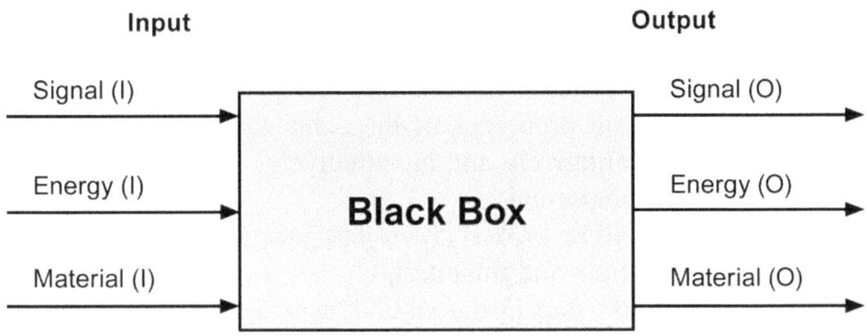


**Fig. 15.15.** Representation of a technical system as a Black Box

The *working structure* represents the combination of working principles from the various sub-functions. At the *working location,* the physical *working principle* is applied by arranging *working surfaces* and selecting *working motions* to carry out the function [15.2].

### c/–h/ Further Design Tasks

Solutions have to be found for the sub-functions (Figure 15.14c). Since there are normally several partial solutions available, they have to be narrowed down by examination and evaluation. The alternatives are then evaluated by carrying out comparisons of different variants, taking into account both technical and economic criteria. The process demands systematic variation of solutions, and critical, formal selection of the solution [15.19].

The principle solutions can be derived systematically, using one of several design methodology tools:

- *Design catalogues* can be used to select solutions. The information for the individual stages in the process is taken from catalogues by applying selection criteria. "Algorithmic selection procedure for designing with design catalogues".
- The systematic search for solutions can make use of classification schemes that represent solution catalogues organised by type and complexity. They provide a means of combining partial solutions to build overall solutions. Such tools are called *morphological matrices* [15.21]. The critical factor is selecting the classifying criteria.
- A more intuitive approach that relies on the dynamics of group interaction is brainstorming. This involves sharing ideas in a group drawn from different disciplines, in a non-judgemental context, to tease out possible solutions. This method is primarily intended for non-technical problems, but can be applied in design engineering.

Having established solution principles for the various sub-functions, a solution field can be compiled. Various solution principles can be derived for the overall solution by alternative combinations (Figure 15.14d). The main problem with the combination method is deciding which principle combinations do not clash. The solution field is constrained by considerations of avoiding clashes.

There then follows an evaluation, selection and decision-making process (Figure 15.14e). The appropriate solutions are selected from the range of solutions available, using systematic, verifiable selection procedures. This process is assisted by compiling selection lists. The main properties of the combinations of principles proposed must be considered qualitatively and quantitatively. Selecting the right evaluation criteria is particularly important.

The subsequent design tasks can be broken down into general form design, detailed form design as well as detailing and final design.

General form design is the first step in the embodiment design phase, and is characterised by large-scale sketches (Figure 15.14f). The main function carriers determining the overall design must be roughly determined, taking into account the given boundary conditions.

The next step is detailed form design (Figure 15.14g). The design of the main and auxiliary function carriers is defined by applying design rules, legal requirements, standards, calculations and test results.

When checking and evaluation have been completed, the layout is finalised and the parts list drawn up to form the basis for detailing (Figure 15.14h). Here, the individual parts are drawn in detail taking into account criteria for optimising shape, material, surfaces, fits and tolerances. The objectives at this stage are good material utilisation and detail design that is cost-effective and efficient to manufacture, taking into account applicable standards, and using as many bought-in parts or existing in-house parts as possible.

The final design involves structuring or organising the production documentation in the form of parts lists and drawings by arranging it function-oriented or production-/assembly-oriented. The production documentation is completed by finalising operating instructions, installation, assembly or transport instructions.

# 16 Transmission Manufacturing Technology

*Economical manufacture of quality products*

Automotive transmissions, like any other product, are in a constant state of competition, both between different transmission manufacturers and between different transmission designs. The product "transmission" must therefore not only satisfy its function but must also be economically producible and reach the desired level of quality. Basic requirements of product development are therefore knowledge of the manufacturing process and consideration of this even in the design phase.

In this chapter, mechanisms of several manufacturing methods in industrial transmission production will be presented in order to provide insight into geometry-generating processes. Several successive manufacturing processes (process chains) are required to produce component properties such as tolerances, surface qualities, strengths etc. If the obtainable accuracies of a method are insufficient to create the required tolerances, a further, more exact method must be connected to the process chain. For this reason, component-specific process chains are also represented, the design and optimisation of which as a whole can lead to much higher economical and technological savings than simply improving individual methods.

In addition, the later sections of the chapter will present organisational and methodical aspects of manufacture, e.g. work preparation, production systems or statistical process control. As a whole, the field of manufacturing technology is very broad, and knowledge thereof can be considerably deepened by pursuing the given literature.

Figure 16.1 shows the classification of manufacturing methods into 6 main groups in accordance with German standard DIN 8580 [16.1]. With respect to manufacturing chains, the components of a vehicle transmission can basically be divided into 5 classes:

1/ *Steel parts*:
e.g. shafts, planetary gear carriers etc.

2/ *Cast parts*:
housings, hydraulic valve plates, small cast components etc.

3/ *Geared components*:
spur gears, geared shafts, bevel gears, sun gears, planetary gears, ring gears etc.

4/ *Sheet metal parts*:
fine-blanked parts, formed parts etc.

5/ *Other parts*:
sealing elements, filters, standardised parts, add-on parts etc.

H. Naunheimer et al., *Automotive Transmissions*, 2nd ed.,
DOI 10.1007/978-3-642-16214-5_16, © Springer-Verlag Berlin Heidelberg 2011

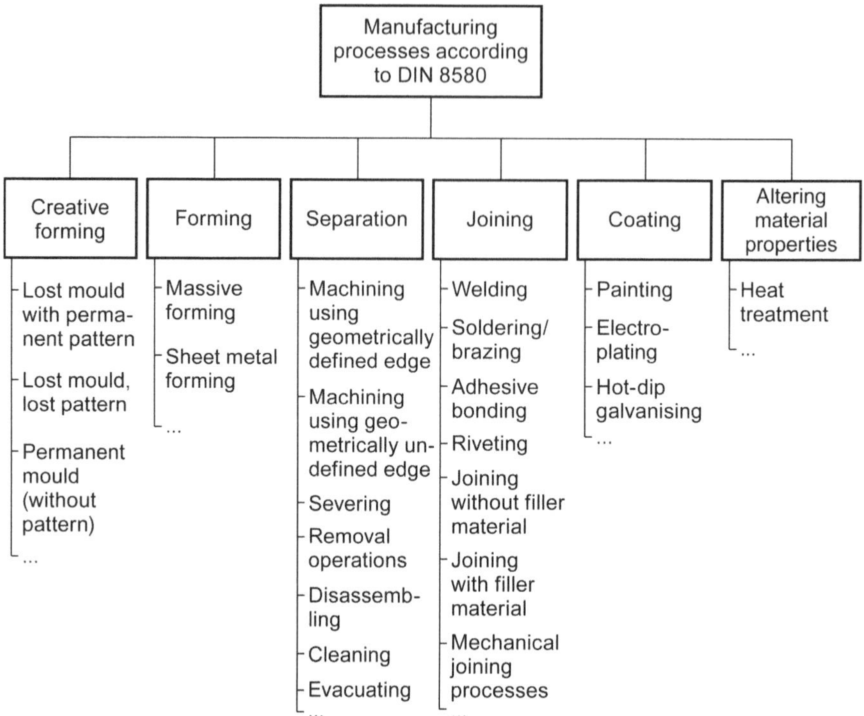


**Fig. 16.1.** Classification of manufacturing processes according to German standard DIN 8580 (example main groups and subgroups)

In the following, process chains of the previously mentioned component classes 1/ to 4/ will be considered using the example of passenger car automatic transmissions.

## 16.1 Process Chains for Steel Part Processing

In the case of steel parts, especially shaft components (Figure 16.2), the process chain is generally composed of the following steps:

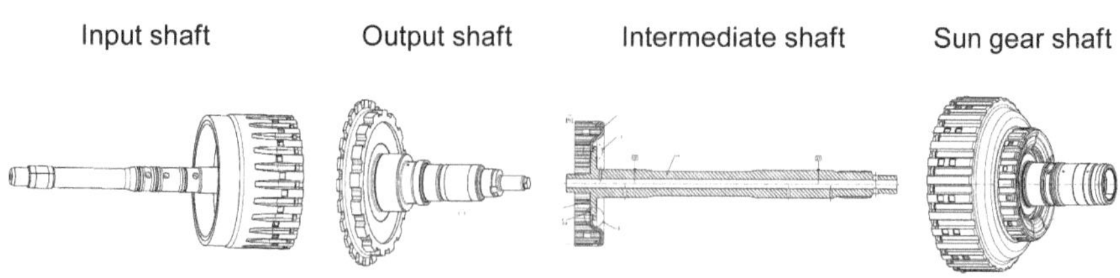


**Fig. 16.2.** Typical shaft parts of an automatic gearbox *(see also Figure 12.25)*

1/ soft machining (of a cast or forging blank),
2/ heat treatment (hardening),
3/ hard machining and, if necessary
4/ joining (joining of shafts with gearwheels, sheet metal parts etc.).

As far as the process chain is concerned, one must bear in mind that, in soft machining, the machine allowance is considered, which is necessary to compensate distortion during hardening and the subsequent hard machining (of functional and mating surfaces).

### 16.1.1 Soft Machining Methods

Machining with geometrically defined cutting edges is used for the soft machining (machining before heat treating) of steel parts. Examples include turning, milling and drilling [16.1, 16.4]. In addition to the main methods of producing "soft" component geometries, the soft machining chain usually also includes deburring and cleaning processes.

### 16.1.2 Heat Treatment Methods

#### *1/ Transformation and Case Hardening*

After the pre-contour is produced, the workpiece is heat-treated, which involves creating the required hard contour either with transformation hardening or case hardening.

In the case of transformation hardening, the material already contains enough carbon (0.3%) to obtain a corresponding increase in hardness by heating and quenching. Hardening of these steels can also be done in an integrated way in the production machine or in the production line (e.g. with laser-beam or induction hardening). In case hardening, carbon has to be added to the workpiece in a furnace atmosphere. The carbon diffuses into the surface layer [16.1].

#### *2/ Tempering*

The martensite structure arising from the hardening process is very brittle. For this reason, a workpiece is usually annealed after hardening, i.e. heated to temperatures between room temperature and annealing temperature (300°C to 600°C depending on the material composition) [16.1].

### 16.1.3 Hard Machining Methods

In the final processing of components, frequently machining methods with geometrically undefined cutting edges are employed, such as grinding, honing, lapping, vibratory grinding and jet machining (hard machining methods, i.e. machining methods after heat treating). Material is removed when more or less irregularly shaped grains made of hard materials come into contact with the material. High levels of accuracy (tolerance ranges) and surface qualities are possible with methods with geometrically undefined cutting edges. But the cutting efficiency is smaller than with methods with geometrically defined cutting edges.

#### *1/ Grinding*

The most important grinding methods are standardised in German standard DIN 8589 and international standard ISO/DP 3002/V. In the case of grinding methods with a grinding wheel, there are also special process variants, such as tooth flank grinding, grinding with continuous dressing, high-speed grinding, cutting-off and high pressure grinding [16.8].

#### *2/ Honing*

Honing (German standard DIN 8589, Part 14) is defined as machining with geometrically undefined cutting edges, whereby the many-edged tools perform a cutting motion consisting of two components. Between the tool and the workpiece, there is a change in the direction of the longitudinal motion, which is overlapped with a second motion (transversal for surfaces and rotational for cylinders or piston surfaces). Obtained surfaces exhibit parallel, intersecting marks.

#### *3/ Lapping*

German standard DIN 8589 defines lapping as machining with loose grain distributed in a fluid or paste (lapping compound), which is applied to a usually shaping counterpart (lapping tool), whereby the cutting paths of the individual grains are as undirected as possible [16.8].

#### *4/ Other Hard Machining Methods*

Besides machining with geometrically undefined cutting edges, there are other methods used in transmission production such as hard turning, spark erosion respectively EDM (Electro Discharge Machining) or ECM (Electro Chemical Machining).

## 16.2 Process Chains for Cast Part Processing

In casting, the component is either created directly and ready-to-install (net-shape) or nearly ready-to-install (near-net-shape). In the latter case, machining methods are then used to create interfaces to other components (e.g. fitting surfaces or fitting bores, sealing faces). One common process chain for manufacturing cast parts consists therefore in casting in combination with machining.

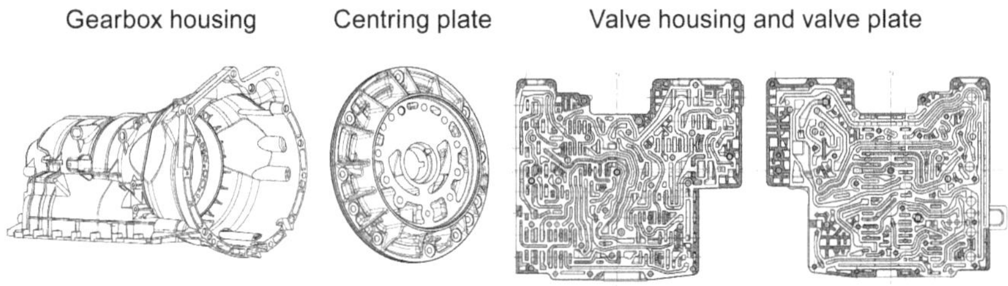


**Fig. 16.3.** Typical cast parts of an automatic transmission *(see also Figure 12.25)*

The advantages of casting consist on the one hand in the broad range of materials, on the other in the complex geometries directly producible in one process step. Figure 16.3 illustrates a typical component spectrum of cast parts for automatic transmissions.

### 16.2.1 Casting Methods

High volume manufacture of transmission components is accomplished with die casting. In die casting, molten metal is injected into a permanent steel mould under high pressure and at high velocity (Figure 16.4).

The resulting components exhibit a comparably high dimensional accuracy (about 0.1–0.4% for a nominal dimension up to 500 mm depending on the material and geometrical shape) as well as smooth and clean surfaces. Subsequent work, often machining, is only required on functional surfaces with high accuracy requirements (fitting surfaces).

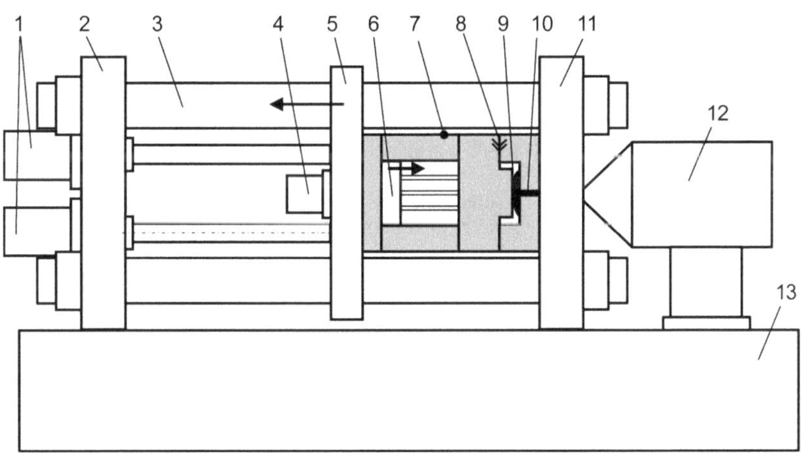


**Fig. 16.4.** Schematic view of a die casting machine. *1* Closing cylinder; *2* head plate; *3* column; *4* ejector cylinder; *5* movable die plate; *6* ejector packet; *7* die cast tool; *8* parting plane; *9* mould cavity; *10* molten material; *11* fixed die plate; *12* injection unit; *13* machine bed

Aluminium and magnesium alloys are the most relevant for vehicle transmissions. With aluminium alloys, cast parts of up to 50 kg can be made. The service life of a mould amounts to up to 80 000 casts [16.6]. See also Section 11.4 "Gearbox Housing".

### 16.2.2 Machining Cast Parts

Figure 16.5 provides an overview of manufacturing systems of various complexities that can be used to machine cast parts. The machining method used for the often very complex transmission cast parts is often milling or drilling. For machining, the workpiece must be fixed on the machine table (set-up). To machine individual locations, different tools are required, which can be changed either manually or automatically (centre) depending on the machine type. The workpiece can be changed either manually or automatically from a storage location and then clamped (cell). Since with one clamping set-up the tool cannot reach all machining locations, processing must take place over the course of several set-ups. Flexible manufacturing systems (FMS) and transfer lines are especially suitable for this. Whether machining on simpler machines (for transmission cast parts usually centres) or on more complex machines (transfer lines or FMS) is more economical depends on the number of pieces of the component required. Flexible manufacturing systems are the interconnection of several cells.

After machining, the workpiece must be cleaned, deburred and examined. After examination of the dimensions and surface tolerances, there is a tightness test to make sure the components do not leak.

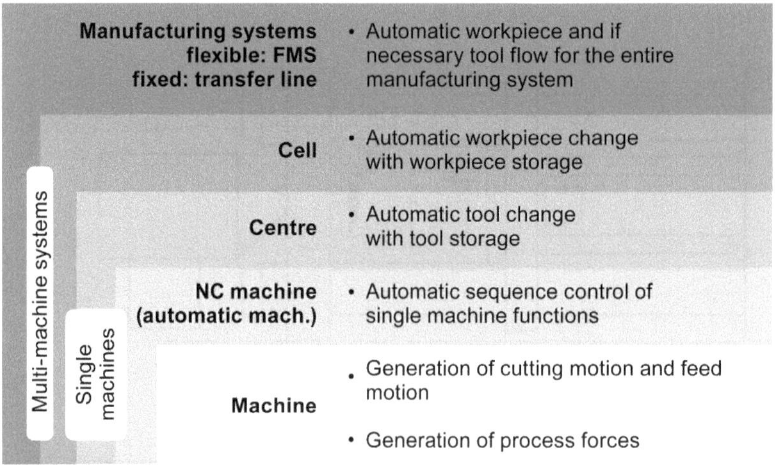


**Fig. 16.5.** Designation of manufacturing systems for cast part processing according to the degree of automation [16.9]

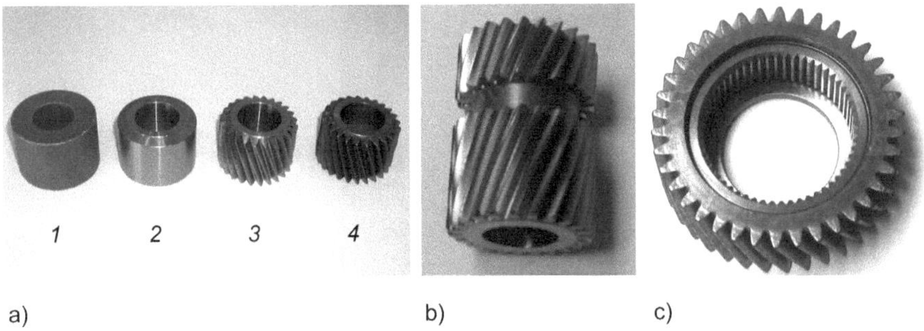


**Fig. 16.6.** Examples of gearwheels [16.10]. *a* Machining steps of a planetary gear: *1* forging, *2* turned blank, *3* soft hobbed gear, *4* hardened and finished (ground) gear; *b* double planetary gear of a Ravigneaux gear set; *c* sun gear (inside: broached spline)

## 16.3 Process Chains for Gear Machining

Gears are an elementary component of vehicle transmissions (see also Section 7.1 "Gearwheel Performance Limits"). The manufacture and machining of gears is therefore a central task in transmission production. There is on the one hand the possibility of giving the tooth contour its final geometry already in the soft state, in which case the gear need only be hardened. The advantage of this is the short process chain and the resultant lower costs. On the other hand, inaccuracies caused by distortion due to hardening are usually unacceptable, in which case hard machining must follow the hardening process. Figure 16.6 shows a planetary gear in various machining steps as well as a planetary and a sun gear.

### 16.3.1 Soft Machining Methods

The dominant machining method in the manufacture of externally toothed, cylindrical gears is hobbing due to its high level of economic efficiency.

#### *1/ Hobbing*

In hobbing, the pairing of a worm with a worm gear is simulated, whereby the tool used is a worm interrupted by chip flutes and the worm gear is the workpiece to be produced (Figure 16.7). The synchronized rotating motions of the hob and the workpiece remove the material. Additionally, the translational motion of the hob along the workpiece axis (axial feed) is superimposed. In this case, we refer to the process as axial hobbing.

The gating technique also varies depending on the machining task. The tool can either be fed directly, or (e.g. when the toothing starts in the middle of a shaft) first there is a radial feed until the depth of immersion of the hob in the workpiece is reached. In industrial production, dry machining has become widespread, i.e. coolant emulsions are not used.

**Fig. 16.7.** Hobbing kinematics [16.10]. *1* Tool axis; *2* hob (tool); *3* hob rotation; *4* axial feed; *5* radial feed; *6* workpiece; *7* workpiece axis; *8* workpiece rotation

### *2/ Broaching*

Broaching is especially important for manufacturing internal gearings in large quantities. The method consists in cutting with a multi-toothed tool, the cutting teeth of which lie in a row and are each staggered by thickness of one cut. This replaces the feed motion (Figure 16.8). The cutting motion is translational (internal broaching, external broaching), in special cases also helical (helical broaching). The advantage of broaching is its high cutting efficiency, since the chip volume per tool tooth is large due to the large width of cut and despite the small thickness of cut. Moreover, usually several teeth are cutting simultaneously.

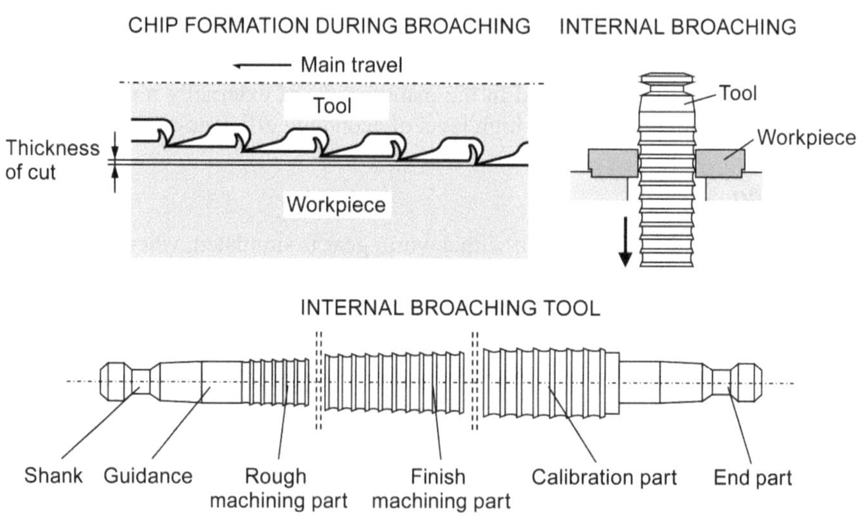


**Fig. 16.8.** Broaching

A further advantage of broaching is that it provides high surface qualities and accuracies, and that IT7 tolerances are complied with. This method is only economical in mass production due to the high costs of tool manufacture and preparation, particularly as every altered workpiece shape necessitates a new tool [16.5].

There are many other methods for the soft machining of gears, such as planing, shaping or shaving, which are not pursued here. Further information can be found in the literature [16.5].

### 16.3.2 Hard Machining Methods

In gear grinding, the shape of the evolvents can be produced either by an exact profiled grinding wheel (form grinding) and/or by means of a relative motion between the workpiece and the tool (hob grinding). In form grinding, the tooth gaps are machined one at a time with a grinding wheel dressed with the specified shape. In hob grinding, the tooth shape is created by simulating the rolling kinematics between the gear rack and the gear with a superimposed cutting motion [16.8].

Power honing of gears is a process variant of form grinding developed in the 1990s. In it, material is removed by rolling the workpiece/gear in a hollow wheel acting as the tool. The tool consists either of grinding wheel ceramics (usually corundum = aluminium oxide) or of metallically bonded cubic boron nitride (CBN). Engagement takes place by means of a relative angle of attack between the workpiece axis and tool axis. Infeed is radial. In this way, rolling creates a relative motion between the tool and the workpiece which leads to chip formation. The advantage of power honing is its short machining times, making it very economical. Since every change in workpiece geometry requires a new tool/dressing tool, this method is especially effective for medium-sized and large production numbers.

In all, there are a large number of process variants in gear machining with grinding. Every method can be used with varying machining strategies (kinematic machining sequence: machining directions, depths of cut etc.), with which the quality of the process results can be influenced. The method is selected with an eye to the required component quality (geometrical accuracy and surface quality) and of course to economical considerations (machining duration as well as costs of machines, tools and machining).

## 16.4 Process Chains for Sheet Metal Machining

Sheet metal parts allow for very low weight design solutions. Figure 16.9 shows a spectrum of typical sheet metal parts for automatic transmissions. See also the description of the automatic transmission example ZF 6 HP 26 in Section 12.1.4 and Figure 12.25.

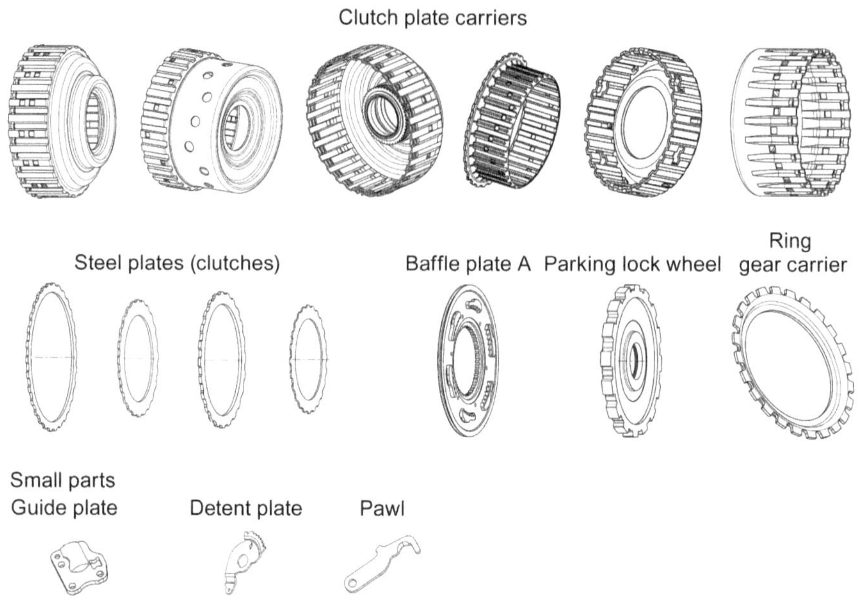


**Fig. 16.9.** Spectrum of sheet metal parts of an automatic transmission *(see also Figure 12.25)*

Sheet metal machining can be subdivided into sheet separation and sheet forming [16.7]. The methods used in the case of the illustrated components include deep drawing, cold rolling (forming) and fine-blanking (separation).

## 16.4.1 Sheet Separation

In sheet processing, separation methods are of great importance, as the initial step of the manufacture of a sheet metal part is almost always associated with separation processes. Separation is necessary both to produce blanks (e.g. a round blank) as well as the final workpiece contour.

According to German standard DIN 8588, cutting methods belong to the "severing" group, whereby one distinguishes between shearing (blanking), knife blade cutting, cutting with approaching blades, slitting and breaking. Blanking is of particular importance here. A tool is required for cutting that consists of the main components cutting punch and blanking die (Figure 16.10). The cutting forces are transferred from the face of the punch and from the blanking die to the workpiece. As a result, the sheet bends through between the punch and the die. This workpiece deformation can be avoided by a process variant, fine-blanking. In fine-blanking, there is an additional press plate (guide plate) that holds the sheet during the cutting process by means of a v-shaped impingement ring. Fine-blanking produces much more accurate components than blanking with not only less deformation but also much smoother edge surfaces.

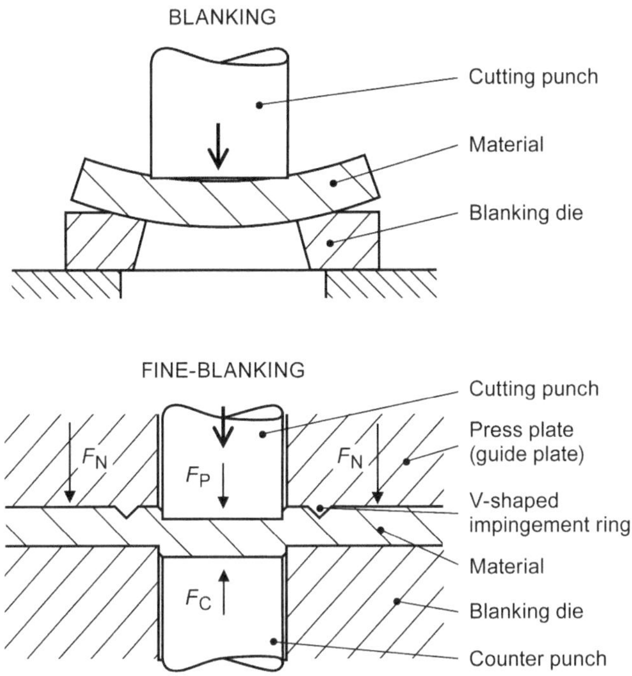


**Fig. 16.10.** Sheet separation: blanking and fine-blanking [16.7]

## 16.4.2 Sheet Forming

Concerning forming techniques, German standard DIN 8582 distinguishes between massive forming and sheet metal forming. The latter can be characterised by the fact that extensive hollow parts are formed without significantly changing the uniform initial wall thickness.

Deep drawing is the most important manufacturing method for producing sheet metal workpieces with general three-dimensional geometries. In the case of the multi plate carriers shown in Figure 16.9, fine-blanking and deep drawing were combined. The complex geometry is then produced by gradual forming, whereby each forming stage has its own tool. The tools are series-connected in a row and are passed by components from beginning to end. Tools that form and cut in one pass are called compound dies.

Sheet metal pots serving as inner or outer multi-plate carriers are connected to shafts in further process steps by welding (see also Figure 16.2). Special care must be taken that the tolerances along the process chain "sheet metal machining – welding" are observed.

## 16.5 Manufacturing and Factory Management

In conclusion, we will look at some organisational and methodical aspects of manufacture.

### 16.5.1 Work Preparation and Planning

Work preparation comprises the totality of all measures required to manufacture the designed product (industrialisation). According to the German Committee for Economical Manufacturing (AWF), production planning is a branch of work preparation. Production planning can be seen as a link between design and production. The range of tasks involved in production planning includes:

- manufacturing consultation of the design department,
- technology and method planning,
- material planning,
- sequence and time planning,
- resource planning and
- cost planning.

The central task of production planning is creating/verifying the production schedule, i.e. the operation sequence with all necessary information on components, machines, processing methods, tools, processing times and so on. It is of decisive importance that individual processes are timed to each other as much as possible in order to avoid losses due to idle times.

### 16.5.2 Production Systems

A production system consists of all the methods and principles that guide a production. A production system is therefore a kind of framework, a directive foundation, on the basis of which a company works. Principles of "lean production" are firmly anchored in production systems. This means that every working step serves to add value, whereby value is always added when the component can be sold at a higher price than before the alteration. One tool for analysing value added within a production process is "value stream mapping".

Standardisation is an important element in production systems. It standardises operations which contributes to quality assurance. Also, the best known operating method is documented, serving as the comparative basis for improvements. Since there is no one perfect process, there is always a need for improvements (Continuous Improvement Process (CIP)). Further information on production systems can be found in the extensive literature on the topic.

### 16.5.3 Statistical Process Control in Manufacture

The goal of every production is to create quality products at the lowest possible cost. Quality is expected, but it does not come about automatically. It must be planned at the beginning of the Product Development Process (PDP) and borne by all line functions of the value chain (sales, development, purchasing, production planning, production and logistics – see also Figure 15.2). Management decisions such as capital budgeting and approval, programme selection, personnel selection and training as well as company culture play a large role. Quality defects, better known as “errors”, resulting from insufficient quality not only cost money when they are removed but also image.

The goal is a quality assurance strategy to prevent failures. Inspection and sorting is not rewarding in the realm of complex products like automotive transmissions. The entire manufacturing process with all its individual processes must be set up and controlled such that no errors occur; if they do occur, they must be recognised immediately and prevented from propagation in the process. Otherwise the stipulated failure rates in ppm (parts per million) are not realisable. The following will present Statistical Process Control (SPC) very simple and generally as a method for preventing errors in manufacture. The reader is also referred to the substantial literature and guidelines on the topic (e.g. [16.2, 16.3]).

Every individual operation is a sub-process and part of the manufacturing process. In vehicle transmission production, many different factors are involved, such as machine, materials, method, man and milieu (Ishikawa’s classic 5 Ms). The resulting variety of influences on the manufacturing process, generally running normally and set for a specified value, are of both random nature (e.g. tool breakage) and systematic nature (e.g. tool wear). Statistical methods are therefore required for process control. The normal distribution, as a type of distribution to explain the SPC by model, can be selected for the statistical analysis of a manufacturing process.

In practice, a number of distribution models are important, all forming the basis for SPC. Non-random disturbances are recognised by statistical process control as errors and discarded by controlling the manufacturing process.

Quality always refers to a demand, to the conformity of target and actual performance. The demand of a feature is specified with nominal value and threshold values (UT = upper tolerance, LT = lower tolerance, Figure 16.11). If the typical distribution of a feature value is known from statistical preproduction investigations, a correspondingly large sample, taken from the manufacturing process in accordance with precisely formulated rules, is sufficient to make a statement regarding the working of the manufacturing process and thus the feature value.

The distribution and position of a manufacturing process is judged with the help of measurement results taken from statistical analyses (short-term: machine capability, long-term: process capability). The following quality indexes show this.

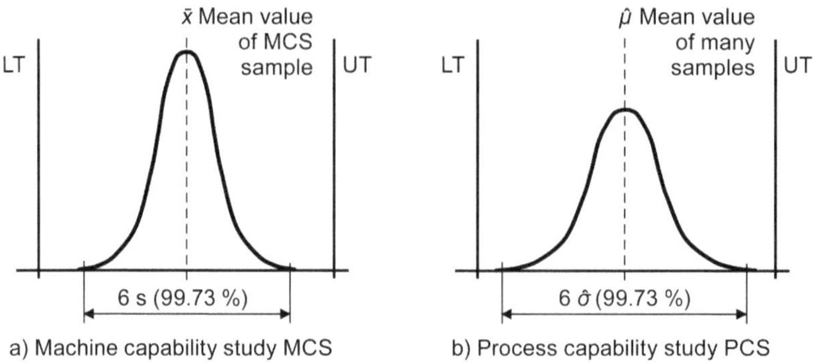


**Fig. 16.11.** Normal distribution and capability indexes. *a* Machine capability $c_m$ determined from short-term study and *b* process capability $c_p$ determined by long-term study

We differentiate between:

- process potential with
  1/ machine capability $c_m$,
  2/ process capability $c_p$,
- process capability with
  1/ critical machine capability $c_{mk}$,
  2/ critical process capability $c_{pk}$.

### *1/ Machine Capability, Indexes for Examining Machine Capability*

Machine capability is determined by the machine capability study (MCS). The MCS is a short-term examination. It is carried out without readjustment of the machine, with the same operator and the same raw material in order to eliminate process influences. It is carried out before the introduction of SPC at the workplace, when a new method has been installed or when a process has proven incapable. The MCS determines the typical distribution produced by the machine in the production of features. It must be carried out for important features that are produced at a part in one operation. The two following, mathematically determined quality indexes provide information on machine capability (Figure 16.11a):

$c_m$ is the machine capability index with respect to spreading. It provides information about the total width of the distribution in proportion to the tolerance width. The factor $c_m$ must be $\geq 1.67$:

$$c_m = \frac{\text{Tolerance width}}{\text{Machine spread width}} = \frac{\text{UT} - \text{LT}}{6s}, \quad c_m \geq 1.67 \,. \tag{16.1}$$

$c_{mk}$ is the position of the mean value of the distribution in proportion to the tolerance limits. It provides information concerning the position of the distribution to the tolerance limits. The factor $c_{mk}$ must be $\geq 1.67$:

$$c_{mk} = \mathrm{MIN}\,(c_{mu}\,; c_{ml}), \quad \text{with } c_{mu} = \frac{\mathrm{UT} - \bar{x}}{3s}, \quad c_{ml} = \frac{\bar{x} - \mathrm{LT}}{3s}. \tag{16.2}$$

### 2/ Process Capability, Indexes for Examining Process Capability

Process capability is determined by a process capability study (PCS). The PCS is a long-term examination. The PCS determines what influences the individual factors (man, machine, materials, method and milieu) that make up the process have on the final result (dimensional accuracy of the manufactured feature). The two following, mathematically determined quality indexes give information about process capability (Figure 16.11b):

$c_p$ is the process capability index with respect to spreading. It gives information about the total width of the distribution (which the process creates) in proportion to the tolerance width. The factor $c_p$ must be $\geq 1.33$:

$$c_p = \frac{\text{Tolerance width}}{\text{Process spread width}} = \frac{\mathrm{UT} - \mathrm{LT}}{6\hat{\sigma}}, \quad c_p \geq 1.33\,. \tag{16.3}$$

$c_{pk}$ is the position of the mean value of the distribution (that the process creates) in proportion to the tolerance limits. It provides information concerning the position of the distribution to the tolerance limits. The factor $c_{pk}$ must be $\geq 1.33$:

$$c_{pk} = \mathrm{MIN}\,(c_{pu}\,; c_{pl}), \quad \text{with } c_{pu} = \frac{\mathrm{UT} - \hat{\mu}}{3\hat{\sigma}}, \quad c_{pl} = \frac{\hat{\mu} - \mathrm{LT}}{3\hat{\sigma}}. \tag{16.4}$$

### Practical Application at the Workplace

If the typical distribution is known, the action limits, important for the machine operator, can be defined (UAL = upper action limit, LAL = lower action limit). The values of the manufacturing process are measured continuously during series production and then compared to these action limits (Figure 16.12). Plotting the check information over time shows the process development, which helps to create a more targeted intervention. It is assumed that the sample withdrawn for process control originates from the typical distribution.

Sometimes, individual processes do not fulfil the abovementioned capability demands ($c_{pk} = 1.33$). The process is then not capable and requires special safeguarding measures such as 100%-testing.

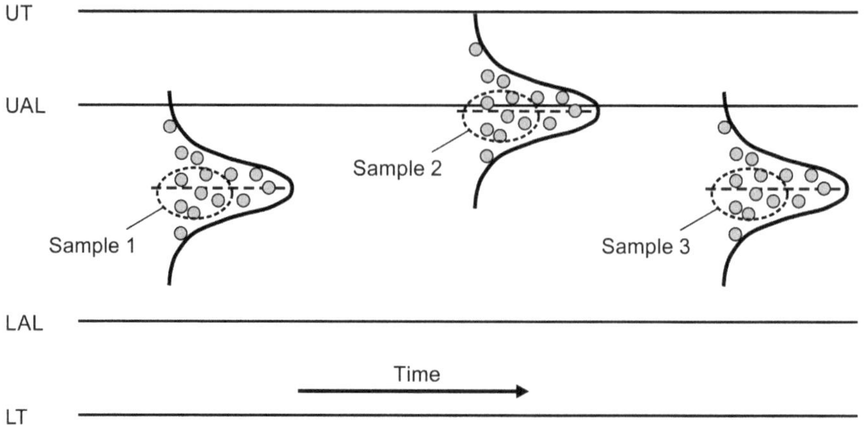


**Fig. 16.12.** The basic idea of process control with SPC

*Example:* Sample 2 is alright with respect to the typical distribution. Two measurements however are above the UAL. There must be an intervention into the manufacturing process, e.g. tool correction.

TS16949 thoroughly describes the requirements of a functioning QM system with respect to structure and organisation of all organisational units and processes in the company. These requirements are the basis for the orientation of QM systems for suppliers in the automotive industry.

# 17 Reliability and Testing of Automotive Transmissions

*The automobile and its components are an outstanding example of complex technologies combined with a high degree of reliability*

Legislation (e.g. that relating to product liability and environment protection), higher product complexity, shorter innovation cycles and increased customer expectations require more and more efforts to produce reliable, safe products. To achieve this, certain basic rules need to be observed, even during the development of the product. The most important basic rules to be observed during the development of vehicle transmissions to ensure reliability are set out below:

- precise specification,
- as few components as possible,
- elimination of risk parts,
- interchangeability of wearing parts,
- computer simulation of practical use,
- investigating the dynamic behaviour of the powertrain,
- early component tests,
- comprehensive test bench and road testing,
- most rigorous quality assurance in-house and with suppliers and
- random inspection of production.

An automotive gearbox cannot be regarded as a single component. It must rather be regarded as a complex system comprising many different components. The individual components are accordingly subjected to the most varied influences and stresses. The reliability of the vehicle transmission system is therefore determined by numerous influencing variables. These influencing variables can basically be divided into two categories, "internal" and "external" (Figure 17.1).

Just as there are numerous influences, so are there also several criteria for defining a reliable vehicle transmission, the most important being:

- *The transmission must have a high average service life expectancy and*
- *The transmission must have practically no premature failures.*

H. Naunheimer et al., *Automotive Transmissions*, 2nd ed.,
DOI 10.1007/978-3-642-16214-5_17, © Springer-Verlag Berlin Heidelberg 2011

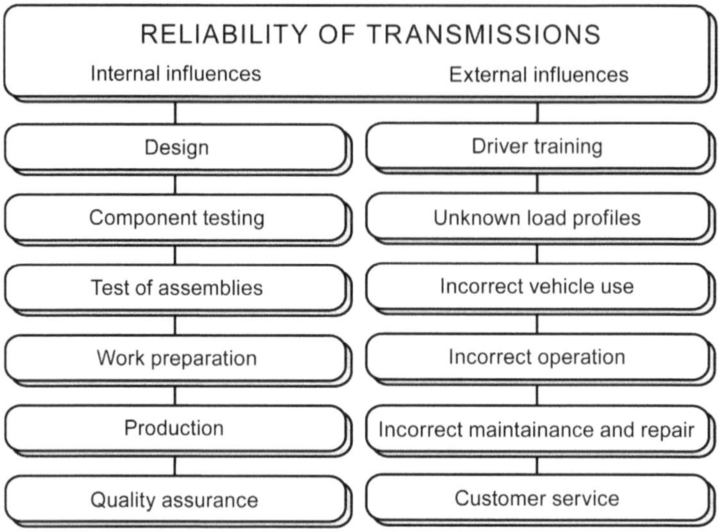


**Fig. 17.1.** "Internal" and "external" influences on the reliability of a transmission

Based on these requirements, two basic measures to improve the reliability of automotive transmissions can be derived:

- *The permissible stress of the weak elements must be increased and the spread of these permissible stresses narrowed down and*
- *Quality assurance systems must be implemented to minimise production and assembly errors, see Section 16.5.*

In order to take account of the above requirements in development, it is necessary for the design engineer to have a number of methods available to him for calculating the reliability of components and complete component systems, or at least estimating them. This chapter discusses the necessary mathematical-statistical principles and principles of reliability theory necessary for this purpose.

## 17.1 Principles of Reliability Theory

The principles of reliability theory are set out below. The reader is referred to the relevant literature for a more detailed treatment of this topic [17.1].

### 17.1.1 Definition of Reliability

It must be possible to define the variable "reliability" qualitatively and quantitatively for the purpose of objective assessment and calculation. The definition of technical reliability in the VDI guideline 4001 [17.15] is:

*RELIABILITY*
*is the probability that a product will not fail during a defined period of time under given functional and environmental conditions.*

### 17.1.2 Statistical Description and Representation of the Failure Behaviour of Components

The service life $t$, describing the failure behaviour of a component is not to be interpreted as a variable to be determined discretely. It is rather a random variable that is subject to a particular spread [17.12].

Figure 17.2a shows a histogram of failure times for a service life test. In the histogram the spread $t_{max} - t_{min}$ is divided into an appropriate number of intervals, and the failures observed allocated to the intervals. The height of the bars then represents the total number of failures occurring in that interval. As the interval width reduces, the contour of the histogram can be approximated by the curve of the density function $f(t)$ (Figure 17.2b).

If the failures noted are added with consecutive interval number, this results in the histogram of cumulative frequency shown in Figure 17.3a. The contour of this histogram can in turn be approximated by a smooth curve with reduction of the interval width. This curve is referred to in statistics as the distribution function $F(t)$, and in reliability theory as the failure probability $F(t)$. Between the density function $f(t)$ and the failure probability $F(t)$, the following relations apply

$$F(t) = \int f(t)\,\mathrm{d}t \qquad \text{or} \qquad f(t) = \frac{\mathrm{d}F(t)}{\mathrm{d}t}. \tag{17.1}$$

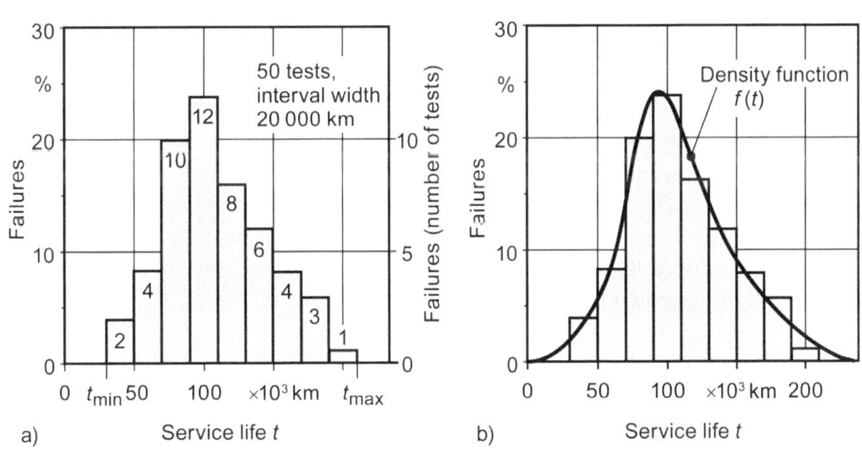


**Fig. 17.2.** *a* Histogram of failure frequency of a service life test; *b* failure frequency and density function $f(t)$

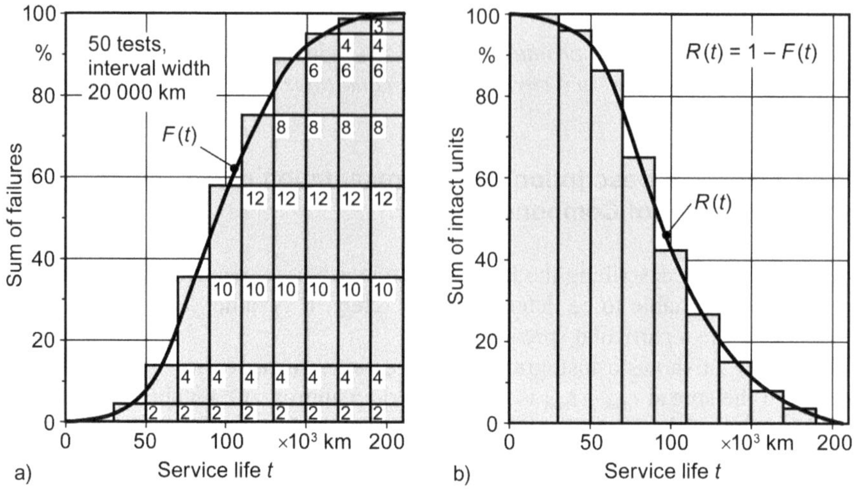


**Fig. 17.3.** *a* Histogram of the cumulative frequency and distribution function or failure probability $F(t)$; *b* histogram of the survival probability or reliability $R(t)$

To represent the units that are still intact, the survival probability $R(t)$ (Reliability) is used (Figure 17.3). Since the failure probability $F(t)$ describes the sum of failed parts, the survival probability $R(t)$ is derived as a complement of $F(t)$ as 1,

$$R(t) = 1 - F(t)\,. \tag{17.2}$$

The survival probability $R(t)$ is sometimes also referred to as reliability $R(t)$ in reliability theory.

A further statistical variable often used to characterise failure behaviour is the failure rate $\lambda(t)$. To determine the failure rate $\lambda(t)$, failures at a given point in time $t$ or in a time interval d$t$ are related to the number of units that are still intact

$$\lambda(t) = \frac{\text{Failures}}{\text{Intact units}}\,. \tag{17.3}$$

Since the density function $f(t)$ describes the failure density and the survival probability $R(t)$ describes the intact units, the failure rate ($\lambda$ rate) $\lambda(t)$ can be derived as the quotient of these two functions

$$\lambda(t) = \frac{f(t)}{R(t)}\,. \tag{17.4}$$

The failure rate $\lambda(t)$ can be determined as a degree of the failure risk of a part if it has already survived up to this point in time $t$.

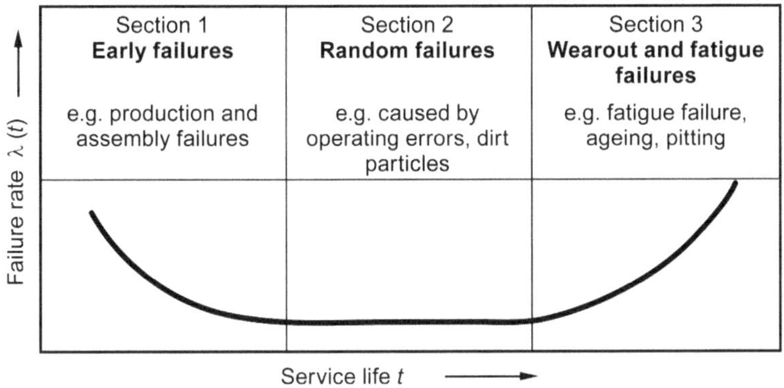


**Fig. 17.4.** Bathtub curve

If the failure behaviour of a product from its production to the end of its life is considered, a typical curve profile emerges (Figure 17.4). Because of its shape, this is called a *bathtub curve*. There are three distinct sections: section 1 relates to early failures, section 2 to random failures, and section 3 to wearout and fatigue failures.

### Section 1: Early Failures

Section 1 is characterised by a declining failure rate. The risk of a part failing decreases with time. These early failures are caused mainly by production and assembly errors. Early failures can be countered by corresponding quality assurance.

### Section 2: Random Failures

Random failures in section 2 have a constant failure rate. The failure risk of a part is thus always the same. The risk is usually relatively low. These random failures are caused e.g. by operating errors or dirt particles.

### Section 3: Wearout and Fatigue Failures

In section 3, failures due to wear and fatigue, the failure rate increases sharply. The failure risk increases for a part as its service life increases. The failures occurring here are, for example, caused by fatigue fracture, ageing, pitting or wear. This section is the most interesting for the design engineer, since the service life of a part is largely determined by this section. It can therefore be substantially improved by taking special account of the possible causes of failure and designing the parts accordingly – service life calculation.

The bathtub curve applies not only to individual components but is also observed in the case of complete systems.

### 17.1.3 Mathematical Description of Failure Behaviour using the Weibull Distribution

In the previous section we saw how failure behaviour can be represented by various statistical functions. Of particular interest, however, is the precise profile of these functions for a specific case, and how the curve can be described analytically. For this purpose, the failure functions derived empirically are to be replaced or approximated by curves that can be described analytically. The service life distributions used for this purpose are considered in this section.

#### *1/ Normal Distribution*

The best-known of these service life distributions is the *normal distribution.* It has as density function $f(t)$ the well known "bell curve", which is completely symmetrical about a mean value. The failure density is highest at the mean value. Thus, basically only one type of failure behaviour can be described. The normal distribution is, however, frequently used in reliability theory.

#### *2/ Exponential Distribution*

The density function $f(t)$ of the *exponential distribution* decreases monotonically as an inverse e-function from an initial value. Thus, only a failure behaviour can be described in which a high number of failures is initially observed, which then continuously decreases. In addition to the continuously decreasing density function, the constant failure rate $\lambda$ is a key characteristic of this distribution. That means the failure risk is unrelated to time. Here too it can be noted that the exponential distribution is mainly only suitable for describing a specific type of failure behaviour.

The exponential distribution is frequently used in electrical and aeronautical engineering, while in mechanical engineering, the service life distributions most frequently used are the *logarithmic normal distribution* and the *Weibull distribution.*

#### *3/ Logarithmic Normal Distribution*

The logarithmic normal distribution, commonly called log normal distribution for short, is derived from the normal distribution. The random variable $t$ is used in the logarithmic form $\ln t$. This means that the logarithmised failure times follow a normal distribution.

With the logarithmic normal distribution, similar curves can be achieved in similar variety as with the Weibull distribution discussed below. The mathematical use of the log normal distribution is more difficult than the Weibull distribution, since it cannot be solved analytically, but only numerically.

### *4/ Weibull Distribution: Basic Concept and Equations*

The Weibull distribution is capable of effectively describing a very different failure behaviour. This is shown most clearly by the density functions represented with the Weibull distribution (Figure 17.5). The formulae and relations of the Weibull distribution are given in Table 17.1.

**Table 17.1.** Formulae and designations of the Weibull distribution [17.1]

| | | |
|---|---|---|
| **Two parametric Weibull Distribution** | | |
| Survival probability | $R(t) = \mathrm{e}^{-\left(\frac{t}{T}\right)^b}$ | (17.5) |
| Failure probability | $F(t) = 1 - \mathrm{e}^{-\left(\frac{t}{T}\right)^b}$ | (17.6) |
| Density function | $f(t) = \frac{\mathrm{d}F}{\mathrm{d}t} = \frac{b}{T}\left(\frac{t}{T}\right)^{b-1} \mathrm{e}^{-\left(\frac{t}{T}\right)^b}$ | (17.7) |
| Failure rate | $\lambda(t) = \frac{f(t)}{R(t)} = \frac{b}{T}\left(\frac{t}{T}\right)^{b-1}$ | (17.8) |
| **Three parametric Weibull Distribution** | | |
| Survival probability | $R(t) = \mathrm{e}^{-\left(\frac{t-t_0}{T-t_0}\right)^b}$ | (17.9) |
| Failure probability | $F(t) = 1 - \mathrm{e}^{-\left(\frac{t-t_0}{T-t_0}\right)^b}$ | (17.10) |
| Density function | $f(t) = \frac{\mathrm{d}F}{\mathrm{d}t} = \frac{b}{T - t_0}\left(\frac{t \; t_0}{T - t_0}\right)^{b-1} \mathrm{e}^{-\left(\frac{t-t_0}{T-t_0}\right)^b}$ | (17.11) |
| Failure rate | $\lambda(t) = \frac{f(t)}{R(t)} = \frac{b}{T - t_0}\left(\frac{t - t_0}{T - t_0}\right)^{b-1}$ | (17.12) |

**Parameter**

$t$: Statistical variable (loading time, load cycles, actuations, ...).

$T$: Characteristic service life, "scale parameter".
For $t = T$, $F(t) = 63.2\%$ or $R(t) = 36.8\%$.

$b$: Shape parameter or failure slope. Determines the shape of the curve.

$t_0$: Failure free time. The parameter $t_0$ determines the point in time from which failures begin to occur. It corresponds to a shifting of the failure behaviour along the $t$-axis.

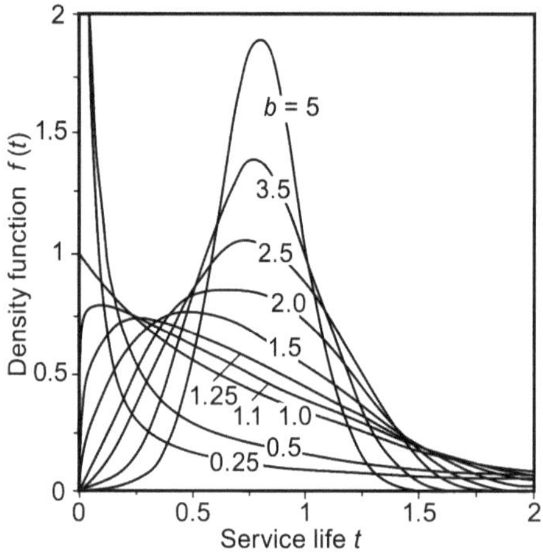


**Fig. 17.5.** Density functions $f(t)$ of the two parametric Weibull distribution for various shape parameters $b$ (characteristic service life $T = 1$)

### *4.1/ Two Parametric Weibull Distribution*

The density function changes distinctly as a function of a parameter of the distribution – shape parameter $b$. The different failure rates of the two parametric Weibull distribution in Figure 17.6 can be divided into three sections identical with the sections of the bathtub curve in Section 17.1.2:

$b < 1$: The failure rates decrease as service life increases. This relates to early failures.

$b = 1$: The failure rate is constant. The shape parameter $b = 1$ is thus suitable for describing random failures in section 2 of the bathtub curve.

$b > 1$: The failure rates increase distinctly as service life increases. Values of $b$ greater than 1 thus relate to failures attributable to wear and fatigue.

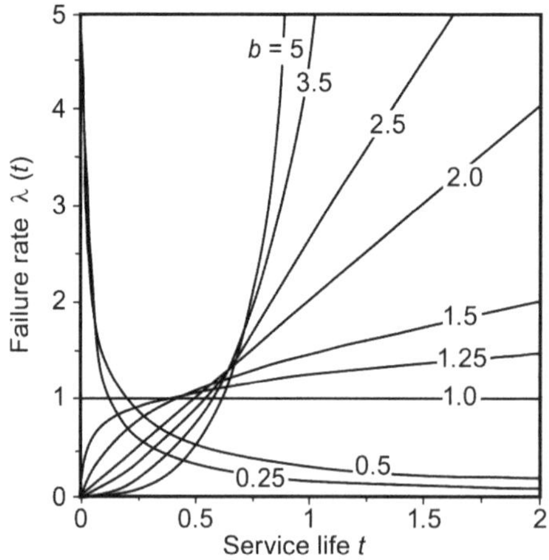


**Fig. 17.6.** Failure rates $\lambda(t)$ of the two parametric Weibull distribution for various shape parameters $b$ (characteristic service life $T = 1$)

The Weibull distribution can be subdivided into a two parametric and a three parametric distribution (Table 17.1). The two parametric Weibull distribution has as its parameters the characteristic service life $T$ (scale parameter) and the shape parameter $b$. The characteristic service life $T$ is assigned the failure probability $F(T) = 63.2\%$ ($R(T) = 36.8\%$). The shape parameter $b$ is a measure of the spreading of failure or for the shape of the failure density. Failures are always described starting from the point in time $t = 0$ in the two parametric Weibull distribution.

### 4.2/ Three Parametric Weibull Distribution

The three parametric Weibull distribution has in addition to the parameters $T$ and $b$, a further parameter, the failure free time $t_0$. In the case of failures due to wear and fatigue, the failure free time $t_0$ is based on a certain time being needed for failures to arise and spread.

A failure probability $F(T) = 63.2\%$ or a survival probability $R(T) = 36.8\%$ is assigned to the characteristic service life $T$. The characteristic service life $T$ can thus be treated as a characteristic value, similar to the median value, for which $F(t) = 50\%$. A further important characteristic is the $B_x$ service life. This is the service life for which the failure probability of the element under investigation is x %.

### Graphic Representation of the Weibull Distribution

The failure probabilities $F(t)$ follow an S-shaped curve. Using special Weibull probability paper it is possible to draw the functions $F(t)$ of the *two parametric Weibull distribution* as *straight lines* (Figure 17.7).

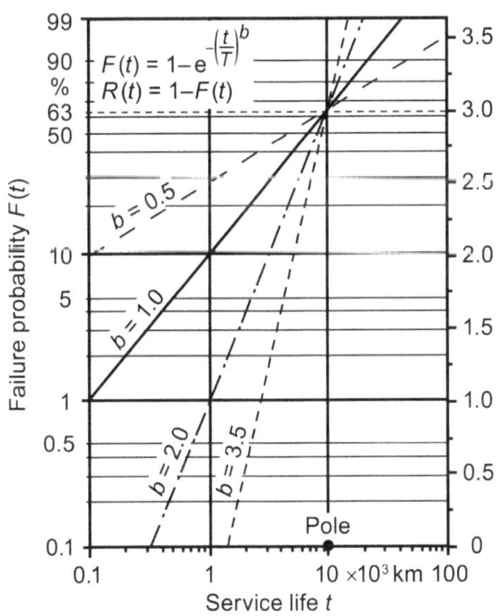


**Fig. 17.7.** Weibull probability paper. Failure curves of the two parametric Weibull distribution (Weibull lines) with different shape parameters $b$

This enables failure behaviour to be shown in a simple graphic form. There are also advantages in evaluating tests, since this enables drawing a straight line through the entered experimental values. The abscissa is logarithmically divided, while the ordinate has a double logarithmic scale:

$$x = \ln t\,, \tag{17.13}$$

$$y = \ln\left[-\ln(1 - F(t))\right] \quad \text{or} \quad y = \ln\left[-\ln R(t)\right]. \tag{17.14}$$

Every two parametric Weibull distribution can thus be represented as a straight line in the Weibull probability network (see Figure 17.7).

The slope of the straight lines in the probability network is a direct measure of the shape parameter $b$. The shape parameter $b$ can be read off on the right ordinate in Figure 17.7, if the straight line is shifted parallel through the pole.

A *three parametric Weibull distribution does not produce a straight line* on Weibull probability paper, but a *curve* (Figure 17.8a), although a three parametric Weibull distribution can also be drawn as a straight line if the corrected failure times ($t - t_0$) are plotted on the abscissa. This time transformation serves to return the three parametric Weibull distribution to a two parametric Weibull distribution (Figure 17.8b).

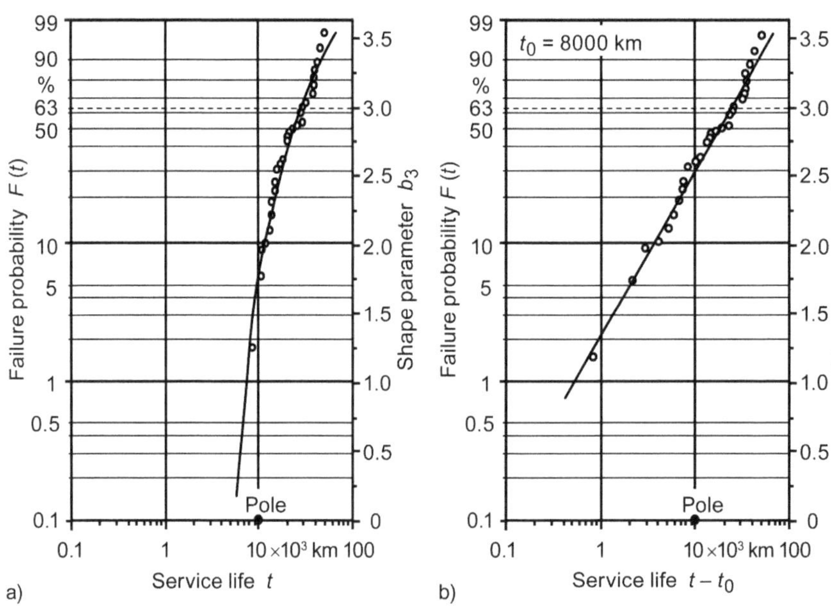


**Fig. 17.8.** Three parametric Weibull distribution on Weibull probability paper. *a* Original values and failure curve (Weibull curve) of the three parametric Weibull distribution; *b* three parametric Weibull distribution with the corrected failure times ($t - t_0$)

### 17.1.4 Reliability with Systems

Engineering products have to be regarded as complex systems consisting of several components. The failure behaviour of the individual components can, as described in the previous chapter, be represented by a Weibull distribution with the parameters $b$, $T$ and $t_0$. The failure behaviour of the total system is derived using a system theory that links the reliabilities of the elements in a suitable manner. One of these system theories is the Boolean theory, the chief premises of which are:

- a system is "non-repairable" (first system failure terminates the system's service life),
- components must be either in a failed or not failed state of condition and
- components are "independent" (the failure behaviour of a component is not affected by the failure behaviour of the other components).

Under these conditions, many engineering products can be treated using the Boolean theory. This theory is used exclusively below.

Reliability schematic diagrams can be constructed with the components, indicating how the failure of an element affects the system as a whole. The connections between the input (I) and output (O) of the schematic diagram (Figure 17.9) represent the possibilities for the system's functionality. The system is thus functioning if there is at least one connection in the reliability schematic diagram between input and output, on which all the components shown are intact.

In a serial structure (Figure 17.9a) the failure of any component leads to failure of the whole system. In a parallel structure (Figure 17.9b) the system only fails when all the elements have failed.

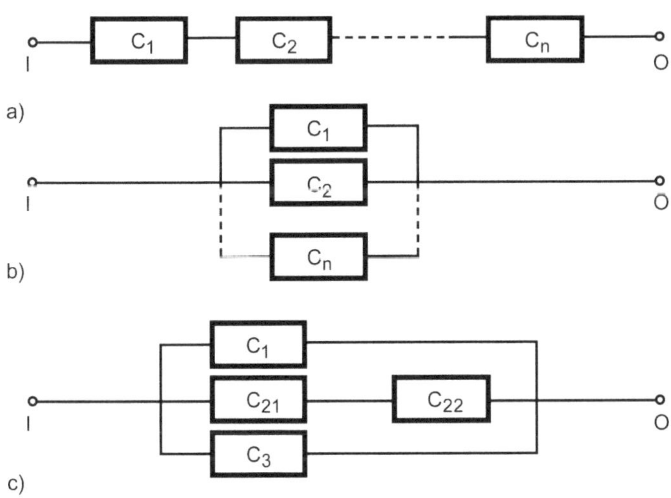


**Fig. 17.9.** Basic structures of reliability schematic diagrams. *a* Serial structure; *b* parallel structure; *c* mixed structure; *I* input; *O* output; *C* component

It should be noted that the structure of the reliability diagram does not relate to the mechanical structure of a design. For example a component can occur at several points in the reliability diagram. Almost all the systems used in mechanical engineering have serial structures, since the construction of parallel redundancies is complex and expensive.

The reliability of a serial system is calculated according to the product law of survival probabilities

$$R_S(t) = R_1(t)\ R_2(t)\ R_3(t)\ldots \quad \text{or} \quad R_S(t) = \prod_{i=1}^{n} R_i\,. \tag{17.15}$$

Since the survival probability of each system element $R_i(t) \leq 1$, the result for system reliability is always a value less than/equal to the reliability of the worst component.

With parallel systems (Figure 17.9b), the system's reliability is derived from the formula

$$R_S(t) = 1-(1-R_1(t))\,(1-R_2(t))\,(1-R_3(t))\ldots$$

$$\text{or} \quad R_S(t) = 1-\prod_{i=1}^{r}(1-R_i)\,, \tag{17.16}$$

with the redundancy level $r$ of the system.

### 17.1.5 Availability of Systems

Reliability describes the survival probability of components or whole systems until the *first* failure. In the case of *repairable systems* the system can be returned to a functional condition by a repair. Failure and subsequent repair can be frequently repeated in the case of repairable systems. The term "availability" was introduced for such systems. Availability $A(t)$ relates to the probability of a system being in a functional condition at a given time. The inherent steady state availability is calculated using the formula

$$A(t) = \frac{MTTF}{MTTF + MTTR}\,. \tag{17.17}$$

where $MTTF$ is the Mean Time To Failures and $MTTR$ is the Mean Time to Repair.

The Markov model is frequently used as a system theory to describe status probability [17.8, 17.11]. This model can capture the status probability of a repairable system at any desired point in time. The status conditions of the system are described by linear differential equations which in most cases can no longer be analytically solved, but have to be solved numerically.

If the reliability or availability of systems cannot be determined analytically, they can be determined using the Monte Carlo method as a simulation model. The Monte Carlo method is sometimes the only practically accessible method for investigating complex systems [17.6, 17.12, 17.16].

## 17.2 Reliability Analysis of Vehicle Transmissions

The main purpose of reliability assurance is to determine the anticipated failure behaviour of a product at the development stage, or to forecast it. Such prognoses are only possible for fatigue and wearout failures, i.e. for section 3 of the bathtub curve (Figure 17.4).

In order to be able to dispense with some of the extensive and time-consuming tests, calculation methods are used which are based on the principles of probability theory described in the previous sections. A reliable prognosis can only be achieved if the failure behaviour of the individual components is known in sufficient detail.

The procedure illustrated in Figure 17.10 [17.3, 17.12] has proven its value in determining system reliability. In system analysis, first all components and their functions are determined. In order to guarantee a complete analysis, it is usually appropriate to divide up the components into groups according to their function or design. In the following second stage, qualitative reliability analysis, the system elements relevant to reliability, and their effect on the functionality of the system are determined and evaluated. In the final stage, the quantitative reliability analysis, the failure behaviour of the system is determined with the principles of probability theory discussed in the preceding stages.

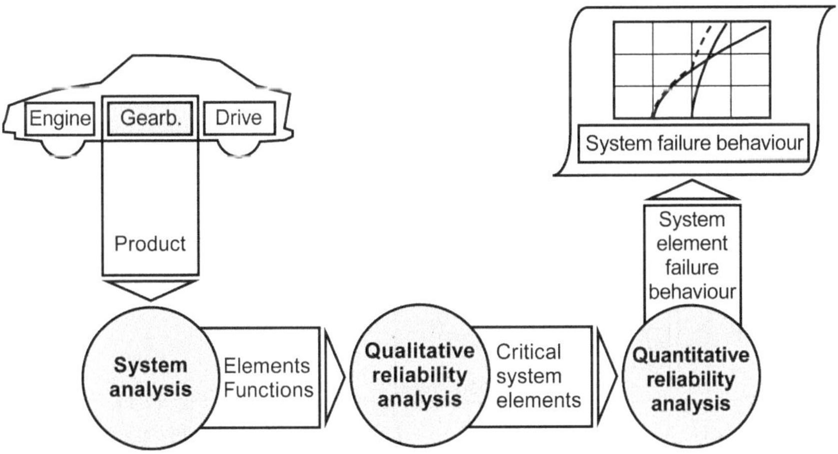


**Fig. 17.10.** Procedure for reliability analysis

These three stages of reliability analysis are examined below using examples from vehicle transmission engineering.

### 17.2.1 System Analysis

For this purpose, the product is initially delineated as a system within its environment or within its superordinate system. In order to gain an overview of the whole system, all the elements arising are then determined.

Elements in this case include both the components and the component interfaces. Component interfaces are, for example, shrink connections, welded connections etc., which also represent reliability-critical elements of a system as well as the components themselves. To illustrate the functions of the system and components, it is helpful to further subdivide complex products into functional groups or assemblies.

#### *1/ System Definition*

To describe the effect of failures of a product on neighbouring systems, it is necessary to establish a system boundary. This then involves determining all interactions across system boundaries. The system boundary relates both to mechanical and hydraulic as well as electrical connections. For a vehicle transmission, for example, this is the input and output, the mounting, the gearshift system and the data link for rotational speed measurement (Figure 17.11).

#### *2/ Interactions between the Components*

The links and interactions of the individual components are shown in the so-called "function block diagram". It is important that this diagram is clearly arranged and complete, since the function block diagram is the starting point and basis for the qualitative reliability analysis described in the following section.

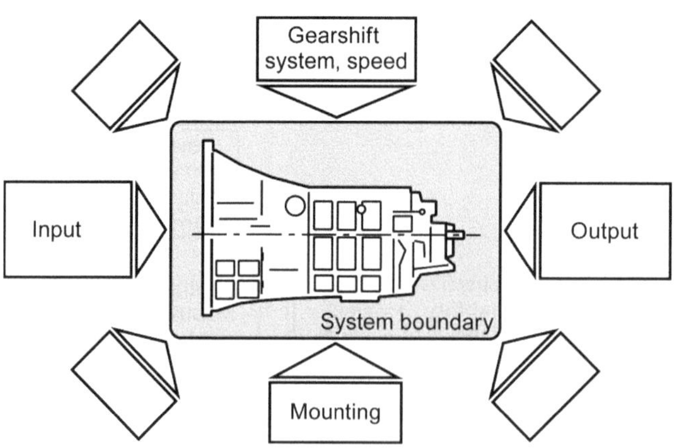


**Fig. 17.11.** System boundary for a passenger car selector gearbox

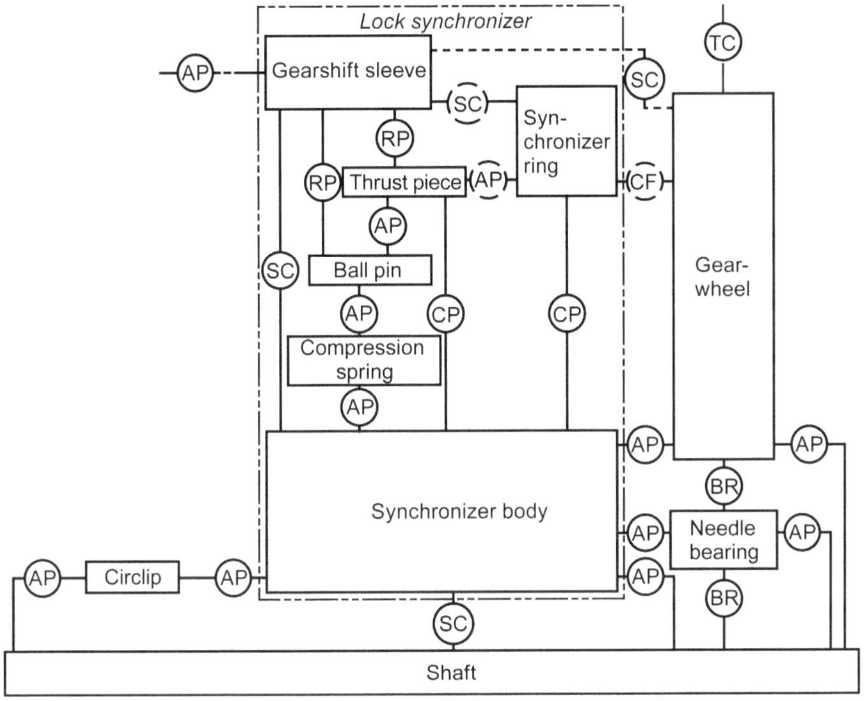


**Fig. 17.12.** Function block diagram of a single-cone synchronizer according to Figure 17.13. Elements of the function block diagram, see Figure 17.14

The arrangement of the components in the function block diagram should correspond as far as possible to the structure of the design, so that the force and power flows can be directly recognised (Figure 17.12).

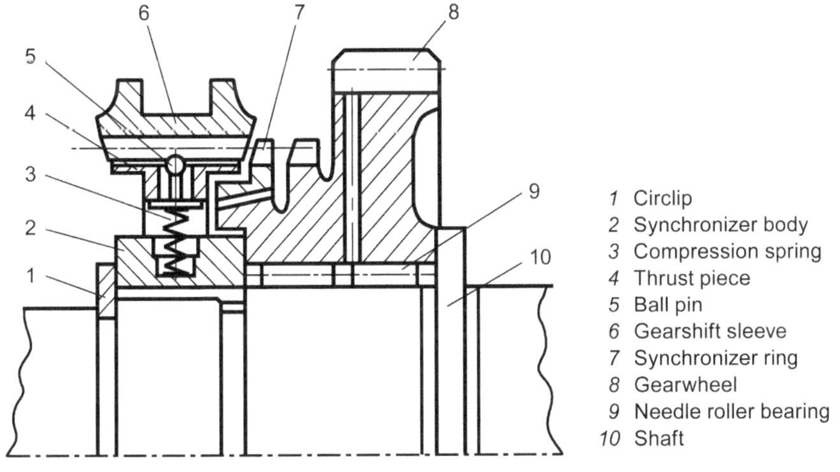


**Fig. 17.13.** Assembly drawing of a single-cone synchronizer

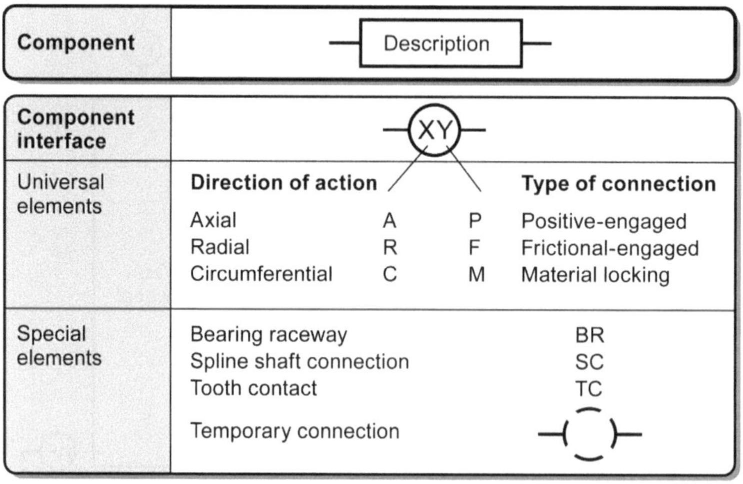


**Fig. 17.14.** Elements of the function block diagram for the single-cone synchronizer

The function block diagram in Figure 17.12 is derived directly from the design drawing in Figure 17.13. The component interfaces should be marked in the function block diagram by indicating the type of link. The links which cannot be described through a linking element, such as the meshing of running gears or rolling contact of rolling bearings, must be introduced as special elements, and marked accordingly or listed separately.

### 17.2.2 Qualitative Reliability Analysis

Qualitative reliability analysis involves investigating the components relevant to reliability in terms of their failure causes, and assessing them in terms of their effect on the functionality of the system as a whole. The assessment should be based on the functions of the product determined in the course of systems analysis, the interactions between components, calculations, results of tests, fault statistics and knowledge gathered from experience. The results of qualitative reliability analysis can be represented as a fault tree (FTA), block diagram (BD) or in the failure mode and effects analysis (FMEA) form sheets (see also German standard DIN 25424 [17.5] and VDA Volume 4 [17.14]).

To document that some components can fail in different ways, the components are broken down into system elements according to the type of fault. In the example described of the single-cone synchronizer, the component, e.g. gearwheel, must be broken down into the system elements tooth failure, pitting or scuffing. The system elements resulting from this breakdown of course fulfil different functions, and thus make a different contribution to system reliability. This means it is neither reasonable nor admissible to regard all system elements as of equal value. It is therefore necessary to preselect elements that are relevant to reliability and those that are not before assessing the elements. This can, for example, be carried out using the so-called ABC analysis.

### 1/ ABC Analysis

A helpful technique for preselecting elements that are relevant to reliability is subdividing them into three categories, as shown in Figure 17.15.

The system elements are classified on the basis of the effect of the system elements on system reliability and calculability of their $B_x$ service life. While the $B_x$ service life of A system elements that are critical to reliability can be calculated, in the case of B system elements which are also critical to reliability, one has to rely on data derived from experience or from test results. The C system elements that are neutral in terms of reliability are not taken into account in subsequent analysis. After this classification, the necessary calculations or tests can be carried out for the rest of the reliability analysis. The ABC analysis can be regarded as a highly simplified form of the failure mode and effects analysis (FMEA) described in the next section.

### 2/ FMEA, FTA, BD

In the case of failure mode and effects analysis (FMEA), potential failures are systematically identified and assessed. It should always be carried out for new and important products in parallel with the design process [17.1, 17.7], in order to be able to take account of the necessary design improvements immediately (see also Section 15.3 “Release Stages in the Product Development Process” and Figures 15.8–15.9).

Fault tree analysis (FTA) [17.5] is an analytical procedure, the result of which is displayed as a fault tree. This involves a deductive procedure in which all associated failure causes are sought, starting from a particular fault.

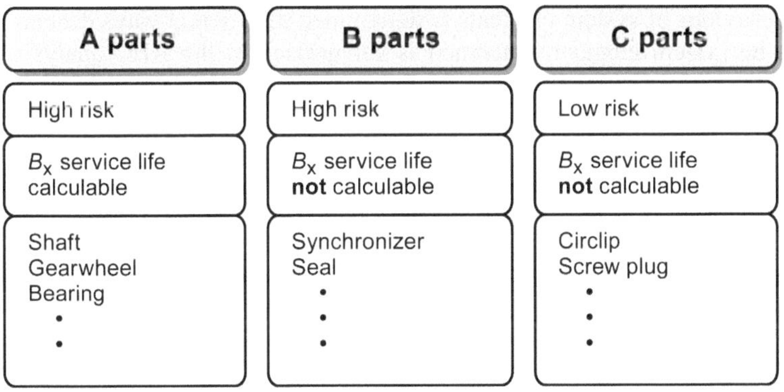


**Fig. 17.15.** ABC analysis

The difference between FMEA and FTA can be described as follows: FMEA describes a completely functioning system that is broken down into its system design elements. Functions and malfunctions are assigned to these elements, and potential effects and failure causes are determined. The result is the risk priority number (RPN). FTA proceeds from a particular fault (primary event). The possible failure causes are determined analytically. With the help of Boolean algebra, both failure frequency density and non-availability can be computed. The fault tree analysis and the failure mode and effects analysis lead basically to the same result.

In the block diagram (BD), the system elements are linked together to form a reliability structure, which can be either a serial or parallel structure, or a combined structure (Figure 17.9).

Generating the block diagram involves an inductive procedure, since conclusions are drawn about system failure behaviour from the failures of individual system elements. As in fault tree analysis, the block diagram is in some way contained in the FMEA. In a serial system, the block diagram is created by setting out the individual system elements (potential failures) arising from the FMEA risk analysis, one behind the other.

The structure encountered in most engineering products in practice is a serial structure. This means that there is no redundancy, and that the first failure of a system element leads to a failure of the whole system.

### 17.2.3 Quantitative Reliability Analysis

The aim of quantitative reliability analysis is to determine the failure behaviour of the system elements identified as critical in the qualitative reliability analysis. On this basis, the failure behaviour of the system is determined in accordance with the principles of reliability theory discussed in the preceding sections.

#### *1/ Failure Behaviour of System Elements*

The failure behaviour of system elements is determined in different ways depending on how the system element concerned is categorised in the ABC analysis. There are relatively precise load profiles for the A system elements and the associated Wöhler curves. This enables the service life of the system elements to be determined by means of an operational fatigue strength calculation. This calculated service life corresponds in most cases to the $B_1$ or $B_{10}$ service life. For the B system elements one has to rely on data derived from experience and test results. If the failure behaviour profile is known, the $B_1$ or $B_{10}$ service life can be converted into the characteristic service life $T$ using appropriate equations, as in [17.1]. To be able to describe the failure behaviour of a system element fully, the associated distribution function is placed in the Weibull chart using the point determined with the $B_x$ service life.

There are different ways of determining the Weibull parameters $t_0$, $T$ and $b$ needed to describe failure behaviour.

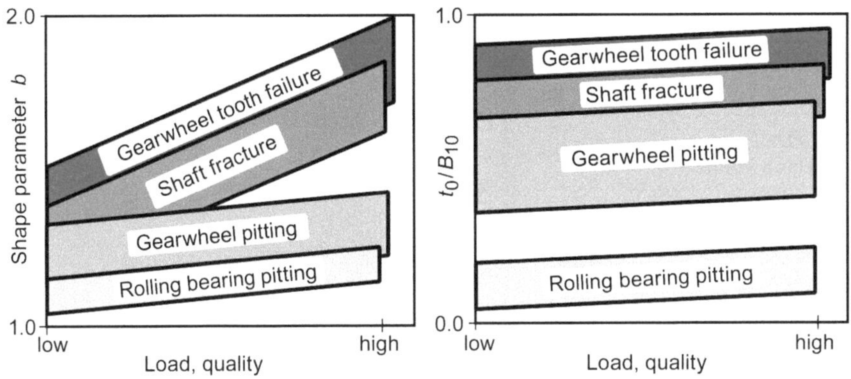


**Fig. 17.16.** Shape parameters $b$ and $t_0/B_{10}$ derived from evaluation of test results

The most reliable is to carry out tests. This is however very costly, since a large number of tests are required to produce representative results. If Weibull parameters are available for the same type of failure under comparable conditions, the Weibull parameters required can be estimated using knowledge derived from experience and calculation. There are now also reliability databases containing the Weibull parameters as a function of load, machining, the material used and the failure mechanism for some machine elements. Using the values stored in the database, and possibly other tests, it is possible to estimate the parameter $b$ (and $t_0$ if necessary) required.

Failure behaviour depends basically on the failure mechanism (e.g. fracture), on the load (fatigue or endurance strength range), the type of machining and the material. Instead of the failure free time $t_0$, the relationship $t_0 / B_{10}$ has proven its value in practice.

The major influence on the failure behaviour of a component is the failure mechanism. For example components which fail as a result of fracture have higher $b$ or $t_0 / B_{10}$ values than components which fail as a result of pitting (Figure 17.16).

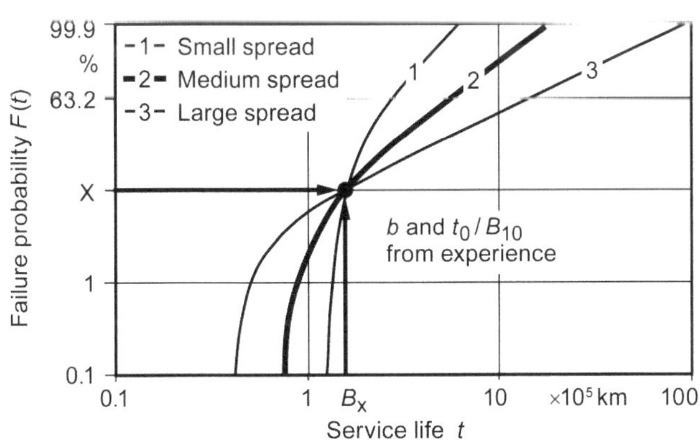


**Fig. 17.17.** Determining failure behaviour from empirical knowledge

**Table 17.2.** Weibull parameters using the example of a lock synchronizer unit

| **System element** | $B_{10}$ ($10^5$ km) | $t_0$ ($10^5$ km) | $t_0 / B_{10}$ | $b$ | $T$ ($10^5$ km) |
|---|---|---|---|---|---|
| Gearwheel<br>Tooth failure | 8.2 | 7.0 | 0.85 | 1.7 | 11.5 |
| Gearwheel<br>Pitting | 1.9 | 1.0 | 0.5 | 1.15 | 7.4 |
| Shaft<br>Fracture | 10.8 | 8.6 | 0.8 | 1.3 | 21.0 |
| Needle roller bearing<br>Pitting | 14.7 | 2.2 | 0.15 | 1.2 | 83.7 |
| Synchronizer ring<br>Wear | 3.7 | 1.9 | 0.5 | 1.1 | 15.8 |

Components with a higher level of machining or material quality also have higher $b$ or $t_0 / B_{10}$ values than components of lower quality. Higher load results in higher $b$ values. More brittle materials usually have a smaller spread than ductile ones [17.3, 17.12]. With the Weibull parameters thus determined, the failure behaviour of the particular element can be determined in accordance with Figure 17.17.

Failure behaviour can be displayed on Weibull paper, and also described in tabular form with the $B_{10}$ service life, the shape parameter $b$ and the value $t_0 / B_{10}$. Table 17.2 shows this for the example of the shaft, idler gear and single-cone synchronizer described above, for comparative purposes.

### *2/ Failure Behaviour of the System*

As discussed in Section 17.1.4, the survival probabilities $R_i(t)$ of the critical system elements are linked in accordance with an appropriate system theory to determine the failure behaviour of the system. In the case of vehicle transmissions, the Boolean theory has proven a good approximation.

If the component failure behaviour can be described with a two parametric Weibull distribution, then starting from $t > 0$ the individual components already have a limited reliability $R_i(t) < 1$. This in turn implies that for calculating system reliability, all components have to be taken into account, and that each additional component in a serial structure further reduces system reliability. Where there are many components in serial, this leads to very low system reliability.

The failure behaviour of many components can however often, like mentioned above, be described better with a three parametric Weibull distribution. In these cases, the components only have to be taken into account for determining system reliability when the time $t > t_0$.

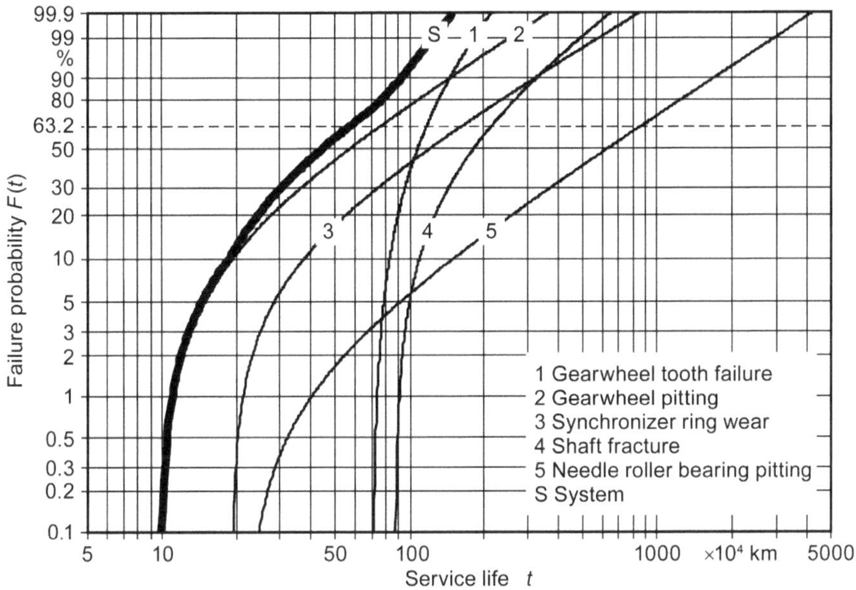


**Fig. 17.18.** System failure behaviour, based on the example of a single-cone synchronizer

Thus, if the early failures (section 1 of the bathtub curve) are prevented by suitable quality assurance measures, the reliability of the system is determined only by the design of these components in terms of fatigue and wear [17.2].

If the failure probabilities of the various system elements are recorded on Weibull paper, it is generally easy to see which elements largely determine system failure behaviour. Figure 17.18 shows the failure probability for the example already mentioned of the single-cone synchronizer. In this example, it is apparent that a minimal improvement in tooth flank load capacity (failure of the gearwheel attributable to pitting) can nearly double the $B_{10}$ service life of the system.

Finally, it should be noted that the analysis is particularly suited to comparing similar products to be operated under exactly the same operating conditions. Reliability analysis is, however, also appropriate for new products if the emphasis is not on absolute figures, but the intention is to investigate the effect of the critical system elements on system failure behaviour by means of parameter variations.

## 17.3 Testing to Ensure Reliability

Despite the availability of modern computational and simulation methods for component dimensioning, practical tests with transmissions and their components is unavoidable. Especially where many parts interact in a complex system under environmental conditions that are not precisely known (dynamics, lubrication), the process is only partially susceptible to numerical simulation. So-called accelerated tests can help to disclose weak points and risk components early on in product development. In many cases comprehensive tests are also prescribed by the customer or legislator, on the basis of which the product is released for series production.

It is essential that the systematic development process and the measures to ensure a product's reliability are interlinked right from the start of the development (Figure 17.19). Component and material tests are necessary even at the planning and conceptual design phase. It is also necessary to determine practical load profiles for dimensioning at this point in time (see also Section 7.4 "Operational Fatigue Strength and Service Life").

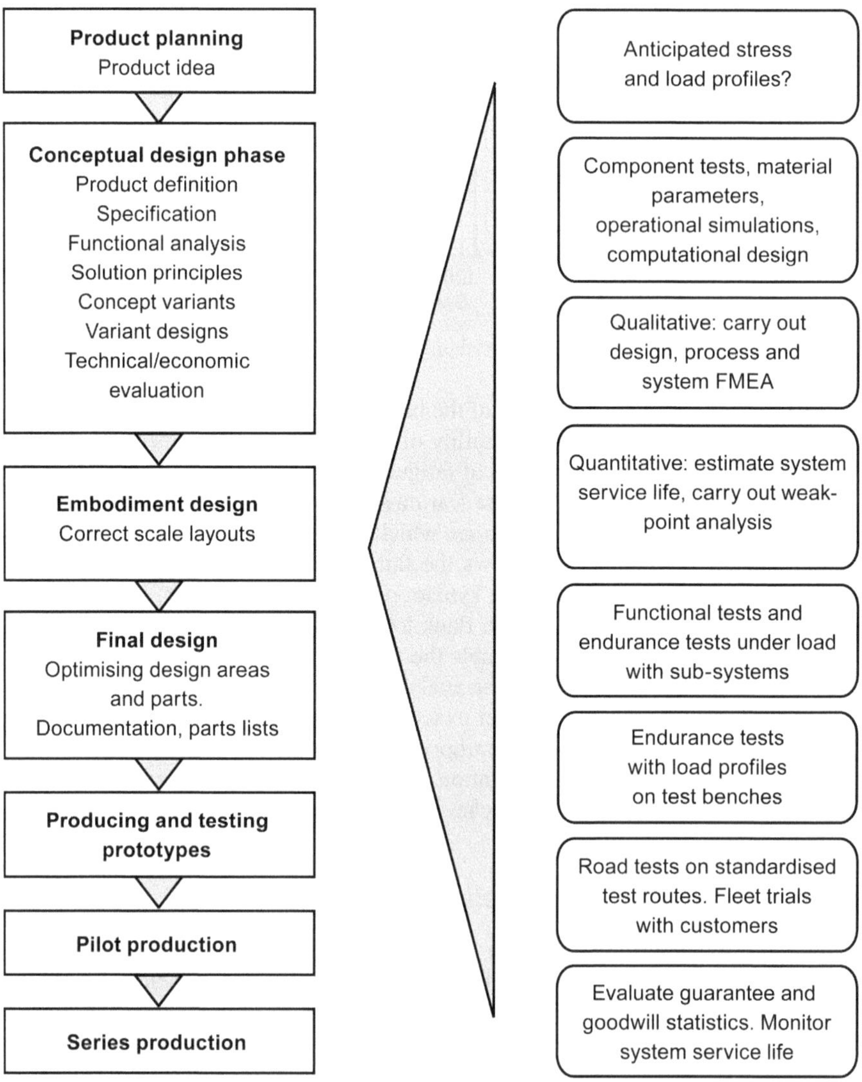


**Fig. 17.19.** Interlinking of the development process and reliability assurance

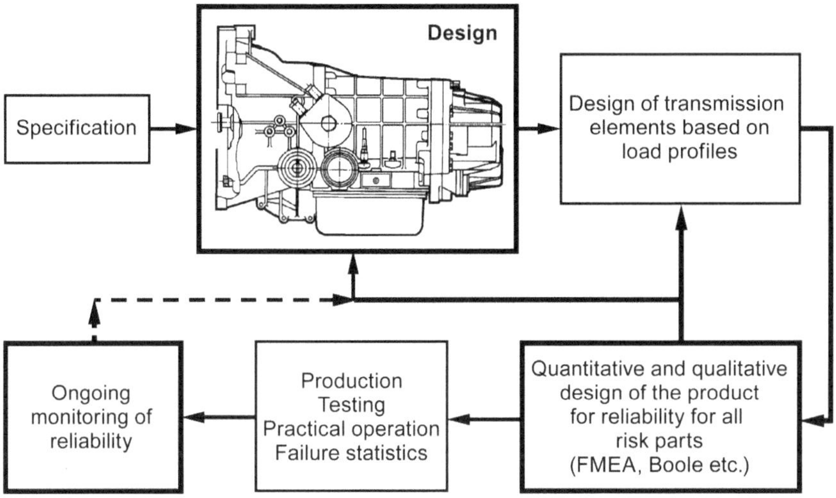


**Fig. 17.20.** Design for reliability as part of a systematic product development process

Due to the growing application of increasingly better simulation and virtual development methods, development loops for cost, durability, function and weight optimisation are used already in the conceptual design phase for the first transmission design stage (see Chapter 14 "Computer-Aided Transmission Development"). After design detailing and prototype production, comprehensive testing is carried out on test benches. With the help of increasingly realistic vehicle simulation on test benches, it is possible to detect faults early on. Information gained in this process can then be exploited in further developments as part of the development loop. In this way, both the time needed for product development and cost-intensive vehicle testing can be reduced.

In order to be able to take account in the development process of the requirements for designing a reliable product, it is necessary to consider reliability in addition to design for operational fatigue strength of the power transmitting and other risk components. The ideal product development process in this regard is one where the individual phases are arranged in a closed loop. Integrating a constant reliability monitoring function at the testing stage and during practical operation enables this ideal case to be nearly realised (Figure 17.20).

### 17.3.1 Classifying Vehicle Transmission Test Programs

The test programs carried out in the last phase of vehicle transmission development can mainly be classified into three main areas:

1/ component testing (component and analogue tests),
2/ prototype bench tests and
3/ vehicle testing.

### 1/ Component Testing

Component testing is carried out with individual components or with "analogue test parts", which is the simplest kind of test. However these analogue test parts only allow for an estimation of the failure behaviour of the components for a particular type of fault.

Precise information on service life or other types of failure are derived from the real component under test. Component behaviour can be calculated more and more exactly thanks to the constant development of mathematical models. A distinction is made in component testing between static and dynamic tests (Figure 17.21). While component testing provides principally information for dimensioning components, the prototype tests described in the next section, provide initial indications of the whole system behaviour.

### 2/ Prototype Bench Tests

For this testing variant, a distinction is made between pure functional testing and endurance testing (Figure 17.22). With functional testing, the power conducting parts are initially tested with low stress. This type of test is commonly used for transmissions, for example to determine their shiftability or oil supply at different lateral and longitudinal inclinations.

With test bench endurance testing, testing is carried out as a function of the possibilities provided by the test bench, e.g. particular load profiles are tested as load runs with precisely defined load/time functions, continuous shifting performance is tested or, with modern simulation-capable test benches, entire driving profiles are simulated. Driving profiles are based on "synthetic routes". When creating routes, shifting frequency, the amount of time in different gears and the load profile are all taken into consideration.

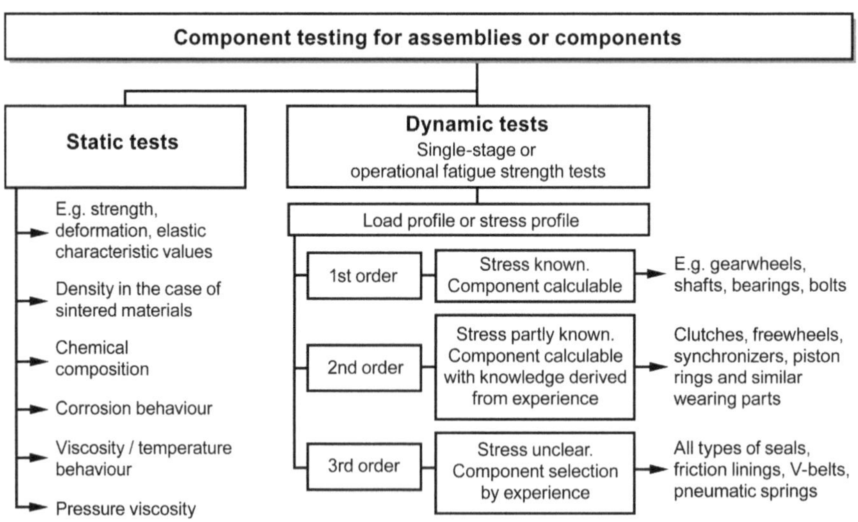


**Fig. 17.21.** Extent of component testing required for assemblies or components

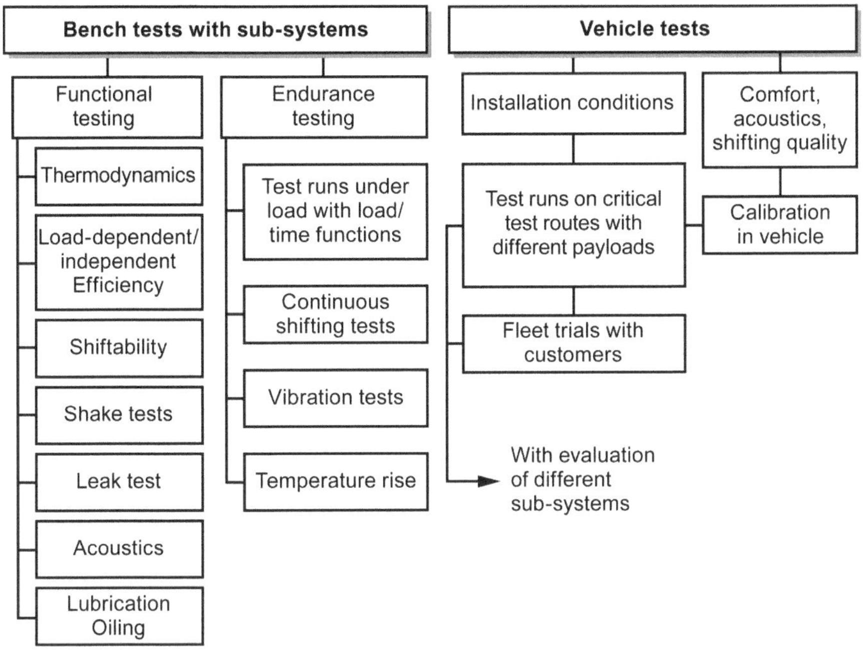


**Fig. 17.22.** Prototype test bench tests with sub-systems and vehicle testing

### *3/ Vehicle Testing*

Vehicle testing involves testing the system as the last part of the testing phase. In this phase the function, service life and reliability of the transmission in the whole vehicle system is determined on various test routes and under different conditions, in addition to testing the installation conditions (Figure 17.22).

The test routes, as representative, customer-oriented load profiles, are characterised by the height profiles, the gradient distribution and the speed distribution. In the case of vehicles, the test routes are often so designed as to include a mixture of motorway, urban road, rural road and considerable mountain roads (see also Tables 2.9 and 2.10).

The test routes are specified at the planning phase of development, since their data, in combination with the vehicle data to be derived, give the load profiles which will be needed for dimensioning right from the development phase. The load/time profiles recorded during the test runs are transferred to the load profiles by a counting procedure (see also Section 7.4.2 "Load Profile and Counting Procedure"). Figure 17.23 shows causes of the load/time function.

In the context of testing for motor vehicles, the trend is now increasingly towards "fleet trials" in co-operation with taxi companies or hauliers. These fleet trials provide an extraordinarily realistic profile in respect of various types of stress. This fact then implies that wear arising in the course of these fleet trials will also occur in actual day-to-day use. Complementing such fleet trials, so-called high-load tests (e.g. driving on race circuits and misuse tests) are also carried out.

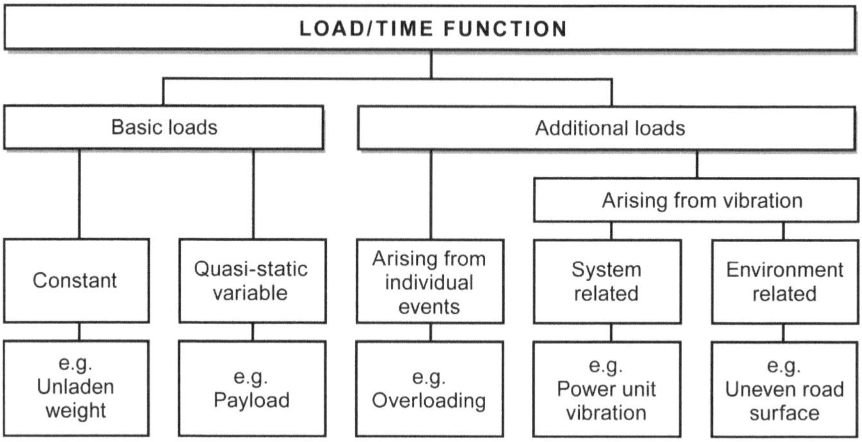


**Fig. 17.23.** Causes of the load/time function. Source: Buxbaum

It should be noted that the mathematical principles of reliability calculation have a bearing on all the testing programs referred to. It is necessary to clarify beforehand how many test units permit a significant result for a given population. It should be mentioned at this point that tests have been carried out to show that when the number of test specimens is increased, their actual test duration of the test specimens can be reduced [17.13]. This is naturally highly desirable with a view to saving time and money when developing new products.

### 17.3.2 Test Benches for the Test Programs

For component and prototype tests, test benches are needed which can be divided into different categories in line with the test programs:

1/ functional test benches,
2/ component test benches,
3/ assembly test benches and
4/ powertrain test benches.

#### *1/ Functional Test Benches*

Typical uses for functional test benches include functional tests after assembly, drag power curves or heating curves for each gear and in varying operating conditions. In addition, oiling tests aimed at improving oil supply are carried out on tilt test benches. Functional test benches usually have only one drive and are used for transmission and axle examinations.

#### *2/ Component Test Benches*

Component test benches test the functionality and endurance of individual components. They examine:

- synchronizers,
- gearshift actuators (external/internal gearshift systems),
- clutch actuators,
- clutches,
- friction coefficients of clutch linings,
- torque converters,
- variators in continuously variable transmissions,
- differential locks and
- pumps.

These test benches have simulation and control structures that can be quite sophisticated.

### 3/ Assembly Test Benches

Assembly test benches are utilised for testing individual assemblies such as gearboxes, all-wheel drive systems, axles or shafts, even the powertrain without the combustion engine. Depending on the test target and test specimen, 2 to 5 electric machine test benches are used with an electric drive machine to replace the combustion engine and up to four output machines.

Assembly test benches are normally used for endurance tests. However, efficiency, running-in behaviour, acoustics, temperature and oil supply are also investigated. Currently, stationary programs or test programs with minimal dynamic requirements are prevalent. Due to advances in propulsion and simulation technology, the trend is increasingly pointing to dynamic test programs. Typical test bench configurations are

- 2 electric machines
  - for gearboxes and shafts.
- 3 electric machines
  - for gearboxes with separate or integrated axle drive. In the case of blocked axles, all-wheel drive powertrains can also be tested.
  - due to the small distance of transverse-mounted gearboxes between input and output, usually an ancillary construct is required, e.g. a belt drive.
- 4 electric machines
  - reduced all-wheel drive, including input, gearbox and one axle drive. For the second axle, output is directly at the cardan shaft.
- 5 electric machines
  - complete all-wheel drive, consisting of an input and 4 outputs.

### 4/ Powertrain Test Benches

The internal combustion engine represents the drive in the case of powertrain test benches. As in the case of assembly test benches, different configurations result for one to four outputs. Electric output machines using direct current, asynchronous and increasingly synchronous technology are predominant. Flywheels with electric machines, eddy-current brakes or hydraulic drives are generally becoming a thing of the past.

For strength analysis, route profiles or synthetic routes are tested on powertrain test benches. Improved simulation possibilities and highly dynamic drives are making functional investigations and the reproduction of misuse tests possible as well [17.9].

### 17.3.3 Simulation during Bench Testing

The use of simulation models is becoming more important in bench testing. They are used both in test specimen control and to improve realistic load conditions.

To operate assemblies such as automatic transmissions, all-wheel drive systems or electrically controlled locks on the test bench, electric control is required. Due to the networking of control units in the vehicle or lacking signals at the test bench (e.g. the transverse or longitudinal acceleration signal from the acceleration sensor), separate assemblies can only rarely be operated on the test bench with the series control unit. Depending on the case at hand, hand-operated units or data sets adjusted for test bench operation can be used. Signals from other control devices required for operation can also be made available by means of rest bus simulation.

Besides the simulation of driving resistances, simulation of vibration behaviour is also necessary for a test bench investigation that is as realistic as possible. Highly dynamic drives with low inertia and constantly increasing simulation capabilities are helping to make significant progress in this field, e.g. by the use of tyre-slip models [17.4] or simulating rotational irregularities of the internal combustion engine [17.10]. This leads to a transfer of testing from the road onto test benches.

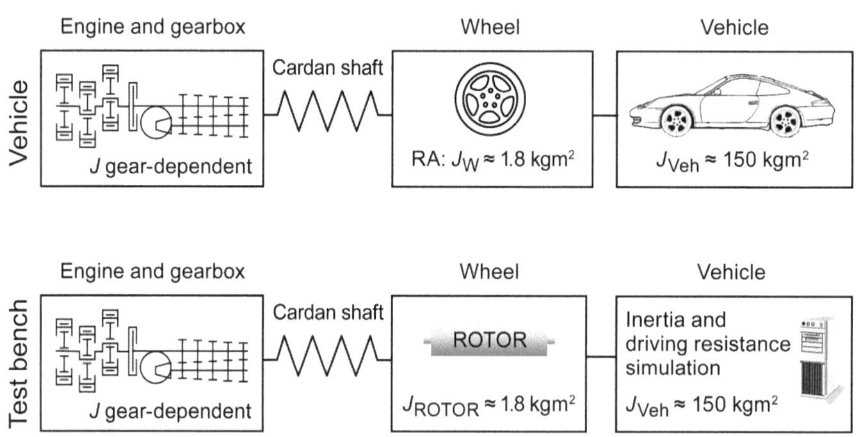


**Fig. 17.24.** Comparison of vibration model vehicle with the test bench

### 1/ Tyre-Slip Simulation

Every vehicle exhibits gear-dependent powertrain vibrations. These are essentially affected by the vehicle's weight and the moments of inertia and stiffness of the powertrain components. Damping and the maximum transferable torque depend mainly on the contact between the tyres and the street.

With the help of electric machines low in inertia, it is possible to adjust the moment of inertia of the rotor to the moment of inertia of the wheel. In this way, the same mechanical vibration model is depicted on the test bench as in the vehicle (see Figure 17.24).

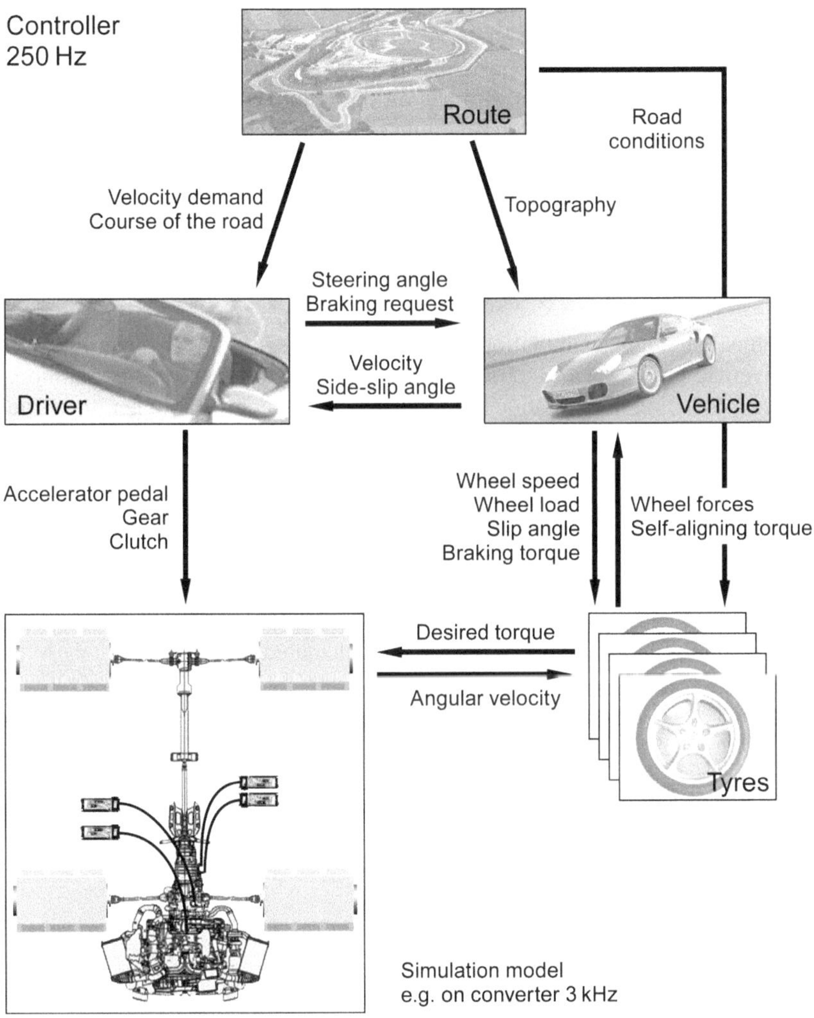


**Fig. 17.25.** Relation between the simulation models for route, driver, vehicle and tyres and the powertrain [17.4]

The maximum torque of the electric machine must be capable of representing the maximum torque transferable from the tyre to the street.

By expanding the vehicle model by a highly dynamic tyre-slip model for every wheel, the vibration behaviour emulates that of the real vehicle. Figure 17.25 shows the structure and interaction of simulation models.

### 2/ Simulation of Rotational Irregularities of the Internal Combustion Engine

In the field of automotive engineering, the powertrain is generally propelled by an internal combustion engine. Gas and inertia forces arising during operation cause a torque pulsation at the crankshaft which is arising in the crankshaft flange as rotational irregularity. These rotational irregularities cause torsional vibrations in powertrain components (e.g. gear rattling) on the one hand and can on the other hand contribute considerably to the damaging of powertrain components.

Under pressure to shorten product development times, powertrain development processes are also being parallelised, i.e. there are often no durable combustion engines with performance data conforming to specifications available yet in initial test bench investigations. The simulation of rotational irregularities in test bench testing makes it possible to evaluate and to optimise powertrain components at an early stage.

In combustion engine simulations, a rotational irregularity is superimposed over the average torque as a function of the cylinder number and load condition. The torque producing the rotational irregularity can be much higher than the maximum torque of the combustion engine. This implies very high demands on the electric machine. The moment of inertia of the rotor must be in the range of the moment of inertia of the combustion engine. Internal combustion engines for passenger cars are generally lower than 0.1 $kgm^2$. At the same time, the required torque of the electric machine, even at higher speeds, is many times that of the maximum torque of the engine to be simulated. The converter must be capable of creating vibrations with a frequency of several hundred Hz as a function of speed and cylinder number. The required frequency is calculated as follows:

$$f = 0.5 \times (\text{Cylinder number}) \times (\text{Engine speed})\,. \tag{17.18}$$

# References

The references are arranged by chapter. Those references marked with " * " are not used directly in the text but amplify and complement the topic under discussion.

## Chapter 1


[1.1] Lohr FW (1989) Marktorientierte Kraftfahrzeugentwicklung (Market-Oriented Motor Vehicle Development). Aachen: 2. Aachener Kolloquium „Fahrzeug- und Motorentechnik": 1–18

[1.2]* Div. authors (1956) Technik. Kleine Enzyklopädie (Technology. Small Encyclopaedia). Verlag Enzyklopädie Leipzig

[1.3]* Div. authors (1966) Genius der Deutschen. Die großen Forscher, Erfinder, Ärzte (Genius of the Germans. The Great Researchers, Inventors, Doctors). Propyläen Verlag, Berlin, Auszug aus dem Sammelwerk „Die Großen Deutschen" von Heimpel, Heuss und Reifenberg. Propyläen Verlag, Berlin

[1.4]* Div. authors (1986) 100 Jahre Automobil. Daimler-Benz 1886–1986 (100 Years of the Automobile. Daimler-Benz 1886-1986). München, Sonderbeilage der Süddeutschen Zeitung (24.01.1986), Nr. 19

[1.5]* Eckermann E (1981) Vom Dampfwagen zum Auto (From Steam Passenger Car to Automobile). Rowohlt-Verlag

[1.6]* Fersen von O (1986) Ein Jahrhundert Automobiltechnik – Personenwagen (One Century of Automotive Engineering – Automobiles). VDI-Verlag, Düsseldorf

[1.7]* Klinckowstroem C Graf v (1959) Knaurs Geschichte der Technik (Knaurs History of Technology). Droemersche Verlagsanstalt TH, Knaur Nachfolger München Zürich

[1.8]* Kühner K (1965) Geschichtliches zum Fahrzeugantrieb (Historical Notes on Vehicle Drive Systems). Friedrichshafen: Eigendruck der Zahnradfabrik Friedrichshafen AG

[1.9]* Lang OR (1982) Geschichte des Gleitlagers (History of the Plain Bearing). Stuttgart: Sonderdruck der Daimler-Benz AG

[1.10]* Lauster E (1980) Untersuchungen und Berechnungen zum Wärmehaushalt mechanischer Getriebe (Investigations and Calculations Relating to Heat Management within Mechanical Transmissions). Dissertation, Universität Stuttgart, Institut für Maschinenelemente

[1.11]* Looman J (1996) Zahnradgetriebe (Gear Mechanisms). 3rd Edition. Springer, Berlin Heidelberg New York



H. Naunheimer et al., *Automotive Transmissions*, 2nd ed.,
DOI 10.1007/978-3-642-16214-5, © Springer-Verlag Berlin Heidelberg 2011

[1.12]* Maier A (1975) Vom Kammrad zur Zahnradfabrik. Eine Dokumentation zur Zahnradgeschichte (From Cogwheel to Gearwheel Factory. A Documentary Record of Gearwheel History). Vol. I–VI. Friedrichshafen: Eigendruck der Zahnradfabrik Friedrichshafen AG

[1.13]* Maier A (1963) Geschichtliche Entwicklung von Fahrzeuggetrieben (Historical Development of Motor Vehicle Transmissions). Friedrichshafen: Sonderdruck der Zahnradfabrik Friedrichshafen AG

[1.14]* Parmee K (1988) Gearing up for the Future – A Story of Commercial Vehicle Transmission Development. Proceedings of the Institution of Mechanical Engineers Vol. 202, Nr. D2: 71–81

[1.15]* Priwitzer G (1982) Geschichte der Fahrzeuggetriebe (History of Automotive Transmissions). Unveröffentlichte Studienarbeit Universität Stuttgart, Institut für Maschinenelemente

[1.16]* Schönhaar P (1984) Beitrag zur Entwicklungsgeschichte der Maschinenelemente, ihrer Berechnung und Konstruktion (Paper on the History of the Development of Machine Elements, their Calculation and Engineering Design). Unveröffentlichte Studienarbeit Universität Stuttgart, Institut für Maschinenelemente

[1.17]* Schweikert H (2005) Voith Antriebstechnik (Voith Drive Technology. 100 Years of the Föttinger Principle). 100 Jahre Föttinger-Prinzip. Springer, Berlin Heidelberg New York

[1.18]* Seherr-Thoss HC Graf v (1965) Die Entwicklung der Zahnradtechnik. Zahnformen und Tragfähigkeitsberechnung (The Development of Gear Engineering. Tooth Formation and Capacity Calculation). Springer, Berlin Heidelberg New York

[1.19]* Sonderheft 100 Jahre ATZ (1998) Geschichte und Zukunft des Automobils (History and Future of the Automobile). ATZ Automobiltechnische Zeitschrift

[1.20]* Sonderheft 100 Jahre VDI-FVT (2004) 100 Jahre Fahrzeugtechnik im VDI (100 Years of Automotive Engineering in the Association of German Engineers (VDI)). ATZ Automobiltechnische Zeitschrift

[1.21]* Sprague de Camp L (1964) Ingenieure der Antike (Engineers of Ancient Times). ECON-Verlag, Düsseldorf Wien

[1.22]* Szabo I (1977) Geschichte der mechanischen Prinzipien (History of Mechanical Principles). Birkhäuser Verlag, Basel Stuttgart

[1.23]* Ungerer H (1986) Die Geschichte der Fahrzeugtechnik in sechs Jahrtausenden (The History of Vehicle Technology over Six Millennia). München: Vortrag auf der 12. Fachtagung (100 Jahre Automobiltechnik) des Münchner Arbeitskreises für Straßenfahrzeuge MAS e.V., 28.02.1986

## Chapter 2

[2.1] Christ H (1989) Strategische Forschung und Entwicklung in einem Unternehmen der Kraftfahrzeug-Zulieferindustrie (Strategic Research and Development in a Company of the Automotive Supply Industry). Konstruktion 41: 389–394

[2.2] DIN 70010 (2001) Systematik der Straßenfahrzeuge. Begriffe für Kraftfahrzeuge, Fahrzeugkombinationen und Anhängefahrzeuge (Classification of Road Vehicles. Terms for Motor Vehicles, Vehicle Combinations and Trailers). DIN Deutsches Institut für Normung e.V., Berlin

[2.3] DIN 70020 (2004) Straßenfahrzeuge. Kraftfahrzeugbau. Teil 1: Personenkraftwagen. Begriffe, Grundlagen, Bestimmungen, Maßkurzzeichen (Road vehicles. Vehicle Manufacturing. Part 1: Passenger Cars. Terms, Fundamentals, Regulations, Measurement Abbreviations). DIN Deutsches Institut für Normung e.V., Berlin

[2.4] DIN ISO 4130 (1979) Straßenfahrzeuge. 3-dimensionales Bezugssystem und primäre Bezugspunkte. Definitionen (Road Vehicles; 3-Dimensional Reference System and Fiducial Marks, Definitions). DIN Deutsches Institut für Normung e.V., Berlin

[2.5] DIW (1990/2005) Verkehr in Zahlen (Traffic in Numbers). Deutsches Institut für Wirtschaftsforschung, Berlin. Hrsg. Bundesministerium für Verkehr, Bau- und Wohnungswesen

[2.6] Drees S, Mayer K (2004) Entwicklung einer sicheren Auslegungsmethode für zukünftige Pkw-Stufenautomaten durch Abgleich von LK-Simulationen mit LK-Messungen (Development of a Reliable Method of Design for Future Automatic Car Transmissions Based on the Comparison of Collective-Load Simulations with Collective-Load Measurements). VDI-Berichte Nr. 1827: 609–625

[2.7] Förster HJ (1974) Transportsystem Straßenverkehr (The Road Traffic Transport System). Automobil-Industrie 2: 35ff

[2.8] Helling J (1972) Umdruck zur Vorlesung Kraftfahrzeuge I (Lecture Notes of the Motor Vehicles I Lecture). RWTH Aachen, Institut für Kraftfahrwesen

[2.9] Helling J (1989) Längsdynamik von Kraftfahrzeugen. Umdruck zur Vorlesung Kraftfahrzeuge I (Longitudinal Dynamics of Motor Vehicles. Lecture Notes of the Motor Vehicles I Lecture). ika Schriftenreihe Automobiltechnik, RWTH Aachen, Institut für Kraftfahrwesen

[2.10] Küçükay F (1990) Rechnerunterstützte Getriebedimensionierung mit repräsentativen Lastkollektiven (Computer-Aided Gear Rating with Representative Load Profiles). ATZ Automobiltechnische Zeitschrift 92: 328–333

[2.11] Looman J (1996) Zahnradgetriebe (Gear Mechanisms). 3rd Edition. Springer, Berlin Heidelberg New York

[2.12] Müller HW (1971) Die Umlaufgetriebe (Epicyclic Gears). Springer, Berlin Heidelberg New York

[2.13] Paul M, Grumbach M, Pelchen C (2004) Der Antriebsstrang als vernetztes System im Fahrzeug (Vehicle Driveline as a Networked System). VDI-Berichte Nr. 1827: 5–19

[2.14] Statistisches Bundesamt (2004) Bevölkerung im früheren Bundesgebiet, neue Länder und Deutschland ab 1950 (Population in Western, Eastern and the Whole of Germany since 1950). Statistisches Bundesamt, Wiesbaden

[2.15] Thyssen Henschel (1991) Transrapid. Produktunterlagen (Transrapid. Product Documentation)

[2.16] VDA (1953–2005) Tatsachen und Zahlen. 17–69. Folge (Facts and Numbers. Volume 17–69). Verband der Automobilindustrie e.V., Frankfurt

[2.17] Wolf A (1968) Die Grundgesetze der Umlaufgetriebe (The Fundamental Principles of Epicyclic Gears). Vieweg-Verlag, Wiesbaden

[2.18] Trendprognosen (Trend Prognoses):
- ZF drive 01/2003
- Ricardo-Analyse 2003
- DRI-WEFA / A.D. Little Studie
- B&D Forecast; Automobil Produktion, Januar 2005

[2.19]* Daimler-Benz AG (1979) Verkehr, Umwelt, Zukunft (Traffic, Environment, Future). Seminar der Forschungsgruppe Berlin, 27.–28.09.1979

[2.20]* Daimler-Benz AG (1979) Die O-Bahn (The O railway). WZ20847/00/00/0479

[2.21]* Daimler-Benz AG (1983) Report 1/83: Perspektiven des zukünftigen Verkehrs (Perspectives on Future Traffic). Schriftenreihe der Daimler-Benz AG. VDI-Verlag, Düsseldorf

[2.22]* Lauster E (1980) Untersuchungen und Berechnungen zum Wärmehaushalt mechanischer Getriebe (Investigations and Calculations Relating to Heat Management within Mechanical Transmissions). Dissertation, Universität Stuttgart, Institut für Maschinenelemente

[2.23]* Leutzbach W (1989) Zukunft des Straßenverkehrs (The Future of Road Traffic). Aachen: 2. Aachener Kolloquium „Fahrzeug- und Motorentechnik": 19–30

[2.24]* Lippl E (1973) Die Zukunft im Schnellverkehr. Eine reale Lösung (The Future for Express Traffic. A Realistic Solution). VDI-Zeitschrift 115: 1132–1135

[2.25]* Lohr FW (1989) Marktorientierte Kraftfahrzeugentwicklung (Market-Oriented Motor Vehicle Development). Aachen: 2. Aachener Kolloquium „Fahrzeug- und Motorentechnik": 1–18

[2.26]* Petri H, Najork R, Eggert U, Krauss C (2004) Systeme/Konzepte/Lösungen – Automatgetriebe 2004 (Systems/Concepts/Solutions – Automatic Transmissions 2004). VDI-Berichte Nr. 1827: 713–748

[2.27]* VDI (1976) Entwicklungslinien in der Kraftfahrzeugtechnik (Lines of Development in Automotive Engineering). Berlin: 2. Jahrestagung der VDI-Gesellschaft Fahrzeugtechnik


## Chapter 3


[3.1] BOSCH (2004) Kraftfahrtechnisches Taschenbuch (Automotive Handbook). 25th Edition. Vieweg-Verlag, Wiesbaden

[3.2] Bussien R (1965) Automobiltechnisches Handbuch (Automobile Engineering Handbook). Vol. 2. 18th Edition. Technischer Verlag Herbert Cram, Berlin

[3.3] Förster HJ (1987) Die Kraftübertragung im Fahrzeug vom Motor bis zu den Rädern. Handgeschaltete Getriebe (Power Transmission from the Engine to the Wheels in Motor Vehicles. Manual Transmissions). Verlag TÜV Rheinland GmbH, Köln

[3.4] Freymann R, Strobl W, Glonner H, Köpf P, Kilian S (2004) Wandlerfreies Automatikgetriebe und Elektromaschinen-Kondensatorsystem, ein neuer Ansatz zur Realisierung einer effizienten Dynamik (Converter-Free Automatic Transmission and E-Motor Capacitor System – A New Approach for the Implementation of Efficient Dynamics). VDI-Berichte Nr. 1827

[3.5] Gebert J, Glonner H, Michael J, Dreibholz R, Kilian S, Reisch M (2004) Neue Funktionspotenziale im Antriebsstrang durch Elektromaschinen im wandlerfreien Automatikgetriebe (New Function Potentials in the Powertrain Using E-Machines in the Converter-Free Automatic Transmission). VDI-Berichte Nr. 1852

[3.6] Jossen A, Weydanz W (2006) Moderne Akkumulatoren richtig einsetzen (Right Application of Modern Accumulators). Print-yourbook-Verlag Inge Reichardt

[3.7] MICHELIN (2002) Betriebsanleitung für Pkw-, Offroad- und Lkw-Reifen (Operating Instructions for Passenger Car, Off-road and Truck Tyres). 31st Edition. Michelin Reifenwerke KGaA, Karlsruhe

[3.8] MICHELIN (2004) Michelin Nutzfahrzeug-Reifen (Michelin Commercial Vehicle Tyres). Michelin Reifenwerke KGaA, Karlsruhe

[3.9] Mitschke M (2004) Dynamik der Kraftfahrzeuge. Band A: Antrieb und Bremsung (Dynamics of Motor Vehicles. Vol. A: Drive and Braking). 4th Edition. Springer, Berlin Heidelberg New York

[3.10] Regar KN, Braess HH, Reister D (1991) Der neue Elektro-3er von BMW – Glied einer langen Entwicklungskette (BMW's New Electric 3 Series – One Link in a Long Chain of Development). Aachen: 3. Aachener Kolloquium „Fahrzeug- und Motorentechnik": 551–598

[3.11] Scheurer K, Braess H, Jungk A, Wolf J (2003) Verkehrswirtschaftliche Energiestrategie (VES), Clean Energy Partnership Berlin (CEP) (Energy Strategy in Transport Economics, Clean Energy Partnership Berlin (CEP)). VDA Technischer Kongress, Wolfsburg

[3.12] VDI-Nachrichten (1997) Ausgabe Nr. 22, 30. Mai 1997 (Newspaper of the Association of German Engineers, Issue 22, 30th May 1997). Verein Deutscher Ingenieure, Düsseldorf

[3.13] Vahlensieck B, Speck FD, Feulner P, Sattler M, Mertinkat R (2005) Hybridantriebe – Module und Systeme von ZF (Hybrid Drive Trains: Modules and Systems by ZF). Aachen: 14. Aachener Kolloquium „Fahrzeug- und Motorentechnik"

[3.14] Wallentowitz H (2005) Längsdynamik von Kraftfahrzeugen. Umdruck zur Vorlesung Kraftfahrzeuge I (Longitudinal Dynamics of Motor Vehicles. Lecture Notes of the Motor Vehicles I Lecture). RWTH Aachen, Institut für Kraftfahrwesen

## Chapter 4

[4.1] Höhn BR (1990) Warum stufenlose Getriebe im Kraftfahrzeug? (Why Continuously Variable Transmissions in Motor Vehicles?) VDI-Berichte Nr. 803: 121–147

[4.2] Stall E (1984) Entwicklung einer elektronischen Regelung für kraftstoffsparendes Motor-Getriebe-Management im Forschungs-Pkw Uni-Car (Development of an Electronic Control for Fuel-Saving Engine/Transmission Management in the Uni-Passenger Car Research Vehicle). Dissertation, RWTH Aachen

[4.3] Theissen M, Petra H, Kern J, Klüting M (1992) Künftige Anforderungen an den Fahrzeugantrieb (Future Requirements to be Met by Vehicle Drives). ATZ Automobiltechnische Zeitschrift 94, Nr. 11: 562–570

## Chapter 5

[5.1] Albers A (1995) Selbsteinstellende Kupplung (SAC) und Zweimassenschwungrad (ZMS) zur Verbesserung des Antriebsstrangkomforts (Self-Adjusting Clutch (SAC) and Dual Mass Flywheel for Improvement of Powertrain Comfort). VDI-Berichte Nr. 1175: 153–168

[5.2] DIN 70020 (2004) Straßenfahrzeuge. Kraftfahrzeugbau. Teil 1: Personenkraftwagen. Begriffe, Grundlagen, Bestimmungen, Maßkurzzeichen (Road vehicles. Vehicle Manufacturing. Part 1: Passenger Cars. Terms, Fundamentals, Regulations, Measurement Abbreviations). DIN Deutsches Institut für Normung e.V., Berlin

[5.3] DIN 70030 (1986) Kraftfahrzeuge. Ermittlung des Kraftstoffverbrauchs. Teil 2: Lastkraftwagen und Kraftomnibusse (Road Vehicles. Determination of Fuel Consumption. Part 2: Goods Vehicles and Buses). DIN Deutsches Institut für Normung e.V., Berlin

[5.4] Fischer R, Jürgens G (1992) Einfluss von Pkw-Automatgetrieben auf den Verbrauch (The Influence of Passenger Car Automatic Transmissions on Consumption). VDI-Berichte Nr. 977: 527–544

[5.5] Heidingsfeld D, Rohs U (2001) Die Entwicklung des Mechanischen Torsions-Dämpfers (MTD) von der Idee bis zum fertigen Produkt (The Development of the Mechanical Torsion Damper, from the Idea to the Product). ATK Antriebstechnisches Kolloquium 2001, RWTH Aachen: 474–496

[5.6] Jörg B (1988) Abbau von Resonanzschwingungen im Antriebsstrang durch Drehschwingungstilger (Reducing Resonance Vibrations in the Powertrain Using Torsional Vibration Absorber Dampers). VDI-Berichte Nr. 697: 159–172

[5.7] Kern M (1990) Fahrstil und Ökonomie (Driving Style and Economy). Lastauto und Omnibus, Nr. 10

[5.8] Küçükay F (1987) Dynamik der Zahnradgetriebe (Dynamics of Gear Mechanisms). Springer, Berlin Heidelberg New York

[5.9] Laschet A (1988) Simulation von Antriebssystemen (Simulating Drive Systems). Springer, Berlin Heidelberg New York

[5.10] Lutz D (1988) Kupplungsmanagement – ein Baustein zur Drehschwingungsdämpfung (Clutch Management – a Component for Damping Torsional Vibrations). VDI-Berichte Nr. 697: 219–231

[5.11] Mitschke M, Wallentowitz H (2004) Dynamik der Kraftfahrzeuge (Dynamics of Motor Vehicles). 4th Edition. Springer, Berlin Heidelberg New York

[5.12] Nissen PJ, Heidingsfeld D, Kranz A (2000) Der MTD – neues Dämpfungssystem für KFZ Antriebsstränge (The MTD – New Damping System for Road Vehicle Powertrains). MTZ Motortechnische Zeitschrift 61: 386–391

[5.13] Peinemann B (2001) Drehzahladaptiver Tilger – eine Alternative für die Schwingungsreduzierung? (Speed-Adaptive Absorber Damper – an Alternative for Reducing Vibrations?) ATZ Automobiltechnische Zeitschrift 103: 290–296

[5.14] Pels T (1998) ISAD – Das Integrierte Starter-Alternator-Dämpfer-System (ISAD – The Integrated Starter Alternator Damper System). VDI-Berichte Nr. 1418: 295–309

[5.15] Petri H, Philipsenburg HJ (1991) Der hydraulische Torsionsschwingungsdämpfer (The Hydraulic Torsional Vibration Damper). VDI-Berichte Nr. 878: 361–379

[5.16] Rohs U, von Dahlen R, Heidingsfeld D (1995) Isolation der Drehschwingungen in einem Kfz-Antriebsstrang durch einen Hydraulischen Torsionsdämpfer (HTD) (Isolation of the Torsional Vibration in a Motor Vehicle Powertrain Using a Hydraulic Torsional Vibration Damper). ATK Antriebstechnisches Kolloquium 1995, RWTH Aachen: 223–242

[5.17] Ryborz J (2003) Klapper- und Rasselgeräuschverhalten von Pkw- und Nkw-Getrieben (Rattle and Clatter Noise Behaviour of Passenger Car and Commercial Vehicle Transmissions). Dissertation, Universität Stuttgart, Institut für Maschinenelemente

[5.18] Weidner G (1991) Klappern und Rasseln von Fahrzeuggetrieben (Rattling and Clattering in Vehicle Transmissions). Dissertation, Universität Stuttgart, Institut für Maschinenelemente

[5.19] Werner K, Eckel HG, Fehrecke H, Jörg B (1999) Motor- und Antriebsstrangberuhigung im System durch einen drehzahladaptiven Tilger (DAT) (Engine and Powertrain Damping Using a Speed-Adaptive Absorber Damper). Aachen: 8. Aachener Kolloquium „Fahrzeug- und Motorentechnik“: 1133–1141

## Chapter 6

[6.1] Allison Transmission (1999) Product Information: Allison Automatic Transmissions – 1000, 2000, 2400 Series

[6.2] Allison Transmission (2004) Operator's Manual: Allison Transmission On-Highway, 1000 and 2000 Product Families, OM3063EN

[6.3] Altmann FG (1940) Ausgleichsgetriebe für Kraftfahrzeuge (Differential Gears for Motor Vehicles). VDI-Zeitung 84, Nr. 31: 545–551

[6.4] Bensinger J, Heißling B (1987) Auslegung und Konzeption des Allradantriebs im Audi 80 quattro (Layout and Design of All-Wheel Drive in the Audi 80 Quattro). ATZ Automobiltechnische Zeitschrift 89: 247–251

[6.5] Blumenstock KU (1990) Von Sonnen und Planeten (Of Suns and Planets). mot Spezial: 72–77

[6.6] Brügel E, Klein H, Willibald K (1986) Die Visko-Kupplung als Bauelement in Antriebssystemen (The Visco-Clutch as a Component in Drive Systems). VDI-Berichte Nr. 579: 321–343

[6.7] Dach H, Gruhle WD, Köpf P (2001) Pkw-Automatgetriebe (Passenger Car Automatic Transmissions). 2nd Edition. Bibliothek der Technik Band 88, Verlag Moderne Industrie, Landsberg/Lech

[6.8] Eaton (1985) Product Information: Eaton Twin Splitter TS/TSO 11612

[6.9] Eaton (1997) Product Information: Eaton S Transmission Series

[6.10] Elser W (1998) Optimierung stufenloser Toroidgetriebe für Pkw (Optimisation of Continuously Variable Toroidal Gearboxes for Passenger Cars). VDI-Berichte Nr. 1393: 513–526

[6.11] Fecht N (2002) Treibende Kraft (Driving Force). Lastauto Omnibus, Nr. 2: 74–75

[6.12] Foth J, Karch G, Speck FD (2005) Getriebesysteme für Busse: Lastschaltgetriebe, automatisierte Getriebe oder Handschaltgetriebe? – Auswahlkriterien heute und morgen (Transmission Systems for Buses: Powershift Transmissions, Automated Transmissions or Manual Transmissions? – Selection Criteria Today and Tomorrow). VDI-Berichte Nr. 1876: 367–388

[6.13] Freymann R, Strobl W, Glonner H, Köpf P, Kilian S (2004) Wandlerfreies Automatikgetriebe und Elektromaschinen-Kondensatorsystem, ein neuer Ansatz zur Realisierung einer effizienten Dynamik (Converter-Free Automatic Transmission and Motor Capacitor System – A New Approach for the Implementation of Efficient Dynamics). VDI-Berichte Nr. 1827

[6.14] Grad K (2003) Getriebetechnologie für Traktoren (Transmission Technology for Tractors). Landtechnik 58, H. 3: 178–180

[6.15] Härdtle W (1997) Ein neues automatisiertes Schaltgetriebe für schwere Nutzfahrzeuge (A New Automated Manual Transmission for Heavy-Duty Commercial Vehicles). ATZ Automobiltechnische Zeitschrift 99: 598–604

[6.16] Härdtle W (1998) Automatisierte Getriebe für Nutzfahrzeuge: Integration von Funktionen und Komponenten (Automated Transmissions for Commercial Vehicles: Integration of Functions and Components). VDI-Berichte Nr. 1993: 193–213

[6.17] Hata H, Kamiya M, Nagamatsu S (2005) Development of a New Hybrid Transmission for FWD Sports Utility Vehicle. Toyota Technical Review Vol. 54 No. 1. Toyota Motor Corporation, Aichi (Japan), ISSN 0916-7501

[6.18] Heißling B, Müller E, Stockmar J (1989) Die Entwicklungsstufen von Pkw-Allradantrieben der Audi AG (The Development Stages in Audi Passenger Car All-Wheel Drives). Aachen: 2. Aachener Kolloquium „Fahrzeug- und Motorentechnik“: 651–685

[6.19] Hoffmann C, Würfel A (2002) Dreiwellengetriebe – Die kurze Art, Sechsganggetriebe zu bauen (Three-Shaft Transmissions – The Short Way to Build 6-Speed-Transmissions). ATZ Automobiltechnische Zeitschrift 104, Nr. 12: 1090–1099

[6.20] James FJ, Wegeng WJ (1999) Leveraging a Commercial Medium-Duty Product for the Heavy-Duty Pickup Truck Market. SAE Technical Paper Series, 1999-01-3743

[6.21] James IB, Greenwood CJ (1998) Pressure Control Issues for Higher Viscosity Traction Fluids in a Torque Controlled IVT. VDI-Berichte Nr. 1393: 695–702

[6.22] Jarchow F (1989) Die Wirkungsweise von Schubgliederketten (The Mode of Operation of Thrust Link Chains). Konstruktion 41: 73–81

[6.23] Jarchow F (1991) Stufenlos wirkende hydrostatisch-mechanische Lastschaltgetriebe (Infinitely Variable Hydrostatic/Mechanical Powershift Transmissions). VDI-Berichte Nr. 878: 623–647

[6.24] Kasauya S, Taniguchi T, Tsukamoto K, Hayabuchi M, Nishida M, Suzuki A (2005) Aisin AW New High Torque Capacity Six-Speed Automatic Transmission for FWD Vehicles. SAE 2005-01-1020

[6.25] Keller W, Härdtle W (2000) Erfahrungen bei Entwicklung und Einsatz automatischer Getriebe für Nutzfahrzeuge (Experiences in Development and Application of Automatic Transmissions for Commercial Vehicles). Aachen: 9. Aachener Kolloquium „Fahrzeug- und Motorentechnik“: 1247–1258

[6.26] Köpf P (1991) Entwicklungstendenzen bei Nkw-Getrieben (Development Trends in Commercial Vehicle Transmissions). VDI-Berichte Nr. 878: 21–40

[6.27] Köpf P (2005) Ausführungsbeispiele automatischer Getriebe und CVT. Lehrgang „Automatische Getriebe“ (Design Examples of Automatic and Continuously Variable Transmissions. Seminar “Automatic Transmissions”). TAE Technische Akademie Esslingen, Ostfildern

[6.28] Kubalczyk R, Ebenhoch M, Schneider HJ (2006) 7-Gang-Doppelkupplungsgetriebe für sportliche Anwendungen (Seven Speed Dual Clutch Transmission System for Sportive Applications). VDI-Berichte Nr. 1943

[6.29] Lechner G (1970) Getriebelehre II. Angewandte Getriebelehre (Theory of Gears II. Applied Theory of Gears). Walter de Gruyter & Co, Berlin

[6.30] Looman J (1996) Zahnradgetriebe (Gear Mechanisms). 3rd Edition. Springer, Berlin Heidelberg New York

[6.31] Möllers W, Müller R (1986) Der Antriebsstrang allradgetriebener Personenkraftwagen (The Powertrain of All-Wheel Drive Passenger Cars). VDI-Berichte Nr. 579: 189–213

[6.32] Ott A (1968) Zur systematischen Synthese mehrgängiger Umlaufräder-Schaltgetriebe (On the Systematic Synthesis of Multi-Speed Planetary Gear Selector Gearboxes). ZF-Sonderdruck aus: ATZ Automobiltechnische Zeitschrift 70, Nr. 1, 3, 4

[6.33] Peschke W (1989) Die Wirkungsweise der Visko-Kupplung und ihr Einfluss auf die Traktion eines Allradfahrzeugs (The Principle of Operation of the Visco-Clutch and its Effect on the Traction of an All-Wheel Drive Vehicle). Dissertation, Universität Hannover

[6.34] Pohlenz J, WD Gruhle (2001) Stufenloses hydrostatisch-mechanisch leistungsverzweigtes Getriebe ECCOM für den Einsatz in Traktoren (Continuously Variable Hydrostatic-Mechanical Power-Split Gearbox ECCOM for the Use in Tractors). VDI-Berichte Nr. 1592: 451–468

[6.35] Raeder M, Rüchardt C, Speck F (2004) Die ZF AS-Tronic-Familie – Automatische Getriebe für alle Nutzfahrzeugklassen (The ZF AS-Tronic Series – Automatic Transmissions for all Commercial Vehicle Classes). ATZ Automobiltechnische Zeitschrift 106: 772–783

[6.36] Renius KT (2003) Hydrostatische Fahrantriebe für mobile Arbeitsmaschinen (Hydrostatic Drive Transmissions for Mobile Machinery). VDI-Berichte Nr. 1793: 65–78

[6.37] Renius KT, Resch R (2005) Continuously variable tractor transmissions. ASAE Distinguished Lecture Series No. 29: 1–37. ASAE Publication Number 913C0305. American Society of Agricultural Engineers, St. Joseph, MI

[6.38] Rösch R, Wagner G (1995) Die elektrohydraulische Steuerung und äußere Schaltung des automatischen Getriebes W5A330/580 von Mercedes Benz – Aufbau und Betriebsverhalten des mechanischen Getriebeteils (The Electrohydraulic Control Unit and External Gearshift System of the Automatic Transmission W5A330/580 by Mercedes-Benz – Design and Operating Behaviour of the Mechanical Gearbox Part). ATZ Automobiltechnische Zeitung 97, Nr. 10: 698–706

[6.39] Rösch R, Zaiser W, Kürschner M (1995) Das neue automatische Getriebe W5A330/580 für Personenwagen von Mercedes-Benz – Aufbau und Betriebsverhalten des mechanischen Getriebeteils (The New Automatic Transmission W5A330/580 for Passenger Cars by Mercedes-Benz – Design and Operating Behaviour of the Mechanical Gearbox Part). ATZ Automobiltechnische Zeitung 97, Nr. 9: 536–545

[6.40] Schaller KV (1997) Automatisiertes Getriebeschaltsystem SAMT B (Semi Automated Manual Transmission SAMT B). VDI-Berichte Nr. 1341

[6.41] Schmidt R, Diessner G, Uhlig G, Otto H (1986) Antriebskonzept, Bauteile und Eigenschaften des VW Golf syncro (Drive Concept, Components and Characteristics of the VW Golf Syncro). ATZ Automobiltechnische Zeitschrift 88: 477–485

[6.42] Tenberge P (1998) Toroidgetriebe mit verbesserten Kennwerten (Toroidal Gearbox with Improved Characteristics). VDI-Berichte Nr. 1393: 703–724

[6.43] Van Doorne´s Transmissie b.v.: Company Publication

[6.44] Wagner G, Bucksch M, Scherer H (2001) Das automatische Getriebe 6 HP 26 von ZF – Getriebesystem, konstruktiver Aufbau und mechanische Bauteile (The Automatic Transmissions 6 HP 26 of ZF – Gearbox System, Engineering Design and Mechanical Components). VDI-Berichte Nr. 1610

[6.45] Witte L, Gorissen W (1987) Beeinflussung der Fahreigenschaften durch den Antrieb (The Influence of the Drive System on Handling Characteristics). VDI-Berichte Nr. 650: 259–282

[6.46] ZF Friedrichshafen Product Information: ZF AS-Tronic

[6.47] ZF Friedrichshafen Product Information: ZF-Ecomat and ZF-Ecomat 2

[6.48] ZF Friedrichshafen (2003) Product Information: ZF-Ecomid

[6.49] ZF Friedrichshafen (2003) Product Information: ZF-Ecosplit

[6.50] ZF Friedrichshafen Product Information: ZF-eTronic

[6.51]* Behles F (1969) Die Triebwerksanordnung beim Frontantrieb (Power Unit Layout with Front-Wheel Drive). ATZ Automobiltechnische Zeitschrift 71: 35–39

[6.52]* Behles F, Barske H (1981) Die Frontantriebsentwicklung bei Audi, NSU, Auto Union und DKW (The Development of Front-Wheel Drive at Audi, NSU, Auto Union and DKW). Part 1. ATZ Automobiltechnische Zeitschrift 83: 487–495; Part 2: ATZ Automobiltechnische Zeitschrift 83: 557–564

[6.53]* Behles FG, Schneeweiss MA (1981) Front Wheel Drive Power Trains and Suspensions. SAE-Paper 810422

[6.54]* Lorenz R (1991) Variable Verteilung von Längs- und Querkräften beim Bremsen mit Allradantrieb (Variable Distribution of Axial and Transverse Forces when Braking with All-Wheel Drive). 20. Mitarbeiterveröffentlichung Daimler-Benz AG: 1–20

## Chapter 7

[7.1] Bellomo P, Cricenti F, De Vito N, Lang CH, Minervini D (2000) Innovative Vehicle Powertrain Systems Engineering: Beating the Noisy Offenders in Vehicle Transmissions. SAE Technical Paper 2000-01-0033

[7.2] Bick R, Britten A, Joachim FJ (1998) Moderne Methoden zur Verzahnungsberechnung (Modern Methods for Tooth Calculation). VDI-Berichte Nr. 1393: 391–416

[7.3] Borenius G (1990) Zur rechnerischen Schädigungsakkumulation in der Erprobung von Kraftfahrzeugteilen bei stochastischer Belastung mit variabler Mittellast (Numerical Damage Accumulation in the Testing of Motor Vehicle Components with Stochastic Loading with Variable Mean Load). Dissertation, Universität Stuttgart, Institut für Maschinenelemente

[7.4] Buxbaum O (1986) Betriebsfestigkeit (Operational Fatigue Strength). Verlag Stahleisen GmbH, Düsseldorf

[7.5] DIN 3990 (1987–1994) Tragfähigkeitsberechnung von Stirnrädern (Capacity Calculation for Spur Gear Drives). Part 1–6, 12, 21, 31, 41. DIN Deutsches Institut für Normung e.V., Berlin

[7.6] DIN 45667 (1969) Klassierverfahren für das Erfassen regelloser Schwingungen (Classification Methods for Evaluation of Random Vibrations). DIN Deutsches Institut für Normung e.V., Berlin

[7.7] DIN 50100 (1978) Dauerschwingversuche (Continuous Vibration Test). DIN Deutsches Institut für Normung e.V., Berlin

[7.8] DIN ISO 362 (2005) Messverfahren der Genauigkeitsklasse 2 für das von beschleunigten Straßenfahrzeugen abgestrahlte Geräusch (Measurement of Noise Emitted by Accelerating Road Vehicles - Engineering Method). Part 1–2. DIN Deutsches Institut für Normung e.V., Berlin

[7.9] DIN ISO 14635 (2005) Zahnräder: FZG-Prüfverfahren (Gears – FZG Test Procedures). DIN Deutsches Institut für Normung e.V., Berlin

[7.10] DISCOM: RotasPro Geräuschanalyse-System (RotasPro Noise Analysis System), Company Publication

[7.11] Fan J, Humberg P (2003) Verdrehspiele im Antriebsstrang, Beispiele zur Optimierung von Schwingungsphänomenen durch Toleranzanalyse in Design und Fertigung (Backlashes in the Powertrain. Examples for the Optimisation of Vibration Phenomena in Design and Manufacturing). Tagung „Dynamisches Gesamtverhalten von Fahrzeugantrieben". Haus der Technik, Essen

[7.12] Finsterhölzl H, Kelle W (1990) Leerlaufqualität in der Fahrzeugbeurteilung (Idling Quality in Vehicle Assessment). ATZ Automobiltechnische Zeitschrift 92, Nr. 5: 268–282

[7.13] Gosdin M (1985) Analyse und Optimierung des dynamischen Verhaltens eines Pkw-Antriebsstrangs (Analysis and Optimisation of the Dynamic Behaviour of a Passenger Car Powertrain). Dissertation, Technische Universität München

[7.14] Goto Y, Yagi Y, Morimoto Y, Kawasaki M (1988) Shift Feel in Manual Transmission – An Analysis of Unsmooth Shifting and Gear Clashing. JSAE Review Vol. 9, No. 4

[7.15] Haibach E (1989) Betriebsfestigkeit (Operational Fatigue Strength). VDI-Verlag, Düsseldorf

[7.16] Hildebrandt M (1994) Schadensfrüherkennung an Wälzkontakten mit Körperschall-Referenzsignalen (Early Damage Detection at Rolling Contacts Using Structur-Borne Noise Reference Signals). Dissertation, Universität Stuttgart, Institut für Maschinenelemente

[7.17] Hirschmann KH (1993) Methodische Lebensdauerberechnung bei Maschinenelementen (Methodical Calculation of Service Life for Machine Elements. TAE Esslingen, Lecture Course No. 17019/60.175: Reliability analysis of systems). TAE Technische Akademie Esslingen, Lehrgang Nr. 17019/60.175: Zuverlässigkeitsanalyse von Systemen

[7.18] Joachim FJ (1992) Schäden an Zahnradgetrieben und ihre Verhütung. TAE-Band 28: Schäden an geschmierten Maschinenelementen: 161–201 (Damages of Gear Transmissions and its Prevention, TAE Volume 28: Damages of Lubricated Machine Components)

[7.19] Joachim FJ, Collenberg HF (1996) Scuffing Resistance of Vehicle Transmission Gears. AGMA Technical Paper, 96FTM10

[7.20] Kanagawa J, Okubo K, Fujii T, Kido R, Takahashi M (2004) Study on Noise Generation Mechanism for Dry Hybrid Type CVT – Influence of Block Motions and Surface Roughness of Pulley on Sound Pressure. SAE Technical Paper Series 2004-01-0477

[7.21] Klätschke H (2002) Standardisierte Lastkollektive und Lastfolgen für Pkw-Antriebsstränge mit Automatgetrieben, CARLOS-PTA (Standardised Load Profiles and Load Effects for Passenger Car Powertrains with Automatic Transmissions, CARLOS-PTA). Fraunhofer Institut für Betriebsfestigkeit mbH (LBF) Darmstadt, LBF-Bericht Nr. 110310/110370(d)

[7.22] Krishnaswami R, DeFore M, Hildebrand D, Metcalf J, Sell D (2001) Gear Whine Improvements for an Automatic Transmission through Design Retargeting and Manufacturing Variability Reduction. SAE Technical Paper 2001-01-1505

[7.23] Kronast M, Hildebrandt M (2000) Vibro-Acoustic Modal Analysis of Automobile Body Cavity Noise. Sound and Vibration, Vol. 34, No. 6

[7.24] Kunz M (2002) Ermittlung des Einflusses fahrzeug-, fahrer- und verkehrsspezifischer Parameter auf die Getriebelastkollektive mittels Fahrsimulation (Determination of the Influence of Vehicle-, Traffic- and Driver-Specific Parameters on the Transmission Load Spectra by Drive-Line Simulation). Dissertation, Universität Stuttgart, Institut für Maschinenelemente

[7.25] Lechner G (1966) Die Fress-Grenzlast bei Stirnrädern aus Stahl (The Scuff-Limited Load for Steel Spur Gears). Dissertation, Technische Universität München

[7.26] Lechner G (1973) Berechnung der Fresstragfähigkeit von Stirn- und Kegelrädern (Calculating the Scuffing Resistance of Spur Gears and Bevel Gears). Technisch-wissenschaftliche Veröffentlichung der ZF AG, Heft 8

[7.27] Miner MA (1937) Cumulative Damage in Fatigue. Trans. New York: ASME (American Society of Mechanical Engineers), Journal of Applied Mechanics 4: A160–A162

[7.28] Müller-Kose JP (2002) Repräsentative Lastkollektive für Fahrzeuggetriebe (Representative Load Profile for Automotive Transmissions). Dissertation, Universität Braunschweig

[7.29] Niemann G, Winter H (2002) Maschinenelemente. Band II: Getriebe allgemein, Zahnradgetriebe – Grundlagen, Stirnradgetriebe (Machine Components. Volume II: Transmissions in General, Gear Transmissions – Fundamentals, Spur Gear Transmissions). Springer, Berlin Heidelberg New York

[7.30] Nill R (1986) Das Schwingungsverhalten loser Bauteile in Fahrzeuggetrieben (The Vibration Behaviour of Loose Components in Vehicle Transmissions). Dissertation, Universität Stuttgart, Institut für Maschinenelemente

[7.31] Palmgren A (1924) Die Lebensdauer von Kugellagern (The Service Life of Ball Bearings). VDI-Zeitschrift 68, Nr. 14: 339–341

[7.32] Reik W (1987) Torsionsschwingungen und Getriebegeräusche (Torsional Vibrations and Transmission Noise). Automobil-Industrie 32, Nr. 1: 37–42

[7.33] Reilhofer KG: EOL-Analyser – Akustische Qualitätssicherung an Getrieben (EOL-Analyser – Acoustic Quality Assurance of Transmissions), Company Publication

[7.34] Directive 92/97/EEC

[7.35] Schöpf HJ, Jürgens G, Fischer R (1989) Optimierung der Komforteigenschaften des Triebstrangs von Mercedes-Benz-Fahrzeugen mit Schaltgetriebe (Optimising the Comfort Aspects of the Powertrain of Mercedes-Benz Manual Transmission Vehicles). ATZ Automobiltechnische Zeitschrift 91, Nr. 10: 568–575

[7.36] Schumacher T, Reitz A, Biermann JW (2001) Lastwechselschwingungen in Kfz-Antriebssträngen – eine Kompromissauslegung zwischen Komfort und Agilität (Load Reversal Vibration in Motor Vehicle Powertrains – a Compromise Design between Comfort and Agility). Tagung „Systemanalyse in der Kfz-Antriebstechnik". Haus der Technik, Essen

[7.37] Weidner G (1991) Klappern und Rasseln von Fahrzeuggetrieben (Rattling and Clattering in Vehicle Transmissions). Dissertation, Universität Stuttgart, Institut für Maschinenelemente

[7.38] Weidner G, Lechner G (1990) Klapper- und Rasselgeräusche in Fahrzeuggetrieben (Rattling and Clattering Noise in Vehicle Transmissions). ATZ Automobiltechnische Zeitschrift 92, Nr. 6: 320–326

[7.39] Westermann-Friedrich, Zenner H (1988) Zählverfahren zur Bildung von Kollektiven aus Zeitfunktionen – Vergleich der verschiedenen Verfahren und Beispiele (Counting Procedures for Creating Load Profiles out of Time Functions – Comparison of Different Methods and Examples). FVA-Merkblatt Kollektive, Nr. 0/14. Forschungsvereinigung Antriebstechnik e.V., Frankfurt

[7.40] ZFN 201 (1990) Zahnradschäden, Begriffsbestimmung (Gearwheel Damage, Definition). Corporate Standard ZF Friedrichshafen AG

[7.41] Zwicker E, Fastl H (1990) Psychoacoustics. Facts and Models. Springer, Berlin Heidelberg New York

[7.42]* Henn H, Sinambari GR, Falle M (1984) Ingenieurakustik (Engineering Acoustics). Vieweg-Verlag, Wiesbaden

[7.43]* James B, Douglas M (2002) Development of a Gear Whine Model for the Complete Transmission System. SAE Technical Paper 2002-01-0700

[7.44]* Küçükay F (1987) Dynamik der Zahnradgetriebe (Dynamics of Gear Mechanisms). Springer, Berlin Heidelberg New York

[7.45]* Schildmeier EF (1998) Gear Noise Reduction Using Design of Experiments. SAE Technical Papers 981194

[7.46]* VDI-Richtlinie 2159 (1985) Emissionskennwerte technischer Schallquellen – Getriebegeräusche (Characteristic Values of Emissions of Technical Sound Sources – Gearing Noise). VDI Verein Deutscher Ingenieure e.V., Düsseldorf

## Chapter 8

[8.1] Dietrich G, Stahl H (1978) Matrizen und Determinanten und ihre Anwendungen in Technik und Ökonomie (Matrices and Determinants and their Applications in Engineering and Economics). VEB, Leipzig

[8.2] DRESP, Schwingungsberechnungsprogramm (Vibration Calculation Program). RWTH Aachen

[8.3] EN 10083 (1996) Vergütungsstähle. Teil 1: Technische Lieferbedingungen für Edelstähle (Steels for Quenching and Tempering – Part 1: General Technical Delivery Conditions for Special Steels). DIN Deutsches Institut für Normung e.V., Berlin

[8.4] Roos E (2005) Festigkeitslehre I. Vorlesungsmanuskript (Strength Theory I. Lecture Notes). Universität Stuttgart

[8.5] Muhs D, Wittel H, Jannasch D, Voßiek J (2005) Roloff/Matek Maschinenelemente (Roloff/Matek Machine Components). 17th Edition. Vieweg-Verlag, Wiesbaden

[8.6] Niemann G, Winter H, Höhn BR (2005) Maschinenelemente. Band 1: Konstruktion und Berechnung von Verbindungen, Lagern, Wellen (Machine Components, Vol. I.: Design and Calculation of Joints, Bearings, Shafts). 4th Edition. Springer, Berlin Heidelberg New York

[8.7] Rühl KH (1952) Die Tragfähigkeit metallischer Baukörper in Bautechnik und Maschinenbau (The Load-Bearing Capacity of Metal Structures in Construction and Mechanical Engineering). Verlag Wilhelm Ernst & Sohn, Berlin

[8.8] Siebel E, Ludwig N (1955) Handbuch der Werkstoffprüfung. Band 2: Die Prüfung der metallischen Werkstoffe (Materials Testing Handbook. Volume 2: Testing Metal Materials). 2nd Edition. Springer, Berlin Heidelberg New York

[8.9] Thum A, Peterson C, Svenson O (1960) Verformung, Spannung und Kerbwirkung (Deformation, Stress and Stress Concentration). VDI-Verlag, Düsseldorf

[8.10] Truong-Canh H (1986) Strukturorientiertes Modellieren, Optimieren und Identifizieren von geometrisch linearen Mehrkörpersystemen (Structure-Oriented Modelling, Optimising and Identifying of Geometrical Linear Multi-Body Systems). Dissertation, Universität Stuttgart, Institut für Maschinenelemente

[8.11] Wellinger K, Dietmann H (1976) Festigkeitsberechnung, Grundlagen und technische Anwendung (Strength Calculation, Basic Elements and Technical Application). Kröner-Verlag, Stuttgart

[8.12]* DIN 3960 (1987) Begriffe und Bestimmungsgrößen für Stirnräder (Zylinderräder) und Stirnradpaare (Zylinderradpaare) mit Evolventenverzahnung (Definitions, Parameters and Equations for Involute Cylindrical Gears and Gear Pairs). DIN Deutsches Institut für Normung e.V., Berlin

[8.13]* Göldner H, Holzweißig F (1967) Leitfaden der Technischen Mechanik. Statik – Festigkeitslehre – Kinematik – Dynamik (Guide to Engineering Mechanics. Statics – Strength Theory – Kinematics – Dynamics). 8th Edition. Steinkopff-Verlag, Darmstadt

[8.14]* Hähnchen R, Decker KH (1967) Neue Festigkeitsberechnung für den Maschinenbau (New Strength Calculation for Mechanical Engineering). Hanser-Verlag, München

[8.15]* Kleinbach K (1989) Qualitätsbeurteilung von Kegelradsätzen durch integrierte Prüfung von Tragbild, Einflankenwälzabweichung und Spielverlauf (Quality Assessment of Bevel Gear Sets by Integrated Inspection and Testing of Contact Pattern, Single Flank Working Devitation and Backlash Pattern). Dissertation, Universität Stuttgart, Institut für Maschinenelemente

[8.16]* Magnus K, Müller HH (2005) Grundlagen der Technischen Mechanik (Basic Elements of Engineering Mechanics). 7th Edition. Teubner-Verlag, Stuttgart

## Chapter 9

[9.1] Bencker R, Walter G (2005) Das neue kleine Sechsgang-Handschaltgetriebe bei BMW (The New Small 6-Speed Manual Transmission of BMW). 4. Internationales CTI-Symposium „Innovative Fahrzeug-Getriebe", Berlin

[9.2] Bencker R, Nussbaumer M, Schlecht B (2004) Schaltkomfortorientierte Getriebe- und Antriebsstrangauslegung (Shifting Comfort Oriented Transmission and Powertrain Design). ATZ Automobiltechnische Zeitschrift 106, Nr. 11: 996–1006

[9.3] Bönning J, Krake F, Krüttgen R, Painter S (2004) Neues Sechsgang-Schaltgetriebe von Opel (New 6-Speed Manual Transmission by Opel). ATZ Automobiltechnische Zeitschrift 106, Nr. 10: 862–872

[9.4] Davis G, Donin R, Findlay M, Harman P, Ingram M, Kelly D (2004) Simulationsgestützte Optimierung der Schaltqualität von Handschaltgetrieben (Simulation Supported Optimisation of Shifting Quality of Manual Transmissions). ATZ Automobiltechnische Zeitschrift 106, Nr. 7/8: 668–675

[9.5] Drott H, Oppertshäuser E (1991) 3-Konus-Synchronisierungen in Opel-PKW-Schaltgetrieben (3-Cone Synchronizers in Opel Passenger Car Manual Transmissions). VDI-Berichte Nr. 878: 293-309

[9.6] Eaton (1997) Company Brochure "Eaton S Transmission Series" and "Eaton Fuller S Transmission with LF Synchronization"

[9.7] Förster HJ (1991) Automatische Fahrzeuggetriebe (Automatic Vehicle Transmissions). Springer, Berlin Heidelberg New York

[9.8] Franke A, Greiner J, Gösele A, Kiesel J, Sperber R, Stenkert J, Wagner B, Weiss P (2004) Triebstrang und Getriebe (Powertrain and Transmission. ATZ/MTZ Special Issue 10/2004: The New Mercedes-Benz A-Class: 150-159). ATZ/MTZ-Sonderausgabe 10/2004: Die neue Mercedes-Benz A-Klasse: 150–159

[9.9] Frey D, Naumann W (1993) Reibbeläge für ölgekühlte Reibschaltelemente (Friction Linings for Oil-Cooled Frictional Shifting Elements). ATK Antriebstechnisches Kolloquium 1993, RWTH Aachen

[9.10] Goll S, Schröper R (1986) Das Schaltelement „Synchronisierung" – eine hochentwickelte Baugruppe bei PKW- und NKW-Getrieben (The "Synchronizer" Shifting Element – a Sophisticated Assembly in Passenger Car and Commercial Vehicle Transmissions). VDI-Berichte, Nr. 579: 287–300

[9.11] Goto Y, Yagi Y, Morimoto Y, Kawasaki M (1988) Shift Feel in Manual Transmission – An Analysis of Unsmooth Shifting and Gear Clashing. JSAE Review Vol. 9, No. 4

[9.12] Greiner J, Alber H, Wagner B, Franke A, Dorfschmid J (2004) Präzises Räderwerk (Precise Gear Train). ATZ/MTZ-Sonderausgabe 4/2004: Die neue SLK-Klasse: 94–99

[9.13] Hämmerl B (1995) Lamellenauslegung, Hitzeflecken, Auslegungsrichtlinien für ölgekühlte Lamellenkupplungen (Multi-Plate Design, Hot Spots, Design Guidelines for Oil-Cooled Multi-Plate Clutches). FVA-Forschungsvorhaben 150, Heft-Nr. 456

[9.14] Hauser C (2004) Einfluss der Ölalterung auf die Reibwertstabilität von Lamellenkupplungen (Influence of Oil Ageing on the Stabilitiy of the Coefficient of Friciton of Multi-Plate Clutches). FVA-Forschungsvorhaben 297, Heft-Nr. 737

[9.15] Holzer N, Frey D, Matthes B (1997) Schleppmomente an nasslaufenden Lamellenbremsen (Drag Torque of Wet Multi-Plate Brakes). VDI-Berichte Nr. 1323: 469–489

[9.16] Köpf P (1980) Die Synchronisierung von Fahrzeuggetrieben (Synchronizing Vehicle Transmissions). Automobil Revue Nr. 18 (24.04.1980): 61–67

[9.17] Looman J (1996) Zahnradgetriebe (Gear Mechanisms). 3rd Edition. Springer, Berlin Heidelberg New York

[9.18] Mosbach C (2001) Untersuchung der Ursachen von Reibschwingungen und Geräuschen an ölgekühlten Reibschaltelementen (Investigation of the Causes of Friction Vibrations and Noises of Oil-Cooled Frictional Shifting Elements). FVA-Forschungsvorhaben 297, Heft-Nr. 650

[9.19] Naumann W, Frey D (1988) Beeinflussung kraftschlüssiger Schaltelemente durch das Öl (Influencing Frictional Shifting Elements by the Oil). VDI-Berichte Nr. 680

[9.20] Oerleke C (2000) Einfluss auf die Schleppmomente schnelllaufender Lamellenkupplungen in Automatgetrieben (Influence on the Drag Torques of Fast-Running Multi-Plate Clutches in Automatic Transmissions). Dissertation, Universität der Bundeswehr Hamburg

[9.21] Oerleke C (2000) Leerlaufverhalten von ölgekühlten Lamellenkupplungen (Open Clutch Behaviour of Oil-Cooled Multi-Plate Clutches). FVA-Forschungsvorhaben 290, Heft-Nr. 607

[9.22] Pfleger F (1998) Einfluss von Lamellenbehandlung und modernen Getriebeölen auf das Lebensdauer- und Schaltverhalten von Lamellenkupplungen (Influence of Multi-Plate Treatment and Modern Transmission Oils on the Service Life and Shifting Behaviour of Multi-Plate Clutches). FVA-Forschungsvorhaben 150, Heft-Nr. 537

[9.23] Rastinger B (1991) Die neuen Handschaltgetriebe in den BMW-Baureihen 8, 5 und 3 (The New Manual Transmissions in the BMW Series 8, 5 and 3). VDI-Berichte Nr. 878: 75–95

[9.24] Rüchardt C, Ebner O, Keller R, Rossmann M, Jäger T, Gansohr M, Erb M, Thüncher G (2004) AS Tronic lite, die neue AS-Getriebebaureihe der ZF für das mittlere Nutzfahrzeugsegment (AS Tronic Lite, the New AS-Transmission Product Range of ZF for Medium Duty Trucks). VDI-Berichte Nr. 1827: 303–326

[9.25] Schröper R (1981) Temperaturmessungen an Synchronisierungen (Temperature Measurements at Synchronizers). Unveröffentlichte Diplomarbeit, Universität Stuttgart, Institut für Maschinenelemente

[9.26] Wagner D (1993) Neue Reibmaterialien und Konzepte für Einfach- und Mehrfachsynchronisierungen (New Friction Materials and Concepts for Single and Multiple Synchronizers). ATZ Automobiltechnische Zeitschrift 95, Nr. 7/8: 380–387

[9.27] Winkelmann S, Harmuth H (1985) Schaltbare Reibkupplungen (Shiftable Friction Clutches). Springer, Berlin Heidelberg New York

[9.28] Winter H, Pflaum H (1991) Synchronisierungen in Kfz-Schaltgetrieben – Einfluss der Werkstoffpaarung auf das Reibungsverhalten (Synchronizers in Motor Vehicle Manual Transmissions – Influence of Material Mating on Friction Behaviour). Antriebstechnik 30, Nr. 8: 50–58

[9.29] ZF Friedrichshafen AG: Calculation Reference Folder

[9.30] ZF Friedrichshafen AG: Graphical Material

## Chapter 10

[10.1] Basedow G (1972) Berechnung der Rutschzeit von Anlaufkupplungen bei veränderlichem Lastmoment (Calculation of the Slipping Time of Starting Clutches under Varying Load Torque). Antriebstechnik 11: 367–370

[10.2] Baumann R, Engel S, Sasse C (2003) Die neue Generation moderner Hochleistungswandlerkupplungen mit patentiertem Ringnutsystem (The New Generation of Modern High Performance Converter Lock-Up Clutches with Patented Annular Groove System). VDI-Berichte Nr. 1786

[10.3] Bausch E (1979) Grenzen der Schaltarbeit an Reibelementen von Kupplungen und Bremsen (Limits of the Shift Energy at Friction Elements of Clutches and Brakes). Antriebstechnik 18: 367–370

[10.4] DIN 2092 (2006) Tellerfedern – Berechnung (Disc springs – Calculation). DIN Deutsches Institut für Normung e.V., Berlin

[10.5] Drexl HJ (1997) Kraftfahrzeugkupplungen – Funktion und Auslegung (Automotive Clutches – Function and Design). Verlag Moderne Industrie, Landsberg/Lech

[10.6] Duminy J (1979) Beurteilung des Betriebsverhaltens schaltbarer Reibkupplungen (Evaluation of the Operating Behaviour of Shiftable Friction Clutches). Dissertation, Technische Univeristät Berlin

[10.7] Förster B, Lindner J, Steinel K, Stürmer W (2004) Kupplungssysteme für schwere Nutzfahrzeuge (Clutch Systems for Heavy-Duty Commercial Vehicles). ATZ Automobiltechnische Zeitschrift 106, Nr. 10

[10.8] Förster HJ (1991) Automatische Fahrzeuggetriebe (Automatic Vehicle Transmissions). Springer, Berlin Heidelberg New York

[10.9] Großpietsch W, Sudau J (2000) Doppelkupplung für lastschaltbare Getriebe – ein altes Schaltelement mit neuer Zukunft? (Dual Clutch for Powershift Transmissions – a Traditional Engaging Element with New Future?) VDI-Berichte Nr. 1565

[10.10] Hasselgruber H (1953) Die Berechnung der Temperaturen an Reibungskupplungen (The Calculation of Temperatures of Friction Clutches). Dissertation, RWTH Aachen

[10.11] Hasselgruber H (1959) Temperaturberechnungen für mechanische Reibkupplungen (Temperature Calculation for Mechanical Friction Clutches). Schriftenreihe Antriebstechnik, Bd. 21

[10.12] Jauch F (1989) Beitrag zur Optimierung des Antriebsstranges von Kraftfahrzeugen unter dem Aspekt von Kraftstoffverbrauch, Transportleistung und Lastkollektiven (Contribution to Optimising the Powertrain of Motor Vehicles in Terms of Fuel Consumption, Transportation Efficiency and Load Profiles). Dissertation, Universität Stuttgart, Institut für Maschinenelemente

[10.13] Krüger H (1965) Das Reibungsverhalten der nassen Lamellenkupplung (The Friction Behaviour of the Wet Multi-Plate Clutch). Konstruktion 17: 54–60

[10.14] Krüger H (1965) Das Reibungsverhalten der nassen Lamellenkupplung (The Friction Behaviour of the Wet Multi-Plate Clutch). Konstruktion 17: 93–99

[10.15] Lebedeu, Bykova, Luck (1981) Zur Theorie der Lamellen-Schaltkupplungen (On the Theory of Multi-Plate Shifting Clutches). Maschinenbautech. 30: 215–221

[10.16] Mc Garth M, Müller B, Maucher E, Marathe B, Bailey G (2002) Der Drehmomentwandler (The Torque Converter). 7. LuK-Kolloquium

[10.17] Middlemann V, Wagner U (1998) The Torque Converter as a System. 6. LuK-Kolloquium

[10.18] Tenbrock F, Roßmann T, Neumann D, Jauch F, Gierer G (1999) Modellbasierte Funktionsentwicklung und Applikation am Beispiel einer Wandlerüberbrückungskupplung (Model-Based Function Development and Application Using a Converter Lock-Up Clutch as Example). Berlin, 2. Symposium Steuersysteme für den Antriebsstrang von Kfz

[10.19] Reik W, Friedmann O, Agner I, Werner O (2004) Die Kupplung – das Herz des Doppelkupplungsgetriebes (The Clutch – the Heart of the Dual Clutch Transmission). VDI-Berichte Nr. 1827

[10.20] SAE J618 (1991) Flywheels for Single-Plate Spring-Loaded Clutches. SAE Society of Automotive Engineers

[10.21] SAE J619 (1993) Flywheels for Two-Plate Spring-Loaded Clutches. SAE Society of Automotive Engineers

[10.22] Sasse C, Frey P, Ackermann J (2004) Hydrodynamischer Drehmomentwandler – „Alter Hut oder Innovationsfaktor im Antriebstrang?“ (Hydrodynamic Torque Converter as Innovative Starting Devices) VDI-Berichte Nr. 1827

[10.23] Schreiber W, Rudolph F, Becker V (2003) Das neue Doppelkupplungsgetriebe von Volkswagen (The New Dual Clutch Transmission of Volkswagen). ATZ Automobiltechnische Zeitschrift 105, Nr. 11

[10.24] Sebulke J (1974) Wärmetechnische Auslegung von Trockenreibungs- und Rutschkupplungen (Thermotechnical Design of Dry Friction and Slipping Clutches). Antriebstechnik 13: 376–382

[10.25] Seybold H (1981) Ein Verfahren zur Ermittlung der Auslegung einer Kraftfahrzeug-Schaltkupplung (A Process for Determining the Design of an Automotive Shifting Clutch). ATZ Automobiltechnische Zeitschrift 83: 355–358

[10.26] Steinhilper W (1962) Der zeitliche Temperaturverlauf in Reibungsbremsen und Reibungskupplungen beim Schaltvorgang (The Time Dependent Temperature Profile in Friction Brakes and Friction Clutches). Dissertation, TH Karlsruhe

[10.27] Steinhilper W (1967) Der Kraftfluss in unter Last geschalteten Lamellenkupplungen und das übertragbare Drehmoment (The Power Flow in Powershift Multi-Plate Clutches and the Transmissible Torque). Konstruktion 19: 262–267

[10.28] Stüper J (1965) Automatische Automobilgetriebe (Automatic Automobile Transmissions). Springer, Berlin Heidelberg New York

[10.29] Sudau J, Sasse C (1998) „ZMSIIi“ – Ein neues, im Drehmomentwandler integriertes Schwingungsdämpfungskonzept für Automatgetriebe (“ZMSIIi” – A new Vibration Damping Concept for Automatic Transmissions Integrated in the Torque Converter). VDI-Berichte Nr. 1393

[10.30] VDI-Richtlinie 2240 (1971) Wellenkupplungen. Systematische Einteilung nach ihren Eigenschaften (Shaft Couplings; Systematic Classification According to Their Properties). VDI Verein Deutscher Ingenieure e.V., Düsseldorf

[10.31] VDI-Richtlinie 2241, Blatt 1 (1982) Schaltbare fremdbetätigte Reibkupplungen und -bremsen, Begriffe, Bauarten, Kennwerte, Berechnungen (Friction Clutches and Brakes; Vocabulary, Types, Characteristic Data, Calculations). VDI Verein Deutscher Ingenieure e.V., Düsseldorf
[10.32] Voith (1987) Hydrodynamik in der Antriebstechnik (Hydrodynamics in Drive Engineering). Vereinigte Fachverlage Krauskopf Ingenieur Digest
[10.33] Winkelmann S, Harmuth H (1985) Schaltbare Reibkupplungen (Shiftable Friction Clutches). Springer, Berlin Heidelberg New York
[10.34] ZF Getriebe GmbH (2004) Interne Information – Hydrodynamischer Drehmomentwandler (Internal Information – Hydrodynamic Torque Converter)
[10.35] ZF-Sachs (2003) Produktinformation: Hydrodynamischer Drehmomentwandler (Product Information: Hydrodynamic Torque Converter)
[10.36] ZF-Sachs (2006) Produktinformation: Antriebskomponenten und -systeme für Pkw (Product Information: Drive Components and Systems for Passenger Cars)
[10.37] ZF-Sachs: Graphical Material
[10.38]* Bauer KH (2002) Nasslaufende Anfahrkupplung in modernen Automatikgetrieben (Wet Starting Clutch in Modern Automatic Transmissions). Offenbach: 1. Internationales IIR-Symposium „Innovative Fahrzeug-Getriebe“
[10.39]* Bauer KH, Heinrich J, Günter F (2005) Kraftstoffeinsparpotentiale durch Doppelkupplungsgetriebe (Potential Fuel Efficiency Improvements by Dual Clutch Transmissions). 26. Internationales Wiener Motorensymposium
[10.40]* Braess HH, Seifert U (2005) Vieweg Handbuch Kraftfahrzeugtechnik (Vieweg Automotive Engineering Handbook). Vieweg-Verlag, Wiesbaden
[10.41]* Götte T, Pape T (2004) DSG – das Direktschaltgetriebe von Volkswagen. Innovative Getriebetechnik mit Doppelkupplung (The Volkswagen DSG Direct Shift Gearbox. Innovative Dual Clutch Transmission Technology). VDI-Berichte Nr. 1827
[10.42]* Kickbusch E (1963) Föttinger-Kupplungen und Föttinger-Getriebe (Föttinger Clutches and Föttinger Transmissions). Springer, Berlin Heidelberg New York
[10.43]* VDI-Richtlinie 2153 (1974) Hydrodynamische Getriebe (Hydrodynamic Transmissions). VDI Verein Deutscher Ingenieure e.V., Düsseldorf
[10.44]* Wolf M (1962) Strömungskupplungen und Strömungswandler (Fluid Clutches and Hydrodynamic Transmissions). Springer, Berlin Heidelberg New York

## Chapter 11

[11.1] Block H (1940) Fundamental mechanical aspects of boundary lubrication. JSAE 46, No. 2
[11.2] BOSCH (2004) Kraftfahrtechnisches Taschenbuch (Automotive Handbook). 25th Edition. Vieweg-Verlag, Wiesbaden
[11.3] Brändlein J, Lorösch HK (1986) Schmutzgeschützte Lager in Kraftfahrzeuggetrieben (Clean Bearings in Vehicle Transmissions). VDI-Berichte Nr. 579: 253–268
[11.4] Collenberg F (1991) Untersuchungen zur Fresstragfähigkeit schnelllaufender Stirnradgetriebe (Investigations Into the Scuffing Resistance of Fast-Running Spur Gear Mechanisms). Dissertation, Technische Universität München, Lehrstuhl Maschinenelemente

[11.5] DIN 3760 (1996) Radial-Wellendichtringe (Rotary Shaft Lip Type Seals). DIN Deutsches Institut für Normung e.V., Berlin

[11.6] DIN 3761 (1984) Radial-Wellendichtringe für Kraftfahrzeuge (Rotary Shaft Lip Type Seals for Automobiles). DIN Deutsches Institut für Normung e.V., Berlin

[11.7] DIN 31651 (1991) Gleitlager (Plain Bearings). DIN Deutsches Institut für Normung e.V., Berlin

[11.8] DIN 31652 (2002) Gleitlager – Hydrodynamische Radial-Gleitlager im stationären Betrieb (Plain Bearings; Hydrodynamic Plain Journal Bearings Designed for Operation under Steady-State Conditions). DIN Deutsches Institut für Normung e.V., Berlin

[11.9] DIN 51509 (1976) Auswahl von Schmierstoffen für Zahnradgetriebe. Teil 1: Schmieröle (Selection of Lubricants for Gears. Part 1: Gear Lubricating Oils). DIN Deutsches Institut für Normung e.V., Berlin

[11.10] DIN ISO 281 Beiblatt 1 (2003) Wälzlager – Dynamische Tragzahlen und nominelle Lebensdauer. Lebensdauerbeiwert $a_{\mathrm{DIN}}$ und Berechnung der erweiterten modifizierten Lebensdauer (Rolling Bearings – Dynamic Load Ratings and Rating Life – Life Modification Factor $a_{\mathrm{DIN}}$ and Calculation of the Modified Rating Life). DIN Deutsches Institut für Normung e.V., Berlin

[11.11] DIN ISO 281 Beiblatt 4 (2003) Wälzlager – Dynamische Tragzahlen und nominelle Lebensdauer. Verfahren zur Berechnung der modifizierten Referenz-Lebensdauer für allgemein belastete Wälzlager (Rolling Bearings – Dynamic Load Ratings and Rating Life – Methods for Calculation of the Modified Reference Rating Life for Universally Loaded Rolling Bearings). DIN Deutsches Institut für Normung e.V., Berlin

[11.12] DIN ISO 2909 (2004) Mineralölerzeugnisse – Berechnung des Viskositätsindex aus der kinematischen Viskosität (Petroleum Products – Calculation of Viscosity Index from Kinematic Viscosity). DIN Deutsches Institut für Normung e.V., Berlin

[11.13] DIN ISO 3547 (2000) Gleitlager – Gerollte Buchsen (Plain Bearings – Wrapped Bushes). DIN Deutsches Institut für Normung e.V., Berlin

[11.14] DIN ISO 12128 (1998) Gleitlager – Schmierlöcher, Schmiernuten und Schmiertaschen. Maße, Formen, Bezeichnung und ihre Anwendung für Lagerbuchsen (Plain Bearings – Lubrication Holes, Grooves and Pockets – Dimensions, Types, Designation and Their Application to Bearing Bushes). DIN Deutsches Institut für Normung e.V., Berlin

[11.15] Dornhöfer G (1994) Lebensdauerschmierung von Kfz-Aggregaten unter Berücksichtigung der Umweltschonung (Life-Time Lubrication of Motor Vehicle Units Taking Into Account Environmental Considerations). Tribologie und Schmierungstechnik, Nr. 2: 86–90

[11.16] Eckhardt F (1985) Stationäre Zahnradgetriebe. Schmierung und Wartung (Stationary Gear Mechanisms. Lubrication and Maintenance). Eigenverlag der Mobil Oil AG, Hamburg

[11.17] Federal Mogul: Product Information

[11.18] Förster HJ (1974) Automatische Fahrzeuggetriebe und Retarder für Nutzfahrzeuge (Automatic Vehicle Transmissions and Retarders for Commercial Vehicles). ATZ Automobiltechnische Zeitschrift 76, Nr. 5: 142–149

[11.19] Freund H, Stubenrauch A (1987) Entwicklungen von Wälzlagerungen in Nutzfahrzeug-Schaltgetrieben (Developments in Rolling Bearings in Commercial Vehicle Manual Transmissions). Antriebstechnik Nr. 4: 28–39

[11.20] Göhring E (1989) Bremsleistung schwerer Nutzfahrzeuge (Braking Performance of Heavy Commercial Vehicles). Aachen: 2. Aachener Kolloquium „Fahrzeug- und Motorentechnik": 397–425

[11.21] Haupt J (2005) Auslegungskriterien der Automatgetriebepumpe (Design Criteria for Automatic Transmission Pumps). ZF Getriebe GmbH

[11.22] Hettich V, v Eiff H, Lechner G (1992) Untersuchung von Flächendichtungen unter dynamischer Scherbeanspruchung (Investigation of Gaskets under Dynamic Shearing Stress). Antriebstechnik 31, Nr. 4: 74–79

[11.23] INA FAG (2006) Wälzlager HR 1. Produktkatalog (Rolling Bearings HR 1. Product Catalogue). Schaeffler KG, Herzogenaurach, Schweinfurt

[11.24] Jäckle M (1994) Entlüftung von Getrieben (Venting Transmissions). Dissertation, Universität Stuttgart, Institut für Maschinenelemente

[11.25] Jäckle M, Lechner G (2000) Mechanisch hochbelastete Flächendichtverbindungen (Mechanically Highly Stressed Surface Seals). VDI-Berichte Nr. 1579

[11.26] Joachim FJ, Thies HE, Brügel E (1992) Lebensdauerschmierung bei Fahrzeuggetrieben (Life-Time Lubrication for Vehicle Transmissions). Tribologie und Schmierungstechnik 39, Nr. 6: 317–324

[11.27] Khosrawi MA (1983) Untersuchung über die Reaktionsschichtbildung bei hochbelastbaren Schmierölen (Investigation Into Reaction Layer Formation in Highly Stressed Lubricants). Dissertation, Universität Stuttgart, Staatliche Materialprüfungsanstalt (MPA)

[11.28] Kreuzer R, Romanos G (2000) Zuverlässigkeit von Flächendichtverbindungen auf der Basis von Flüssigdichtmitteln unter dynamischer Beanspruchung (Reliability of Surface Seals Based on Fluid Seals under Dynamic Load). VDI-Berichte Nr. 1579

[11.29] Krieg WE (1993) Untersuchungen an Gehäuseabdichtungen von hochbelasteten Getrieben (Investigations into Housing Seals for Highly Stressed Transmissions). Dissertation, Universität Stuttgart, Institut für Maschinenelemente

[11.30] Kubalczyk R (2000) Gehäusegestaltung von Fahrzeuggetrieben im Abdichtbereich (Housing Design of Automotive Transmissions in the Sealing Area). Dissertation, Universität Stuttgart, Institut für Maschinenelemente

[11.31] Kürzeder H (1987) Verlängerung der Lebensdauer neuer und bereits beschädigter Getriebe durch Spezialschmierstoffe (Extending the Service Life of New and Damaged Transmissions by Using Special Lubricants). Antriebstechnik 26, Nr. 7: 18–25

[11.32] Lane TB, Hughes JK (1951) A Practical Application of the Flash Temperature Hypotesis of Gear Lubrication. Third World Petrol Congr.: 7

[11.33] Lauth HJ, Webert D, Scholz I, Agner I (2001) Bedarfsorientiert ansteuerbare Pumpen (Demand-Oriented Controllable Pumps). 7. LuK-Kolloquium: 119–131

[11.34] Lechner G (1969) Untersuchungen zur Schmierfilmbildung an Zahnrädern (Investigations into Lubricant Film Formation on Gearwheels). VDI-Zeitschrift 111, Nr. 4: 269–274

[11.35] Lechner G (1973) Berechnung der Fresstragfähigkeit von Stirn- und Kegelrädern (Calculating the Scuffing Resistance of Spur Gears and Bevel Gears). Technisch wissenschaftliche Veröffentlichung der Zahnradfabrik Friedrichshafen AG, Heft 8

[11.36] Looman J (1996) Zahnradgetriebe (Gear Mechanisms). 3rd Edition. Springer, Berlin Heidelberg New York

[11.37] Lorösch HK (1993) Wirtschaftliche, umweltschonende Wälzlagerungen durch Vermeiden von Überdimensionierung (Economic, Environmentally Acceptable Rolling Bearings by Avoiding Over-Dimensioning). Antriebstechnik 32, Nr. 3: 89–93

[11.38] Lundberg S (1977) Anstellen von Kegelrollenlagern in Pkw-Getrieben mit Leichtmetallgehäuse (Adjusting Tapered Roller Bearings in Passenger Car Transmissions with Light-Alloy Housings). SKF-Kugellager-Zeitschrift 193, Jahrgang 52

[11.39] Meyer K, Kloss H, Reichelt F (1990–91) Übertragbarkeit von Fressverschleißergebnissen aus Modellapparaturen auf praxisnahe Bedingungen (Transferability of Gear Scuffing Results from Model Apparatus to Practical Conditions). Tribologie und Schmierungstechnik. Part I: 21 (1990), Nr. 10: 299–302; Part II: 22 (1991), Nr. 3: 68–71; Part III: 22 (1991), Nr. 4: 105–107

[11.40] Molykote (1991) Molykote-Handbuch (Molykote Handbook)

[11.41] Müller HK (1990) Abdichtung bewegter Maschinenteile (Fluid Sealing Technology). Medienverlag Müller, Waiblingen

[11.42] Niemann G (1955) Schmierfilmbildung, Verlustleistung und Schadensgrenzen bei Zahnrädern mit Evolventenverzahnung (Lubricant Film Formation, Power Loss and Damage Limits in Gearwheels with Involute Toothing). VDI-Zeitschrift 97, Nr. 10

[11.43] Niemann G, Lechner G (1967) Die Fressgrenzlast bei Stirnrädern aus Stahl (Maximum Scuffing Load for Steel Spur Gears). Erdöl und Kohle 20, Nr. 2: 96–106

[11.44] Oster P (1982) Beanspruchung der Zahnflanken unter Bedingungen der Elastohydrodynamik (Stressing Tooth Faces Under Conditions of Elasto-Hydrodynamics). Dissertation, Technische Universität München, Lehrstuhl Maschinenelemente

[11.45] Pittroff H (1961) Riffelbildung bei Wälzlagern infolge Stillstandserschütterungen (Grooving in Rolling Bearings as a Result of Shocks at Rest). Eigenverlag SKF-Kugellagerfabriken GmbH, Schweinfurt

[11.46] Schiberna P, Gesenhaus R, Nowatschin K, Fleischmann HP, Hommes G (2004) Audi multitronic® – leistungsfähig und sportlich (Audi Multitronic – Powerful and Sporty). VDI-Berichte Nr. 1827: 447–459

[11.47] Schwuchow S (1996) Sonderverzahnungen für Zahnradpumpen mit minimaler Volumenstrompulsation (Special Gearings for Gear Pumps with Minimal Volumetric Flow Pulsation). Dissertation, Universität Stuttgart, Institut für Maschinenelemente

[11.48] Sonntag W (1988) Vermeidung von Zahnradschäden durch geeignete Schmieröle (Avoiding Gearwheel Damage by Using Suitable Lubricants). Tribologie und Schmierungstechnik 19, Nr. 10: 292–295

[11.49] Technische Akademie Esslingen (1984) Additive für Schmierstoffe. Reihe Tribotechnik, Band 2 (Additives for Lubricants. Tribology Series, Vol. 2). Curt R. Vincentz-Verlag, Hannover

[11.50] Uetz H, Föhl J, Sommer K (1989) Einfluss v. Reaktionsschichten auf das tribologische Verhalten langsamlaufender Wälzpaarungen (The Influence of Reaction Layers on the Tribological Behaviour of Slow-Running Roller Pairs). Antriebstechnik 28, Nr. 10: 57–62

[11.51] VDA-Empfehlung 263 (1999) Abnahmeprüfung von Pkw-Getrieben (Final Inspection of Passenger Car Transmissions). VDA Verband der Automobilindustrie e. V., Frankfurt

[11.52] VDI-Richtlinie 2204 (1992) Auslegung von Gleitlagerungen (Design of Plain Bearings). VDI Verein Deutscher Ingenieure e.V., Düsseldorf

[10.53] WABCO (2004) Fahrzeug-Vorschriften (Automotive Specifications). 22. Ausgabe. WABCO Vehicle Control Systems

[11.54] Wagner G, Remmlinger U, Fischer M (2004) Das stufenlose Getriebe CFT 30 von ZF – Ein CVT mit Kettenvariator für 6-Zylinder Motoren in Fahrzeugen mit Front-Quer-Antrieb (The CFT30 of ZF – a Continuously Variable Automatic Transmission with a Chain Driven Variator for 6-Cylinder Engines in Front-Transverse Installations). VDI-Berichte Nr. 1827: 461–497

[11.55] Wender B, Lecong S (1991) Gestaltung von Versteifungsrippen und Bohrungen an Getriebegehäusen (Designing Bracing Ribs and Drill Holes in Transmission Housings). Konstruktion 43: 365–370

[11.56] Winter H, Michaelis K (1975) Fresstragfähigkeit von Stirnradgetrieben. Teil I: Grundlagen zum Ansatz einer Integraltemperatur (Scuffing Resistance of Spur Gear Mechanisms. Part I: Basic Principles for Estimating an Integral Temperature). Antriebstechnik 14, Nr. 7: 405–409

[11.57]* DIN 51512 (1988) Schmierstoffe. SAE-Viskositätsklassen für Schmieröle für Kraftfahrzeuggetriebe (Lubricants; SAE Viscosity Grades for Lubricating Oils for Automotive Gears). DIN Deutsches Institut für Normung e.V., Berlin

[11.58]* Michaelis K (2003) Tribologie des Zahnrades. Vorlesungsmanuskript (Gearwheel Transmission Tribology. Lecture Notes). Technische Universität München, Lehrstuhl für Maschinenelemente

[11.59]* Müller HK, Nau BS: www.fachwissen-dichtungstechnik.de

[11.60]* Schmid E (1981) Handbuch der Dichtungstechnik (Handbook of Sealing Technology). Expert-Verlag, Grafenau/ Württemberg

[11.61]* Tietze W (2003) Handbuch Dichtungspraxis (Handbook Sealing in Practice). Vulkan-Verlag, Essen

## Chapter 12

[12.1] Allison Transmission (1999) Product Information: Allison Automatic Transmissions – 1000, 2000, 2400 Series

[12.2] Allison Transmission (2004) Operator's Manual: Allison Transmission On-Highway, 1000 and 2000 Product Families, OM3063EN

[12.3] Barnbeck A, Drott, H, Helms G (1993) Das neue Opel-Sechsgang-Getriebe (The New Opel 6-Speed Transmission). ATZ Automobiltechnische Zeitschrift 95, Nr. 6: 288–292

[12.4] Boch P (1998) EE DRIVE: Der elektrische Radnabenantrieb der ZF (EE DRIVE: The Electric Wheel Hub Drive from ZF). Nahverkehrspraxis 01/1998

[12.5] Böning J, Krake F, Krüttgen R, Painter S (2004) Das Neue Sechsgang-Schaltgetriebe von Opel (The New Opel 6-Speed Manual Transmission). ATZ Automobiltechnische Zeitschrift 106, Nr. 10: 862–872

[12.6] Braess HH, Seifert U (2005) Vieweg Handbuch Kraftfahrzeugtechnik (Vieweg Automotive Engineering Handbook). Vieweg-Verlag, Wiesbaden

[12.7] Collenberg HF, Fischer D, Führer G, Müller W (1998) Möglichkeiten zur Leistungssteigerung, Geräuschoptimierung und Schaltkraftreduzierung von Nkw-Schaltgetrieben (Possibilities to Increase Performance, Optimise Noise and Reduce Shifting Force of Commercial Vehicle Manual Transmissions). VDI-Berichte Nr. 1393

[12.8] Duckstein H, Leinfellner H, Sommer HD (1985) Der VW-Transporter Syncro (The VW Transporter Syncro). ATZ Automobiltechnische Zeitschrift 87, Nr. 9: 405–417

[12.9] Eaton (1985) Product Information: Eaton Twin Splitter TS/TSO 11612

[12.10] Eaton (1997) Product Information: Eaton S Transmission Series

[12.11] Ebenhoch M, Krieg WE, Kubalczyk R, Drabek M (2004) Getriebebaukasten für ein Doppelkupplungs-, automatisiertes Handschalt- und Handschaltgetriebe auf Basis eines gemeinsamen Radsatzes (Modular Transmission System for a MT, AMT and DCT Based on a Common Wheelset). VDI-Berichte Nr. 1827

[12.12] Eggert U (1990) CVT – Elektronische Regelung und Fahrdynamik (CVT – Electronic Control and Vehicle Dynamics). VDI-Berichte Nr. 803: 197–234

[12.13] Eschrich G, Schwarz J, Rüchardt C (2001) Funktionen automatischer Nkw-Getriebe im Verbund elektronischer Systeme (Features of Automatic Commercial Vehicle Transmissions in Combination with Electronic Systems). VDI-Berichte Nr. 1610: 101–115

[12.14] Fleischmann H, Gutz H, Kumpf G, Martin F, Schöffmann M (2004) Die neuen Getriebe im Audi A6 (The New Transmissions in the Audi A6). ATZ/MTZ-Sonderausgabe 3/2004: Der neue Audi A6: 128–138

[12.15] Foth J, Karch G, Speck FD (2005) Getriebesysteme für Busse: Lastschaltgetriebe, automatisierte Getriebe oder Handschaltgetriebe? – Auswahlkriterien heute und morgen (Transmission Systems for Buses: Powershift Transmissions, Automated Transmissions or Manual Transmissions? – Selection Criteria Now and in the Future). VDI-Berichte Nr. 1876: 367–388

[12.16] Franke A, Greiner J, Gösele A, Kiesel J, Sperber R, Stenkert J, Wagner B, Weiss P (2004) Triebstrang und Getriebe (Powertrain and Transmission). ATZ/MTZ-Sonderausgabe 10/2004: Die neue Mercedes-Benz A-Klasse: 150–159

[12.17] Freymann R, Strobl W, Glonner H, Köpf P, Kilian S (2004) Wandlerfreies Automatikgetriebe und Elektromaschinen-Kondensatorsystem, ein neuer Ansatz zur Realisierung einer effizienten Dynamik (Converter-Free Automatic Transmission and Motor Capacitor System – A New Approach for the Implementation of Efficient Dynamics). VDI-Berichte Nr. 1827

[12.18] v Glasner EC, Povel R, Holloh KD (1997) Der intelligente Antriebsstrang (The Intelligent Powertrain). Berlin: Symposium „Steuerungssysteme für den Antriebsstrang von Kraftfahrzeugen“ 18.–19. September 1997

[12.19] Göddel T, Hillenbrand H, Hopff C, Jud M (1998) Das neue Fünfgang-Automatgetriebe W5A180 (The New W5A 180 5-Speed Automatic Transmission). ATZ/MTZ-Sonderausgabe 10/1998: Die neue A-Klasse von Daimler-Benz: 96–101

[12.20] Grad K (2005) Development Trends for Tractor Powertrains. 3rd AVL International Commercial Powertrain Conference, 20th–21st April 2005, Graz, Austria.

[12.21] Greiner J, Dörr C, Nauerz H, Graeve M (2004) The new „7G-TRONIC" of Mercedes-Benz. Innovative Transmission Technology for Better Driving Performance, Comfort and Fuel Economy. SAE 2004-01-0649

[12.22] Greiner J, Indlekofer G, Nauerz H, Dorfschmid J, Gödecke T, Dörr C (2003) Das neue Automatikgetriebe 7G-TRONIC von Mercedes-Benz (The New Automatic Transmission 7G-TRONIC of Mercedes-Benz). ATZ Automobiltechnische Zeitung 105, Nr. 10: 920–930

[12.23] Hall W, Bock C (2001) Sechsgang-Stufenautomatikgetriebe für den neuen 7er BMW (Six-Speed Automatic Transmission for the New 7 Series BMW). ATZ Automobiltechnische Zeitung 103, Nr. 9: 714–724

[12.24] Härdtle W (1997) Ein neues automatisiertes Schaltgetriebe für schwere Nutzfahrzeuge (A New Automated Manual Gearbox for Heavy Commercial Vehicles). ATZ Automobiltechnische Zeitschrift 99: 598–604

[12.25] Härdtle W (1998) Automatisierte Getriebe für Nutzfahrzeuge: Integration von Funktionen und Komponenten (Automated Transmissions for Commercial Vehicles: Integration of Functions and Components). VDI-Berichte Nr. 1993: 193–213

[12.26] Hata H, Kamiya M, Nagamatsu S (2005) Development of a New Hybrid Transmission for FWD Sports Utility Vehicle. Toyota Technical Review Vol. 54 No. 1. Toyota Motor Corporation, Aichi (Japan), ISSN 0916-7501

[12.27] Hoffmann C, Würfel A (2002) Dreiwellengetriebe – Die kurze Art, Sechsganggetriebe zu bauen (Three-Shaft Transmissions – The Short Way to Build 6-Speed-Transmissions). ATZ Automobiltechnische Zeitschrift 104, Nr. 12: 1090–1099

[12.28] Hunger A (2002) Porsche Cayenne. Delius-Klasing Verlag, Bielefeld

[12.29] James FJ, Wegeng WJ (1999) Leveraging a Commercial Medium-Duty Product for the Heavy-Duty Pickup Truck Market. SAE Technical Paper Series 1999-01-3743

[12.30] Jaskiewicz Z: Mosty Napedowe

[12.31] Kasauya S, Taniguchi T, Tsukamoto K, Hayabuchi M, Nishida M, Suzuki A (2005) Aisin AW New High Torque Capacity Six-Speed Automatic Transmission for FWD vehicles. SAE 2005-01-1020

[12.32] Katou N, Taniguchi T, Tsukamoto K, Hayabuchi M, Nishida M (2004) Aisin AW new six-speed automatic transmission for FWD vehicles. SAE 2004-01-0651

[12.33] Keller W, Härdtle W (2000) Erfahrungen bei Entwicklung und Einsatz automatischer Getriebe für Nutzfahrzeuge (Experiences in Development and Application of Automatic Transmissions for Commercial Vehicles). Aachen: 9. Aachener Kolloquium „Fahrzeug- und Motorentechnik": 1247–1258

[12.34] Kiesel J, Greiner J, Veil A, Strenkert J (2005) Das neue CVT-Getriebe AUTOTRONIC von Mercedes Benz (The New Front Wheel Drive CVT AUTOTRONIC from Mercedes Benz). Wien: 26. Wiener Motorensymposium

[12.35] Knorra U (2005) Der Lexus RX400h (The Lexus RX400h). ATZ Automobiltechnische Zeitung 107, Nr. 9

[12.36] König RJ, Hylto JD (1999) Vocational and Body Builder Controls Features of the Allison 1000/2000/2400 Series TM Automatic Transmission. SAE Technical Paper Series, 1999-01-3744

[12.37] Kubalczyk R, Ebenhoch M, Schneider HJ (2006) 7-Gang-Doppelkupplungsgetriebe für sportliche Anwendungen (Seven Speed Dual Clutch Transmission System for Sportive Applications). VDI-Berichte Nr. 1943

[12.38] Lechner G, Naunheimer H (1994) Fahrzeuggetriebe (Automotive Transmissions). 1st Edition. Springer, Berlin Heidelberg New York

[12.39] Ledwinka E (1972) Der Steyr-Puch-„Pinzgauer"-Geländewagen (The Steyr-Puch "Pinzgauer" Off-Road Vehicle). ATZ Automobiltechnische Zeitschrift 74: 228–235

[12.40] Leitermann W (2004) Das neue 7-Gang-AMT im BMW M5 (The New 7-Speed AMT – BMW M5). Würzburg: 3. Internationales CTI-Symposium „Innovative Fahrzeuggetriebe"

[12.41] Looman J (1996) Zahnradgetriebe (Gear Mechanisms). 3rd Edition. Springer, Berlin Heidelberg New York

[12.42] Magna Drivetrain: Graphical Material. Magna Drivetrain, Graz/Lannach, Austria

[12.43] Manthey D, Fleischmann O (1989) Das Allradsystem des Opel Vectra (The All-Wheel Drive System of the Opel Vectra). Aachen: 2. Aachener Kolloquium „Fahrzeug- und Motorentechnik": 687–702

[12.44] Marx P, Schmidtmann M, Wieland D, Greiner J (2004) Verbrauchsoptimierung des Flottenverbrauchs mittels FE-HAG (Fuel-Economy Hinterachsgetriebe) (Fuel Consumption Optimized Vehicle Fleet by Means of FE-HAG (Fuel Economy Rear Axle Differential)). VDI-Berichte Nr. 1827: 289–301

[12.45] Mohan SK (2005) Torque Vectoring Systems: Architecture, Stability Performance and Efficiency Considerations. Graz: 6. Grazer Allradkongress

[12.46] Mozer H, Piepenbrink A, Sommer S (2001) The Technology of the ZF CVT – CFT 23. SAE 2001-01-08

[12.47] Nett HP (2005) Der modulare Winkeltrieb – die preiswerte Möglichkeit unterschiedlicher Allradkonzepte für eine Plattform (The Modular Bevel Gear – the Inexpensive Possibility for Different All-Wheel Concepts for One Platform). Graz: 6. Grazer Allradkongress

[12.48] Pape E (2002) VW in 4Motion – Vortrieb in allen Lebenslagen (VW in 4Motion – Propulsion for All Purposes). Graz: 3. Grazer Allradkongress

[12.49] v Petery G (2004) Kraftstoffersparnis durch maßgeschneiderte Lager für Achsgetriebe von BMW (Fuel Savings by Means of Custom-Made Bearings for Axle Drives of BMW). ATZ Automobiltechnische Zeitschrift 106, Nr. 12: 1096–1100

[12.50] Pichler P, Erjawetz K (2005) 4x4 Hybridmodul mit elektrischer Vorderachse (4x4 Hybrid Module with Electric Front Axle). Graz: 6. Grazer Allradkongress

[12.51] Pilgrim R, Rohardt H, Srock R (1988) 911 Carrera 4, der Allrad-Porsche. Teil 1 (911 Carrera 4, the All-Wheel Drive Porsche. Part 1). ATZ Automobiltechnische Zeitschrift 90, Nr. 11: 615–625

[12.52] Pohlenz J, Gruhle WD (2001) Stufenloses hydrostatisch-mechanisch leistungsverzweigtes Getriebe ECCOM für den Einsatz in Traktoren (Continuously Variable Hydrostatic-Mechanical Power-Split Gearbox ECCOM for the Use in Tractors). VDI-Berichte Nr. 1592: 451–468

[12.53] Raeder M, Rüchardt C, Speck FD (2004) Die ZF AS-Tronic-Familie – Automatische Getriebe für alle Nutzfahrzeugklassen (The ZF AS-Tronic Series – Automatic Transmissions for All Commercial Vehicle Classes). ATZ Automobiltechnische Zeitschrift 106, Nr. 9: 772–783

[12.54] Rastinger B (1991) Die neuen Schaltgetriebe für die BMW-Baureihen 8, 5 und 3 (The New Manual Transmissions for the BMW Series 8, 5 and 3). ATZ Automobiltechnische Zeitschrift 93, Nr. 11: 698–705

[12.55] Richardson BR, Wegeng WJ (1997) Allison Transmission – New Family of Transmissions: the 1000/2000 Series. SAE Technical Paper Series 973278

[12.56] Rösch R, Zaiser W, Kürschner M (1995) Das neue automatische Getriebe W5A330/580 für Personenwagen von Mercedes-Benz – Aufbau und Betriebsverhalten des mechanischen Getriebeteils (The New Automatic Transmission W5A330/580 for Passenger Cars by Mercedes-Benz – Design and Operating Behaviour of the Mechanical Gearbox Part). ATZ Automobiltechnische Zeitung 97, Nr. 9: 536–545

[12.57] Schaller KV (1997) Automatisiertes Getriebeschaltsystem SAMT B (SAMT B Automated Gearshift System). Berlin: Symposium „Steuerungssysteme für den Antriebsstrang von Kraftfahrzeugen“

[12.58] Schiberna P, Gesenhaus R, Nowatschin K, Fleischmann HP, Hommes G (2004) Audi multitronic® – leistungsfähig und sportlich (Audi Multitronic – Powerful and Sporty). VDI-Berichte Nr. 1827

[12.59] Schmidt R, Diessner G, Uhlig G, Otto H (1988) Antriebskonzept, Bauteile und Eigenschaften des VW Golf Syncro (Drive Concept, Components and Characteristics of the VW Golf Syncro). ATZ Automobiltechnische Zeitschrift 90, Nr. 10: 535–542

[12.60] Söffge F, Dietz M, Bofinger G, Lawitzki S (1989) 911 Carrera 4, der Allrad-Porsche. Teil 2 (911 Carrera 4, the All-Wheel Drive Porsche. Part 2). ATZ Automobiltechnische Zeitschrift 91, Nr. 1: 15–24

[12.61] Schöpf HJ, Jürgens G, Pickard J (1989) Das neue Fünfgang-Automatgetriebe von Mercedes-Benz (The New Mercedes-Benz 5-Speed Automatic Transmission). ATZ Automobiltechnische Zeitschrift 91, Nr. 9: 444–453

[12.62] Schreiber W, Rudolph F, Becker V (2003) Das neue Doppelkupplungsgetriebe von Volkswagen (The New Dual Clutch Transmission by Volkswagen). ATZ Automobiltechnische Zeitung 105, Nr. 11: 1022–1039

[12.63] Schröder J, Gabler R, Pfau W, Nistler G (2004) Das Verteilergetriebe im BMW xDrive. Der neue Allradantrieb im BMW X3 und BMW X5 (The Transfer Gearbox in the BMW xDrive. The New All-Wheel Drive in the BMW X3 and X5). VDI-Berichte Nr. 1827: 237–252

[12.64] Stockmar J (2004) Das große Buch der Allradtechnik (The Big Book of All-Wheel Drive Technology). Pietsch Verlag, Stuttgart

[12.65] Wagner G, Bucksch M, Scherer H (2001) Das automatische Getriebe 6 HP 26 von ZF – Getriebesystem, konstruktiver Aufbau und mechanische Bauteile (The Automatic Transmission 6 HP 26 from ZF – Gearbox System, Engineering Design and Mechanical Components). VDI-Berichte Nr. 1610: 631–654

[12.66] Wagner G, Remmlinger U, Fischer M (2004) ZF-Ecotronic CFT 30. VDI-Berichte Nr. 1827

[12.67] ZF Product Information: ZF-AS Tronic – Das automatische Getriebesystem für den modernen Lkw (ZF AS-Tronic – The Automatic Transmission System for the Modern Truck)

[12.68] ZF Product Information (2003) ZF-Ecomid

[12.69] ZF Product Information (2003) ZF-Ecosplit

[12.70] ZF Product Information (2003) ZF-Ecomat and ZF-Ecomat 2

[12.71] ZF Product Information (2004) Das automatische Getriebesystem in Van und Pickup: ZF-eTronic (The Automatic Transmission System in Vans and Pickups: ZF-eTronic)

[12.72] Product Information: ZF TC-Tronic

[12.73]* Hipp F (2005) ITC – Das Verteilergetriebe im neuen Land Rover Discovery (ITC – The Transfer Gearbox in the New Land Rover Discovery). Graz: 6. Grazer Allradkongress

## Chapter 13

[13.1] Oberhauser M, Vetter H (2003) Mechatronische Getriebesysteme (Mechatronic Transmission Systems). Expert Verlag, Renningen

[13.2] Robert Bosch GmbH (2002) Autoelektrik, Autoelektronik (Automotive Electrics, Automotive Electronics). Vieweg-Verlag, Wiesbaden

[13.3] Tapia C, Bauer G, Dornseiff M (2003) Automatisierte Softwaretests (Automated Software Tests). ATZ/MTZ-Sonderausgabe: Automotive Electronics II/2003: 16–23

[13.4] Zurawka T, Schäuffele J (2003) Automotive-Software-Engineering – Grundlagen, Prozesse, Methoden und Werkzeuge (Automotive Software Engineerung – Basics, Processes, Methods and Tools). Vieweg-Verlag, Wiesbaden

## Chapter 14

[14.1] Bathe KJ (2002) Finite-Elemente-Methoden (Finite Elements Methods). Springer, Berlin Heidelberg New York

[14.2] Butscher K, Kötzle H, Müller K (1988) Fahrzeug-Messsystem zur Erfassung topographischer Straßendaten. (Vehicle Measuring System for Acquiring Topographical Road Data). VDI-Berichte Nr. 687: 511–520

[14.3] David W (1989) Neuartiges, modular aufgebautes Rechenprogramm zur Längsdynamiksimulation (Innovative, Modular Computer Program for Longitudinal Dynamics Simulation). Aachen: 2. Aachener Kolloquium „Fahrzeug- und Motorentechnik“: 497–519

[14.4] Demmerer S, Graswald C, Kwittner G (2003) Simulationsunterstützte Versuchsplanung zur Optimierung des Schaltkomforts eines Kfz-Getriebes (Simulation Based Design of Experiments for the Optimisation of Gear Change Comfort of Automotive Gearboxes). VDI-Berichte Nr. 1749: 109–124

[14.5] Dirschmid W (1992) Einsatz der Simulationstechnik unter dem Gesichtspunkt des Nutzen-Kosten-Aspektes (Cost/Benefit Aspects of the Use of Simulation Techniques). ATZ Automobiltechnische Zeitschrift 94, Nr. 9: 488–491

[14.6] Drees S, Mayer K (2004) Entwicklung einer sicheren Auslegungsmethode für zukünftige Pkw-Stufenautomaten durch den Abgleich von LK-Simulation mit LK-Messung (Development of a Reliable Method for the Design of Future Automatic Car Transmissions based on the Comparison of Collective-Load Simulations with Collective-Load Measurements). VDI-Berichte Nr. 1827: 609–625

[14.7] DRESP, FVA-Schwingungsberechnungsprogramm (DRESP, FVA Vibration Calculation Program). RWTH Aachen, Institut für Maschinenelemente und Maschinengestaltung

[14.8] Hagedorn P, Otterbein S (1987) Technische Schwingungslehre (Applied Vibration Theory). Springer, Berlin Heidelberg New York

[14.9] Heitz U (2005) Entwurf mechatronischer Getriebesysteme mit modernen CAE-Methoden (Development of Mechatronic Transmission Systems with Modern CAE-Methods). Beitrag in: Oberhauser M, Vetter H: Mechatronische Getriebesysteme. Expert Verlag

[14.10] Jauch F (1989) Beitrag zur Optimierung des Antriebsstranges von Kraftfahrzeugen unter dem Aspekt von Kraftstoffverbrauch, Transportleistung und Lastkollektiven (Contribution to Optimising the Powertrain of Motor Vehicles in Terms of Fuel Consumption, Transportation Efficiency and Load Profiles). Dissertation, Universität Stuttgart, Institut für Maschinenelemente

[14.11] Kollmann F (2000) Maschinenakustik (Machine Acoustics). Springer, Berlin Heidelberg New York

[14.12] Kücükay F (1987) Dynamik der Fahrzeuggetriebe (Automotive Transmission Dynamics). Springer, Berlin Heidelberg New York

[14.13] Kunz M (2002) Ermittlung des Einflusses fahrzeug-, fahrer- und verkehrsspezifischer Parameter auf die Getriebelastkollektive mittels Fahrsimulation (Determination of the Influence of Vehicle-, Traffic- and Driver-Specific Parameters on the Transmission Load Spectra by Drive-Line Simulation). Dissertation, Universität Stuttgart, Institut für Maschinenelemente

[14.14] Kunz M, Weidler A, Niemann K, Lechner G, Bertsche B (2001) Simulationsrechnungen zur Ermittlung des Streckenverbrauchs eines Hybridbusses unter realistischen Bedingungen (Simulations to Determine the Fuel Consumption of a Hybrid Bus under Realistic Conditions). VDI-Berichte Nr. 1610: 703–726

[14.15] Ludmann J, Diekamp R, Lerner G (1992) PELOPS – Ein Programmsystem zur Untersuchung neuer Längsdynamikkonzepte im Verkehrsfluss (PELOPS – A Program System for Investigating New Longitudinal Dynamics Concepts in Traffic Flow). VDI-Berichte Nr. 1007: 192–209

[14.16] Naunheimer H (1995) Beitrag zur Entwicklung von Stufenlosgetrieben mittels Fahrsimulation (Contribution to the Development of Continuously Variable Transmissions Using Driving Simulation). Dissertation, Universität Stuttgart, Institut für Maschinenelemente

[14.17] Naunheimer H, Zelßmann H, Lechner, G (1993) Satellitengestützte Aufzeichnung von Streckendaten für Fahrsimulationsprogramme (Satellite-Based Recording of Test Run Data for Driving Simulation Programs). ATZ Automobiltechnische Zeitschrift 95, Nr. 12: 650–654

[14.18] Piehler J, Zschiesche HU (2002) Simulationsmethoden (Simulation Methods). Harri Deutsch Verlag

[14.19] Plieske M, Gierlich K, Rohrer A (2004) Das ZF-Berechnungsportal – eine konzernweite Plattform zur breiteren Nutzung der Berechnung (The ZF Computational Portal – A Group Wide Platform for the Wide Use of Computational Methods). VDI-Berichte Nr. 1846: 655–666

[14.20] Plieske M, LeCong S, Weber W (1995) Einsatz der Finite-Element- und der Boundary-Element-Methode zur Entwicklung lärmarmer Fahrzeuggetriebe (Using Finite Elements and Boundary Elements Method for Developing Low-Noise Automotive Transmissions). ATK Antriebstechnisches Kolloquium 1995, RWTH Aachen

[14.21] Schiberna P (1998) Geschwindigkeitsvorgabe für Fahrsimulationen mittels Verkehrssimulation (Velocity Specification for Driving Simulation Using Traffic Simulation). Dissertation, Universität Stuttgart, Institut für Maschinenelemente

[14.22] Schöttl C, Hintereder J, Breitling U (2004) CAD-integrierte Berechnung, Anwendererfahrung beim Vergleich mit klassischer FEA (CAD-Based Structural Analysis, Practical Experience in Comparison to Standalone FEA). VDI-Berichte Nr. 1846: 711–734

[14.23] Spörl T (1996) Modulares Fahrsimulationsprogramm für beliebig aufgebaute Fahrzeugtriebstränge und Anwendung auf Hybridantriebe (Modular Driving Simulation Program for Arbitrarily Designed Vehicle Powertrains and Application to Hybrid Drives). Dissertation, Universität Stuttgart, Institut für Maschinenelemente

[14.24] Steinbuch R (2004) Simulation im konstruktiven Maschinenbau (Simulation in Mechanical Design Engineering). Fachbuchverlag Leipzig

[14.25] Tasche J, Dettmar M (2003) Optimierung des Schwingungsverhaltens von Fahrzeug-Antriebssträngen durch Simulation (Vibrational Optimisation of Automotive Drive Trains Using Simulation). VDI-Berichte Nr. 1749: 241–254

[14.26] Vollmer D, Balasubramanian B, Siegert E (1992) Fahrsimulation unter realistischen Umfeldbedingungen (Driving Simulation Under Realistic Conditions). VDI-Berichte Nr. 1007: 263–280

[14.27] Wagner G (1992) Berechnung der Verlustleistung von Kfz-Vorgelegegetrieben (Calculation of Power Loss of Automotive Countershaft Transmissions). VDI-Berichte Nr. 977: 175–198

[14.28] Wiedemann R, Schwerdtfeger T (1987) Makroskopisches Simulationsmodell für Schnellstraßennetze mit Berücksichtigung von Einzelfahrzeugen (Macroscopic Simulation Model for Highway Networks Taking Into Account Individual Vehicles). Forschung Straßenbau und Straßenverkehrstechnik, Heft 500

[14.29] Willmerding G (1992) Simulationssystem zur Optimierung des Antriebsstrangs unter Einbeziehung des Verkehrsflusses (Simulation System for Optimising the Powertrain Taking Account of Traffic Flow). ATZ Automobiltechnische Zeitschrift. Part 1: ATZ 94, Nr. 5: 286–291; Part 2: ATZ 94, Nr. 6: 342–349

[14.30] Wolz U (1992) Die Einbindung der technischen Berechnung bei einem Fahrzeugzulieferer in den Entwicklungsprozess (Incorporating Engineering Calculation in the Development Process at a Vehicle Supplier). VDI-Berichte Nr. 1007: S. 59–79
[14.31] www.avl.com: CRUISE
[14.32] www.dynasim.se: Dymola (Dynamic Modeling Laboratory)
[14.33] www.mathworks.de: Tutorial for Matlab

## Chapter 15

[15.1] Adams J (1976) Conceptual Blockbusting. Norton & Company, New York London
[15.2] Beitz W (1983) Neue Arbeitstechniken beim Konstruieren (New Techniques for Design Engineering). VDI-Berichte Nr. 492
[15.3] Brankamp K (1971) Planung und Entwicklung neuer Produkte (Planning and Developing New Products). De Gruyter & Co, Berlin
[15.4] Conrad P, Schiemann H, Vömel G (1978) Erfolg durch Methodisches Konstruieren (Success Through Systematic Design). Lexika-Verlag, Grafenau/Württ.
[15.5] Ewald O (1975) Lösungssammlungen für das Methodische Konstruieren (Solution Compendiums for Systematic Design). VDI-Verlag, Düsseldorf
[15.6] Glegg G (1971) The Design of Design. University Press, Cambridge
[15.7] Hansen F (1966) Konstruktionssystematik (Design Methodology Systematics). VEB-Verlag Technik, Berlin
[15.8] Hohmann K (1977) Methodisches Konstruieren (Systematic Design). Verlag W. Girardet, Essen
[15.9] Hubka V (1976) Theorie der Konstruktionsprozesse (Theory of Engineering Design Processes). Springer, Berlin Heidelberg New York
[15.10] Koller R (1979) Konstruktionsmethode für den Maschinen-, Geräte- und Apparatebau (Design Engineering Methods for Mechanical, Equipment and Apparatus Engineering). Springer, Berlin Heidelberg New York
[15.11] Pahl G, Beitz W (2005) Konstruktionslehre (Engineering Design Theory). Springer, Berlin Heidelberg New York
[15.12] Rodenacker W (1970) Methodisches Konstruieren (Systematic Design). Springer, Berlin Heidelberg New York
[15.13] Rodenacker W, Claussen U (1973) Regeln des Methodischen Konstruierens (Rules of Systematic Design). Krausskopf-Verlag GmbH, Mainz
[15.14] Roth K (1974) Aufbau und Handhabung von Konstruktionskatalogen (Compiling and Managing Design Engineering Catalogues). VDI-Berichte Nr. 219
[15.15] Steinwachs H (1976) Praktische Konstruktionsmethode (Practical Design Engineering Methods). Vogel-Verlag, Würzburg
[15.16] Storp HJ, Christof S, Hackh D, Kuhr M (2005) Projekt- und Wissensmanagement (Project and Knowledge Management). ATZ/MTZ-Sonderausgabe 10/2005: Die neue S-Klasse von Mercedes-Benz: 192–195
[15.17] VDA-QMC (1998) Sicherung der Qualität vor Serieneinsatz – Projektplanung (Quality Assurance prior to Serial Application – Project Planning). VDA Band 4, Teil 3. Verband der Automobilindustrie; Qualitäts-Management-Center

[15.18] VDI-Richtlinie 2221 (1985) Methodik zum Entwickeln und Konstruieren technischer Systeme und Produkte (Systematic Approach to the Development and Design of Technical Systems and Products). VDI Verein Deutscher Ingenieure e.V., Düsseldorf

[15.19] VDI-Richtlinie 2222 Teil 1 (1977) Konstruktionsmethodik: Konzipieren technischer Produkte (Design Engineering Methodics; Conceptioning of Industrial Products). VDI Verein Deutscher Ingenieure e.V., Düsseldorf

[15.20] Wolff J (1976) Kreatives Konstruieren (Creative Engineering Design). Verlag W. Girardet, Essen

[15.21] Zwicky F (1971) Entdecken, Erfinden, Forschen im Morphologischen Weltbild (Discovering, Inventing, Researching in the Morphological World Picture). Verlag Droemer-Knaur, München Zürich

## Chapter 16

[16.1] Beitz W, Grote KH (1997) Dubbel – Taschenbuch für den Maschinenbau (Dubbel – Handbook of Mechanical Engineering). 19th Edition. Springer, Berlin Heidelberg New York

[16.2] DGQ-Band 16–31 (1990) SPC 1 – Statistische Prozesslenkung (SPC 1 – Statistical Process Control). Deutsche Gesellschaft für Qualität

[16.3] Does RJ, Roes CB, Trip A (1999) Statistical Process Control in Industry Implementation and Assurance of SPC. Series: Mathematical Modelling: Theory and Applications, Vol. 5. Springer, Berlin Heidelberg New York

[16.4] König W, Klocke F (1997) Fertigungsverfahren. Band 3: Abtragen und Generieren (Manufacturing Processes. Volume 3: Eroding and Generating). Springer, Berlin Heidelberg New York

[16.5] König W, Klocke F (1995) Fertigungsverfahren. Band 1: Drehen, Fräsen, Bohren (Manufacturing Processes. Volume 1: Turning, Milling, Drilling). VDI-Verlag, Düsseldorf

[16.6] König W, Klocke F (1995) Fertigungsverfahren. Band 4: Massivumformung (Manufacturing Processes. Volume 4: Forming). VDI-Verlag, Düsseldorf

[16.7] König W, Klocke F (1995) Fertigungsverfahren. Band 5: Blechbearbeitung (Manufacturing Processes. Volume 5: Sheet-Metal Forming). VDI-Verlag, Düsseldorf

[16.8] König W, Klocke F (1989) Fertigungsverfahren. Band 2: Schleifen, Honen, Läppen (Manufacturing Processes. Volume 2: Grinding, Honing, Lapping). VDI-Verlag, Düsseldorf

[16.9] Weck M (1991) Werkzeugmaschinen – Fertigungssysteme. Band 1: Maschinenarten, Bauformen und Anwendungsbereiche (Machine Tools – Manufacturing System. Volume 1: Machine Types, Layouts and Fields of Application). 4th Edition. VDI-Verlag, Düsseldorf

[16.10] ZF Getriebe GmbH: Graphical Material

## Chapter 17

[17.1] Bertsche B, Lechner G (2004) Zuverlässigkeit im Fahrzeug- und Maschinenbau. Ermittlung von Bauteil- und System-Zuverlässigkeiten (Reliability in Automotive and Mechanical Engineering. Determination of Component and System Reliability). Springer, Berlin Heidelberg New York

[17.2] Bertsche B, Lechner G (1987) Einfluss der Teileanzahl auf die Systemzuverlässigkeit (The Effect of the Number of Parts on System Reliability). Antriebstechnik 26, Nr. 7

[17.3] Brodbeck P, v Eiff H, Lechner G (1991) Zuverlässigkeitsanalyse von Getrieben (Reliability Analysis of Gear Systems). VDI-Berichte Nr. 878

[17.4] Brodbeck P, Pfeiffer M, Germann S, Schyr C, Ludemann S (2001) Verbesserung der Simulationsgüte von Antriebsstrangprüfständen mittels Reifenschlupfsimulation (Improving the Simulation Quality of Powertrain Test Benches Using Tyre-Slip Simulation). VDI-Berichte Nr. 1610

[17.5] DIN 25424 (1981) Fehlerbaumanalyse (Fault Tree Analysis). DIN Deutsches Institut für Normung e.V., Berlin

[17.6] Dubi A (1990) Stochastic Modeling of Realistic Systems. Ben-Gurion University of the Negev

[17.7] Kersten G (1987) FMEA – Fehlermöglichkeits- und -einflussanalyse in der industriellen Praxis (FMEA Failure Mode and Effect Analysis in Industrial Practice). Verlag Moderne Industrie

[17.8] Koslow BA, Uschakow IA (1979) Handbuch zur Berechnung der Zuverlässigkeit für Ingenieure (Handbook for Calculating Reliability for Engineers). Carl Hanser Verlag, München Wien

[17.9] Müller-Kose JP, Patzer J, Brodbeck P, Merkt T (2004) Transfer von Missbrauchstests an manuellen Schaltgetrieben vom Fahrzeug auf den Prüfstand (Transfer of Misuse Tests on Manual Transmissions from the Prototype Vehicle to the Test Rig). VDI-Berichte Nr. 1827

[17.10] Nicola A, Rosenberger J, Sauer B, Nemeth G (2004) Verbrennungsmotoren hochdynamisch simuliert (Internal Combustion Engines Simulated Highly Dynamically). Antriebstechnik 11

[17.11] Schäfer E (1979) Zuverlässigkeit, Verfügbarkeit und Sicherheit in der Elektronik (Reliability, Availability and Safety in Electronics). Vogel-Verlag, Würzburg

[17.12] TAE-Lehrgang Nr. 1267/60.175 (1993)

1/ Lechner G: Zuverlässigkeit und Lebensdauer von Systemen (Reliability and Service Life of Systems);

2/ Bertsche B: Mathematische Behandlung der Zuverlässigkeit von Komponenten und Systemen (Mathematical Treatment of Reliability of Components and Systems);

3/ Hirschmann KH: Methodische Lebensdauerberechnung bei Maschinenelementen (Systematic Service Life Calculation for Machine Elements);

4/ v Eiff H: Vorgehensweise bei der Zuverlässigkeitsanalyse (Procedure for Reliability Analysis);

5/ Brodbeck P: Vorgehensweise bei der Zuverlässigkeitsanalyse (Procedure for Reliability Analysis);

6/ Wu Z: Vorgehensweise bei der Zuverlässigkeitsanalyse (Procedure for Reliability Analysis);

7/ Herchenröder P: Systemanalyse durch Monte-Carlo-Simulation (Systems Analysis Using the Monte Carlo Simulation);
8/ Lechner G: Zuverlässigkeitssicherung bei Systemen der Fahrzeugtechnik (Reliability Assurance in Automotive Engineering Systems).
Technische Akademie Esslingen

[17.13] VDA (1984) Qualitätskontrolle in der Automobilindustrie – Zuverlässigkeitssicherung bei Automobilherstellern und Lieferanten (Quality Control in the Automobile Industry. Reliability Assurance for Automobile Manufacturers and Suppliers). 2nd Edition. Verband der Automobilindustrie e.V., Frankfurt

[17.14] VDA-Band 4 (2003) Sicherung der Qualität vor Serieneinsatz (Quality Assurance prior to Serial Application). Verband der Automobilindustrie e.V., Frankfurt

[17.15] VDI-Richtlinie 4001, Blatt 2 (2005) Terminologie der Zuverlässigkeit (Reliability Terminology). VDI Verein Deutscher Ingenieure e.V., Düsseldorf

[17.16] Wu Z (1992) Vergleich und Entwicklung von Methoden zur Zuverlässigkeitsanalyse von Systemen (Comparison and Development of Methods for Reliability Analysis of Systems). Dissertation, Universität Stuttgart, Institut für Maschinenelemente

# Index of Companies/Transmissions

# Index of Names

The following index contains the names of persons appearing in the text of the book. The names of persons appearing in Tables 1.2 to 1.7 "History of Automotive Transmissions" are not listed here.

# Subject Index

# About the Authors

**Harald Naunheimer** studied at the Universität Stuttgart. He got his degree in mechanical engineering in 1988. Afterwards, he worked as a research assistant in the field of automotive transmissions at the Universität Stuttgart. He earned his doctorate in mechanical engineering in 1995. Harald Naunheimer is author of the English edition of "Automotive Transmissions". From 1995-2001 he worked in the vehicle transmissions division of RENK AG in Augsburg, Germany. He joined ZF in 2001 as a senior manager for application engineering in the automatic transmissions business unit at ZF Getriebe GmbH in Saarbruecken, Germany. From 2006–2009 he was Director of Product Development, Automatic Transmissions for ZF Getriebe GmbH, in Friedrichshafen. Harald Naunheimer was named Vice President of Corporate Research and Development at ZF Friedrichshafen AG during July 2009, headquartered in Friedrichshafen, Germany.

**Professor Bernd Bertsche** earned his degree and doctorate in mechanical engineering at the Universität Stuttgart. Afterwards, he worked for Mercedes-Benz AG in the passenger car development unit followed by a position as professor at the University of Applied Sciences in Albstadt-Sigmaringen. Later, he was appointed at the Universität Stuttgart. Since 2001, Professor Bertsche is the ordinary and director for the Institute of Machine Components at the Universität Stuttgart.

**Joachim Ryborz** studied mechanical engineering and worked as research assistant of Professor Gisbert Lechner in the field of automotive transmissions at the Universität Stuttgart. He earned his doctorate in mechanical engineering in 2003. He is co-author of the 1st English edition of this book as well as co-author of the 2nd German edition. Since 2002, he has been working in the transmission development for light commercial vehicle at ZF Friedrichshafen AG, amongst others as project manager for manual, automated and hybrid transmissions.

**Wolfgang Novak** studied mechanical engineering at the Universität Stuttgart. From 2002–2003 he worked at McLaren Cars Limited as a development engineer for sports car transmissions. Afterwards, he worked as a research assistant of Professor Bernd Bertsche at the drive technology department and he is co-author of the 2nd German edition. He earned his doctorate in mechanical engineering in 2010. Since 2008, he has been working in the transmission development center at Daimler AG in Stuttgart, Germany.

MIX
Papier aus verantwortungsvollen Quellen
Paper from responsible sources
FSC® C105338

If you have any concerns about our products,
you can contact us on
**ProductSafety@springernature.com**

In case Publisher is established outside the EU,
the EU authorized representative is:
**Springer Nature Customer Service Center GmbH**
**Europaplatz 3, 69115 Heidelberg, Germany**

Printed by Libri Plureos GmbH
in Hamburg, Germany